European Nuclear Physics Conference (EuNPC 2022)

EPJ Web of Conferences Volume 290 (2023)

Santiago de Compostela, Spain
24-28 October 2022

Editor:

D. Cortina-Gil

ISBN: 978-1-7138-8493-4

Printed from e-media with permission by:

Curran Associates, Inc.
57 Morehouse Lane
Red Hook, NY 12571

Some format issues inherent in the e-media version may also appear in this print version.

For additional information, please contact EDP Sciences – Web of Conferences
at the address below.

EDP Sciences – Web of Conferences
17, Avenue du Hoggar
Parc d'Activité de Courtabœuf
BP 112
F-91944 Les Ulis Cedex A
France

Phone: +33 (0) 1 69 18 75 75
Fax: +33 (0) 1 69 28 84 91

contact-edps@webofconferences.org

Additional copies of this publication are available from:

Curran Associates, Inc.
57 Morehouse Lane
Red Hook, NY 12571 USA
Phone: 845-758-0400
Fax: 845-758-2633
Email: curran@proceedings.com
Web: www.proceedings.com

TABLE OF CONTENTS

P1 ACCELERATORS AND INSTRUMENTATION

P2 NUCLEAR STRUCTURE, SPECTROSCOPY AND DYNAMICS

P3 NUCLEAR ASTROPHYSICS

P4 ASTROPARTICLE PHYSICS

P5 HEAVY ION COLLISIONS AND QCD PHASES

P6 HADRON STRUCTURE, SPECTROSCOPY AND DYNAMICS

P7 FUNDAMENTAL SYMMETRIES AND INTERACTIONS

P8 NUCLEAR PHYSICS APPLICATIONS

P9 FEW-BODY SYSTEMS

PLENARY TALK

EUNPC 2022

European Nuclear Physics Conference

Santiago de Compostela | 24-28 October

Sessions

Nuclear Astrophysics
Astroparticle Physics
Accelerators and Instrumentation
Nuclear Structure, Spectroscopy, and Dynamics
Heavy Ion Collisions and QCD Phases
Hadron Structure, Spectroscopy, and Dynamics
Fundamental Symmetries and Interactions
Nuclear Physics Applications
Few-Body Systems

EuNPC

bit.ly/eunpc2022
eunpc2022@igfae.usc.es
@igfae_hep

Scientific Committee

(Board of the Nuclear Physics Division (EPS))

R. Burku Cakirli
Alessandra Fantoni (co-Chair)
Elena Gonzalez
Zsofia Kertesz
Ann-Cecile Larsen
Silvia Leoni
Constantin Mihai
Silvia Niccolai
Zsolt Podolyak
Stephan A. Pomp
Rene Reifarth
Guy Ron
Timo Sajavaara
Eugenio Scapparone
Giuseppe Verde

Local Committee

Héctor Álvarez
Yassid Ayyad
Saúl Beceiro
Fabio Bellini
José Benlliure
Dolores Cortina (Chair)
Beatriz Fernández
Luis M. Fraile
Heduardo Garrido
Elena González (Chair)
Carlos Guerrero
Jose A. Hernando
Jose L. Rodríguez
Tomás Rodriguez
Laura Tolos
Silvia Viñals

edp sciences

EPJ Web of Conferences

Statement of Peer review

In submitting conference proceedings to EPJ Web of Conferences, I certify to the Publisher that I adhere to the **Policy on Publishing Integrity** of the journal in order to safeguard good scientific practice in publishing.

1. All articles have been subjected to peer review administered by the proceedings editors.
2. Reviews have been conducted by expert referees, who have been requested to provide unbiased and constructive comments aimed, whenever possible, at improving the work.
3. Proceedings editors have taken all reasonable steps to ensure the quality of the materials they publish and their decision to accept or reject a paper for publication has been based only on the merits of the work and the relevance to the journal.

Title, date and place of the conference

European Nuclear Physics Conference EnNPC 2022, October 24th -28 th 2022 Santiago de Compostela (Spain)

Proceedings editor(s):

Dolores Cortina-Gil

Date and editor's signature

20th November 2023

EPJ Web of Conferences **290**, 01001 (2023) https://doi.org/10.1051/epjconf/202329001001
EuNPC 2022

Standardisation of ^{18}F using new portable $4\pi\beta(LS)-\gamma$ coincidence detection system at ENEA-INMRI

Farmesk Abubaker [1,2*], *Marco* Capogni[3], *Francesco* Tortorici [1,2], *Concetta* Sutera [1,2], *Francesco* Pepe[4], *Matteo* Corbo[4], *Pierino* De Felice[3], and *Vincenzo* Bellini[1]

[1] University of Catania – Department of Physics, Catania, Italy
[2] INFN, Istituto Nazionale di Fisica Nucleare - Sezione di Catania, Italy
[3] ENEA, Istituto Nazionale di Metrologia delle Radiazioni Ionizzanti (INMRI), C.R Casaccia, Rome, Italy
[4] CAEN, Costruzioni Apparecchiature Elettroniche Nucleari, Viareggio, Italy

Abstract. This project concerned with the development at ENEA-INMRI of new in-situ $4\pi\beta(LS)-\gamma$ coincidence detection system for activity measurement of the short half-life radionuclides used in Nuclear Medicine. The hardware of the new portable $4\pi\beta(LS)-\gamma$ coincidence detection system was implemented at ENEA-INMRI, in collaboration with Catania University and INFN, by adding a gamma channel on the existing TDCR portable detector available at ENEA-INMRI. A new data analysis software was developed at CAEN, independently from an existing one elaborated at ENEA.INMRI, in order to analyse the data recorded in list-mode by the new detector equipped with the CAEN desktop digitizer DT5720. The activity for short half-life radionuclides used in nuclear medicine can be then computed. Two primary activity measurement TDCR and $4\pi\beta(LS)-\gamma$ coincidence methods were then used to determine the activity of ^{18}F at ENEA-INMRI. The TDCR parameter is measured for the ^{18}F standard solution using both CAEN and ENEA-INMRI data analysis software.

1 Introduction The $4\pi\beta-\gamma$ coincidence counting technique has been a powerful and widely used method for radionuclide standardisation and to measure the absolute activity concentration of radionuclides [1, 2, 3]. Radionuclide activity can be measured using a direct activity measurement method, which doesn't require any quenching-indicating parameters or reference standards to determine detection efficiency. [4]. Among the currently available methods of radionuclides activity measurement, the coincidence counting technique is one of the most widely utilised in any of its variants [5, 2, 3]. The direct measurement of activity concentration can be performed for radionuclides decaying via pure β, electron capture (EC), $\beta-\gamma$, $EC-\gamma$, and $X-\gamma$ without making any assumptions about the values of the detector efficiencies, branching ratios, or decay scheme parameters [6]. The principles and theory of the $4\pi\beta-\gamma$ coincidence counting method are well-established and thoroughly understood. Nevertheless, it's difficult to apply these principles to actual measurements. Attempts to improve the accuracy and the precision of $4\pi\beta-\gamma$ coincidence counting has been ongoing for decades. Grigorescu studied the accurate fulfilment of the $\gamma-$ray channel linearity conditions [7]. Campion authored the first and most renowned paper on the $4\pi\beta-\gamma$ coincidence counting [2]. Campion showed that a $4\pi\beta-\gamma$ coincidence system is principally made up of two detectors; beta and gamma detectors and an electronic unit to generate a coincidence pulse by detecting synchronised gamma and beta pulses which come from the same nuclear disintegration.

He demonstrated that precise measurements of $\beta-\gamma$ emitter radionuclides can be accomplished by employing a $\beta-$detector with 4π geometry for the coincidence method. In this configuration, many of the necessary corrections are minimal, ensuring accurate results. It is expected that the $\gamma-$detector responds only to $\gamma-$rays, the $\beta-$detector responds only to $\beta-$particles, and the coincidence channel records only the simultaneous pulses come from the two detectors (true coincidences) [2]. This technique has been the focus of many papers and comprehensive reports [8, 6] since the publications of Dunworth [1] and Campion [2]. The fundamental principle of coincidence counting can be illustrated through the prototypical case of the activity measurement of a radionuclide with a simple decay-scheme consisting of $\beta-$emission followed directly by a $\gamma-$ray from the de-excitation of the daughter radionuclide or annihilation process. The basis of the coincidence technique comes from the additional coincidence channel. This additional channel records disintegration pulses that emanate from both $\beta-$ and $\gamma-$counters. The standardisation of a wide range of radionuclides can be achieved using the original $4\pi\beta-\gamma$ coincidence counting and its extended techniques [9]. Most laboratories have used a proportional counter (*PC*) in the $\beta-$channel for a long time to standardise radionuclides using the $4\pi\beta(PC)-\gamma$ coincidence method such as Laboratoire National Henri Becquerel (LNHB) and Physikalisch-Technische Bundesanstalt: National Metrology Institute (PTB) [10, 11]. Afterwards, $4\pi\beta-\gamma$ coincidence system has been updated by replacing the proportional counter with the liquid scintillation (LS) counter as a beta counter. This new coincidence counting system has been developed and improved in metrology laboratories. $4\pi\beta(LS)-\gamma$ coincidence counting systems had already been implemented [12, 13, 14], but the goal was to have a complete TDCR (Triple to Double Coincidence System) counter as the $\beta-$channel [15, 16]. The

*Corresponding author: farmesk.abubaker@ct.infn.it

https://doi.org/10.1051/epjconf/202329001001

TDCR system has been specially developed for the purpose of standardising pure beta and pure electron capture emitters, and it can be extended to standardise simple $\beta-\gamma$ and $EC-\gamma$ emitter radionuclides.

2 Experimental setup of ENEA $4\pi\beta(LS)$ - γ coincidence counting system

At the Italian National Institute of Ionizing Radiation Metrology (INMRI), belonging to ENEA and located in Casaccia Research Center, the TDCR method is used for absolute activity measurements and standardisation of pure electron capture (EC) or pure beta emitting radionuclides [17, 18]. A portable TDCR detector consists of three photomultipliers (PMTs) operating in coincidence. These PMTs are symmetrically positioned around a specially designed optical chamber [19]. It allows for the measurement of radioactive material in solution of liquid scintillator, contained in a vial which is put inside the optical chamber [17].
ENEA-INMRI in collaboration with Catania University, Istituto Nazionale di Fisica Nucleare (INFN)- Sezione di Catania section and Costruzioni Apparecchiature Elettroniche Nucleari (CAEN) developed a new in-situ coincidence system for activity measurement for short half-life radionuclides used in nuclear medicine which is called $4\pi\beta(LS)-\gamma$ coincidence detection system. The portable TDCR counter [19] at ENEA-INMRI is extended to $4\pi\beta(LS)-\gamma$ coincidence system by implementing a cylindrical-type Scionix Standard 51B51/2M NaI(Tl) scintillation detector as a gamma counter. The NaI(Tl) scintillation detector consists of a scintillation material integrally coupled to a photomultiplier tube. The NaI(Tl) scintillation detector with the (3"×3") crystal dimension includes a 14 pins photomultiplier (0.4 mm) thick aluminium housing inserted into the bottom of the TDCR counter as shown in Fig.1.

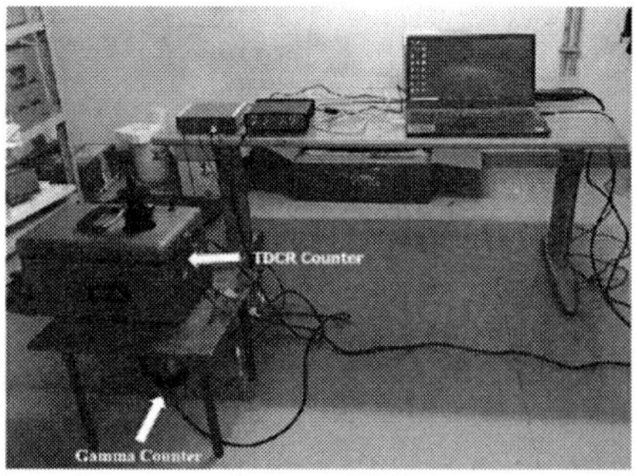

Fig. 1. Experimental setup of ENEA-INMRI portable $4\pi\beta$ (LS) $-\gamma$ coincidence system.

The new ENEA-INMRI portable $4\pi\beta(LS)-\gamma$ coincidence system can be used in environmental applications and in nuclear medicine laboratories of the hospitals to measure the activity of a radiopharmaceutical injected into patients, such as ^{18}F,^{11}C or ^{131}I. The schematic configuration of the new ENEA-INMRI portable $4\pi\beta\,(LS)-\gamma$ coincidence counting system is shown in Fig. 2.

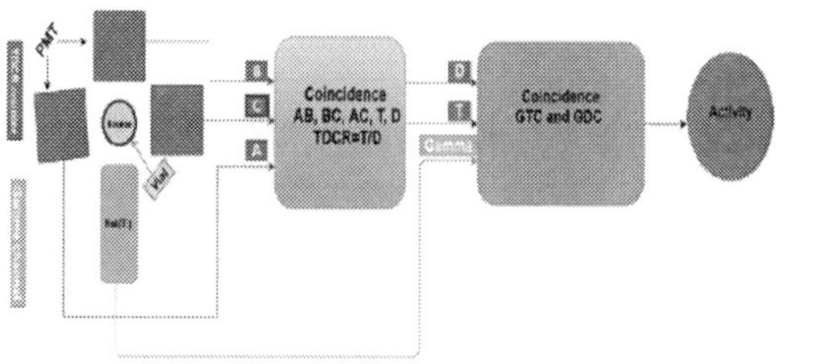

Fig. 2. The schematic configuration of $4\pi\beta(LS)$ - γ coincidence system available at ENEA-INMRI.

3 $4\pi\beta(LS)-\gamma$ data analysis software

The $4\pi\beta(LS)-\gamma$ coincidence system is directly linked to the CAEN DT5720 digitizer. The CAEN DPP-CI Control Software remotely controls the digitizer [20]. The acquisition parameters such as gate length, DC offset, pulse polarity, threshold level, etc. can be set independently for each channel to configure the hardware, and perform the data readout. In order to perform a $4\pi\beta(LS)-\gamma$ analysis on the collected event lists coming from the TDCR detector and the gamma detector, a customised TDCR DAQ software has been updated and modified. The software is written in C language and designed for off-line implementation of $4\pi\beta-\gamma$ coincidence counting method. Since all the acquisition channels of a digitizer are synchronised, it is possible to correlate events from different input channels comparing their Trigger Time Tag (TTT)s. Once the run is over, the CAEN TDCR DAQ software scans the resulting event lists looking for coincidences between the events come from the β −channels in the TDCR counter. Afterwards, the two output files are created for both Triple N_T and Double N_D coincidences data inside the CAEN TDCR code and they are saved as a ASCII file or binary file. These two output files are used as input files for the CAEN TDCRGamma code. Next, the $4\pi\beta(LS)$–γ (CAEN TDCRGamma) DAQ software scans the event lists generated, searching for coincidences between the gamma channel and beta channel. It specifically looks for coincidences between gamma event comes form gamma counter and the triple coincidences and the logical sum of double coincidences output files within the TDCR counter, as illustrated in Fig. 3. This data subset from TDCR counter is used together with the set of data coming from the γ −channel. The set of events in the β −channels which are time correlated with the gamma event are taken into account. One can select the γ − energy window close to the full-energy peak of the γ −ray for the γ −channel. Furthermore, the code will look for the event coming from the beta counter and gamma counter by looking to the minimum TTT of each event in double or triple coincidences data output files in β −counter and γ −counter to create a new coincidence window. In other word, let's assume τ_γ and τ_β are the minimum TTT of gamma and beta events respectively, and ω is the coincidence window. Moreover, if $\tau_\gamma < (\tau_\beta+\omega)$, this means that there is a coincidence between γ −channel and β −channel, otherwise there is no coincidence between the events coming from the both detectors.

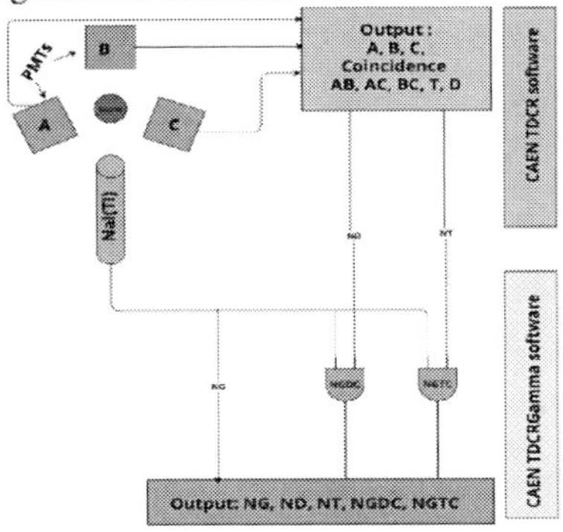

Fig. 3. $4\pi\beta\ (LS)\ -\ \gamma$ coincidence system configuration.

Following this, the TDCRGamma software includes the minimum TTT event as a dataset within the beta coincidence window in the triple and double coincidence data files. Thus, it starts looking for coincidences with events in the gamma counter and extending the coincidence window until the event is found in the gamma channel. Finally, TDCRGamma software will compare the TTT for all the acquisition from the three input channels (γ −channel and the two output files from β −channel). Subsequently, the TTT for each selected event in coincidence in the TDCR counter and gamma counter will be saved. This intimates that the coincidence window for the new $4\pi\beta(LS)-\gamma$ coincidence system starts from the minimum time of measuring the first event in files in the β −channel until the event in the γ −channel is measured. The new time coincidence window for $4\pi\beta(LS)\ -\gamma$ coincidence system is longer than the time coincidence window of the TDCR system. The dead-time of the TDCR counter is longer than the selected dead time for the gamma channel, in order to eliminate the after-pulses in the liquid scintillation counter. The recorded data stream is then analysed by the CAEN TDCRGamma analysis software. Moreover, the DAQ software implements the same algorithms of the well-known French MAC3 analogic module emulating its operations in terms of coincidence definition and extendable-type dead time management [21]. In particular, the CAEN TDCRGamma software ignores all those events found in the lists whose TTTs lie in a dead time window; the dead time is then extended according to the MAC3 logic [20]. As a result of this analysis, the new CAEN TDCRGamma DAQ software can provide the counting from the three channels: The single count rate for each channel, gamma count rate N_G, from γ −channel. The triple coincidence N_T and the logical sum of the double coincidence N_D, from β −channels. Moreover, the coincidences between gamma channel with double coincidence in beta channel N_{GDC}, and the coincidences between gamma channel with triple coincidence in beta channel N_{TDC} will be measured. In addition, real-, dead-, and live-times can be evaluated as shown in Fig. 3. The activity of the radionuclides can be measured, at the "zero order approximation", by using the conventional formula of the $4\pi\beta(LS)-\gamma$:

$$A=\frac{N_G\cdot N_D}{N_{GDC}}=\frac{N_G\cdot N_T}{N_{TDC}} \qquad (1)$$

Independently from CAEN, another TDCR analysis software was developed at ENEA-INMRI by implementing the MAC3 philosophy in the CERN ROOT framework [20], an object-oriented package for physics analysis. The algorithm for the TDCR coincidence analysis was then developed by following the main idea described by Bouchard and Cassette [20] and by taking into account the powerful CERN ROOT resources for analysing data coming from complex detectors. Both codes can run either on Windows or Linux machines. The results obtained by the two software have been compared by measuring a standard solution of ^{18}F with TDCR parameters by using CAEN software and ENEA software. In Table 1, one can observe the typical TDCR value acquired from these measurements along with the corresponding efficiencies for the logical sum of double coincidences and triple coincidences, as analysed using both CAEN and ENEA software. A coincidence window of t_c = 140 ns and a dead-time 50 μs were applied on the set of recorded data to perform TDCR analysis for ^{18}F. The deviation in percent between the TDCR parameter TDCRCAEN computed by CAEN code and TDCRENEA computed by ENEA code for ^{18}F sources are also reported in Table 1.

Table 1. TDCR values for radionuclide ^{18}F by using CAEN software and ENEA software code.

Radionuclide	Software	N_D	N_T	TDCR
^{18}F	CAEN	2693230	2673821	0.9928±0.03
	ENEA	2693096	2673695	0.9928±0.03
	Δ%	0.7126	0.005	0.00

At the same time, the activity of ^{18}F is calculated using conventional Eq. 1 without any knowledge of the detection efficiency of the detectors. The γ −energy window close to the full-energy peak about 511 keV of ^{18}F for the gamma channel is selected. A coincidence window of t_c = 2000 ns and a dead-time 10 μs were applied on the set of recorded data to perform the 4πβ (LS) − γ analysis for ^{18}F. Table 2 shows that the ^{18}F activity obtained by both CAEN TDCR and $4\pi\beta(LS)-\gamma$ DQA methods, and the minimum deviation less than 0.1% between the two methods is measured. In the TDCR method one can determine the activity of ^{18}F source by using the following equation:

$$A=\frac{R_cD}{\varepsilon_D}=\frac{R_cT}{\varepsilon_T} \qquad (2)$$

Where R_cD and R_cT are logical sum of double and triple corrected count rates for ^{18}F respectively. Corrected count rate can be calculated by multiply the net count rate of logical sum of double count R_nD and triple count R_nT by the correct Factor. Another important parameters are the logical sum of double efficiency ε_D and triple efficiency ε_T which can be computed theoretically by using Monte Carlo Simulation GEANT4 code [22]. The measurement conducted with TDCR achieves a level of uncertainty at 0.5%, considering all aspects of uncertainty. This includes not only statistical uncertainty but also uncertainties associated with factors such as the decay model, background measurements, coincidence resolving time, dead-time, and decay during measurement, as well as decay at the reference date. The good agreements between the two methods confirms that CAEN $4\pi\beta(LS)-\gamma$ (TDCRGamma) data analysis software developed for in-situ $4\pi\beta(LS)-\gamma$ coincidence system at ENEA operates correctly.

Table 2. Activity measurements of radionuclides ^{18}F using TDCR and $4\pi\beta(LS)-\gamma$ methods.

Radionuclide	Method	Activity (Bq)
^{18}F	TDCR	41379.69 $\pm$ 0.15
	$4\pi\beta(LS)-\gamma$	41777.06 $\pm$ 0.13
	Δ%	0.1

4 Conclusion

The $4\pi\beta(LS)-\gamma$ coincidence counting technique is the most powerful technique for the absolute activity measurements of $\beta-\gamma$ or $EC-\gamma$ emitting nuclides. The $4\pi\beta(LS)$–γ coincidence method allows for the direct standardisation of short-lived pure β, α, EC, β –γ, and EC –γ radionuclides directly on-site. Furthermore, it can be used to calibrate the devices used in the Nuclear Medicine Departments, such as activimeters, PET, SPECT, Gamma camera, etc, without moving the radioactive sources between the Hospitals and the Metrology Institutes. The activity of ^{18}F is determined using two different methods, such as TDCR and $4\pi\beta\ (LS)\ -\gamma$ coincidence methods. The standardisation in situ of short half-lived radionuclides is advisable to avoid the transport of radioactive sources between the hospital and metrology institutes.

ENEA-INMRI and CAEN developed TDCR software independently, to perform the TDCR analysis by implementing the MAC3 philosophy. Then, the TDCR parameters are measured for a ^{18}F standard solution by using CAEN and ENEA software. Both codes are compatible and work properly. The activity measurement of the radionuclides is independent on the detection efficiency for either detectors using the $4\pi\beta\ (LS)-\gamma$ method. On the other hand, in TDCR method the detection efficiency has a crucial role in the activity measurement of the radionuclides. For that reason, one can compute the detection efficiency of the detectors by using the Monte Carlo Simulation; in this project GEANT4 Code is used. Another important feature of DAQ TDCRGamma software is that the new software is able to run the $4\pi\beta\ (LS)-\gamma$ detection analysis both online or off real-time measurements and previously acquired data. This makes it possible to perform a parameter sweep analysis repeating the $4\pi\beta\ (LS)-\gamma$ detection technique on the same data set changing automatically for example the coincidence resolving time and/or dead time. It is possible, therefore, to study the behaviour of the results obtained with different parameter values applying them on the same set of data, without repeating the measurements several times and just by changing the parameter values in the software configuration, with no hardware operations.

This work was funded by the University of Catania cooperation with ENEA-INMRI, INFN and the CAEN S.P.A in the framework of the National Operational Program on Research and Innovation PON 2014-2020. PON was carried out with the funding of the European Union. The authors are grateful to the radiochemistry laboratory in ENEA-INMRI for supplying the radioactive source in the LS solution.

References

1. J. V. Dunworth, Rev. Sci. Instr. **11**, 167 (1940).
2. P. J. Campion, Int J. Appl. Radiat. Isotopes **4**, 232 (1959).
3. A. P. Baerg, Metrologia **2**, 23 (1966).
4. C. Bobin, J. Bouchard, S. Pierre, and et al., Appl. Radiat. Isot. **70**, 2012 (2012).
5. J. V. Dunworth, Rev. Sci. Instr. **11**, 167 (1940).
6. Particle Counting in Radioactivity Measurements: International Commission on Radiation Units and Measurements (ICRU), ISBN 0-913394-51-3 **Report 52**, (1994).
7. L. Grigorescu, Int. J. Appl. Radiat. Isot **34:8**, p1151 (1982).
8. R. Resnick, D. Halliday, and K. S. Krane, A handbook of radioactivity measurements procedures (National Council on Radiation Protection and Measurements (PUBLISHER, Washington, 1985), No. 58.
9. P. J. Campion, J. G. V. Taylor, and J. S. Merritt, Int. J. Appl. Radiat. Isot. **8**, 8 (1960).
10. Ch. Bobin and J. Bouchard, Appl. Radiat. Isot. **64**, 124 (2006).
11. E. Sch¨ofeld, H. Janssen, R. Klein, JC. Hardy, V. Iacob, M. Sanchezvega, HC. Griffin, and Ludington MA., Appl. Radiat. Isot. **56**, 215 (2002).
12. B. R. S. Simpson and B. R. Meyer, Nucl. Instr. and Meth. A **14**, 339 (1994).
13. P. Cassette, M. M. Be, F. Jaubert, and M. C. Lepy, Appl Radiat. Isot. **60**, 439 (2004).
14. A. Chylinski and T. Radoszewski, Nucl. Instrum. Meth. A **369**, 336 (1996).
15. K. Pochwalski, R. Broda, and T.Radoszewski, Appl. Radiat. Isot. **39**, 165–172 (1988).
16. J. Bouchard and P. Cassette, Nucl. Instrum. Meth **A 505**, 72–75 (2003).
17. R. Vatin, Mesure absolue de l'activite des radionucleides emetteurs beta purs par scintillateur liquide a l'aide d'un detecteur a 3 photomultiplicateurs. In: Magana, J.F. (Ed.), Mesure de l'activite des radionucleides emetteurs beta purs (Bulletin BNM, Paris, 1991), No. 85 BNM, p. 13–21.
18. Ziemek, T., Broda, R., Listkowska, A. et al. Standardization of an 55Fe solution using the TDCR method in POLATOM as part of the CCRI (II)-K2.Fe-55.2019 key comparison. J Radioanal Nucl Chem **331**, 3241–3248 (2022).
19. Capogni M, De Felice P. A prototype of a portable TDCR system at ENEA. *Appl Radiat Isot.* **93**:45-51(2014)
20. Giuliano Mini, Francesco Pepe, Carlo Tintori, and Marco Capogni, Applied Radiation and Isotopes **87**, 166–170 (2014).
21. J. Bouchard and P. Cassette, Applied Radiation and Isotopes **52**, 669 (2000).

 https://doi.org/10.1051/epjconf/202329001001

22. C. Thiam, C. Bobin, B. Chauvenet, and J. Bouchard, Applied Radiation and Isotopes **70**, 2195 (2012).

EPJ Web of Conferences **290**, 01002 (2023)
EuNPC 2022
https://doi.org/10.1051/epjconf/202329001002

Investigation of Electron Capture in ^{176}Lu with a LYSO crystal scintillator

Francesco Nozzoli[1,2,*], *Luigi Ernesto* Ghezzer[2], *Riccardo* Nicolaidis[2,1], *Roberto* Iuppa[2,1], and *Paolo* Zuccon[2,1]

[1]Trento Institute for Fundamental Physics and Applications, INFN-TIFPA
[2]Physics Dep. University of Trento, Via Sommarive 14 I-38123 Trento, It.

Abstract. The nuclide ^{176}Lu is one of the few naturally occurring isotopes that are potentially unstable with respect to electron capture (EC). Although experimental evidence for ^{176}Lu EC decay is still missing, this isotope is instead well known to β^- decay into ^{176}Hf with an half-life of about 38 Gyr. Previous searches for the ^{176}Lu EC decay were performed by using a passive Lutetium source coupled with an HP-Ge spectrometer. Our approach uses a LYSO crystal both as Lutetium source and as an active detector and the obtained ^{176}Lu EC branching ratio limits are 7×10^{-4} and 4.4×10^{-4} for the capture of the 1s and 2s electron, respectively. Previously known limits for the EC in ^{176}Lu are improved by a factor 3 to 15 depending on the considered EC channel.

1 Introduction

The naturally occurring isotope ^{176}Lu (2.6% abundance) it is known to β^- decays to ^{176}Hf with a half-life of ~38 Gyr. However, ^{176}Lu is also one of the six naturally occurring isotopes that are potentially unstable with respect to electron capture (EC). In particular, evidence for EC decay was found for ^{40}K [1], ^{50}V [2, 3] and ^{138}La [1], but is still missing for ^{123}Te [4], ^{176}Lu [5] and ^{180m}Ta [6]. Figure 1 shows the decay scheme of ^{176}Lu: on the right side the dominant decay chain to ^{176}Hf is depicted, while on the left side there is the expected EC decay process to ^{176}Yb.

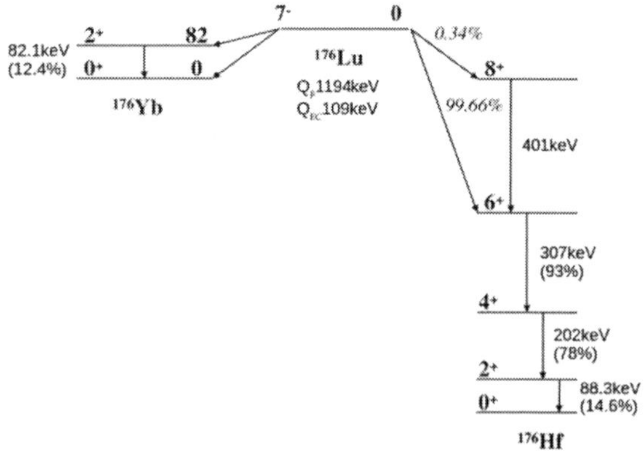

Figure 1. Decay scheme of ^{176}Lu [7, 8]. The reaction Q-value for EC decay of ^{176}Lu to the ^{176}Yb ground state is $\approx$ 109 keV. The percentages in the brackets are the γ-ray emission yields.

Previous searches for the ^{176}Lu EC decay were performed by using a passive Lutetium source coupled to a HP-Ge detector searching for the ^{176}Yb* 82 keV γ-ray or the characteristic Yb x-rays [5]. Our "active source" approach uses a 7.9g, 2mm thick, Lutetium–Yttrium OxyorthoSilicate (LYSO [9]) scintillator coupled to a R5946 PMT both as a ^{176}Lu source and as a detector placed on top of a GC2020 [10] HP-Ge detector. A passive shield made of 2.5mm of Cu and 5cm of Pb surrounds both detectors to minimize the environmental γ background. The measured energy resolution of HP-Ge detector was 1.65keV FWHM at 75keV and the measured energy resolution of the LYSO crystal was 20keV FWHM at 60keV. The energy scale of the LYSO crystal was calibrated with external ^{241}Am and ^{136}Cs sources, the known effect of the non-linearity of the LYSO Light Yield was taken into account in the evaluation of the measured LYSO energy. Further details on the experimental set-up are summarized in [11].

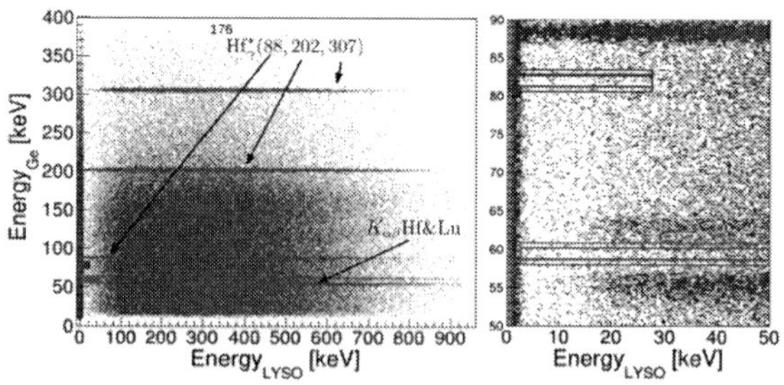

Figure 2. Energy$_{Ge}$ vs Energy$_{LYSO}$ spectrum and a zoom in the region of interest for EC. Red and magenta boxes are the regions of interest for ^{176}Lu EC to ^{176}Yb$^{2^+}$ and to ^{176}Yb$^{0^+}$ respectively. The zoom shows the $\pm1\sigma$ "Bulk" and $\pm(1\sigma$ to $2\sigma)$ "Sideband" regions considered in the analysis.

A total number of 1.5 Mevents with Energy$_{Ge}$ >15keV were acquired during 90h exposure time, most of the events collected by the HP-Ge detector come from the ^{176}Lu β-decay and the intrinsic/environmental back-

*e-mail: Francesco.nozzoli@unitn.it

ground. These two classes of events are well recognized in the bidimensional spectrum of FIG. 2. In particular, the population of events characterized by $Energy_{LYSO}$ <3keV (vertical strip in FIG. 2) are mostly due to intrinsic or environmental background. On the other hand, events characterized by $Energy_{LYSO}$ >3keV are mostly due to the continuous electron distribution emitted by ^{176}Lu β-decay. For this last class of events, in FIG. 2 are also visible the ^{176}Hf* γ-lines characterized by a fixed energy detected in the HP-Ge and the relative continuous distributions due to the Compton scattering. Similarly, at lower energy, the Hf x-rays due to internal conversion of ^{176}Hf* levels and the Lu x-rays due to photoelectric absorption, within the crystal, of ^{176}Hf* γ-rays, are visible. The rejection capability of the "active source" technique can be noticed in FIG. 2 where the red and magenta boxes are the regions of interest for ^{176}Lu EC to $^{176}Yb^{2^+}$ and to $^{176}Yb^{0^+}$ respectively.

In Figure 3 the energy distribution of the events measured by the HP-Ge is shown (magenta line). These data are plotted without any requirement on the LYSO measured energy, and as a consequence this plot is dominated by the background from the ^{176}Lu β-decay. This corresponds to typical landscape of the passive source measurement approach used in [5]. Profiting of the "active source" technique, most of ^{176}Lu β-decay events are rejected by the request E_{LYSO} <27keV (blue points).

This approach provides a factor ~20 in background suppression, and the resulting measured spectrum mostly corresponds to the intrinsic/environmental background characteristic of our HP-Ge detector. The black points on FIG. 3, are the dedicated measurement of the background events collected during 90h exposure time with the same set-up once the LYSO crystal has been removed. The x-rays of Pb and the characteristics γ-lines of ^{234}Pa* are identified in the background spectrum[1].

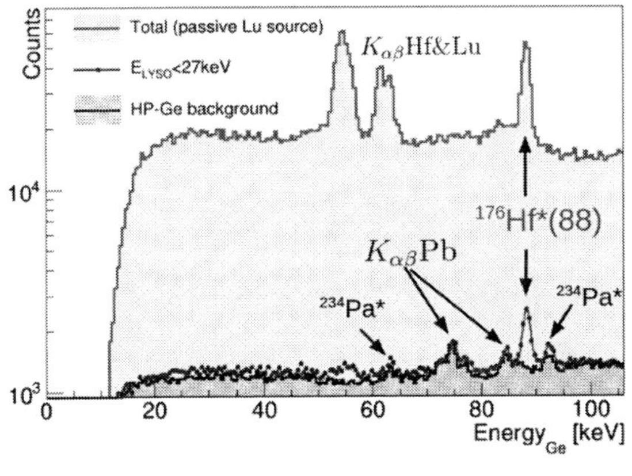

Figure 3. Energy distribution measured by the HP-Ge with no selections (magenta line) and selecting E_{LYSO} <27keV (blue points). A strong suppression of β decay background is achieved and the residual spectrum is dominated by the intrinsic/environmental background characteristic of our HP-Ge detector (black points).

2 Search for ^{176}Lu EC/REC

The Q-value for EC decay of ^{176}Lu (J^π= 7$^-$) to the ^{176}Yb ground state (J^π= 0$^+$) is ≃ 109 keV [8, 13, 14], and the one to the ^{176}Yb first excited state (J^π= 2$^+$) is ≃ 27 keV. These EC decay branches, however, would be forbidden transitions of 7th and 5th order respectively. The relative branching ratios are then expected to be very small, also considering that forbidden transitions of order >4th have never been experimentally observed so far [15]. In these cases of highly forbidden transitions, that are suppressed by a large variation of the nucleus spin as compared with the sum of the angular momentum of the captured electron and of the emitted neutrino, an important channel for the decay is provided by the Radiative Electron Capture (REC) [16]. In this process an x-ray can be radiated either by the captured electron (Internal Bremsstrahlung) or by the father/daughter nucleus (Detour Transitions). This photon contributes with his spin to the overall angular momentum conservation. The REC decay has been experimentally observed for the forbidden decay of: ^{41}Ca, ^{59}Ni, ^{81}Kr,^{137}La and ^{204}Tl [17–21].

2.1 Search for ^{176}Lu EC to $^{176}Yb^{2^+}$

Considering the 5th forbidden EC transition of ^{176}Lu to $^{176}Yb^{2^+}$, a strong identification signature is the 82.1keV de-excitation γ-ray observed in the HP-Ge detector. The maximum neutrino energy for this decay is $E_{82}^{\nu_{max}} = Q_{EC} - E_b - 82.1$keV, where E_b is the Yb atomic binding energy of the captured electron [22]. $E_{82}^{\nu_{max}} \geq 0$ therefore the capture of the 1s electron is not energetically allowed for this detection channel. The energy emitted in the Yb atomic shell rearrangement along with a possible REC x-ray are typically detected by the LYSO crystal.

The search for EC transition of ^{176}Lu can proceed by rejecting a large fraction of ^{176}Lu β-decay background by selecting an $Energy_{LYSO}$ window. This analysis procedure and results are summarized in [11]. In the following a complementary analysis approach is adopted. A LYSO energy distribution that is abundant in EC signals can be obtained by selecting E_{Ge} in the "Bulk" of the expected signal region [82.1keV$-1\sigma < E_{Ge} <$ 82.1keV$+1\sigma$] (see FIG. 2). By construction we expect that 68.3% of the total EC signal events belong to this [81.4 keV - 82.8 keV] energy range. We compare this last LYSO energy distribution with the one obtained by selecting E_{Ge} in the two "Sidebands" [82.1keV$-2\sigma < E_{Ge} <$ 82.1keV-1σ] and [82.1keV$+1\sigma < E_{Ge} <$ 82.1keV$+2\sigma$]. By construction only 13.6% of EC signal events lies in each sideband. The difference of the LYSO energy distributions obtained with "Bulk" and "Sidebands" events, will still contains 41% of the total number of the EC signal events, however the background contribution will be strongly suppressed[2].

Figure 4 shows the energy distributions measured by the LYSO detector for events selected in the "Bulk" range and in the "Sidebands", and also the residuals of the difference; no evidence for a statistically significant excess

[1]Natural radioactive chain of ^{238}U contains the ^{234}Pa isotope [12].

[2]A complete background rejection occurs when the energy distribution of the background is a linear function in this (narrow) region.

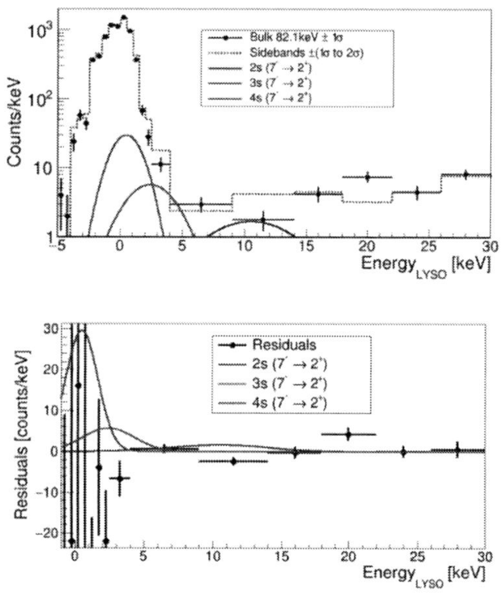

Figure 4. Top panel: Energy distributions measured by the LYSO detector for events selected in the [81.4keV< E_{Ge} < 82.8keV] range ($\pm 1\sigma$ "Bulk", data points) and in the two "Sidebands" (dashed lines). Bottom: residuals of the "Bulk"-"Sidebands" difference. The red, green and blue lines are the upper limits for the possible contribution of EC of the 4s, 3s and 2s electron of ^{176}Lu, respectively.

was found. Red, green and blue distributions shown in FIG. 4 are the 95% upper limits for the possible contribution of the Electron Capture of the 4s, 3s and 2s electrons, respectively; in particular the energy is emitted in the atomic rearrangement of the electron vacancy of the daughter, Yb, atom. Taking into account the 41% efficiency of this approach, the upper limits on the number of the ^{176}Yb$^{2^+}$ 82.1keV de-excitation γ-rays, N_{82}, are 210; 65 and 37 for EC of 4s; 3s and 2s electrons, respectively.

Following the approach of [5] an upper limit on the EC branching fraction can be obtained by comparing the upper limit on N_{82} with the number of events related with the β-decay branching fraction, thus evaluating the number of ^{176}Hf* events in the 88.3keV γ-peak in the HP-Ge event distribution (N_{88}=119.9 ± 0.5 × 10^3 events):

$$B_{82} = \frac{(1+\alpha_T^{82})N_{82}/\varepsilon_{88}}{(1+\alpha_T^{88})N_{88}} \tag{1}$$

where α_T^{82}=7.06 and α_T^{88}=5.86 are the total electron conversion coefficients for first excited levels of ^{176}Yb and ^{176}Hf, respectively [7], and $\varepsilon_{88}(82) \simeq 80\%$ is the detection efficiency loss of 82keV γ-ray with respect to 88keV γ-ray evaluated by Geant4 simulation and verified in this experiment by comparing the measured K_α and K_β peak amplitudes of Lu and Hf with the 88keV peak [11].

On the other hand, with a probability 18.1% the 82.1 keV level can decay with a K-shell internal conversion [23] and with a probability 15.1% a 59.3 keV K_β x-ray is emitted [1]. This is a different channel to search for the same EC decay to ^{176}Yb$^{2^+}$, providing additional (22.8 keV) energy released in the LYSO (see FIG. 5).

Table 1 is summarizing the upper limits obtained in the EC branching fraction by applying this analysis approach to the different decay channels. The expected distribution for the REC was evaluated by considering the approximations of [24] and assuming for ^{176}Lu a negligible contribution of Detour Transitions[3] [20, 21].

2.2 Search for ^{176}Lu EC to ^{176}Yb$^{0^+}$

The EC decay of ^{176}Lu to the fundamental state of ^{176}Yb is a 7th degree forbidden transition, thus it is expected to be suppressed with respect to the 5th degree forbidden EC decay in ^{176}Yb*; however if a ^{176}Lu nucleus captures a K-shell electron, the result must be a ^{176}Yb in the ground state. Considering that the K-shell atomic orbital is the one with the largest superposition with the nucleus, this possibility deserves a dedicated investigation.

A signature of the K-shell capture is provided by the x-rays emitted when the Yb atomic vacancy is filled. In particular, $K_{\alpha 2}$=51.35keV, $K_{\alpha 1}$=52.4keV and K_β=59.3keV are be emitted with a probability of 27.2%, 48.1% and 15.1%, respectively [1]. By comparing these x-ray energies with the measured energy distribution (see e.g. FIG. 3) it is clear that the Yb K_β line is the only one distant enough from the nearby Lu and Hf K-shell lines, to be easily identified. Moreover, due to self absorption in the LYSO crystal and in the HP-Ge dead layers, the expected detection efficiency for Yb K_α lines is quite small, whereas the Yb K_β line has $\varepsilon_{88} \simeq 100\%$ [11].

Figure 5 shows the Energy distributions measured by the LYSO detector for events selected in the "Bulk" range of Yb K_β and in the relative "Sidebands", and also the residuals of the difference; no evidence for a statistically significant excess was found. Thus, following the approach previously described, the upper limits on the number of the Yb K_β 59.3keV x-rays, N_{59}, are 90 and 50 for the possible contribution of EC or REC of the 1s electron of ^{176}Lu, respectively. The upper limit to the corresponding branching fraction is:

$$B_{59} = \frac{N_{59}/(\varepsilon_{59}\varepsilon_{88})}{(1+\alpha_T^{88})N_{88}} \tag{2}$$

where ε_{59}=15.1% is the probability of a K_β x-ray emission filling an Yb K-shell vacancy and $\varepsilon_{88} \simeq 100\%$ is the detection efficiency of 59keV x-ray with respect to 88keV.

3 Discussion and conclusions

The search for Electron Capture in ^{176}Lu is an opportunity for a laboratory measurement of 5th degree forbidden processes that may provide valuable information for nuclear theory models. Our active source technique, that uses a LYSO crystal scintillator also as detector, allows a powerful reduction of the background from ^{176}Lu β-decay and the HP-Ge intrinsic/external contamination.

[3]This cautiously minimizes the emitted energy in the REC signal distribution [11].

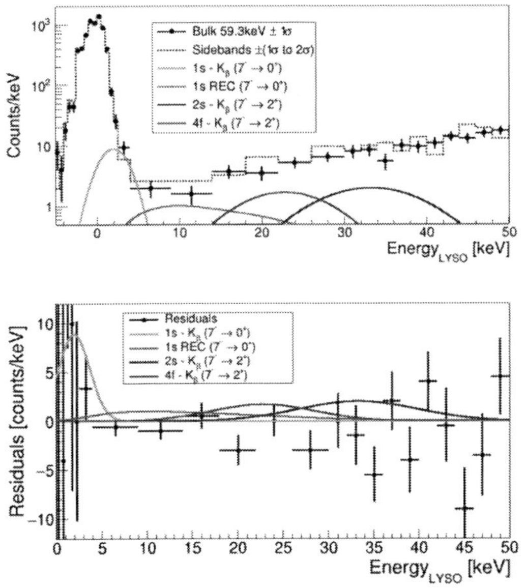

Figure 5. Top panel: Energy distributions measured by the LYSO detector for events selected in the [58.6keV< E_{Ge} < 60keV] range ($\pm 1\sigma$ "Bulk", data points) and in the two "Sidebands" (dashed lines). Bottom: residuals of the "Bulk"-"Sidebands" difference. The green and magenta lines are the upper limits for the possible contribution of EC or REC of the 1s electron of ^{176}Lu, whereas red and blue lines are the upper limits for the contribution of EC to $^{176}Yb^{2+}$ for 4f and 2s, respectively.

Table 1. Upper limits on the branching ratio (B.R.) for specific Electron Capture channels of ^{176}Lu.

EC/REC channel	B.R. Limit (95% C.L.)	Previous Limit (68% C.L.) [5]
1s EC + 59keV	0.07%	0.36%
1s REC + 59keV	0.04%	
2s/2p EC + 82keV	0.044%	0.45%
3s/3p EC + 82keV	0.083%	
3d EC + 82keV	0.11%	
4s EC + 82keV	0.26%	
4f EC + 59keV	0.26%	
2s REC + 82keV	0.06%	
2p REC + 82keV	0.05%	
3/4s REC + 82keV	0.055%	
3p REC + 82keV	0.075%	

No evidence for the EC process was found. Table 1 reports our estimation of the upper limits for the different EC channels. Improvements in sensitivity of 2-3 order of magnitude could be possible, in principle, by adopting the techniques of low-background experiments. In particular, by considering the possibility to operate small LYSO crystals as cryogenic bolometers, a very good signature of EC peaks in the LYSO spectrum is also expected [25].

References

[1] S. Chu, L. Ekström, R. Firestone, *Table of radioactive isotopes, http://nucleardata.nuclear.lu.se/toi/*

[2] M. Laubenstein, B. Lehnert, S.S. Nagorny, S. Nisi, K. Zuber, Phys. Rev. C **99**, 045501 (2019)

[3] F.A. Danevich, M. Hult, D.V. Kasperovych, V.R. Klavdiienko, G. Lutter, G. Marissens, O.G. Polischuk, V.I. Tretyak, Phys. Rev. C **102**, 024319 (2020)

[4] D. Münstermann, K. Zuber, Journal of Physics G: Nuclear and Particle Physics **29**, B1 (2003)

[5] E. Norman, E. Browne, I. Goldman, P. Renne, Applied Radiation and Isotopes **60**, 767 (2004)

[6] B. Lehnert, M. Hult, G. Lutter, K. Zuber, Phys. Rev. C **95**, 044306 (2017)

[7] IAEA, *Nuclear data https://www-nds.iaea.org* (2022)

[8] BNL, *https://www.nndc.bnl.gov/qcalc/* (2023)

[9] Saint-Gobain, LYSO *https://www.crystals.saint-gobain.com/radiation-detection-scintillators/crystal-scintillators/lyso-scintillation-crystals*

[10] Mirion, GC2020 *https://www.mirion.com/products/sege-standard-electrode-coaxial-ge-detectors*

[11] L.E. Ghezzer, F. Nozzoli, R. Nicolaidis, R. Iuppa, P. Zuccon, C. De Santis, Phys. Rev. C **107**, 045504 (2023)

[12] B. Mauz, P.J. Nolan, P.G. Appleby, Geochronology **4**, 213 (2022)

[13] W. Huang, M. Wang, F. Kondev, G. Audi, S. Naimi, Chinese Physics C **45**, 030002 (2021)

[14] M. Wang, W. Huang, F. Kondev, G. Audi, S. Naimi, Chinese Physics C **45**, 030003 (2021)

[15] B. Singh, J. Rodriguez, S. Wong, J. Tuli, Nuclear Data Sheets **84**, 487 (1998)

[16] L. Kalinowski, Z. Janas, M. Pfützner, A. Płochocki, P. Hornshøj, H. Nielsen, Nuclear Physics A **537**, 1 (1992)

[17] B. Mysłek, Z. Sujkowski, J. Żylicz, Nuclear Physics A **215**, 79 (1973)

[18] Z. Janas, M. Pfützner, A. Płochocki, P. Hornshøj, H. Nielsen, Nuclear Physics A **524**, 391 (1991)

[19] S. Mianowski, E. Werner-Malento, A. Korgul, M. Pomorski, K. Pachucki, M. Pfützner, B. Szweryn, J. Żylicz, P. Hornshøj, T. Nilsson et al., Phys. Rev. C **82**, 044308 (2010)

[20] M. Pfützner, P. Hornshøj, H. Nielsen, J. Kurpeta, A. Płochocki, Nuclear Physics A **611**, 270 (1996)

[21] J. Kurcewicz, M. Pfützner, P. Hornshøj, H. Nielsen, B. Szweryn, Nuclear Physics A **728**, 3 (2003)

[22] J.A. BEARDEN, A.F. BURR, Rev. Mod. Phys. **39**, 125 (1967)

[23] T. Kibédi, T. Burrows, M. Trzhaskovskaya, P. Davidson, C. Nestor, Nucl. Instr. and Meth. A **589**, 202 (2008)

[24] R.J. Glauber, P.C. Martin, Phys. Rev. **104**, 158 (1956)

[25] S. Fu, et al., *https://ltd20.genimice.com//paper/ preview.do?idx=1237* (LTD20 *conference, 2023)*

Commissioning of miniBELEN-10A, a moderated neutron counter with a flat efficiency for thick-target neutron yields measurements

N Mont-Geli[1,*], *A* Tarifeño-Saldivia[2,**], *L M* Fraile[3], *S* Viñals[4], *A* Perea[5], *M* Pallàs[1], *G* Cortés[1], *G* Garcia[4], *E* Nácher[2], *J L* Tain[2], *V* Alcayne[6], *O* Alonso-Sañudo[3], *A* Algora[2], *J* Balibrea-Correa[2], *J* Benito[3], *M J G* Borge[5], *J A* Briz[5], *F* Calviño[1], *D* Cano-Ott[6], *A* De Blas[1], *C* Domingo-Pardo[2], *B* Fernández[7,8], *R* Garcia[1], *J* Gómez-Camacho[7,8], *E M* González-Romero[6], *C* Guerrero[7,8]*J* Lerendegui-Marco[2], *M* Llanos[3], *T* Martínez[6], *V* Martínez-Nouvilas[3], *E* Mendoza[6], *J R* Murias[3], *S E A* Orrigo[2], *A* Pérez de Rada[6], *V* Pesudo[6], *J* Plaza[6], *J M* Quesada[7], *A* Sánchez[6], *V* Sánchez-Tembleque[3], *R* Santorelli[6], *O* Tengblad[6], *J M* Udías[3], and *D* Villamarín[6]

[1]Institut de Tècniques Energètiques (INTE), Universitat Politècnica de Catalunya (UPC), E-08028, Barcelona, Spain
[2]Instituto de Física Corpuscular (IFIC), CSIC – Univ. Valencia (UV), E-46071, Valencia, Spain
[3]Grupo de Física Nuclear (GFN) and IPARCOS, Universidad Complutense de Madrid (UCM), E-28040, Madrid, Spain
[4]Centro de Micro-Análisis de Materiales (CMAM), Universidad Autónoma de Madrid (UAM), E-28049, Madrid, Spain
[5]Instituto de Estructura de la Materia (IEM), CSIC, E-28006 Madrid, Spain
[6]Centro de Investigaciones Energéticas, Medioambientales y Tecnológicas (CIEMAT), E-28040, Madrid, Spain
[7]Departamento de Física Atómica, Molecular y Nuclear, Universidad de Sevilla (US), E-41012 Sevilla, Spain
[8]Centro Nacional de Aceleradores (CNA), Universidad de Sevilla (US) - J. Andalucía - CSIC, E-41092, Sevilla, Spain

Abstract. miniBELEN-10A is a modular and transportable moderated neutron counter with a nearly flat detection efficiency up to 8 MeV. The detector was designed to carry out measurements of (α, n) reactions in the context of the Measurement of Alpha Neutron Yields (MANY) project. In this work we present the results of the commissioning of miniBELEN-10A using the relatively well-known thick-target neutron yields from $^{27}Al(\alpha, n)^{30}P$.

1 Introduction

The production of neutrons through α-induced reactions plays a crucial role in many different fields. Specifically, in nuclear astrophysics, (α, n) reactions are the major source of neutrons for the slow neutron-capture nucleosynthesis process (the s-process) [1] and are involved in the weak rapid neutron-capture process (named the weak r-process, also known as the α-process) [2]. Moreover, (α, n) reactions are important in underground physics, as they are one of the main sources of the neutron-induced background in underground facilities, which is typically a crucial issue for the experiments carried out there [3]. Other fields of interest include nuclear technologies such as fission [4] and fusion reactors [5] and non-destructive assays for non-proliferation and spent fuel management applications [6, 7].

All these applications rely on accurate and precise experimental data. However, most of the available data on (α, n) reactions was measured decades ago, is incomplete or presents large discrepancies not compatible with the declared uncertainties. Furthermore, in many cases the uncertainties on the cross-sections and neutron energy spectra are rather large. To address such problems new measurements are required [3]. To that end the Measurement of Alpha Neutron Yields (MANY) collaboration was formed. The project relies on the employment of the currently existing infrastructure in Spain, in particular the α-particle beams produced by the accelerators at CMAM (Madrid) [8] and CNA (Sevilla) [9]; and on the use of neutron detection systems such as miniBELEN, which is a moderated neutron counter based on the use of ^{3}He-filled detectors, and MONSTER, which is a time-of-flight neutron spectrometer based on the use of BC501/EJ301 liquid scintillation modules [10]. Both systems are complemented by γ-spectroscopy measurements using an array of fast $LaBr_3(Ce)$ scintillation detectors of the FATIMA type [11] with angular resolution capabilities and a high-purity germanium detector.

2 miniBELEN

The miniBELEN detector builds upon the experience gained from previous experiments involving other detectors of the BELEN-type [12–14]. It is a 4π moderated neutron counter which is based on the use of a set of ^{3}He-filled thermal neutrons proportional counters embedded in a modular block of High-Density PolyEthylene (HDPE) which acts as moderator. Modularity means that the HDPE moderator is formed by smaller blocks which can be re-assembled in different ways in order to provide the detector with different types of response. In practice, in the case

*e-mail: nil.mont@upc.edu
**e-mail: atarisal@ific.uv.es

https://doi.org/10.1051/epjconf/202329001003

of miniBELEN, this means that we have three detectors or configurations for the prize of one. These configurations have been designed in order to present a neutron detection efficiency nearly independent from the neutron energy, namely a flat efficiency, up to 8 MeV with a slightly negative slope up to 10 MeV.

The basic components of miniBELEN are the 7x10x70 cm^3 HDPE blocks with a central hole where the neutron counters (60 cm length and 1 inch diameter) are embedded. Each of these blocks is made of seven smaller 7x10x10 cm^3 blocks which are assembled using two stainless-steel rods as it is shown in figure 1.

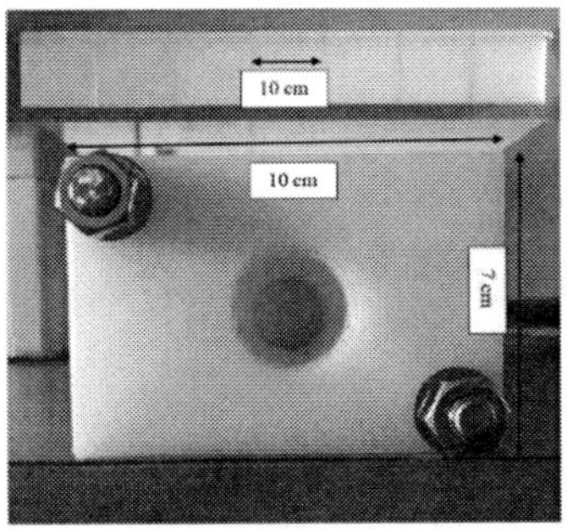

Figure 1. HDPE block from miniBELEN (7x10x70 cm^3) with central hole to be filled with a neutron counter. Each of these blocks is made of seven smaller blocks (7x10x10 cm^3) which are assembled using two stain-less steel rods.

In miniBELEN the flat efficiency is achieved by partially covering some of the neutron counters with cadmium filters in order to optimize the contribution of each ring (i.e., a group of counters which are all placed at the same distance from the centre of the detector) to the total detection efficiency (see figure 2). This is what we call the composition method.

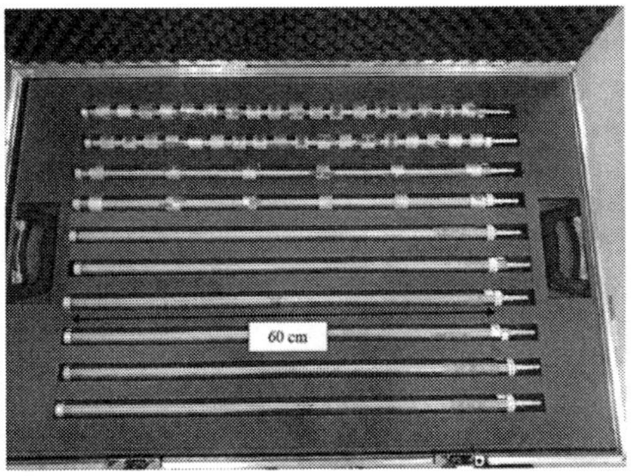

Figure 2. The ten ^{3}He-filled neutron proportional counters of miniBELEN-10A (60 cm length and 1 inch diameter). Some of the counters are partially covered with cadmium filters (2 cm length and 0.5 mm thickness) in order to provide the detector with a flat neutron efficiency.

In his work we present the results from the commissioning of miniBELEN-10A, which is a configuration of the miniBELEN detector using ten cylindrical ^{3}He-filled proportional counters manufactured by LND [15]. The gas pressure is generally 10 atm although in three counters of the outer ring the gas pressure is 4, 8 and 20 atm. Figure 3 shows a schematic plot of the modular HDPE moderator used in miniBELEN-10A.

3 Detector efficiency

Monte Carlo calculations were carried out using *Particle Counter* [16], which is a Geant4 application [17], in order to determine the neutron detection efficiency of miniBELEN-10A. These calculations were carried out using a point-like and isotropic neutron source placed at the center of the detector. Results are shown in figure 4 in the form of a 0.5 MeV bin-size histogram.

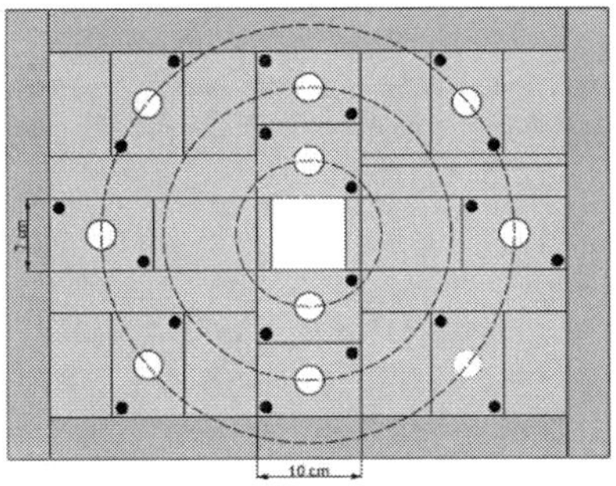

Figure 3. Schematic plot of the HDPE moderator of miniBELEN-10A. The total volume is 58x43x70 cm^3. The length of the red blocks is 50 cm while the length of the blue blocks is 70 cm. The area of the central hole (white) is 7x7 cm^2. The white circles represent the position of the proportional counters and the black circles represent the stainless-steel rods. Dashed circles: rings of counters.

The nominal efficiency of miniBELEN-10A (ε) is defined as the average efficiency within a certain energy range (from 0 up to E_{Max}),

$$\varepsilon = \frac{1}{M}\sum_{k=0}^{E_{Max}} \varepsilon_k = 0.0705 \qquad (1)$$

Being ε_k the detection efficiency of bin k and M the total number of bins. In the case of miniBELEN-10A the nominal efficiency is calculated up to E_{Max} = 8 MeV. It is worth noting that the alpha-energies involved in this experiment were not larger than 9.3 MeV and that the Q-value of ^{27}Al$(\alpha,n)^{30}$P is 2.64241(8) MeV. Consequently, the maximum energy reached by the emitted neutrons will always be lower than 8 MeV.

In order to take into account the efficiency fluctuations, the uncertainty of the nominal efficiency ($\delta\varepsilon$) is defined in the following way,

$$\delta\varepsilon = \max(\varepsilon_k^{max} - \varepsilon, \varepsilon - \varepsilon_k^{min}) = 0.0052 \qquad (2)$$

Being ε_k^{max} and ε_k^{min}, respectively, the maximum and minimum values of the neutron detection efficiency between 0 and E_{Max}.

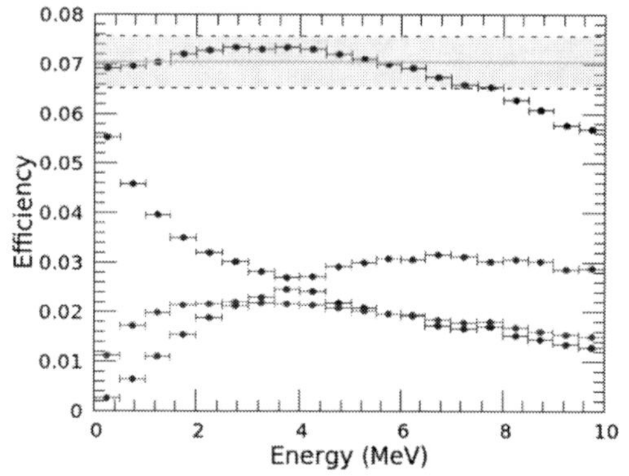

Figure 4. Monte Carlo calculated neutron detection efficiency of miniBELEN-10A. The y-axis error-bars are smaller than the size of the data points. Black: total efficiency. Red: efficiency of ring 1. Blue: ring 2. Green: ring3. Orange line: nominal efficiency calculated according to equation 1. Grey band (limited by the black dashed lines): nominal error-band calculated according to equation 2.

An experimental characterization of miniBELEN-10A was carried out using several ^{252}Cf neutron sources. Due to the lack of a well-calibrated neutron source the detection efficiency was determined using the Neutron Multiplicity Counting (NMC) technique [18, 19]. Relative discrepancies between the experimental efficiencies and Monte Carlo simulations using a point-like and isotropic neutron source were not larger than 5%. In these simulations the ^{252}Cf neutron emission energy spectrum from reference [20] was used.

4 Commissioning of miniBELEN-10A

The commissioning was carried out by measuring the relatively well-known ^{27}Al$(\alpha, n)^{30}$P thick-target neutron yields at Centro de Micro-Análisis de Materiales (CMAM, Madrid, Spain). The alpha-particles beam (charge +2e) was accelerated using the 5 MV tandem accelerator at CMAM and was then passed through a stainless steel cylindrical collimator which also acted as secondary electrons suppression system after applying a -100 V bias voltage. The collimator was covered with a natural tantalum (^{181}Ta) thick foil in order to avoid the production of neutrons. It must be taken into account that the (α, n) threshold for natural tantalum is 10.009(5) MeV.

The target consisted on a high-purity (99.9995%) natural aluminium (^{27}Al) thick foil which was placed at the center of the detector immediately after the collimator. The beam-current was measured by collecting the charge reaching the target and integrating it with an ORTEC-439 digital current integrator. The data acquisition system (DACQ) is based on the use of a SIS3316 sample digitizer (16 channels and 250 MHz sampling frequency) from Struck Innovative Systeme. The digitizer is controlled by the *Gasific 7.0* software [21].

Figure 5. Experimental setup at CMAM with miniBELEN-10A on the foreground and the 5MV tandem accelerator behind it.

4.1 Thick-target yields

At a given beam-energy E_α, the thick-target neutron production yields $Y(E_\alpha)$ can be determined in the following way,

$$Y(E_\alpha) = \frac{\sum_{i=1}^{10} r_i}{\varepsilon I} \tag{3}$$

Being r_i the neutron counting rate in counter i, ε the nominal neutron detection efficiency and I the beam-current expressed in particles per time unit.

The neutron background was estimated by carrying out measurements with a dummy target (natural tantalum) at 3.6, 3.7, 5.0, 5.5, 6.0, 7.4 and 8.5 MeV. A cubic spline interpolation of the data points was used in order to get the background yields at each energy relevant for the experiment. For measurements at beam-energies larger than 5 MeV the background contribution was found to be negligible (lower than 1%). For energies between 3.6 and 3.7 MeV the contribution of the background yields ranged from 16% up to 30%.

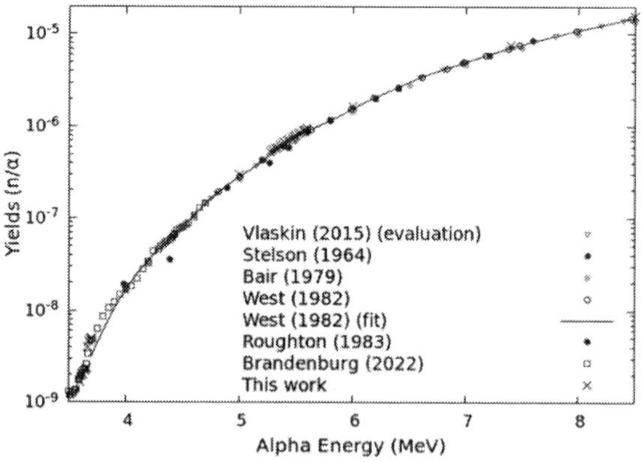

Figure 6. ^{27}Al$(\alpha, n)^{30}$P thick-target neutron yields. Blue-crosses: measurement with miniBELEN-10A. Circles: data from references [22–26]. Grey triangles: evaluation from reference [27]. Black line: cubic spline fit to data from reference [24].

When the production yields are large enough, dead-time effects become relevant and the counting rates in miniBELEN-10A, unless corrected, are underestimated. Our experience suggests that for counting rates below 10 kilo-counts per second and channel, the behavior of the DACQ can be modeled as a nonparalyzable system so that the correction model for such type of acquisition systems can be applied.

The thick-target neutron yields from ^{27}Al$(\alpha,n)^{30}$P measured with miniBELEN-10A are shown in figure 6. The relative errors are about 7% and are due to the uncertainty of the nominal efficiency. The plot also includes data from previous measurements carried out by other groups (via direct neutron counting [22–24, 26] or activation [25]) and the evaluation from Vlaskin *et al.* [27]. Our measurements are consistent with most of the previously existing experimental data within the margin of uncertainty.

5 Summary and final remarks

miniBELEN is a modular moderated neutron counter based on the use of ^{3}He-filled detectors with a nearly flat response up to 8 MeV with a slightly negative slope up to 10 MeV. Modularity means that the moderator can be reassembled in different ways in order to obtain different types of response. In this work we have presented the results of the commissioning of configuration miniBELEN-10A through the measurement of the relatively well-known thick-target neutron yields from ^{27}Al$(\alpha,n)^{30}$P. Assuming a flat nominal efficiency determined through Monte Carlo calculations, we have been able to reproduce the results from previous experiments carried out by other groups.

This work has been supported by the Spanish Ministerio de Economía y Competitividad under grants FPA2017-83946-C2-1 and C2-2, PID2019-104714GB-C21 and C22, PID2021-126998OB-I00 and RTI2018-098868-B-I00, the Generalitat Valenciana Grant PROMETEO/2019/007, both cofounded by FEDER (EU), and the SANDA project funded under H2020-EURATOM-1.1 Grant No. 847552. The authors acknowledge the support from Centro de Micro-Análisis de Materiales (CMAM) - Universidad Autónoma de Madrid, for the beam time proposal (*Commissioning of neutron detection systems for (α,n) reaction measurements*) with code P01156, and its technical staff for their contribution to the operation.

References

[1] F. Käppeler, R. Gallino, S. Bisterzo, W. Aoki, Rev. Mod. Phys. **83**, 157 (2011)
[2] J. Bliss, A. Arcones, F. Montes, J. Pereira, J. Phys. G Nucl. Part. Phys. **44**, 054003 (2017)
[3] S. Westerdale, A. Junghans, R. de Boer et al., Tech. Rep. INDC(NDS)-0836, International Atomic Energy Agency (2022)
[4] S.V. Bedenko, N. Ghal-Eh, I.O. Lutsik, I.V. Shamanin, Appl. Radiat. Isot. **147**, 189 (2019)
[5] C.J. Cerjan, L. Bernstein, L.B. Hopkins et al., J. Phys. G Nucl. Part. Phys. **45**, 033003 (2018)
[6] D.P. Broughton, S. Croft, C. Romano, A. Favalli, Nucl. Instrum. Methods. Phys. Res. A **1009**, 165485 (2021)
[7] C.E. Romano, D.A. Brown, S. Croft et al., Tech. Rep. ORNL/TM-2020/1789, Oak Ridge National Laboratory (2020)
[8] A. Redondo-Cubero, M.J.G. Borge, N. Gordillo et al., Eur. Phys. J. Plus **136**, 1 (2021)
[9] J. Gómez-Camacho, J.G. López, C. Guerrero et al., Eur. Phys. J. Plus **136**, 273 (2021)
[10] A.R. Garcia, T. Martínez, D. Cano-Ott et al., JINST **7**, C05012 (2012)
[11] V. Vedia, M. Carmona-Gallardo, L.M. Fraile et al., Nucl. Instrum. Methods. Phys. Res. A **857**, 98 (2017)
[12] M.B. Gomez-Hornillos, J. Rissanen, J.L. Taín et al., J. Phys. Conf. Ser. **312**, 052008 (2011)
[13] J. Agramunt, A.R. Garcia, A. Algora et al., Nucl. Data Sheets **120**, 74 (2014)
[14] A. Tarifeño-Saldivia, J.L. Tain, C. Domingo-Pardo et al., JINST **12**, P04006 (2017)
[15] https://www.lndinc.com/
[16] https://www.particlecounter.net//
[17] S. Agostinelli, J. Allison, K. Amako et al., Nucl. Instrum. Methods. Phys. Res. A **506**, 250 (2003)
[18] D. Henzlova, A. Favalli, S. Croft, Metrologia **56**, 015017 (2019)
[19] D.G. Langner, J.E. Stewart, M.M. Pickrell et al., Tech. Rep. LA-13422-M, Los Alamos National Laboratory (1998)
[20] R. Radev, T. McLean, Tech. Rep. LLNL-TR-664160, Lawrence Livermore National Laboratory (2014)
[21] J. Agramunt, J.L. Tain, M.B. Gómez-Hornillos et al., Nucl. Instrum. Methods Phys. Res. A **807**, 69 (2016)
[22] P.H. Stelson, F.K. McGowan, Phys. Rev. **133**, B911 (1964)
[23] J.K. Bair, J. Gomez del Campo, Nucl. Sci. Eng. **71**, 18 (1979)
[24] D. West, A.C. Sherwood, Ann. Nucl. Energy **9**, 551 (1982)
[25] N.A. Roughton, T.P. Intrator, R.J. Peterson et al., At. Data Nucl. Data Tables **28**, 341 (1983)
[26] K. Brandenburg, G. Hamad, Z. Meisel et al., Nucl. Sci. Eng. **197**, 1 (2022)
[27] G.N. Vlaskin, Y.S. Khomyakov, V.I. Bulanenko, At. Energy **117**, 357 (2015)

Warsaw Active-Target TPC: a new detector for photonuclear reactions studies at astrophysical energies

Mikołaj Ćwiok[1,*], *Wojciech* Dominik[1], *Aleksandra* Fijałkowska[1], *Mateusz* Fila[1], *Zenon* Janas[1], *Artur* Kalinowski[1], *Krzysztof* Kierzkowski[1], *Magdalena* Kuich[1], *Chiara* Mazzocchi[1], *Wojciech* Okliński[1], *Piotr* Podlaski[1], *Marcin* Zaremba[1], *Moshe* Gai[2], *Sarah R.* Stern[2], *Deran K.* Schweitzer[2], and *Dimiter L.* Balabanski[3]

[1]Faculty of Physics, University of Warsaw, Poland
[2]LNS at Avery Point, University of Connecticut, CT, USA
[3]Extreme Light Infrastructure - Nuclear Physics, IFIN-HH, Magurele, Romania

Abstract. An active-target time-projection chamber (TPC) optimised for studying nuclear reactions of astrophysical interest was developed by the University of Warsaw for studying photo-disintegration reactions using intense, monochromatic γ-ray beams. Different reactions can be studied by tuning composition and density of the gaseous target for particular energy of the gamma beam. The Warsaw TPC detector, with its active volume of about $33\times20\times20\,\mathrm{cm}^3$ centred around the beam axis and micro-pattern structures (GEMs) employed to amplify the primary ionisation induced by charged particles produced in reactions in the gaseous target, is characterised by a readout based on signal strips, arranged into 3-coordinate redundant system. The 3D kinematics of charged particles in the event are reconstructed from a total of about 10^3 channels read out by digitising front-end electronics based on the General Electronics for TPCs (GET). In this paper the principles of detector operation and basic track reconstruction methods are discussed and illustrated by means of preliminary results from pilot measurements conducted in 2021-22.

1 Introduction

Modelling of the process that govern stars evolution requires as input, among others, the reaction rate per particle pair for the thermonuclear reactions that take place in the stars core. These rates are energy (temperature) dependent and directly proportional to the reaction cross section, which decreases exponentially with decreasing energy. The cross section has to be measured at the same energies at which the reactions happen in the stars (i.e. at the so-called Gamow peak) [1]. Since the relevant energies are well below the Coulomb barrier, these thermonuclear reactions are characterised by extremely low cross sections, often beyond the reach of modern-day experiments. Typically, experimental knowledge is gathered for as low energies as possibly achieved and extrapolations by means of R-matrix models are employed to extrapolate the cross section at the Gamow peak. It is therefore necessary to measure the cross section for astrophysically-relevant reactions at as low energies and as accurately as possible in order to constrain the R-matrix fit. Among the most important reactions that take place in the core of a star are direct-capture reactions, namely (α,γ) and (p,γ). From the experimental point of view, the tiny rates that characterise these reactions can be increased significantly by taking advantage of the principle of detailed balance [2]. One of the key problems in nucleosynthesis is how to understand or explain the ratio between oxygen and carbon in the Universe. This ratio is governed by one reaction in particular, $^{12}\mathrm{C}(\alpha,\gamma)^{16}\mathrm{O}$, at centre-of-mass energies below $\sim$2 MeV, with the Gamow peak for helium-burning temperatures being at $\sim$300 keV.

In this work, cross sections for nuclear reactions at astrophysical energies are studied by means of the inverse photo-disintegration reactions, with the flagship being $^{16}\mathrm{O}(\gamma,\alpha)^{12}\mathrm{C}$. The goal is to investigate it at the lowest possible energies, i.e. to about 1 MeV in the centre-of-mass. For this purpose, intense, monochromatic γ ray beams are needed, as well as a detection system able to measure the momenta of the charged reaction products. In comparison with direct capture measurements the reactions induced by γ-rays are characterised by inherently low experimental backgrounds, different systematic uncertainties and larger cross sections due to detailed balance principle of nuclear reactions.

An active-target Time Projection Chamber (TPC) detection technique is employed for momenta measurements and particle identification. In TPCs of this kind a gaseous active volume acts as both the target for nuclear reactions and as the medium for charge transport and amplification. Nowadays active target detectors with tracking capabilities of low-energy charged particles are adopted in more than a dozen experiments in nuclear and particle physics [3–5].

The Warsaw TPC detector was specifically designed to operate with lower-than-atmospheric gas pressures and is optimised for studying reactions induced by narrow neutral beams, such as γ-rays and neutrons. The experimental

*e-mail: Mikolaj.Cwiok@fuw.edu.pl

technique presented in this work results from a decade-long R&D programme carried out by the University of Warsaw (UW) in collaboration with the University of Connecticut, CT, USA and the Extreme Light Infrastructure Nuclear Physics / IFIN-HH, Bucharest-Magurele, Romania, to build and use in experiment a full-scale demonstrator of the TPC dedicated to ELI-NP [6–8].

2 Detector and data acquisition system

The apparatus consists of a low-pressure vessel, an internal TPC structure and external front-end electronics attached on top of the vessel (see Figure 1).

The internal detector structure is composed of: a drift cage, micro-pattern amplification structures and a segmented readout anode (see Figure 2). The electron drift cage consists of: a 4 mm-thick cathode plate made of aluminium and 12 field shaping 2 mm-thick aluminium electrodes having 16 mm pitch, for a total drift length of 196 mm, and with uniformity of the drift electric field better than 1% in the entire active volume. Primary ionisation charge is amplified by a stack of three 50 μm-thick Gas Electron Multiplier (GEM) foils [9] interlaced with 3 mm-wide transfer gaps. The induced signals are collected by a segmented multi-layer PCB anode. The active area is a 330×200 mm^2 rectangle with chamfered corners at 30° and consists of many tessellated diamond-shaped gold-plated pads (1 mm edge, *Via in Pad* technology). The pads are arranged along three main axes at 60° with respect to each other, and which form three independent linear sets of strips (U, V, W) with 1.5 mm pitch (see insert in Figure 2). Such a redundant planar 3-coordinate system allows for unambiguous creation of virtual pixels on a 2D plane for a given instant of time provided that measured track topologies are relatively simple, e.g. several straight tracks in a single recorded event. Most of the strips are split in half and read out by two independent electronic channels in order to further reduce ambiguity for more complex event topologies. The U-strips are parallel to the nominal gamma beam trajectory. In total 1018 electronics channels are being read out: 264 for U, 376 for V and 378 for W strips, respectively. The third cartesian coordinate along the drift electric field, $\vec{E}_{\rm drift}$, is determined from drift time assuming known electron drift velocity.

The TPC internal structure is immersed in a stainless steel vacuum vessel of 170 dm^3 volume with horizontally-oriented UVW readout PCB. The top lid and side walls of the vessel are equipped with several vacuum ports for installing high voltage and analogue signal feedthroughs, gamma beam windows and radioactive calibration sources. The vessel is connected to a low-pressure gas system consisting of: a high-pressure gas cylinder with a double-stage pressure regulator, an upstream mass-flow controller unit, two precise gas dosing valves at inlet and outlet ports, a set of vacuum and pressure gauges, a programmable Proportional-Integral-Derivative (PID) pressure control unit, a downstream dry four-stage diaphragm pump and a turbo-molecular pump. After initial evacuation of the vessel down to vacuum levels of 10^{-5} – 10^{-4} mbar, the detector is operated in an open loop flow mode with gas exchange rates varying between 1 and 4 detector volumes per day.

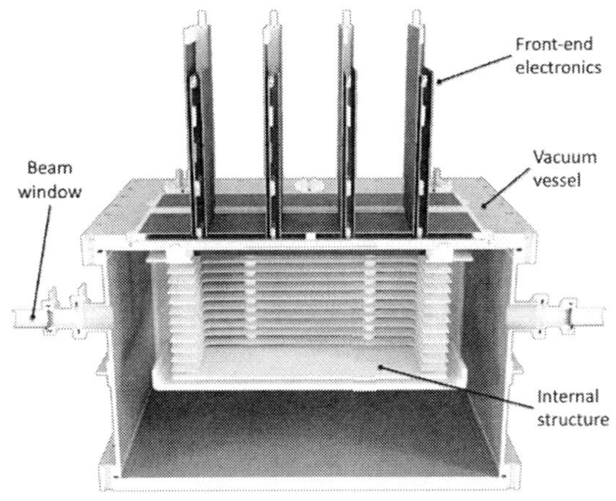

Figure 1. Half-section view of the Warsaw TPC detector.

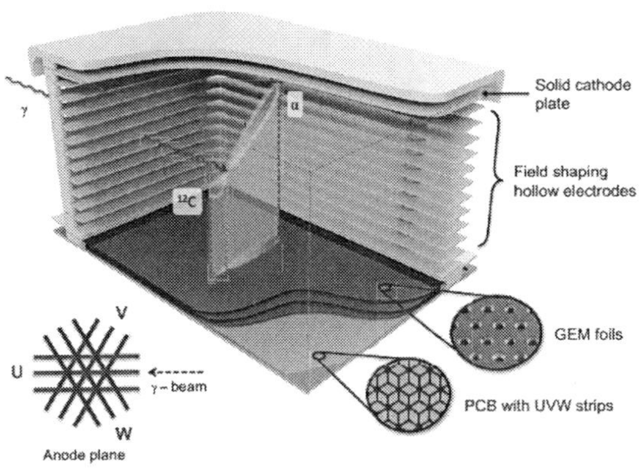

Figure 2. Details of internal structure of the Warsaw TPC with superimposed artistic rendering of ^{16}O photo-disintegration event. The concept of UVW readout strips is illustrated in the bottom-left insert.

The data acquisition (DAQ) system is based on the General Electronics for TPCs (GET) [10]. Analogue signals from the readout anode are extracted via 4 vacuum feedthroughs and routed to 4 commercial *ASIC and ADC* (AsAd) front-end cards equipped with custom-designed protection boards and electromagnetic shielding. Each AsAd card contains 4 *ASIC for GET* (AGET) chips with 64 active channels and 4 blinded channels for evaluating characteristic fixed-pattern noise. Inside each AGET chip signals undergo amplification and shaping stages, which are followed by a Switched Capacitor Array (SCA) acting as an analogue memory buffer with 512 time cells. Signal sampling frequency can be adjusted from 1 to 100 MS/s (in 7 steps), dynamic range from 120 fC to 10 pC (in 4 steps) and characteristic shaping time from 70 ns to 1 μs (in 16 steps). Upon receiving the event accept signal all stored charges are digitised by a 12-bit ADC and transmitted outside of the AsAd board to the main 1024-channel DAQ module for further processing. This module, referred as *z-CoBo*, was custom-designed and built at the Faculty of Physics, UW. It employs *System on Chip* architecture that combines the Field Programmable Gate Array (FPGA) part and the Hard Processor System (HPS) part in a single Xilinx Zynq-7045 chip. This novel solution offers similar functionality to the original Concentrator Board (CoBo) of GET electronics, which was based on the legacy Xilinx Virtex-5 chip. The DAQ module also provides event time-stamping and optional high-level triggering. The raw-data are transmitted by one (or more) 1 Gbit/s Ethernet link(s) from the z-CoBo to a dedicated DAQ server PC and then to a Network Attached Storage disk matrix. The maximal sustained trigger rate is limited to 80 Hz for a single Ethernet link and without zero-suppression of the data.

In case of semi-continuous beams or radioactive calibration sources the Warsaw TPC is operated in self-trigger mode, in which a collective analogue signal from the last GEM amplification stage is routed to the fast NIM electronics for further amplification, shaping and discrimination in order to form a $100\,\mu$s-wide gate signal for the DAQ module. The acquisition is then stopped after certain programmable delay so that the arrival time of the first charged particle track corresponds to 10% of the full time scale. In this mode the position of a track along detector's axis parallel to the drift electric field is determined only up to an additive constant. The discriminator's threshold is adjusted for a given TPC high-voltage working point for efficient detection of particles of the reaction of interest.

A dedicated Detector Control System (DCS) software framework allows for remote operation of the apparatus, including: setting TPC high-voltage working points, changing working gas pressure and periodical recording of various working conditions, such as: gas flow, pressures, voltages, currents and temperatures [11].

3 Experimental setup

The detector was assembled at the Faculty of Physics, UW in 2020. It was initially commissioned with two collimated ^{241}Am radioactive sources installed in close proximity to the active volume, namely: the first one located 6 mm behind the cathode plate and emitting α-particles parallel to $\vec{E}_{\text{drift}}$ through a tiny hole, and the second one located on the side, about 73 mm away from the active volume, that emitted α-particles perpendicular to $\vec{E}_{\text{drift}}$. These calibration measurements with weak radioactive sources allowed us to validate simulations of electron drift velocity and particle ranges.

The first measurement with a gamma beam was conducted in June 2021 at the Van der Graaf accelerator at the Institute of Nuclear Physics, Polish Academy of Sciences (IFJ-PAN), Cracow, Poland. In this experiment 13 MeV gamma photons were created in the ^{15}N(p, γ)^{16}O reaction by bombarding ^{15}NCr target deposited on a tantalum backing with a 1.03 MeV proton beam. The resulting gamma beam was shaped by means of a lead absorber inserted between the solid target and the brass beam entrance window to illuminate only the middle part of the drift cage. The detector was operated with a pure CO_2 gas kept at 250 mbar absolute pressure. Candidate events for ^{16}O(γ, p)^{15}N and ^{16}O(γ, α)^{12}C reactions were clearly observed [12].

Between April and September 2022 the second ^{16}O and ^{12}C photo-disintegration experiment was conducted at the High Intensity γ-Ray Source (HIγS) facility at the Triangle University Nuclear Laboratory (TUNL), Durham, NC, USA [13]. The γ-rays were produced in Compton back-scattering of a free-electron laser beam with a relativistic electron beam. A lead collimator with 10.5 mm aperture located about 60 m away from the γ production point provided almost collinear beam that fit inside the 14 mm-wide gap between two adjacent field shaping electrodes in the middle plane of the TPC drift cage. The detector was equipped with Kapton windows of 25 mm diameter at beam entrance and exit ports ($50\,\mu$m and $125\,\mu$m-thick, respectively).

The TPC vessel was aligned using a dedicated visible laser beam that was collinear with the nominal gamma beam trajectory. A refined detector alignment was done with the gamma beam using a set of lead plugs placed in front and rear ports of the detector and a gamma camera located downstream of the TPC vessel. The remaining residual beam tilt and offset are determined for each beam energy in the offline analysis step on the basis of vertex distribution on the UVW readout plane for candidate photo-disintegration events.

In total 15 beam energy points were measured during 275 h of the allocated beam time, ranging from 8.51 MeV to 13.9 MeV (nominal E_{CM} from 1.35 MeV to 6.7 MeV for ^{12}C(α, γ) reaction, respectively) [2]. For each energy the TPC operating conditions were optimised for charged particle ranges in detector's active volume. In particular, three pressure settings of pure CO_2 gaseous target were used during the measurements: 130, 190 and 250 mbar. The readout electronics was operated with: dynamic range of 120 fC, characteristic shaping time of 232 ns and two settings of the sampling frequency: 12.5 MHz or 25 MHz.

During data-taking the relative beam intensity was constantly monitored by a set of scintillation counters positioned before the TPC. For most of the scanned energy points the (γ, n) activation measurements of thin ^{197}Au discs placed at the exit window of the TPC were performed in order to determine absolute photon fluxes and to establish correlation with the scintillator-based method. For one beam energy a small photo-fission ionisation chamber with ^{238}U target was used at the rear TPC window as an alternative method to measure the absolute photon flux over time. Preliminary analysis of partial data yields beam intensities between $1 \cdot 10^8$ γ/s and $5 \cdot 10^8$ γ/s depending on the beam energy. The energy spectra of γ-rays were measured at 8-20 h intervals by means of a high-purity germanium detector located behind the TPC. The FWHM energy spread for the lowest studied energy point of E_γ=8.51 MeV was about 350 keV.

4 Event selection and reconstruction

All recorded raw-data waveforms were corrected for pedestals of individual electronic channels and for fixed-pattern-noise shape common for 64 channels of a given

https://doi.org/10.1051/epjconf/202329001004

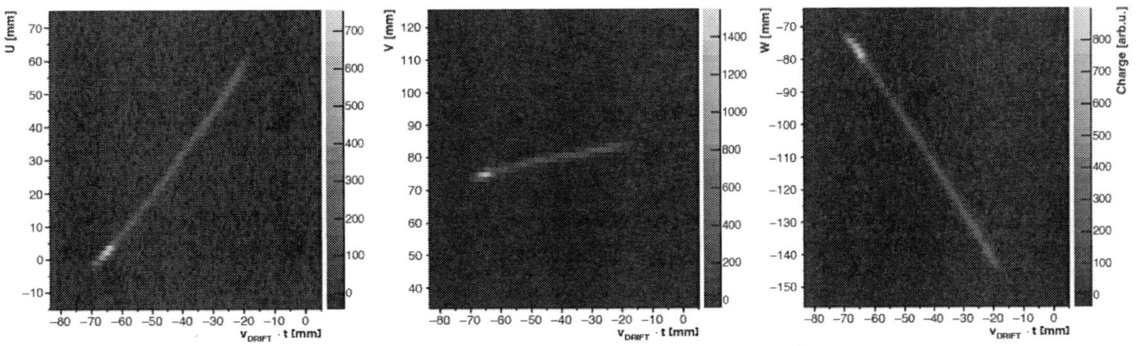

Figure 3. Example of a ^{16}O photo-disintegration candidate event collected at E_γ=13.9 MeV. Positions in strips on the anode plane and time bins along the drift direction are represented in millimetres assuming known strip pitch and electron drift velocity. A 90 mm × 90 mm region of interest is enlarged for each of U, V and W projections (p=250 mbar, 12.5 MHz sampling rate).

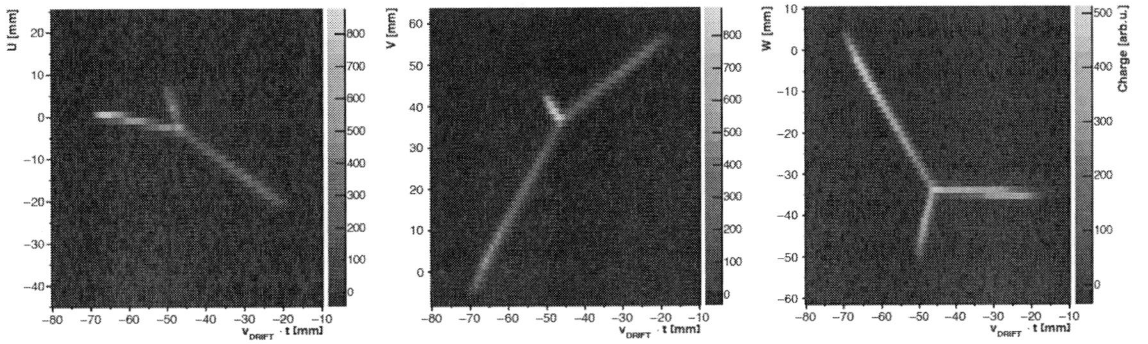

Figure 4. Example of a ^{12}C photo-disintegration candidate event collected at E_γ=13.9 MeV. Positions in strips on the anode plane and time bins along the drift direction are represented in millimetres assuming known strip pitch and electron drift velocity. A 70 mm × 70 mm region of interest is enlarged for each of U, V and W projections (p=250 mbar, 12.5 MHz sampling rate).

AGET chip in the front-end electronics. Afterwards 2D charge distributions of strip position vs. drift length were created separately for U, V and W strip arrays. Such distributions were visually inspected and collected events were assigned to categories according to their respective topology: 2- and 3-particles, background/noise. In Figures 3 and 4 the raw-data distributions from U, V and W strip arrays are depicted for a typical event from 2- and 3-particle category, respectively.

Background consisted of α particles from radon generated in the decay of ^{232}Th and ^{238}U isotopes that are present in welded seams of the vacuum vessel (amounting to ~0.15 Hz in total trigger rate), and of recoils following beam-induced (γ, n) reactions on the nuclei of CO_2 molecules. Such events appear as long bright tracks and short bright tracks ("dots"), respectively, and they are easily separated from the remaining events. Those events classified as 2- or 3-particles were analysed further to apply detector fiducial cuts and to determine the momenta of the particles in 3D. The fiducial cuts ensured that: (i) positions of the vertices on the UVW plane were inside a band defined by the gamma beam collimator with 2 mm safety margin; (ii) positions of the endpoints of all tracks were fully contained within the UVW active area with 5 mm safety margin; (iii) tracks were fully contained within the SCA memory buffer with a safety margin of at least 2 time cells; (iv) tracks did not interfere with the first 25 time cells of the SCA memory buffer used for calculating pedestals on event-by-event basis. Last but not least, the rate of observed pile-up effects with more than one gamma interaction in the same event was negligible.

Due to the presence of natural isotopes of oxygen and carbon in the CO_2 gaseous target, the 2-particle event category also contains admixture of $^{17,18}O(\gamma, \alpha)^{13,14}C$, $^{16}O(\gamma, p)^{15}N$ and $^{13}C(\gamma, \alpha)^{9}Be$ reactions. Such an isotopic contaminant background can be discriminated based on reaction Q-values. Separation energies for beam-induced background reactions relevant to the studied energy range are listed in Table 1. The 3-particle event category contains triple-alpha final states stemming from $^{12}C(\gamma, 3\alpha)$ reaction.

Two methods of track finding were used for preliminary analysis presented in this paper. The first one is based on human-based determination of track segments and the second one on automatic reconstruction, which is currently optimised only for identifying 2-particle events with back-to-back topology. In case of automatic procedure the charge deposits, *hits*, are grouped into clusters around *seed hits* that passed certain threshold level. The clustered hits are used to find the *approximate axis* of a 2-particle event in 3D space based on three 2D projections from U, V and W strip arrays. Afterwards all clustered hits are projected onto this approximate axis to form a 1D charge distribution. The resulting distribution is then fitted for a given reaction hypothesis with the sum of two templates of energy loss along the track, dE/dx, in order to fine-tune endpoints of the tracks and position of the common vertex in 3D space (see Figure 5). In the final step a parameterised dependence of the average range as a function of the kinetic energy for a given isotope and gas density is employed to determine scalar momenta of the particles. The dE/dx templates and particle ranges are obtained from Monte Carlo simulations [15, 16]. In order to overcome imperfections of the modelling of particles' ranges we plan to fine-tune the absolute energy scale *in situ* using known resonances in ^{16}O for the studied energy range, namely: a narrow 2^+ resonance at excitation energy of 9.845 MeV and three broader ones at excitation energies of 11.51 MeV (2^+), 12.45 MeV (1^-) and 13.09 MeV (1^-) [17].

5 Preliminary results

In this paper we present preliminary results from partial statistics of the collected data sample during the HIγS experiment, which demonstrate various technical capabilities of the Warsaw TPC, such as discriminating between reaction channels and measuring angular distributions. The beam enters the detector horizontally parallel to the anode plane and perpendicular to the $\vec{E}_{\text{drift}}$ vector (vertical). The coordinate system origin is the centre of the TPC drift cage. For momenta representation, a spherical coordinate system (BEAM) is used, with Z axis parallel to the beam, and with θ and ϕ being the polar and azimuthal angles, respectively. Assignment of the 2-particle event to the $^{16}\text{O}(\gamma,\alpha)^{12}\text{C}$ or to the $^{17,18}\text{O}(\gamma,\alpha)^{13,14}\text{C}$ reactions can be done by comparing the track lengths, since the reactions have Q-values that differ by about 1 MeV (see Table 1). In Figure 6 the histogram of the sum of measured α and carbon track lengths is shown for E_γ=8.66 MeV. The group of events below 50 mm corresponds to the $^{16}\text{O}(\gamma,\alpha)^{12}\text{C}$ decay channel, while the group of events above this threshold to the $^{17,18}\text{O}(\gamma,\alpha)^{13,14}\text{C}$. It can be seen that the $^{16}\text{O}(\gamma,\alpha)$ decay channel is well distinguished from the background reactions that dominate the rate of measured 2-particle events at this γ beam energy.

The angular distributions of particles emitted in the reactions can be reconstructed as well and used to improve reaction identification. An example of two-dimensional

Table 1. Separation energies in MeV for: neutrons, protons and alpha particles in ^{12,13}C and ^{16,17,18}O isotopes [14].

	^{12}C	^{13}C	^{16}O	^{17}O	^{18}O
S_n	18.722	4.946	15.664	4.144	8.044
S_p	15.957	17.533	12.127	13.780	15.942
S_α	7.367	10.648	7.162	6.359	6.227

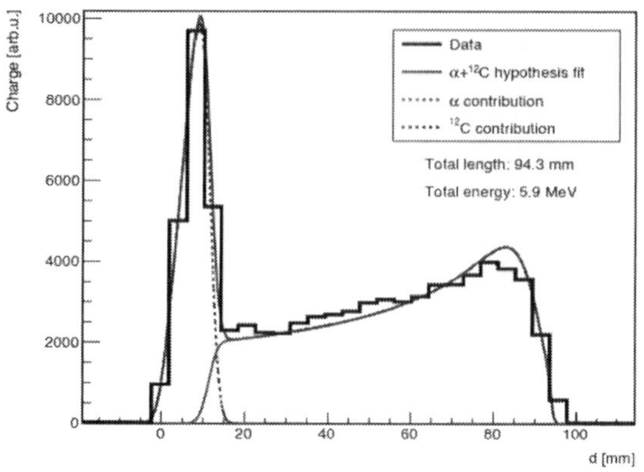

Figure 5. Fitted $\frac{dE}{dx}$ templates superimposed on the observed charge distributions along candidate tracks for 2-particle event shown in Figure 3 (E_γ=13.9 MeV, p=250 mbar, automatic reconstruction).

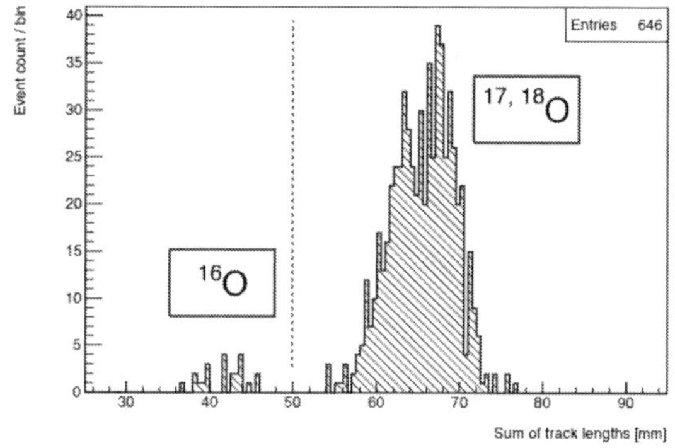

Figure 6. Example of 1D identification plot based on the sum of α and carbon track lengths (E_γ=8.66 MeV, p=130 mbar, manual reconstruction, partial statistics).

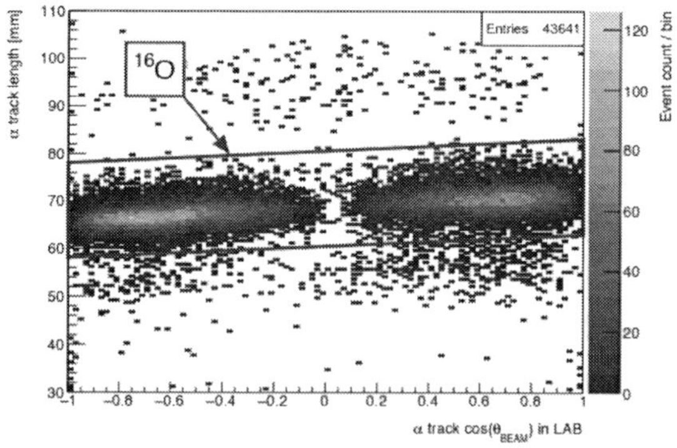

Figure 7. Example of 2D identification plot based on α-particle track length and its cosine of the polar angle θ (E_γ=11.5 MeV, p=190 mbar, automatic reconstruction, LAB reference frame).

selection cuts for ^{16}O photo-disintegration candidates at E_γ=11.5 MeV is shown in Figure 7, in which the track length of the α particle is plotted as a function of cosine of the polar angle θ in the laboratory reference frame. Events corresponding to $^{16}\text{O}(\gamma,\alpha)$ decay channel are contained in-

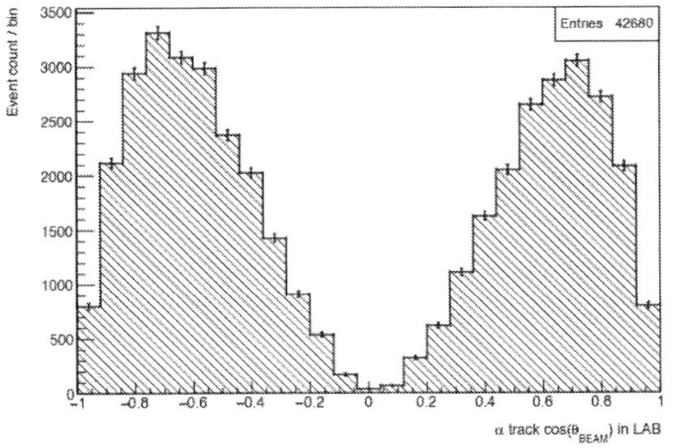

Figure 8. Distribution of cosine of the polar angle θ of α particles from $^{16}O(\gamma,\alpha)^{12}C$ candidate events after selection cuts shown in Figure 7 (E_γ=11.5 MeV, p=190 mbar, automatic reconstruction, LAB reference frame).

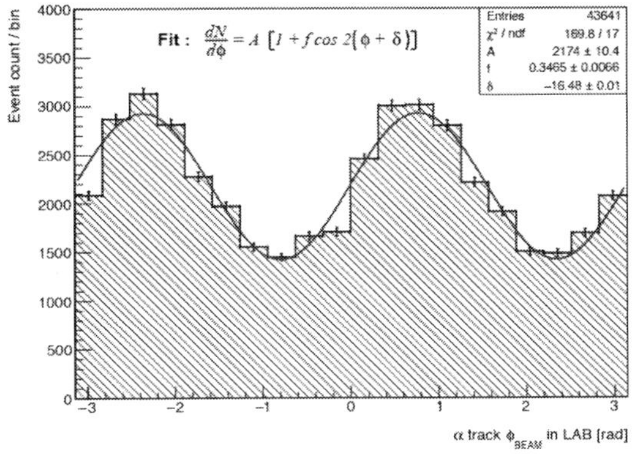

Figure 9. Distribution of azimuthal angle ϕ of α particles in 2-particle event category. The fit equation and parameters are shown in the figure (E_γ=11.5 MeV, p=190 mbar, automatic reconstruction, LAB reference frame).

side the indicated selection band. Such a cut results in better purity than the one based on α-particle track length alone. In Figure 8 the resulting distribution of cosine of the polar angle θ is shown for events tagged as $^{16}O(\gamma,\alpha)$ using 2D selection cut from Figure 7. The characteristic shape of E2 multipolarity can be observed as expected for the 2^+ resonance in ^{16}O at the excitation energy of 11.51 MeV [17].

The distributions of the azimuthal angle ϕ were used to characterise the gamma beam delivered to the TPC detector. An example of such distribution from the reconstructed momenta of 2-particle events is depicted in Figure 9 for E_γ=11.5 MeV. The observed shape corresponds to a gamma beam circularly polarised more than 90%, as expected from independent measurements of the polarisation of the laser beam [18].

6 Summary and outlooks

At the University of Warsaw an R&D program to design and construct active target time-projection chambers optimised for reactions induced by neutral beam (γ-rays and neutrons) has been developed over the last 10 years. The physics goals driving the project consist in the study of nuclear reactions of astrophysical interest at the relevant energies. The main features characterising the detector are illustrated as well as data processing. The first experiments with the full-scale demonstrator were conducted in 2021 and 2022 at IFJ-PAN, Cracow, Poland and HIγS at TUNL, Durham, NC, USA, respectively. The highlight of the experiments was the study of the $^{16}O(\gamma,\alpha)^{12}C$ reaction at energies below 2 MeV in the centre-of-mass. Full statistics of the collected data is under evaluation. New experiments are planned at HIγS and ELI-NP facilities after 2024 to explore even lower centre-of-mass regions using γ-ray beams of higher intensity.

The authors would like to thank the staffs of HIγS and IFJ-PAN for providing high-quality beams during the experiments, as well as engineers H. Czyrkowski and R. Dąbrowski (UW) for their excellent technical support. This scientific work is supported by the National Science Centre, Poland (contract no. UMO-2019/33/B/ST2/02176), by the University of Warsaw, Poland (Interdisciplinary Centre for Mathematical and Computational Modelling – computational allocation no. G89-1286 and Excellence Initiative Research University – IDUB program), by the US Department of Energy, Office of Science, Office of Nuclear Physics (grant no. DE-FG02-94ER40870) and by the Romanian Ministry of Research, Innovation and Digitalization (contract no. PN 23.21.01.06).

References

[1] W.A Fowler, Rev. Mod. Phys. **56**, 149 (1984)
[2] M. Ćwiok *et al.*, EPJ Web Conf. **279**, 04002 (2023)
[3] D. Bazin *et al.*, Progress in Particle and Nucl. Phys. **114**, 103790 (2020)
[4] P. Egelhof *et al.*, ACTAF TDR, FAIR/GSI (2016)
[5] B. Adams *et al.*, arXiv:1808.00848 [hep-ex]
[6] M. Ćwiok, Acta Phys. Pol. B **47**, 707 (2016)
[7] O. Tesileanu *et al.*, Romanian Rep. Phys. (Suppl.) **68**, S699 (2016)
[8] M. Ćwiok *et al.*, Acta Phys. Pol. B **49**, 509 (2018)
[9] F. Sauli, Nucl. Inst. and Methods A **386**, 531 (1997)
[10] E.C. Pollacco *et al.*, Nucl. Inst. and Methods A **887**, 81 (2018)
[11] M. Fila, PhD thesis, Univ. of Warsaw (in preparation)
[12] M. Kuich *et al.*, Acta Phys. Pol. B (Proc. Suppl.) **16**, 4-A17 (2023)
[13] H.R. Weller *et al.*, Progress in Particle and Nucl. Phys. **62**, 257 (2009)
[14] M. Wang *et al.*, Chin. Phys. C **45**, 030003 (2020)
[15] J.F. Ziegler, M.D. Ziegler and J.P. Biersack, Nucl. Inst. and Methods B **268**, 1818 (2010)
[16] J. Allison *et al.*, IEEE Trans. Nucl. Sci. **53**, 270 (2006)
[17] R.J. deBoer *et al.*, Rev. Mod. Phys. **89**, 035007 (2017)
[18] Y.K. Wu, priv. comm. (2022)

Integration of the scattering chamber of the NUMEN experiment

Diego Sartirana[1*], *Daniela* Calvo[1], *Clementina* Agodi[2], *Francesco* Cappuzzello[2,3], *Manuela* Cavallaro[2], *Carlo* Ferraresi[4,1], *Jose Roberto* Brandão de Oliveira[5], and *Antonio Domenico* Russo[2] for NUMEN collaboration

[1]INFN, Sezione di Torino, Torino, Italy
[2]INFN, Laboratori Nazionali del Sud, Catania, Italy
[3]Dipartimento di Fisica e Astronomia "Ettore Majorana", Università di Catania, Catania, Italy
[4]DIMEAS, Politecnico di Torino, Torino, Italy
[5]Instituto de Fisica, Universidade de São Paulo, São Paulo, Brazil

Abstract. The NUMEN experiment points to measure Double Charge Exchange reaction (DCE) cross sections with heavy-ion beams interacting with specific isotopes of interest for Neutrinoless Double Beta Decay process ($0\nu\beta\beta$). To achieve this, NUMEN requires a complete upgrade to MAGNEX spectrometer in INFN-LNS and a high intensity ion beam. This work focuses on the design of a new scattering chamber where the ion beam will interact with the target precisely positioned by an automatic system. The target will be cooled by a custom system based on HOPG and a cryo-cooler. The fist integration of the chamber with all its components has been done at INFN section of Turin to check compliance with the requirements and to study the procedures for the target replacement. Some results of this tests will be presented.

1 Introduction

The NUMEN experiment points to measure Double Charge Exchange reaction (DCE) cross sections with heavy-ion beams of unprecedented intensity interacting with isotopes of interest for the Neutrinoless Double Beta Decay ($0\nu\beta\beta$) [1-3]. DCE proved to be useful for getting information on the nuclear matrix elements of the Neutrinoless Double Beta Decay due to similarities between these two processes. The proposed technique of measuring was tested with pilot runs (until 2020 [4-6]) at *Laboratori Nazionali del Sud* (INFN-LNS), in Catania, with ion beams provided by pre-existing cyclotron and the magnetic spectrometer MAGNEX. This is a large acceptance spectrometer featuring a Scattering Chamber (SC) with a target positioned inside, a quadrupole, a dipole and a Focal Plane Detector [7]. MAGNEX is mounted on a rotating platform that could modifies angle between the target, always orthogonal to the beam, and the optical axis of the spectrometer allowing the selection of specific physics channels to study. [8].

However, to get statistical significance from DCE low cross sections, higher intensity ion beams are required. The super conducting cyclotron is being fully refurbished featuring ion beams with energy from 15 up to 70 MeV/u and intensity up to 10^{13} pps [9] this will require a complete upgrade of MAGNEX [10, 13-15] including the SC, where beam hit the target, and the installation of a beam dump. Since the design and the realisation of the chamber is already completed, it has been assembled for the first time at INFN section of Turin with all requested elements. This will be useful to check compliance to the requirements and to define best procedures in automatic and remote-controlled systems.

In this article the system will be presented in detail with first results of this testing.

1.1 Scattering chamber integration requests

To better understand the new design of the SC it is important to describe the requests for the complete system and the constraints dictated by the new working conditions. Here the main features requested to integration of this system are listed:

- Adaptation of the target of the experiment to the new experimental conditions (higher intensity beam);
- Integration to both the new High Intensity (HI) and the old Low Intensity (LI) beam line with attention to the relative positions between chamber and the beams;
- Implementation of an external manipulation system for replacement of the target due to high radiation level. The upper estimated limit is about 10^6 Gy/month for a target made of 2,5 μm of Ge and 100 μm of graphite as considered by the *INFN-LNS Radioprotection Office*. However, the current target being designed by NUMEN is thinner and therefore the expected dose reduces to a lower value;
- Compliance with high vacuum requirements inside the chamber and connected systems (10^{-6} mbar);
- Positioning of diagnostic and calibration systems (infrared camera, alumina, Faraday cup and others);
- Positioning of 110 Gamma ray Detectors (G-NUMEN) with a spherical geometry radially around the SC pointing to the target.

* e-mail: diego.sartirana@to.infn.it

https://doi.org/10.1051/epjconf/202329001005

2 Integration of the Scattering Chamber

2.1 Target and target holder design

The present target of the experiment consists of a thin isotope film (few hundreds of nanometres) evaporated on a Highly Oriented Pyrolytic Graphite (HOPG) thin foil of about 2 μm featuring in-plane high thermal conductivity [10]. This layer is clamped in a Target Holder (TH) made of OFHC (Oxygen-Free High Conductivity) copper which holds it in position and connect it to the Cold Finger (CF) of a cryo-cooler to dissipate the heat generated by the interaction of the ion beam with target material (lower than 10 Watt for the presented target design [11-12]).

TH has different apertures on the height to accommodate not only the target but also other tools for beam monitoring (Figure 1 on the left). To correctly align the different apertures on the beam, the cryo-cooler will be vertically shifted by a suited high precision motorized jack with a stroke of 100 mm.

TH and CF connection is a custom designed bayonet coupling that can be handled by a manipulation system (Figure 1 on the right). The target always offers its surface orthogonally to the incident beam no matter the rotation of the spectrometer platform because the support of cryocooler is anchored to the floor.

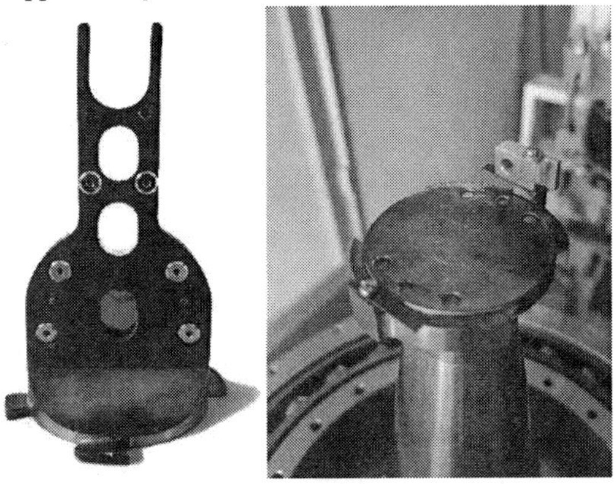

Fig. 1. On the left the TH with all different apertures, which will host starting from the top: alumina (for beam tuning), a free hole (to measure background from the frame), a graphite layer (for reference measure), the target. In the lowest part of the TH there is the connection elements to CF. The total heigh of the target holder is 98 mm. On the right the CF top where the TH will be connected through the bayonet coupling.

2.2 Scattering chamber and integration to the High Intensity beam lines

The shape of the SC is spherical 460 mm in diameter, 6 mm in thickness, due to the presence of G-NUMEN (Figure 2 on the left). The material of the body is aluminium (series 5000) to reach maximum efficiency for these detectors. There are different openings all around the chamber, seven in total, two of them are rectangular shaped. All flanges are made in stainless steel (SS 304L) to accommodate metallic sealing in OFHC copper. The connection of the main body to the flanges is obtained with the explosion bonding technique (to weld different materials together).

A rectangular flange present in the frontal part of the chamber allows the connection alternatively of the LI and HI beam lines. For connection to the HI beams (Ne or O) two plates can be alternatively mounted here, these have a surface inclined of +3° or -3° to compensate rotation of the whole system depending on the beam entering (in Figure 2 on the right there is a rendering of this connection, reference *a* indicates the plate). These plates also house a window in germanium used to inspect the target with an infrared camera (*b*). This camera will map the thermal field of the target signalling degradation of graphite, when not in use it will be protected by a remote-controlled lead screen. Beam diagnostic (*c*) is positioned upstream the chamber, connected through a bellow (d) to the plate. This bellows will allow little corrections of angles of ±1° in positioning the MAGNEX platform.

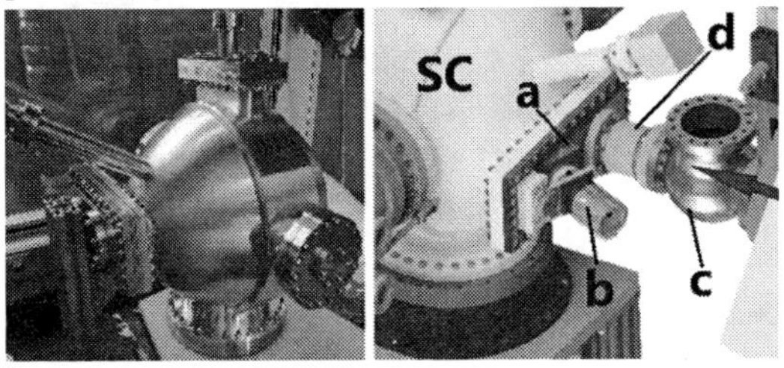

Fig. 2. On the left the SC body seen from outside. On the right a 3D rendering of the SC connection to the HI beam line, where the direction of the beam line is indicated with the arrow right above the structure of the last quadrupole of the beam line before SC.

2.3 Scattering chamber integration

Figure 3 shows the complete integration of the SC on the MAGNEX platform. Here we can spot the automatic system (*e*) designed to manage the replacement of TH inside the chamber [16]. This system is designed to work by remote control when experimental run is stopped and replacing of target is required. It is composed by a main pneumatic gripper directed by two axis servo-mechanical wrist, both introduced in SC by a pneumatic cylinder with 800 mm stroke. This manipulator could enter in the chamber exclusively when air is returned and a specific gate valve with 200 mm diameter is opened (*f*). The automatic system also includes a temporary storage where the new THs and the ones that are outgoing the chamber are stocked. Here a second gripper moves the activated ones to a vertical storage, wrapped with lead screen all around for radiation confinement; it can accommodate up to 6 THs before being removed from experimental area. Under the manipulator valve there is the vacuum pumping system (*g*) for the SC composed by a dry scroll pump and a turbomolecular pump with a speed of 1050 l/s.

The SC support (*h*) accommodates the cryocooler inside and the lifting actuator for it, a bellows guarantees the necessary stroke for the vertical movement. This support is constrained to the rotating platform of the spectrometer, so the SC rotates together with the platform. Since the TH and cryocooler are anchored to the floor (as seen in subsec. 2.1) a specific flanges system is designed to allow this rotation keeping the vacuum once in position. On the top of the SC the actuator for the Faraday Cup (FC) (*i*) is also visible.

Fig. 3. 3D rendering of the SC complete integration (red arrow indicates beam direction).

3 Testing of the system

3.1 Testing setup of the integration

The SC and the target system were designed and assembled in the *Laboratorio Tecnologico of INFN* in Turin. For this purpose, a specific test bench was built. This simulates all positions and working conditions of MAGNEX at INFN-LNS, thus allowing to test all the designed features of the integration. A specific rotary table positioned in the lowest part of the structure simulates the rotation of the platform rotating, for simplicity, only the cryocooler and TH. A picture of the whole setup is shown in Figure 4, with the same references used previously in Figure 3.

Fig. 4. Picture of the test bench with the same references of Figure 3 indicated (the arrow again indicates beam direction).

This test bench allows also to evaluate the positioning repeatability of the target manipulation system. Ad hoc control programs for the movement were developed using software of commercial motor drivers. Their integration with pneumatic lines, limit switches and safety systems are obtained programming a PLC by Crouzet using the SFC language (Sequential Function Chart). This made it possible to write down automatic procedures for each step of TH handling and optimize mechanics. Following this way, 100% repeatability for the target holder positioning and disassembly cycle is achieved.

An example of automatic sequence tested is mounting of the TH above the CF inside the SC, it can be divided in nine different steps shown in Figure 5:

• I the starting condition of the sequence, TH is not present on CF;

• II the manipulator wrist enters the chamber holding the TH horizontally with gripper (G), the motion of the cylinder that allow this movement is stopped in a calibrated position;

• III the first axis of the wrist rotates vertically the gripper (G) aligning TH with CF;

• IV the second axis of the wrist rotates TH around central axis of the cryocooler to the engage correct position for bayonet coupling;

• V the CF is lifted 8 mm to make contact with TH;

• VI the second axis of the wrist rotates again TH around central axis of the cryocooler in order to complete the lock the bayonet coupling;

• VII the gripper opens and disengages TH;

• VIII the CF with the TH coupled is lowered;

• IX the first axis of the wrist rotates the gripper horizontally.

This represents only the part inside the scattering chamber of the complete automatic procedure for placing a TH in position.

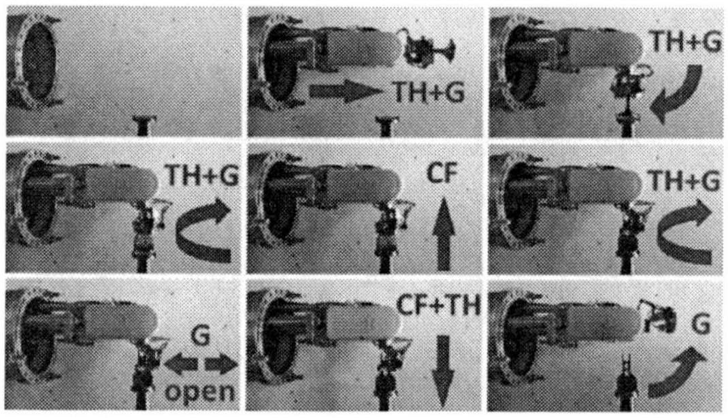

Fig. 5. Sequence for mounting the TH in position above the CF. For this test the SC is removed from its position to have the full view of the procedure ongoing.

3.2 Testing of the Faraday Cup actuator

The FC is designed to be positioned inside the scattering chamber for measure the beam current on request, to be operated only with LI beam lines. Since the positioning precision will be very important for functionality (in-beam maximum admitted errors must be ±0,05 mm from nominal position) it is connected to a specific system to align in correct position and to move it on-line and off-line (Figure 6 on the left). The actuator of the FC is a harmonic geared rotational motor positioned coaxial with cryocooler on the top of the chamber (*l*), it transmits motion through a magnetic coupling (*m*) inside the vacuum volume of SC. Here the motion is collected by an arc shaped frame in aluminium (one-piece structure to reduce mechanical backlash) (*o*) that sustains the FC and positions it with its opening at 148 mm from the centre of the SC. The connection between the arc and the FC is mediated by a shim made in boron nitride (BN) to electrical insulate Cup main body still allowing the heat to be transmitted. In this way the arc will be rotated by the actuator by 90° to move in-beam and off-beam the diagnostic element. The adjustment for working position is mediated by a port aligner (*n*) that moves relative to the SC the motor and all elements connected to it (consequently also the same FC).

The alignment of that with the beam line will be performed before the experimental runs and then there will be no possibility to replicate at each rotation, so it is important to evaluate its positioning repeatability.

For this purpose, a setup to test this system was realized at *Laboratorio Tecnologico of INFN* in Turin (Figure 6 on the right). The test bench for the FC actuator positions the system in the same final working condition which will be used on MAGNEX, the only difference here is that an aluminium block (*p*) simulates the FC with the same weight (150 g). On the frame that supports the actuator two comparators with resolution of 0,01 mm are mounted, one on the left side, one on the right side of the frame. The 90° rotation will be performed symmetrically to the support, touching alternatively the two comparators.

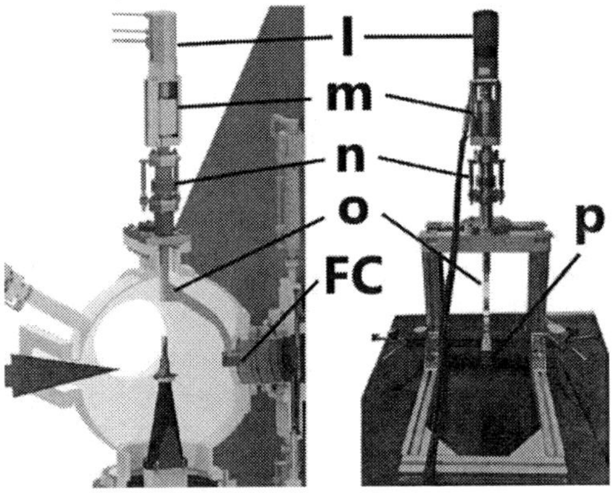

Fig. 6. On the left rendering 3D of a section of the SC with FC system in-beam (the arrow indicates beam direction). On the right picture of the testing setup for the FC actuator.

The test procedure is as follows: at the beginning of each test both the comparators are set to zero at the ends of a 90° arc described by the FC block during first rotation. Then back-and-forth rotation is performed by actuator, so the deviation from the starting position is measured by each comparator. A positive value of the deviation indicates an overcoming of the initial set position, a negative value indicates the opposite.

Sets of ten rotations at different speed of the actuator (1°/s, 2°/s and 5°/s) were studied getting the values for the left comparator showed in Figure 7. Both for the left and right comparators results are similar and under the required threshold (±0,05 mm) also proving no influences on repeatability of the rotation velocity.

Then the velocity was set to 1°/s and a group of 50 back-and-forth rotations is performed, giving similar results from those obtained previously. This means that also with high number of rotations the repeatability of the system allows not to cumulate errors in positioning above the desired value.

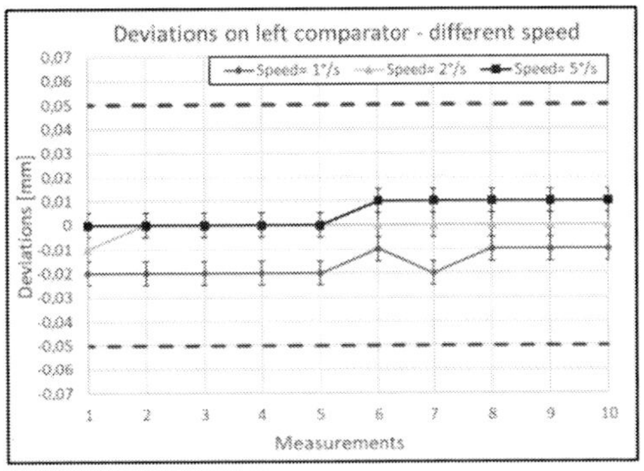

Fig. 7. Test results on left comparator at different rotational speeds, the dashed lines indicate maximum allowable error.

4 Conclusions and future works

The mechanical work of integration of the new MAGNEX SC is concluded and now the design of the G-NUMEN supports is in progress. The testing activity is now focused on analyse manipulator cooperation with the other remote-controlled or automated systems around the chamber. For example, it is important to test the interlocks of the manipulator sequences with the vacuum procedures or with the FC movements.

Future tests will allow to characterize temperature trends of the target in experimental run conditions. These tests will be conducted with a laser beam to simulate heat generated by HI beam-target interaction.

Acknowledgement

Special thanks go to *Riccardo Panero* for his skilful work and support for the mechanics practical integration and to Bruno Romano for collaboration on tests developing on the manipulator and the FC during his master degrees at the Turin Polytechnic.

References

1. F. Cappuzzello et al., Eur. Phys. J. A **54**, 72 (2018)
2. H. Lenske et al., PPNP **109**, 103716 (2019)
3. F. Cappuzzello et al., PPNP **128**, 103999 (2023)
4. D. Carbone et al., Phys. Rev. C **102**, 044606 (2020)
5. A. Spatafora et al., Phys. Rev.C **100**, 034620 (2019)
6. V. Soukeras et al., Res. in Phys. **28**, 104691 (2021)
7. D. Torresi et al., NIM A **989**, 164918 (2021)
8. F. Cappuzzello et al., Eur. Phys. J. A **52**, 167 (2016)
9. C. Agodi et al., Universe **7**(3), 72 (2021)
10. F. Cappuzzello et al., IJMP A **36**, 30 (2021)
11. V. Capirossi et al., Acta Phys. Polon. B **51**, 3 (2020)
12. V. Capirossi et al., NIM A **954**, 161122 (2020)
13. D. Calvo et al., NIM A **1041**, 167336 (2022)
14. D Calvo et al., JPCS **2374**, 012029 (2022)
15. P. Finocchiaro Universe **06**, 129 (2020)
16. D. Sartirana et al., Adv. in Serv. and Industrial Robotics (RAAD 2020) **84**, pp. 535-543 (2020)

EPJ Web of Conferences **290**, 01006 (2023)
EuNPC 2022
https://doi.org/10.1051/epjconf/202329001006

Event reconstruction in ACTAR TPC: Adaptation for transfer experiments

Miguel Lozano-González[1,*], *Beatriz* Fernández-Domínguez[1], *Juan* Lois-Fuentes[1], and *Thomas* Roger[2]

[1]IGFAE and Dpt. de Física de Partículas, Univ. of Santiago de Compostela, E-15758, Santiago de Compostela, Spain
[2]GANIL, CEA/DRF-CNRS/IN2P3, Bd. Henri Becquerel, BP 55027, F-14076 Caen, France

Abstract. Preliminary results of the performance of the RANSAC algorithm over data measured with the ACTAR TPC detector are herewith presented. With the aim of improving track reconstruction in highly contaminated events, an initial assessment of the algorithm is given for elastic reactions involving p, d with a beam of ^{20}O. Results show great possibilities for data/noise discrimination, which in the near future will be extended to small angle and more energetic events, which are key for transfer reactions.

1 A prime example of an active target

The ACtive TARget and Time Projection Chamber (ACTAR TPC) [1] is part of a new generation of detectors aimed to overcome the limitations of traditional devices, more prominent when performing transfer experiments with exotic nuclei, where the low-intensity beams require thicker targets.

An active (gaseous) time-projection chamber (TPC) allows us to tackle this by working in inverse kinematics, obtaining the vertex and the tracks with high resolution, provided a highly segmented pad, and correcting the energy lost by the particles during their propagation inside the gas.

Major advances in R&D, mainly in the field of gaseous detectors and data acquisition, combined with the experience gained with other successful active targets such as MAYA at GANIL or IKAR at GSI, have led to the development of ACTAR TPC.

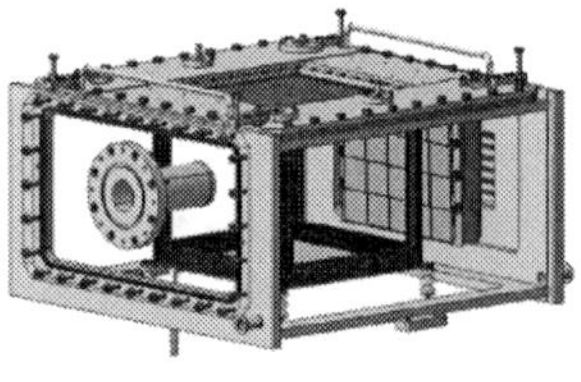

Figure 1. ACTAR TPC layout, from [1].

2 Track reconstruction with RANSAC

Detector electronics provide us with a collection of *hits*: XYZ position and collected charge. A fitting procedure (*algorithm*) has to be chosen so that we can associate those hits to *tracks*.

RANSAC (RANdom SAmple Consensus) fulfils our requirement: extracting a *model* (track) from a cloud of hits full of *outliers* (noise), which in our case is originated mainly from $\delta-$electrons [2, 3]. One can sum up the algorithm in three steps:

1. Generate N models, lines in our setup, by sampling iteratively 2 hits.
2. Given a *distance threshold*, one can rank each sampled model by counting the number of hits whose distance is below the said threshold (*inliers*).
3. Then, in descending order of inliers, a 3D linear regression is applied to the points belonging to the model, obtaining a track. Next, the points from the track are deleted from the cloud, and the sequence is halted when no more fits can be done.

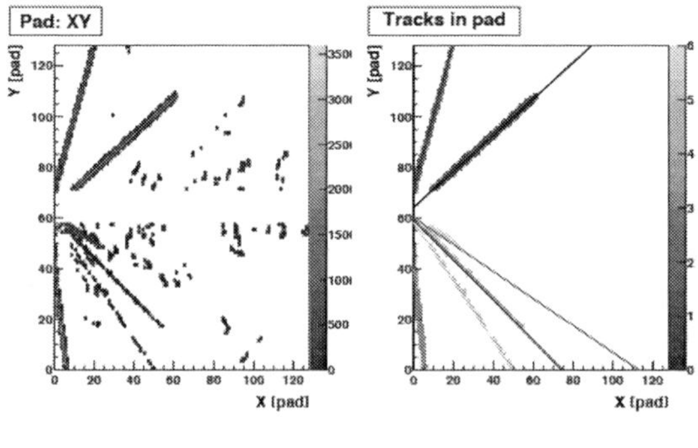

Figure 2. An example of RANSAC in our dataset.

3 Testing on data

The data for this test was gathered using a setup with ACTAR TPC and two silicon arrays placed at the left side and in front of the beam direction.

*e-mail: miguellozano.gonzalez@usc.es

https://doi.org/10.1051/epjconf/202329001006

Employing a ^{20}O beam at 35 MeV u^{-1} and a chamber filled with a mixture of 90 % D_2 and 10 % C_4H_{10}, we can measure elastic, inelastic and transfer reactions, such as:

$$^{20}\mathrm{O(p,p)}^{20}\mathrm{O}, \quad ^{20}\mathrm{O(d,d)}^{20}\mathrm{O},$$
$$^{20}\mathrm{O(p,d)}^{19}\mathrm{O} \quad \text{and} \quad ^{20}\mathrm{O(d,t)}^{19}\mathrm{O}$$

Note: during execution of the algorithm, beam hits were masked, since further development has yet to be done to adapt RANSAC to this intricate region.

4 Preliminary results

Once tracks have been reconstructed, the reaction vertex and the silicon impact point can be obtained directly from our fits, given the detector layout. Afterwards, we can extract the light recoil ID, which constitutes an excellent measurement of our algorithm performance. Figure 3 depicts an example of this identification, where we plot the energy measured in the silicon detectors against the *averaged* (per unit of track length) deposited charge in the pad plane.

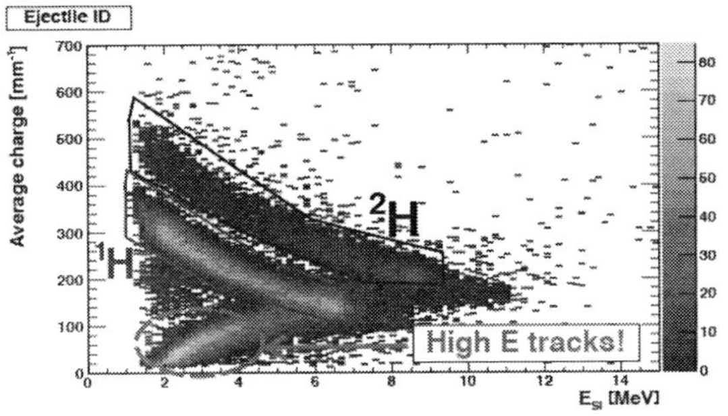

Figure 3. Light recoil ID for events triggering the side arrays.

Unlike the previous clustering algorithm still in use, this RANSAC implementation shows its great potential to reconstruct extremely energetic tracks, whose energy deposit is lower, both in the pad plane and in the silicon layer, as they suffer punchthrough.

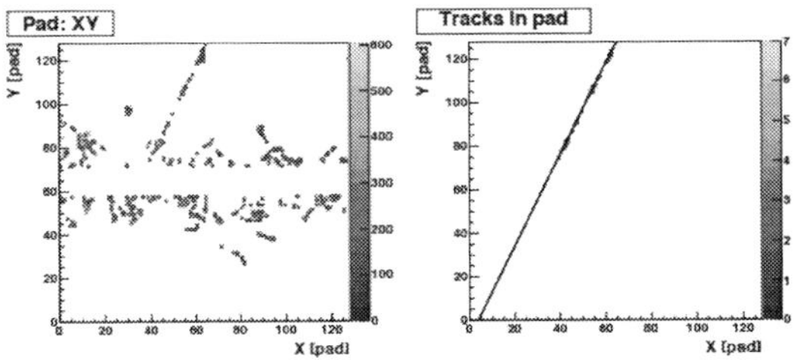

Figure 4. ID of a high E event that punched through the silicon layer.

Besides, more data from other experiments will be needed to perform in-depth studies concerning the adaptability of the algorithm under variations in the number of outgoing particles or the amount of background pollution. A final assessment of the algorithm is expected soon.

5 Current status and future prospects

After evaluating our results, RANSAC exhibits great capabilities when it comes to identifying low-charge depositing tracks and finding multiple track events. The cleaning of δ−electron contamination achieved with this algorithm, a major matter of concern for some experiments, is also noteworthy.

However, a few drawbacks still require major modifications. Firstly, there is the issue of the time consumption: the sampling step represents most of the CPU time consumption per event. A pre-clustering stage or a *filter* could be added to reduce the hit matrix dimensionality and improve execution times.

Secondly, very thick and long tracks may be split or identified by RANSAC as different particles. A post-clustering step could be implemented, merging aligned clusters or tracks with similar fit parameters.

Finally, δe^- rejection should be further improved, as it is crucial for front silicon detectors, where tracks are shorter and emerge in a narrower angular window.

Acknowledgements

M.L.G. and J.L.F. wish to acknowledge financial support from Xunta de Galicia (Spain) grants number ED481A-2022/419 and ED481A-2020/069. This work was supported by the Spanish MINECO through the project PGC2018-096717-B-C22 and PID2021-128487NB-I00. This work was also partially supported by the Xunta de Galicia under project ED431B 2018/015, 2021-PG045 and the Maria de Maeztu Unit of Excellence MDM-2016-0692.

References

[1] B. Mauss, P. Morfouace, T. Roger, J. Pancin, G. Grinyer, J. Giovinazzo, V. Alcindor, H. Álvarez Pol, A. Arokiaraj, M. Babo et al., Nucl. Instrum. Methods Phys. Res. A **940**, 498 (2019)
[2] Y. Ayyad, W. Mittig, D. Bazin, S. Beceiro-Novo, M. Cortesi, Nucl. Instrum. Methods Phys. Res. A **880**, 166 (2018)
[3] J. Zamora, G. Fortino, Nucl. Instrum. Methods Phys. Res. A **988**, 164899 (2021)

EPJ Web of Conferences **290**, 02001 (2023) https://doi.org/10.1051/epjconf/202329002001
EuNPC 2022

Study of multinucleon knockout reactions of exotic nuclei in the region of Sn

M. Feijoo-Fontán[1]*, *E.* Kudaibergenova[2], *I.* Lihtar[3], *I.* Gasparic[3], *A.* Horvat[2,3], *J.L.* Rodriguez-Sanchez[1], *V.* Panin[4], *H.* Alvarez-Pol[1], *L.* Atar[2], *T.* Aumann[2,4,5], *J.* Benlliure[1], *K.* Boretzky[4], *M.J.G.* Borge[6], *L.T.* Bott[7], *C.* Caesar[4], *E.* Casarejos[8], *J.* Cederkall[9], *A.* Chatillon[10], *D.* Cortina-Gil[1], *E.* De Filippo[11], *T.* Dickel[4], *M.* Duer[2], *A.* Falduto[2], *D.* Galaviz[12], *G.* García-Jiménez[1], *Z.* Ge[4], *E.I.* Geraci[11,13], *R.* Gernhäuser[14], *B.* Gnoffo[11,13], *A.* Graña-González[1], *K.* Göbel[7], *E.* Haettner[4], *A.L.* Hartig[2], *M.* Heil[4], *A.* Heinz[15], *T.* Hensel[16], *M.* Holl[15], *C.* Hornung[4], *A.* Jedele[2], *D.* Jelavic-Malenica[2], *T.* Jenegger[14], *L.* Ji[2], *H.T.* Johansson[15], *N.* Kalantar-Nayestanaki[17], *E.* Kazantseva[4], *A.* Kelić-Heil[4], *O.A.* Kiselev[4], *P.* Klenze[4,14], *R.* Knoebel[4], *D.* Körper[4], *T.* Kröll[2], *B.* Löher[4], *N.S.* Martorana[18], *L.* Milhomens da Fonseca[2], *P.* Morfouace[10], *I.* Mukha[4], *S.* Murillo-Morales[19], *E.* Nácher[6], *C.* Nociforo[4], *A.* Obertelli[2], *S.* Paschalis[19], *A.* Perea[6], *M.* Petri[19], *S.* Pietri[4], *S.* Pirrone[11], *L.* Ponnath[14], *H.B.* Rhee[2], *L.* Rose[19], *D.M.* Rossi[2,4], *P.* Russotto[18], *D.* Savran[4], *C.* Scheidenberger[4], *H.* Simon[4], *J.P.* Simon[2], *S.* Storck-Dunite[2], *A.* Stott[19], *Y.* Sun[2], *D.* Symochko[2], *C.* Sürder[2], *J.* Taïeb[10], *R.* Taniuchi[19], *O.* Tengblad[6], *H.T.* Törnqvist[2,4], *S.* Velardita[2], *F.* Wamers[4], *H.* Weick[4], *J.* Zhao[4], and the R^{3}B collaboration

[1]IGFAE, Universidade de Santiago de Compostela, E-15782 Santiago de Compostela, Spain
[2]Technische Universität Darmstadt, Fachbereich Physik, Institut für Kernphysik, 64289 Darmstadt, Germany
[3]RBI Zagreb, Bijenicka cesta 54, HR10000, Zagreb, Croatia
[4]GSI Helmholtzzentrum für Schwerionenforschung, Planckstraße 1, 64291, Darmstadt, Germany
[5]Helmholtz Research Academy Hesse for FAIR, 64289 Darmstadt, Germany
[6]Instituto de Estructura de la Materia, CSIC, E-28006 Madrid, Spain
[7]Goethe-Universität Frankfurt, Max-von-Laue Str. 1, 60438, Frankfurt am Main, Germany
[8]CINTECX, Universidade de Vigo, E-36200 Vigo, Spain
[9]Department of Physics, Lund University, P.O. box 118, 221 00 Lund, Sweden
[10]CEA, DAM, DIF, F-91297 Arpajon, France
[11]INFN Sezione di Catania, Via Santa Sofia 64, 95123, Catania, Italy
[12]Laboratório de Instrumentação e Física Experimental de Partículas, LIP, Av. Prof. Gama Pinto 2, 1649-003 Lisbon, Portugal
[13]Università di Catania, Dipartimento di Fisica e Astronomia "Ettore Majorana", Catania, Italy
[14]Technische Universität München, James-Franck-Str 1, 85748, Garching, Germany
[15]Institutionen för Fysik, Chalmers Tekniska Högskola, 412 96 Göteborg, Sweden
[16]Helmholtz-Zentrum Dresden-Rossendorf, Institute of Radiation Physics, Bautzner Landstraße 400, 01328, Dresden, Germany
[17]ESRIG, University of Groningen, Groningen, Netherlands
[18]INFN Laboratori Nazionali del Sud, Via Santa Sofia 62, 95133, Catania, Italy
[19]School of Physics, Engineering and Technology, University of York, YO10 5DD York, UK

Abstract. The experimental data collected during the S515 experiment performed by the R^{3}B collaboration at GSI/FAIR represent a great opportunity to investigate nucleon knockout reactions of exotic nuclei in the region of Sn using complete kinematics measurements. These cross sections can be used in the future to investigate the quenching in the knockout of the minority species (neutrons or protons) in nuclei far from stability. Some of the arguments put forward are the underestimation of the knockout of deeply bound nucleons, final state interactions or the role of short-range correlations (SRC). Recently, several works based on inclusive measurements have shown that these SRCs could reduce the single nucleon knockout cross sections by around 50%, depending on the neutron excess (N/Z) of the initial projectile. The S515 data can help us to go further in this investigation because it allows to correlate the knockout cross sections of one, two or more nucleons with the number of protons and neutrons emitted from the target and which can be detected by the CALIFA and NeuLAND detectors, respectively, and perform complete kinematical studies on the nature of the event (SRC, evaporation, emission of clusters, final-state interactions...). Here the results obtained for the charge distribution of reaction residues are presented, which is one of the first steps of the still on-going analysis.

1 Introduction

Knockout reactions offer a powerful tool for identifying the single-particle structure of nuclei since they are direct nucleon-nucleon collisions of the probe particles with the shell nucleons. In particular, one of the objects of study can be the determination of spectroscopic factors, i. e., how much the independent particle model (IPM) is fulfilled in reality. A significant deviation from IPM model that seemed independent from the asymmetry parameter ΔS defined as $S_n - S_p$ ($S_p - S_n$) for neutron (proton)

*email: *martina.feijoo.fontan@usc.es*

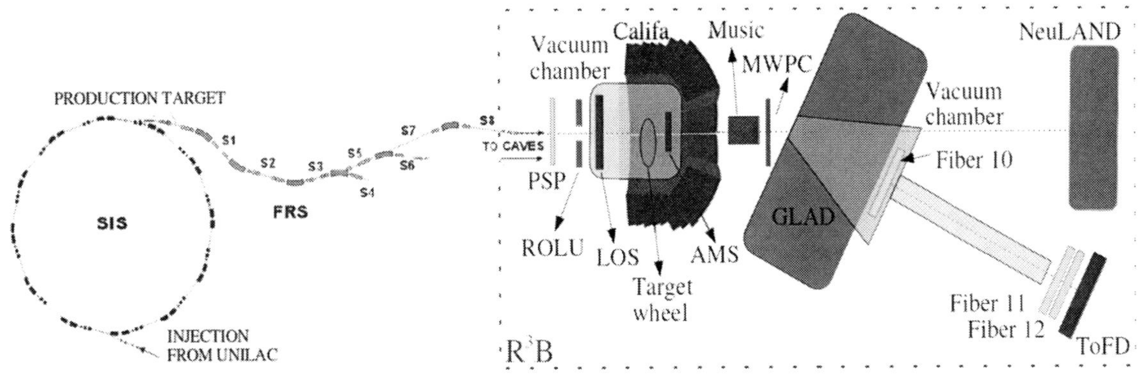

Figure 1. Schematic representation of the experimental setup used in the present work. Sizes are not to scale.

knockout was found [1, 2] both when studying transfer reactions as well as quasi-elastic electron scattering $(e, e'p)$. Nevertheless, experiments performed in inverse kinematics and with composite nuclear targets, such as beryllium and carbon, found a clear dependency between ΔS and the reduction of the experimental cross sections of single nucleon knockout compared to the shell model predictions [3]. Several works, based on inclusive measurements [4, 5] focused on the study of neutron-rich-medium-mass nuclei around Sn, showed that this reduction expressed as dependence on the neutron excess (N/Z) (which can be translated into ΔS) of the initial projectile is more noticeable for proton-removal reactions.

The quenching can be partially explained as an effect of SRC, which are pairs of nucleons having large relative momentum, proton-neutron being the dominant ones [6]. Because of this, the fraction of paired protons is larger than the neutron pairs, in the case of neutron-rich nuclei. In addition, investigations in Refs. [3, 4] indicate that the excitation energy of the remnant nuclei calculated with the current models is underestimated, resulting in a higher probability for particle evaporation after the reaction and reducing the measured fragments associated with the single-knockout channel. Another factor contributing to the overestimation of the core survival probability could be the omission of statistical cascade-model-like processes [7], mainly for weakly-bound cores. Finally, the emission of clusters could also be playing a significant role.

In this experiment the single nucleon knockout (A-1) channel was extended to the multiple ones (A-2, A-3,...). Moreover, the evaporated particles can also be detected, hence this experiment provides for the first time access to complete kinematic information for such reactions. This work will try to study the different contributions of the mentioned factors to the observed quenching in the knockout cross sections. In addition, gamma emission of excited states of Sn isotopes can be measured as well, since an unexpected favored gamma-deexcitation was observed above neutron evaporation threshold [8].

The following outlines the experimental setup briefly. Moreover, the results obtained for the identification of the incoming projectiles, delivered via the fragment separator FRS, and the reaction residues are presented as a function of the atomic number (Z), as well as their production cross sections. These are the first steps of the analysis, which will be continued in order to link the produced fragments and their knocked-out particles with the effects described above.

2 Experiment

The S515 experiment was performed in spring 2021 at GSI, using two different production mechanisms, cocktail beams produced at the entrance of the FRS and guided through the spectrometer to the experimental cave (see Figure 1). The first type of beam was produced from the fragmentation of ^{136}Xe on a Be target, resulting in a secondary beam centered on the isotope ^{124}Sn. This was done at two different beam energies in order to get the secondary beam at 405 and 900 MeV/u in the experimental hall. The second cocktail beam was produced via fission of ^{238}U on two different Pb targets (2386 mg/cm^2, 1542 mg/cm^2), resulting in secondary cocktail beams containing ^{124}Sn at 872 MeV/u and ^{132}Sn 678 MeV/u. With this combination of settings, a wide range of Sn isotopes can be studied using the inverse kinematics technique. Using this method a large momentum in the center of mass is reached and complete kinematic measurements of all the outgoing fragments can be accomplished.

This kind of measurement is performed with the R^{3}B experimental setup shown in the Figure 1. It consists of several detectors. A Position Sensitive Pin diode (PSP) to measure the charge of the incoming particles, a plastic scintillator (ROLU) that works as a veto to make sure only collimated events are recorded. Another plastic scintillator (LOS) is used to determine the Time-of-Flight (ToF) between the central focal plane of the FRS (S2) and the experimental cave before the target wheel. The different materials and thicknesses used as target were: C(1g/cm^2, 2g/cm^2), CH_2 (1g/cm^2, 2g/cm^2) and Pb (980mg/cm^2). Surrounding the target-area the highly segmented (1504 CsI(Tl) crystals) calorimeter CALIFA (CAlorimeter for In-Flight detection of γ-rays and high energy charged pArticles) is placed to measure gammas and light charged particles [9]. The MUSIC detector (MUlti-Sampling Ionization Chamber) is used for the charge identification of the reaction residues. Mass identification of both incoming and outgoing isotopes is performed using the ΔE-Bρ-ToF technique. Fiber detectors are used to determine the positions for tracking after the target, while ToF is determined from timing information of the TOFD wall [10] and the start detector LOS. The large acceptance superconducting dipole GLAD is placed behind the R^3B target, bending the trajectories of the reaction residues depending on their A/Q. The forward-emitted neutrons are detected in the NeuLAND detector [11].

3 Results

The first step of the analysis is the identification of the isotopes in the cocktail beam as shown in Figure 2 for the fragmentation setting at 900 MeV/u. In the present work only this FRS setting is considered, focusing the study on the case of ^{124}Sn which is indicated in Figure 2. In the future other isotopes will be studied, as well as the fission settings.

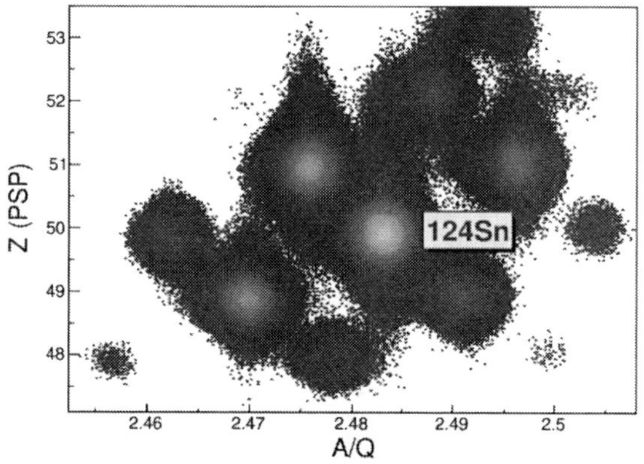

Figure 2. (Color online) Cluster plot displaying the identification of secondary beams in atomic number (Z) and mass-over-charge (A/Q) for an FRS magnetic setting centered on ^{124}Sn.

The charge identification after the target is made, being able to measure a broad range of products with a resolution of 0.48 charge units (FWHM). In Figure 3 the charge distribution is shown for the three different targets when selecting ^{124}Sn as the incoming isotope. The distributions are scaled to equal height at for Z=48, where the electromagnetic excitation contribution is assumed to be negligible. The main difference arises for the Pb target, since the electromagnetic field induced by this element is much larger than for the other two. Because of this, some reaction channels that do not modify the nuclear charge of the projectile and which are not accesible for C, CH_2 targets are open for reaction with a Pb target. Coulomb excitation results, e.g., in an increase of the neutron emission due to the excitation of giant dipole and quadrupole resonances [12]. Consequently, the amount of projectiles that remain with the same charge (Z=50) after the reaction is larger in the case of the lead target. In Figure 3 it can be seen that the difference is of almost one order of magnitude.

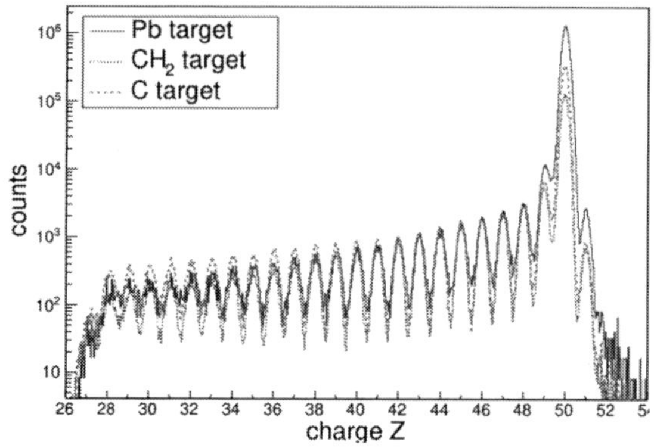

Figure 3. (Color online). Charge distribution comparison of the reaction residues produced by projectiles of ^{124}Sn at 900 MeV/u impinging on C, CH_2 and Pb targets.

Once both identifications are performed, it is possible to determine the charge production cross sections σ_Z for the different targets as follows

$$\sigma_Z = \left(\frac{N_{\text{out}}^{\text{target}}}{N_{\text{in}}^{\text{target}}} - \frac{N_{\text{out}}^{\text{empty}}}{N_{\text{in}}^{\text{empty}}} \right) \frac{M(t)}{d(t) \cdot N_A} \quad (1)$$

where $M(t), d(t)$ are the molar mass and the thickness of the target, respectively, and N_A is Avogadro's number. N_{out} refers to the number of outgoing fragments and N_{in} to the incoming projectiles. Subtracting the number of measured particles from an empty-target measurement ensures that the background is removed, consisting mainly of reactions that occured in matter layers of the setup, such as the scintillators. Finally, a correction regarding secondary reactions in the target is needed. A factor for each target and outgoing charge is calculated utilizing two concatenated simulations: the first one, based on the use of the intranuclear cascade and abrasion reaction models, a distribution of reacted fragments is obtained for ^{124}Sn+p and ^{124}Sn+C. This is the input for the second one, based on Geant4 where the fragments pass through the corresponding material (C, Pb) but considering only half of the length of the target in the experiment. In this way, all secondary reactions are approximated to happen in the center of the target. Afterwards, the correction factors for secondary reactions are calculated as the ratio of incoming over outgoing fragments for each charge. Finally, the cross section obtained from Eq. 1 for each charge is multiplied by the corresponding correction factor. For the measurement with the CH2 target the correction factor is assumed to be the same and is applied to the cross section accordingly.

Preliminary results are shown in Figure 4 for the three targets, considering only statistical uncertainties so far. The proton contribution is obtained as $\sigma_Z^p = 1/2 \cdot (\sigma_Z^{CH_2} - \sigma_Z^C)$. When comparing the results with model calculations, for the two light targets, the INCL (Liège Intranuclear Cascade) model [13, 14] is used, whereas for Pb the calculation is performed using ABRABLA07 [15].

 https://doi.org/10.1051/epjconf/202329002001

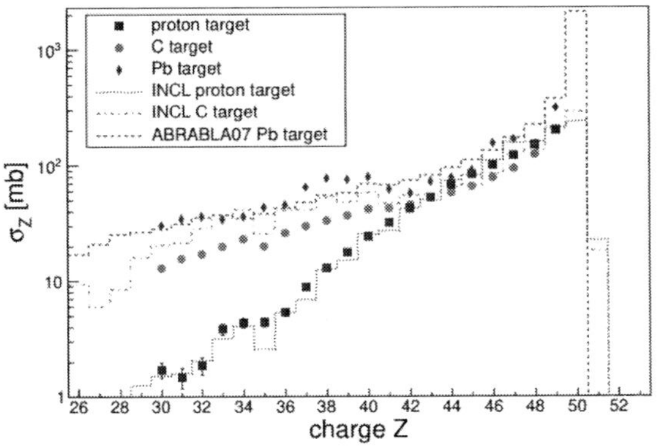

Figure 4. (Color online) Production cross section as a function of the atomic number (Z) measured for ^{124}Sn on p, C and Pb targets. The data are compared to different model calculations.

In general, it can be seen that the obtained experimental production cross sections follow the same trend as the calculated ones. Nevertheless, a further analysis is needed to check for possible mismatches, especially the behaviour found for σ_Z^{Pb} around $Z = 38$ and the sudden decrease visible for $Z = 48$. For the case of σ_Z^{C}, the reason why the measured values are systematically below those calculated is still not clear.

4 Conclusions and outlook

The fragment separator FRS is combined with the R^{3}B experimental setup to study multi-nucleon knockout reactions of exotic nuclei around Sn. This complex experiment requires the calibration of many detectors in order to identify the incoming and outgoing fragments as well as the emitted particles. In this work we present the first results concerning the charge distribution of reaction residues produced with projectiles of ^{124}Sn impinging onto different targets at kinetic energies of 900 MeV/u. After the detector calibrations, the charge distributions are obtained with a resolution of 0.48 (FWHM). The number of counts for each charge is then corrected by dead times and secondary reactions and normalized to the number of incoming projectiles to obtain the production cross sections. After applying all the corrections the cross sections are obtained with uncertainties smaller than 4% for the direct measurements with the C, CH_2 and Pb targets. For the hydrogen, this uncertainty reaches 19% for the lighter atomic numbers but it is less than 9% for $Z \geq 34$. As expected, the production cross sections of lighter reaction residues increase with the target size, being larger for the Pb target. The data are also compared to INCL+ABLA and ABRABLA07 calculations, showing a reasonable agreement. The next step will be to obtain the cross sections for multi-nucleon knockout channels and extend this investigation to other incoming isotopes.

Additionally, total neutron-removal cross sections could be used to determine the neutron skin thickness and to constrain the slope parameter L of the symmetry energy down to ± 10 MeV [16]. A similar constrain for the L parameter can be achieved studying the dipole polarizability [17] thanks to the measurements performed with the Pb target. The results of these promising investigations are still under analysis and will be presented in forthcoming publications.

Acknowledgments

This experiment was performed within the FAIR Phase-0 beam-time. The work has been supported by the Spanish Ministry for Science and Innovation under grants No. PGC2018-099746-B-C21, PGC2018-099746-B-C22, PID2019-104390GB-I00, PID2021-125771NB-C21, and PID2021-125771NB-C22, by the Regional Government of Galicia under the program "Grupos de Referencia Competitiva" ED431C-2021-38, by the "María de Maeztu" Units of Excellence program MDM-2016-0692, by the German Federal Ministry of Education and Research (BMBF, project 05P2015RDFN1) and by the GSI-TU Darmstadt cooperation agreement. J.L.R.-S. is thankful for support from Xunta de Galicia grant No. ED481D-2021-018. I.G, I.L., D.J.M. have been supported by Croatian Science Foundation (HRZZ) under porject numbers IP-2018-01-1257 and CEMS at RBI.

References

[1] L. Lapikás et al., Nuclear Physics A **553**, 297 (1993)
[2] B.P. Kay et al., Phys. Rev. Lett. **111**, 043502 (2013)
[3] A. Gade et al., Phys Rev. C **90**, 057602 (2014)
[4] J. Díaz-Cortés et al., Phys. Lett. B **811**, 135962 (2020)
[5] V. Vaquero et al., Physics Letters B **795**, 356 (2019)
[6] M. Duer et al., Nature **560**, 617 (2018)
[7] C. Louchart et al., Phys. Rev. C **83**, 011601 (2011)
[8] V. Vaquero et al., Phys. Rev. Lett. **118** 202502 (2017)
[9] H. Alvarez-Pol et al., Nucl. Instrum. Methods A **767**, 453 (2014)
[10] M. Heil et al., Eur. Phys. J. A **58**, 248 (2022)
[11] K. Boretzky et al., Nucl. Instrum. Methods Phys. Res. A **1014**, 165701 (2021)
[12] X. Roca-Maza et al., Phys. Rev. Lett. **106**, 252501 (2011)
[13] J. Hirtz et al., Phys. Rev. C **101**, 014608 (2020)
[14] J. L. Rodríguez-Sánchez et al., Phys. Rev. C **105**, 014623 (2022)
[15] J.-J. Gaimard et al., Nucl. Phys. A **531**, 709 (1991)
[16] T. Aumann et al., Phys. Rev. Lett. **119**, 262501 (2017)
[17] X. Roca-Maza et al., Phys. Rev. C **92**, 064304 (2015)

Broadband storage-ring mass and lifetime spectrometry

Yuri A. Litvinov[1,*] and *M. Shahab* Sanjari[1,2]

[1]GSI Helmholtzzentrum für Schwerionenforschung GmbH, Planckstrasse 1, 64291 Darmstadt, Germany
[2]Aachen University of Applied Sciences, Bayernallee 11, 52066 Aachen, Germany

Abstract. The mass and half-life of a nucleus are its basic properties which reflect its structure and decay properties. These data are among the most important nuclear physics inputs to astrophysical nucleosynthesis modelling. Tremendous experimental efforts are carried out to obtain yet unknown quantities, which are as a rule belong to short-lived nuclei far away from stability. Storage ring mass spectrometry is a technique ideally suited for addressing many different nuclear species at the same time. In the last few years it went through major developments regarding nuclear mass determinations, thereby boosting its efficiency, sensitivity and precision. A non-destructive detection capability is being presently implemented aiming at simultaneous mass, lifetime and decay branching measurements. Recent developments and future perspectives are briefly discussed.

1 Introduction

Atomic nuclei are composed of two kinds of quantum particles, protons and neutrons, which are kept together by a highly complicated nuclear force. This force defines the structure of nuclei, their shapes and moments, decay properties and correlations among nucleons [1]. Total energies, $E(Z, N)$, of various nuclei with different numbers of protons, Z, and neutrons, N, spanning over the entire chart of the nuclides, contain answers on the result of this nuclear interaction at work. $E(Z, N)$, or the rest mass, $m(Z, N)$, is directly connected to the binding energy, $B(Z, N)$, via:

$$m(Z, N) = E(Z, N)/c^2 = Z \cdot m_p + N \cdot m_n - B(Z, N), \quad (1)$$

where m_p, m_n are the rest masses of proton and neutron, respectively. The mass of the corresponding atom can be obtained by adding the masses of electrons and removing their binding energies [2]. Nuclear masses plotted on a nuclear chart form the *nuclear mass surface*. This surface is in general very smooth and changes in nuclear structure appear on it as irregularities [3].

Masses of the nuclei define whether and which radioactive transformation between them is possible. Of particular interest are weak decays, which dominate throughout the nuclear chart. Both quantities, masses and beta-decay lifetimes, are required in nuclear astrophysics, where the former ones define the pathways of nucleosynthesis processes on the nuclear chart and the latter ones affect the accumulated element abundances [4, 5].

In the past 100 years masses of about 2550 nuclei have been measured [6], whereas about 7000 bound nuclei are expected to exist [7]. These extensive data are employed to benchmark nuclear theories [8, 9]. However, their predictive powers are still insufficient for many applications in nuclear structure and astrophysics. Therefore, the measurements of still unknown masses is highly desired, which will again be used to further constrain theory. Taken their importance, mass measurements belong to research programs at basically all present and planned radioactive-ion beam facilities. Mass spectrometry methods are exhaustively reviewed in literature [10–17].

The experimental masses are evaluated in the Atomic Mass Evaluation (AME) [6], which also contains short-range extrapolations to yet unmeasured nuclides. The masses and half-lives are conveniently presented in the NUBASE database [18]. By inspecting NUBASE, one can notice that half-lives are typically known for more exotic nuclei than those with measured masses, which can be explained by the fact that many particles are required in modern mass spectrometry methods. Measurements of yet unknown masses are extremely challenging. To significantly advance further, a technique is required which would be capable to determine precisely the mass from ideally a single particle. At the same time, such a technique needs to be very fast since the characteristic half-lives of the nuclei of interest are well below one second. Furthermore, a dream would turn into reality, if the lifetime determination is conducted at the same time. In this context, storage ring mass spectrometry is a promising approach.

There are presently three heavy-ion storage rings coupled to radioactive ion-beam facilities [16, 17]. The facility at GSI Helmholtz Center in Germany comprises heavy-ion synchrotron SIS, projectile fragment separator FRS and the experimental storage ring ESR [19, 20]. At the Institute of Modern Physics in China, the experimental cooler-storage ring CSRe is connected to the synchrotron CSRm via the fragment separator RIBLL2 [21, 22]. Quasi-DC beams from a high-power cyclotron are employed at RIKEN Nishina Center in Japan, where the rare-ion storage ring R3 is placed behind the high-acceptance separator Big-RIPS [23]. At all these facili-

*e-mail: y.litvinov@gsi.de

ties, the secondary ions of interest are produced at high energies of a few hundreds of AMeV, utilizing mainly the projectile fragmentation or fission nuclear reactions [24]. At these energies, fragments are as a rule fully-ionized or have just a very few bound electrons [25]. At the ESR and CSRe, all fragments fitting into the acceptance of the machines are transported, injected and stored. The peculiarity of the R3 is that the injection of individual pre-identified particles is implemented [26].

This short contribution does not aim at providing details of specific facilities and/or experiments. Instead it focuses on recent developments of the storage ring mass spectrometry as well as on somewhat speculative future perspectives.

2 Storage ring mass spectrometry

Motion of a relativistic particle with mass-over-charge ratio m/q and velocity v in a heavy-ion storage ring is analogous to the one in a synchrotron and is given by [27]:

$$\frac{\Delta f}{f} = \eta \frac{\Delta p}{p}, \tag{2}$$

where f and $p = mv\gamma$ are, respectively, the revolution frequency and momentum of the particle. The parameter η is defined as:

$$\eta = \frac{1}{\gamma_t^2} - \alpha_p = \frac{1}{\gamma^2} - \frac{1}{\gamma_t^2}, \tag{3}$$

with γ and γ_t being the relativistic Lorentz factor corresponding to particle energy in general and at a specific value known as *transition energy*, which is defined by the ion optics of the ring. In accelerator physics, $\alpha_p = 1/\gamma_t^2$ is termed *momentum compaction factor*. It describes the variation of the orbit length $\Delta C/C$ corresponding to a change of the magnetic rigidity $\Delta B\rho/B\rho$, with $B\rho = mv\gamma/q$. In order to enable acceleration or deceleration in a synchrotron, the condition $\eta \neq 0$ or $\gamma_t \neq \gamma$ must hold. However, beam can be stored also for $\eta = 0$ or $\gamma_t = \gamma$. These optical settings are called *standard* and *isochronous* modes, respectively [28].

Assuming a closed trajectory, the revolution frequency is given as $f = v/C$. By combining the above equations, the basic formula for *storage ring mass spectrometry* can be derived [20]:

$$\frac{\Delta f}{f} = -\frac{\Delta T}{T} = -\frac{1}{\gamma_t}\frac{\Delta(m/q)}{m/q} + \frac{\Delta v}{v}\left(1 - \frac{\gamma^2}{\gamma_t^2}\right), \tag{4}$$

where T is the revolution time.

According to Eq. (4), the revolution frequencies of the stored particles reflect their mass-over-charge ratios, provided the second term on the right hand side vanishes. The revolution frequency scale is calibrated utilizing numerous different nuclear species with experimentally known masses that are transmitted and injected in the same machine setting. The remaining challenge is that the secondary particles are produced in a nuclear reaction and their velocity spread $\Delta v/v$ in the present context is huge [25].

In the standard mode, $\Delta v/v$ is reduced by cooling the beam utilizing stochastic and/or electron cooling [29, 30]. For low-intensity fragment beams, small velocity spreads of $\Delta v/v \approx 10^{-7}$ can be achieved. However, the cooling process requires a few seconds to complete. Therefore, in spite of very successful usage in the past [31–36], this approach has a limited applicability for mass measurements of short-lived nuclei. Nevertheless, beam cooling will remain inevitable for investigations of exotic decay modes [37, 38].

In the isochronous mode, ion optics of the ring is changed so that the transition energy can be matched to the energy of the particle. In this mode, the differences in particle momenta for a given particle species are compensated by the lengths of their orbits, leading to the same revolution frequencies [39, 40]. The time consuming cooling is not needed and, apart from the negligible flight time from the production target until the injection into the ring, the measurement can start without any further delay. Therefore, *isochronous mass spectrometry* (IMS) is the technique of choice for short lived nuclei.

To be stored, $B\rho$ of the ion must match the range of $B\rho$ values accepted by the machine. Due to the broad momentum distributions, particles with very different m/q can be transmitted. On the one hand, this enables us to cover a large ($\Delta(m/q)/(m/q)$ of several percent) region of the nuclidic chart in a single setting [41, 42]. On the other hand, the required condition $\gamma = \gamma_t$ is not fulfilled for all ions. According to Eq. (4), the spread $\Delta f/f$ or $\Delta T/T$ increases with γ deviating from the fixed γ_t. Black squares in Fig. 1 illustrate the growth of revolution time widths, σ_T, as a function of m/q for ^{58}Ni projectile fragments measured in the CSRe [43, 44]. The range of m/q values corresponding to the smallest peak widths of $\sigma_T \approx 2$ ps is termed *isochronous window*. The nuclei outside of the isochronous window are typically disregarded in data analyses [41].

3 Revolution frequency determination

There are two ways to measure revolution frequencies or, equivalently, revolution times of the stored ions.

The first one employs a very thin foil inserted into the ring aperture. When a relativistic ion punches through the foil, secondary electrons are released on both sides of the foil. These electrons are guided by perpendicularly arranged electrostatic and weak magnetic fields to fast avalanche detectors, typically micro-channel plates. The signal from the anode is continuously sampled with high rate, reaching nowadays 50 giga-samples per second. As a result, each particle causes a train of repeating signals, which are unambiguously identified in the offline analysis. Procedures to extract mean revolution times take into account variations due to field inhomogeneities, energy losses in the foil, etc. The number of secondary electrons depends on the charge of the ion, which can be utilized as an additional identification criterion [47, 48].

Several implementations of such time-of-flight detectors have been in use at all three facilities in operation [49–53]. Also new-generation detectors are being prepared and

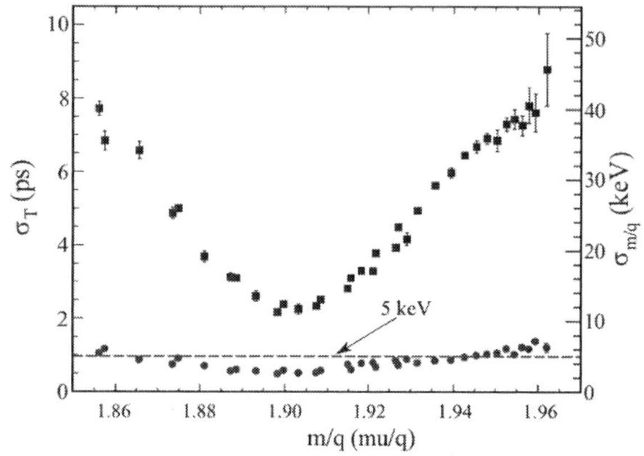

Figure 1. Standard deviations of the revolution-time peaks (left scale) for ^{58}Ni projectile fragments measured in the CSRe. The widths obtained with and without using the information on particle velocities are shown, respectively, with blue circles and black squares. The right scale provides the corresponding absolute accuracy of mass-over-charge ratios. Taken from [43, 44].

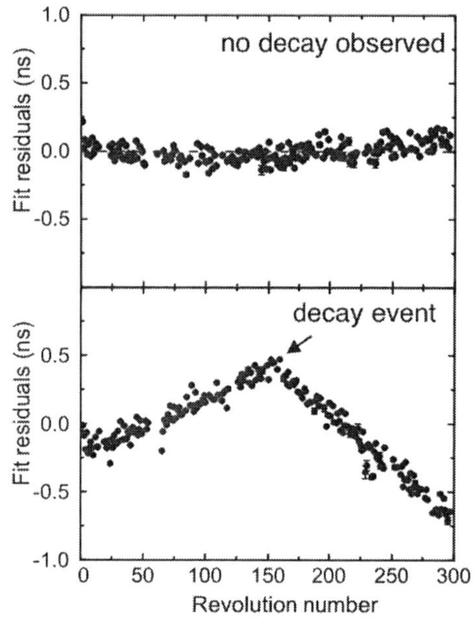

Figure 2. Residuals to the fits of timing signals for two ^{94m}Ru^{44+} ions measured by a time-of-flight detector in the CSRe. No changes in the slope, which would indicate a change of the m/q value, are seen for the ion in the top panel. A clear kink is observed in the lower panel. This kink manifests the de-excitation of ^{94m}Ru^{44+} isomer to the corresponding ground state ^{94g}Ru^{44+}. Modified from [56].

are optimized for the future storage rings [54]. The advantage of this approach is that only a few ten turns are sufficient to determine the revolution time, where one revolution takes merely a fraction of microsecond. An obvious disadvantage is that the overall method is destructive. Particles are lost after a few hundred revolutions. Therefore, in the context of the present discussion, no simultaneous lifetime measurements is possible. Exceptions are microsecond isomers which decay within these few hundred revolutions [55, 56], as illustrated in Fig. 2.

The second method employs Schottky detectors. Revolving particles periodically pass by these detectors. Highly charged particles induce an equal amount of charge on the conductive surface of the vacuum pipe, and hence also on the surface of the detector electrodes. Over the years, detectors have evolved from capacitive copper plate electrodes to sophisticated cavity-based resonators [57]. The Fourier analysis of the amplified output from the detector yields the noise power spectrum or *Schottky frequency spectrum*. Typically, higher harmonics of the ion revolution frequencies are analyzed, reaching $h \approx 200$ in the latest detector in the ESR [58]. The main advantage of these devices is that the detection of revolution frequencies is non-destructive. Whereas the frequencies in the Schottky spectrum are used for mass determinations, the amplitudes of the peaks are proportional to the corresponding numbers of the stored ions [59, 60]. By monitoring the development of peak areas in time, decay constants can be obtained [61, 62]. Furthermore, since the storage ring has a large acceptance, the decay daughter ions often stay in the machine. In such cases, combined analysis of the decrease of the peak area of the parent ions and the increase of the area(s) of the daughter-ion species enables one to establish also the branching ratios [63–66]. In addition, the detectors have no dynamic range limitations and are capable of simultaneously addressing single stored ions and milliampere beams.

Until very recently, Schottky-noise detectors were employed in experiments in standard operation mode with cooled ions. This technique is broadly known as time-resolved *Schottky mass spectrometry* (SMS) [67]. The

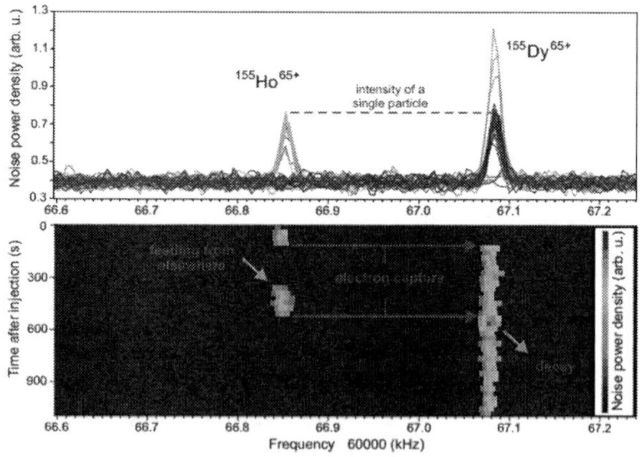

Figure 3. Time-resolved Schottky frequency spectra of $A = 165$ isobars produced in fragmentation of ^{209}Bi measured in the ESR. The time resolution is 30 s. Displayed is the 30th harmonic of the revolution frequency. Top panel illustrates individual spectra which indicate the constancy of the single-particle areas. The bottom panel shows the development of peak areas in time. See labels in the figure for more details. Redrawn from [67].

first generations of Schottky detectors required integration over several ten seconds to observe revolution frequencies of single ions. One such example is shown in Fig. 3. However, the detectors based on resonant cavities allowed for a considerable increase of time resolution. Presently, several milliseconds are sufficient for the frequency measurement [68, 69]. With this development, a straightforward next step is to employ such detectors in the isochronous mode. The first application of such combined time-resolved Schottky plus isochronous mass spectrometry (S+IMS) was done in the CSRe where half-lives of fully ionized $^{49}Cr^{24+}$ and $^{53}Fe^{26+}$ ions were measured [70].

The S+IMS technique has successfully been applied in the ESR to measure two-photon de-excitation of the first 0^+ state in ^{72}Ge [71]. The corresponding half-life is just a few ten milliseconds which confirms the power of the method. The data analysis is in its final stage and the results will be reported soon.

4 Velocity determination

The limitation set by the isochronous window severely reduces the efficiency of the method, where only a part of measured data is used. It has been suggested in [72] that an additional information on $B\rho$ or v of each stored ion is needed to expand the isochronous window ideally over the entire ring acceptance. In a test experiment in the ESR, $B\rho$ was defined by mechanical slits in the FRS to $\Delta B\rho/(B\rho) \approx 1.5 \times 10^{-4}$ [73]. Measurements were conducted employing a time-of-flight detector. A part of the revolution time spectrum in a region far away from the isochronous window is illustrated in Fig. 4. By applying this so-called *Bρ-tagging* method, improvement of the resolving power by up to a factor of eight was reported, though it is evident that the transmission efficiency is dramatically reduced.

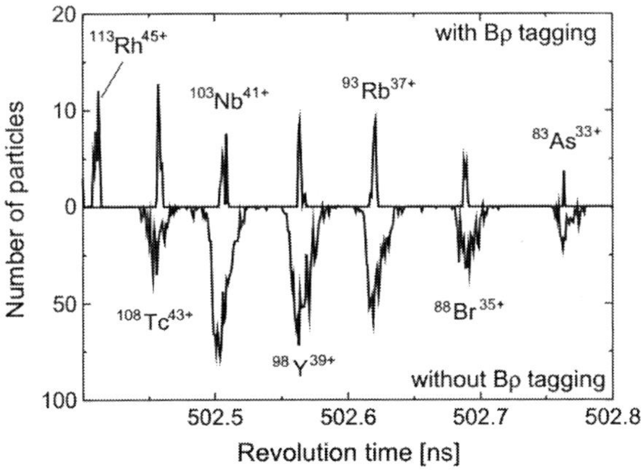

Figure 4. Parts of the revolution time spectra of ^{238}U fission fragments measured in the ESR by applying a time-of-flight detector. The shown region is far away from the isochronous window. Upper panel: The magnetic rigidities were limited in the FRS to $\Delta B\rho/(B\rho) \approx 1.5 \times 10^{-4}$. Lower panel: No cuts have been applied to the transmitted beam. Taken from [73].

The determination of $B\rho$ or v for every ion without limiting the transmission became the major quest of the storage ring mass spectrometry. At the R3, as mentioned earlier, the necessary identification of every ion includes the measurement of its time-of-flight in the Big-RIPS, which delivers information on the ion velocity prior to its injection into the ring [74]. At the CSRe, two identical time-of-flight detectors were installed in a straight section [75, 76]. Two trains of timing signals are obtained from these detectors, which provides information v and in turn on C and $B\rho$ for every ion [77]. Since the ions with the same $B\rho$ must have the same mean C, a very powerful analysis has been developed which enabled correction for non-isochronicity throughout the entire CSRe acceptance [43]. The improvement of the resolving power can be followed through the reduced peak widths shown with blue circles in Fig. 1. The dashed line indicated the absolute $m/q = 5$ keV limit. This means that by detecting just a single ion, its m/q value can be determined with this mass uncertainty, which is remarkable. This has been demonstrated through mass measurements of exotic ^{78}Kr projectile fragments in the CSRe. Only 4 particles were observed for ^{70}Kr and for ^{75}Sr, corresponding to the rate of two ions per week [78, 79]. This technique was named $B\rho$-defined IMS or just $B\rho$-IMS. By combining this sensitivity, speed of the measurement, efficiency and achieved mass precision, $B\rho$-IMS became one of the most powerful mass measurement methods.

However, the destructive time-of-flight detectors do not allow for addressing lifetimes and this remains an unsolved challenge. The pursued solution is to employ position sensitive Schottky cavities installed in a dispersive location of the ring [80–82]. The center of the cavity is placed at an offset to the beam pipe. Two cavities for two transverse directions are needed. The amplification gradient falls from the middle of the cavity towards its edge. Particles of the same charge passing the cavity at different locations will thus produce signals with different amplitudes. A high resolution Schottky cavity will be used for high precision frequency measurement which in turn will provide information on the charge of every particle. Combined information will be used to determine $B\rho$ of every ion. The prototype cavities for the proof-of-principle experiments are being constructed for the ESR and the R3 [83]. Dependent on the position sensitivity, a similar resolving power as shown in Fig. 1 could be envisaged.

5 Outlook

The mass measurement programs at the present ESR, CSRe and R3 storage rings have delivered important results, see e.g. [84–96]. They will continue in the future and even further expanded at the next-generation facilities FAIR, Germany and HIAF, China. The collector ring (CR) at FAIR [97] and spectroscopy ring (SRing) at HIAF [98] have been dedicatedly designed for IMS experiments. In comparison to IMS with a single time-of-flight detector and $B\rho$-tagging option, the $B\rho$-IMS technique has proven its superior efficiency. Therefore, both rings will have setups with two high-aperture time-of-flight detectors as required by the $B\rho$-IMS.

Implementation of longitudinal-sensitive Schottky cavities in the IMS enabled the S+IMS, which has been applied to half-life determinations. Further improvement of the time resolution of the S+IMS will be possible with higher quality cavities, though the price to pay is the lower bandwidth. Hence, several cavities tuned at different overlapping frequency ranges will be needed to cover the ring acceptance [99]. The most promising development is related to cavities with transverse-sensitivity. The non-destructive position determination will boost the mass resolving power. If successful, this will enable non-destructive mass measurements of short-lived nuclides. Simultaneously, the information on their lifetimes will be acquired as well.

Employing multiple Schottky cavities is a challenging task extending beyond the design on the detectors. Major advances in data handling and data analysis methods have been made in recent years, contributing to an unprecedented understanding of the devices themselves, but also allowing larger amounts of data to be processed for increased accuracy. For offline analysis, software frameworks have been developed to run on larger computer clusters to process large amounts of raw data [100]. The operation of Schottky detectors requires dedicated data acquisition systems. Data from multiple narrow-band and broadband Schottky detectors acquired by separate DAQ systems can be combined for online and offline analysis in order to combine the advantages of narrow-band data acquisition with a super-wide spectrum covering multiple revolution harmonics. To this end, open source and open hardware DAQ systems based on Software Defined Radio SDR for continuous and distributed data acquisition are planned [101].

In conclusion, taken that in the IMS: (i) a big region of the nuclidic chart is addressed at the same time, (ii) a single ion is sufficient for a successful measurement, and (iii) the time resolution is in the ten millisecond range, storage ring mass and lifetime spectrometry has the potential to deliver properties of nuclides in ground and isomeric states, which are simply inaccessible by other techniques.

Acknowledgements

This contribution would not be possible without successful collaboration with colleagues working at the ESR, CSRe and R3. This research received support from the European Research Council (ERC) under the EU Horizon 2020 research and innovation programme (ERC-CoG "ASTRUm"), and the State of Hesse within the Research Cluster ELEMENTS (Project ID 500/10.006).

References

[1] A. Bohr and B.R. Mottelson, *Nuclear Structure*, World Scientific Publ., Singapore, 1998.

[2] G. C. Rodrigues, *et al.*, At. Data Nucl. Data Tab. **86**, 117–233 (2004).

[3] Yu. N. Novikov, *et al.*, Nucl. Phys. A **697**, 92–106 (2002).

[4] E. M. Burbidge, *et al.*, Rev. Mod. Phys. **29**, 547–650 (1957).

[5] J. J. Cowan, *et al.*, Rev. Mod. Phys. **93**, 015002 (2021).

[6] W. J. Huang, *et al.*, Chin. Phys. C **45**, 030002 (2021).

[7] J. Erler, *et al.*, Nature **486**, 509 (2012).

[8] A. Sobiczewski, *et al.*, Phys. Rev. C **89**, 024311 (2014).

[9] A. Sobiczewski, *et al.*, At. Data Nucl. Data Tables **119**, 1–32 (2018).

[10] D. Lunney *et al.*, Rev. Mod. Phys. **75**, 1021–1082 (2003).

[11] K. Blaum, Phys. Rep. **425**, 1–78 (2006).

[12] F. Bosch, *et al.*, Int. J. Mass Spectr. **251**, 212–219 (2006).

[13] F. Bosch, *et al.*, Int. J. Mass Spectr. **349-350**, 151–161 (2013).

[14] Y. Litvinov, *et al.*, Nucl. Instr. Meth. B **317**, 603–616 (2013).

[15] T. Eronen, *et al.*, Prog. Part. Nucl. Phys. **91**, 259–293 (2016).

[16] Y. H. Zhang, *et al.*, Phys. Scripta **91**, 073002 (2016).

[17] T. Yamaguchi, *et al.*, Prog. Part. Nucl. Phys. **120**, 103882 (2021).

[18] F. G. Kondev, *et al.*, Chin. Phys. C **45**, 030001 (2021).

[19] H. Geissel, *et al.*, Phys. Rev. Lett. **68**, 3412-3415 (1992).

[20] B. Franzke, H. Geissel and G. Münzenberg, Mass Spectr. Rev. **27**, 428–469 (2008).

[21] J. Xia, *et al.*, Nucl. Instr. Meth. A **488**, 11–25 (2002).

[22] M. Z. Sun, *et al.*, Front. Phys. **13**, 132112 (2018).

[23] A. Ozawa, *et al.*, Prog. Theor. Exp. Phys. **2012**, 03C009 (2012).

[24] C. A. Bertulani and P. Danielewicz, *Introduction to Nuclear Reactions*, CRC Press, Boca Raton, 2021.

[25] H. Geissel, G. Münzenberg and K. Riisager, Ann. Rev. Nucl. Part. Sci. **45**, 163-203 (1995).

[26] T. Yamaguchi, *et al.*, Int. J. Mass Spectr. **349-350**, 240–246 (2013).

[27] F. Hinterberger, *Physik der Teilchenbeschleuniger und Ionenoptik*, Springer-Verlag, Berlin, Heidelberg, 2008.

[28] M. Steck and Yu. A. Litvinov, Prog. Part. Nucl. Phys. **115**, 103811 (2020).

[29] H. Geissel, *et al.*, Nucl. Phys. A **746**, 150–155 (2004).

[30] H. Geissel, *et al.*, Eur. J. Special Topics **150**, 109–115 (2007).

[31] T. Radon *et al.*, Phys. Rev. Lett. **78**, 4701–4704 (1997).

[32] T. Radon, *et al.*, Nucl. Phys. A **677**, 75-99 (2000).

[33] Y. Litvinov, *et al.*, Nucl. Phys. A **756**, 3–38 (2005).

[34] L. Chen *et al.*, Phys. Rev. Lett. **102**, 122503 (2009).

[35] L. Chen *et al.*, Phys. Lett. B **691**, 234–237 (2010).

[36] D. Shubina, *et al.*, Phys. Rev. C **88**, 024310 (2003).

[37] Y. Litvinov, *et al.*, Rep. Prog. Phys. **74**, 016301 (2011).
[38] F. Bosch, *et al.*, Prog. Part. Nucl. Phys. **73**, 84–140 (2013).
[39] M. Hausmann, *et al.*, Nucl. Instr. Meth. A **446**, 569–580 (2000).
[40] M. Hausmann, *et al.*, Hyperf. Interactions **132**, 289–295 (2001).
[41] B. Sun, *et al.*, Nucl. Phys. A **812**, 1–12 (2008).
[42] X. Tu, *et al.*, Nucl. Instr. Meth. A **654**, 213–218 (2011).
[43] M. Wang *et al.*, Phys. Rev. C **106**, L051301 (2022).
[44] M. Zhang *et al.*, Eur. Phys. J. A **59**, 27 (2023).
[45] M. Zhang *et al.*, Nucl. Instr. Meth. A **1027**, 166329 (2022).
[46] Y. M. Xing *et al.*, Phys. Rev. C **107**, 014304 (2023).
[47] P. Shuai, *et al.*, Phys. Lett. B **735**, 327–331 (2014).
[48] X. Zhou, *et al.*, Nucl. Sci. Techniq. **32**, 37 (2021).
[49] J. Trötscher *et al.*, Nucl. Instr. Meth. B **70**, 455–458 (1992).
[50] B. Mei *et al.*, Nucl. Instr. Meth. A **624**, 109–113 (2009).
[51] W. Zhang *et al.*, Nucl. Instr. Meth. A **755**, 38–43 (2014).
[52] W. Zhang *et al.*, Nucl. Instr. Meth. A **756**, 1–5 (2014).
[53] D. Nagae *et al.*, Nucl. Instr. Meth. A **986**, 164713 (2021).
[54] N. Kuzminchuk-Feuerstein *et al.*, Nucl. Instr. Meth. A **821**, 160–168 (2016).
[55] B. Sun *et al.*, Phys. Lett. B **688**, 294–297 (2010).
[56] Q. Zeng *et al.*, Phys. Rev. C **96**, 031303 (2017).
[57] F. Nolden, *et al.*, Nucl. Instr. Meth. A **659**, 69–77 (2011).
[58] M. S. Sanjari, *et al.*, Rev. Sci. Instr. **91**, 083303 (2020).
[59] H. Irnich, *et al.*, Phys. Rev. Lett. **75**, 4182–4185 (1995).
[60] Y. Litvinov, *et al.*, Phys. Lett. B **573**, 80–85 (2003).
[61] Y. Litvinov, *et al.*, Phys. Lett. B **664**, 162–168 (2008).
[62] D. Atanasov, *et al.*, J. Phys. B **48**, 144024 (2015).
[63] T. Ohtsubo, *et al.*, Phys. Rev. Lett. **95**, 052501 (2005).
[64] Y. Litvinov, *et al.*, Phys. Rev. Lett. **99**, 262501 (2007).
[65] N. Winckler, *et al.*, Phys. Lett. B **679**, 36–40 (2009).
[66] D. R. Atanasov, *et al.*, Eur. Phys. J. A **48**, 22 (2012).
[67] Y. Litvinov, *et al.*, Nucl. Phys. A **734**, 473-476 (2004).
[68] P. Kienle, *et al.*, Phys. Lett. B **726**, 638–645 (2013).
[69] F. Ozturk, *et al.*, Phys. Lett. B **797**, 134800 (2019).
[70] X. L. Tu, *et al.*, Phys. Rev. C **97**, 014321 (2018).
[71] W. Korten, *et al.*, Proposal, GSI Program Advisory Committee (2022).
[72] H. Geissel, *et al.*, J. Phys. G **31**, S1779–S1783 (2005).
[73] H. Geissel, *et al.*, Hyperf. Interactions **173**, 49–54 (2006).
[74] H. F. Li, *et al.*, Phys. Rev. Lett. **128**, 152701 (2022).
[75] P. Shuai, *et al.*, Nucl. Instr. Meth. B **376**, 311–315 (2016).
[76] X. L. Yan, *et al.*, Nucl. Instr. Meth. A **931**, 52–59 (2019).
[77] X. Zhou, *et al.*, Phys. Rev. Accel. Beams **24**, 042802 (2021).
[78] X. Zhou, *et al.*, Nat. Phys., in press (2023).
[79] M. Wang, *et al.*, Phys. Rev. Lett., in press (2023).
[80] M. S. Sanjari, *et al.*, Phys. Scripta **T166**, 014060 (2015).
[81] X. Chen, *et al.*, Hyperf. Interactions **235**, 51–59 (2015).
[82] X. Chen, *et al.*, Nucl. Instr. Meth. A **826**, 39–47 (2016).
[83] D. Dmytriiev, *et al.*, Nucl. Instr. Meth. B **463**, 320–323 (2020).
[84] J. Stadlmann, *et al.*, Phys. Lett. B **586**, 27–33 (2004).
[85] Y. Litvinov, *et al.*, Phys. Rev. Lett. **95**, 042501 (2005).
[86] M. W. Reed, *et al.*, Phys. Rev. Lett. **105**, 172501 (2010).
[87] X. Tu, *et al.*, Phys. Rev. Lett. **106**, 112501 (2011).
[88] Y. Zhang, *et al.*, Phys. Rev. Lett. **109**, 102501 (2012).
[89] X. L. Yan, *et al.*, Astroph. J. Lett. **766**, L8 (2013).
[90] L. Chen, *et al.*, Phys. Rev. Lett. **110**, 122502 (2013).
[91] X. Xu, *et al.*, Phys. Rev. Lett. **117**, 182503 (2016).
[92] R. Knöbel, *et al.*, Eur. Phys. J. A **52**, 138 (2016).
[93] R. Knöbel, *et al.*, Phys. Lett. B **754**, 288–293 (2016).
[94] P. Zhang, *et al.*, Phys. Lett. B **767**, 20–24 (2017).
[95] Y. H. Zhang, *et al.*, Phys. Rev. C **98**, 014319 (2018).
[96] Y. M. Xing, *et al.*, Phys. Lett. B **781**, 358–363 (2018).
[97] P. Walker, *et al.*, Int. J. Mass Spectr. **349-350**, 247–254 (2013).
[98] J. Yang, *et al.*, Nucl. Instr. Meth. B **317**, 263–265 (2013).
[99] M. S. Sanjari, *et al.*, *Technical Design Report for the Schottky Detector System of ILIMA* (2018).
[100] M. S. Sanjari, Zenodo, **7615693**, (2023), https://zenodo.org/record/7615693
[101] D. Dmytriiev, *et al.*, J. Phys. Conf. Ser. 1668 **012014** (2020).

Alpha emission from fast neutron interactions with ^{64}Zn nuclei

Cristiana Oprea[1,*], *Alexandru* Mihul[2,3], *Ioan* Oprea[1], *Ion* Gruia[1,2] and *Sorin* Zgura[4]

[1]Romanian Scientific Research Agency, Bucharest (ANCS), County Department for Education Bihor, Mihai Eminescu 11, Oradea 410010, Bihor, Romania
[2]University of Bucharest, Faculty of Physics, Magurele 077125, Ilfov, Romania
[3]European Center for Nuclear Research (CERN), 1211 Geneva 23, Switzerland
[4]Institute for Space Sciences, 077124 Magurele, Romania

Abstract. In this study, the nuclear process $^{64}Zn(n, \alpha)^{61}Ni$ induced by fast neutrons in the ^{64}Zn nucleus was investigated. With the authors' computer programs and Talys codes, cross-sections, angular correlations, and forward-backward asymmetry effects were investigated for incident neutrons with energies ranging from 0.5 MeV to 25 MeV. ^{61}Ni cross-sections were determined through an investigation of nuclear reaction mechanisms (direct, compound, pre-equilibrium) and discrete and continuum states within the ^{61}Ni residual nucleus. Our analysis of the neutron incident and alpha emergent channels included the extraction of optical potential parameters (with real and imaginary parts) with volume, surface, and spin-orbit components. The cross-sections calculated theoretically and found in our experiments confirm earlier published data. For neutrons with energies of a few MeV, the $^{64}Zn(n, \alpha)^{61}Ni$ reaction produced by fast neutrons exhibits an unusual forward-backward asymmetry. At this energy, the observed effect cannot be explained by direct mechanisms with orders of magnitude lower; compound processes that predominate must be used. To understand the observed forward-backward asymmetry effect, the spectra of emitted alpha particles were modelled by the direct Monte Carlo method while taking into consideration the energy loss of alpha particles on Zn targets with finite dimensions.

*Corresponding author: coprea2007@yahoo.co.uk

1. Introduction

Nuclear reactions caused by fast neutrons that are followed by the emission of charged particles (protons and alphas) are traditionally studied. [1-5]. Fast neutrons processes are of interest for basic and applied researches. In the field of fundamental physics fast neutrons reactions give new data on the structure of atomic nuclei and the mechanisms of nuclear reactions. Accurate information is provided by these processes for nuclear fission and future fusion reactors, handling long-life nuclear waste, transmutation, energy, and accelerated driven systems (ADS) programs [6,7].
There are five naturally occurring isotopes of the chemical element zinc: ^{64}Zn, ^{66}Zn, ^{67}Zn, ^{68}Zn, and ^{70}Zn, with abundances of 49.2%, 27.7%, 4%, 18.5%, and 0.6%, respectively [8]. Nuclear reaction ^{64}Zn(n,α)^{61}Ni induced fast neutrons with energies starting from 0.5 MeV up to 25 MeV was investigated. Using programs written by the authors and also using Talys code, cross-sections, angle distributions, alpha spectra, and parameters of Wood-Saxon potential in incident and emergent channels were produced [9]. Theoretical evaluations and conclusions are in good agreement with the available experimental data.

2. Elements of theory

Cross-sections, angular distributions and alpha spectra were obtained using Talys code [9]. All nuclear reaction mechanisms are taken into account in the analyses. Distorted Wave Born Approximation (DWBA) [9,10] is used to explain direct processes, a statistical model of nuclear reaction using Hauser-Feshbach formalism [9,11] is used to describe compound processes, and a two-component exciton model is used to characterize pre-equilibrium processes. [9,12]. In the framework of the Fermi gas model with constant temperature, the nuclear states density was determined [9].

3. Results and discussions

Cross-section of ^{64}Zn(n, α)^{61}Ni reaction (heat of reaction Q = 3.86 MeV) was evaluated with Talys. Contributions to the cross-section of nuclear reaction mechanisms as well as the discrete and continuum states of residual nuclei were obtained. Results are shown in Fig. 1.
Compound processes are dominant in the whole energy interval (Fig. 1.a curve 2). Pre-equilibrium mechanisms are activated after 7-8 MeV (Fig. 1.a curve 3). Direct mechanisms can be neglected (Fig. 1.a curve 4). Compound processes for neutrons with energies more than 15 MeV originate from a pre-equilibrium mechanism.

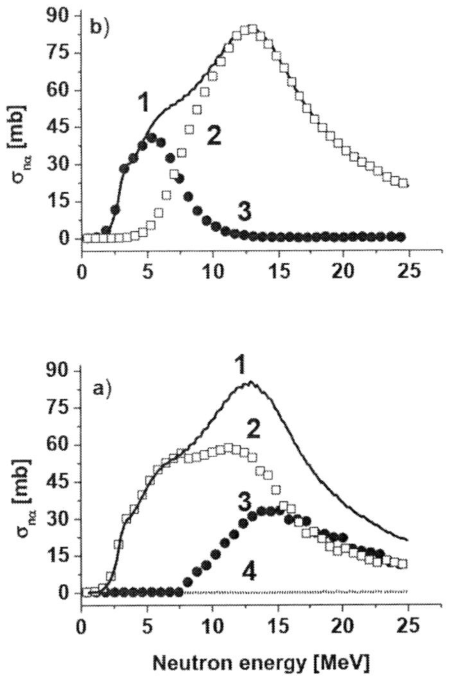

Fig. 1. Cross-section of ^{64}Zn(n, α)^{61}Ni reaction. Contribution of: a) Nuclear reaction mechanisms; 1) Total (1 -> 2 + 3 + 4) 2) Compound 3) Pre-equilibrium 4) Direct. b) States of residual nuclei; 1) Total (1 -> 2 + 3) 2) Continuum 3) Discrete

Contributions to the cross-section of the discrete and continuum states of the residual nucleus are shown in Fig. 1.b. Cross-section is primarily provided by discrete states for neutron energies up to 7 - 8 MeV (Fig. 1.b curve 3). There is a rising contribution of continuum states to the cross-section at incident energy higher than 8 MeV. After 10 MeV neutron energy, the cross-section can only be produced by continuum states (Fig. 1.b curve 2).

The theoretical assessment of the ^{64}Zn(n,α)^{61}Ni cross-section for incident neutron energies ranging from 0.5 MeV to 25 MeV is shown in Fig. 1.ab curve 1. Figure 2 compares estimated cross-sections to experimental data from [13–16]. There are two sets of experimental data that can be seen. Two alternative sets of parameters can be obtained primarily by adjusting the optical potential's characteristics. In Fig. 2 a set of parameters which is passing between both groups of data is represented. Because different cross-section values were measured in the experiment for the same incident neutron energy, the authors decided to choose the above approach.
For example, the experimental cross-section had three distinct values at 4 MeV incident energies: 18, 41, and 60 mb, respectively [13–15]. At the same energy, theoretical cross-section is 33 mb. Taking into account the above considerations, experimental data are in good agreement with theoretical evaluations.

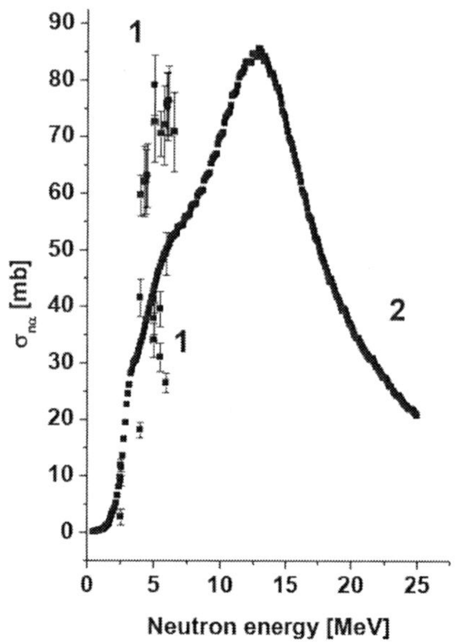

Fig. 2. Comparison between experimental and theoretical data. Points – experiment. Line – Talys evaluation (Curve 2 from Fig. 1.ab)

Differential cross-section was evaluated for a large incident neutrons energy range. For incident neutron energies E_n = 4 and 5.5 MeV, respectively, theoretical calculations and experimental results of ^{64}Zn(n, α)^{61}Ni differential cross-section as a function of polar angle θ are shown in Fig. 3.ab. Experimental information is sourced from [14].

The computed cross-sections from Fig. 1 are $\sigma_{n\alpha}$ = 33 mb and 47 mb for incident neutron energies E_n = 4 and 5.5 MeV, respectively. The experimental cross-sections for the same energies in Fig. 3 are shown as $\sigma_{n\alpha}(E_n = 4$ MeV$)$ = (59.6±3.3) mb and $\sigma_{n\alpha}(E_n = 5.5$ MeV$)$ = (70.5±4.0) mb, respectively [14]. Therefore, a new set of Wood-Saxon potential parameters had to be discovered in order to have the same cross-section value in both theory and experiment. If this assumption is fulfilled then the measured differential cross-section is described very well by theoretical evaluation for neutron energy E_n = 4 MeV, and satisfactorily for E_n = 5.5 MeV. Theoretical analysis reveals that there is only a compound mechanism present for neutron energy of 4 and 5.5 MeV and no direct process. Computer simulations were carried out using the results of the differential cross-section from Fig. 3.ab. Starting from differential cross section, angular correlations were calculated and applying direct Monte-Carlo method, it was possible to determine the angular distribution of the polar angle θ [17].

The forward-backward effect was evaluated using simulated angular distribution for targets with different dimensions. In any event, simulations have not identified any forward-backward effect for either a point-like target or a target with a thickness of 266.3 g/cm^2 [14]. In the simulations, the forward-backward asymmetry effect is defined as the ratio between all forward events (polar angle θ from 0 to π/2) and all backward events (polar angle θ from π/2 to π) [2, 14]. The alpha particle loss in the target was assessed for targets with finite dimensions [18].

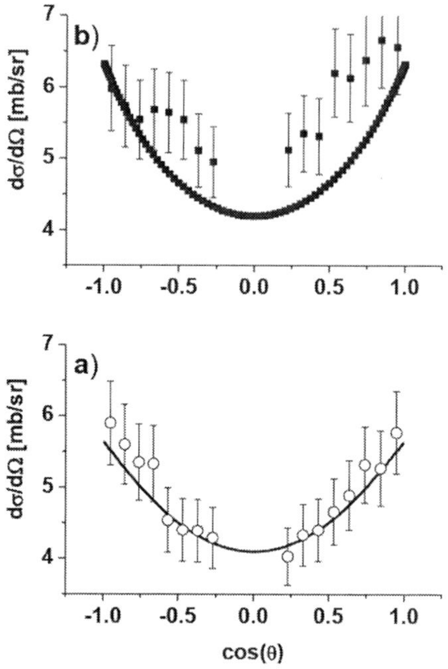

Fig. 3. ^{64}Zn(n, α)^{61}Ni differential cross-section for neutron energy, a) E_n = 4 MeV; b) 5.5 MeV. Line – Talys evaluation. Points – experimental data

Table 1. Parameters of Wood Saxon potential with components: volume, surface and spin-orbit with real and imaginary part

Channel	Volume WS- Real		
	V [MeV]	r_V [fm]	a_V [fm^{-1}]
n	52.69	1.203	0.668
α	169.37	1.184	0.676
	Spin – orbit – Real		
	V_{SO} [MeV]	r_{VSO} [fm]	a_{VSO} [fm^{-1}]
n	5.08	1.024	0.590
α	0	0	0
Channel	Volume WS - Imaginary		
	W [MeV]	r_W [fm]	a_W [fm^{-1}]
n	0.19	1.279	0.668
α	25.70	1.340	0.500
	Spin – orbit - Imaginary		
	W_{SO} [MeV]	r_{WSO} [fm]	a_{WSO} [fm^{-1}]
n	0.01	1.024	0.590
α	0	0	0

The absence of the forward-backward effect can be attributed to the lack of direct processes in the cross-sections The forward-backward effect is suggested by experimental results of differential cross-section for En

= 5.5 MeV (see Fig. 3b) and some higher energies [13-16], although the observed asymmetry is not brought on by the target thickness. If this effect does indeed exist, it can be accounted for by the presence of other open channels with the participation of alpha particles. All open exit channels were taken into account during the analyses. 30 distinct states of the residual nucleus were considered for both elastic and inelastic channels, but only 10 levels were considered for reactions. Table 1 lists the optical potential parameters for the incident and emerging channels according to [9].

4. Conclusions

The nuclear reaction ^{64}Zn(n, α)^{61}Ni, with neutron energy ranging from 0.5 to 25 MeV, was under investigation. Using Talys code and the author's programs, cross-sections, angle distributions, and alpha spectra were examined. Theoretical assessments and available experimental data are in good agreement. The largest contribution to the cross-sections is given by compound processes. Practically, direct processes can be neglected. Up to 7-8 MeV contribution of the discrete states of the residual nucleus in the cross-sections are dominant but as input neutron energy increases, continuum states become important and gradually replace the discrete ones. Contrary to what differential cross-sections for neutrons with energy higher than 5 MeV anticipated, forward-backward effects were not visible in the modelled alpha spectra starting from angular distribution. In the absence of direct mechanisms, the observed forward-backward asymmetry effect can be explained by the existence of additional open emergent channels with the participation of alpha particles.

References

1. B. Huaiyong, J. Haoyu, L. Yi, C. Zengqi, C. Jinxiang, Z. Guohui, Yu. M. Gledenov, M. V. Sedysheva, G. Khuukhenkhuu, R. Xichao, H. Hanxiong, R. Jie, F. Qiwen, Phys. Rev. **C 99**, 024619 (2019)
2. C. Oprea, A. I. Oprea, Bulletin of the Russian Academy of Sciences: Physics, Allerton Press Inc. and Springer Nature Switzerland AG, 85, **12**, 1471 (2021)
3. J. Han, J. Liu, X. Liu, X. Fan, Zh. Wang, J. Chen, G. Zhang, Yu. M. Gledenov, M. V. Sedysheva, L. Krupa, G. Khuukhenkhuu, P. J. Szalanski, European Physical Journal **A 51**, 12 (2015)
4. G. Khuukhenkhuu, Yu. M. Gledenov, M. V. Sedysheva, M. Odsuren, J. Munkhsaikhan, T. Delgersaikhan, Journal Physics of Elementary Particles and Atomic Nuclei (PEPAN) Letters, 11 **6**, 1159 (2014)
5. Yu. M. Gledenov, P. E. Koehler, J. Andrzejewski, K. H. Guber, Phys. Rev. **C 62** 042801(R)
6. M. Salvatores, I. Slessarev, A. Tchistiakov, Nucl. Sci. Eng. 309, 130, (1998)
7. M. Salvatores, A. Zaetta, C. Girard, M. Delpech, I. Slessarev, J. Tomassi, Appl. Radiat. Isot. 46 **6**, 681 (1995)
8. J. Meija, T. B. Coplen, M. Berglund, W. A. Brand, P. D. Bièvre, M. Gröning, N. E. Holden, J. Irrgeher, R. D. Loss, T. Walczyk, T. Prohaska, Pure Appl.Chem, 88 **3**, 265 (2016)
9. A. J. Koning, S. Hilaire and M.C. Duijvestijn, TALYS-1.0., Proceedings of the International Conference on Nuclear Data for Science and Technology, April 22-27, 2007, Nice, France, editors O. Bersillon, F. Gunsing, E. Bauge, R. Jacqmin, S. Leray, EDP Sciences, 211 (2008)
10. G. R. Satchler, Direct Nuclear Reactions, Oxford University Press, New York (1983)
11. G. W. Hauser, H. Feshbach, Phys. Rev., 87 **1**, 366 (1952)
12. A.J. Koning, M.C. Duijvestijn, Nucl. Phys. A, 744, p. 15 (2004)
13. G. Zhang, R. Cao, J. Chen, G. Tang, Yu. M. Gledenov, M. Sedysheva, G. Khuukheenkhuu, Nuclear Science and Engineering, 156, 115 (2007)
14. G. Zhang, J. Zhang, R. Cao, L. Guo, J. Chen, Yu. M. Gledenov, M. V. Sedysheva, P. J. Szalanski, Nuclear Science and Engineering 160, 123 (2008)
15. J. Yuan, Z. Chen, G. Tang, G. Zhang, J. Chen, Z. Shi, Yu. M. Gledenov, M. Sedysheva, G. Khuukheenkhuu, Nuclear Science and Engineering, 144, 108 (2003)
16. Yu. M. Gledenov, M. V. Sedysheva, P. V. Sedyshev, A. Oprea, G. Khuukheenkhuu, Z. Chen, Y. Chen, J. Yuan, G. Tang, G. Zhang, J. Andrzejevski, P. J. Szalanski, Bulletin of Russian Academy of Science: Physics Series, 67, 1784 (2003)
17. C. Oprea, A. I. Oprea, Bulletin of Russian Academy of Sciences: Physics, Allerton Press Inc. and Springer Nature Switzerland AG, 86 **11**, 1410 (2022)
18. J. F. Ziegler, SRIM & TRIM Software – http://www.srim.org – accessed at 15.02.2023

Recent results from the DESPEC campaign at GSI

Giovanna Benzoni[1,*], for the HISPEC-DESPEC collaboration

[1]Istituto Nazionale di Fisica Nucleare, Sezione di Milano, Via Celoria 16, I-20133 Milano, Italy

Abstract. HISPEC-DESPEC is a collaboration, part of NUSTAR, aiming at studying the structure of nuclei produced by projectile fragmentation reactions exploiting the beams delivered by the accelerator system at present and future facilities GSI and FAIR. In the early stages of the experimental activity at FAIR Phase-0, started in 2020, the collaboration focused the attention on studies following the decay of long-living isomeric states and β-decays with lifetimes in the millisecond-to-second range. A description of the set-up and of first results obtained in the experimental campaigns covering years 2020-2022 are the subject of this talk.

1 Introduction

The HISPEC-DESPEC (H/D) collaboration, part of the NUSTAR research program, exploits the actual and future high energy and intensity beams delivered by the GSI-FAIR accelerator complex, with two main approaches: HISPEC (High-resolution In-flight SPECtroscopy) aims at performing detailed spectroscopic studies by measuring outgoing γ radiation via nuclear and Coulomb excitation techniques, while DESPEC (DEcay SPECtroscopy) experiments aim at a detailed study of the decay mechanism in the most exotic nuclear species at reach at GSI-FAIR, thanks to the measurement of decay half-lives, competing decay modes and isomeric states.

The core of the HISPEC setup is the Advanced GAmma Tracking Array (AGATA) [1], latest generation array of segmented HPGe detectors, able to reconstruct the initial energy of the emitted γ radiation also in case of large Doppler displacement. To track the secondary reaction products AGATA is coupled to the LYCCA (Lund York Cologne Calorimeter) array [2], based on time of flight (TOF), energy loss and total energy measurements, in conjunction with a magnetic spectrometer. Lifetimes of specific nuclear states can be measured using plunger devices, or the active detectors of the LISA project [3]. For the DESPEC experimental program, ions are stopped in the Advanced Implantation Detector Array (AIDA) and their subsequent decays measured. AIDA is surrounded by a modular and compact high-resolution γ-detection array, DEGAS (DEspec Germanium Array Spectrometer), as well as the neutron arrays MONSTER (MOdular Neutron time of flight SpectromeTER) and Beta-dELayEd Neutron detector (BELEN). In addition, the Decay Total Absorption Spectrometer (DTAS) and the FAst TIMing Array (FATIMA) can be employed.

At the restart of operations at the GSI-FAIR facility, the H/D collaboration initiated an experimental campaign exploiting mainly the DESPEC suite of detectors, with the aim of addressing the nuclear structure of exotic nuclei my means of isomeric and radioactive decays.

The study of isomeric states and of the radioactive decays can provide the first glimpses of the internal structure via its γ-ray 'fingerprints'. A systematic study of key experimental signatures, such as the energy of the first excited state and/or of the ratio of the excitation energies of the lowest-lying levels in even-even nuclei can, e.g., demonstrate the erosion of the established magic numbers or reveal the emergence of nuclear collective excitations.

The experimental campaign targeted several areas in the chart of nuclei, ranging from the p-rich nuclei south of ^{100}Sn, the rare-earth region around ^{190}W and the heavy systems above ^{208}Pb.

The set-ups and experiments performed using the so-called Phase-0 campaign at the restart of the GSI-FAIR facility are described in these proceedings, together with an overview of future plans.

2 The DESPEC setup for the GSI-FAIR Phase-0 campaigns

The exotic nuclei under investigation are produced by fragmentation reaction of the relativistic beams, provided by the LINAC+SIS18 accelerator complex. The beams imping onto a primary target of ^{9}Be, located at the entrance of the FRagment Seprator (FRS) [4]. The zero-degree spectrometer, FRS, is used to select and transport the ions of interest towards the final detection set-up, located, for the campaigns in 2020-2022, in the S4 area at the end of the FRS, corresponding to its 4th focal point. The separation is achieved thanks to the $B\rho - \Delta E - B\rho$ technique, where the analysing capabilities of FRS dipoles are coupled to energy losses in passive degraders, to select and focus the wanted species at the final focus.

Postion, time of flight (TOF) and energy loss in detectors located along the flying path in the intermediate and

*e-mail: giovanna.benzoni@mi.infn.it

final focal planes of FRS, serve to reconstruct, on an event-by-event basis, the ions of interest [5].

During the Phase-0 campaign, several detector combinations have been commissioned and used. The common piece of instrumentation is the AIDA set-up, based on Double Sided Silicon Strip Detectors (DSSSD) characterised by a high degree of pixelation. Each unit is a 8x8 cm^2 DSSSD tile with thickness of 1 mm, composed of 128x128 strips (16384 pixels), with a 0.560 mm interstrip pitch. The DSSSDs are purchased as single (8×8cm^2) or triple (24×8cm^2) wafer devices. AIDA is used both to detect the implantation of the energetic fragments, depositing energies up to several GeVs, and the particles resulting from the radioactive decay of the parent nucleus, on the MeV energy scale.

Thanks to the high degree of pixelation, one can define the position (X, Y, Z, depth) where the nucleus has been implanted, and correlate it to the subsequently emitted decay particles. The measurement of the time elapsed between the implantation and the decay returns the half-life of the radioactive decay itself, ranging between tens of ms to several tens of seconds.

In the Phase-0 campaign AIDA detectors were used in both the narrow (8 cm wide) and wide (24 cm) combinations, with detectors placed in a stack to cover a large implantation range. AIDA is sandwiched between plastic scintillator detectors, also capable of measuring implants and decay particles. The fast plastic detector, named βPlastic, consists of a tile of BC-400 scintillator material, covering the same area as the AIDA detectors readout by a series of SiPM glued to each side. The βPlast detectors work together with AIDA in order to identify ion-β correlations resulting in an overall efficiency of 12% for events implanted in AIDA and β particles released in both the DSSSD and the plastic detector. AIDA itself is measured to have an efficiency of 35% for ion-β correlations.

In order to get access to various experimental quantities, several detector set-ups have been deployed in the Phase-0 campaign.

A combination of high-resolution, high-efficiency HPGe detectors and the FAst TIMing Array (FATIMA) set-up of $LaBr_3$:Ce scintillators is used to measure the internal γ-deexcitation following internal deexcitation of the nuclei, while the DESPEC Beta Decay Total Absorption Gamma-Ray Spectrometer (DTAS) provides access to the full B(GT).

The DESPEC Germanium Array Spectrometer (DEGAS) high-resolution high-efficiency HPGe array exploits the EUROBALL Cluster detectors, rearranged into clusters of three detectors, so-called triple clusters. DEGAS is designed to cover an implantation area of about 24×8 cm^2, demanded by the wider focal plane of the Super-FRS at FAIR, and well suited for the S4 focal plane of the FRS. The on-board electronics includes low-noise preamplifiers with overload recovery, high-voltage generators, power management and temperature controls. GSI-made (FEBEX) modules have been successfully employed in the experimental campaign performed in 2020-2022. The digitizer data are used to obtain both energy and time information [6].

FATIMA is a modular system, where the detector number and size, as well as angular coverage, could be changed according to the needs of the experiment. In the standard configuration, it consists of 36 $LaBr_3$:Ce crystals with a diameter size of 1.5×2 $inch^2$ length. FATIMA can be used to determine lifetimes by measuring the time difference between two detected γ rays, or, alternatively, it can be used to measure lifetimes of excited states populated by β decay. Details of the implementation of the FATIMA set-up at GSI are described in Ref, [7, 8], while a further description of the DESPEC detetcors implementation at GSI-FAIR is given in [9].

The high-resolution set-up here described can be replaced by the DESPEC Beta Decay Total Absorption Gamma-Ray Spectrometer (DTAS) [10–12]. A modular design based on NaI(Tl) scintillators was chosen to take advantage of the information on γ-ray cascade multiplicity in data analysis. Sixteen modules with crystal dimensions 15×15×25 cm^3 constitute DTAS providing the following figures of merit: measured energy resolution of 6.0% at 1.33 MeV, measured time resolution of 10 ns, expected spectrometer efficiency (from MC simulations) 86% (total) and 58% (peak) at the 1.33 MeV ^{60}Co line.

Figure 1 shows, in the top panel, the layout of the HPGe+FATIMA high-resolution, while the bottom panel presents the coupling to the DTAS array, both realised and used in the Phase-0 campaign.

The experimental approach used in the campaign in 2020-2022 is well established, having the DESPEC collaboration a long tradition in such studies. Novelties and latest improvements are focused on advances in the detector set-up and the use of digital electronics. Specifically, the newly developed DESPEC array, based on the coupling of the AIDA highly-segmented Silicon array and the high-efficiency DEGAS HPGe array, provides a step forward in the study of decay radiation in a fragmentation facility. The high-granularity of the DSSSDs of AIDA allows to sustain a high-count rate, making it feasible to study several species at the same time, and reducing the background induced by non-correlated implantations. In the case of isomeric decay this is particularly important, since it allows to accept a higher number of incoming ions. The increased γ-detection efficiency provided by the DEGAS array allows to detect weak transitions and low-branching ratios from new weakly populated states. In addition, the improved efficiency helps to reconstruct level schemes by allowing the analysis of high-fold coincidences and improved particle-γ correlations.

With comparable primary beam intensities to previous years, the improved sensitivity of the DESPEC array allows to reach further out into new the proton and neutron rich isotopes, with the GSI-FAIR accelerator complex coupled to the FRS.

3 DESPEC Experimental Campaigns in 2020-2022

The DESPEC collaboration successfully commissioned the majority of its subsystem during the runs in 2020-2022,

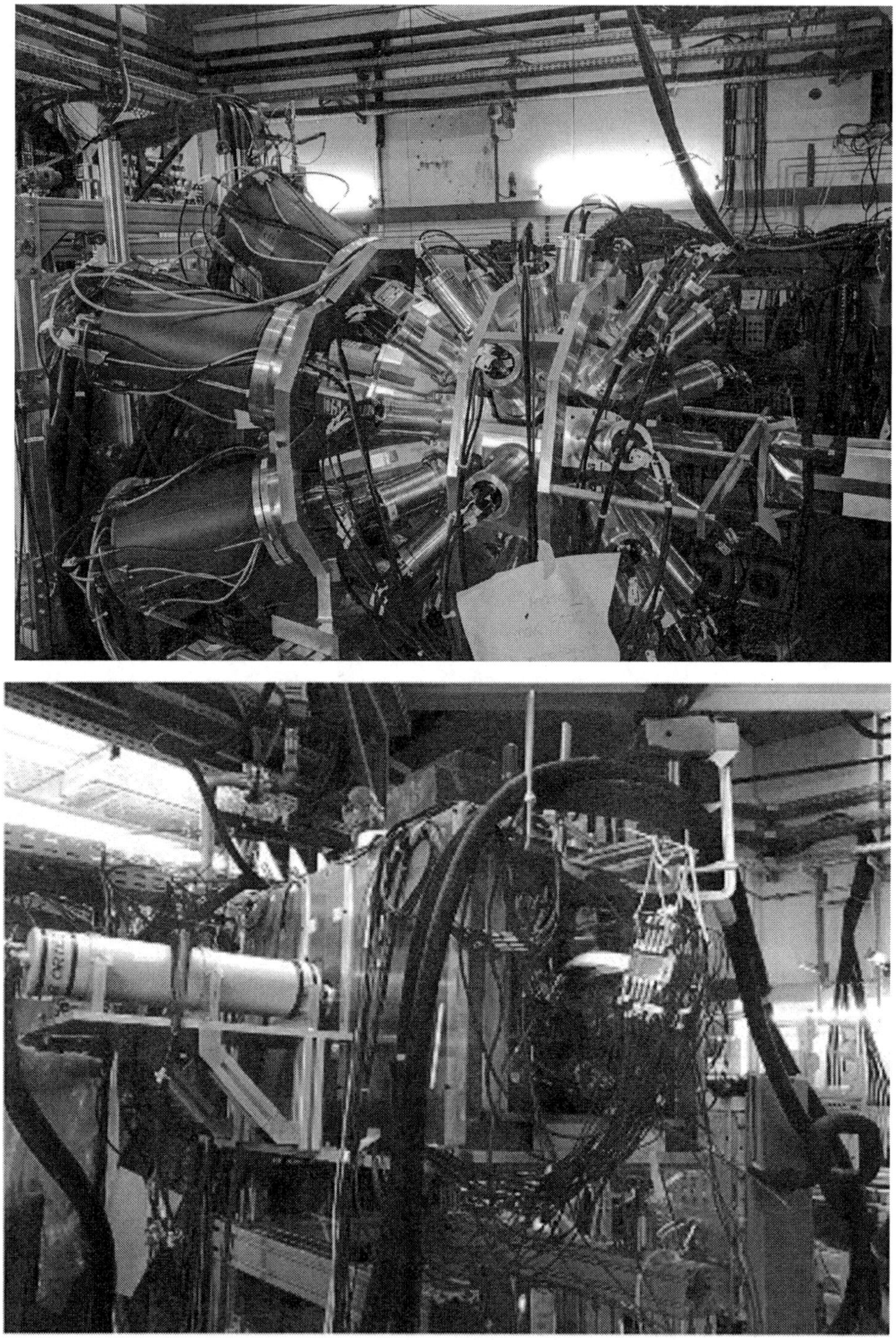

Figure 1. Top Panel: High-resolution DESPEC set-up consisting of HPGe detectors (left side) and FATIMA ones (right side), as employed during the Phase-0 campaign in 2020-2021. Bottom Panel: DTAS implementation at S4 FRS focal plane.

performing a series of experiments using AIDA, FATIMA, DEGAS and DTAS, arranging them in various configurations.

In 2020 the collaboration completed an experiment approaching the heaviest N=Z nuclei by studying seniority and electromagnetic transition rates in ^{94}Pd, populated by fragmentation of a ^{124}Xe beam. Thanks to the Fast-Timing technique, the E2 strength for the $2^+ \to 0^+$, $4^+ \to 2^+$ transitions were measured in ^{94}Ru, exhibiting a drastic enhancement of transition strength in comparison with pure-seniority predictions as well as standard shell model calculations in the fpg proton hole space with respect to doubly-magic ^{100}Sn. This anomalous behavior is ascribed to a subtle interference between the wave function of the lowest seniority, ν=2, $J^\pi = 4^+$ state and that of a closely-lying ν=4 state, that exhibits partial dynamic symmetry, resulting in a quantum phase transition from the seniority conserved regime to a seniority mixed regime. Results of the successful measurement are published in Physics Review C Letters [13].

The FATIMA set-up was also employed during the campaign in 2021, where three experiments were performed, covering three different regions in the nuclide chart: the first experiment focused on shape evolution in the A∼190 region, in particular focusing on n-rich W isotopes. The second experiment performed a study of octupole deformation in the region of mass around A∼225, where little to no experimental information on the internal structure and decay half-lives is available. Further details are found in the talk by M. Polettini (this issue). The last experiment of the 2021 campaign focused on the ^{100}Sn region with the aim of studying the isomeric states in the Cd chain and providing access to E2 strengths in n-poor Sn isotopes. In this last run the wide layout of AIDA, covering the full focal plane, was employed.

The two experiments, performed in the campaign in 2022, aimed at attacking the same region in the chart of nuclides, around N=126, accessing the isomeric and β-delayed γ transitions in Os, Ir and Au nuclei, via two complementary approaches, exploiting the high resolution of DEGAS HPGe dectors to study the detailed decay scheme, and the overall B(GT) response thanks to the total absorption spectroscopy made available by the DTAS scintillators.

The analysis of the collected data is on-going for all the mentioned runs, with preliminary results already presented at several conferences.

4 Future perspectives

After a short campaign at Riken (J), the FATIMA array will be back to GSI-FAIR in 2024, to be used together with the DEGAS and AIDA detectors, for additional experimental runs within the FAIR Phase-0 framework.

Beams from SIS18 will further increase in intensity and allow to continue map the neutron and proton rich sides of the chart of nuclides. Additional detection equipment, in particular for neutron detection, can be coupled to this array to further extend the experimental information at reach.

With the advent of Super-FRS, the DESPEC set-up, thanks to its easy installation and quick feedback on the populated nuclei, will be exploited to help in the commissioning phases.

Later, when SIS-100 beams will become available, the HISPEC-DESPEC collaboration will move to the HISPEC part of its experimental program, in particular exploiting the AGATA spectrometer.

The collaboration plans to exploit the DESPEC suite of detectors for campaign with first beams from SIS-100 and Super-FRS at the Early Start of FAIR. Later the collaboration aims at starting its in-beam program deploying the AGATA segmented γ array to study secondary-fragmentation, knock-out and relativistic Coulomb excitation reactions. For such campaigns AGATA will be followed by a magnetic transport system in order to help mass and charge identification of the outgoing ions in the LYCCA array.

The HISPEC-DESPEC collaboration is actively commissioning next generation implantation detectors, capable of sustaining higher implantation rates while presenting higher radiation hardness and faster responses. This will help the measurement of β-delayed neutron emission, where a low-threshold and fast response is required.

A dedicated campaign to measure nuclear moments, the g-SPEC campaign, is also under development with the design of a custom electromagnet and a linear version fo the DEGAS triple detectors.

In addition new analysis techniques and novel approaches to the analysis of the data are being developed in order to handle the increasing volume of acquired data and fully exploit the details of the digitally readout signals.

5 Acknowledgements

The excellent work of the GSI accelerator department, the fruitful discussions and collaborative efforts of the FRS team, the ingenuous solutions of the mechanics and electronic workshops in GSI are here acknowledged. This work is carried out as part of FAIR-Phase-0 at the GSI Helmholtzzentrum für Schwerionenforschung, Darmstadt, Germany. The contribution of INFN is also acknowledged.

References

[1] S. Akkoyun et al., Nuclear Instruments and Methods in Physics Research Section A: Accelerators, Spectrometers, Detectors and Associated Equipment **668**, 26 (2012)

[2] P. Golubev et al., Nuclear Instruments and Methods in Physics Research Section A: Accelerators, Spectrometers, Detectors and Associated Equipment **723**, 55 (2013)

[3] *Lifetime measurements with solid active targets*, https://web-docs.gsi.de/~kwimmer/

[4] H. Geissel, P. Armbruster, K. Behr, A. Brünle, K. Burkard, M. Chen, H. Folger, B. Franczak, H. Keller, O. Klepper et al., Nuclear Instruments and

Methods in Physics Research Section B: Beam Interactions with Materials and Atoms **70**, 286 (1992)

[5] G. Münzenberg, Nuclear Instruments and Methods in Physics Research Section B: Beam Interactions with Materials and Atoms **70**, 265 (1992)

[6] G. Li, R. Lozeva, I. Kojouharov, J. Gerl, M. Górska, Nuclear Instruments and Methods in Physics Research Section A: Accelerators, Spectrometers, Detectors and Associated Equipment **987**, 164806 (2021)

[7] M. Rudigier, Z. Podolyák, P. Regan, A. Bruce, S. Lalkovski, R. Canavan, E. Gamba, O. Roberts, I. Burrows, D. Cullen et al., Nuclear Instruments and Methods in Physics Research Section A: Accelerators, Spectrometers, Detectors and Associated Equipment **969**, 163967 (2020)

[8] A. Banerjee, M. Wiebusch, M. Polettini, A. Mistry, H. Heggen, H. Schaffner, H.M. Albers, N. Kurz, M. Górska, J. Gerl, Nuclear Instruments and Methods in Physics Research Section A: Accelerators, Spectrometers, Detectors and Associated Equipment **1028**, 166357 (2022)

[9] A. Mistry, H. Albers, T. Arıcı, A. Banerjee, G. Benzoni, B. Cederwall, J. Gerl, M. Górska, O. Hall, N. Hubbard et al., Nuclear Instruments and Methods in Physics Research Section A: Accelerators, Spectrometers, Detectors and Associated Equipment **1033**, 166662 (2022)

[10] J. Tain, A. Algora, J. Agramunt, V. Guadilla, M. Jordan, A. Montaner-Pizá, B. Rubio, E. Valencia, D. Cano-Ott, W. Gelletly et al., Nuclear Instruments and Methods in Physics Research Section A: Accelerators, Spectrometers, Detectors and Associated Equipment **803**, 36 (2015)

[11] V. Guadilla, J. Tain, A. Algora, J. Agramunt, J. Äystö, J. Briz, A. Cucoanes, T. Eronen, M. Estienne, M. Fallot et al., Nuclear Instruments and Methods in Physics Research Section A: Accelerators, Spectrometers, Detectors and Associated Equipment **910**, 79 (2018)

[12] V. Guadilla, J. Tain, A. Algora, J. Agramunt, J. Äystö, J. Briz, A. Cucoanes, T. Eronen, M. Estienne, M. Fallot et al., Nuclear Instruments and Methods in Physics Research Section A: Accelerators, Spectrometers, Detectors and Associated Equipment **910**, 79 (2018)

[13] B. Das, B. Cederwall, C. Qi, M. Górska, P.H. Regan, O. Aktas, H.M. Albers, A. Banerjee, M.M.R. Chishti, J. Gerl et al., Phys. Rev. C **105**, L031304 (2022)

Scattering of ^{15}C on ^{208}Pb at energies near the Coulomb barrier: Study of the experimental device response via the ^{12}C+^{208}Pb scattering

*V.G.*Távora[1,*], *J.D.* Ovejas[1], *I.* Martel[2], *O.* Tengblad[1], *M.J.G.* Borge[1], *L.* Acosta[5], *T.* Kurtukian-Nieto[1], *A.M.* Sánchez-Benítez[2], *A.* di Pietro[3], *P.* Figuera[3], *N.* Keeley[4], *K.* Rusek[4], and the IS619 collaboration.

[1]Instituto de Estructura de la Materia, CISC, Serrano 113bis, E-28006 Madrid, Spain
[2]Departamento de Física Aplicada, Universidad de Huelva, E-21071, Huelva, Spain
[3]INFN, Laboratori Nazionali del Sud, via S. Sofia 62, 1-95123, Catania, Italy
[4]National Center for Nuclear Research, ul Andrzeja Soltana 7, 05-400 Otwock, Poland
[5]Instituto de Física, Universidad Nacional Autónoma de México, Mexico City, Mexico

Abstract. There are evidences that the ^{15}C nucleus has an extended neutron distribution or a "neutron halo", but this situation is far from clear. If ^{15}C has 1n-halo, the scattering dynamics should be affected and the angular distribution of the elastic channels should be sensitive to coupling effects due to the halo configuration. The objetive of this study is to understand the role of the halo in ^{15}C by investigating its dynamical response in intense electric fields at energies close to the Coulomb barrier. For this purpose experiment IS619 was conducted performing the ^{15}C + ^{208}Pb elastic scattering reaction at HIE-ISOLDE, CERN. The beam energy was 4.37 *MeV/u*, which is very close to the Coulomb barrier of the system. The experimental setup used was the global detection system GLORIA, a six silicon telescopes array enable to measure the energy and angular distributions of the scattered particles. During the experiment, the ^{12}C+^{208}Pb scattering at 4.37 *MeV/u* was measured for calibration. The results of the ^{12}C+^{208}Pb measurement were used to fine-tune the geometry of the experimental setup and it is presented in this contribution.

1 Introduction: neutron halos

Nuclear systems such as ^{6}He, ^{11}Li, ^{11}Be, ^{14}Be are known to have an extended neutron distributions: the so-called neutron halos [1, 2]. This feature occurs when the separation energy of valence neutrons is very low and certainly much smaller than the average binding energy per nucleon in a nucleus, so the valence neutrons can tunnel out of the nuclear potential to large distances with sizable probability. It has been an intense experimental and theoretical activity dedicated to study the existence of halos and their dynamics in reaction processes. The neutron halos produces a pronounced maxima at low excitation energies in the Coulomb dipole strength B(E1), very narrow transverse momentum distributions and large interaction cross-sections when measured at high energies [3].

The dynamics of the halo nuclei scattering at low energies, around the Coulomb barrier, is dominated by the coupling between the elastic channel and collective excitations, neutron transfer and breakup. The angular distributions of the elastic cross section and the core fragments present large sensitivity to these coupling effects, due to the halo configuration. This has been demonstrated in previous studies with light exotic beams of ^{6}He, ^{11}Li and ^{11}Be scattered on heavy targets [4–6]. The angular distribution of the elastic channel shows strong absorption patterns where the nuclear and Coulomb interference completely disappears.

The ^{15}C nucleus ($T_{1/2}$ = 2.449(5) s) has a low single neutron separation energy S_n = 1218.1(8) keV in comparison to the two-neutron separation energy S_{2n} = 9394.5(8) keV [7] and the neighbors ^{14}C (S_n = 8176.43 keV) ^{16}C (S_n = 4250 keV). The spins and parities of the ground and first excited state at E = 740 keV of ^{15}C are known to be $I_\pi = 1/2^+, 5/2^+$, respectively.

The halo structure of ^{15}C has been investigated at relatively high energies in several experiments. The reaction cross section at 83 *MeV/u* shows an enhancement respect to the neighboring ^{14,16}C isotopes [8], and the longitudinal momenta of the ^{14}C fragments after 1n-breakup present a FWHM distribution between 64-70 *MeV/c* depending of the target that it is narrower than that of the neighbour ^{14,16}C isotopes, ~ 200 *MeV/c*, but wider than the typical ~ 40-50 *MeV/c* found for the archetype cases [3]. These properties have hinted the presence of a 1n-halo configuration in the ^{15}C nucleus that would be unique in the sense that it can be described with an almost pure $s_{1/2}$ ground state wavefunction [8].

On the other hand, the enhancement of the cross section observed at 83 *MeV/u* is not seen at 950 *MeV/u* [8]. Also nuclear matter radii and density were revisited in [9] obtaining a value of $\kappa = R_v/R_c = 1.81$ for the ratio between radius of the valence neutron and core radius distribution, where $\kappa > 2$ for typical halo nucleus.

To complete the understanding of the role of the halo in ^{15}C, its dynamical response at energies close to the Coulomb barrier has been studied in IS619 experiment,

*e-mail: vicente.garcia@csic.es

where, according with [10, 11], a scattering dominated by the competition of 1n-stripping and breakup is expected.

A ^{12}C beam at the same energy per nucleon was measured with the same ^{208}Pb targets for calibration of the experimental device.

2 Experimental Methodology

The study of ^{15}C +^{208}Pb elastic scattering was performed at HIE-ISOLDE, CERN. Where the ^{15}C beam was produced from the impact of 1.4 GeV proton pulses on a CaO production target. The isotope of interest was extracted, mass-separated and post-accelerated to 4.37 *MeV/u*, which is very close to the Coulomb barrier of the system.

The radioactive beam impacted on a target of ^{208}Pb with purity higher than 98% and two different thicknesses; one of 2.1 and the other of 1.2 mg/cm^2. Using the thicker target, the energy losses increased but favors a higher amount of reactions and more statistics. Using the thin target the statistics decreased but there was less straggling, and better energy resolution.

The experimental setup for IS619 was the global reaction array, called GLORIA [12]. It consisted of 6 silicon telescopes, all tangent to a 6 cm radius sphere in which center the reaction target was placed. All telescopes were composed of a 40 μm ΔE detector and a 1 mm E detector. Both detectors had 16x16 strips, which means that they worked as a set of 256 pixels each with an angular resolution between 2 and 3°. Dynamic ranges were adjusted to around 50 MeV to detect the scattered ^{15}C ions at ~ 65 MeV lab. energy. The system had an angular coverage from 15 to 165° in a continous way with some overlapping areas between pairs of telescopes. The geometric efficiency was 25% of 4π. The detector resolution was in the order of 30 keV FWHM.

A scheme of this configuration is presented in Fig. 1 as well as the nominal angular range covered by each detector in Tab. 1. The target ladder was placed with a 30° tilt respect to the beam direction, this avoids shadowing at 90° but introduces an asymmetry in the energy losses through the azimutal angle. The telescope C did not work during the experiment.

Table 1. The angular coverage of the six telescopes is given. Center of the detectors, geometric angle range for each telescope is given assuming that beam and target were centered and the final effective angular coverage.

Telescope	Center (°)	Geometric angular range (°)	Effective angular range (°)
A	38	15-62	23-58
B	38	15-62	22-57
F	75	52-97	59-95
E	105	82-128	91-124
D	142	117-165	123-154
C	142	117-165	

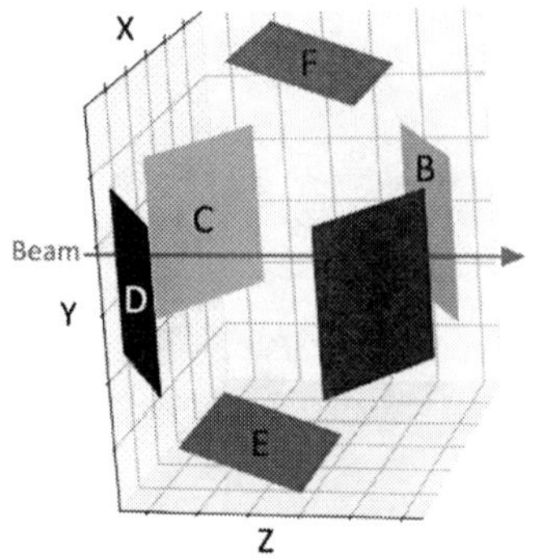

Figure 1. Schematic representation of the telescope configuration used during the experiment.

3 Analysis procedure

For single particle detection in double-sided detectors, the juntion (P-side) and the ohmic (N-side) are read separately. For good events, these signals correspond to the same deposited energy. Experimentally a certain deviation between both sides is expected, so the $|E_p - E_n| \leq 2\sigma$ condition has been included in the data processing, where σ value is obtained from the gaussian fit applied to $|E_p - E_n|$ distribution for each detector, either front detector, ΔE, or back detector, E.

As it is explained in [13], the beam is not polarised so the reaction fragments follow axial symmetry and the same physics is expected to happen over the same dispersion angle θ. Due to that symmetry, the scattering for each θ-angle follows a conical surface. This property allow the grouping of all the pixels that belong to a determined angular sector ($\Delta\theta$) (Fig. 2 left).

The ΔE-E telescope configuration makes possible ion identification (Fig. 2 right), hence, those events can be selected by a gate in the corresponding range of the ^{15}C elastic scattering events. The border strips of both detectors of each telescope were removed due to the lower coincidence efficiency.

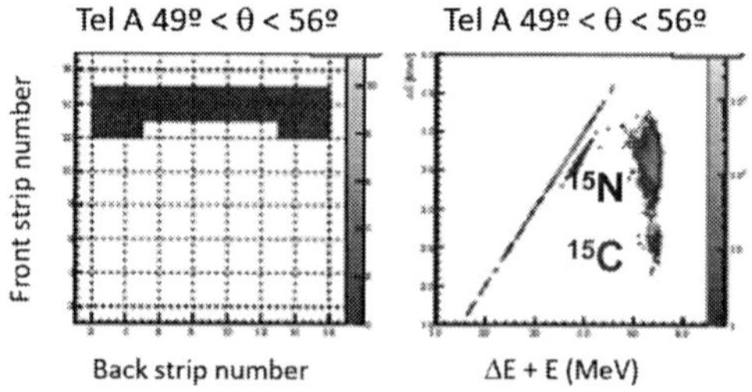

Figure 2. On the left hand side, the pixels selected in the angular coverage 49°< θ < 56° are shown. On the right hand side, the ΔE-E plot where the contaminant ^{15}N and the beam ^{15}C are clearly separated.

3.1 $^{12}C+^{208}Pb$ elastic scattering

The comparison performed between ^{15}C Rutherford cross section and the experimental cross section of ^{15}C elastic scattering on lead given in [13] shows large error bars hampering any conclusion about the dynamic behaviour of ^{15}C beam on heavy targets. In [13], the geometry optimization was done using the stable ^{15}N present in the beam.

In addition, the $^{12}C+^{208}Pb$ scattering was measured under the same conditions described in section 2. At these energies, a pure elastic scattering behavior is expected for ^{12}C. This can be used as an additional set of data to check the optimization of the geometry made in [13] and to normalize the distribution of the cross section obtained for $^{15}C+^{208}Pb$ by the elastic scattering of the stable isotope. $^{12}C+^{208}Pb$ data were treated with the same procedure explained in sections 3.

The angular distribution of the ratio between $\sigma_{^{12}C+^{208}Pb}$ and σ_{Ruth} is presented in Fig. 3. With $\sigma_{^{12}C} = I_{pixel}/\Omega_{pixel}$ and $\sigma_{Rutherford} = N/\sin^4(\theta_{pixel}/2)$, where I_{pixel} is the integral of ^{12}C ions for each pixel, N is the normalization constant, Ω_{pixel} and θ_{pixel} are the solid angle and scattering angle subtended for each pixel respectively.

Despite the efforts dedicated to the alignment of the chamber and detectors, a small shift ($\sim mm$) on the reaction point or a tilt of the beam direction is always expected. This results on small deviations of the solid angles Ω and angles (θ, ϕ) of each pixel of the detectors due to the very close geometry of the setup so, a cloud of points with small deviations around 1 was expected due to the pure elastic scattering reaction of ^{12}C. In this case, a wide distribution of the points is shown in Fig. 3 where a clear upward trend can be observed in the data-sets for each telescope, in addition to the lack of continuity for the intersection zones between telescopes and the lack of symmetry for mirror telescopes A and B. This behavior indicates the need for further geometry optimization to be performed.

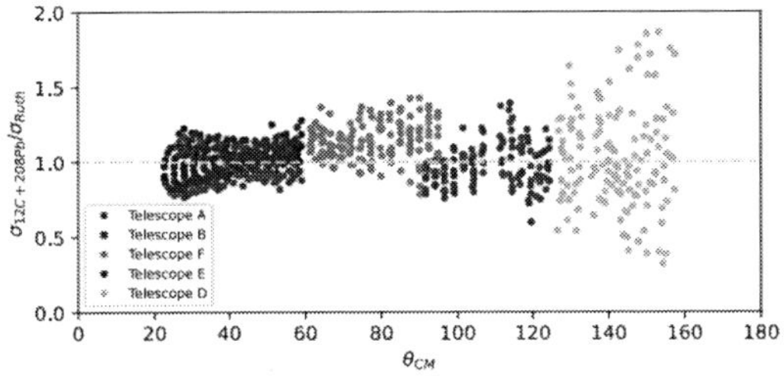

Figure 3. Differential elastic scattering of $^{12}C+^{208}Pb$. The data per pixel shows large dispersion angular distribution that indicates thar a geometrical optimization is needed.

4 Geometry optimization

To carry out the geometric optimization of the experimental setup, a program generating the 3D geometry has been built in *python*. It takes the spherical coordinates of the center of each detector, the position (X, Y, Z) of the reaction point and the beam orientation as inputs. From these data, it computes the central angle and the solid angle subtended by each pixel of each telescope.

4.1 Solid angle (Ω) optimization

In order to compute the solid angle (Ω), it is necessary to calculate the coordinates of three corners (A, B, C) for each pixel, using a coordinate system centered on the target. Each pixel is then divided into $m \times n$ smaller pixels with coordinates (A_{ij}, B_{ij}, C_{ij}), where $i \in 0, ..., n$ and $j \in 0, ..., m$ and $m = n = 100$, and these smaller pixels are treated as differential areas. The coordinates of each smaller pixel can be calculated using equation 1.

$$Pixel_{ij} = \left(A_{ij}, B_{ij}, C_{ij}\right) = \begin{cases} A_{ij} = A + i\frac{\overrightarrow{d_{AB}}}{n} + j\frac{\overrightarrow{d_{AC}}}{m} \\ B_{ij} = A + (i+1)\frac{\overrightarrow{d_{AB}}}{n} + j\frac{\overrightarrow{d_{AC}}}{m} \\ C_{ij} = A + i\frac{\overrightarrow{d_{AB}}}{n} + (j+1)\frac{\overrightarrow{d_{AC}}}{m} \end{cases} \tag{1}$$

The normal vector (Eq. 2) and center (Eq. 3) of each smaller pixel is computed using the coordinates obtained above. The vector going from the pixel center to the target center ($\overrightarrow{r_{ij}} = C_{ij} - Target$) is then calculated, along with the modulus of that vector ($r_{ij} = |\overrightarrow{r_{ij}}|$), which is the distance between the target and pixel center.

$$\overrightarrow{n_{ij}} = \overrightarrow{A_{ij}B_{ij}} \times \overrightarrow{A_{ij}C_{ij}} \tag{2}$$

$$Center_{ij} = \left(\frac{C_{x,ij} + B_{x,ij}}{2}, \frac{C_{y,ij} + B_{y,ij}}{2}, \frac{C_{z,ij} + B_{z,ij}}{2}\right) \tag{3}$$

To compute the area of each smaller pixel, the pixel area is multiplied by the cosine of the angle between the normal vector ($\overrightarrow{n_{ij}}$) and the vector going from the pixel center to the target center ($\overrightarrow{r_{ij}}$), as $\overrightarrow{n_{ij}}$ is not parallel to $\overrightarrow{r_{ij}}$. The effective pixel area is given by equation 4:

$$S_{ij} = \left|\overrightarrow{A_{ij}B_{ij}}\right| \left|\overrightarrow{A_{ij}C_{ij}}\right| \cos\varphi_{ij} \tag{4}$$

where φ_{ij} is the angle between $\overrightarrow{n_{ij}}$ and $\overrightarrow{r_{ij}}$, and can be calculated using the dot product of these two vectors.

$$\overrightarrow{n_{ij}} \cdot \overrightarrow{r_{ij}} = \left|\overrightarrow{n_{ij}}\right| \left|\overrightarrow{r_{ij}}\right| \cos\varphi_{ij} \; ; \; \cos\varphi_{ij} = \frac{\overrightarrow{n_{ij}} \cdot \overrightarrow{r_{ij}}}{\left|\overrightarrow{n_{ij}}\right| \left|\overrightarrow{r_{ij}}\right|} \tag{5}$$

The solid angle for each smaller pixel is calculated using equation 6. This equation involves computing α_{ij} and β_{ij}, which are the half-lengths of the sides of the smaller pixel divided by the distance between the target and the pixel center, $\alpha_{ij} = \frac{|\overrightarrow{A_{ij}B_{ij}}|}{2r_{ij}}$ and $\beta_{ij} = \frac{|\overrightarrow{A_{ij}C_{ij}}|}{2r_{ij}}$. Since detector pixels are very small, they are treated as surface differentials, and so equation 6 can be approximated using only the first order term (equation 7).

$$\Omega_{ij} = 4\arctan\left(\frac{\alpha_{ij}\beta_{ij}}{\sqrt{1 + \alpha_{ij}^2 + \beta_{ij}^2}}\right) \tag{6}$$

$$\Omega_{ij} = 4\alpha_{ij}\beta_{ij} = \frac{S_{ij}}{r_{ij}^2} = \frac{\left|\overrightarrow{A_{ij}B_{ij}}\right|\left|\overrightarrow{A_{ij}C_{ij}}\right|\cos\varphi_{ij}}{\left|\overrightarrow{r_{ij}}\right|^2} \quad (7)$$

The total solid angle of the detector is the sum the solid angle of each pixel (Eq. 8):

$$\Omega = \sum_{ij}\Omega_{ij} = \sum_{ij}\frac{S_{ij}}{r_{ij}^2} = \sum_{ij}\frac{\left|\overrightarrow{A_{ij}B_{ij}}\right|\left|\overrightarrow{A_{ij}C_{ij}}\right|\cos\varphi_{ij}}{\left|\overrightarrow{r_{ij}}\right|^2} \quad (8)$$

4.2 Optimizer

Once all the variables are computed the program returns θ, ϕ and Ω associated to the index of each pixel. From these data, the cross section and the ratio with respect to Rutherford are recalculated. The optimizer is based on Gradient Descent method with a loss function of normalized χ^2 to optimize the computing time.

The hit patterns of telescope A has been used to optimize the geometry due to the fact that the largest statistic was present in the two front telescopes (A and B) and telescope A seems the most centered one, as shown in Fig. 4. On the other hand, telescope B seems off-centered so its position has to be optimized independendly.

The parameters obtained for the reaction point position and beam orientation are presented in Tab. 2. And the effective angles are given in Tab. 1. The parameter space was constrained between ±5 *mm* for X and Y, and ±3° for α and β.

Table 2. Parameters obtained after optimization of the geometry. Coordinates X and Y of the reaction point and rotations of the beam around X axis (α) and Y axis (β) using the right-hand rule.

Telescope	X (*mm*)	Y (*mm*)	α (°)	β (°)
A, F, E, D	-2.5	-2	0.5	1.5
B	4.5	-4.5	-1.5	3

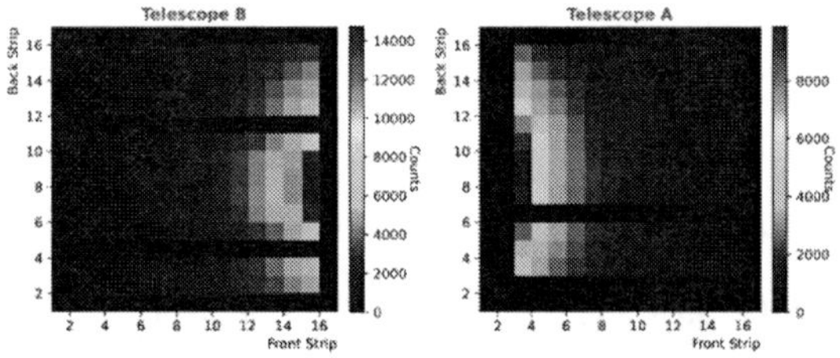

Figure 4. Hit maps of two front-side telescopes, where the beam comes in the middle between them. It can be seen how Telescope A is well centered around back strip 9 while Telescope B needs a geometric adjustment.

4.3 Electronic efficiency correction

One of the consequences of the high counting rate in front telescopes A and B was that telescopes F, E and D count less as expected due to the low trigger rate for detectors placed backwards. To calculate the electronic efficiency factor, the angular region that matches telescopes A and B with F (1.126), F with E (1.174) and E with D (1.426) has been used. The application of those factors in addition to the geometry optimization give the final distribution for the elastic scattering of ^{12}C on ^{208}Pb presented in Fig. 5.

5 Conclusions

We have optimized the GLORIA device for the present $^{15}C+^{208}Pb$ study using the $^{12}C+^{208}Pb$ scattering at 4.37 *MeV/u* energy well below the ^{12}C Coulomb barrier. We have determine the effective position of the ^{208}Pb target in comparison with the geometrical center of GLORIA. We have also determine the electronic response that make the distribution typical of Rutherford scattering. This normalization parameters will be used to extract the differential $^{15}C+^{208}Pb$ elastic cross section.

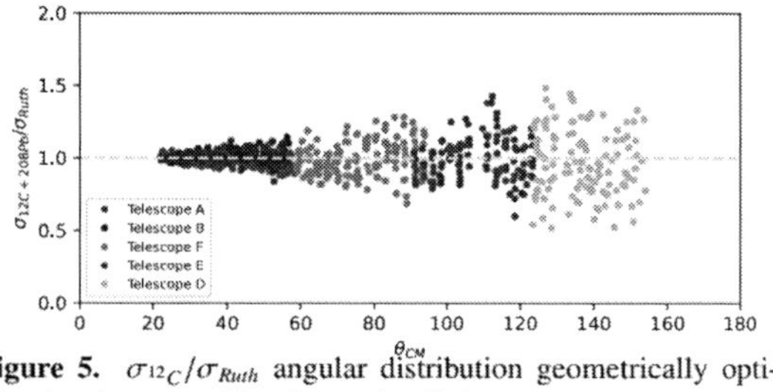

Figure 5. $\sigma_{^{12}C}/\sigma_{Ruth}$ angular distribution geometrically optimized and corrected by electronic efficiency.

Acknowledgements

We acknowledge the funding through the grant MCIN/AEI PID2019-104390GB-I00 and GIFMAN associated unit between CSIC and Universidad de Huelva.

References

[1] I. Tanihata et al., Physical Review Letters **55**, 2676 (1985)
[2] K. Riisager, Physica Scripta **2013**, 014001 (2013)
[3] T. Aumann, The European Physical Journal A-Hadrons and Nuclei **26**, 441 (2005)
[4] A. Sánchez-Benítez et al., Nuclear Physics A **803**, 30 (2008)
[5] M. Cubero et al., Physical review letters **109**, 262701 (2012)
[6] V. Pesudo et al., Phys. Rev. Lett. **118**, 152502 (2017)
[7] M. Wang et al., Chinese Physics C **36**, 1603 (2012)
[8] A. Ozawa, Nuclear Physics A **738**, 38 (2004), proceedings of the 8th International Conference on Clustering Aspects of Nuclear Structure and Dynamics
[9] A. Dobrovolsky et al., Nuclear Physics A **1008**, 122154 (2021)
[10] C. Beck et al., Physical Review C **75**, 054605 (2007)
[11] N. Keeley et al., The European Physical Journal A **50**, 1 (2014)
[12] G. Marquínez-Durán et al., Nuclear Instruments and Methods in Physics Research Section A **755**, 69 (2014)
[13] J. Díaz Ovejas, Thesis (2021)

 https://doi.org/10.1051/epjconf/202329002006

FIRST FULL BETA-STRENGTH MEASUREMENT WITH DTAS ACROSS N=126 AT FAIR PHASE-0

*D.*Rodriguez-Garcia[1,*], *J. L.*Tain[1], *A. I.*Morales[1], *G.*Aggez[2,3], *J.*Agramunt[1], *M.*Alaqeel[4,5], *B.*Alayed[5,6], *H. M.*Albers[2], *G.*Alcala[1], *A.*Algora[1], *A.*Alharbi[5], *S.*Alhomaidhi[2,7], *F.*Amjad[2], *T.*Arici[2], *M.*Armstrong[8], *M.*Bajzek[2], *A.*Banerjee[2], *G.*Bartram[2], *G.*Benzoni[9], *Z.*Chen[2], *B.*Das[2], *T.*Davinson[10], *T.*Dickel[2], *I.*Dillmann[11], *C.*Domingo-Pardo[1], *H.*Ekawa[12], *Z.*Ge[2], *W.*Gelletly[13], *J.*Gerl[2], *M.*Gorska[2], *E.*Haettner[2], *O.*Hall[10], *P.*Herrmann[2], *C.*Hornung[2], *N.*Hubbard[2,7], *C.*Jones[2], *E.*Kazantseva[2], *R.*Knoebel[2], *I.*Kojouharov[2], *G.*Kosir[21], *D.*Kostyleva[2], *T.*Kurtukian-Nieto[15], *N.*Kurz[2], *M.*Labiche[16], *A.*Mccarter[5], *M.*Mikolajczuk[2], *A. K.*Mistry[2,7], *I.*Mukha[2], *E.*Nacher[1], *M.*Nakagawa[12], *B.S.*Nara-Singh[20], *S.*Nishimura[12], *S. E. A.* Orrigo[1], *P.*Papadakis[13], *S.*Pietri[2], *W.*Plass[2], *Z.*Podolyak[13], *M.*Polettini[9,17], *R.*Prajapat[2,18], *E.*Rocco[2], *B.*Rubio[1], *E.*Sahin[2,7], *M.*Satrazani[5], *H.*Schaffner[2], *C.*Scheidenberger[2], *A.*Sharma[19], *Y.K.*Tanaka[12], *A.*Tolosa-Delgado[21], *M.*Vencelj[14], *J.*Vesic[14], *P.*Vi[12], *J. A.*Victoria[1], *H.*Weick[2], *K.*Wimmer[2], *H. J.*Wollersheim[2], *A.*Yaneva[2,8], and *J.*Zhao[2,22]

[1]Instituto de Fisica Corpuscular, CSIC-Universidad de Valencia, E-46980 Paterna, Spain
[2]GSI Helmholtzzentrum für Schwerionenforschung, 64291 Darmstadt, Germany
[3]Institute of Graduate Studies in Sciences, Istanbul University, 34452 Istanbul, Turkey
[4]Physics Department, Imam Mohammad Ibn Saud Islamic University (IMISU), P.O. Box 90950, Riyadh, 11623, Saudi Arabia
[5]Department of Physics, University of Liverpool, Liverpool, L69 7ZE, UK
[6]Department of Physics, ArRass College of Sciences and Art, Qassim University, Saudi Arabia
[7]Institut für Kernphysik, Technische Universität Darmstadt, D-64289 Darmstadt, Germany
[8]IKP, University of Cologne, D-50937 Cologne, Germany
[9]INFN Sezione di Milano, I-20133 Milano, Italy
[10]School of Physics and Astronomy, University of Edinburgh, Edinburgh, EH9 3FD, UK
[11]TRIUMF, 4004 Wesbrook Mall, Vancouver, BC V6T 2A3, Canada
[12]RIKEN Cluster for Pioneering Research, RIKEN, Saitama 351-0198 Japan
[13]Department of Physics, University of Surrey, Guildford, GU2 7XH, UK
[14]Faculty of mathematics and physics of the University of Ljubljana, SI-1000 Ljubljana, Slovenia
[15]Instituto de Estructura de la Materia, CSIC, E-28006, Madrid, Spain
[16]Science and Technology Facilities Council, Daresbury Laboratory, Daresbury, WA4 4AD, UK
[17]Dipartimento di Fisica, Università degli Studi di Milano - Milano, Italy
[18]Department of Physics, Indian Institute of Technology Roorkee, Roorkee-247667, Uttarakhand, India
[19]Department of Physics, Indian Institute of Technology Ropar, Rupnagar - 140001, India
[20]Paisley, University of the West of Scotland, High St, Paisley PA1 2BE, UK
[21]Faculty of Mathematics and Science, University of Jyväskylä, FI-40014Jyväskylä, Finlandia
[22]School of Physics, Peking University, Beijing 100871, China

Abstract. An experiment was performed at GSI with the objective of measuring the β-intensity distribution in the decay of Hg, Au and Pt isotopes around N=126 using the total absorption gamma-ray spectroscopy technique. The aim is to benchmark theoretical models used to make predictions of half-life and neutron emission probabilities of exotic nuclei involved in the rapid neutron capture process, leading to the synthesis of very heavy elements. This paper presents some experimental details and the current status of the analysis.

1 Introduction

Our current understanding of the production of the heaviest elements in the Universe remains incomplete, especially with regards to the contribution of the rapid neutron capture (r-) process to the observed abundances of elements with A>180 and the astrophysical site responsible for this process. To shed light on this puzzle, a combination of astronomical observations, including gravitational wave detection, nuclear physics laboratory experiments and theoretical modelling is necessary.

From the point of view of the nuclear data, the situation is challenging, as the abundance distribution in the heavy mass region peaks at A ~ 195, the so-called 3rd abundance peak, which is related to the N=126 shell closure in very neutron-rich nuclei. There is a lack of experimental information in this remote region, and it is impossible at current facilities to fill the gap. As a result, we must rely on theoretical predictions, particularly for the important macroscopic quantities of half-life ($T_{1/2}$), neu-

*e-mail: David.Rodriguez@ific.uv.es

tron emission probability (P_n), and mass. P_n and $T_{1/2}$ are extracted from the calculated β-strength distribution, which depends on nuclear structure. However, the existence of a number of theoretical calculations providing largely conflicting results for both quantities when crossing N=126, see for example [1, 2], calls for additional studies to understand the origin of these differences.

To address this issue, we aim to contrast calculated and measured β-strength distributions, rather than $T_{1/2}$ and P_n, for the first time in this region. Total Absorption Gamma-ray Spectroscopy (TAGS) [3] is the most suitable technique to provide β-strength distributions in the decay energy window. This technique relies on the use of large calorimeters that are sensitive to γ cascades rather than individual γ rays, and a suitable analysis method [4] to extract feeding intensities as a function of daughter excitation energy E_x.

2 Experiment

The experiment, the first with the Decay Total Absorption Spectrometer (DTAS)[5] at the GSI facility, was performed in June 2022 as part of the FAIR Phase-0 DESPEC campaign. We measured the decay of ^{207}Hg, $^{204-206}$Au and ^{203,204}Pt produced in fragmentation and relativistic charge-exchange reactions [6, 7] of a 1 GeV/u ^{208}Pb beam on a 1625 mg/cm^2 ^{9}Be target. The beam intensity was 10^9 pps with a time structure of 1.6 s spill every 3.8 s. The isotopes were selected using the FRagment Separator (FRS) and identified by the $B\rho - \Delta E - B\rho$ method [8].

Figure 1. Experimental set-up installed at the end of the FRS.

The Advanced Implantation Detector Array (AIDA) [9] was used to measure ion implantation and decay particles. In our experiment we employed two 1-mm thick silicon Double-Sided Strip Detectors (DSSD) separated 10 mm, each with 128 strips in X and Y direction and a strip pitch of 0.56 mm. Two position sensitive plastic scintillation detectors (βPlast [10]) were placed within the AIDA snout ~15 mm upstream and downstream from the DSSDs. With an area of 80 mm × 80 mm and a thickness of 3 mm, they were used as subsidiary ion/β detectors. Isomeric and β-delayed γ-ray cascades were measured with DTAS surrounding AIDA. It consisted of 16 independent NaI(Tl) modules with crystals of size 150 mm × 150 mm × 250 mm. The setup was completed with two X-ray sensitive HPGe detectors (LOAX), each 30 mm thick and 70-mm diameter. They were inserted in DTAS to the left and right of the beam.

Up to five independent Data ACQuisition (DACQ) systems were used (FRS, AIDA, DTAS, βPlast, LOAX) all integrated into a Multi Branch System (MBS) using the White Rabbit (WR) common synchronization signal. For DTAS DACQ [11] this required encoding WR time information following developments made at Chalmers Technical University and GSI [12]. The progress of the measurement was controlled via Go4 based near-line analysis. Further details and references can be found in [10].

3 Status and outlook

The analysis of the data is at an early stage. Currently, we are working on the calibration of the FRS detectors in order to improve the identification of the nuclei of interest. At the same time, we are analysing DTAS data obtained from the measurement of several radioactive sources and comparing the results with detailed Monte Carlo simulations to benchmark the calculated spectrometer response. Next we will address the calibration of AIDA data and the reconstruction of ion and β events. Then the information from the three systems will be merged to obtain ion-γ and ion-β-γ correlations from which we will obtain the spectra to be analyzed.

Acknowledgements

Work partially supported by Spanish Ministry of Science grant PID2019-104714GB-C21 and Generalitat Valenciana grant PROMETEO/2019/007. D.Rodriguez-Garcia acknowledges FPI fellowship PRE2020-904341.

References

[1] A.I. Morales et al., Phys. Rev. Lett. 113, 022702 (2014)
[2] R. Caballero-Folch et al., Phys. Rev. Lett. 117, 012501 (2016)
[3] B. Rubio et al., J. Phys. G 31, S1477 (2005)
[4] J.L. Tain et al., Nucl. Instrum. Meth. Phys. Res. Sect. A 571, 728 (2007)
[5] V. Guadilla et al., Nucl. Instrum. Meth. Phys. Res. Sect. A 910, 79 (2018)
[6] T. Kurtukian-Nieto et al., Phys. Rev. C. 89, 024616 (2014)
[7] A. I. Morales et al., Phys. Rev. C. 84, 011601(R) (2011)
[8] H. Geissel et al., Nuclear Instruments and Methods in Physics Research Section B 70, 286, (1992)
[9] O. Hall et al., Nucl. Instrum. Meth. Phys. Res. Sect. A 1050, 168166 (2023)
[10] A.K. Mistry et al., Nucl. Instrum. Methods Phys. Res. Sect. A 1033, (2022) 166662
[11] J. Agramunt, et al., Nucl. Instrum. Methods Phys. Res., Sect. A 807, 69 (2016)
[12] H. Johansson and S. Löeher, private communication http://fy.chalmers.se/ f96hajo/rataser/

Isomeric Lifetime Measurement in the Neutron-rich ^{189}Ta

Sultan Alhomaidhi[1,2,3,4*], *Elif* Sahin[1,2,3], *Volker* Werner[1,3], *Jan* Jolie[5], *Patrick* Regan[12], *Norbert* Pietralla[1], *Jürgen* Gerl[2], *Magdalena* Górska[2], *Giovanna* Benzoni[7,8], *Andrew K.* Mistry[2], *Nicolas* Hubbard[1,2,3], *Guler* Aggez[2], *Helena M.* Albers[2], *Tugba* Arici[2], *Michael* Armstrong[2,5], *Akashrup* Banerjee[2], *Biswarup* Das[2], *Thomas* Davinson[6], *Oscar* Hall[6], *Ivan* Kojouharov[2], *Pavlos* Koseoglou[1,3], *Michal* Mikolajczuk[2,23], *Zsolt* Podolyák[12], *Marta* Polettini[7,8], *Matthias* Rudigier[1], *Henning* Schaffner[2], *Hans-Jürgen* Wollersheim[2], *Aleksandrina* Yaneva[2,5], *Plamen* Boutachkov[2], *Timo* Dickel[2], *Emma* Haettner[2], *Henning* Heggen[2], *Christine* Hornung[2], *Ronja* Knöbel[2], *Daria* Kostyleva[2], *Nikolaus* Kurz[2], *Natalia* Kuzminchuk[2], *Ivan* Mukha[2], *Stephane* Pietri[2], *Wolfgang* Plaß[2], *Christoph* Scheidenberger[2], *Yoshiki* Tanaka[2], *Helmut* Weick[2], *Usama* Ahmed[1,3], *Özge* Aktas[13], *Alejandro* Algora[14], *Corrigan* Appleton[6], *Andrey* Blazhev[5], *Alison* Bruce[10], *Bo* Cederwall[13], *Muhammad M. R.* Chishti[12,24], *Martha L.* Cortes[2], *Fabio* Crespi[7,8], *Jose J.* Dobon[18], *Maria* Doncel[20], *Aysegül* Ertoprak[13], *Arwin* Esmaylzadeh[5], *Marcos L.* Expósito[11], *Luis M.* Fraile[11], *Jaime B.* García[11], *Andrea* Gottardo[18], *Jeongsu* Ha[9], *Shaheen* Jazrawi[12], *Philipp R.* John[1], *Calum* Jones[2], *Vasil* Karayonchev[5], *Ralph* Kern[2], *Lukas* Knafla[5], *Gregor* Kosir[16,19], *Radomira* Lozeva[21], *Daniele* Mengoni[9], *Theo J.* Mertzimekis[17], *Bénédicte* Million[8], *Anabel* Morales[14], *Javier R.* Murias[11], *Sonja E. A.* Orrigo[14], *Julgen* Pellumaj[18], *Stefanos* Pelonis[17], *Jana* Petrović[13], *Sara* Pigliapoco[9], *Begoña* Quintana[22], *Francesco* Recchia[9], *Kseniia* Rezynkina[9], *Lewis* Sexton[6], *Nara* Singh[15], *Pär-Anders* Söderström[25], *Arshiya* Sood[13], *Polytimos* Vasileiou[17], *Jelena* Vesic[16], *Johannes* Wiederhold[1], *guangxin* zhang[9], *Radostina* Zidarova[1,3] and *Aikaterini* Zyriliou[17].

[1]Institut für Kernphysik, Technische Universität Darmstadt, 64289 Darmstadt, Germany
[2]GSI Helmholtzzentrum für Schwerionenforschung GmbH, 64291 Darmstadt, Germany
[3]Helmholtz Forschungsakademie Hessen für FAIR (HFHF)
[4]King Abdulaziz City for Science and Technology, P.O Box 6080, Riyadh 11442, Saudi Arabia
[5]Institut für Kernphysik, Universität zu Köln, 50937 Köln, Germany
[6]School of Physics and Astronomy, University of Edinburgh, EH9 3FD Edinburgh, UK
[7]Universita degli Studi di Milano, Via Celoria 16, 20133, Milano, Italy
[8]INFN, sez. di Milano, Via Celoria 16, 20133, Milano, Italy
[9]Dipartimento di Fisica e Astronomia dell'Universita di Padova and INFN Padova, 35131, Padova, Italy
[10]School of Computing, Engineering, and Mathematics, University of Brighton, BN2 4GJ Brighton, UK
[11]Grupo de Física Nuclear and IPARCOS, Universidad Complutense de Madrid, CEI Moncloa, E-28040 Madrid, Spain
[12]Department of Physics, University of Surrey, GU2 7XH Guildford, UK
[13]KTH Royal Institute of Technology, Stockholm, Sweden
[14]Instituto de Física Corpuscular, CSIC-Universidad de Valencia, E-46071 Valencia, Spain
[15]School of Computing, Engineering and Physical Sciences, University of the West of Scotland, PA1 2BE Paisley, UK
[16]Jozef Stefan Institute, Jamova cesta 39, 1000 Ljubljana, Slovenia
[17]National and Kapodistrian University of Athens, Athens 157 72, Greece
[18]INFN Laboratori Nazionali di Legnaro, Viale dell'Università 2, 35020 Legnaro PD, Italy
[19]Faculty of Mathematics and Physics, University of Ljubljana, Ljubljana, Slovenia
[20]Stockholm University, SE-106 91 Stockholm, Sweden
[21]Université Paris-Saclay, IJCLab, CNRS/IN2P3, F-91405 Orsay, France
[22]Universidad de Salamanca, 37008 Salamanca, Spain
[23]Faculty of Physics, University of Warsaw, 00-681 Warsaw, Poland
[24]Department of Physics, Lund University, SE-221 00 Lund, Sweden
[25]Extreme Light Infrastructure-Nuclear Physics (ELI-NP)/Horia Hulubei National Institute for Physics and Nuclear Engineering (IFIN-HH), Str. Reactorului 30, Magurele 077125, Romania

Abstract. Isomeric states of the neutron-rich isotope $^{189}_{73}\mathrm{Ta}_{116}$ were populated via fragmentation of a primary beam of ^{208}Pb ions at 1 GeV/u impinging on a ^{9}Be target at GSI, Darmstadt, Germany. The isotopes of interest were separated, identified and delivered to the DESPEC setup. Two isomers were deduced in $^{189}\mathrm{Ta}_{116}$ and their lifetimes were measured based on the γ-ray time distributions.

1 Introduction

Neutron-rich nuclei in the A ~ 190 mass region exhibit a variety of nuclear structural properties. For instance, well deformed prolate, triaxial, oblate and spherical shapes (at N = 126) could be characterized by the properties of their respective ground-state shapes [1-4]. The evolution from prolate to oblate nuclei passing

* Corresponding author: s.alhomaidhi@gsi.de

through the γ-soft nuclei have been described as the prolate-oblate phase transitional systems [5]. In odd-mass systems, the shape of the atomic nucleus is influenced by the addition of an unpaired particle or hole. A study of odd-A neutron-rich Rhenium isotopes (Z =75) in the A ~ 190 region indicated increasing γ deformation and the evolution of triaxiality with increasing neutron number [6]. The $^{187}Ta_{114}$ nucleus was studied recently using multinucleon transfer (MNT) reactions by Walker et al. The γ rays depopulating the isomeric states of $^{187}Ta_{114}$ were associated with a perturbed rotational band, showing the prolate-oblate shape transition effect [7].

However, with increasing neutron number, it becomes difficult to synthesize and select isotopes using MNT reactions and the isotope separation method. To overcome this obstacle, projectile fragmentation reactions at relativistic energies proved to be an efficient method for populating states in neutron-rich isotopes. For $^{189}Ta_{116}$, two data sets were obtained from the RISING campaigns in 2006 and 2007, which reported different isomeric half-lives of $T_{1/2}$ = 1.6(2) μs [8] and $T_{1/2}$ = 0.58(22) μs [9], respectively. The involvement of two different isomeric states were assumed to explain the discrepancy in half-lives.

The $^{189}Ta_{116}$ nucleus and other neutron-rich nuclei in the A~190 region were populated for the study of shape evolution, using the projectile fragmentation reactions at GSI Helmholtzzentrium für Schwerionenforschung. The spectroscopic study was carried out using the DEcay SPECtroscopy (DESPEC) [10] setup within the FAIR Phase-0 campaign in March 2021.

2 Experimental Details

The nuclei of interest were produced by the fragmentation of a 1 GeV/u primary beam of ^{208}Pb, impinging on a 2.7 g/cm^2 ^{9}Be target. The primary beam was delivered by the UNILAC and SIS-18 synchrotron with a beam intensity of up to 10^9 ions/s. The fragments were separated and identified on an event-by-event basis in the FRagment Separator (FRS) [11], operated in the standard achromatic mode with an Al degrader placed in its intermediate focal plane. The FRS was tuned to transmit fully-stripped $^{190}Ta_{117}$ ions in the central trajectory of the FRS to the final focal plane. A summary of the FRS setting is shown in Table 1.

Table 1. Summary of FRS setting parameters.

Parameter	Value
Magnetic rigidity $B\rho_{12}$	12.5166 Tm
Magnetic rigidity $B\rho_{34}$	10.5367 Tm
S2 degrader thickness	2500 mg/cm^2
S4 degrader thickness	4838 mg/cm^2
Spill length	1.5 s

A series of detectors at the intermediate and final focal planes of the FRS were used for the secondary beam identification. The time of flight (ToF) was determined by measuring the time difference between the two scintillation detectors placed at the intermediate and the final focal planes of the FRS. Time projection chambers (TPCs) were used for position measurements. Two multi-sampling ionization chambers (MUSICs), placed at the final focal plane, were used for the energy loss (ΔE) measurements.

An Al degrader was used to decelerate the ions arriving at the final focal plane and allowed for their implantation into an active stopper, the Advanced Implantation Detector Array (AIDA) [12]. AIDA consisted of 3 layers of (8 x 8) cm^2 Double-Sided Silicon Strip Detectors (DSSSDs), for the identification of the implanted ions and the subsequent β decays. Two fast plastic scintillators were mounted upstream and downstream of AIDA with 10 mm distance for the β-decay timing. The downstream plastic was also used for vetoing light ions passing through the detectors. Surrounding AIDA, two high-purity Ge cluster detectors of the EUROBALL array [13], each with 7 segments, and 36 $LaBr_3$(Ce) detectors of the FAst TIMing Array (FATIMA) [14] were used to detect γ radiation. The $LaBr_3$(Ce) detectors were used for fast-timing spectroscopy, while HPGe detectors provided precise energy information.

All the subsystems of the DESPEC setup were triggered independently, in which the White Rabbit common time clock with ~ 1 ns precision was employed to correlate the events (for more details about the setup, the reader is referred to Ref. [10]).

3 Analysis and Results

As the particle identification is based on the mass-over-charge ratio (A/Q) as a function of the atomic number (Z), the charge states of the ions traversing the FRS must be selected first. Previous studies in the A ~ 190 mass region have shown how the different charge states can overlap in the particle ID plot [8, 9]. In particular, isotopes with atomic mass equal to (A – 3), Z and charge (Q = Z – 1), have comparable magnetic rigidity to fully-stripped ions (A, Q = Z). Fully-stripped ions transmitted through the FRS can pick up an electron when they interact with matter and detectors at the intermediate focal plane. In order to overcome this, the energy loss ΔE_{deg} in the degrader in the intermediate focal plane can be deduced from the difference in the magnetic rigidity, Bρ, between the first and the second halves of the FRS. The charge states of the ions were determined by correlating the energy loss, ΔE_{deg}, with the atomic number (Z) [15].

As shown in Fig. 1, two main isotopic regions (1, 2) in the two-dimensional plot can be seen, one is for the H-like ions from Platinum (Z = 78) to Osmium (Z = 76), and the second region corresponds to the fully-stripped ions from Rhenium (Z = 75) to Hafnium (Z = 72). The isotopes from Z = 76 to 78 acquire an electron at the mid-focal plane and retain it to the final focal plane.

 https://doi.org/10.1051/epjconf/202329002007

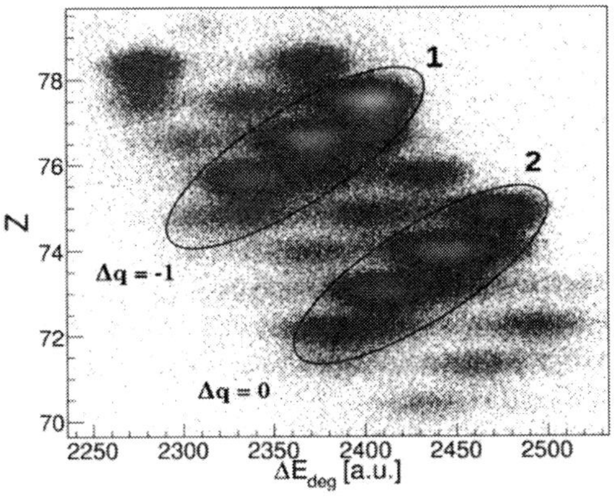

Fig. 1. (color online) Two-dimensional plot of the effective energy loss ΔE_{deg} as function of Z. Region (1): H-like charge state ions from (Z = 78 to 76). Region (2): fully-stripped ions from (Z = 75 to 72).

After the particle identification, the fully-stripped ions (Q = Z), shown in Fig. 2, were used for the γ-spectroscopic analysis. The presence of the delayed 292 keV γ-ray transition in the $^{188}Ta_{115}$ isotope [16], confirms the validity of the used method.

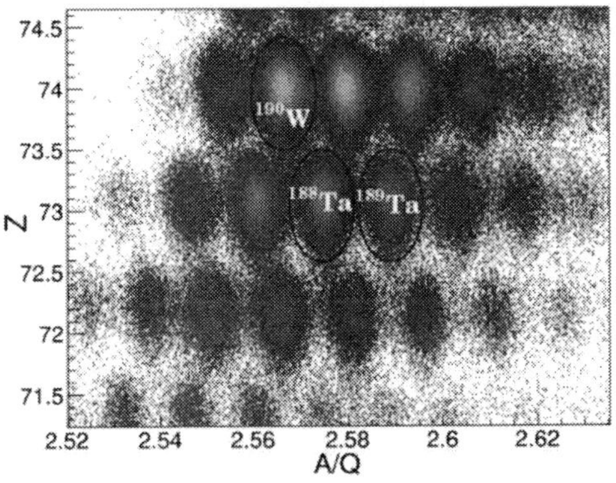

Fig. 2. (color online) Particle identification plot. The main isotopes of interest in the experiment are labelled (^{190}W, ^{188}Ta, ^{189}Ta).

The γ rays decaying from the isomeric states in ^{189}Ta were detected within a time widow of 10 μs for HPGe clusters and 7 μs for the $LaBr_3$(Ce) detectors after the implantation. The presence of previously-reported 83, 134, 154, 199, 246, 283, 389 and 481 keV γ-ray transitions have been confirmed in the present data set. As shown in Fig. 3, the delayed γ-ray transitions spectrum, where the energy peaks are labelled, as recorded in HPGe detectors.

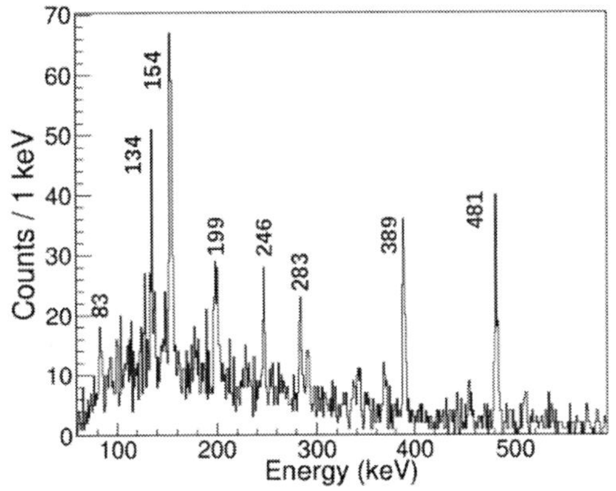

Fig. 3. Delayed energy spectrum of the γ rays known in ^{189}Ta as registered in HPGe detectors.

In the previous measurements by Alkhomashi et al. and Steer et al., one isomeric state was identified. The half-life of the isomer was reported to be $T_{1/2}$ = 1.6(2) μs and $T_{1/2}$ = 0.58(22) μs in [8] and [9], respectively. In this experiment, two isomeric states have been identified. The γ-ray emission time was registered with respect to the time of implantation. The time behaviour of the 134 keV transition, as shown in Fig. 4, shows a shorter decay constant in comparison with time behaviour of the 154, 283, 389 and 481 keV transitions.

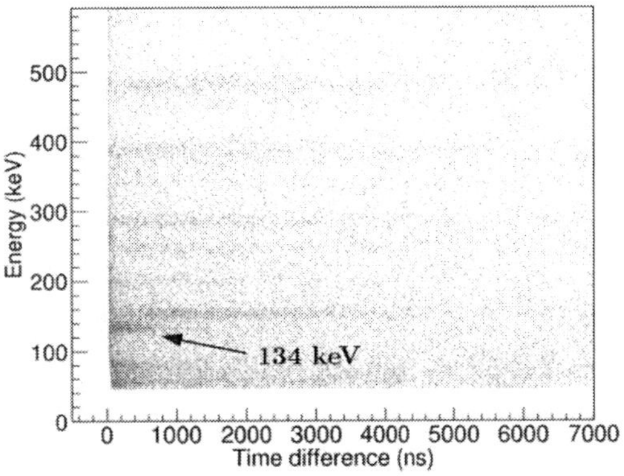

Fig. 4. (color online) The energy-time matrix for the $LaBr_3$(Ce) detectors. The 134 keV γ-ray transition is indicated by the black arrow.

The 154, 199, 246, 283, 389 and 481 keV γ-ray transitions were found to be in coincidence within a ~ 10 ns time window of the γ-γ-ΔT cube. The 134 keV transition is in coincidence with the other transitions mentioned above, when the ΔT range is long (10 ns < ΔT < 5 μs). The time difference between transitions for example: ΔT (154 keV – 134 keV) > 10 ns suggests that the first isomer is decaying from a higher-lying energy state via a 134 keV γ-ray, feeding the second isomer which then depopulates to the ground state via other γ-ray transitions.

With a single-component exponential decay fit to the time difference distribution of the 134 keV transition with respect to the implant time, the half-life of the first isomer was estimated to be $T_{1/2}$ ~ 200 ns. For the second isomer, fitting to the exponential decay of the sum of the time difference distributions of the 154, 283, 389 and 481 keV γ-ray transitions yields a half-life of $T_{1/2} \sim 1.2$ μs. A comparison with theoretical calculations, interpretation of the isomeric states and a suggestion for the level scheme for the neutron-rich $^{189}Ta_{116}$ isotope will be reported elsewhere [17].

4 Conclusions

The isomeric states of $^{189}Ta_{116}$ were produced by fragmentation reactions at the GSI accelerator facility, Darmstadt. The study of γ rays via spectroscopy using the DESPEC setup shows that two isomers are present in this isotope. The half-life of the shorter isomer was estimated to be $T_{1/2} \sim 200$ ns, and $T_{1/2} \sim 1.2$ μs for the longer one. The γ-γ-ΔT analysis has shown that the shorter isomer is decaying from a higher-lying energy state and feeding the second isomer.

The authors would like to thank the staff of the FRS and the GSI accelerator for their support. The results were obtained in the context of FAIR Phase-0 Darmstadt, Germany. This work was supported by BMBF under Verbundprojekt 05P2021 (ErUM-FSPT07) grants 05P21PKFN1 and 05P21RDFN1; by STFC(UK); by the Helmholtz Research Academy Hesse for FAIR (HFHF). PAS was supported by contract PN 23.21.01.06 sponsored by the Romanian Ministry of Research, Innovation and Digitalization. S.A. acknowledges support from King Abdulaziz City for Science and Technology for the Ph.D. funding.

References

1. R. F. Casten, Nucl. Phys. **A443**, 1 (1985).
2. P. D. Stevenson, M. P. Brine, Z. Podolyak, P. H. Regan, P. M. Walker, and J. R. Stone, Phys. Rev. C **72**, 047303 (2005).
3. P. Sarriguren, R. Rodriguez-Guzman, and L. M. Robledo, Phys. Rev. C **77**, 064322 (2008).
4. L. M. Robledo, R. Rodríguez-Guzmán, and P. Sarriguren, J. Phys. G: Nucl. Part. Phys. **36**, 115104 (2009).
5. J. Jolie and A. Linnemann, Phys. Rev. C **68**, 031301(R) (2003).
6. M. W. Reed, et al., Phys. Lett. B **752**, 311 (2016).
7. P. M. Walker, et al., Phys. Rev. Lett. **125**, 192505 (2020).
8. N. Alkhomashi, et al., Phys. Rev. C **80**, 064308 (2009).
9. S. J. Steer, et al., Phys. Rev. C **84**, 044313 (2011).
10. A.K. Mistry et al., Nucl. Instrum. Meth. Phys. Res. A **1033** (2022) 166662.
11. H. Geissel et al., Nucl. Instrum. Methods Phys. Res., Sect. B **70**, 286 (1992).
12. Available: https://edms.cern.ch/ui/file/1865809/2/TDR_HISPEC_DESPEC_AIDA_public.pdf
13. J. Simpson, The Euroball Spectrometer, Z. Phys. A **358** (1997).
14. M. Rudigier et al., Nucl. Instrum. Methods Phys. Res. A **969**, 163967 (2020).
15. J. Benlliure,K.-H. Schmidt,D.Cortina-Gil, T. Enqvist, F. Farget, A. Heinz, A. R. Junghans, J. Pereira, and J. Taieb, Nucl. Phys. A**660**, 87 (1997).
16. M. Caamaño et al., Eur. Phys. J. A **23**, 201 (2005).
17. S. Alhomaidhi et al., (to be published).

EPJ Web of Conferences **290**, 02008 (2023) https://doi.org/10.1051/epjconf/202329002008
EuNPC 2022

R-Matrix study of the β^+decay of ^{8}B to the highly excited states of ^{8}Be

D. Fernandez Ruiz[1*], *S.* Viñals[2,4], *O.* Tengblad[1], *M.J.G.* Borge[1], *H.O.U.* Fynbo[3], *K.* Riisager[3] *and E.* Nácher[5]

[1] *Instituto de Estructura de la Materia, CSIC, E-28006 Madrid, Spain*
[2] *Centro de Micro-Análisis de Materiales (CMAM-UAM), Madrid, Spain*
[3] *Department of Physics and Astronomy, University of Aarhus, DK-8000 Aarhus C, Denmark*
[4] *Universidad Complutense de Madrid, Madrid, Spain*
[5] *Instituto de Física Corpuscular, CSIC, Valencia, Spain*

Abstract. Experiment IS633 was conducted at the ISOLDE facility with the aim of studying the 2^+ isospin doublet of ^{8}Be, the only expected case of nearly equal isospin configuration mixing. The doublet was previously probed by reaction studies where the feeding does not depend on the isospin. Beta decay studies are sensitive to the isospin composition of the doublet since the Fermi and Gamow-Teller strengths are heavily dependent of isospin. In this experiment the doublet was probed through the EC/β^+ feeding from ^{8}B. A four-particle telescope setup with a C-foil in the centre was employed to stop the ^{8}B beam, the implanted nucleus would decay and populate the doublet that breaks up in two alpha particles that are detected. The statistics achieved in this experiment were two orders of magnitude higher than that of any previous experiment, enabling the first experimental observation of both contributions to the doublet.

1 Introduction

The ^{8}B nucleus is a 1-proton halo nucleus that undergoes β^+ decay and electron capture to ^{8}Be. This EC/ β^+ decay is the main source of high-energy solar neutrinos above 2 MeV and has therefore been the subject of extensive research in astrophysics, particularly regarding the so-called "solar neutrino problem" [1]. At the same time, the decay also offers a unique opportunity to investigate the structure of the 2^+ isospin doublet in ^{8}Be. This doublet, formed by two narrow levels at 16.6 and 16.9 MeV with dominant configurations of ^{7}Li+p and ^{7}Be+n, respectively, is the best-known case of fully mixed isospin states [2]. However, the proposed total isospin mixture has yet to be determined experimentally.

Within the Q-value window, Q_{EC} = 17.9798(1) MeV, the following states in ^{8}Be can be fed through allowed beta transitions as shown in Fig 1.

Primarily, the beta decay feeds a broad state at 3 MeV ($J^\pi = 2^+$, Γ=1513(13) keV) which takes most of the feeding, (>88%) and it is the primary source of high-energy solar neutrinos [1].

Secondly, the ^{8}B decay feeds by allowed EC/β^+ transitions the 2^+ isospin doublet. Decay through this doublet is modelled assuming that the Fermi strength goes to the T=1 component and Gamow-Teller strength only to T=0. As the states are isospin mixed, the wave function of the doublet states can be expressed as:

$$|a\rangle = \alpha\ |T=0\rangle + \beta|T=1\rangle \qquad (1.a)$$
$$|b\rangle = \beta\ |T=0\rangle - \alpha|T=1\rangle \qquad (1.b)$$

Where α and β are the mixture coefficients of the pure states. Therefore, if the states are fully mixed then $\alpha^2/\beta^2=1$.

The α and β mixture coefficients can be related with the resonances through the following expressions:

$$\frac{\alpha^2}{\beta^2} = \frac{B_{GT-16,6}}{B_{GT-16,9}};\ \frac{\alpha^2}{\beta^2} = \frac{B_{F-16,9}}{B_{F-16,6}} \qquad (2)$$

$$\alpha^2 = \frac{\Gamma_{16,6}}{\Gamma_{16,6}+\Gamma_{16,9}};\ \beta^2 = \frac{\Gamma_{16,9}}{\Gamma_{16,6}+\Gamma_{16,9}} \qquad (3)$$

Where B_{GT} and B_F are Gamow-Teller and Fermi strength respectively, and Γ the decay width of each level. To determine the α and β coefficients the relevant information must be extracted from the experimentally obtained alpha spectrum.

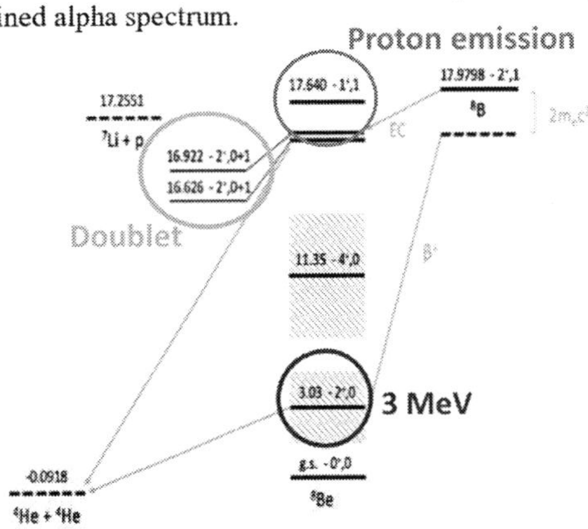

Fig. 1. Decay scheme of ^{8}B to ^{8}Be.

* Corresponding author : daniel.fernandez@csic.es

Considering that the ^{8}Be is unbounded with all its states breaking up into two α particles, and the broad nature of the 3 MeV 2^{+} state, an α-continuum spectrum is measured as observed in previous works [5-10].

Thirdly, in the Q_{EC}-window there is a highly excited 1^{+} level at 17,640 MeV, that decays into ^{7}Li by emitting a low energy (330 keV) proton. This EC-decay can be modelled as occurring in the ^{7}Li core with the halo proton as a spectator [3]. The strength of this branch is estimated from the β-decay of ^{7}Li to be B_{GT} of 1.83 [4].

The first EC/β^{+} decay experiments of 8B were performed by Gilbert in 1954 [5] and improved by Farmer [6] et al, six years later. Its purpose was to test the mirror symmetry with the already studied case of the β^{-}-decay of the ^{8}Li mirror nucleus and to search for a possible new level in the excited ^{8}Be nucleus. The results confirmed the symmetry in the decay schemes of the mirror nuclei (^{8}Li and ^{8}B) and proposed for the first time the existence of two excited states at 3 MeV and 16.7 MeV. In 1964 Matt et al. [7] measured the single α spectrum emphasising the region above 7.5 MeV, the results pointed towards the existence of a 16.6 MeV state with spin and parity 2^{+} and an isospin of T=1 with a small contribution of T=0. Later, Brown et al. [8] fitted the experimental data via a reaction mechanism that assumed a 2^{+} doublet [9] strongly mixed by the Coulomb interaction.

In summary, the 16.626(3) MeV state has been observed by several groups in EC/β^{+}-decay studies. The mainly EC-decay to the 16.922(3) MeV state was first hinted at the previous JYFL08 experiment [10], where 5 events were attributed to the breakup of this state.

2 Experiment IS633

The MAGISOL (Madrid-Aarhus-Goteborg at ISOL)-collaboration has conducted several experiments to study the structure of ^{8}Be. The latest of them, experiment IS633, was performed at the ISOLDE facility at CERN to resolve the doublet in ^{8}Be via a β^{-}-feeding process and from that try to determine the isospin mixing coefficients.

The experimental setup, depicted in Fig. 2, comprised four particle telescopes, each consisting of a Double-Sided Silicon strip Detector (DSSD) with a thickness of 40-60 μm, stacked with a 1500 μm thick Si-PAD detector. A 30 μg/cm² carbon foil catcher was positioned at the centre of the setup. To increase the angular coverage of beta particles, a thick 500 μm DSSD was placed beneath the carbon foil. Overall, the system has an angular coverage of Ω = 38% of 4π. For further details, refer to [11].

A 50 keV $^{8}BF_2$ beam produced online at ISOLDE [12] was implanted in the 138 nm C-foil at a depth of 26 nm [11]. The implanted ^{8}B nucleus decays in the foil, feeding the excited states of ^{8}Be, which breaks up into two alpha particles. The experimental setup surrounding the C-foil detects the emitted alpha particles, yielding information on the excited states of the ^{8}Be nucleus.

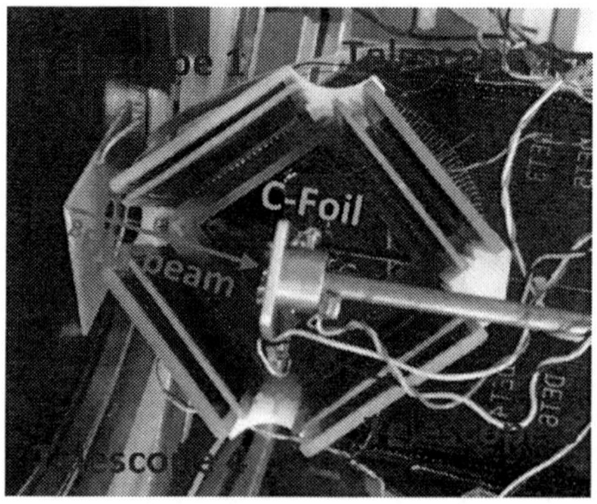

Fig. 2. Top view of the IS633 set up. The four particle telescopes surround the central C-foil catcher, the $^{8}BF_2$ beam impacts from the left side of the set up.

In the analysis procedure, for an event to be considered valid, it must fulfil coincidence conditions: two α particles must be detected in opposite detectors. The line connecting the points where the α-particles are detected in each DSSD should cross the area of the beam spot. This beam spot creates a cone in the DSSD that includes the maximum deviation angle between the two particles, therefore, an event must be detected within this cone. The high statistics obtained in experiment IS633 allowed to resolve the doublet for the first time in an EC/β+ decay study as it can be seen in the lower part of Fig 3

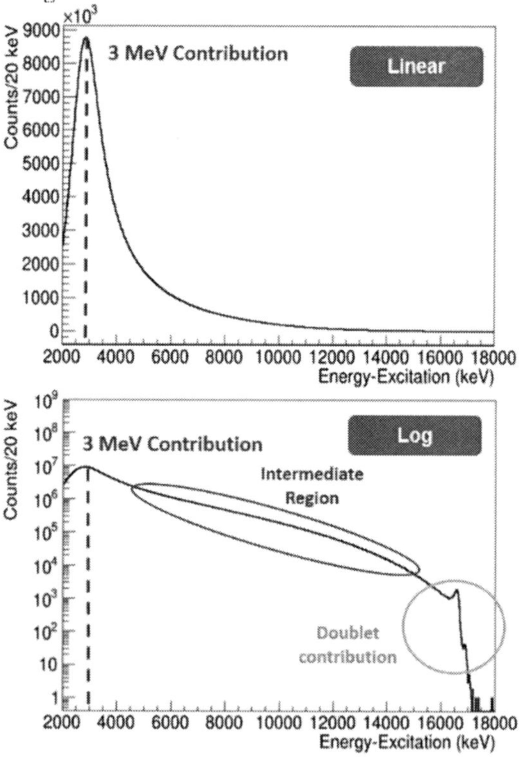

Fig. 3. α-α coincidence spectrum as function of the ^{8}Be excitation energy [$E_{\alpha1}$+$E_{\alpha2}$-92 keV]. Upper figure: linear scale. Lower figure: logarithmic scale with three main regions: the 3 MeV peak, the intermediate region associated with the transition to the continuum, and the 2^{+} doublet (resolved for the first time).

The spectrum shown in Fig. 3 exhibit three distinct regions: the dominant 3 MeV peak, the region of the doublet, and an intermediate region, which is difficult to explain only due to the tail contributions of the 2^+ states, for this region, one should consider whether it corresponds to a direct transition to the continuum.

The α-α coincidence spectrum obtained is corrected for energy losses in the C-foil and for the breakup energy of the alphas. With the corrections applied, a formalism to fit the spectrum is required. Due to all the states being merged into a continuum, a good candidate for fitting the spectrum is the R-Matrix formalism [9].

The R-Matrix formalism was originally developed to study nuclear reactions [13]. This approach divides the configuration space into two radial regions: internal and external, separated by a constant called the channel radius r_o. Each region contains a different part of the reaction process and is dominated by a specific potential. The external region ($r_o < r < \infty$) is dominated by the Coulomb force, while the nuclear interaction dominates the internal region ($0 < r < r_o$). Imposing continuity of the wave function and its derivative at the border between these regions, an expression for a matrix that relates the eigenstates of both regions is obtained, the Reaction Matrix.

$$R_{cc'} = \sum_{\lambda} \frac{\gamma_{\lambda c}\gamma_{\lambda c'}}{E_\lambda - E} \quad (4)$$

R-Matrix is a parametrization of the configuration space in terms of individual resonances, each associated with a reaction channel. These resonances are characterized by specific parameters related to the actual physical reaction, most notably the cross-section. Baker et al. [12] extended this formalism to nuclear β decays.

From an experimental point of view, a spectrum can be decomposed into a series of nuclear resonances, each characterized by specific parameters simplifying a complex fit into a series of individual resonances.

3 R-Matrix analysis

The R-Matrix formalism was employed to analyse the excitation spectrum of ^{8}Be shown in Fig 4. The spectrum is decomposed into four resonances: the 3 MeV, 16.6 MeV, 16.9 MeV levels and a background (BKG) level that models the effects of the combination of the tails of higher energy resonances. Each level is characterized by the energy (E), decay width ($\Gamma_{\alpha\alpha}$), Fermi (B_F) and/or Gamow-Teller (B_{GT}) strengths, the R-Matrix fit must be folded with the response function of the detector [14].

The algorithm to perform an R-Matrix fit has been implemented into the ORM_FIT program developed at Aarhus University [10].

The ORM_FIT code includes a feature that selects the fitting range for the R-Matrix. For this analysis, two main fits were performed: First, a local fit to the doublet, in which only the parameters associated with the 16.6 MeV and 16.9 MeV resonances were allowed to change. Second, was a global fit to the entire spectrum (see Fig. 4), where the initial doublet parameters were set to the ones obtained from the local fit and therefore let them evolve to properly fit the full range. The parameters of the other resonances (3 MeV and BKG) where set free. In all cases the channel radius was fixed to r_0=1.35 fm.

While results of the local fit were consistent with literature [2], in the case of the global fit, the decay width ($\Gamma_{\alpha\alpha}$) of the 3 MeV resonance deviates from the literature value by 400 keV ($\Gamma_{\alpha\alpha}$=1513 (15) keV vs $\Gamma_{\alpha\alpha}$=1957 (18) keV).

Since the global R-Matrix fit gives values that do not align well with those of the literature, a series of internal and external consistency tests were performed. First, the global fits were repeated imposing certain restrictions over the fitting parameters (internal consistency checks), the objective of these checks was to ensure that the algorithm always gives similar results.

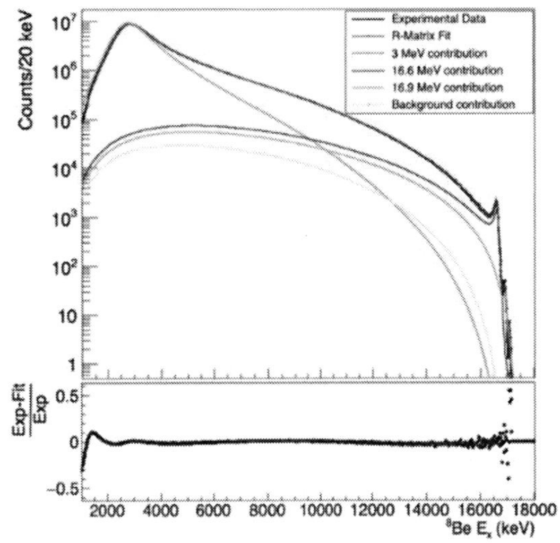

Fig. 4. Upper figure: Global R-Matrix fit (Red) to α-α coincidence spectrum as function of the ^{8}Be excitation energy (black), the individual contributions included in the R-Matrix are those of the 3 MeV (purple), 16.6 MeV (blue), 16.9 MeV (green) and BKG, 21 MeV (yellow) resonances. Lower figure: residue function between the R-Matrix and the data.

A first fit was performed fixing the R-Matrix parameters to those found in the literature, resulting in a fit with a χ^2 of 3991 denoting the low fit quality. Secondly, the literature values were used as a starting point for the fit and allowed to change. The resulting R-Matrix values converged to those of the previous global fit ($\Gamma_{\alpha\alpha}$=1949 (18) keV) with a χ^2 of 14.3. Finally, the fit was repeated with the BKG contribution fixed to 37 MeV. This choice of the background level was motivated by S.Hyldegaar's work [15], who found that a 37 MeV was the optimal energy for the background level. In this case, the fit had a χ^2 of 12.2, better than the initial fit but the width of the 3 MeV level was $\Gamma_{\alpha\alpha}$=1886 (12) keV, still far from the standard literature value.

None of these tests managed to significantly improve the previous results. Fixing the fitting parameters to literature values produced an extremely bad fit, while the other two fits converged to the values obtained in the initial global fit, with only a slight improvement if the BKG level was set to 37 MeV.

Apart from the internal consistency checks of the R-matrix fitting programme, the α-α coincidence spectrum was compared with that of the previous experiment JYFL08 [9] to explore possible contributions from pile-up that should mainly affect the spectrum at around 4.5-6 MeV. To compensate the difference in statistics the IS633 data set was normalized to the 2-4 MeV region of JYFL08 by a scaling factor of 0,0162.

The α-α spectrum of JYFL08 (red solid line) and the IS633 α-α spectrum normalized to the intermediate region (dashed black line) are displayed in Fig. 5. It can be appreciated that both spectra are almost identical.

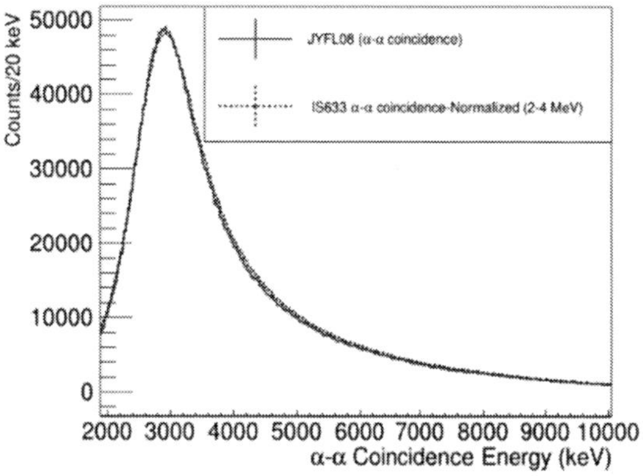

Fig. 5. The α-α coincidence spectrum Red- JYFL08 data; Black- IS633 normalized data.

From this analysis, one can calculate the decay width of the 3 MeV level for both data sets (IS633 and JYFL08) and compare them to see if there is any significant difference. To obtain the width, two methods where proposed: a direct determination of the width at half-maximum (FWHM) and the the R-Matrix fit. The latter was repeated varying the fitting ranges to test the influence of the fitting over the R-Matrix. The results obtained are presented in Table 1.

Table 1: FWHM and R-Matrix (R-Mat) fit for the 3 MeV region for multiple fitting ranges.

R-Mat Fit Range (MeV)	JYFL08		IS633	
	E (keV)	$\Gamma_{\alpha\alpha}^{3\,MeV}$ (keV)	E (keV)	$\Gamma_{\alpha\alpha}^{3\,MeV}$ (keV)
2-4	3034(34)	1488(12)	2997(36)	1470(15)
2-5	3036(28)	1475(17)	3006(32)	1588(18)
2-6	3047(31)	1516(20)	3020(38)	1655(16)
2-7	3060(26)	1565(13)	3030(37)	1706(18)
FWHM	2980(20)	1510(30)	2980(20)	1525(20)
Lit	**E (keV)**	3030(10)	$\Gamma_{\alpha\alpha}^{3\,MeV}$**(keV)**	1513(15)

The decay width for the 3 MeV level obtained from the R-Matrix fit for both experimental data stets coincides with the literature value of Γ=1513 keV when the fitting range is reduced to the 3 MeV resonance (2-5 MeV).

However, as the fitting range increases and enters into the intermediate region, the value of the decay width increases and starts to diverge from the literature value. This effect is more pronounced when the contributions from borth members of the doublets are included. The explanation for the dependency of the results with the fitting range might lie within the limitations of the R-Matrix method, whose core is the decomposition of the spectrum into well-defined resonances. This approximation is valid for the low and high-energy regions. However, the intermediate region could be caused by a direct non-resonant transition to the continuum. In that case, R-Matrix cannot resolve these non-resonant contributions. One could think that adding another background level could improved the fitting of the intermidiate region, nevertheless the results obtained from this approach are not clear to interpret.

4 Conclusion

Experiment IS633 constitutes an improvement over previous attempts to measure, by beta-decay, the 2^+ doublet of ^{8}Be, resolving the both doublet contributions in the excitation energy spectrum fot the first time.

The R-Matrix formalism is used to study the very complex nature of the spectrum. However, the R-Matrix algorithm could not fit the whole spectrum accurately and the resulting parameters for the well-known level at 3 MeV diverge from the values of the literature. Moreover, the parameters of the 2^+ doublet obtained are dependent of the fitting range. The contribution of a nonresonant direct transition to the continuum could be a possible explanation of this phenomenom.

Acknowledgements

We acknowledge the funding through the grants MCIN/AEI PID2019-104390GB-I00; PRE2020-094522 "ESF Investing in your future", CT31/21-UCM and fundings Next Generation of the European Union, and the MAGISOL collaboration.

References

1. J. N. Bahcall et al., New J. Phys **6** (2004) 63.
2. D.R.Tilley et al., Nucl. Phys. A **745** (2004)155.
3. T. Nilsson et al., Hyperfine Int **129** (2000) 67.
4. M.J.G. Borge et al., J. Phys. G **40** (2013) 035109.
5. F.C.Gilbert, Phys. Rev. **93** (1954) 499.
6. B.J. Farmer et al., Nucl. Phys **15** (1960) 626.
7. E. Matt et al., Phys. Lett. **9** (1964) 174.
8. C.P. Browne et al., Phys. Lett. **23** (1966) 371
9. F.C.Barker, Aust. J. Phys. **22** (1969)
10. O. Kirsebom et al., Phys. Rev. C **83** (2011) 065802.
11. S.Viñals, PhD thesis. Complutense University of Madrid (2020).
12. J.Ballof et al., Eur. Phys. J.A **55** (2018) 65.
13. A. M. Lane and R. G. Thomas, Rev. Mod. Phys **30** (1958) 257.
14. S. Viñals et al., EPJA **57** (2021) 1.
15. S.Hyldegaar, PhD thesis. Aarhus University (2010).

EPJ Web of Conferences **290**, 02009 (2023) https://doi.org/10.1051/epjconf/202329002009
EuNPC 2022

Study of (p,2p) fission reactions in inverse kinematics using the R3B set-up.

G. García-Jiménez[1,*], H. Alvarez-Pol[1], A. Graña-González[1], J.L. Rodríguez-Sánchez[1], J. Benlliure[1], D. Cortina-Gil[1], L. Atar[2], L. Audouin[3], G. Authelet[4], A. Besteiro[5], G. Blanchon[6], K. Boretzky[7], M.J.G. Borge[8], P. Cabanelas[1], E. Casarejos[5], J. Cederkall[9], A. Chatillon[6], A. Corsi[4], M. Feijoo[1], E. De Filippo[10], D. Galaviz[11], I. Gasparic[12], E. Geraci[10], R. Gernhäuser[13], B. Gnoffo[10], M. Heil[7], A. Heinz[14], T. Hensel[15], M. Holl[14], D. Jelavic Malenica[12], T. Jenegger[13], L. Ji[2], H.T. Johansson[14], O.A. Kiselev[7], P. Klenze[13], A. Knyazev[9], D. Körper[7], T. Kröll[2], I. Lithar[12], Y.A. Litvinov[7], B. Löher[7], N. S. Martorana[10], P. Morfouace[6], D. Mücher[16,17], S. Murillo-Morales[18], A. Obertelli[2], V. Panin[7], J. Park[19], S. Paschalis[18], A. Perea[8], M. Petri[18], S. Pirrone[10], L. Ponnath[13], A. Revel[4], H. Rhee[2], L. Rose[18], D.M. Rossi[2], P. Russotto[10], H. Simon[7], A. Stott[18], Y. Sun[2], C. Sürder[2], J. Taïeb[6], R. Taniuchi[18], O. Tengblad[8], H.T. Törnqvist[2,7], M. Trimarchi[10], S. Velardita[2], J. Vesic[20], B. Voss[7], and the R³B Collaboration

[1]IGFAE, University of Santiago de Compostela, E-15782 Santiago de Compostela, Spain
[2]Technische Universität Darmstadt, Fachbereich Physik, Institut für Kernphysik, 64289 Darmstadt, Germany
[3]IPN Orsay, 15 rue Georges Clemenceau, 91406, Orsay, France
[4]CEA Saclay, IRFU/DPhN, Centre de Saclay, 91191, Gif-sur-Yvette, France
[5]CINTECX, Universidade de Vigo, DSN, 36310 Vigo, Spain
[6]CEA, DAM, DIF, F-91297 Arpajon, France
[7]GSI Helmholtzzentrum für Schwerionenforschung, Planckstraße 1, D-64291 Darmstadt, Germany
[8]Instituto de Estructura de la Materia, CSIC, E-28006 Madrid, Spain
[9]Department of Physics, Lund University, P.O. box 118, 221 00 Lund, Sweden
[10]INFN Laboratori Nazionali del Sud, Via Santa Sofia 62, 95123, Catania, Italy
[11]LIP and Faculty os Sciences, University of Lisbon, 1000-149 Lisbon, Portugal
[12]RBI Zagreb, Bijenicka cesta 54, HR10000, Zagreb, Croatia
[13]Technische Universität München, James-Franck-Str 1, 85748, Garching, Germany
[14]Institutionen för Fysik, Chalmers Tekniska Högskola, 412 96 Göteborg, Sweden
[15]Institut für Kern- und Teilchenphysik, Technische Universität Dresden, 01069 Dresden, Germany
[16]University of Guelph, 50 Stone Road E, N1G 2W1, Guelph, ON, Canada
[17]Institut für Kernphysik der Universität zu Köln, Zülpicher Strasse 77, D-50937 Köln, Germany
[18]School of Physics, Engineering and Technology, University of York, YO10 5DD York, UK
[19]Institute for Basic Science, Center for Exotic Nuclear Studies, 34126, Daejeon, Republic of Korea
[20]Jozef Stefan Institute, Jamova cesta 39, 1000 Ljubljana, Slovenia

Abstract. A new experimental fission approach is presented in the context of the R³B (Reactions with Relativistic Radioactive Beams) collaboration, at the GSI/FAIR facility, in which knockout reactions in inverse kinematics are used to induce fission of ^{238}U that will allow to characterise the excitation energy of the fission process and all the fission products. The CALIFA (CALorimeter for In-Flight detection of γ-rays and high energy charged pArticles) calorimeter, a key part of the R³B set-up, is used to reconstruct the momenta of the two protons from the (p, 2p) reactions. Preliminary results show that kinematic variables and first estimates for nucleon-removal cross sections are well reconstructed and in good agreement with other experimental measurements.

1 Motivation

Nuclear fission has been used as a tool for the study of nuclear properties since its discovery in 1939 by O. Hahn, F. Strassmann and L. Meitner after the irradiation of uranium nuclei by neutrons [1, 2]. The process can be understood as a complex mechanism in which a heavy nucleus splits into two smaller fragments, releasing a large amount of energy (compared with another nuclear reactions) and lighter particles, such as γ-rays and neutrons. The fission mechanism is governed by intrinsic and collective degrees of freedom. The latter can be divided into macroscopic and microscopic processes. The fission path followed by the system until it reaches the scission point is lead by a collective motion of its constituents and deformations, such as the elongation and mass asymmetry. In particular, at low excitation energies, the way the nucleus splits after the saddle point is strongly affected by its nuclear shell structure [3].

With the advance of nuclear reactions in inverse kinematics it was possible for the first time to measure

*e-mail: gabrielgarcia.jimenez@usc.es

EPJ Web of Conferences **290**, 02009 (2023) https://doi.org/10.1051/epjconf/202329002009
EuNPC 2022

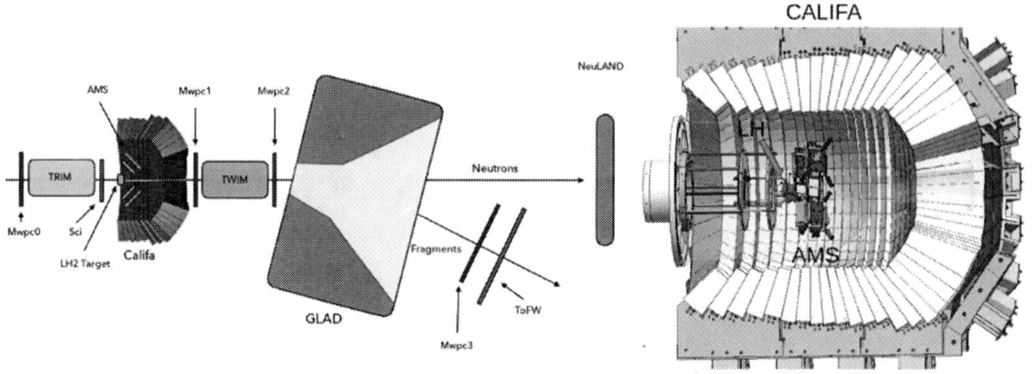

Figure 1: **(Left)** Top schematic view of the experimental set-up, in which the beam enters from left and impinges on the LH_2 target. Sizes are not to scale. **(Right)** CAD drawing of the CALIFA calorimeter, together with the AMS-type detectors and the LH_2 target.

simultaneously both fission fragments in terms of their charge and mass numbers [4, 5], which allowed for a complete kinematics measurement of the fission process. Such experiments made it possible to extract correlations between fission observables sensitive to the dynamics of the fission process at high excitation energies [6, 7] and to the nuclear shell structure at low excitation energies [8, 9]. With the development of new radioactive beam accelerator facilities more exotic nuclei are at the disposal of the experimentalists but, as they are typically short-lived rare isotopes, new methods are needed to measure fission barriers heights and fission yields. Coulomb-induced fission can be used to investigate fission of exotic nuclei at low excitation energies taking advantage of inverse kinematics, but the lack of reaction products prior to fission and the absence of information about the impact parameter makes not possible to study and characterize nascent fragments as a function of the excitation energy, which is assumed to be in a certain range. Transfer-induced fission allows a precise reconstruction of the excitation energy, but the cross section is much lower [10] compared with knockout reactions and the reaction is more peripheral, exploring only the outer shells of the nuclei.

Single-nucleon knockout induced fission reactions would overcome some of these limitations, allowing the study of this fission process at low excitation energies. At sufficiently high beam energies, quasifree scattering reactions become possible, with a strong localized interaction in which the knockout proton leaves a single hole state, producing an excited nuclear remnant, being this excitation released in the subsequent fission. The measurement of the two proton momenta allows to determine the excitation energy of the remnant prior to fission on an event-by-event basis and therefore a study of key fission observables, such as the mass and charge of both fission fragments, evaporated neutrons and fission probabilities, as function of the excitation energy.

2 Experiment

The experiment was carried out at GSI on March 2021 using the R^3B experimental set-up located at Cave-C, as shown in Fig. 1 (left). A primary beam of ^{238}U at 560 AMeV was delivered to the Cave from SIS18 impinging onto a 15-mm liquid hidrogen target. The CALIFA calorimeter [11, 12] is used together with a silicon tracker based on AMS-type detectors [13] to provide angular, energy and vertex reconstruction of the two scattered protons. The charge and the horizontal angle of the two nascent fragments is then measured by a double multi-sampling ionization chamber (Twin-MUSIC) detector [14]. Several multiwire proportional chambers (MWPC's) [15] are placed along the set-up, so a precise tracking of the beam and the fragments is possible. The large-acceptance superconducting dipole magnet GLAD bends the trajectory of each fragment, which in combination with a Time-of-flight wall [16] allows a mass spectroscopy study of the fragments. Finally, the NeuLAND (Large Area Neutron Detector) [17] can be used to reconstruct the neutron kinematics and to obtain the neutron multiplicities.

The CALIFA detector is composed of 2432 CsI(Tl) detection units (1504 currently installed) with right-pyramid frustum shape, and lenghts varying from 14 to 22 cm. A carbon-fibre structure acts as a honeycomb-like holder for the crystals. Each crystal is coupled on its back to a Large Avalanche Photo-Diode (APD) that collects the light output in the material. The detector is divided into two different parts. The barrel part has a cylindrical shape and is made of 1952 crystals, covering from 43° to 143° (right now up to 90°) in polar angle. The end-cap part comprises

480 crystals with more irregular shapes, covering the range from 21° to 43°. This part is equiped with double-ranged preamplifiers, so both γ-rays and protons can be reconstructed using the same electronics and analysis software, whereas the barrel electronics only allows for a single-gain range that, for the present experiment, was set to the proton range.

3 Data analysis and results

After a proper calibration of the TWIM Music [18] and the CALIFA detector, the preliminary analysis of the reaction confirms that the proton pair can be properly correlated with the fission fragment charge distribution. Each proton is measured in coincidence in the CALIFA detector, so a kinematic line and an opening angle can be extracted.

The analysis comprises two reaction channels:

1. The knockout of a proton before fission: $^{238}U(p,2p)^{237}Pa$ + Fission.

2. The knockout of a proton and evaporation of neutrons, without fission: $^{238}U(p,2p)^{X}Pa$.

The observation of the two channels results in two very different opening angle distributions, as it can be seen in Figs. 2a and 2b. No fission events are more likely to come after a clean knockout of a proton from ^{238}U (i.e quasifree) with lower excitation energy, while fission events are produced mostly after a rescattering inside the incoming nucleus, so more energy is at disposal of the remanent making fission more probable. Selecting events with the quasifree condition, as illustrated in Fig. 3(a), in the fission channel and correlating then with the fragment charge distribution gives the evolution presented in Fig. 3(b) and 3(c) for quasifree and multi-knockout reactions, respectively. This proves that quasifree knockouts produce an asymmetric distribution of the fragments, with a larger contribution from the standard I and standard II fission modes, while a cut in the rescattered events shows a more symmetric distribution, with more contribution from the "super-long" (SL) fission mode. The next step would be then to reconstruct the reaction vertex and fine momenta of the scattered proton pair, so a direct calculation of the excitation energy can be performed using an invariant mass approach, as described in Ref [19].

3.1 Cross sections

Preliminary cross sections have been obtained by imposing the condition of two proton hits in the CALIFA calorimeter and, in the case of fission reactions, of two fission fragments in the spectrometer. For the fission channel selection we require fissioning systems with charge 91, which corresponds to fission fragments that fulfill the condition Z_1+Z_2=91. Additionally, we also request two hits in the segmented ToF wall located at the end of the R³B setup since it provides the fission trigger. The results obtained for the cross sections are listed in Table 1, which also includes the cross section measured in previous experiments performed with the GSI fragment separator.

Reaction	Cross sections (mb)
$^{238}U(p,2p)^{X}Pa$ (This work)	78 ±13
$^{238}U(p,2p)^{X}Pa$ ([20])	82.6 ±5.6
$^{238}U(p,2p) + Fission + QF$	24 ±8

Table 1: Preliminary cross section estimation for the two channels of interest. The uncertainties given for this work are purely statistical.

4 Conclusions

The first (p,2p)-fission experiment has been carried out at the GSI/FAIR facility with projectiles of ^{238}U at 560A MeV impinging on a liquid hydrogen target. The R³B experimental set-up is used to perform complete kinematics measurements of the reaction.

The reconstruction of the (p,2p) reactions with the CALIFA calorimeter shows the presence of two kinematical

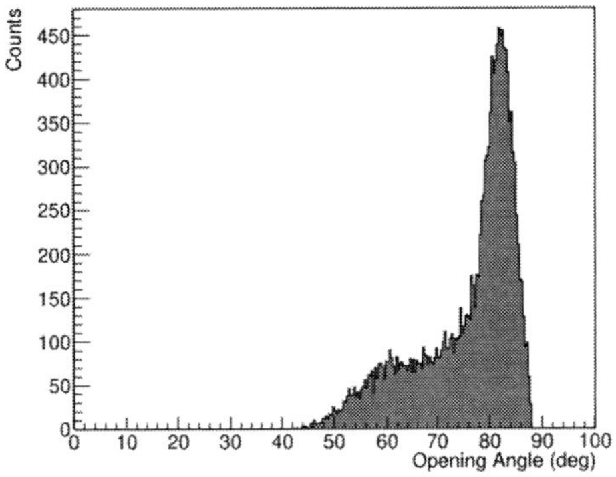

(a)

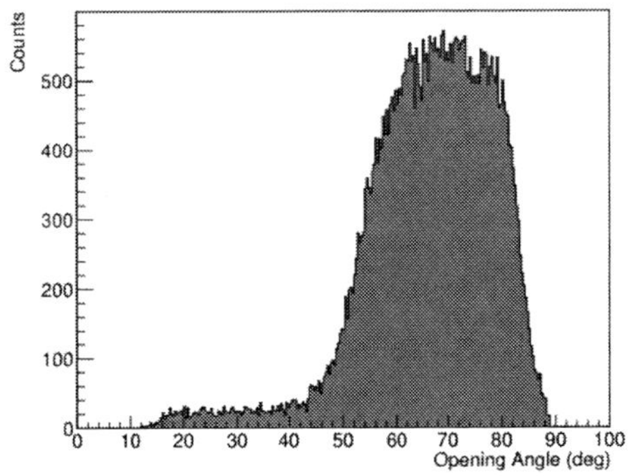

(b)

Figure 2: (a): Opening angle distribution for the two scattered protons in the case of a knockout without fission. (b): Opening angle distribution for the two scattered protons in the case of a knockout-induced fission.

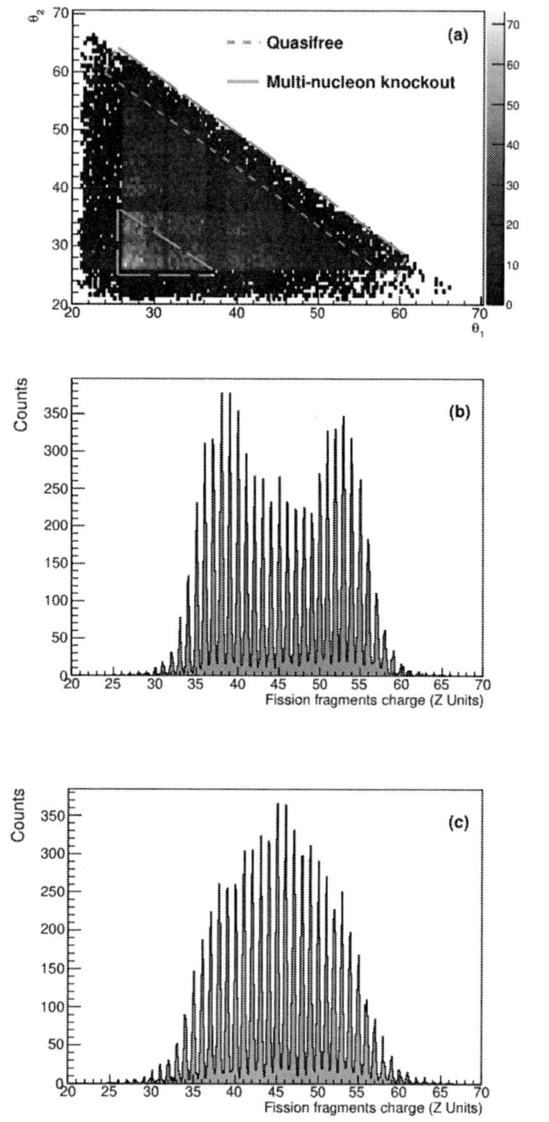

Figure 3: (a) Polar angle correlation of outgoing protons with the angular ranges selected to investigate the fission yields evolution with the opening angle. (b) Charge distribution for quasi-free (p,2p) fission reactions. (c) Charge distribution for multi-nucleon knockout reactions.

regions related to quasi-free scattering reactions close to opening angles of 80° and re-scattering at smaller angles. The correlation of these two kinematic regions with the charge distribution of the fission fragments shows a clear evolution from asymmetric to symmetric fission related to low and high excitation energies, respectively. Finally, inclusive cross sections for the two channels of interest have been extracted for the first time deploying the large acceptance and complete-kinematics capabilities of the R³B set-up. This analysis proves the validity of the method for future studies with more exotic beams, allowing the study of fission observables as a function of the excitation energy.

Acknowledgments

This work was partially supported by the Spanish Ministry for Science and Innovation under Grants PGC2018-099746-B-C21 and PGC2018-099746-B-C22, by the Regional Government of Galicia under the program "Grupos de Referencia Competitiva" ED431C-2021-38 and by the "María de Maeztu" Units of Excellence program MDM-2016-0692. J.L.R.-S. is thankful for support from Xunta de Galicia under the postdoctoral fellowship Grant ED481D-2021-018. J. P. Acknowledges the support by the Institute for Basic Science (IBS-R031-D1). G. García-Jiménez thanks the support by the Spanish Ministry for Science and Innovation under the Grant PRE2019-087415.

References

[1] L. Meitner and O. R. Frisch, Nature (London) **143**, 239 (1939).
[2] O. Hahn and F. Strassmann, Naturwissenschaften **27**, 11 (1939).
[3] P. Möller, D. Madland, A. Sierk, and A. Iwamoto, Nature **409**, 785 (2001).
[4] J. L. Rodríguez-Sánchez et al. Phys. Rev. C **91**, 064616 (2015)
[5] E. Pellereau et al. Phys. Rev. C **95**, 054603 (2017)
[6] J. L. Rodríguez-Sánchez et al., Phys. Rev. C **94**, 034605 (2016)
[7] J. L. Rodríguez-Sánchez et al., Phys. Rev. C**94**, 061601(R) (2016)
[8] A. Chatillon et al., Phys. Rev. Lett. **102**, 202502 (2020)
[9] A. Chatillon et al., Phys. Rev. C **106**, 024618 (2022)
[10] F. Farget et al., Eur. Phys. J. A , **175** (2015)
[11] D. Cortina-Gil et al, Nuclear Data Sheets **120**, 99 (2014)
[12] H. Alvarez Pol et al. Nucl. Instrum. Methods Phys. Res. A **767**, 453 (2014)
[13] J. Alcaraz et al., Nucl. Instrum. Methods Phys. Res. A **593**, 376 (2008)
[14] E. Pellereau et al, EPJ Web of Conferences **62**, 06005 (2013)
[15] ALICE Collaboration, J. Phys. G: Nucl. Part. Phys. **30**, 1517 (2004)
[16] A. Ebran et al., Nucl. Instr. Meths. A **728**, 40 (2013)
[17] K. Boretzky et al., Nucl. Instrum. Methods Phys. Res. A **1014**, 165701 (2021)
[18] A. Graña-González et al., arxiv.org/abs/2210.06054
[19] T. Aumann, Eur. Phys. J. A **26**, 441 (2005)
[20] J. Taieb et al, Nuclear Phys. A **724**, 413 (2003)

 https://doi.org/10.1051/epjconf/202329002010

Reaction studies with neutron-rich light nuclei at the upgraded SEC Device: Exploring the excited structure of ^{11}Li

D. Fernandez Ruiz[1*], *O.* Tengblad[1], *M.J.G.* Borge[1], *H.O.U.* Fynbo[2] and *K.* Riisager[2]

[1] *Instituto de Estructura de la Materia, CSIC, E-28006 Madrid, Spain*
[2] *Department of Physics and Astronomy, University of Aarhus, DK-8000 Aarhus C, Denmark*

Abstract. The ^{11}Li nucleus is considered the best example of a two-neutron halo nucleus. While the structure of the ground state is well established there is still a significant debate regarding its excited states despite of the multiple experiments realised mainly from the excitation of the ground state of ^{11}Li that is rather complex. We propose a novel approach, to directly feed the different excited states of ^{11}Li by the (t,p) reaction as ^{9}Li has a simpler structure. We will employ a ^{9}Li beam produced at ISOLDE and post-accelerated by HIE-ISOLDE to energies of 7 MeV/u to impinge on a ^{3}H-target located at the upgraded SEC station, this station contains a new particle detection set up: composed of 5 Si-particle telescopes in a pentagon configuration with end caps consisting of 2 CD Si detectors downstream and 1 CD upstream has been prepared to offer the maximum angular coverage and resolution. Geant4 Monte Carlo simulations with the efficiency of the setup is presented in the following proceeding.

1 Introduction: Halo Nuclei

The term halo nucleus was coined to describe nuclei with a low biding energy for the last nucleons and an unusually large spatial extension, diverging from the standard $r=r_0A^{1/3}$. The first empirical observation of this behaviour came from scattering experiments of accelerated Lithium isotopes [1]. The interaction cross-section for neutron-rich nuclei was measured, and it was found that the cross-section of ^{11}Li was unexpectedly larger than that of ^{9}Li. The drastic increase of cross-section pointed towards a nuclear radius larger than the theoretical prediction. This discovery was interpreted as a new type of nuclear structure [3], formed by a compact core and an external set of nucleons. A few years later [4], the ^{9}Li momentum distribution obtained from ^{11}Li break-up experiments confirmed this hypothesis.

The ^{11}Li nucleus is the archetype of a two-neutron halo: a three-body system formed by two somehow correlated neutrons loosely bound to the ^{9}Li ground state (g.s) [5]. While the gs structure of ^{11}Li is known to be a mixture of p(59(1)%), s(35(4)%) and d(6(4)%) waves [6], knowledge of higher energy levels is not well settle as the different reaction studies give dispersed results.

^{11}Li has no bound excited state. The low-lying continuum spectrum is dominated by broad dipole structures observed in several experiments, while narrower resonances have been proposed up to 6.2 MeV However, multiple experiments do not agree in the energies and width of the observed resonances.

Recent results on the low-lying continuum structure in ^{11}Li have been obtained from inelastic p and d scattering at TRIUMF [6,7]. Both measurements gave consistent elastic cross sections. However, the inelastic scattering results indicated a resonant state at 0.80(4) MeV, Γ=1.15(6) MeV for the proton inelastic scattering [7]. While the determinate resonance was characterized to be at 1.03(4) MeV, Γ= 0.51(11) MeV for the deuteron case [6]. Although the values are not far apart neither the energy nor the width coincide. It has been argued [8] that the difference could be due to low statistic in the (d,d') case, or problems with the angular determination that is particularly difficult in a cryogenic target. However, there is a more relevant question concerning the physics process involved: excitation to a resonance or excitation directly to the continuum.

Most of the discussed experiments explore the excited structure of ^{11}Li by promoting a ^{11}Li nucleus in the g.s to the excited levels. The only exception being the study of the (very complex) ^{14}C(π^-,p+d) reaction [9]. Whose low resolution did not allow a detailed characterization of the resonances. Instead, we propose [10] to use the two-neutron transfer reaction ^{9}Li(t,p)^{11}Li to probe the resonant structure of ^{11}Li, see Fig. 1.

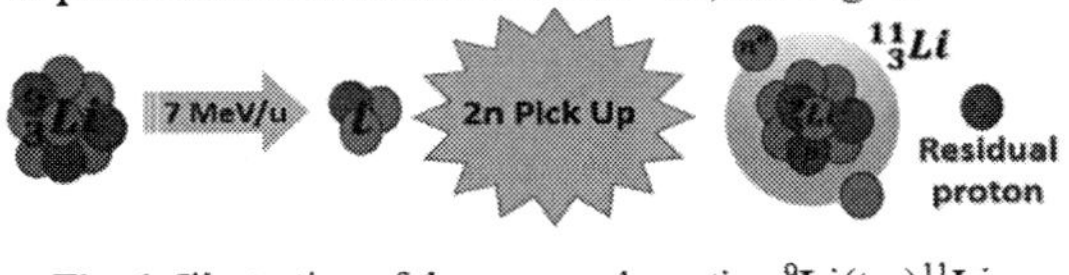

Fig. 1. Illustration of the proposed reaction ^{9}Li(t,p)^{11}Li.

The dominant reaction channel is a one-step neutron transfer process, the DWBA [11] calculations indicate a possible L=0 transmission to the ^{11}Li g.s or three L=1 excitations whose differential cross section peak for inverse kinematics at backward angles [12].

* Corresponding author : daniel.fernandez@csic.es

Knowledge of the excited states will come from the energy and momentum distribution of the residual protons, detected using forward and backward angles.

This experiment would complement the ^{11}Li(p,t)^{9}Li experiment carried at TRIUMF [13] and could also study other interesting reaction channels such as the elastic scattering channel which is essential to fix optical potentials in the theoretical models.

2 Experiment IS690

The experiment, IS690, will be carried at the HIE ISOLDE facility at CERN in 2023. A post-accelerated ^{9}Li beam at 7 MeV/u will impinge on a ^{3}H-target (^{3}H absorbed in a thin Ti-foil to a ~1/1 ratio). The energy of the incoming ^{9}Li beam is chosen, to facilitate the 2n transfer while reducing the number of additional open channels. Identifying the influence of background reactions, especially the ^{9}Li(p,d)^{10}Li and elastic channels, requires a setup that can differentiate between very similar reaction products and offer an optimal angular coverage.

The upgraded detection set-up at SEC (Scattering Experimental Chamber) at ISOLDE will consist of three detector structures: a) the charged particle detectors, b) eight GAGG scintillators for gamma detection, and c) the SAND neutron time-of-flight detector.

The charged particle detector set up has been designed based on our previous experience from reaction experiments. Its objective is to offer a maximum angular coverage of the reaction target while allowing for an accurate differentiation of the multiple reaction products. It's formed by the following parts (detectors from Micron Semiconductor Ltd), see fig 2 for the different coloured parts.

I. 5 particle telescopes formed by W1-60 μm DSSDs backed by MSX25-1500 μm Si-PAD and forming a pentagon ($44^{O} - 104^{O}$)

II. 2 Silicon Circular Detectors (CD), (S5-60 μm + S3-1000 μm) forming a telescope and covering the forward direction ($6^{O} - 32^{O}$).

III. An additional CD (S5-300 μm) covering the backward direction. ($100^{O} - 145^{O}$).

The target will be placed between the pentagon and the backwards CD detector, the incoming ^{9}Li beam will enter via the central hole. The heavy reaction products will move in the forward direction, while the lighter ones will be detected by the surrounding detectors.

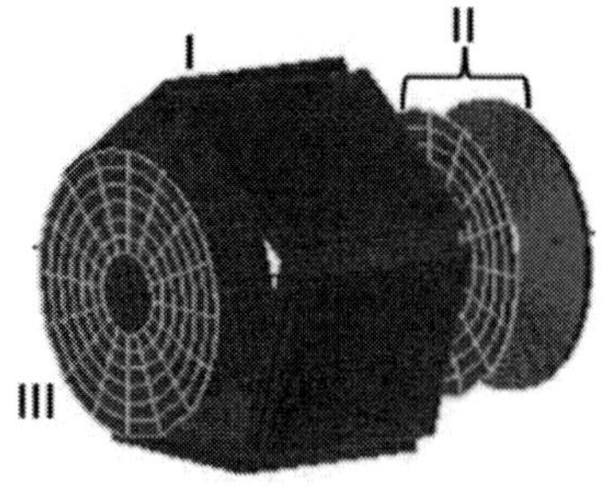

Fig. 2. Geant4 simulation of the experimental set-up with the three parts shown in different colours.

Geant4 simulations of the Si-device, see Fig. 2, were used to fine tune the thickness and positions of the detector, optimizing angular resolution and coverage. These simulations where also employed to model the effect of the background channels identified by ΔE- E plots, see Fig 3. The ratio of the (t,p) and (t,d) reaction was 0,3 and 0,7 respectively.

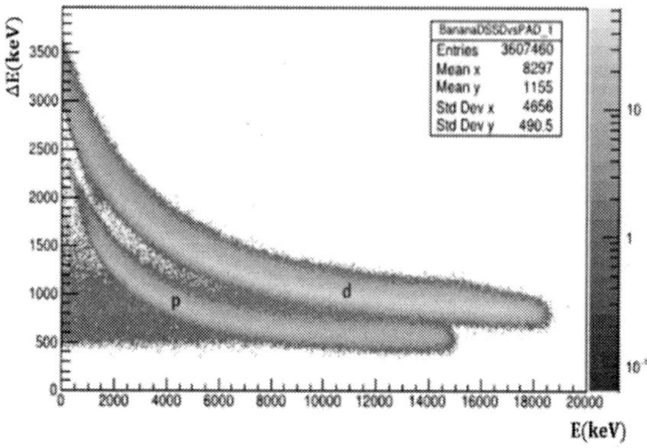

Fig. 3. Simulated ΔE-E plot used for particle identification. Lower trace: proton from the main ^{9}Li(t,p)^{11}Li reaction. upper deuterium ions from the ^{9}Li(t,d)^{10}Li background reaction.

Acknowledgements

We acknowledge the funding through the grant MICIN/Feder PID2019-104390GB-I00 & PRE2020-094522 "ESF Investing in your future", and the MAGISOL collaboration.

References

1. I.Tanihata et al., Phys. Rev.Lett **55** (1985) 2676.
2. I.Tanihata et al., Phys Lett B 160 (1985) 380.
3. P.G. Hansen and B. Jonson, Europhys. Lett **4** (1987) 409.
4. T.Kobayashi et al., Phys. Rev.Lett **60** (1988) 2599.
5. M.V. Zhukov et al., Phys Rep 231 (1995) 151.
6. J. Tanaka et al., Phys. Lett. B **774** (2017) 268.
7. R. Kanungo et al., Phys. Rev. Lett. **114** (2015) 192502.
8. I. Tanihata and K. Ogata, Eur. Phys. J. A **55** (2019) 239.
9. M.G.Gornov et al., Phys. Rev. Lett. **81** (1998) 766.
10. M.J.G. Borge and J. Cederkäl, Proposal to the ISOLDE and Neutron Time-of-Flight Committee (2021) European organization for nuclear research
11. E.Garrido and A.S.Jensen, Phys. Rev. C **101** (2019) 034003
12. Y.Kubota et al., Phys. Rev.Lett **125** (2020) 252501
13. T. Roger et al., Phys. Rev. C **79** (2009) 031603(R).

EPJ Web of Conferences 290, 02011 (2023) https://doi.org/10.1051/epjconf/202329002011
EuNPC 2022

Fission decay modes of ^{254}Fm* compound nucleus formed in ^{16}O+^{238}U reaction

Amandeep Kaur[1,*] and *Manoj* K. Sharma[2]

[1]Department of Physics, Faculty of Science, University of Zagreb, Bijenička c. 32, 10000 Zagreb, Croatia
[2]School of Physics and Materials Science, Thapar Institute of Engineering and Technology, 147004 Patiala, Punjab, India

Abstract. The quantum mechanical fragmentation theory (QMFT) based dynamical cluster-decay model (DCM) is applied to analyze the probable fission decay modes of ^{254}Fm* compound nucleus produced in ^{16}O+^{238}U nuclear reaction at excitation energy E^*_{CN}=45.9 MeV. The fission valley of collective fragmentation potential and the multi-humped peaks of preformation probability P_0 profile are analyzed by considering compact as well as elongated configurations of quadrupole (β_2) deformed fragments. The competitive emergence of different symmetric [symmetric superlong (SL), symmetric supershort (SS)] and asymmetric [standard 1 (S1), standard 2 (S2), standard 3 (S3)] fission modes have been observed for the case of elongated configuration. The division of mass and charge in nuclear fission of ^{254}Fm* depicts the importance of spherical and deformed magic shell closures. The most energetic light (A_L and heavy (A_H) decay fragments of aforementioned fission modes are identified. Moreover, the DCM-calculated fission cross-sections ($\sigma_{fission}$) show reasonable agreement with the experimental measurements [24].

1 Introduction

The nuclear fission process is crucial not only for the experimental and theoretical nuclear/astrophysics studies, but also for the production of energy and exotic nuclear isotopes that are useful in the medical and industrial applications. In a broader scenario, a fissioning nucleus disintegrates in either a symmetric or asymmetric fission mode. However, there may be a possibility of multimodal fission, during which a fissioning nucleus may choose different decay paths such as symmetric superlong (SL), symmetric supershort (SS), asymmetric standard 1 (S1), asymmetric standard 2 (S2), asymmetric standard 3 (S3) and superasymmetric (SA), for description see [1]. The proton (Z) and neutron (N) number of such fission fragments are determined by the shell effects. The predominance of asymmetric fission mode has been seen in the nuclear elements of actinide region [2]. The primary observation of a transition from asymmetric to symmetric fission has been measured in the region of mass A=254-258 of Fm isotopes [3, 4]. Similar kind of results have been predicted lately in the fission decay of ^{254}Es nucleus [5]. It is indeed highly desirable to investigate the nuclear fission near mass 254 due to the interplay among symmetric and asymmetric fission decay paths.

It is well known that macroscopic and microscopic effects impart a substantial impact on the emergence of symmetric and asymmetric fission of a nucleus, but in addition, the deformation and orientation of decay fragments also influence the fission fragment mass distribution [1, 6]. The side-to-side (polar or compact) and tip-to-tip (equatorial or elongated) configurations of nuclei have great impact on the formation and decay dynamics of a nucleus [7]. Recently, we have made an effort to see the effect of compact and elongated configurations of quadrupole (β_2) deformed decay fragments on the spontaneous fission of $^{242-260}$Fm isotopes using preformed cluster model [8]. It has been observed that tip-to-tip (elongated) configuration results in the production of double-peaked (asymmetric) to triple-humped (multimodal) fission fragment mass distribution with increase in neutron number of Fm isotopes.

In the present work, quantum mechanical fragmentation theory (QMFT) [9] based dynamical cluster-decay model (DCM) [10, 11] is applied to analyze the possibility of multimodal fission mode of excited ^{254}Fm* compound nucleus (CN) formed in ^{16}O+^{238}U nuclear reaction. The calculations are made at center-of-mass energy $E_{c.m.}$≈84 MeV (excitation energy E^*_{CN}=45.9 MeV) near the Coulomb barrier by considering quadrupole β_2-deformed nuclei having compact/elongated configurations with optimum orientations ($\theta_i^{opt.}$) [7]. The temperature (T) dependence is included in β_2-deformations (called as β_2-dynamic deformation), as explained in section 2. The objectives of this work is (i) to examine the different fission decay paths of ^{254}Fm* nucleus by analyzing the valleys of collective fragmentation potential $V_R(\eta,T)$ and peaks of preformation probability P_0 of fission fragments; (ii) to study the role of compact and elongated orientations of β_2-deformed fragments on the structure of fission fragment mass distribution; (iii) to predict the energetically favorable light (A_L) and heavy (A_H) fission fragments in order to see the role of spherical and deformed magic shell

*Corresponding author: akaur.phy@pmf.hr

closure; and (iv) a comparison of DCM-calculated results with the recent experiment measurements [12].

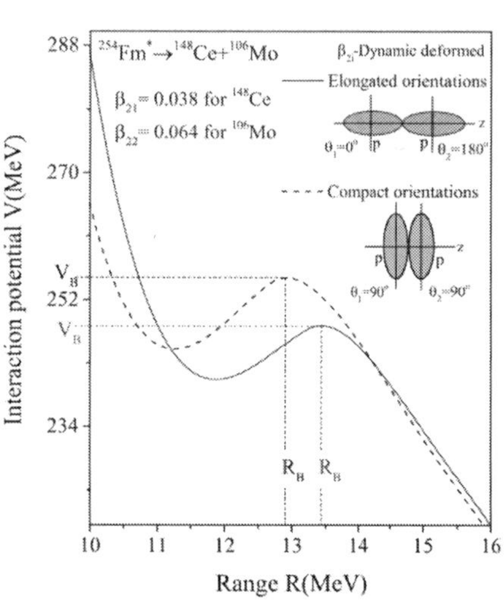

Figure 1. The interaction *or* scattering potential V(R) is plotted for asymmetric fission channel $^{254}Fm^* \rightarrow ^{148}Ce + ^{106}Mo$ using both β_2-dynamic elongated and compact configurations.

2 Methodology or Formalism

Various fission channels of $^{254}Fm^*$ CN are studied using dynamical cluster-decay model (DCM) [10, 11], which in terms of collective coordinates of mass (and charge) asymmetries $\eta_A = (A_1 - A_2)/(A_1 + A_2)$ (and $\eta_Z = (Z_1 - Z_2)/(Z_1 + Z_2)$) (where 1 and 2 stand for heavy (A_H) and light (A_L) fragments), relative separation R, the multipole deformations $\beta_{\lambda i}$ (λ=2, 3, 4; i=1, 2) and orientations θ_i (i=1, 2). Using these collective coordinates, the combine effect of attractive and repulsive interactions on a decay process is explored in terms of fragmentation potential $V_R(\eta,T)$ as

$$\begin{aligned} V_R(\eta,T) &= \sum_{i=1}^{2}[V_{LDM}(A_i,Z_i,T)] \\ &+ \sum_{i=1}^{2}[\delta U_i]exp(-T^2/T_0^2) \\ &+ V_C(R,Z_i,\beta_{\lambda i},\theta_i,T) \\ &+ V_P(R,A_i,\beta_{\lambda i},\theta_i,T) \\ &+ V_\ell(R,A_i,\beta_{\lambda i},\theta_i,T). \end{aligned} \quad (1)$$

The V_C, V_P and V_ℓ are respectively, the T-dependent Coulomb, the nuclear proximity and centrifugal potentials for deformed and oriented nuclei [13]. The prox1977 version of the V_P [14] is considered for the calculations. V_{LDM} is T-dependent liquid drop energy of Davdison et al. [15] and δU, the "empirical" shell correction, as given in [16], made T-dependent as per [17]. The constants of V_{LDM} at T=0 are refitted [13] to give the experimental binding energies of Audi et al. [18] or the theoretical estimates of Möller et al. [19] if not available in [18]. The quadrupole deformations (β_{2i}) are taken from the theoretical estimates of Möller et al. [19], and the "optimum" orientations $\theta_i^{opt.}$ for hot-compact and cold-elongated configurations are taken from Table 1 of Ref. [7]. The temperature effects are included in deformation parameters using following relation [20, 21], $\beta_{\lambda i}(T) = exp(-T/T_0)\beta_{\lambda i}(0)$, where $\beta_{\lambda i}(0)$ represents the static deformation and T_0 is the temperature of the nucleus at which shell effects start to vanish (T_0 = 1.5 MeV) [17]. The nuclear temperature (T) is related to CN excitation energy as $E^*_{CN} = E_{c.m.} + Q_{in} = (A_{CN}/9)T^2 - T$, where Q_{in} is the Q-value of entrance channel. To study the fragment mass distribution and energetically favoured nascent fragments, the preformation probability P_0 is calculated using the fragmentation potential $V_R(\eta,T)$ by solving the Schrödinger equation in η-coordinate at fixed $R = R_a$,

$$\left\{-\frac{\hbar^2}{2\sqrt{B_{\eta\eta}}}\frac{\partial}{\partial\eta}\frac{1}{\sqrt{B_{\eta\eta}}}\frac{\partial}{\partial\eta} + V(\eta,R,T)\right\}\psi^\nu(\eta) = E^\nu\psi^\nu(\eta), \quad (2)$$

with $\nu = 0, 1, 2, 3...$ referring to ground state (ν = 0) and excited state solutions, with the ground state P_0 given as

$$P_0 = |\psi(\eta(A_i))|^2 \sqrt{B_{\eta\eta}}\frac{2}{A_{CN}}, \quad (3)$$

and $|\psi|^2 = \sum_{\nu=0}^{\infty}|\psi^\nu|^2 exp(\frac{-E^\nu}{T})$ is Boltzmann-like function. $B_{\eta\eta}$ in Eq. (3) represent the smooth hydrodynamical mass parameter [22]. The penetration probability P of decaying fragments is calculated using Wenzel-Kramers-Brillouin integral. The first turning point of penetration path is defined as $R_a(T) = R_1(\alpha_1,T) + R_2(\alpha_2,T) + \Delta R(T) = R_t(\alpha,T) + \Delta R(T)$, with radius vectors R_i (i = 1, 2)

$$R_i(\alpha_i,T) = R_{0i}(T)\left[1 + \sum_\lambda \beta_{\lambda i}Y_\lambda^{(0)}(\alpha_i)\right] \quad (4)$$

and T-dependent nuclear radii R_{0i} of the equivalent spherical nuclei [23], $R_{0i}(T) = [1.28A_i^{\frac{1}{3}} - 0.76 + 0.8A_i^{\frac{-1}{3}}](1 + 0.0007T^2)fm$. ΔR is the only parameter of the model known as neck-length parameter. Using P_0 and P, for ℓ partial waves, the decay and fragment's production cross-section is given by

$$\sigma(A_1,A_2) = \frac{\pi}{k^2}\sum_{\ell=0}^{\ell_{max}}(2\ell+1)P_0P; \; k = \sqrt{\frac{2\mu E_{c.m.}}{\hbar^2}}, \quad (5)$$

where $\mu = m[A_1A_2/(A_1 + A_2)]$ is the reduced mass, and ℓ_{max} is maximum angular momentum fixed for the vanishing of light particle cross sections.

3 Results and Discussions

Figure 1 shows the scattering potential V(R) for the asymmetric fission channel $^{254}Fm^* \rightarrow ^{148}Ce + ^{106}Mo$ for β_2-dynamic elongated and compact configurations at ℓ=0$\hbar$. The heavy A_H and light A_L fission fragments ^{148}Ce and ^{106}Mo are prolate in shape. Depending on the prolate (p) and oblate (o) shapes (+ and - signs) of β_{2i} deformations, the orientation angles are uniquely fixed, which result in the "hot-compact" and "cold-elongated" configurations [7], a pictorial representation is shown in the figure. It is observed from the figure that barrier properties

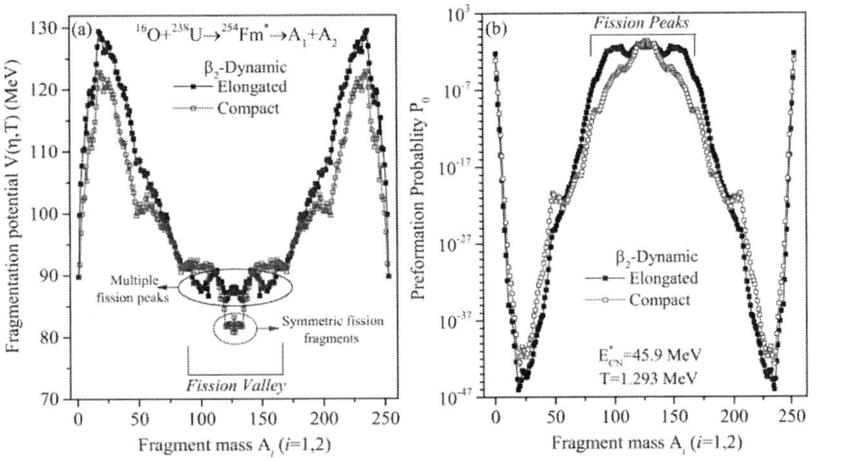

Figure 2. DCM-calculated (a) collective fragmentation potential $V_R(\eta, T)$ and (b) preformation probability P_0 as a function of fragment mass A_i (i=1,2) of ^{254}Fm* at E^*_{CN}=45.9 MeV are plotted for β_2-dynamic deformations for elongated and compact configurations.

Table 1. DCM-calculated and experimental [24] $\sigma_{fission}$ of ^{254}Fm* formed in ^{16}O+^{238}U reaction at E^*_{CN}=45.9 MeV for β_2-deformed compact and elongated configurations.

Orientation	ΔR (fm)	ℓ_{max} ($\hbar$)	Fragments (A_2)	$\sigma^{DCM}_{fission}$ (mb)	$\sigma^{Expt.}_{fission}$ (mb)
Elongated	0.854	156	85-127	115.4	118.5
Compact	0.900	142	95-127	120.8	

are significantly modified with change in orientation criteria. The barrier height (V_B) is small and barrier position (R_B) is large for the case of elongated configuration as compared to the compact one. This suggests that using distinct methods of orientations the fragmentation potential $V_R(\eta$,T) and preformation probability P_0 get modified, and hence elongated and compact orientations impact on the fission dynamics significantly. Further, table 1 shows the DCM calculated fission cross-sections ($\sigma_{fission}$) along with other related quantities for the most probable fission fragments (A_1+A_2) of ^{254}Fm* nucleus at E^*_{CN}=45.9 MeV, which are calculated by fixing neck-length parameter ΔR for both elongated and compact fragment configurations. The calculated $\sigma_{fission}$ show nice agreement with experimental data given in [24].

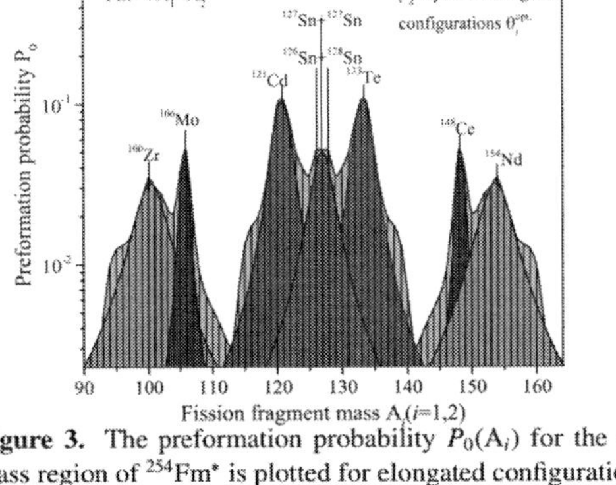

Figure 3. The preformation probability $P_0(A_i)$ for the fission mass region of ^{254}Fm* is plotted for elongated configurations.

For the analysis of fission valleys, the fragmentation potential $V_R(\eta$,T) of ^{254}Fm* is plotted as a function of fragment mass A_i(i=1,2) at E^*_{CN}=45.9 MeV and ℓ_{max}-value in figure 2(a) for β_2-deformed elongated as well as compact approach. The first look of fragmentation potential suggests relatively more structural effects in the fission valley of tip-to-tip elongated configurations in contrast to side-to-side compact configuration. The fission dips of compact interactions show the dominance of symmetric fission fragments whereas the elongated configurations depict the multiple dips, which point towards the possibility of the co-existence of various fission decay channels for ^{254}Fm* CN. Next, figure 2(b) represents the preformation probability $P_0(A_i)$ of ^{254}Fm* at E^*_{CN}=45.9 MeV and ℓ_{max}-value for both elongated and compact orientations. Note that the fragmentation potential goes as an input in the Schrödinger equation used to calculate the preformation probability P_0, and the minima of the fragmentation potential correspond to the maxima of the preformation probability and vice versa, as evident from figures. 2(a) and 2(b). The symmetric peak becomes sharp for the compact orientations while multi-humped fission fragment mass distribution is observed for the case of elongated configurations. This means β_2-deformed compact and elongated configurations follow different fission decay paths. According to recent experimental and theoretical observation of [12], the fragment mass distribution of ^{254}Fm* at excitation energy $E^*_{CN}\approx$45 MeV may decompose into different fission paths. Therefore, it would be interesting to analyze the energetically favoured fission peaks of preformation probability for elongated oriented fission fragments in DCM calculations, which indicates the possibility of competing fission decay paths.

To explore the multiple fission decay paths of ^{254}Fm* at E^*_{CN}=45.9 MeV and ℓ_{max}-value, a broader representation of preformation probability P_0 as a function of fission fragment mass is represented in figure 3 for the elongated fragment configurations. Several symmetric and asymmetric fission decay modes as described in [1] are marked in figure and represented in table 2 along with experimentally observed heavy fission mass fragments A_H. Note that

Table 2. DCM-predicted light A_L and heavy A_H fission fragments in several fission decay channels of ^{254}Fm* nucleus at E^*_{CN}=45.9 MeV are presented along with the experimental observations of [12].

Fission mode	Heavy mass fission fragment A_H	Light mass fission fragment A_L	A_H [12]
Symmetric superlong (SL)	^{127}Sn(Z_H=50, N_H=77); (β_2=0.0)	^{127}Sn(Z_L=50, N_L=77); (β_2=0.0)	127.00
	^{128}Sn(Z_H=50, N_H=78); (β_2=0.0)	^{126}Sn(Z_L=50, N_L=76); (β_2=0.0)	
Asymmetric Standard 1 (S1)	^{133}Te(Z_H=52, N_H=81); (β_2=0.001)	^{121}Cd(Z_L=48, N_L=73); (β_2=0.024)	135.00
Asymmetric Standard 2 (S2)	^{148}Ce(Z_H=58, N_H=90); (β_2=0.038)	^{106}Mo(Z_L=42, N_L=64); (β_2=0.064)	143.00
	^{154}Nd(Z_H=60, N_H=76); (β_2=0.048)	^{100}Zr(Z_L=40, N_L=60); (β_2=0.064)	
Asymmetric Standard 3 (S3)	^{159}Sm(Z_H=62, N_H=97); (β_2=0.052);	^{95}Sr(Z_L=38, N_L=57); (β_2=0.055)	163.61±1.46

deformations mentioned in the table are T-dependent. The symmetric superlong SL fission channel is mainly situated near $A_{CN}/2$ i.e. A_H=127, 128 and both fission decay fragments are spherical in shape. This fission mode is mainly defined by the macroscopic liquid drop part of fragmentation potential. Standard S1 and S2 are asymmetric fission paths which can be found at either side of the symmetric hump. Generally, the S1 and S2 modes are situated at A_H=132-135 and A_H=138-144, respectively and at neutron shell closures N_H=82 (spherical) and $N_H\approx$88 (deformed), respectively. Similar results we have observed in DCM calculations. S1 mode has one spherical heavy fragment and a deformed light fragment and for S2 mode we have observed two deformed prolate fragments. For asymmetric S3 fission mode, the role of deformed magic shell closures of Z_L=38 and N_L=60 can be seen. The DCM predicted heavy and light fission fragment mass numbers show reasonable agreement with the previous experimental and theoretical observations [1, 2, 12]. Hence, the deformations, oreintations and shell effects play significant role in the splitting of a nucleus. However, for the present study the deformations are used up to quadrupole only and it would be interesting to see the effect of octupole deformations of fragments in the fission process as stabilized octupole configuration around Z=52 and 56 is supposed to play important role in the splitting of a fissioning nucleus.

4 Summary

Summarizing, the ^{16}O-induced fission of ^{254}Fm* is studied within the framework of dynamical cluster-decay model (DCM) at excitation energy E^*_{CN}=45.9 MeV. The scattering potential V(R), collective fragmentation potential $V_R(\eta,T)$ and preformation probability P_0 are computed for two type of decay fragment orientations: elongated and compact. The barrier characteristics have been considerably modified in terms of barrier height (V_B) and barrier position (R_B) with change in orientation degree of freedom. The fragmentation potential represents more structure in elongated interaction as compared to compact one. Hence, multiple peaks are observed in preformation probability distribution for elongated case, which indicates the possibility of different fission decay modes for ^{254}Fm* nucleus. These results are in agreement with the available data. Spherical/deformed magic shell closures play significant role in the division of fissioning nuclei. It would be interesting to include the pear shaped deformations in the decaying fragments for the case of elongated orientations, to analyse the role of deformed shell closures in fission modes of ^{254}Fm* and nearby nuclei.

5 Acknowledgements

This work is supported by the QuantiXLie Centre of Excellence, a project co-financed by the Croatian Government and European Union through the European Regional Development Fund, the Competitiveness and Cohesion Operational Programme (KK.01.1.1.01.0004). MKS is thankful to DST-SERB project grant CRG/2021/001144.

References

[1] U. Brosa, S. Grossmann, A. Müller, Phys. Rep. **197**, 167 (1990)

[2] K. Mahata, C. Schmitt, Shilpi Gupta, A. Shrivastava, G. Scamps, K.-H. Schmidt, Phys. Lett. B **825**, 136859 (2022)

[3] E. K. Hulet et al., Phys. Rev. C **40**, 770 (1989)

[4] J. E. Gindler, K. F. Flynn, L. E. Glendenin, R. K. Sjoblom, Phys. Rev. C **16**, 1483 (1977)

[5] M. D. Usang, F. A. Ivanyuk, C. Ishizuka, S. Chiba, Sci. Rep. **9**, 1525 (2019)

[6] K. Nishio, H. Ikezoe, S. Mitsuoka, I. Nishinaka, Y. Nagame, Y. Watanabe, T. Ohtsuki, K. Hirose, S. Hofmann, Phys. Rev. C **77**, 064607 (2008)

[7] R. K. Gupta, M. Balasubramaniam, R. Kumar, N. Singh, M. Manhas, W. Greiner, J. Phys. G: Nucl. Part. Phys. **31**, 631 (2005)

[8] A. Kaur, N. Sharma, M. K. Sharma, Phys. Rev. C **103**, 034618 (2021)

[9] R. K. Gupta, W. Scheid, W. Greiner, Phys. Rev. Lett. **35**, 353 (1975)

[10] A. Kaur, M. K. Sharma, Phys. Rev. C **99**, 044611 (2019)

[11] A. Kaur, M. K. Sharma, Eur. Phys. J. A **55**, 89 (2019)

[12] T. Banerjee et al., Phys. Rev. C **105**, 044614 (2022)

[13] B. B. Singh, M. K. Sharma, R. K. Gupta, Phys. Rev. C **77**, 054613 (2008)

[14] R. Kumar, M. K. Sharma, Phys. Rev. C **85**, 054612 (2012)

[15] N. J. Davidson, S. S. Hsiao, J. Markram, H. G. Miller, Y. Tzeng, Nucl. Phys. A **570**, 61 (1994)

[16] W. Myers, W. J. Swiatecki, Nucl. Phys. **81**, 1 (1966)

[17] A. S. Jensen, J. Damgaard, Nucl. Phys. A **203**, 578 (1973)

[18] G. Audi, A. H. Wapstra, C. Thibault, Nucl. Phys. A **729**, 337 (2003)

[19] P. Möller, J. R. Nix, W. D. Myers, W. J. Swiatecki, At. Data Nucl. Data Tables **59**, 185 (1995)

[20] M. Rashdan, A. Faessler, W. Waida, J. Phys. G: Nucl. Part. Phys. **17**, 1401 (1991)

[21] M. Muenchow, W. Scheid, Phys. Lett. B **162**, 265 (1985); Nucl. Phys. A **468**, 59 (1987)

[22] H. Kröger, W. Scheid, J. Phys. G: Nucl. Part. Phys. **6**, L85 (1980)

[23] G. Royer, J. Mignen, J. Phys. G: Nucl. Part. Phys. **18**, 1781 (1992)

[24] H. Zhang, Z. Liu, J. Xu, X. Qian, Y. Qiao, C. Lin, K. Xu, Phys. Rev. C **49**(2), 926 (1994)

Experimental study of high-energy fission and quasi-fission with fusion-induced fission reactions at VAMOS++

D. Fernández[1,*], *M. Caamaño*[1,**], *D. Ramos*[2,***], *A. Lemasson*[2], *M. Rejmund*[2], *H. Álvarez-Pol*[1], *L. Audouin*[3], *J. D. Frankland*[2], *B. Fernández-Domínguez*[1], *E. Galiana-Baldó*[4], *J. Piot*[2], *C. Schmitt*[5], *D. Ackermann*[2], *S. Biswas*[2], *E. Clement*[2], *D. Durand*[6], *F. Farget*[6], *M. O. Fregeau*[2], *D. Galaviz*[4], *A. Heinz*[7], *A. Henriques*[8], *B. Jacquot*[2], *B. Jurado*[8], *Y. H. Kim*[2], *P. Morfouace*[2], *D. Ralet*[9], *T. Roger*[2], *P. Teubig*[4], and *I. Tsekhanovich*[8]

[1]IGFAE – Universidad de Santiago de Compostela, E–15706 Santiago de Compostela, Spain
[2]GANIL, CEA/DRF-CNRS/IN2P3, BP 55027, F-14076 Caen Cedex 5, France
[3]IPN Orsay, Université de Paris-Saclay, CNRS/IN2P3, F-91406 Orsay Cedex, France
[4]LIP Lisboa, 1649-003 Lisbon, Portugal
[5]IPHC Strasbourg, Université de Strasbourg-CNRS/IN2P3, F-67037 Strasbourg Cedex 2, France
[6]LPC Caen, Université de Caen Basse-Normandie-ENSICAEN-CNRS/IN2P3, F-14050 Caen Cedex, France
[7]Department of Physics, Chalmers University of Technology, SE-41296 Göteborg, Sweden
[8]CENBG, IN2P3/CNRS-Université de Bordeaux, F-33175 Gradignan Cedex, France
[9]CSNSM, CNRS/IN2P3, Université de Paris-Saclay, F-91405 Orsay, France

Abstract. Over the past decade, inverse kinematics has been increasingly employed in experimental studies of fission. This approach has yielded a wealth of new observables that can be obtained in single measurements, enabling their analysis and correlations. One ongoing application of this technique involves a series of experiments performed at GANIL using the variable-mode, large-acceptance VAMOS++ spectrometer. A recent experiment focused on examining the survival of nuclear structure effects at high excitation energy in both fission and quasi-fission. The results of the study involved a full isotopic identification of fragments, as well as an analysis of the elemental yields their relation to fission dynamics. The results indicate that fission and quasi-fission involve different mechanisms, which could be exploited to distinguish between the two phenomena.

1 Introduction

Although both fission and quasi-fission lead to similar final products, there are notable distinctions between these processes. Firstly, fission comes from an equilibrated compound nucleus, while quasi-fission does not. Additionally, quasi-fission is significantly faster than fission [1]. Researchers aim to identify observables that can differentiate between the two phenomena, and the complete identification of fragments offers new opportunities.

A prior experiment [2] demonstrated that in fusion-induced fission of ^{250}Cf, structural effects may persist in both the neutron-to-proton ratio of fragments and the total kinetic energy, even at high excitation energies of approximately 40 MeV.

In 2017, the E753 experiment was performed at GANIL using the VAMOS++ spectrometer to study the fission and quasi-fission processes also at high excitation energy, and analyse the survival of nuclear structure effects in both processes. The study of fusion-fission reactions is critical for advancing our knowledge of nuclear physics and has important practical applications such as nuclear energy production and nuclear waste management. In this paper, preliminary results of ^{265}Db high-energy fission from ^{238}U + ^{27}Al reaction are presented and discussed. The isotopic-fission yields $Y(Z, A)$, the elemental yields $Y(Z)$, the neutron excess and the velocity in center of mass are presented.

2 Experimental setup

VAMOS++ is a variable mode spectrometer [3, 4] composed of a large magnetic dipole and two quadrupoles, and a set of detectors at the focal plane that measure the energy, energy loss, angles and positions before and after the magnets, and time of flight of those particles that reach the end of the focal plane. The magnetic rigidity is reconstructed using the positions and the angles at focal plane, while the emission vector is measured with those at the target position, before the spectrometer. Regarding the fission fragments, the measured observables are: atomic number (Z), mass number (A), and velocity vector. For the fissioning system, Z, A, and its excitation energy (E^*) are obtained [5].

The VAMOS++ angular acceptance is $\Delta\theta=\pm7°$ and $\Delta\phi=\pm10°$. In this experiment, the spectrometer was used in two settings of central magnetic rigidity ($B\rho_0$) and an-

*e-mail: dani.fernandez@usc.es
**e-mail: manuel.fresco@usc.es
***e-mail: diego.ramos@ganil.fr

gle (θ_{lab}): $B\rho_0$=1.24 Tm, θ_{lab}=14°; and $B\rho_0$=1.1 Tm, θ_{lab} = 21.5°. Further details on VAMOS++ along with typical performances for fission-fragment detection are given in Refs. [6–8].

In this experiment, a ^{238}U beam at 5.9*A* MeV impinged on four different light targets (0.5 mg/cm^2 of ^{9}Be, 0.1 mg/cm^2 of ^{nat}B, 0.5 mg/cm^2 of ^{24}Mg, and 0.2 mg/cm^2 of ^{27}Al) to generate different fissioning systems (FS) through transfer and fusion reactions. In the case of transfer or inelastic reactions, the target-like recoil is measured with a silicon telescope placed around the target. Fusion reactions are assumed when no recoil is detected [9]. Once a FS is formed, it splits into two fission fragments (FF) and one of them may be detected in VAMOS++.

Using inverse kinematics in fission experiments offers several advantages. One of the key benefits is the higher velocity of the FF in the laboratory frame, which facilitates their traversal of multiple detector layers. This, in turn, provides access to a broader range of observables that cannot be achieved through direct kinematics. Additionally, inverse kinematics allows for better control over the reaction products, leading to more accurate measurements and increased understanding of the fission process. Finally, the use of inverse kinematics can also reduce the effects of background noise, which can improve the overall quality of the experimental results.

3 Results with the aluminium target

The compound nucleus produced in the reaction with the Al target is ^{265}Db with E^* = 61.2 MeV. The properties of this compound and its subsequent fission products provide valuable insights into the fundamental physics of nuclear fusion and fission reactions.

3.1 Fission yields

Isotopic-fission yields are derived using the procedure presented in Refs. [10, 11]. In this procedure, the fission events are counted and weighted by various factors, including the spectrometer acceptance, the angular and intrinsic efficiencies, and the relative normalisation between settings. The procedure also includes the subtraction of contamination from transfer-fission events. The resulting isotopic-fission yields provide information about the distribution of fission products and can be used to study the underlying process.

Figure 1 displays the isotopic yields of ^{265}Db as a function of the fragment mass, with each color representing a different atomic number (Z) between Z = 30 and Z = 66. In this and the rest of the figures of this document, the error bars represent the statistical uncertainties; systematic uncertainties range from 2% in the heavier fragments up to 10% in the lighter ones. The resolution in the identification of atomic number is $\Delta Z/Z \sim 1/80$, while the resolution in mass identification is $\Delta A/A \sim 1/200$.

The elemental yields *Y(Z)* can be calculated by summing the contributions of each mass *A*: $Y(Z) = \sum_A Y(Z, A)$. The resulting elemental yields are displayed in Fig. 2 and are compared to the predictions of the semi-empirical code GEF [12] as well as a previous measurement of a similar reaction [13]. In addition, an approximation of the quasi-fission distribution is shown in a solid red line, which is obtained from the difference between the yield and its complementary using the symmetry of the elemental fission yields. This approximation assumes that the yield for a given element *Y(Z)* is equal to the yield of its complementary element at $Z_{comp} = Z_{FS} - Z$, where Z_{FS} is the nuclear charge of the compound nucleus.

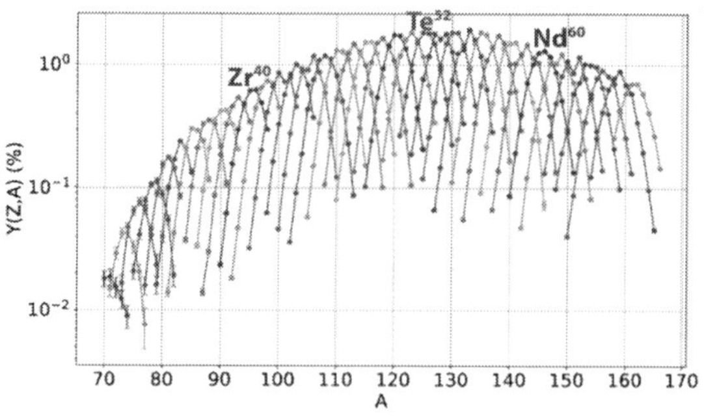

Figure 1. Isotopic fission yields *Y(Z, A)* of fusion-induced fission of ^{265}Db at E^* = 61.2 MeV. Each colour line corresponds to one element. Zr^{40}, Te^{52}, and Nd^{60} are shown for reference.

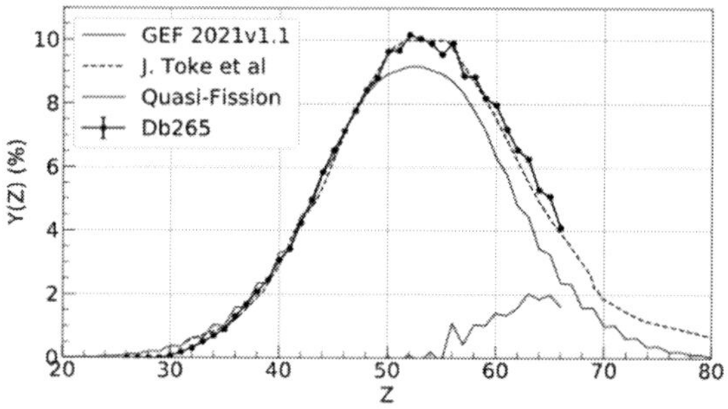

Figure 2. Elemental yield distribution *Y(Z)* of fragments from ^{265}Db fusion-fission reactions (black points and solid black line) compared with the GEF prediction (solid green line) and previous measurement from [13] (dashed blue line). The minimum quasi-fission contribution is shown with solid red line.

The results reveal a non-symmetric distribution, which is produced by to the restricted coverage of the centre-of-mass (c.m.) angle of the experiment. This limited coverage is illustrated in Fig. 3, where the blue solid lines indicate the angle restriction, and the dashed blue lines shows the limits of mass identification (before neutron evaporation). As mentioned earlier, there is a strong correlation between the fragment mass and the c.m. angle. In this case, the experimental coverage only includes the heavy-fragment mass of quasi-fission, represented by the solid red line circle in Fig. 3, while the corresponding light fragments would be in the complementary c.m. angle, indicated by the dashed red line circle, which is not covered in this experiment.

The $Y(Z)$ distribution is in agreement with the previous measurement [13], once the same restrictions are applied. The comparison with the GEF model [12] shows an agreement on the light-fragment part of the distribution. This is because the fission component of the reaction is not affected by the restriction in angular coverage.

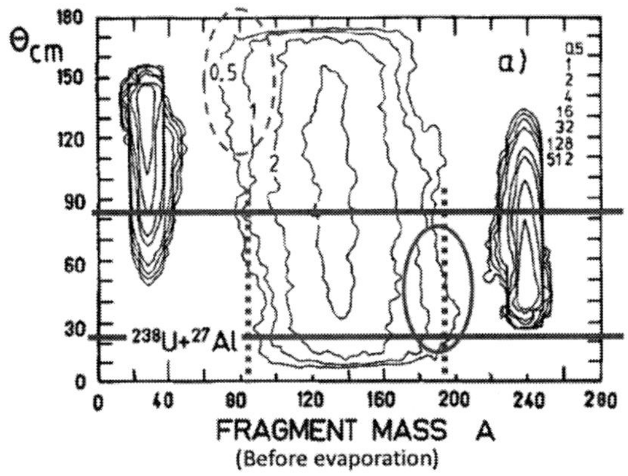

Figure 3. Experimental measurement [13] of fragment masses before evaporation versus c.m. angle for the reaction $^{238}U + ^{27}Al$ with the VAMOS++ limits in angle and mass (blue solid and dashed lines, respectively). Red circles show approximately the quasi-fission contribution to the reaction. The solid red line corresponds to the contribution accessible with VAMOS++.

3.2 Neutron excess

The fissioning system is excited before the saddle point and will emit neutrons. Since the information on these neutrons cannot be accessed directly in this experiment, an estimation with the GEF code is used, resulting in a FS average mass of $\langle A_{FS}\rangle \sim 262.4$. Assuming a negligible proton emission, the pre-evaporation masses of the fragments $\langle A_i^*\rangle$ can be evaluated with this estimated mass $\langle A_{FS}\rangle$. The nuclear structure effects would appear in the pre-evaporation step (at scission). The information about the fissioning system at the scission point can be obtained from the neutron excess after evaporation and the fragment velocities in c.m. The neutron excess $\langle N\rangle/Z = [(\sum_A A \cdot Y(Z,A))/(\sum_A Y(Z,A)) - Z]/Z$ is shown in Fig. 4 compared to GEF.

Since the fissioning system is highly excited, it can emit neutrons before reaching the saddle point, leading to a decrease in the effective excitation energy before the start of the fission process. The calculations with GEF show a smaller $\langle N\rangle/Z$ than what is measured; although the evolution as a function of Z is very similar. GEF predicts a large number of emitted neutrons before scission, including 2.4 pre-saddle neutrons, which could explain the excess. However, there is no way to access information on these neutrons due to current experimental limitations. Further studies on neutron evaporation are necessary to explain this behaviour.

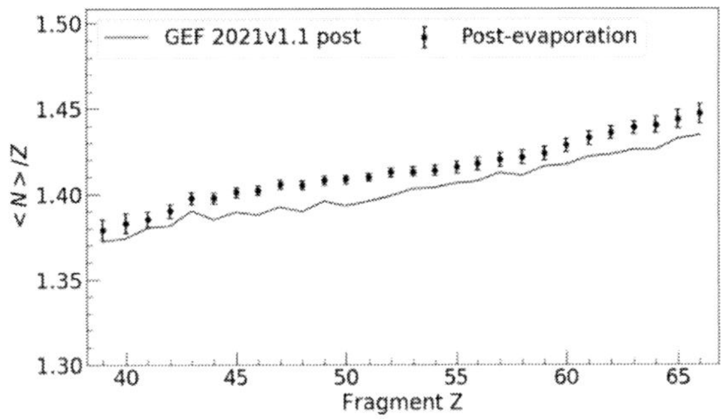

Figure 4. Post-evaporation average neutron-excess $\langle N\rangle/Z$ of ^{265}Db fission fragments as a function of fragment Z (black symbols) and corresponding GEF predictions (solid red line).

3.3 Isotopic velocities

As both the laboratory frame (V_{lab}, θ_{lab}) and the fissioning system frame are measured, the c.m. velocity of the fission fragments $V^{FF}_{c.m.}$ can be determined on an event-by-event basis, enabling access to fission dynamics. Figure 5 displays $V^{FF}_{c.m.}$ as a function of the fragment A with each panel corresponding to one Z. Although the experimental data agree well with GEF for $Z > 40$, there is a deviation observed for the light fragments. The figure exhibits a slight overestimation in GEF prediction, with the largest deviations occurring in more neutron-rich isotopes.

4 Conclusions

The preliminary results presented were obtained with the VAMOS++ setup, which allows for the measurement of a variety of observables, including complete isotopic and elemental yield distributions of the ^{265}Db fissioning system. The data analysis reveals a contribution from the quasi-fission process in the elemental fission yields, which is identified with the heaviest contribution of the quasi-fission. This process can contribute to the observed fragment mass distributions, and it is important to account for it when studying heavy-ion reactions. Furthermore, the neutron excess and isotopic velocity in the centre-of-mass frame of the fission fragments are also reported.

References

[1] D. J. Hinde, R. du Rietz, C. Simenel, M. Dasgupta, A. Wakhle, M. Evers, and D. H. Luong, AIP Conf. Proc. 1423, 65 (2012).

[2] M. Caamaño, F. Farget, O. Delaune, K. H. Schmidt, C. Schmitt, L. Audouin, C. O. Bacri, J. Benlliure, E. Casarejos, X. Derkx, B. Fernández-Domínguez, L. Gaudefroy, C. Golabek, B. Jurado, A. Lemasson, D. Ramos, C. Rodríguez-Tajes, T. Roger, and A. Shrivastava. Phys. Rev. C 92, 034606 (2015).

[3] S. Pullanhiotan, M. Rejmund, A. Navin, W. Mittig, S. Bhattacharyya, Nucl. Instr. and Meth. A 593, 343 (2008).

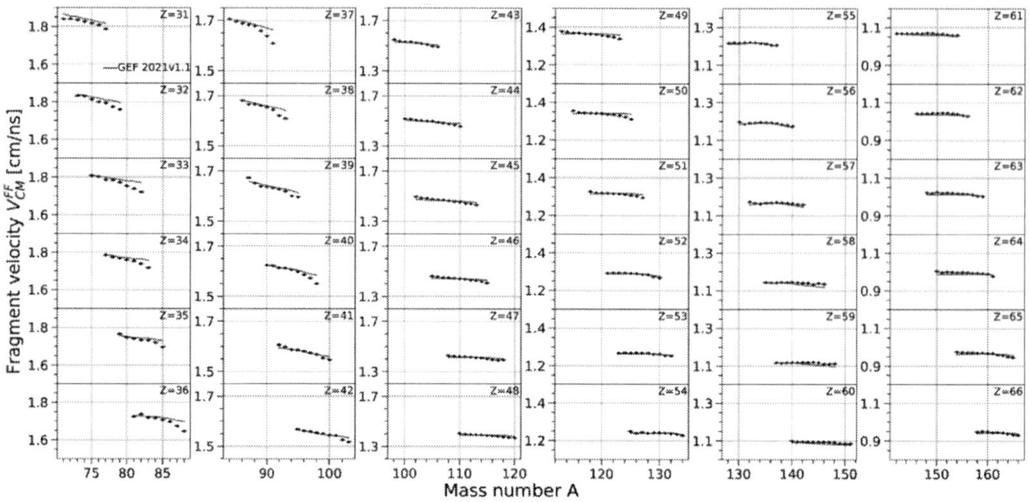

Figure 5. Each panel shows the centre of mass velocity as a function of *A* for the different elements produced in fusion-fission of ^{265}Db (black points). The GEF prediction is shown with red lines.

[4] M. Rejmund, B. Lecornu, A. Navin, C. Schmitt, S. Damoy, O. Delaune, J.M. Enguerrand, G. Fremont, P. Gangnant, L. Gaudefroy, et al., Nucl. Instr. and Meth. A 646, 184 (2008).

[5] C. Rodríguez-Tajes, F. Farget, X. Derkx, M. Caamaño, O. Delaune, K.-H. Schmidt, E. Clement, A. Dijon, A. Heinz, T. Roger, L. Audouin, J. Benlliure et al., Phys. Rev. C 89, 024614 (2014).

[6] M. Rejmund, B. Lecornu, A. Navin, C. Schmitt, S. Damoy, O. Delaune, J.M. Enguerrand, G. Fremont, P. Gangnant, L. Gaudefroy, B. Jacquot, J. Pancin et al., Nucl. Instr. and Meth. A 646, 184 (2011).

[7] M. Vandebrouck, A. Lemasson, M. Rejmund, G. Fremont, J. Pancin, A. Navin, C. Michelagnoli, J. Goupil, C. Spitaels, and B. Jacquot. Nucl. Instr. and Meth. A 812,112 (2016).

[8] Y. Kim, A. Lemasson, M. Rejmund, A. Navin, S. Biswas, C. Michelagnoli, I. Stefan, R. Banik, P. Bednarczyk, S. Bhattacharya, S. Bhattacharyya, E. Clement et al., Eur. Phys. J. A 53, 162 (2017).

[9] D. Ramos, M. Caamaño, A. Lemasson, M. Rejmund, L. Audouin, H. Álvarez-Pol, J. D. Frankland, B. Fernández-Domínguez, E. Galiana-Baldó, J. Piot, D. Ackermann, S. Biswas, E. Clement, D. Durand, F. Farget, M. O. Fregeau, D. Galaviz, A. Heinz, A. I. Henriques, B. Jacquot, B. Jurado, Y. H. Kim, P. Morfouace, D. Ralet, T. Roger, C. Schmitt, P. Teubig, and I. Tsekhanovich, Phys. Rev. L 123, 092503 (2019).

[10] M. Caamaño, O. Delaune, F. Farget, X. Derkx, K.-H. Schmidt, L. Audouin, C.-O. Bacri, G. Barreau, J. Benlliure, E. Casarejos, A. Chbihi, B. Fernández-Domínguez et al., Phys. Rev. C 88, 024605 (2013).

[11] D. Ramos, M. Caamaño, F. Farget, C. Rodríguez-Tajes, L. Audouin, J. Benlliure, E. Casarejos, E. Clement, D. Cortina, O. Delaune, X. Derkx, A. Dijon et al., Phys. Rev. C 97, 054612 (2018).

[12] K.-H. Schmidt, B. Jurado, C. Amouroux, and C. Schmitt, Nucl. Data Sheets 131, 107 (2016).

[13] J. Toke, R. Bock, G. X. Dai, A. Gobbi, S. Gralla, K. D. Hildenbrand, J. Kuzminski, W. F. J. Müller, A. Olmi, and H. Stelzer Nucl. Phys. A 440, 327 (1985).

Torque acting on ^{252}Cf fission fragments

Guillaume Scamps[1,*]

[1]Department of Physics, University of Washington, Seattle, Washington 98195-1560, USA

Abstract.
The generation of angular momentum in the fission fragments is the subject of a renewal of interest since new experimental data [Wilson, et al.] shows that the magnitudes of the spins of the fragments are almost uncorrelated. In this proceeding, the collective potential perceived by the fission fragments is studied using static frozen Hartree-Fock calculation and the dynamic Time-Dependent Hartree-Fock. The results are in qualitative agreement with previous studies done in the case of the ^{240}Pu.

1 Introduction

Fission is the process in which one nucleus separates into two nuclei of comparable mass. An important goal of the nuclear theory [1] is to determine the properties of the fission fragments, mass, charge, energy, and spins. Experimentally the fission fragments have a large spin between 5 and 10 $\hbar$ [2]. The source of spin of the fission fragment is still an open question with different possible explanations, i) the build-up of stochastic fluctuation during the descent of the potential from the saddle to scission [3–6] ii) the quantum effect at scission [7–9] iii) the Coulomb repulsion that creates a torque on the fragments [10–12].

The two fission reactions studied here are, ^{252}Cf $\rightarrow$ ^{132}Sn + ^{120}Cd and ^{252}Cf $\rightarrow$ ^{144}Ba + ^{108}Mo. In the present approach, It is assumed that the two fragments are already formed and correspond to the ground state or a deformed configuration of the corresponding nuclei.

2 Frozen Hartree-Fock

In the Frozen-Hartree-Fock (FHF) [13, 14] calculation, the fission fragment are assumed rigid. Both fragments are put in the lattice at a distance D between their center of mass. The heavy and light fragments are oriented to form respectively an angle θ_H and θ_L with the fission axis. This calculation neglects more complicated shapes that are described in dynamical models such as the time-dependent Hartree-Fock plus BCS pairing model (TDHF+BCS) [15, 16] or Time-Dependent Superfluid Local Density Approximation (TDSLDA) [17]. In these models, a neck is present between the two fragments at the scission point. Nevertheless, these models respect symmetries, and so the deformed fragments at the scission are aligned with the fission axis. As we will observe with the FHF calculations, this corresponds to the arrangement that minimizes the energy. That configuration is then expected to be the most probable but fluctuations due to thermal and/or quantum effects should populate fission configurations where the fragments principle axis deviates slightly from the fission axis.

The fragments of the output channel are chosen to i) describe the main populated states ii) have different shapes to study the impact of different types of deformation on the potential. As can be seen in Table 1, the chosen fission fragment have different shapes: spherical for the ^{132}Sn, quadrupole for the ^{120}Cd and ^{108}Mo and an octupole ^{144}Ba. The ^{108}Mo also shows a hexadecapole deformation. These two output channels belong to the channels that are enough populated to be experimentally studied [18].

Table 1. The deformations and rigid moments of inertia of the four nuclei examined in this study. The rigid moment of inertia in unit of [$\hbar^2$/MeV] is obtained on an axis perpendicular to the main deformation axis of the nucleus. The excitation energy of the deformed state is also shown in the case of the ^{120}Cd which is not in its ground state.

Nuc.	β_2	β_3	β_4	E^* [MeV]	I_{Rigid}
^{132}Sn	0.	0.	0.	0	50.0
^{120}Cd	0.42	0.	0.08	4.4	51.5
^{144}Ba	0.22	0.16	0.15	0	63.1
^{108}Mo	0.58	0.	0.25	0	46.1

The calculations are done in a lattice with a mesh constant dx = 0.8 fm. All calculations of the proceeding are obtained with the Sly4d functional [19]. The fragments are obtained with the Sky3d code [20] on a cubic lattice with $L_x = L_y = L_z$=24 fm. The fragments are then placed on an extended lattice with Lx = 60 fm. The energy of the total system is then computed, and the radial potential is defined as,

$$V = E_{FHF}(\theta, D) - E_{FHF}(\theta = 0, D). \quad (1)$$

In the present static study, the calculation are limited to only one non-zero orientation angle at a time. The angle of

*e-mail: gscamps@uw.edu

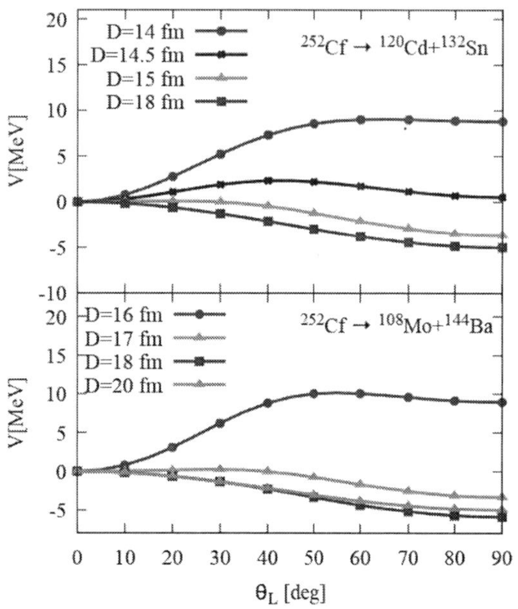

Figure 1. Frozen Hartree-Fock radial potential as a function of the orientation angle of the light fragment for different separation between the fission fragments. The heavy fragment is aligned with the fission axis θ_H=0.

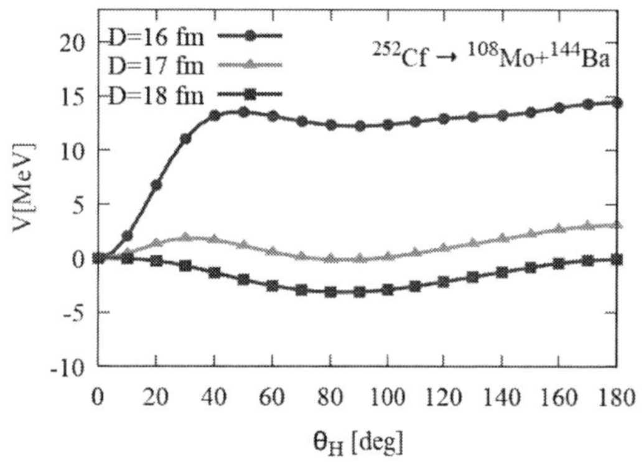

Figure 2. Same as Fig. 1 for the heavy fragment. This fragment being pear-shaped deformed the x-axis of the figure vary from 0 (the heavy fragment point to the other fragment) to 180 degrees. The light fragment is aligned with the fission axis θ_L=0.

the other fragment is assumed to be zero. More complex effects of the correlation between the angles have been studied in Ref. [21] and show a small impact of the correlations. At small angles, the principal impact of the orientation on the energy is then the sum of each fragment contribution. In Fig. 1, the angular potential is shown for the heavy fragment. In both case, the potential show similar behavior, at small distances the fragments can touch each other, and the attractive nucleus-nucleus potential favors the θ=0 configuration. When the distance increases, the nuclear effect drop-off, and the Coulomb potential dominates. The Coulomb energy favors configurations at 90 degrees, then the lowest configuration changes during the scission process from 0 degrees to 90 degrees.

From this static picture, we can estimate the evolution of the system, the fragments are first oriented in the 0 degrees configuration. Beyond the classical approximation, it is expected that statistical or quantum fluctuations will make the fragments deviate from the exact 0 degrees configuration. Then once the system passes the scission point, the 0 degrees configuration become unstable. This lead to a rotational acceleration of the fission fragments due to the Coulomb torque.

3 Time-dependent Hartree-Fock

To describe microscopically that evolution, we solve the TDHF equations with the initial configuration described previously i.e the center of mass of the fragments are at a distance D and oriented with an angle $\theta_{L,R}$. The evolution of the one-body density is shown by a contour plot, in Fig. 3 and 4 panel (a), (b), and (c). Initially oriented at an angle of 25 degrees and at a distance for which there is a nucleus-nucleus interaction (14 fm and 16 fm respectively for the ^{252}Cf → ^{132}Sn + ^{120}Cd and ^{252}Cf → ^{144}Ba + ^{108}Mo channels). The static potential described in the previous section and the additional contribution due to the neck, makes the fragment rotate.

During the evolution, for each fragment, the local total angular momentum is calculated as,

$$\mathbf{J}(\mathbf{r}) = \hbar \sum_{i,\mathbf{F}} \langle \Phi_i(\mathbf{r})| \left((\hat{\mathbf{r}} - \mathbf{r}_{\rm cm}) \times (\hat{\mathbf{p}} - \mathbf{p}_{\rm cm}) + \hat{\mathbf{s}}\right) \; \mathbf{\Theta_F} |\Phi_i(\mathbf{r})\rangle, \tag{2}$$

With $\mathbf{r}_{\rm cm}$ and $\mathbf{p}_{\rm cm}$ respectively the position and impulsion of the fragments. The sum over i includes all occupied protons and neutrons wave function $|\Phi_i(\mathbf{r})\rangle$. The Heaviside function $\mathbf{\Theta_F}$ is the projector operator on the half-space containing the fragment $\mathbf{F}$. That operator reveals how the angular momentum is distributed inside the fragment. The distribution of a rigid rotor would behave as $r\rho(\mathbf{r})$.

That quantity is shown on the three first panels of Fig. 3 and 4. It shows that the fragment deviates from the rigid rotor model. The fragments being excited, different modes of vibrations are populated.

The integral of $\mathbf{J}(\mathbf{r})$ leads to the total angular momentum of the fragments. its evolution as well as the contribution of the protons and neutrons are shown on panel (d) of Fig. 3 and 4. The evolution is strongly dependent on the shapes of the fragments. In the initially spherical ^{132}Sn, the interaction with the other fragment breaks the spherical symmetry and so enables the generation of a small angular momentum. Both ^{120}Cd and ^{108}Mo have a large quadrupole deformation. They see their angular momentum increase rapidly due to the restoring torque. After separation, the angular momentum decreases slightly due to the Coulomb torque. The Coulomb torque can either increase or decrease the angular momentum depending on

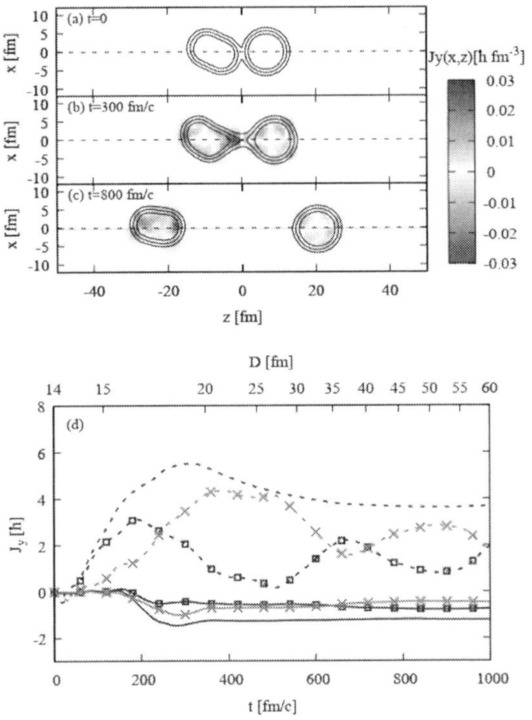

Figure 3. Panel (a-c) Snapshot of the contour density (solid lines) as a function of time for the fission reaction $^{252}Cf \rightarrow {}^{132}Sn + {}^{120}Cd$. The colors show the local angular momentum. The calculation starts at 14 Fm. The light fragment is oriented at an angle θ_L=25°. Panel (d) shows the evolution as a function of time and as a function of the distance D the total spin of the heavy (solid line) and light (dashed line) fragments. For both fragments, the contribution of protons and neutrons are shown respectively by blue squares and green crosses.

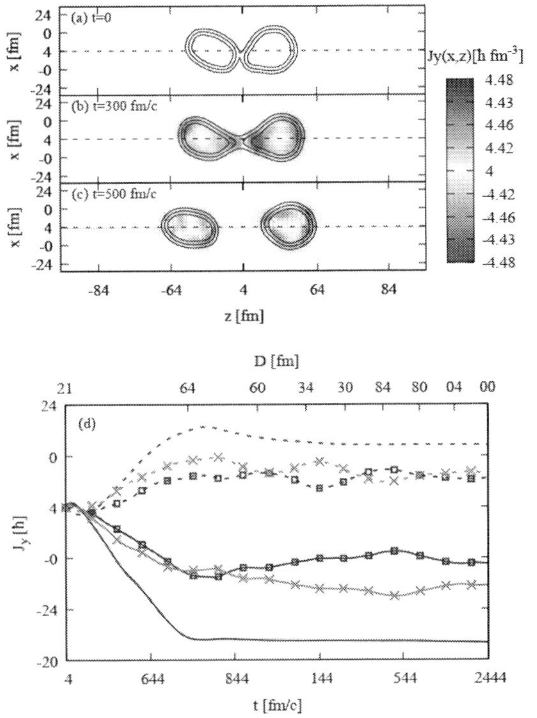

Figure 4. Same as Fig. 3 for the reaction $^{252}Cf \rightarrow {}^{144}Ba + {}^{108}Mo$. The calculation starts at a distance D=16 fm.

the orientation and direction of the angular momentum of the fragments. The pear-shaped ^{144}Ba has the largest final angular momentum. This result is coherent with the static picture since the potential for the heavy fragment (see Fig. 2) is stiffer than for the light fragment (See Fig. 1(b)). This confirms the finding in Ref. [22] that the octupole deformation increases substantially the generation of angular momentum.

A surprising effect, already shown in Ref. [21], is the large transfer of angular momentum between protons and neutrons visible in panel (d) of Fig. 3 and 4. An extreme case is found in the ^{120}Cd fragment. The angular momentum is generated initially in the protons but at t = 500 fm/c the angular momentum is completely dominated by the neutrons.

4 Conclusion

The present results concerning the fission of ^{252}Cf are in qualitative agreement with Ref. [21] which was done for two ^{240}Pu fission channels. A strong restoring force takes place at scission, which creates an angular momentum in the fission fragments if the orientation deviates from θ=0. The goal of the present study is not to obtain results that can be compared to experimental results since the final angular momentum in the fragment depends strongly on the initial angle. Nevertheless, the present FHF results can be used to determine the final angular momentum using the collective hamiltonian model [22].

Acknowledgements

The funding from the US DOE, Office of Science, Grant No. DE-FG02-97ER41014 is greatly appreciated. This research used resources of the Oak Ridge Leadership Computing Facility, which is a U.S. DOE Office of Science User Facility supported under Contract No. DE-AC05-00OR22725.

References

[1] M. Bender, *et. al.*, J. Phys. G: Nucl. Part. Phys. **47** 113002 (2020).

[2] J. B. Wilhelmy, E. Cheifetz, R. C. Jared, S. G. Thompson, H. R. Bowman, and J. O. Rasmussen, Phys. Rev. C **5**, 2041 (1972).

[3] L. G. Moretto and R. P. Schmitt, Phys. Rev. C **21**, 204 (1980).

[4] T. Døssing and J. Randrup, Nuclear Physics A **433**, 215-279 (1985).

[5] J. Randrup and R. Vogt, Phys. Rev. Lett. **127**, 062502 (2021).

[6] J. Randrup, T. Døssing, and R. Vogt, Phys. Rev. C **106**, 014609 (2022).

[7] I. N. Mikhailov and P. Quentin, Physics Letters B **462**, 7–13 (1999).

[8] L. Bonneau, P. Quentin, and I. N. Mikhailov, Phys. Rev. C **75**, 064313 (2007).

[9] T. M. Shneidman, G. G. Adamian, N. V. Antonenko, S. P. Ivanova, R. V. Jolos, and W. Scheid, Phys. Rev. C **65**, 064302 (2002).

[10] Marvin M. Hoffman, Phys. Rev. **133**, B714 (1964).

[11] K. Alder, and A. Winther, *Electromagnetic excitation: Theory of Coulomb excitation with heavy ions* (North-Holland, Amsterdam) (1975).

[12] G.F. Bertsch, arXiv:1901.00928 [nucl-th] (2019).

[13] C. Simenel, A. S. Umar, K. Godbey, M. Dasgupta, and D. J. Hinde, Phys. Rev. C **95**, 031601(R) (2017).

[14] A. S. Umar, C. Simenel, and K. Godbey, Phys. Rev. C **104**, 034619 (2021).

[15] G. Scamps, C. Simenel, D. Lacroix, Phys. Rev. C **92**, 011602(R) (2015).

[16] G. Scamps and C. Simenel, Nature volume **564**, 382-385 (2018).

[17] A. Bulgac, P. Magierski, K. J. Roche, and I. Stetcu, Phys. Rev. Lett. **116**, 122504 (2016).

[18] J.N. Wilson, D. Thisse, M. Lebois, et al. Nature, **590**, 566–570 (2021).

[19] K.-H. Kim, T. Otsuka, and P. Bonche, J. Phys. G **23**, 1267 (1997).

[20] B. Schuetrumpf, P.-G. Reinhard, P.D. Stevenson, A.S. Umar, and J.A. Maruhn, Comp. Phys. Comm. **229**, 211-213 (2018).

[21] G. Scamps, Phys. Rev. C **106**, 054614 (2022).

[22] G. Scamps and G. Bertsch, arXiv:2302.07792 [nucl-th] (2023).

EPJ Web of Conferences **290**, 02014 (2023) https://doi.org/10.1051/epjconf/202329002014
EuNPC 2022

Collective excitations in even-even nuclei with a stepped infinite square well potential

Andreea-Ioana Budaca[1,*] and *Radu* Budaca[1,2,**]

[1]"Horia Hulubei" National Institute for Physics and Nuclear Engineering, Str. Reactorului 30, RO-077125, POB-MG6 Bucharest-Măgurele, Romania
[2]Academy of Romanian Scientists, Splaiul Independenţei 54, 050044, Bucharest, Romania

Abstract.
A solution is worked out for a γ-unstable Bohr Hamiltonian with an infinite square well potential having an adjustable step. The evolution of the model's spectral characteristics are investigated with varying height and width of the additional step, revealing shape coexisting an mixing features. The description of the dynamical shape phase transition in the collective bands of the Te even-even nuclei is offered as a numerical application.

1 Introduction

The introduction of the critical point solution $E(5)$ [1] gave the start for a variety of studies regarding algebraic and phenomenological description of critical phenomena in the collective excitations of atomic nuclei. Basically, the $E(5)$ model is constructed on the solution of the five-dimensional γ-unstable Bohr Hamiltonian (BH) [2] with an infinite square well (ISW) potential only in the variable associated to axial deformation. The flat potential serves as good phenomenological picture for the critical point of the transition between deformed γ-soft rotor and the spherical harmonic vibrator pictures of the nuclear surface fluctuation. It's solution can also be used as a suitable basis state for diagonalization of double well collective potentials [3]. Here we present a new extension of the $E(5)$ by introducing a finite step in the ISW [4].

2 Theoretical framework

For the γ unstable case, the general five-dimensional Bohr Hamiltonian can be reduced, after integration on the $SO(5)$ coordinates, to the following equation for the axial variable β:

$$\left[-\frac{\partial^2}{\partial\beta^2}+\frac{(\tau+1)(\tau+2)}{\beta^2}+v(\beta)\right]F(\beta)=\epsilon F(\beta),$$

where $v(\beta)=2BV(\beta)/\hbar^2$ and $\epsilon=2BE/\hbar^2$ are reduced potential and energies scaled by the mass parameter B, while τ is the so called seniority [5]. For the states considered in this study, the $L=2\tau$ correspondence to angular momentum holds. We solve the above equation for a ISW potential with a step:

$$v(\beta)=\begin{cases}0, & \beta\leq\beta_s,\\ v_0, & \beta_s<\beta\leq 1,\\ \infty, & \beta>1,\end{cases} \tag{1}$$

with $v_0>0$ and $\beta_s\in[0,1]$ being the height and the starting position of the step. The general solution is a combination of Bessel functions of the first and second kind, with distinct behavior in specific regions of β values depending on the energy relative to the height v_0. The wave function must vanish at external walls and match at the position of the step. These boundary and continuity conditions provide determining equations for the total energy and the wave functions. The phenomenological interpretation of the obtained solution is easily extracted from its correlation with the shape of the effective potential

$$v^{\tau}_{eff}(\beta)=\frac{(\tau+1)(\tau+2)}{\beta^2}+v(\beta).$$

As can be seen in Fig.1, the inclusion of the centrifugal effect induces a double-pocket shape even in the ground state. Such double well potentials are well known to be associated with shape coexisting and mixing phenomena [3, 4, 6–9].

3 Model application

The energy is determined from a transcendental equation which is solved numerically. The $E(5)$ critical point model [1] serves as an upper and lower limit for the energy spectrum when $v_0\rightarrow 0$ or ∞ and $\beta_s\rightarrow 0$ or 1. This property is used as a test for the numerical results. The $E(5)$ low-lying states are most perturbed for wider steps and when are positioned near its height. The presence of the additional step usually compress the ground state band spectrum to

*e-mail: abudaca@theory.nipne.ro
**e-mail: rbudaca@theory.nipne.ro

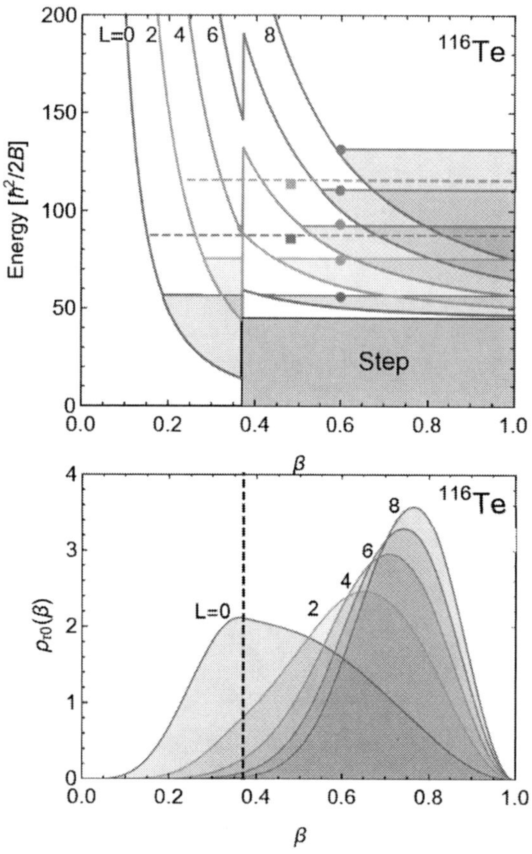

Figure 1. Theoretical effective potential v^{τ}_{eff} with its isolated step (top), as well as the probability distribution (bottom) $\rho_{\tau 0}(\beta) = |F(\beta)|^2$ corresponding to the $L = 2\tau = 0, 2, 4, 6, 8$ ground band states of the ^{116}Te nucleus. The solid and dashed lines denote the theoretical ground band and β excited band states, while circles and squares represent experimental data [10] for the ground and respectively β excited states within the same reference.

resemble the vibrational level sequence, while simultaneously lowering more significantly the β excited states.

The even-even Te isotopes present both vibrational-like ground band and low-lying excited states and are therefore a perfect ground for testing the model. The model fits on few Te nuclei predicted a step at half ISW width with an increasing height with mass. Moreover, the model reproduces with great accuracy the very low β excited states and their rotational excitation. The low 0^+ excited states is considered as a possible signature of shape coexistence [11]. An example in this sense is provided in Fig.1 for the ^{116}Te which has the lowest observed 0^+_2 state. As can be seen from the deformation probability distribution, there is a transition in this nucleus happening from a low deformed ground state to more deformed rotational states. The ground state also has a more extended profile, suggesting a sizable mixing between low and high deformation pockets of the effective potential.

4 Summary

The Bohr Hamiltonian for a γ-unstable ISW potential with an adjustable step is solved analytically. The resulted formalism constitutes an extension of the critical point model $E(5)$ which is recovered at extreme limits of the step's width and height. The centrifugal contribution transforms the stepped potential into an effective double-pocket potential, making it suitable for treating shape coexistence and mixing phenomena. This aspect is state-dependent and can be used to describe the dynamical evolution of the rotational states from low to high deformation configurations. Moreover, due to its special construction, the model can attain very low excited β states. These features are demonstrated for the low-lying states of ^{116}Te nucleus, which was found to exhibit a shape coexisting ground state.

Acknowledgments

This work was supported by project PN-23-21-01-01/2023 of the Ministry of Research, Innovation and Digitalization, Romania.

References

[1] F. Iachello, Phys. Rev. Lett. **85**, 3580 (2000).
[2] A. Bohr, B. R. Mottelson, *Nuclear Structure*, vol. 2 (Benjamin, Reading, Massachusetts, 1975).
[3] R. Budaca, A. I. Budaca, P. Buganu, J. Phys. G Nucl. Part. Phys. **46**, 125102 (2019).
[4] R. Budaca, A. I. Budaca, Eur. Phys. J. Plus **136**, 983 (2021).
[5] L. Fortunato, Eur. Phys. J. A **26**, s01 (2005).
[6] R. Budaca, A. I. Budaca, EPL **123**, 42001 (2018).
[7] R. Budaca, P. Buganu, A. I. Budaca, Nucl. Phys. A **990**, 137 (2019).
[8] A. Ait Ben Mennana, R. Benjedi, R. Budaca, P. Buganu,Y. EL Bassem, A. Lahbas, M. Oulne, Phys. Scr. **96**, 125306 (2021).
[9] A. Ait Ben Mennana, R. Benjedi, R. Budaca, P. Buganu,Y. EL Bassem, A. Lahbas, M. Oulne, Phys. Rev. C **105**, 034347 (2022).
[10] J. Blachot, Nucl. Data Sheets **111**, 717 (2010).
[11] K. Heyde, J. L. Wood, Rev. Mod. Phys. **83**, 1467 (2011).

EPJ Web of Conferences **290**, 02015 (2023) https://doi.org/10.1051/epjconf/202329002015
EuNPC 2022

Fission studies in inverse kinematics with the R^{3}B setup

A. Graña-González[1]*, J. L. Rodríguez-Sánchez[1], J. Benlliure[1], G. García-Jiménez[1], H. Alvarez-Pol[1], D. Cortina-Gil[1], L. Atar[2], L. Audouin[3], G. Authelet[4], A. Besteiro[5], G. Blanchon[6], K. Boretzky[7], M. J. G. Borge[8], P. Cabanelas[1], E. Casarejos[5], J. Cederkall[9], A. Chatillon[6], A. Corsi[4], M. Feijoo[1], E. De Filippo[10], D. Galaviz[11], I. Gasparic[12], E. Geraci[10], R. Gernhäuser[13], B. Gnoffo[10], M. Heil[7], A. Heinz[14], T. Hensel[15], M. Holl[14], D. Jelavic Malenica[12]; T. Jenegger[13], L. Ji[2], H. T. Johansson[14], O. A. Kiselev[7], P. Klenze[13], A. Knyazev[9], D. Körper[7], T. Kröll[2], I. Lihtar[12], Y. A. Litvinov[7], B. Löher[7], N. S. Martorana[10], P. Morfouace[6], D. Mücher[16,17], S. Murillo Morales[18], A. Obertelli[2], V. Panin[7], J. Park[19], S. Paschalis[18], A. Perea[8], M. Petri[18], S. Pirrone[10], L. Ponnath[13], A. Revel[4], H.-B. Rhee[2], L. Rose[18], D. M. Rossi[2], P. Russotto[10], H. Simon[7], A. Stott[18], Y. Sun[2], C. Sürder[2], J. Taïeb[6], R. Taniuchi[18], O. Tengblad[8], H. T. Törnqvist[2,7], M. Trimarchi[10], S. Velardita[2], J. Vesic[20], B. Voss[7], and the R^{3}B Collaboration

[1]IGFAE, University of Santiago de Compostela, E-15782 Santiago de Compostela, Spain
[2]Technische Universität Darmstadt, Fachbereich Physik, Institut für Kernphysik, 64289 Darmstadt, Germany
[3]IPN Orsay, 15 rue Georges Clemenceau, 91406, Orsay, France
[4]CEA Saclay, IRFU/DPhN, Centre de Saclay, 91191, Gif-sur-Yvette, France
[5]CINTECX, Universidade de Vigo, DSN, 36310 Vigo, Spain
[6]CEA, DAM, DIF, F-91297 Arpajon, France
[7]GSI Helmholtzzentrum für Schwerionenforschung, Planckstraße 1, D-64291 Darmstadt, Germany
[8]Instituto de Estructura de la Materia, CSIC, E-28006 Madrid, Spain
[9]Department of Physics, Lund University, P.O. box 118, 221 00 Lund, Sweden
[10]INFN Laboratori Nazionali del Sud, Via Santa Sofia 62, 95123, Catania, Italy
[11]LIP and Faculty os Sciences, University of Lisbon, 1000-149 Lisbon, Portugal
[12]RBI Zagreb, Bijenicka cesta 54, HR10000, Zagreb, Croatia
[13]Technische Universität München, James-Franck-Str 1, 85748, Garching, Germany
[14]Institutionen för Fysik, Chalmers Tekniska Högskola, 412 96 Göteborg, Sweden
[15]Institut für Kern-und Teilchenphysik, Technische Universität Dresden, 01069 Dresden, Germany
[16]University of Guelph, 50 Stone Road E, N1G 2W1, Guelph, ON, Canada
[17]Institut für Kernphysik der Universität zu Köln, Zülpicher Strasse 77, D-50937 Köln, Germany
[18]School of Physics, Engineering and Technology, University of York, YO10 5DD York, UK
[19]Institute for Basic Science, Center for Exotic Nuclear Studies, 34126, Daejeon, Republic of Korea
[20]Jozef Stefan Institute, Jamova cesta 39, 1000 Ljubljana, Slovenia

Abstract. Nuclear fission is a complex dynamical process, whose description involves the coupling between intrinsic and collective degrees of freedom, as well as different quantum-mechanical phenomena. For this reason, to this day it still lacks a satisfactory and complete microscopic description. In addition to the importance of describing fission itself, studies of the r-process in astrophysics depend on fission observables to constrain the theoretical models that explain the isotopic abundances in the Universe. To improve on the existing data, fission reactions of heavy nuclei in inverse kinematics are produced in quasi-free (p,2p) scattering reactions, which induce fission through particle-hole excitations that can range from few to tens of MeV. In order to study the evolution of the fission yields with temperature, the excitation energy of the fissioning system must be reconstructed, which is possible by measuring the four-momenta of the two outgoing protons. Performing this kind of experiment requires a complex experimental setup, providing full isotopic identification of both fission fragments and an accurate measurement of the momenta of the two outgoing protons. This was realized recently at the GSI/FAIR facility and some of the results obtained for the charge distributions are presented in this work.

1 Introduction

The nucleosynthesis r-process is responsible for the synthesis of half of the elements heavier than Fe in the Universe. It is associated with high neutron flux scenarios such as supernovae or neutron star mergers, where rapid neutron captures followed by beta decays lead to the heavy neutron rich nuclear chart region where fission takes place. It was demonstrated that nuclear fission plays an important role in determining the final abundances of the r-process due to the so-called "fission cycling" phenomena, which limits the mass range of the r-process. Therefore, fission studies in the neutron-rich region are required in order to measure fission yields and fission barrier heights to im-

*Corresponding author: antia.grana.gonzalez@usc.es

https://doi.org/10.1051/epjconf/202329002015

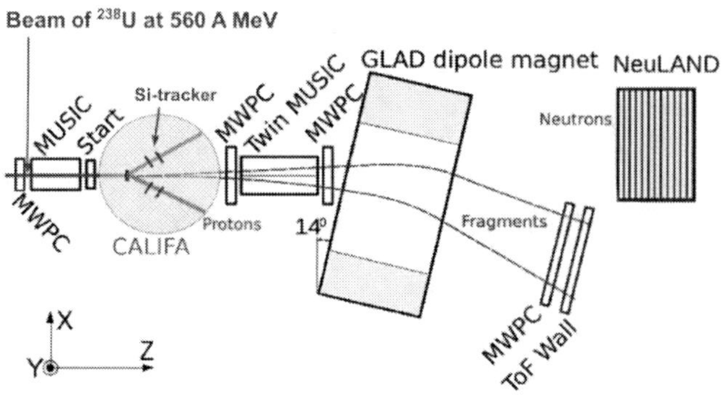

Figure 1. Schematic view of the experimental R^3B setup used in the present work to study quasi-free (p,2p) induced fission reactions.

prove r-process calculations [1, 2] and to obtain more accurate models for the observed isotopic abundances in the Universe.

In the last decade, unprecedented fission experiments have been carried out at the GSI/FAIR facility using the inverse-kinematics technique in combination with state-of-the-art detectors developed by the R^3B (Reactions with Relativistic Radioactive Beams) collaboration. For the first time in the history of fission, it was possible to simultaneously measure and identify both fission fragments in terms of their mass and atomic numbers [3, 4] and to extract correlations between fission observables sensitive to the dynamics of the fission process [5] and the nuclear structure at the scission point [6, 7].

However, previous studies based on either Coulomb excitation, transfer or fusion-fission reactions, were limited in terms of the reachable excitation energy and often limited to fissioning nuclei close to stablity. The excitation energy plays a fundamental role in the description of the process. Depending on the amount of available energy, different regions of the potential energy surface can be populated, leading to different fission paths and consequently to different fission yields distributions. The gradual weakening of the influence of shell structure on the fission yields with the increase of excitation energy can be quantified with the so-called suppression function [8], which minimizes the microscopic term of the energy-dependent effective potential. However, model parameters used in this approach are still not constrained due to the lack of data at different excitation energies. The experiment described in this work combined fission with quasi-free (p,2p) reactions, which allows to determine the excitation energy with stable and exotic nuclei. Thus, the new dataset will provide very valuable data to constrain the suppression function parameters. Another interesting question is the possible dependence of the energy sorting of the fission fragments on the excitation energy of the compound nucleus, since in direct kinematics studies with neutrons, the well known saw-tooth shape shows higher neutron multiplicities for higher incident neutron energies [9].

The ultimate scientific goal of this work is to investigate the energy dependence of the fission yields by using quasi-free (p,2p) reactions. Such data can be obtained combining the previous experimental setup with a silicon tracker based on AMS-type detectors [10] and the calorimeter CALIFA (CALorimeter for In-Flight detection of γ rays and high energy charged pArticles) [11].

2 Experiment

This report is based on the first results obtained in an experiment performed at GSI in March 2021, by using a primary beam of ^{238}U delivered by the SIS18 synchrotron, which was guided to the experimental area and impinged on a liquid hydrogen target to produce quasi-free (p,2p) reactions at 560 MeV/u. The employed technique was the so-called inverse-kinematics, in which the heavy nucleus undergoing fission takes the role of projectile while the target is constituted by lighter nuclei. This methodology was developed at GSI and represents a major breakthrough in the characterization of fission fragments. The major advantage is that the fission fragments are emitted in a narrow cone in the forward direction with very high velocities, permitting to detect both fragments at the same time and to measure the isotopic yields. Using this methodology, previous fission experiments at GSI already provided complete isotopic identification of both fission fragments simultaneously [3, 4].

Since the main goal is to study the excitation energy dependence of the fission yields, the excitation energy is obtained via the aforementioned quasi-free (p, 2p) reactions, which are knockout reactions between protons of the target and protons bound in the heavy projectile nucleus. The products of the reaction are the two outgoing protons and the excited compound nucleus characterized by the particle-hole excitations created by the removed proton. This excited compound nucleus can then de-excite via fission or neutron evaporation. From the measurement of the four momenta of the two outgoing protons the excitation energy of the compound nucleus can be reconstructed using the missing energy method [12].

The experimental setup is displayed in the Fig 1. The detectors named as "MWPCs" in the scheme are multi-wire proportional chambers, which serve to track the trajectories of the particles along the setup. The "MUSIC" detectors are Multi-Sampling Ionization Chambers, used to obtain the charge of the ions by measuring the energy loss when they pass trough the gas filling the chambers. The time of flight is measured by two plastic scintillator detectors, one located before the target (named as "Star" in the scheme) and the ToF-Wall at the end of the setup. After the start scintillator, the ^{238}U beam enters the vaccum chamber and impinges onto the LH_2 target, where the (p, 2p) reaction takes place. To reconstruct the trajectory of the two protons, a silicon tracker is located inside of the chamber right after the target. The tracker includes two arms, each of them consisting of an array of three AMS-type [10] 0.3 mm thick double-sided silicon-strip detectors. To measure the energy of the protons, the CALIFA calorimeter [11] is placed surrounding the vaccum chamber. CALIFA consists of 1504 CsI(Tl) crystal scintillators covering a polar angle range between 22 and 90 degrees. After fission, the two fragments continue moving in the forward direction passing through the rest of the setup. The Twin-MUSIC ionization chamber is designed with two independent ionization chambers, and those two are divided into two sections, each one segmented into 16 anodes. This design allows to measure the atomic number of both fission fragments in coincidence. The two MWPCs, located upstream and downstream from the Twin MUSIC, serve to determine the horizontal angle of the fission fragments inside the Twin. Once inside the GLAD dipole, the fragments will follow different trajectories according to their magnetic rigidities. After GLAD, another MWPC measures the final positions of the fragments and the ToF-Wall detector composed by 28 plastic scintillators provides the time of flight with respect to the start signal. Neutrons emitted from the fission fragments are detected in NeuLAND (New Large-Area Neutron Detector) [13], located 15m downstream from the target. Using the measurements of charge, time of flight and trajectory, the mass of the fission fragments can be obtained employing the $\Delta E - B\rho - ToF$ technique, allowing therefore for full isotopic identification.

3 Results

Data analysis begins with the charge identification of both fission fragments given by the Twin-MUSIC. The correlation of the derived charges for both fission fragments Z_1 and Z_2 is shown in the Fig. 2 (top) in a two-dimensional cluster plot. Each diagonal of clusters in this figure corresponds to a different fissioning system produced in the reaction, and each cluster on the diagonal corresponds to a different pair of resulting fission fragments. Since the beam is $^{238}_{92}$U, when the desired quasi-free (p,2p) reaction takes place the excited compound nucleus is $^{237}_{91}$Pa, which corresponds to the diagonal of clusters enclosed by the red line. Nevertheless, it is important to notice that this diagonal does not only consist of ^{237}Pa systems, it contains also the contributions from other lighter Pa isotopes as well, since the ^{237}Pa can also de-excite via neutron emission and undergo fission after. It is also possible to obtain different excited Pa isotopes when the reaction inducing fission is not a quasi-free process. A selection of 80 degrees in the opening angle between the two protons serves to select and differentiate the quasi-free events.

If the charges Z_1 and Z_2 are added for each diagonal in this plot, the result is the distribution of the fissioning systems, which is shown in the Fig. 2 (top) inset. The peak with the highest statistics in the plot corresponds to $Z_1 + Z_2 = 91$ (Pa), and the peaks of the distribution are clearly observed for fissioning systems charges ranging from 80 to 94 (both included). The two peaks to the right of the 91 peak correspond to reactions in which neutrons might have

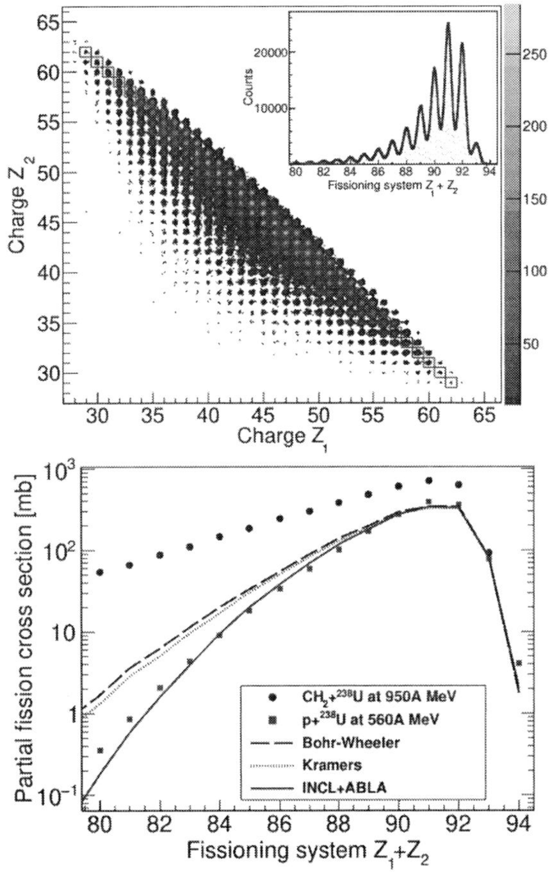

Figure 2. Top: Two-dimensional cluster plot of the charges Z_1 and Z_2 of both fission fragments for all fissioning systems. The inset shows the fissioning system distribution. Bottom: Partial fission cross sections for the different fissioning systems produced in this experiment (red squares, p+^{238}U) and comparison with theoretical models and fragmentation experimental data for the reaction CH_2 + ^{238}U at 950 MeV/u. .

been kicked out of the projectile nucleus, but no proton was removed. In particular, the systems with charge 93 correspond to the less likely case in which the projectile picks up one proton from the target. The peaks to the left of 91 originates from fissioning systems in which more than one proton was removed from the projectile when the (p, 2p) reaction took place.

The Fig. 2 (bottom) shows the fission cross sections for each fissioning system. To obtain this information, the statistics of each peak in Fig. 2 (top) inset is normalized to the total fission cross section measured in the present experiment, derived from an integration over all events where two fission fragments hit the ToF-Wall. The results obtained in this work are represented by red squares and compared with several theoretical models. The dashed line corresponds to the Bohr-Wheeler statistical model [14], the earliest attempt to describe fission. The dotted line corresponds to the Kramer's stationary solution [15], who pointed out that the fission rates provided by the statistical model might be overestimated by not taking into account the energy dissipation from intrinsic to collective degrees of freedom. Thus, a description of fission as a diffusion process was proposed. The experimental data from this work are compatible with the theoretical preditions by INCL+ABLA [16, 17] (solid line). This calculation describes the reaction and the following de-excitation processes as fission, by using a dynamical description based on the Fokker-Planck equation. This kind of dynamical model introduces a transient time at high excitation energies that delays the fission and thus lowers the fission rates in comparison with the precedent models. In addition, the data from this work are also compared with the data for the reaction CH_2+^{238}U at 950 MeV/u [18] (dots). In this case, fission is induced mainly after abrasion processes, which are more violent than the spallation ones used in this work, and thus it comes more likely that more nucleons are removed from the compound nucleus, resulting in larger partial fission cross sections for the lighter fissioning systems.

4 Conclusions

The first (p,2p)-fission experiment has been carried out at GSI/FAIR with projectiles of ^{238}U at energies of 560 MeV/u impinging on a liquid hydrogen target. The R^{3}B experimental setup allows to measure all the reaction products event by event. These complete kinematics measurements enable to correlate the fission yields with the excitation energy of the fissioning compound nucleus and to study the energy sharing between the two fission fragments. Additionally, this experiment permits to develop the data analysis and validate the methodology to use this approach in future experiments with exotic nuclei. At the current stage of the analysis the charge identification is completed and the fission cross sections for each fissioning system have been deduced and found to be compatible with theoretical calculations. The next steps will be the identification in mass number and the reconstruction of the excitation energy spectrum. In the near future further (p,2p)-fission experiments are planned to be carried out with exotic neutron-rich projectiles close to the neutron shell N=152 to obtain their fission yields and fission barrier heights for constraining r-process calculations.

Acknowledgments

This work was partially supported by the Spanish Ministry for Science and Innovation under grants PGC2018-099746-B-C21, PGC2018-099746-B-C22 and PID2019-104390GB-I00, by the Regional Government of Galicia under the program "Grupos de Referencia Competitiva" ED431C-2021-38 and by the "María de Maeztu" Units of Excellence program MDM-2016-0692. J.L.R.S. thanks the support from Xunta de Galicia under the program of postdoctoral fellowship ED481D-2021-018. J. P. acknowledges the support by the Institute for Basic Science (IBS-R031-D1).

References

[1] N. Vassh et al. J. Phys. G: Nucl. Part. Phys. **46** 065202 2019

[2] Samuel A. Giuliani et al. Phys. Rev. C **97** 034323 2018

[3] J. L. Rodríguez-Sánchez et al. Phys. Rev. C **91** 064616 2015

[4] E. Pellereau et al. Phys. Rev. C **95** 054603 2017

[5] J. L. Rodríguez-Sánchez et al. Phys. Rev. C **94** 061601(R) 2016

[6] A. Chatillon et al. Phys. Rev. C **99** 054628 2019

[7] A. Chatillon et al. Phys. Rev. Lett. **102** 202502 2020

[8] J. Randrup et al. Phys. Rev. C **88** 064606 2013

[9] A.N. Andreyev et al. Reports Prog. Phys. **81** 016301 2017

[10] J. Alcaraz et al. Nucl. Instrum. Methods Phys. Res. A **593** 376 2008

[11] H. Alvarez Pol et al. Nucl. Instrum. Methods Phys. Res. A **767** 453 2014

[12] T. Noro et al. Prog. Theor. Exp. Phys. **093D02** 2020

[13] K. Boretzky et al. Nucl. Instrum. Method Phys. Res. A **1014** 165701 2021

[14] N. Bohr and J.A. Wheeler. Phys. Rev. **56.5** 426 1939

[15] H.A. Kramers. Physica. **7.4** 284-304 1940

[16] J. Hirtz et al. Phys. Rev. C **101** 014608 2020

[17] J. L. Rodríguez-Sánchez et al. Phys. Rev. C **105** 014623 2022

[18] K.H. Schmidt and B. Jurado et al. Phys. Rev. Lett. **104** 212501 2010

Fragmentation-induced fission reactions of ^{236}U in inverse kinematics to investigate the pre-fragment angular momentum parameterizations

J. L. Rodríguez-Sánchez[1,2a], M. Feijoo[1], J. Benlliure[1], A. Chatillon[3], J. Taïeb[3], L. Audouin[4], H. Alvarez-Pol[1], Y. Ayyad[1], G. Bélier[3], G. Boutoux[3], M. Caamaño[1], E. Casarejos[5], D. Cortina-Gil[1], F. Farget[6], B. Fernández-Domínguez[1], T. Gorbinet[3], L. Grente[3], A. Heinz[7], H. T. Johansson[7], B. Jurado[8], A. Kelić-Heil[9], D. Körper[9], N. Kurz[9], B. Laurent[3], S. Lindberg[7], B. Löher[9], J.-F. Martin[3], C. Nociforo[9], C. Paradela[1], S. Pietri[9], D. Ramos[1], C. Rodríguez-Tajes[6], H. Simon[9], H. T. Törnqvist[9,10], B. Voss[9], H. Weick[9] and Y. Yan[4]

[1] *IGFAE, University of Santiago de Compostela, E-15782 Santiago de Compostela, Spain*
[2] *CITENI, Campus Industrial de Ferrol, Universidade da Coruña, E-15403 Ferrol, Spain*
[3] *CEA, DAM, DIF, F-91297 Arpajon, France*
[4] *IPN Orsay, 15 rue Georges Clemenceau, F-91406 Orsay, France*
[5] *CINTECX, Universidade de Vigo, DSN, E-36310 Vigo, Spain*
[6] *CNRS, GANIL, Bd H. Becquerel, F-14076 Caen, France*
[7] *Institutionen för Fysik, Chalmers Tekniska Högskola, 412 96 Göteborg, Sweden*
[8] *CNRS, CENBG, F-33175 Gradignan, France*
[9] *GSI Helmholtzzentrum für Schwerionenforschung, Planckstraße 1, D-64291 Darmstadt, Germany*
[10] *Technische Universität Darmstadt, Fachbereich Physik, Institut für Kernphysik, D-64289 Darmstadt, Germany*

Abstract. In the last decades, measurements of spallation, fragmentation and Coulex induced fission reactions in inverse kinematics have provided valuable data to accurately investigate the fission dynamics and nuclear structure at large deformations of a large variety of stable and non-stable heavy nuclei. The collected data were used to constrain dynamic and nuclear structure parameters of different de-excitation models, such as ABLA and GEF, but the data can also be used to constrain the parameterizations describing the pre-fragment properties after the nuclear collision, such as the angular momentum gained by the pre-fragment. In this work, the fissioning system yields are compared to calculations assuming different parameterizations for modeling the angular momentum gained by the compound nuclei. Our findings indicate that the parameterizations utilized by abrasion models clearly underestimate the angular momentum, resulting in the underestimation of the production of lighter fissioning systems.

1 Introduction

Nuclear fission is the clearest example of large-scale collective excitations in nuclei. Since its discovery by Hahn, Meitner, Strassmann and Frisch in 1939 [1, 2], the progress in the understanding of the fission process has been driven by new experimental results. Despite the recent theoretical progress in the investigation of fission, a complete description still represents a challenge in nuclear physics because it is a very complex dynamical process, whose description involves the coupling between intrinsic and collective degrees of freedom, emission of light particles and γ-rays, as well as different quantum-mechanical phenomena [3]. Therefore, its investigation requires complex experimental setups that allow for complete kinematics measurements of the fission products.

In the late 90's, with the advance of heavy-ion accelerators, a new generation of experimental approaches for fission studies was developed. The use of the inverse kinematics technique permited for the first time an in-flight identification of fission fragments in charge and mass number. The first measurements based on this technique were performed at the GSI facility in Darmstadt (Germany) using the fragment spectrometer FRS [4] to detect and identify one of the two fission fragments in charge and mass number [5]. The data provided relevant information on the fission process dynamics [6], resulted in the discovery of new isotopes and isomeric states [7], as well as providing production cross sections of more than 1000 nuclear fission residues [8–11]. The FRS spectrometer was also used to produce secondary radioactive beams of neutron-deficient actinides and preactinides between At and U elements [12] that impinged onto an active target to induce fission through Coulomb-excitation and fragmentation reactions. The fission fragments were identified in charge by a double ionization chamber. The charge distribution measurements provided relevant information on the transition from symmetric to asymmetric fission [12] and on the presaddle fission dynamics [13, 14].

[a] *Corresponding author: j.l.rodriguez.sanchez@udc.es*

Recently, a great effort was made by the SOFIA (Studies On FIssion with Aladin) collaboration at GSI [15, 16] to overcome the restrictions of conventional fission experiments and to provide complete isotopic measurements of both fission fragments inducing fission through spallation, fragmentation, and Coulomb-excitation reactions. A state-of-the-art experimental setup specially designed to investigate fission permits, by combining position, angle, energy loss and time-of-flight (ToF) measurements, to apply the ΔE-Bρ-ToF method to identify in coincidence both fission fragments in terms of their mass and atomic numbers. It became possible to extract correlations between fission observables sensitive to the dynamics of the fission process [17–20] and the nuclear structure at the scission point [21–24].

In this work, fission observables obtained from fragmentation-induced fission reactions of ^{236}U in inverse kinematics are used for the first time to investigate the effects of the existing pre-fragment angular momentum prescriptions. Recently, it has been pointed out that the parameterizations used for modeling the angular momentum gained by the pre-fragments after a nucleus-nucleus collision underestimate by a factor of ten the measurements of high angular-momentum isomeric states populated in medium-mass and heavy nuclei [25–27]. These investigations were performed with medium-mass and heavy fragmentation residues at a few hundred MeV/u, in which one would expect low angular momenta because such residues are produced in peripheral collisions. In this case, we use fission reactions covering a large range of impact parameters to study the effects of the angular momentum from peripheral to more central collisions.

2 Experimental setup

The realization of this kind of experiment requires a complex experimental setup, providing an unambiguous identification of the secondary beams and a complete kinematics measurement of fission fragments. Such a measurement was performed in 2014 at the GSI facility by combining the fragment spectrometer FRS [4] and the SOFIA experimental setup. In the following, a detailed description of the experimental setup and some preliminary results are presented.

The SIS18-synchrotron was utilized to accelerate heavy ions of ^{238}U at relativistic energies around 720*A* MeV. The secondary beams were produced by fragmentation reactions on a 1032 mg/cm^2 Be target mounted together with a 223 mg/cm^2 Nb stripper located at the entrance of the fragment separator FRS. The secondary beam of ^{236}U was selected by the FRS operated as a momentum-loss achromatic spectrometer [28] and then guided to the experimental area *Cave-C* to induce fission reactions in an active target.

The SOFIA experimental setup, shown in Fig. 1, is divided into two parts, one to characterize the incoming projectile nuclei and another to measure the fission products. The first part consists of a triple multiple-sampling ionization chamber (MUSIC), a multi-wire proportional counter

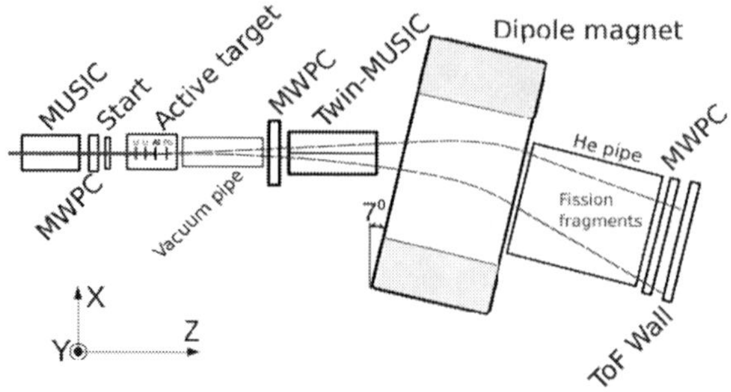

Figure 1. (Color online) Top schematic view of the experimental setup used in this work to identify unambiguously both fission fragments. Sizes are not to scale.

(MWPC) [29], and a plastic scintillator detector used to measure the time-of-flight (ToF) of the incoming projectiles and outgoing fission fragments. The combination of these detectors is used for the beam identification and to determine the beam position on the target.

The second part consists of two MWPC detectors, a double multi-sampling ionization chamber (Twin MUSIC), a large acceptance dipole magnet (ALADIN), and a large ToF wall. The Twin MUSIC chamber has a central vertical cathode that divides its volume into two active regions (left and right) both of which are divided into two sections (up and down). Each section is then segmented in 16 anodes that provide 16 independent energy-loss and drift-time measurements. This segmentation allows us to obtain the atomic number of our fission fragments with a resolution better than 0.38 charge units full width at half maximum (FWHM) and the angles on the *X*-*Z* plane with a resolution around 1 mrad (FWHM). MWPCs, situated in front and behind the dipole magnet, provide the horizontal (*X*) and vertical (*Y*) positions of the fission fragments. The MWPC situated in front of the dipole magnet provides the *X* and *Y* positions with a resolution of around 200 μm and 1.5 mm (FWHM), respectively, while the MWPC situated behind the dipole magnet provides those positions with a resolution of around 300 μm and 2 mm (FWHM), respectively. The ToF wall consists of 28 plastic scintillators that allow to measure the ToF of the fission fragments with respect to the start signal provided by the plastic scintillator located at the entrance of the experimental setup with a resolution of around 40 ps (FWHM) [30]. The ALADIN magnet was set to a magnetic field of around 1.6 T and its gap was filled with helium gas at atmospheric pressure to reduce the energy and angular stragglings of the fission fragments.

The atomic number of the fission fragments is deduced based on the fact that the energy loss is proportional to the atomic number squared. In Fig. 2 we show the measured atomic-number correlation plot in the twin-MUSIC detector, where the energy lost by the fission fragments was corrected by the corresponding ToF measurements. The achieved resolution is better than 0.38 charge units

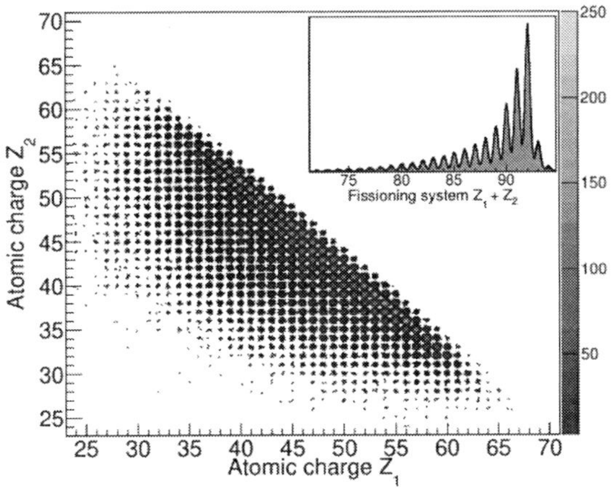

Figure 2. (Color online) Charge distribution correlation of fission fragments produced in the reaction ^{236}U + Al at 650*A* MeV. The inset shows the charge distribution for the fissioning system.

(FWHM) over the whole range of fission fragments. In the inset we display the charge distribution for the fissioning systems produced in the reaction ^{236}U + Al at 650*A* MeV, with the maximum at $Z_1 + Z_2 = 92$.

3 Results

The fission yields as a function of fissioning system Z_1+Z_2 are displayed in Fig. 3(a), which were obtained by normalizing the counts of each fissioning system to the total number of fission events. The data are compared to model calculations based on the intranuclear cascade code INCL++ [31] coupled to the de-excitation model ABLA++ [32], which have been benchmarked with different observables obtained from spallation-induced fission reactions [35]. In this case we consider different parameterizations for the description of the angular momentum. First, we compare the data to standard INCL calculations where the angular momentum is obtained from energy and momentum conservation applyed during the nucleus-nucleus collision, as explained in Refs. [33, 34]. Second, we calculate the angular momentum by using the phenomelogical parameterization introduced in the abrasion-ablation model ABRABLA07 [36], which is based on Goldhaber's prescription [37].

One can see clearly in the Fig. 3(a) that the calculation based on the abrasion parameterization (solid line) underestimates the fission yields for the lighter fissioning systems, whereas the standard INCL calculation (dashed line) gives a better description. This discrepancy can be explained in terms of the angular momentum gained by the fissioning compound nuclei, as shown by the INCL predictions (dashed line) in the Fig. 3(b). Whereas the fragmentation residues are produced with an average angular momentum around $16\hbar$ and a tail that can populate values up to $70\hbar$, the fissioning systems can reach higher angular momenta up to $140\hbar$, with an average angular momentum around $42\hbar$. This fact is in contradiction with the abrasion

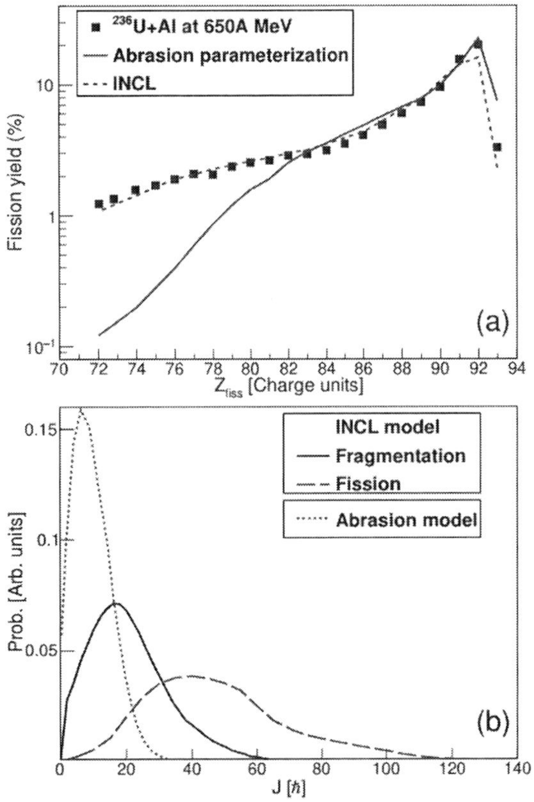

Figure 3. (Color online) (a) Fission yields as a function of fissioning system $Z_1 + Z_2$ for the reaction ^{236}U + Al at 650*A* MeV compared to model calculations based on different parameterizations for the angular momentum. (b) Angular momentum predictions for fragmentation and fission nuclear residues produced in the reaction ^{236}U + Al at 650*A* MeV.

model. The latter predicts that fragmentation residues and fission pre-fragments are produced with the same angular momentum distribution (red dotted line).

Finally, according to these calculations, one can interpret that in fragmentation reactions induced by light target nuclei the pre-fragments could cover a large range in angular momentum, reaching values up to $140\hbar$. This large angular momentum favours the fission channel since it reduces the fission barrier height down to zero, as shown in Ref. [38], increasing the fission probabilities.

4 Conclusions

The fragment spectrometer FRS was combined with the SOFIA experimental setup to study fragmentation-induced fission reactions of ^{236}U in inverse kinematics. A secondary beam of ^{236}U at 650*A* MeV was produced and selected with the fragment spectrometer FRS and guided to the experimental area to induce fission reactions of ^{236}U in an active target. The SOFIA experimental setup was then used to perform complete kinematics measurements of the two fission fragments. In this work, we focus on the charge distributions of the fission fragments and their correlations to obtain the atomic number of the fissioning systems. These yields are compared to intranuclear cascade calculations based on INCL model using the standard description for the pre-fragment angular momentum distributions and also to phenomelogical prescriptions used in abrasion models like ABRABLA07. The comparison reveals that the abrasion parameterization underestimates the fission yields for ligher fissioning systems produced with large angular momentum, whereas the INCL standard parameterization based on energy and angular momentum conservations gives a better agreement.

Further experimental studies on Coulex, spallation and fragmentation induced fission reactions within the R^3B (Reactions with Relativistic Radioactive Beams) collaboration have been carried out recently, measuring the charge and mass distributions for many short-lived neutron-deficient nuclei in the region between Re and Th. The new set of data could be used to systematically investigate the fissioning system yields as a function of the neutron-to-proton asymmetry or isospin. This investigation could also be combined with new approaches based on quasi-free scattering reactions to induce fission reactions of neutron-deficient and neutron-rich nuclei, which will provide direct access to the excitation energy of the fissioning system. Therefore the current and future experiments look very promising to accurately measure fission yields for many exotic nuclei and to study their evolution with the excitation energy.

Acknowledgments

This work was partially supported by the Spanish Ministry for Science and Innovation under Grants PGC2018-099746-B-C21 and PGC2018-099746-B-C22, by the Regional Government of Galicia under the program "Grupos de Referencia Competitiva" ED431C-2021-38 and by the "María de Maeztu" Units of Excellence program MDM-2016-0692. J.L.R.-S. is thankful for the support from Xunta de Galicia under the postdoctoral fellowship Grant ED481D-2021-018 and from the "Ramón y Cajal" program under the Grant RYC2021-031989-I, funded by MCIN/AEI/10.13039/501100011033 and by "European Union NextGenerationEU/PRTR".

References

[1] L. Meitner and O. R. Frisch, Nature (London) **143**, 239 (1939).
[2] O. Hahn and F. Strassmann, Naturwissenschaften **27**, 11 (1939).
[3] P. Möller et al., Nature **409**, 785 (2001).
[4] H. Geissel et al., Nucl. Instrum. Methods Phys. Res. B **70**, 286 (1992).
[5] P. Armbruster et al., Phys. Rev. Lett. **93**, 212701 (2004).
[6] J. Benlliure et al., Phys. Rev. C **74**, 014609 (2006).
[7] A. Jungclaus et al., Phys. Rev. Lett. **99**, 132501 (2007).
[8] T. Enqvist et al., Nucl. Phys. A **686**, 481 (2001).
[9] M. Bernas et al., Nucl. Phys. A **725**, 213 (2003).
[10] E. Casarejos et al., Phys. Rev. C **74**, 044612 (2006).
[11] D. Pérez-Loureiro et al., Phys. Rev. C **99**, 054606 (2019).
[12] K.-H. Schmidt et al., Nucl. Phys. A **665**, 221 (2000).
[13] B. Jurado et al., Phys. Rev. Lett. **93**, 072501 (2004).
[14] C. Schmitt et al., Phys. Rev. C **81**, 064602 (2010).
[15] J. L. Rodríguez-Sánchez et al., Phys. Rev. C **91**, 064616 (2015).
[16] E. Pellereau et al., Phys. Rev. C **95**, 054603 (2017).
[17] J. L. Rodríguez-Sánchez et al., Phys. Rev. C **90**, 064606 (2014).
[18] J. L. Rodríguez-Sánchez et al., Phys. Rev. C **92**, 044612 (2015).
[19] J. L. Rodríguez-Sánchez et al., Phys. Rev. C **94**, 034605 (2016).
[20] J. L. Rodríguez-Sánchez et al., Phys. Rev. C **94** 061601(R) (2016).
[21] A. Chatillon et al., Phys. Rev. C **99**, 054628 (2019).
[22] A. Chatillon et al., Phys. Rev. Lett. **102**, 202502 (2020).
[23] J.-F. Martin et al., Phys. Rev. C **104**, 044602 (2021).
[24] A. Chatillon et al., Phys. Rev. C **106**, 024618 (2022).
[25] Zs. Podolyák et al., Phys. Lett. B **632**, 203 (2006).
[26] S. Pal et al., Phys. Lett. B **665**, 164 (2008).
[27] X. L. Tu et al., Phys. Rev. C **95**, 014610 (2017).
[28] K.-H. Schmidt et al., Nucl. Instrum. Methods Phys. Res. A **260**, 287 (1987).
[29] ALICE Collaboration, J. Phys. G: Nucl. Part. Phys. **30**, 1517 (2004).
[30] A. Ebran et al., Nucl. Instr. Meths. A **728**, 40 (2013).
[31] J. Hirtz et al., Phys. Rev. C **101**, 014608 (2020).
[32] J. L. Rodríguez-Sánchez et al., Phys. Rev. C **105**, 014623 (2022).
[33] A. Boudard et al., Phys. Rev. C **87**, 014606 (2013).
[34] D. Mancusi et al., Phys. Rev. C **90**, 054602 (2014).
[35] J. Benlliure et al., Eur. Phys. J. Plus **132**, 120 (2017).
[36] J.-J. Gaimard et al., Nucl. Phys. A **531**, 709 (1991).
[37] A. S. Goldhaber et al., Phys. Lett. B **53**, 306 (1974).
[38] A. J. Sierk, Phys. Rev. C **33**, 6 (1986).

Universal Models for Heavy-Ion Fusion Cross Section Above-Barrier

Ivano Lombardo[1,2,*], *Daniele* Dell'Aquila[3,4,**], *Brunilde* Gnoffo[1,2], *Luigi* Redigolo[1,2], *Francesco* Porto[1,2], and *Marco* Russo[1,2]

[1]Dipartimento di Fisica e Astronomia, University of Catania, Catania, Italy
[2]INFN-Sezione di Catania, Catania, Italy
[3]Dipartimento di Fisica "Ettore Pancini", University of Naples "Federico II", Naples, Italy
[4]INFN-Sezione di Napoli, Napoli, Italy

Abstract. The paper discusses a recent re-investigation of a large body of heavy-ion fusion cross section data with the aim of deriving a simple phenomenological model able to describe data from the Coulomb barrier up to the onset of nuclear multifragmentation. To this end, we adopted two complementary approaches: a first universal phenomenological model was derived exploiting a novel artificial intelligence tool for the formal modelling of large datasets. This tool is capable of advanced feature selection and is ideal to drive the discovery process even using traditional methods. A second phenomenological model was derived using a sum-of-difference approach and achieved an unprecedented accuracy in describing above-barrier fusion excitation functions data. Future perspectives and opportunities arising from the present models are also discussed in the text.

1 Introduction

During the past decades, extensive experimental campaigns were performed to measure the cross section of the fusion between heavy ions at relatively low energies [1–5]. This gave a crucial boost also to many experimental techniques conventionally adopted in nuclear physics (gamma-ray analysis, time-of-flight and magnetic spectrometers, charged particle detection with telescope arrays). Data were collected especially above the Coulomb barrier and up to the onset of multi-fragmentation phenomena.

Understanding the fusion between heavy ions at energies above-barrier is anything but trivial. A number of models, mainly semi-classical, have been developed to attempt to describe large systematics of data [6]. For light-to-medium-mass systems, i.e. $A_{tot} \simeq 20-140$, the fusion cross section, as a function of the inverse center-of-mass energy $1/E_{cm}$, exhibits peculiar trends that can be schematically described in three main regions[2, 7]. Region I starts approximately at the Coulomb barrier, and shows a quasi linear increase of the cross section for decreasing $1/E_{cm}$ values [1, 2, 7–9]. In this region, the fusion mechanisms is in competition with multi-nucleon-transfer processes [10–12] and quasi-fission (especially for heavy systems) [13–15]. Region II is instead characterized by an almost stagnant trend of the fusion cross section. This is due to the competition with deep inelastic phenomena [16, 17] and quasi-fission. Finally, region III is the highest energy region, where the fusion cross section has a sharp decrease for decreasing values $1/E_{cm}$ giving rise to incomplete fusion phenomena, [18–20], multi-fragmentation [21–23], and other complicated reaction scenarios [5]. At higher energies, the fusion cross section is negligibly small, due to mechanical and thermodynamical instabilities of the transient system formed in the collisions [22–28]. From a theoretical point of view, the fusion cross section between heavy ions has been deeply investigated. To interpret experimental data in regions II and III, where the existence of a limiting angular momentum was previously suggested [29], two main families of macroscopic models have been developed: critical distance models [7, 30], and models based on limitations to compound nucleus [31]. In addition to macroscopic models, also microscopical (as Time-Dependent Hartree-Fock, TDHF, see e.g. [32, 33]), molecular dynamics (see e.g. [34, 35]), and phenomenological models (see, e.g., [1, 27, 28, 36–38]) have been proposed. For example, Refs. [28, 36–39] report phenomenological models capable to describe complete fusion data from several collision systems in the I-II regions, while the model of Ref. [27] is capable to describe the data in the highest energy region (III). Despite the relevant experimental and theoretical efforts done so far, there are yet unanswered questions, especially regarding possible effects of nuclear structure parameters (i.e. characteristics of projectiles, targets and compound nuclei) on fusion cross sections, or the occurrence of shell effects in the fusion between light-to-medium-mass systems. To try to contribute in the field, we analyzed a very large dataset of complete fusion cross section data, with the aim to derive universal models to describe the entire dataset and to inspect the correlation of the fusion cross section on nuclear structure variables. In this paper, we describe the derivation of new computationally simple models, for the description of the fusion cross section between heavy

*e-mail: ivano.lombardo@ct.infn.it
**e-mail: daniele.dellaquila@unina.it

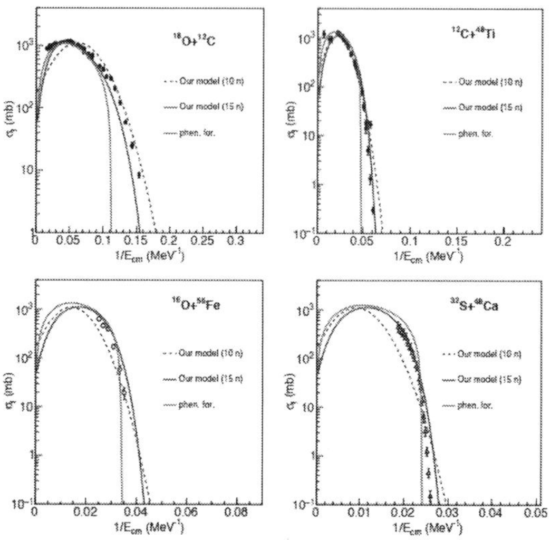

Figure 1. Results of the models of eq. (1), red dashed line, and (2), red solid line, for some of the collisions systems analyzed in this work. The results of the model from Ref. [39] is shown with the green line.

ions, with two complementary approaches: (1) the first exploits novel artificial intelligence algorithms for symbolic regression and advanced feature selection techniques; (2) the second is a modified sum-of-differences (mSOD) approach, which exploits the dataset and the variables refined with the artificial intelligence analysis.

2 Dataset and Modeling

The data used to derive the models were originally extracted from the Nuclear Reaction Video Project (NRV) database [40, 41], taking into account the results of Ref. [27]. We gathered about 4500 data points for 124 light-to-medium-mass collision systems ($Z_p \cdot Z_t < 250$), and additional data from 63 medium-mass systems ($Z_p \cdot Z_t > 250$). We considered systems with $Z_{p,t} \geq 6$, and with $A_{tot} \leq 130$, thus excluding from the dataset too heavy systems, for which fusion-fission and quasi-fission can be the dominant reaction modes, and also too light systems, where the presence of break-up and transfer reactions complicates the analysis. In addition, we included in the dataset only fusion data at incident energies from approximately 1.2 times the Coulomb barrier [2].

2.1 Modeling with advanced artificial intelligence tools

The models derived using artificial intelligence techniques exploited a novel hybridization of genetic programming and artificial neural networks for the formal modelling of data called Brain Project (BP) [42, 43]. BP was previously used in a number of investigations involving different research fields (see, e.g., [44–47]) and adopts a well-known programming scheme inspired by the natural selection in biological systems [48, 49]. In the scheme implemented by BP, analytical models for the description of a set of experimental data points form a *population* of individuals. Each individual is associated with a *fitness* value, which measures the degree to which an individual is a good solution for the modelling problem to solve. At each iteration, new individuals are generated by a suitable crossover of other existing individuals, and the new individuals are inserted into the population replacing previous individuals. Individual selection and replacement criteria usually account for the individual fitness, simulating natural selection rules. In a similar scheme, the average fitness of the population is typically improved at each iteration. In the end, after a suitably large number of iterations, the best individual of the population is taken to represent the model to reproduce the experimental data.

A crucial point of BP is the *feature selection*, i.e. the automatic selection of the variables required to describe the data, neglecting poorly relevant variables. This is typically a complicated task with traditional statistical analyses, especially in the presence of a large number of variables. In the present application, in order to inspect possible dependencies of the fusion cross section on nuclear structure variables, we used 31 variables, described in Ref. [50], to represent colliding nuclei, their structure, and the collision energy of each given data point to model. Following the procedure discussed in detail in Ref. [50], we derived two models for the description of the fusion cross section between heavy ions, with different trade-offs between accuracy and complexity:

$$\sigma_{fus}^{(simple)}(E_{cm}) = 1103 \cdot \exp\left[-\left(1.387-0.468 \cdot \frac{Z_2 \cdot Z_1}{E_{cm}}\right)^2\right] \quad (1)$$

$$\sigma_{fus}^{(complex)}(E_{cm}) = 1116 \cdot$$
$$\cdot \exp\left[-\sinh^2\left(-1.359 + \mathrm{erf}\left(\frac{S_{2n}}{E_{cm}}\right) + 0.061 \cdot \frac{A_1 \cdot A_2}{E_{cm}}\right)\right] \quad (2)$$

The simplest model, eq. (1), is found to exploit exclusively 3 variables, namely the center-of-mass energy, and the charges of projectile and target. This is a very important fact, because it indicates that, given the presently published experimental data, E_{cm}, Z_1, and Z_2 suitably contain the whole informative content of the entire set of 31 variables required for their description. The most complex model, eq. (2), foresees 4 variables, namely E_{cm}, A_1, A_2, and the two-neutron separation energy of the compound system S_{2n}. In this case, a fourth variable is needed to slightly enhance the accuracy. It is found that further enhancing the complexity of the model does not help to further improve the accuracy in a statistically significant way.

The newly derived models help to improve the description of the entire set of fusion cross section data compared to former phenomenological models. Fig. 1 shows the results of our models (red lines) against some typical data, alongside the results of the phenomenological model from Ref. [39] (green line). Compared to the previously published model, our new models crucially improve the de-

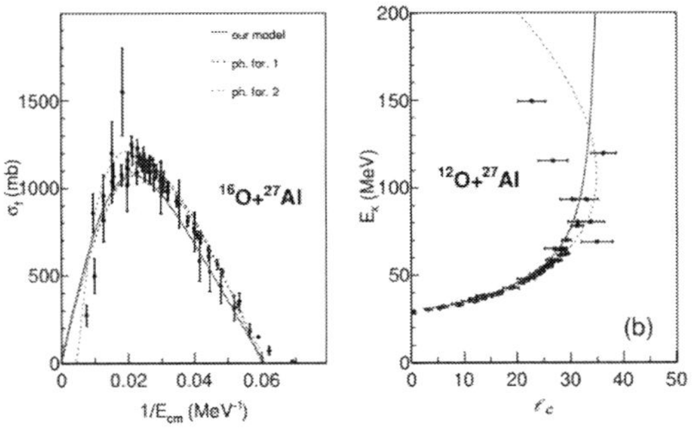

Figure 2. Phenomenological model descriptions of the $^{16}O+^{27}Al$ fusion cross section and critical angular momentum (red lines). For comparisons, analogous results obtained with models of Refs. [39] (blue) and [38] (green) are reported.

scription of the data in the vicinity of the Coulomb barrier. Close to the maximum of the cross section, our models achieve a reasonable description of data, comparably to the results of Ref. [39].

2.2 The improved SOD model

The artificial intelligence analysis described in the previous section helps to suggest the relevant variables required to describe the complete systematics of fusion cross section data, and to improve the dataset itself by identifying mislabelled data. Guided by these suggestions, and exploiting the improved dataset, we derived a phenomenological model starting from a sum-of-difference (SOD) approach.

In the SOD approach, the fusion cross section can be simplified as $\sigma_f \approx 2\pi \int_{\theta_f}^{\pi} \sigma_R(\theta) sin\theta d\theta$, which gives:

$$\sigma_f = \pi \left(\frac{\eta}{k}\right)^2 \cot^2 \frac{\theta_f}{2} = \frac{\pi}{k^2}(kD)^2 \left(1 - \frac{2\eta}{kD}\right) \tag{3}$$

where η is the Sommerfeld parameter, k the wave number, and D the distance of closest approach of the Coulomb trajectory linked to θ_f. By factorizing D as $D = d(\sqrt[3]{A_1} + \sqrt[3]{A_2})$, d being dependent on the energy and other characteristics of the collision, as discussed in Ref. [51], and introducing the quantity $F(\varepsilon) = \frac{kd}{A_{tot}}$, $\varepsilon \equiv \frac{E}{A_{tot}}$, one obtains:

$$\sigma_f = \frac{C_1}{E} F(\varepsilon) \left[C_2 F(\varepsilon) - \frac{C_3}{\sqrt{E}} \right] \tag{4}$$

where the constants C_1, C_2, C_3 depends only on the mass and atomic numbers of the reagents, as discussed in [50]. By inverting eq. (4), one finds an expression for kd as a function of σ_f. This allowed to build $(kd)_{exp}$ values for all data points in the dataset. The modeling procedure, described in detail in Ref. [51], consisted thus in finding the proper expression of kd to describe $(kd)_{exp}$ values for the entire dataset. We find the following expression for $\frac{kd}{A_{tot}}$:

$$\frac{kd}{A_{tot}} = F(\varepsilon, \Delta, n) = a \cdot f_A(\Delta, n) \cdot g_v(\varepsilon) \cdot h_C(\varepsilon) \tag{5}$$

where $f_A = 1 - \Delta^3 \cdot n$ is a factor related to the neutron richness of the total system formed by fusion, being $n \equiv N_{fin}/N_{tot}$, and $N_{fin} = 1.072 \cdot Z + 2.32 \cdot 10^{-3} Z^2$ [52]; $g_v(\varepsilon) = 1 - e^{-b\varepsilon}$ is a term related to friction-like forces occurring during the collision ($b = 2.26 \pm 0.01 MeV^{-1}$ [39]); and $h_C(\varepsilon) = 1 + \kappa_1 e^{-\kappa_2 \varepsilon}$ is used to reproduce the trend of data in the proximity of the Coulomb barrier. $\kappa_1 = 4.22 \pm 0.04$ and $\kappa_2 = 9.40 \pm 0.04$ are free parameters, whose value is determined from a fit to the experimental data.

The improved SOD model described in the present section achieves an unprecedented accuracy in describing previously published fusion cross section data. Fig. 2 (left panel) shows the results of the model (red line) compared to the experimental points for one of the collisions systems, alongside the results of the phenomenological models from Refs. [38, 39]. An interesting feature of the newly derived model is the presence of limiting values of the critical angular momentum ℓ_c, as visible in the right panel of Fig. 2, where we show the excitation energy versus angular momentum as derived from data in the *sharp cut-off* approximation [2, 53] (black dots) and from our model (red lines). This is a feature not commonly shown by phenomenological models (see, e.g. the dashed green line, obtained with the model of Ref. [38]). The present model leads to a saturation of ℓ_c values that is, at least for not too heavy systems ($A_{tot} < 90$), in reasonable agreement with the trend of experimental data and also with the prediction of the liquid drop model [29].

3 Conclusions and Perspectives

In this manuscript, we describe recent efforts to model the fusion cross section between heavy ions from the Coulomb barrier to the onset of multifragmentation phenomena.

Universal data-driven models, with different trade-offs between complexity and accuracy, have been derived using a novel hybridization of genetic programming and artificial neural networks, with advanced feature selection capabilities. The underlying analysis shows that a few variables allow to describe the entire dataset with analytically simple formulas.

A more refined model is derived with a SOD approach, exploiting the entire dataset. The new SOD model is capable to improve the state-of-the-art and shows saturation of the critical angular momentum in agreement with the liquid drop model.

The present analysis evidences the need for new accurate experiments in the regions II and III of the fusion cross section. These studies could benefit from the availability of modern arrays for charged particles being developed in the past decades [55–62].

References

[1] D. Glas, U. Mosel, Nucl. Phys. A **237**, 429 (1975)
[2] R. Bass, *Nuclear Reactions with heavy Ions*, 1st edn. (Springer-Verlag, Sunderland, MA, 1980), ISBN 978-3-540-09611-5
[3] S.J. Sanders, A.S.d. Toledo, C. Beck, Phys. Rep. **311**, 487 (1999)
[4] B.B. Back et al., Rev. Mod. Phys. **86**, 317 (2014)
[5] V. Jha, V.V. Parkar, S. Kailas, Phys. Rep. **845**, 1 (2020)
[6] P. Fröbrich, Phys. Rep. **116**, 337 (1984)
[7] S.M. Lee, T. Matsuse, A. Arima, Phys. Rev. Lett. **45**, 165 (1980)
[8] A. Pakou et al., Eur. Phys. J. A **51**, 55 (2015)
[9] M. Mazzocco et al., Phys. Rev. C **92**, 024615 (2015)
[10] L. Corradi, G. Pollarolo, S. Szilner, J. Phys. G.: Nucl. Part. Phys. **36**, 113101 (2009)
[11] W. von Oertzen, A. Vitturi, Rep. Prog. Phys. **64**, 1247 (2001)
[12] T. Mijatović, Frontiers in Physics **10**, 965198 (2022)
[13] J. Tõke, R. Bock, G.X. Dai, A. Gobbi, S. Gralla, K.D. Hildenbrand, J. Kuzminski, W.F.J. Müller, A. Olmi, H. Stelzer et al., Nucl. Phys. A **440**, 327 (1985)
[14] D.J. Hinde, D.Y. Jeung, E. Prasad, A. Wakhle, M. Dasgupta, M. Evers, D.H. Luong, R. du Rietz, C. Simenel, E.C. Simpson et al., Phys. Rev. C **97**, 024616 (2018)
[15] E.M. Kozulin et al., Eur. Phys. J. A **56**, 6 (2020)
[16] L.G. Moretto, R.P. Schmitt, Rep. Prog. Phys. **44**, 533 (1981)
[17] C. Beck, B. Djerroud, F. Haas, R.M. Freeman, A. Hachem, B. Heusch, A. Morsad, M. Youlal, Y. Abe, A. Dayras et al., Zeitschrift fur Physik A Hadrons and Nuclei **343**, 309 (1992)
[18] K. Hanold, L.G. Moretto, G.F. Peaslee, G.J. Wozniak, D.R. Bowman, M.F. Mohar, D.J. Morrissey, Phys. Rev. C **48**, 723 (1993)
[19] C. Gerschel, Nucl. Phys. A **387**, 297 (1982)
[20] V.V. Samarin, Phys. Atom. Nuc. **81**, 486 (2018)
[21] R. Bougault et al., Phys. Rev. C **97**, 024612 (2018)
[22] B. Borderie et al., Phys. Lett. B **782**, 291 (2018)
[23] E. De Filippo et al., Acta Phys. Pol. B **40**, 1199 (2009)
[24] F. Amorini et al., Phys. Rev. Lett. **102**, 112701 (2009)
[25] G. Cardella et al., Phys. Rev. C **85**, 064609 (2012)
[26] L. Manduci et al., Phys. Rev. C **94**, 044611 (2016)
[27] P. Eudes et al., Phys. Rev. C **90**, 034609 (2014)
[28] R. Giordano, S. Sambataro, A. Scalia, F. Porto, P, Figuera, S. Pirrone, Nuov. Cim. A **103**, 47 (1990)
[29] C. Beck, A. Szanto de Toledo, Phys. Rev. C **53**, 1989 (1996)
[30] T. Matsuse, A. Arima, S.M. Lee, Phys. Rev. C **26**, 2338 (1982)
[31] R. Bass, Phys. Rev. Lett. **39**, 265 (1977)
[32] P.G. Reinhard et al., Phys. Rev. C **93**, 044618 (2016)
[33] H. Zheng et al., Phys. Rev. C **98**, 024622 (2018)
[34] T. Maruyama et al., J. Nucl. Rad. Sci. **3**, 77 (2002)
[35] Z.Q. Feng, G.M. Jin, F.S. Zhang, Nucl. Phys. A **802**, 91 (2008)
[36] D. Horn, A.J. Ferguson, Phys. Rev. Lett. **41**, 1529 (1978)
[37] M. Lozano, G. Madurga, Phys. Lett. B **90**, 50 (1980)
[38] S. Kailas, S.K. Gupta, Z. Phys. A **302**, 355 (1981)
[39] F. Porto, S. Sambataro, Nuov. Cim. **83**, 339 (1984)
[40] A.V. Karpov et al., Nucl. Instrum. Meth. Phys. Res. A **859**, 112 (2017)
[41] V.I. Zagrebaev, A.S. Denikin, A.V. Karpov, A.P. Alekseev, M.A. Naumenko, V.A. Rachkov, V.V. Samarin, V.V. Saiko, *NRV web knowledge base on low-energy nuclear physics* (1999), `http://nrv.jinr.ru/`
[42] M. Russo, Swarm Evo. Comput. **27**, 145 (2016)
[43] M. Russo, Soft Comput. **24**, 16885 (2020)
[44] M. Russo, G. Leotta, P.M. Pugliatti, G. Gigliucci, Solar Energy **105**, 264 (2014)
[45] G. Campobello, D. Dell'Aquila, M. Russo, A. Segreto, Appl. Soft Comput. **94**, 106488 (2020)
[46] E. Buccheri, D. Dell'Aquila, M. Russo, Diabetes Res. Clin. Pract. **174**, 108722 (2021)
[47] E. Buccheri, D. Dell'Aquila, M. Russo, Obes. Med. **31**, 100398 (2022)
[48] J.R. Koza, *Genetic Programming: On the Programming of Computers by Natural Selection* (MIT Press, Cambridge, MA, USA, 1992)
[49] D. Dell'Aquila, M. Russo, Comput. Phys. Commun. **259**, 107667 (2021)
[50] D. Dell'Aquila, B. Gnoffo, I. Lombardo, F. Porto, M. Russo, Jour. Phys. G.: Nucl. Part. Phys. **535**, 88 (2023)
[51] D. Dell'Aquila, B. Gnoffo, I. Lombardo, F. Porto, Phys. Lett. B **837**, 137642 (2023)
[52] R. Charity, Phys. Rev. C **58**, 1073 (1998)
[53] D.A. Bromley, *Treatise on Heavy-Ion Science*, Vol. 2 (Plenum Press, 1984)
[54] S. Kailas, S.K. Gupta, Phys. Rev. C **34**, 357 (1986)
[55] D. Dell'Aquila et al., Nucl. Instr. Meth. Phys. Res. A **877**, 227 (2018)
[56] L. Acosta et al., J. Phys.: Conf. Ser. **730**, 012001 (2016)
[57] G. Pastore et al., Nucl. Instr. Meth. Phys. Res. A **860**, 42 (2017)
[58] S. Valdré et al., Nucl. Instr. Meth. Phys. Res. A **930**, 27 (2019)
[59] D. Dell'Aquila et al., Nucl. Instr. Meth. Phys. Res. A **929**, 162 (2019)
[60] D. Dell'Aquila et al., EPJ Web of Conf. **117**, 06011 (2016)
[61] J. Bishop et al., Phys. Rev. C **100**, 034320 (2019)
[62] G. Cardella et al., Nucl. Instr. Meth. Phys. Res. A **799**, 64 (2015)

https://doi.org/10.1051/epjconf/202329002018

A walk along the proton drip-line by β-decay spectroscopy

Sonja E. A. Orrigo[1,*], *Berta* Rubio[1], and *William* Gelletly[2]

[1]Instituto de Física Corpuscular, CSIC-Universidad de Valencia, E-46071 Valencia, Spain
[2]Department of Physics, University of Surrey, Guildford GU2 7XH, Surrey, UK

Abstract. During the last decade we have carried out a systematic study of the β decay of neutron-deficient nuclei, providing rich spectroscopic information of importance for both nuclear structure and nuclear astrophysics. We present an overview of the most relevant achievements, including the discovery of a new exotic decay mode in the *fp*-shell, the β-delayed γ-proton decay in ^{56}Zn, the first observation of the 2^+ isomer in ^{52}Co and the latest results on the heavier systems ^{60}Ge and ^{62}Ge. We also report on our deduced mass excesses in comparison with systematics and a recent measurement. Finally, we summarise our results on the half-lives of T_z = -1/2, -1 and -2 neutron-deficient nuclides, and analyse their trend.

1 Introduction

The investigation of nuclei close to the limits of nuclear stability, *exotic nuclei*, is at the frontier of modern nuclear physics. Their study is challenging but it is becoming more possible thanks to new-generation facilities for the production and acceleration of radioactive ion beams (RIBs). Neutron-deficient nuclei can be populated even up to the boundaries of nuclear stability, the *proton drip-line*, thus enabling one to perform detailed decay studies.

Decay-spectroscopy experiments provide us with structure information of paramount importance [1–6], such as the half-life of the unstable nucleus and the energies and branching ratios for β-delayed γ or particle emission, e.g. protons, α particles or neutrons, depending on whether the β-decaying nucleus is either neutron-deficient or neutron rich. This is important since it allows us to determine the energy levels populated in the daughter nucleus together with their β-feedings and thus reconstruct the partial decay scheme. From all this information one can finally determine the absolute values of the Fermi B(F) and Gamow-Teller B(GT) transition strengths and, in favourable cases, also deduce the mass excesses of the daughter and parent nuclei.

If isospin were a good quantum number in nuclear physics then mirror nuclei, where the number of neutrons and protons are exchanged, would be identical. In reality their properties, such as the structure of the levels, quantum numbers, half-lives, β-decay strengths, etc., are very similar but not identical. There can be differences which break the ideal isospin symmetry. Charge-exchange (CE) reactions are the mirror strong interaction process of β decay [7, 8]. The comparison between β-decay data and CE reactions carried out on the stable mirror target allows us to investigate fundamental questions related to the role of isospin in atomic nuclei.

Moreover, many exotic nuclei lie on the reaction pathways involved in various processes of nucleosynthesis, responsible for the production of the chemical elements in the Universe. Nuclear-structure properties such as half-lives, masses and β-strengths are of great significance for nuclear astrophysics. One such nucleosynthesis processes is the rapid proton-capture, *rp-process*, happening in explosive stellar environments [9]. It acts on the neutron-deficient side of the nuclear chart and produces many medium-mass/heavy proton-rich elements, passing through neutron-deficient nuclei in the *fp*-shell and above.

During the last decade we have performed a systematic study of neutron-deficient nuclei along or close to the proton drip-line by β-decay spectroscopy experiments with implanted RIBs. Figure 1 shows the nuclei studied, spanning the *fp*-shell and above: the nuclei marked with blue/yellow/purple circle where produced at the GSI/GANIL/RIKEN laboratories, respectively, while a blue or red circumference line indicates a value of -1 or -2 for the third component of the isospin quantum number, T_z. The textbox indicates the primary beam that was used to produce the secondary RIB of interest in the various laboratories, where in the case of GANIL a ^{58}Ni primary beam was used for the experiment focused on the production of the T_z = -2, ^{56}Zn nucleus [1, 2], while a ^{64}Zn beam was used for the experiment devoted to the study of the T_z = -1, ^{58}Zn nucleus [4]. The results from the GSI experiment are published in Ref. [10]. Results from the GANIL experiments are published in Refs. [1–5]. The most recent study of the β decay of ^{60}Ge and ^{62}Ge, performed at RIKEN, is reported in Ref. [6]. Results on ^{64}Se and ^{66}Se will be published soon.

In the following we focus especially on the GANIL and RIKEN experiments. The experimental setups and the results obtained are discussed in Sect. 2. Our results on the mass excesses are summarised in Sect. 3 and compared with the mass evaluation systematics and a recent

*e-mail: Sonja.Orrigo@ific.uv.es

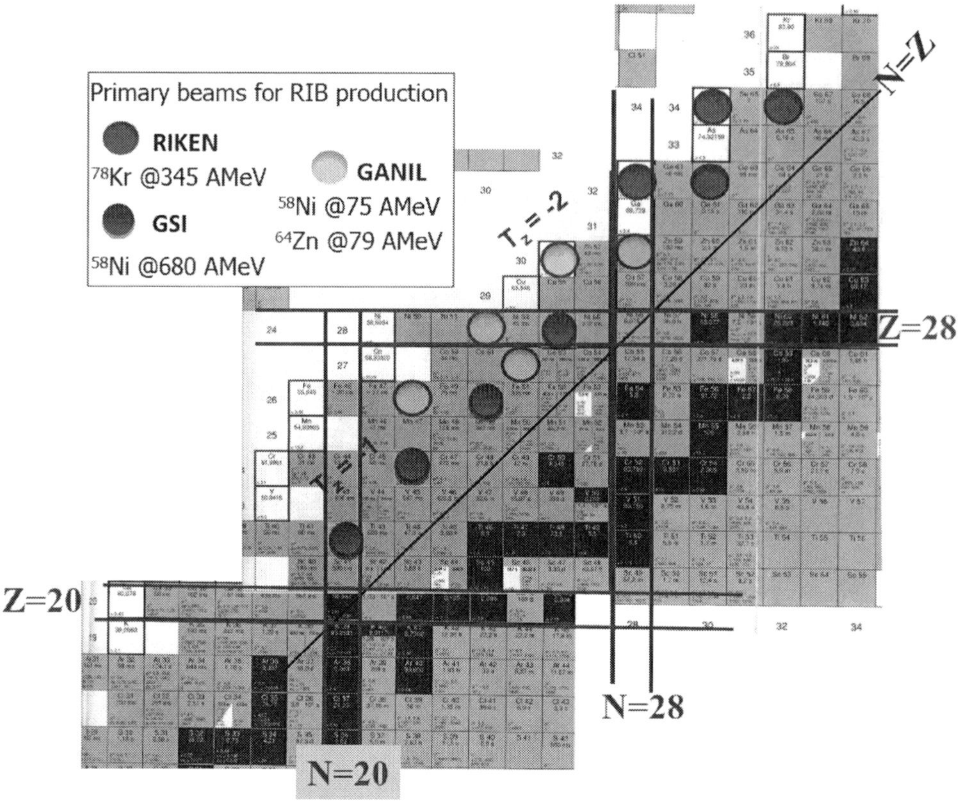

Figure 1. Summary of the β-decay spectroscopy experiments carried out along the proton drip-line, in the *fp*-shell and above [1–6, 10]. Neutron-deficient nuclei produced at the GSI/GANIL/RIKEN laboratories are marked by a blue/yellow/purple circle, where a blue or red circumference line indicates a value of T_z = -1 or -2. The primary beams used in the different laboratories and their energies are reported in the textbox.

Table 1. Nuclei produced at GANIL and RIKEN: reference, isotope of interest, its T_z value and the experimental results for the number of implants N_{imp} and half-life $T_{1/2}$. The last two columns specify the laboratory and primary beam used for their production.

Ref.	Isotope	T_z	N_{imp}	$T_{1/2}$(ms)	Laboratory	Beam, energy(AMeV)
[1]	^{56}Zn	-2	8.9×10^3	32.9(8)	GANIL	^{58}Ni, 75
[2]	^{48}Fe	-2	5.0×10^4	51(3)		
[2]	^{52}Ni	-2	5.3×10^5	42.8(3)		
[3]	^{52}Co^m	-1	2.3×10^5 $^{(a)}$	102(6)		
[3]	^{52}Co$_{g.s.}$	-1	4.1×10^6 $^{(b)}$	112(3)		
[4]	^{58}Zn	-1	1.8×10^5	86(2)		^{64}Zn, 79
[6]	^{62}Ge	-1	2.1×10^6	73.5(1)	RIKEN	^{78}Kr, 345
[6]	^{60}Ge	-2	1.5×10^4	25.0(3)		

$^{(a)}$ Estimated based on production as a by-product of ^{52}Ni implantation [2, 3].
$^{(b)}$ Calculated as 83% of the total ^{52}Co implants (5.0×10^6) including both g.s. and isomeric state [3].

measurement [11]. Finally, half-life trends for T_z = -1/2, -1 and -2 nuclei, including all the newly measured values, are presented in Sect. 4.

2 β-decay spectroscopy experiments and results

As shown in figure 1, our β-decay spectroscopy experiments were performed worldwide at different RIB facilities, but conceptually they are the same. A primary beam is accelerated and fragmented on a thick target: for example beams of ^{58}Ni or ^{64}Zn are fragmented on a natural Ni target at GANIL, while ^{78}Kr is fragmented on a Be target at RIKEN. The fragments are then selected by the LISE3 separator [12] at GANIL, or by the BigRIPS separator [13] at RIKEN. Thereafter they are implanted into Double-Sided Silicon Strip Detectors (DSSSD), used to detect both the implanted heavy ions and any subsequent charged-particle decay: a 300-μm thick DSSSD at GANIL, or the WAS3ABi setup [14] comprising three 1-mm thick DSSSDs at RIKEN. Finally, arrays of high-purity Ge detectors surround the DSSSD setup to detect the β-delayed γ-rays: four EXOGAM clovers [15] at GANIL, or the EURICA array [16] consisting of 12 clusters at RIKEN. More details on the setups, the procedures employed for particle identification and data analysis are available in Refs. [2, 6].

Table 1 shows the number of implanted ions (N_{imp}) together with their measured half-lives $T_{1/2}$ and details, such as the laboratory and primary beam, of their production. The unprecedented statistics available at the RIKEN Nishina Center allowed us to extend the systematic exploration of neutron-deficient nuclei to higher masses, along the proton drip-line.

Several nuclei have been studied in our experimental campaign. We have measured β-delayed proton and γ emissions and the related branching ratios. Decay schemes and absolute B(F) and B(GT) strengths have been determined. For some of the cases the mass excesses have also been deduced. For some of the $T_z = -2$ nuclides under study [1, 2, 5] it was possible to enrich the β-decay data by comparison with complementary (^{3}He,t) CE reactions performed on their stable mirror target at RCNP Osaka [8, 17, 18].

Some common features emerge when looking at nuclei with the same T_z value. The decay of the $T_z = -2$ nuclei proceeds by both β-delayed proton emission and β-delayed γ de-excitation. An exotic feature that we have observed in all the $T_z = -2$ systems studied (^{56}Zn, ^{48}Fe, ^{52}Ni and ^{60}Ge) is the competition between the γ de-excitation and the proton emission from the $T = 2$, 0^+ isobaric analogue state (IAS) populated in the daughter nucleus [2, 19]. The β-delayed proton emission from the IAS is isospin-forbidden, but it is observed and this is attributed to a $T = 1$ isospin impurity in the IAS wave function. In the cases of ^{48}Fe and ^{52}Ni the β-proton component constitutes only 14% and 25% of the respective IAS total decays. A reason for this is the relatively low energy of the protons emitted by the daughter nuclei, corresponding to calculated proton half-lives of the same order-of-magnitude as the γ-decay Weisskopf transition probabilities [2].

In the decay of ^{56}Zn (shown in figure 2), the β-proton component de-exciting the IAS is 44% of the total, i.e., much larger. The ^{56}Zn daughter, ^{56}Cu, has another 0^+ state within 100 keV of the IAS and, since the mixing depends on how close the two states are, the strong isospin mixing of 33% favours proton decay [1]. At this point one would expect that the much faster proton decay ($t_{1/2} \sim 10^{-18}$ s) from the ^{56}Cu IAS should dominate over the γ de-excitation ($t_{1/2} \sim 10^{-14}$ s in the mirror). However the competing γ decay from the IAS is still observed, being 56% of the total. The reason for such behaviour lies in the nuclear structure. Two independent shell model calculations [20, 21] found that the proton decay of the $T = 1$ IAS component is hindered by a factor of 10^3. Finally, in the case of ^{60}Ge the β-delayed proton emission from the IAS is estimated to be within 74.5% - 95% of the total [6] which may again be due to nuclear structure.

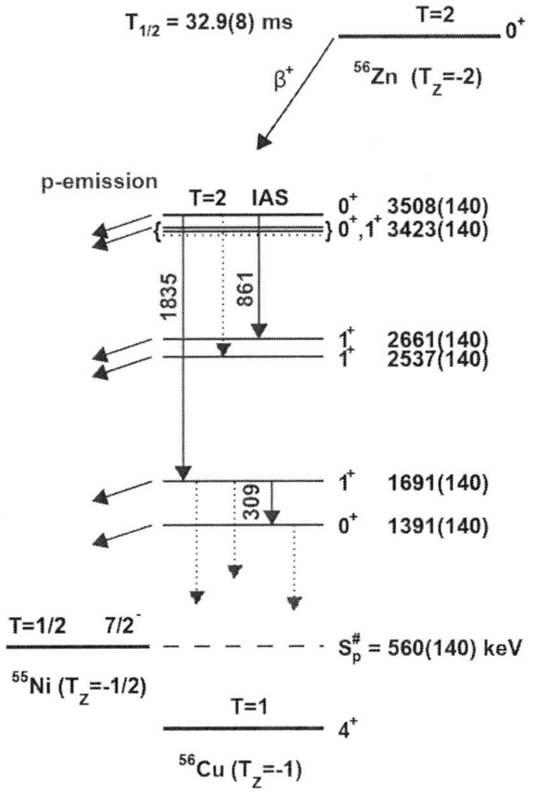

Figure 2. Partial decay scheme of ^{56}Zn (reprinted without changes from Ref. [5] under CC BY 3.0). Transitions corresponding to those observed in the mirror ^{56}Co nucleus are represented by dotted lines.

^{56}Zn, lying at the proton drip-line, has been one of the most intriguing cases since it presents many interesting and unusual features [1]. As mentioned above, the IAS is divided between two mixed levels. These levels are both fed by the Fermi transition, so that the Fermi strength is shared between them. Moreover, in ^{56}Zn we have discovered a new exotic decay mode in the fp-shell: the β-delayed γ-proton decay. This exotic decay pattern is possible because in ^{56}Zn the β-delayed γ rays populate levels in the daughter nucleus which are located above the proton separation energy and hence are unbound. The levels then decay by proton emission to the ground state (g.s.) of ^{55}Ni. Therefore the sequence is β-delayed γ-proton decay and we have observed three such sequences. The comparison between the β decay of ^{56}Zn to ^{56}Cu [1] and the mirror CE process, the ^{56}Fe(^{3}He,t)^{56}Co reaction [17], shows a re-

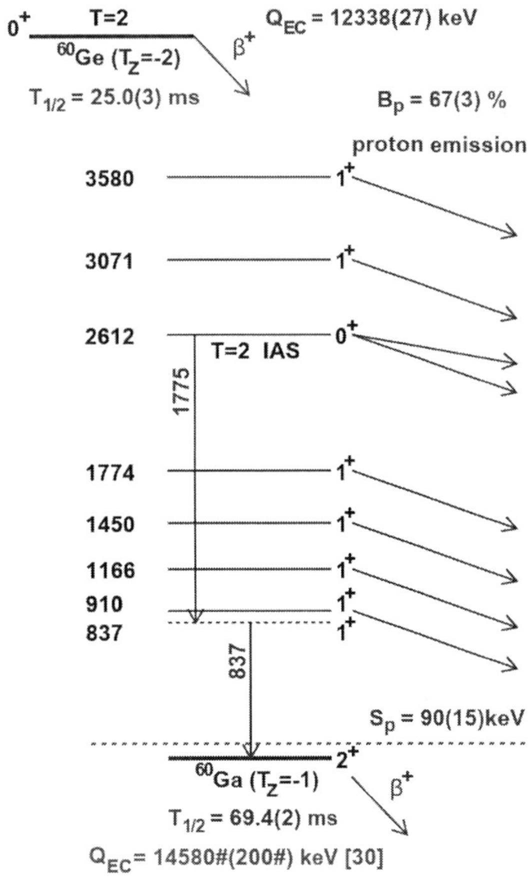

Figure 3. Partial decay scheme of ^{60}Ge. The transitions populating levels in ^{60}Ga are shown. Observed β-delayed protons and β-delayed γ rays are indicated [6].

markable isospin symmetry: all the dominant transitions are observed in both cases and with very similar strengths. Such a comparison allowed us to clarify some aspects of the level structure in ^{56}Cu which would have remained unclear otherwise.

^{60}Ge is another fascinating case. Its decay was almost unknown before the present experiment, where we obtained the first experimental information on both the β-delayed proton and γ emissions, reconstructing a complex decay scheme involving five different nuclei [6]. The partial decay scheme showing the transitions populating energy levels in the daughter nucleus ^{60}Ga, which were unknown before, is reported in figure 3. ^{60}Ga lies right at the proton drip-line, thus its structural properties are of relevance for the rp-process [11].

A common feature that we have found in the T_z = -1 nuclei [6, 10] is the suppression of isoscalar γ transitions between $J^\pi = 1^+$, $T = 0$ states (Warburton and Weneser *quasi-rule* [22, 23]). We have verified experimentally that it also holds in the heavier ^{62}Ge nucleus. In addition, we did not find evidence of enhanced low-lying Gamow-Teller strength in ^{62}Ga due to isoscalar proton-neutron pairing, confirming the findings of a previous measurement [24].

Finally, among the many results we emphasise the first observation of the 2^+ isomer in ^{52}Co [3]. There had been speculation that such an isomeric state exists but it was not observed earlier because, when one attempted to populate ^{52}Co directly, both the g.s. and isomeric state of very similar half-lives are produced and implanted together. We have succeeded disentangling the two decays, measuring for the first time the isomer half-life [102(6) ms] and the g.s. half-life without contamination [112(3) ms]. The trick was to look at the implantation of ^{52}Ni, whose decay process directly populates the $0^+, T = 2$ IAS in ^{52}Co. The latter then de-excites by γ-ray emission populating in a selective way the 2^+ isomeric state.

3 Mass excesses

The mass excesses of the β-decaying neutron-deficient nucleus and its daughter nucleus are important to determine key quantities such as the Q_β value of the decay, needed to deduce the B(F) and B(GT) strengths. The proton separation energy S_p in the β-daughter nucleus can be determined by knowing its mass excess and that of the β-proton daughter nucleus.

There is a lack of mass measurements in this mass region. When the mass excesses are not known, the atomic mass evaluation (AME) systematics can be used. Another option is to determine the mass excess of the nucleus of interest from the Isobaric Multiplet Mass Equation (IMME) [25–27]:

$$\Delta m(\alpha, T, T_z) = a + b\,T_z + c\,T_z^2 . \qquad (1)$$

This can be done when at least three other members of the isospin multiplet are known. In equation 1, T_z is the third component of the isospin T and α stands for all the other quantum numbers. From our β-decay data we have determined the g.s. mass excesses of the T_z = -2 nuclei ^{48}Fe, ^{52}Ni and ^{56}Zn in Ref. [2] and ^{60}Ge in Ref. [6] from the IMME, knowing four members of each quintuplet. We have also deduced the g.s. mass excesses of ^{62}Ge and ^{60}Ga, and S_p in ^{60}Ga [6]. Our deduced mass excesses (red filled circles and diamonds) are shown in figure 4 and compared with the values obtained from the 2003 [28], 2012 [29] and 2020 [30] AME systematics. In the figure open and filled symbols represent the measured and deduced values in the AME systematics.

We have observed since our first study of ^{56}Zn [1] that, for proton-rich nuclei in this region of the mass chart, the AMEs published subsequent to that in 2003 are in poorer agreement for the mass excess in comparison with our IMME or deduced values. After us, other authors reported similar issues [11, 31]. As shown in figure 4, the values from the 2003 AME lie closer to our estimates than the values from the 2012 AME. The values from the 2016 AME [32] (not shown in the figure) have a very similar behaviour. The 2020 AME also behaves similarly for the unmeasured nuclei, but includes new measured values for

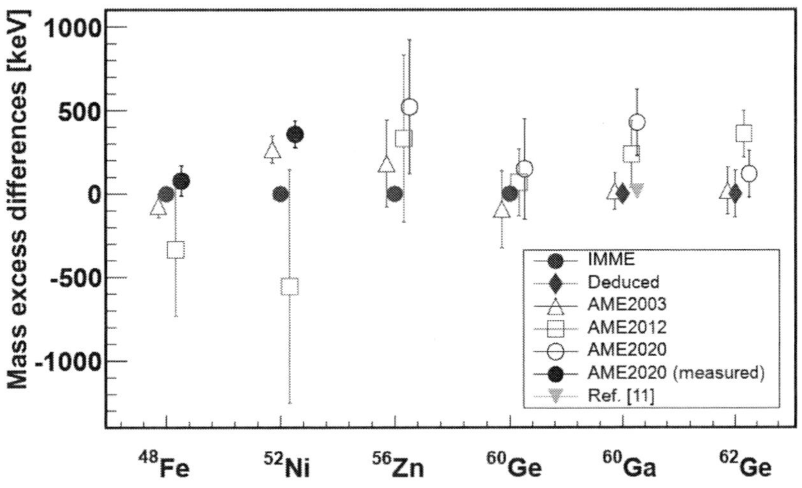

Figure 4. Mass excesses of ^{48}Fe, ^{52}Ni, ^{56}Zn, ^{60}Ge, ^{60}Ga and ^{62}Ge. The difference is shown between the values we obtain (red filled circles and diamond) [2, 6] and values from the 2003, 2012 and 2020 AMEs [28–30]. Open symbols represent values from systematics, while filled symbols are experimental or deduced values. The cyan filled triangle is the measurement from Ref. [11]. The data points belonging to each nucleus are slightly displaced to show the error bars better.

^{48}Fe and ^{52}Ni (black filled circles). The AME systematic evaluation does not include isobaric multiplets because the IAS might be mixed and thus its energy might deviate from the IMME formula. An IAS-mixed case is ^{56}Zn [1]. We think that, with caution, this knowledge can help to calculate extrapolated values. New mass measurements in this region are important to better constrain the future AME.

A recent measurement of the mass excess in ^{60}Ga is also shown in figure 4 (cyan filled triangle) [11], which is in excellent agreement with our indirect determination from the β-decay data [6]. Paul *et al.* have also determined S_p= 78(30) keV in ^{60}Ga, in agreement with our value of 90(15) keV [6]. The 2020 AME systematic value is -340(200) keV [30]. By combining the two experimental values, $S_p(^{60}\mathrm{Ga})$ = 88(18) keV is obtained, establishing the proton-bound nature of ^{60}Ga. This value, together with the fact that ^{59}Ga was not observed in fragmentation reactions at NSCL, provides strong evidence that ^{60}Ga is the last proton-bound gallium isotope [11]. ^{59}Ga was also not observed in our recent experiment at RIKEN, strengthening the conclusion that ^{60}Ga marks the location of the proton drip-line for Z = 31.

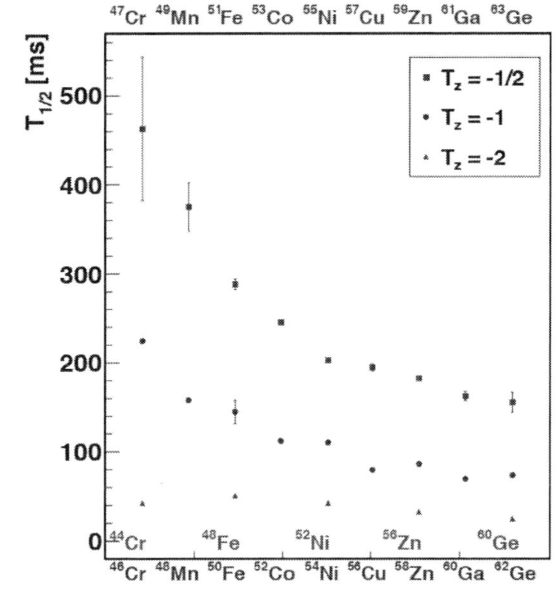

Figure 5. $T_{1/2}$ values measured for the T_z = -1/2 (red squares), -1 (blue dots) and -2 (green triangles) nuclei as a function of atomic number.

4 Half-life trends

In the present Section we summarise the results concerning the half-lives of the nuclei studied and analyse their trends. Figure 5 represents all the measured half-lives as a function of the atomic number Z and for the T_z = -1/2, -1 and -2 nuclei. It is an updated version of figure 7 from Ref. [4], where we have included/updated the half-lives of the following nuclides: ^{56}Zn [1]; ^{48}Fe and ^{52}Ni [2]; ^{52}Co [3]; ^{60}Ge, ^{60}Ga, ^{62}Ge and ^{59}Zn [6] and ^{44}Cr [19].

Three curves are obtained, corresponding to the different T_z values, because the B(F) value driving the decay is the same for all the nuclei with the same T_z. As discussed in Ref. [4], the systematic decrease of the $T_{1/2}$ values with the mass, seen in the T_z = -1/2 nuclei, reflects the increase in the Q_β value. A similar decreasing pattern is observed in the half-lives of the T_z = -1 nuclei and, on top of this behaviour, a typical odd-odd and even-even effect is found. The latter is due to the fact that in the odd-odd nuclei there exist other excited states below the IAS that receive a significant amount of β feeding, which makes their half-lives slightly shorter in comparison with their even-even neighbours [4]. In the T_z = -2 nuclei, which are all even-even, a smoother decreasing trend is observed. The half-life of the most exotic T_z = -2 nucleus, ^{60}Ge, is only 25.0(3) ms.

5 Conclusions

We have given an overview of the most relevant achievements from our β-decay spectroscopy experiments at GANIL and RIKEN. Detailed spectroscopic information has been obtained for several neutron-deficient nuclei, starting from lighter to heavier systems along the proton drip-line. Half-lives, decay schemes, β-decay transition strengths, mass excesses have been determined, many of them for the first time. These results are relevant for both nuclear structure and nuclear astrophysics.

Our deduced mass excesses have been compared with different AME systematics, indicating the need for more mass measurement in this region of the nuclear chart. The half-life values as a function of the mass number have been analysed, providing a comprehensive understanding of the half-life trends in terms of the Fermi strength and the Q_β value.

We have shown that valuable spectroscopic information can be obtained from this kind of experiment, thus improving our knowledge of the properties of neutron-deficient nuclei.

Acknowledgments. This work was supported by the Generalitat Valenciana Grant No. PROMETEO/2019/007, Spanish Grants No. PID2019-104714GB-C21, FPA2017-83946-C2-1-P and FPA2014-52823-C2-1-P (MCIN,MINECO/AEI/FEDER), Centro de Excelencia Severo Ochoa del IFIC SEV-2014-0398; *Junta para la Ampliación de Estudios* Programme (CSIC JAE-Doc) co-financed by FSE.

References

[1] S. E. A. Orrigo *et al.*, Phys. Rev. Lett. **112**, 222501 (2014)
[2] S. E. A. Orrigo *et al.*, Phys. Rev. C **93**, 044336 (2016)
[3] S. E. A. Orrigo *et al.*, Phys. Rev. C **94**, 044315 (2016)
[4] L. Kucuk, S. E. A. Orrigo *et al.*, Eur. Phys. J. A **53**, 134 (2017)
[5] S. E. A. Orrigo *et al.*, J. Phys.: Conf. Ser. **1056**, 012041 (2018)
[6] S. E. A. Orrigo *et al.*, Phys. Rev. C **103**, 014324 (2021)
[7] Y. Fujita *et al.* , Phys. Rev. Lett. **95**, 212501 (2005)
[8] Y. Fujita, B. Rubio, W. Gelletly, Prog. Part. Nucl. Phys. **66**, 549 (2011)
[9] H. Schatz *et al.*, Phys. Rep. **294**, 167 (1998)
[10] F. Molina *et al.*, Phys. Rev. C **91**, 014301 (2015)
[11] S. F. Paul *et al.*, Phys. Rev. C **104**, 065803 (2021)
[12] R. Anne, A. C. Mueller, Nucl. Instrum. Methods B **70**, 276 (1992)
[13] N. Fukuda *et al.*, Nucl. Instrum. Methods B **317**, 323 (2013)
[14] S. Nishimura, Progr. Theor. Exp. Phys. **2012**, 03C006 (2012)
[15] J. Simpson *et al.*, Acta Phys. Hung. A **11**, 159 (2000)
[16] P.-A. Söderström *et al.*, Nucl. Instrum. Methods B **317**, 649 (2013)
[17] H. Fujita *et al.*, Phys. Rev. C **88**, 054329 (2013)
[18] E. Ganioğlu *et al.*, Phys. Rev. C **93**, 064326 (2016)
[19] C. Dossat *et al.*, Nucl. Phys. A **792**, 18 (2007)
[20] B. Rubio *et al.*, Nucl. Phys. Review **33**, 225 (2016)
[21] N. A. Smirnova *et al.*, Phys. Rev. C **93**, 044305 (2016)
[22] G. Morpurgo, Phys. Rev. **110**, 721 (1958)
[23] D. H. Wilkinson, *Isospin in Nuclear Physics* (1969)
[24] E. Grodner *et al.*, Phys. Rev. Lett. **113**, 092501 (2014)
[25] E. Wigner, *Proc. Robert A. Welch Found. Conf.* (Robert A. Welch Foundation, Houston, Texas, 1957) 67
[26] W. Benenson, E. Kashy, Rev. Mod. Phys. **51**, 527 (1979)
[27] M. MacCormick, G. Audi, Nucl. Phys. A **925**, 61 (2014)
[28] G. Audi *et al.*, Nucl. Phys. A **729**, 3 (2003)
[29] G. Audi *et al.*, Chin. Phys. C **36**, 1157 (2012)
[30] M. Wang *et al.*, Chin. Phys. C **45**, 030003 (2021)
[31] M. Del Santo *et al.*, Phys. Lett. B **738**, 453 (2014)
[32] M. Wang *et al.*, Chin. Phys. C **41**, 030003 (2017)

Core-valence absorption in breakup and stripping reactions and its isospin dependence

Mario Gómez-Ramos[1,*], *Joaquín* Gómez-Camacho[1,2,], and *Antonio Matías* Moro Muñoz[1,3,]

[1]Departamento de Física Atómica, Molecular y Nuclear, Facultad de Física, Universidad de Sevilla, Apartado 1065, E-41080 Sevilla, Spain
[2]Centro Nacional de Aceleradores (U. Sevilla, J. Andalucía, CSIC), Tomás Alva Edison, 7, 41092 Sevilla, Spain
[3]Instituto Interuniversitario Carlos I de Física Teórica y Computacional (iC1), Apdo. 1065, E-41080 Sevilla, Spain

Abstract. In this work, the effect of nucleon-core absorption on nucleon removal reaction is explored through the use of complex nucleon-core interactions. Results are presented for exclusive breakup reactions, where absorption is explored through the use of a binormal basis in the continuum-discretized coupled-channel method, and for stripping nucleon knockout reactions, where absorption is considered through the application of an effective density in the eikonal approximation. Both methods show an increased effect of nucleon-core absorption when removing a deeply-bound nucleon, which leads to smaller cross sections, a reduction that is larger than in the weakly-bound case.

1 Introduction

Nucleon-removal reactions with medium-mass targets (^{9}Be, ^{12}C) have a long and successful history in the study of the single-particle properties of nuclei [1, 2]. In this kind of reactions, a projectile P is made to collide with a target T so that a nucleon N is removed from the projectile, and a residual core C is detected. If the nucleon N and the target T (in its ground state) are detected, $P(N+C)+T \rightarrow N+C+T$, the reaction corresponds to elastic (sometimes called diffractive) breakup, while the reaction process is called stripping if the nucleon is absorbed by the target and only the core is detected $P(N+C)+T \rightarrow C+X$. Elastic breakup is the main nucleon-removal process for the removal of halo and weakly-bound nucleons, while for more deeply-bound nucleons the importance of stripping increases with the binding energy of the removed nucleon, being the dominant contribution for removal from stable nuclei and for deeply-bound nucleons.

For the study of elastic breakup for weakly-bound nuclei, the state-of-the-art method is the Continuum-Discretized Coupled Channels method (CDCC) [3] while stripping reactions, which are usually measured at an energy of $\sim$ 100 MeV per nucleon, are usually analyzed using an eikonal sudden description [4]. Both methods, in their standard form, require a real interaction between the nucleon and the core in the final state, in order to ensure the properties of orthogonality and closure for the nucleon-core eigenstates. However, the final-state-interaction between removed nucleon and core can lead to excitation and breakup of the latter, leading to a loss of flux (and therefore cross section) that is usually described by an imaginary part in the interaction, so an extension in these methods to consider complex interactions is required to describe this processes.

In this work we describe briefly two such extensions for the CDCC and eikonal methods and present results for cross sections corresponding to the removal of weakly- and deeply-bound nucleons.

2 Elastic breakup

2.1 Theoretical framework: CDCC

In the CDCC method, the full three-body wavefunction for the $C+N+T$ system $\Psi^{(+)}$ is expanded in eigenstates of the $C+N$ Hamiltonian ϕ_j, with a discretization procedure (usually a binning procedure) to express the infinite continuum eigenstates in a finite discrete basis [3].

$$\Psi(\mathbf{R},\mathbf{r}) = \sum_j \chi_j(\mathbf{R})\phi_j(\mathbf{r}), \tag{1}$$

where $\mathbf{R}$ is the relative coordinate between projectile and target, $\mathbf{r}$ is the one between nucleon and core and $\chi_j(\mathbf{R})$ is the coefficient of state $\phi_j(\mathbf{r})$ in $\Psi(\mathbf{R},\mathbf{r})$. Provided the eigenstates ϕ_j are orthogonal, one can obtain from the Schrödinger equation a set of coupled equations for $\chi_j(\mathbf{R})$,

$$\begin{aligned}&\sum_j \left((T_R - E_i)\left\langle\phi_i|\phi_j\right\rangle + \left\langle\phi_i|U_{CT}+U_{NT}|\phi_j\right\rangle\right)\chi_j(\mathbf{R}) = \\ &\sum_j \left((T_R - E_i)\delta_{ij} + U_{ij}\right)\chi_j(\mathbf{R}) = 0,\end{aligned} \tag{2}$$

where $U_{ij} = \left\langle\phi_i|U_{CT}+U_{NT}|\phi_j\right\rangle$ is the coupling potential and U_{CT}, U_{NT} are the core-target and nucleon-target

*e-mail: mgomez40@us.es

https://doi.org/10.1051/epjconf/202329002019

interactions respectively. This method relies on the orthogonality of states $\phi_{i,j}$, which is no longer fulfilled when the nucleon-core interaction is complex and energy-dependent, as required to describe nucleon-core absorption. This can fortunately be solved by the introduction of the binormal basis $\tilde{\varphi}_j$, which, for a set of non-orthogonal states φ_j verifies $\langle\tilde{\varphi}_i|\varphi_j\rangle = \langle\varphi_i|\tilde{\varphi}_j\rangle = \delta_{ij}$ and for a finite basis can be computed analytically [5]. The use of the binormal basis leads to a set of coupled equations analogous to the standard CDCC which can be solved through standard methods, where the only change is that the coupling potential is substituted by $\tilde{U}_{ij} = \langle\tilde{\varphi}_i|U_{CT} + U_{NT}|\varphi_j\rangle$.

$$\sum_j \Big((T_R - E_i)\langle\tilde{\varphi}_i|\varphi_j\rangle + \langle\tilde{\varphi}_i|U_{CT} + U_{NT}|\varphi_j\rangle\Big)\chi_j(\mathbf{R}) = \sum_j \Big((T_R - E_i)\delta_{ij} + \tilde{U}_{ij}\Big)\chi_j(\mathbf{R}) = 0. \quad (3)$$

It should be noted that the introduction of complex potentials doubles the basis of eigenstates, as states with outgoing $\varphi_i^{(+)}$ and incoming $\varphi_i^{(-)}$ boundary conditions are no longer equivalent. For the description of knockout reactions the states with incoming boundary conditions $\varphi_i^{(-)}$ is the adequate basis, which may come as a surprise, as they have larger-than-one norms for absorptive potentials. However, we note that for weakly-energy-dependent potentials one can verify that $\tilde{\varphi}_i^{(-)} \simeq \varphi_i^{(+)*}$ [6], as is shown in Fig. 1 for a reference bin, so the norm for $\tilde{\varphi}_i^{(-)}$ is smaller than unity and the DWBA matrix element $\tilde{U}_{i0} = \langle\tilde{\varphi}_i^{(-)}|U_{CT} + U_{NT}|\varphi_{gs}\rangle$ between bound and continuum states leads to smaller cross sections, as expected of absorptive processes.

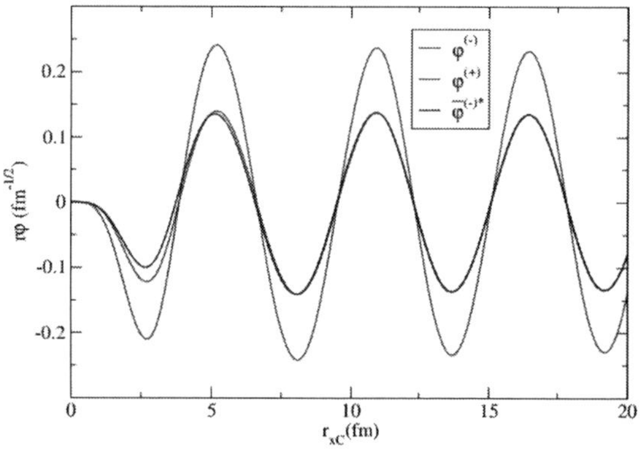

Figure 1. Real part for a bin computed for an absorptive potential. The red and green curves correspond to bins with incoming and outgoing boundary conditions respectively. The blue curve corresponds to the conjugate of the binormal of the state with incoming boundary conditions.

2.2 Results

We have applied this formalism to the neutron removal reaction with a ^{12}C target at 70 MeV per nucleon for the nuclei ^{11}Be (weakly-bound nucleon) and ^{41}Ca (deeply-bound nucleon). For the interaction between neutron and core, the Köning-Delaroche[7] interaction was used for ^{41}Ca, while for ^{11}Be the real part was taken from [8] and the imaginary part was adjusted to reproduce the systematics from [9]. Further details on the interactions and calculations can be found in [10]. The results of the calculations are presented in Fig. 2.

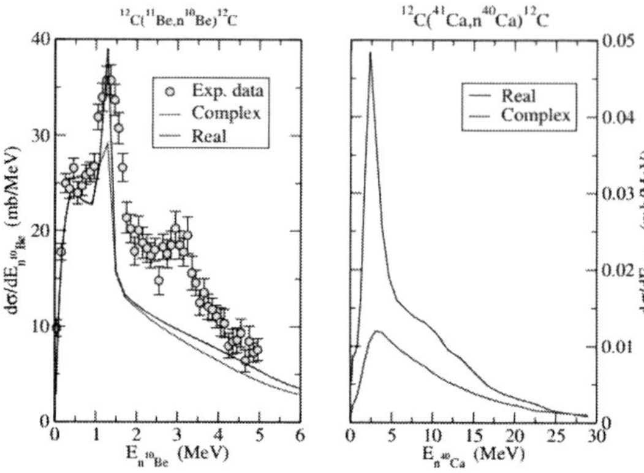

Figure 2. Energy distribution for the elastic breakup of a neutron for ^{11}Be (left panel) and ^{41}Ca (right panel) at 70 MeV per nucleon. The black curve corresponds to standard CDCC calculations while the red curve corresponds to CDCC extended to include neutron-core absorption. For ^{11}Be, experimental data from [11] are presented.

As can be seen in the figure, the reduction due to the absorption is moderate in the weakly-bound ^{11}Be case, reducing the overall cross section by a factor of ~10% and affecting mostly the contribution of the $5/2^+$ resonance while for the more deeply-bound ^{41}Ca nucleus the reduction is much more significant, of ~50%. It should be noted that in [12] it was found that the cross section for elastic neutron breakup in ^{11}Be has a significant contribution associated to the excitation of the ^{10}Be core, which is not considered in this calculation. Therefore it is not surprising that the calculations do not describe well the experimental data, particularly at higher excitation energies. The larger reduction in ^{41}Ca can be understood as the energies explored when removing the more deeply-bound neutron are larger, as seen in Fig. 2, so the imaginary potential is larger at these higher energies. Also, for the weakly-bound neutron the number of open channels is smaller, particularly at low energies, so the loss of flux, and hence the imaginary potential. is smaller.

3 Stripping reactions

3.1 Theoretical framework: Eikonal model

For the description of nucleon removal reactions at intermediate beam energies the standard description is an eikonal one [4], where the stripping part of the cross section can be computed through a simple formula

$$\sigma_{\rm str} = \int d\mathbf{b} \int d\mathbf{b_{NC}} \rho(\mathbf{b_{NC}}) |S_{CT}(\mathbf{b_{CT}})|^2 (1 - |S_{NT}(\mathbf{b_{NT}})|^2), \quad (4)$$

where S_{NT} and S_{CT} are the nucleon-target and core-target S-matrices respectively and $\mathbf{b_{NT}}$ and $\mathbf{b_{VT}}$ the corresponding impact parameters, with $\mathbf{b_{NC}}$ and $\mathbf{b}$ being the nucleon-core and projectile-target impact parameters. $\rho(\mathbf{b_{NC}})$ is a density function which in the standard eikonal model can be obtained as $\rho(\mathbf{b_{NC}}) = |\phi_{gs}(\mathbf{b_{NC}})|^2$, where ϕ_{gs} is the bound state nucleon-core wavefunction. This approximation of the density results from the application of closure over the continuum eigenstates of the V_{NC} interaction, which is no longer applicable when considering absorptive complex interactions between core and nucleon. Therefore, it is necessary to modify the formulas to include these effects. This modification results in a non-local density, which can be approximated to a local form by a judicious choice of the average local values of the coordinates:

$$\rho(\mathbf{r_1}, \mathbf{r_2}) = \phi^*_{gs}(\mathbf{r_1})\phi_{gs}(\mathbf{r_2}) \int d\mathbf{k} \varphi^{(+)}_{NC}(\mathbf{k}, \mathbf{r_1}) \varphi^{*(+)}_{NC}(\mathbf{k}, \mathbf{r_2}) \quad (5)$$

$$\rho^{\rm eff}(x, y) = \int d\mathbf{r_1} d\mathbf{r_2} \delta(x - \frac{x_1 + x_2}{2}) \delta(y - \sqrt{\frac{y_1^2 + y_2^2}{2}}) \rho(\mathbf{r_1}, \mathbf{r_2}), \quad (6)$$

where $\varphi^{(+)}_{NC}$ are the continuum eigenstates of the complex U_{NC} interaction, x is the projection of $\mathbf{b_{NC}}$ in the direction of $\mathbf{b}$ and y the perpendicular component of $\mathbf{b_{NC}}$. This effective density $\rho^{\rm eff}$ can be used in Eq. (4) to obtain stripping cross sections with the effects of nucleon-core absorption. More information can be found in [13].

3.2 Results

As an application of this formalism, we have considered a number of the reactions presented in [14], namely neutron and proton removal from ^{40}Si, ^{12}C and ^{24}Si, where the S-matrices and real bound-state potential have been taken from the original references [1, 15, 16] and the absorptive interaction has been modelled as the imaginary part of the global dispersive potential by Morillon *et al.* [17] with a null real part, as the real part is expected to have a reduced effect in absorption processes. More details can be found in [13]. In Fig. 3 the effective density as a function of x is presented for the neutron and proton in ^{40}Si in the f-wave and d-wave respectively. Note the difference in binding energies (4.72 and 23.1 MeV for neutron and proton respectively).

In the figure the green curve corresponds to the standard eikonal calculation. In order to test the effective density, we performed a calculation using no nucleon-core interaction, which is presented in red, and which should reproduce the eikonal result. The agreement is very good for both proton and neutron, except for some oscillations at small x, which can be associated to the cutoff of the integral in $\mathbf{k}$ in Eq. (5). In black the results for the effective density using the Morillon imaginary potential are presented. The inclusion of the imaginary potential leads to a significant reduction in the effective density, particularly in the interior, a reduction that is more pronounced in the case of the (more bound) proton, which can be understood, as for the more deeply-bound nucleon there are more non-elastic open channels so it suffers from a larger absorption.

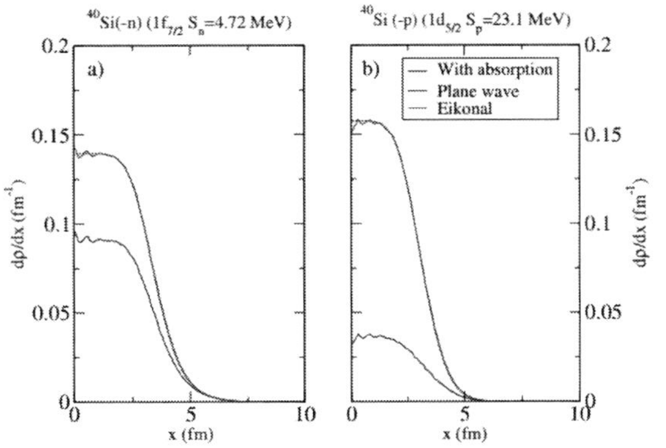

Figure 3. Effective density as a function of x for neutron (left panel) and proton (right panel) removal from ^{40}Si. The green curve corresponds to the standard eikonal calculation and the red curve to one using the effective density with no nucleon-core interaction. The black curve corresponds to densities using the imaginary part of the Morillon interaction.

We then compute the stripping cross section σ_{str}, as per Eq. (4) and compare it to the standard eikonal calculation $\sigma_{\rm eik,str}$. Their ratio is presented in Fig. 4 as a function of ΔS, the difference of the binding energy of the removed nucleon and that of its isospin partner, with weakly-bound nucleons appearing to the left and deeply-bound nucleons to the right. The red squares correspond to a test calculation with no potential, showing a difference to the standard eikonal calculation of less than 1.5%. Black circles correspond to calculations with the Morillon imaginary potential. The results for the Morillon calculation present a ratio of $\sim$ 0.6, with a negative slope. The peripherality of the reaction results in the reduction being less significant than the reduction in effective density in Fig. 3, as the surface, being sensitive to smaller nucleon-core energies, suffers less absorption than the nuclear interior.

The larger absorption for more deeply-bound nucleons yields smaller theoretical cross sections, which would result in larger spectroscopic factors when comparing to experimental results. This would soften the dependence on ΔS of the so-called "quenching factors", the ratio between the theoretical and experimental cross sections $R_s = \sigma_{\rm exp}/\sigma_{\rm theor}$. This dependence was found to be rather steep in nucleon knockout reactions with ^{9}Be and ^{12}C targets [1, 14] but later analyses found the dependence to be flatter for transfer [18] and $(p, 2p)$ [19] reactions, so the inclusion of this effect would help reduce this discrepancy. However, first tests show an overestimation of absorption

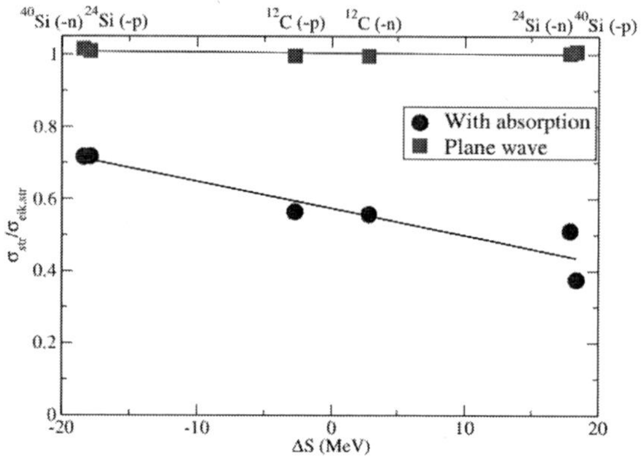

Figure 4. Ratio between computed cross sections and standard eikonal calculations as a function of the difference in binding energies $\Delta S = S_{n(p)} - S_{p(n)}$. Red squares correspond to an effective density with no potential and black circles to one with the Morillon imaginary potential(See text). Linear fits of the trends of the calculations with and without absorption are shown.

in the weakly-bound cases, which results in unphysical $R_s > 1$ values for the "quenching factors" for weakly-bound-nucleon removal. The reason for this overestimation could be related to the fact that the potential by Morillon (as is usual for global potentials) presents a finite imaginary part even for weakly-bound nucleons at low energies, where the lack of open channels should lead to no absorption. This imaginary part is meant to describe compound-nucleus processes which are not well described in optical model calculations but that ultimately lead to the original elastic channel and thus do not result in absorption and therefore should not be considered in the present calculations. Methods to estimate these contributions to remove them from the calculations would help reduce this problem at low binding energies and are currently in consideration [13].

4 Acknowledgements

The authors are grateful to A. Di Pietro for her help and insights on compound-nucleus calculations. This work has been partially funded by MCIN/AEI/10.13039/501100011033 under I+D+i project No. PID2020-114687GB-I00 and under grant IJC2020-043878-I (also funded by "European Union NextGenerationEU/PRTR"), by the Consejería de Economía, Conocimiento, Empresas y Universidad, Junta de Andalucía (Spain) and "ERDF-A Way of Making Europe" under PAIDI 2020 project No. P20_01247, and by the European Social Fund and Junta de Andalucía (PAIDI 2020) under grant number DOC-01006.

References

[1] A. Gade, P. Adrich, D. Bazin, M.D. Bowen, B.A. Brown, C.M. Campbell, J.M. Cook, T. Glasmacher, P.G. Hansen, K. Hosier et al., Phys. Rev. C **77**, 044306 (2008)

[2] R. Chatterjee, R. Shyam, Progress in Particle and Nuclear Physics **103**, 67 (2018)

[3] N. Austern, Y. Iseri, M. Kamimura, M. Kawai, G. Rawitscher, M. Yahiro, Physics Reports **154**, 125 (1987)

[4] P. Hansen, J. Tostevin, Annual Review of Nuclear and Particle Science **53**, 219 (2003)

[5] B.H. McKellar, C.M. McKay, Australian Journal of Physics **36**, 607 (1983)

[6] D.C. Brody, Journal of Physics A: Mathematical and Theoretical **47**, 035305 (2013)

[7] A. Koning, J. Delaroche, Nuclear Physics A **713**, 231 (2003)

[8] P. Capel, G. Goldstein, D. Baye, Phys. Rev. C **70**, 064605 (2004)

[9] A. Bonaccorso, R.J. Charity, Phys. Rev. C **89**, 024619 (2014)

[10] M. Gómez-Ramos, J. Gómez-Camacho, A. Moro, Physics Letters B **832**, 137252 (2022)

[11] N. Fukuda, T. Nakamura, N. Aoi, N. Imai, M. Ishihara, T. Kobayashi, H. Iwasaki, T. Kubo, A. Mengoni, M. Notani et al., Phys. Rev. C **70**, 054606 (2004)

[12] A.M. Moro, J.A. Lay, Phys. Rev. Lett. **109**, 232502 (2012)

[13] M. Gomez-Ramos, J. Gomez-Camacho, A.M. Moro, *Isospin dependence in single-nucleon removal cross sections explained through valence-core destruction effects* (2023), `https://arxiv.org/abs/2303.00426`

[14] J.A. Tostevin, A. Gade, Phys. Rev. C **103**, 054610 (2021)

[15] B.A. Brown, P.G. Hansen, B.M. Sherrill, J.A. Tostevin, Phys. Rev. C **65**, 061601 (2002)

[16] S.R. Stroberg, Ph.D. thesis, Michigan State University, Michigan State University, East Lansing (2016)

[17] B. Morillon, P. Romain, Phys. Rev. C **76**, 044601 (2007)

[18] F. Flavigny et al., Phys. Rev. Lett. **110**, 122503 (2013)

[19] L. Atar et al. (R^3B Collaboration), Phys. Rev. Lett. **120**, 052501 (2018)

 https://doi.org/10.1051/epjconf/202329002020

Proton single-particle energy gaps in Sc isotopes

Pawan kumar[1,2*], *Gabriel* Martínez-Pinedo[2,1], and *Riccardo* Mancino[1,2]

[1]Institut für Kernphysik (Theoriezentrum), Fachbereich Physik, Technische Universität Darmstadt, Darmstadt 64298, Germany
[2]GSI Helmholtzzentrum für Schwerionenforschung, Darmstadt 64291, Germany

Abstract. Variation in the proton single-particle energy gaps is investigated within the framework of nuclear shell model. The change is identified to originate mainly from the central force of the proton-neutron interaction. The relationship between the nuclear state and the single-particle energy gap is discussed. The first $3/2^-$ state of ^{55}Sc reveals the fragile character of $N = 34$ semi-magic shell gap.

1 Introduction

Since the beginning of the twenty-first century, the physics of the neutron-rich nuclei has gained a lot of interest. Shell evolution is one of the intriguing phenomena observed in the neutron-rich nuclei. It reveals that the shell structure in nuclei changes depending on the *N/Z* ratio [1, 2]. Following are some examples of the shell evolution: a) the canonical $N = 20$ and 28 magic shell gaps disappear for $Z = 8$–12 nuclei [3–9] and $Z < 16$ nuclei [10, 11], respectively, b) the new semi-magic shell gaps emerge at $N = 14$, 16 in oxygen [12, 13], and at $N = 32$ and 34 in calcium [14, 15].

To uncover the origin of shell evolution, many efforts have been made by theoretical nuclear physicists in the past. In the nuclear shell model framework, Otsuka and his collaborators have unveiled that the bare $\pi + \rho$ meson exchange tensor force

$$V(1,2) = f(r)[Y^2.(\sigma_1 \otimes \sigma_2)^2]^0 \quad (1)$$

is a predominant source of the shell evolution [16]. The robust effects of the tensor force on the single-particle energy gaps has been demonstrated in Refs. [16, 17].

As phenomenological nucleon-nucleon interactions are used in the nuclear shell model, a more rigorous method was put forward by Umeya *et al.*, [18] and Smirnova *et al.*, [19, 20] to elucidate the origin of the shell evolution. This method relies on the spin-tensor decomposition [21, 22] that allows to decompose a shell model interaction into its central, spin-orbit and tensor force components and peruse their role in the shell evolution. It has been demonstrated by Umeya *et al.*, [18] that the proton-neutron central force is crucial for the evolution of the N = 8 shell gap. On the other hand, it has been shown by Smirnova *et al.*, [19] that both proton-neutron central and tensor forces are important for the evolution of N = 20 and 28 shell gaps.

In this article, we intend to discuss the effects of proton-neutron force and its different components on the proton single-particle energy gaps in Sc isotopes. Recently, the low-energy states of neutron-rich ^{53,55}Sc has been populated in the experiments [23, 24]. Among the observed states, their first $3/2^-$ state have been found near 2.1 MeV and 0.7 MeV, respectively. As such these isotopes have $N = 32$ and 34 semi-magic shell gaps in their neutron space, so, if the excitation energy of their $3/2^-$ state is compared with that of ^{49}Sc which is about 3.1 MeV, it can be put forward that the proton $1p_{3/2} - 0f_{7/2}$ magic shell gap present at $N = 28$ reduces at $N = 32$ and 34, in particular, substantially at the latter. With motivation to obtain comprehensive information about this point, we have carried out the theoretical investigation within the framework of nuclear shell model. The calculations were performed in the *pf* harmonic oscillator shell using the GX1N interaction [25]. Here we discuss the results of those calculations.

This article is organized as follows. In Sec. 2, we have given a brief introduction of the effective single-particle energy and the spin-tensor decomposition. In Sec. 3.1, we have discussed the effects of the proton-neutron interaction and its different components on proton-single particle energy gaps. In Sec. 3.2, we have perused the relationship between the $1p_{3/2} - 0f_{7/2}$ proton single-particle energy gap and the first $3/2^-$ state of ^{49,53,55}Sc. The summary of this work in presented in Sec. 4.

2 Effective single particle energy and spin-tensor decomposition

In the nuclear shell model, the effective single particle energy *ESPE* gauges the shift in the single-particle energy of an orbit j caused by valence nucleons. The *ESPE* is determined from the monopole component of shell model Hamiltonian $\hat{H}$. In proton-neutron formalism,

*e-mail: p.kumar@gsi.de

the monopole Hamiltonian $\hat{H}_m$ is expressed as [26, 27]

$$\hat{H}_m = \sum_i \varepsilon_{\pi_i}\hat{n}_{\pi_i} + \sum_i \varepsilon_{\nu_i}\hat{n}_{\nu_i} + \sum_{i\leq j} \frac{\hat{n}_{\pi_i}(\hat{n}_{\pi_j}-\delta_{ij})}{1+\delta_{ij}}\bar{V}^{\pi\pi}_{ij} + \sum_{i\leq j} \frac{\hat{n}_{\nu_i}(\hat{n}_{\nu_j}-\delta_{ij})}{1+\delta_{ij}}\bar{V}^{\nu\nu}_{ij} + \sum_{ij}\hat{n}_{\pi_i}\hat{n}_{\nu_j}\bar{V}^{\pi\nu}_{ij}, \quad (2)$$

where, $\hat{n}_{\rho_i}$ is number operator for particle type ρ and orbit i. The index i consists of all quantum numbers (n, l, and j) to define the orbit, and ε is the unperturbed single-particle energy of that orbit. $\bar{V}$ is total angular momentum (J) averaged NN matrix elements

$$\bar{V}^{\rho\rho'}_{ij} = \frac{\sum_J (2J+1)(1+(-1)^J\delta_{\rho\rho'}\delta_{ij})V^{\rho\rho'}_{ij;J}}{(2i+1)(2j+1-\delta_{\rho\rho'}\delta_{ij})}, \quad (3)$$

where, summation runs only over the Pauli principle allowed J values.

In a simple approximation, the *ESPE* of orbit j for particle ρ can be obtained from the differentiation of the monopole Hamiltonian with respect to particle number n [28]

$$\epsilon'^{\rho}_j \equiv \frac{\partial\langle H_m\rangle}{\partial n^{\rho}_j} \approx \epsilon^{\rho}_j + \sum_{j'} n^{\rho'}_{j'}\bar{V}^{\rho\rho'}_{jj'}. \quad (4)$$

For the proton orbit j in Sc isotopes, the above expression of *ESPE* can be rewritten as

$$\epsilon'^{\pi}_j = \epsilon^{\pi}_j + \sum_{j'} n^{\nu}_{j'}\bar{V}^{\pi\nu}_{jj'}. \quad (5)$$

Here, the sum runs over the neutron orbits.

At any time, the information about the single-particle energy gaps can be obtained from the effective single-particle energy of the orbits.

The spin-tensor decomposition [21, 22] has remained a handful method to gain knowledge about the different forces of an effective interaction. Since nucleons are spin 1/2 fermions, the interaction between two nucleons can be written as the linear sum of the scalar product of configuration space operator Q and spin space operator S

$$V = \sum_{k=0}^{2} V(k) = \sum_{k=0}^{2} Q^k.S^k, \quad (6)$$

where, rank k = 0, 1 and 2 represent central, spin-orbit and tensor force, respectively. To obtain the matrix elements of these force in jj coupling, one needs to transform the jj coupled state into to *LS* coupled state in the standard way. In the *LS* coupled state, the matrix of $V(k)$ can be obtained from V as

$$\langle (ab), LS; JM|V(k)|(cd), L'S'; JM\rangle = (2k+1)(-1)^J \times \begin{Bmatrix} L & S & J \\ S' & L' & k \end{Bmatrix} \sum_{J'} (-1)^{J'}(2J'+1) \begin{Bmatrix} L & S & J' \\ S' & L' & k \end{Bmatrix} \times\langle (ab), LS; J'M|V|(cd), L'S'; J'M\rangle, \quad (7)$$

where, a is shorthand notation for the quantum numbers n_a and l_a. We have followed this method to obtain the central spin-orbit and tensor force matrix elements of GX1N interaction.

3 Results and discussion

3.1 Proton single-particle energy gaps

Figure 1 presents the single particle energy of proton pf orbits in Sc isotopes. Although these orbits evolve more or less similarly, they build unique patterns for the proton single-particle energy gaps. When neutrons occupy $0f_{7/2}$ and $0f_{5/2}$ orbits, the $1p_{3/2} - 0f_{7/2}$ and $1p_{1/2} - 0f_{7/2}$ proton energy gaps enhance. However, when neutrons occupy $1p_{3/2}$ and $1p_{1/2}$ orbits, the same single-particle energy gaps reduce. Particularly, the $\pi 1p_{3/2} - \pi 0f_{7/2}$ energy gap reduces so much at $N = 34$ relative to $N = 28$, which in a way supports the low excitation energy of $3/2^-$ state of ^{55}Sc. There is an interesting pattern in the $0f_{7/2} - 0f_{5/2}$ energy gap. Unlike the others two gaps, this energy gap remain approximately constant, which manifest that the splitting of spin-orbit partners are independent from the condition that which orbit is occupied by neutrons.

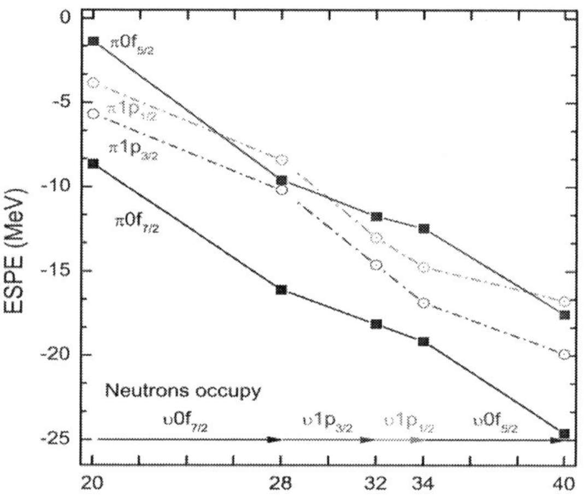

Figure 1. Effective single-particle energy ESPE of proton – pf orbits in Sc isotopes.

In Fig. 2, we show the effects of the central, spin-orbit, and tensor forces components of the proton-neutron force on the proton single-particle gaps. It can be easily followed that when neutrons occupy $0f_{7/2}$ and $0f_{5/2}$ orbits, the central force predominantly contributes to enhance the $1p_{3/2}-0f_{7/2}$ and $1p_{1/2}-0f_{7/2}$ energy gaps. The same is observed when neutrons occupy $1p_{3/2}$ and $1p_{1/2}$ orbits, however, this time central force acts with the opposite effect, which cause narrowing of gaps. It is worth mentioning here that tensor force tries to widen the $1p_{1/2} - 0f_{7/2}$ energy gap when the $1p_{1/2}$ orbit is filled with neutrons. But, its effect is negated by the central force which contributes with nearly twice of its strength. Further, it can be observed that the central force enhances the $1p_{3/2}-0f_{7/2}$ and $1p_{1/2} - 0f_{7/2}$ energy gaps by about 0.5 MeV irrespective of j of neutron $0f$ orbits, and similarly reduce the same

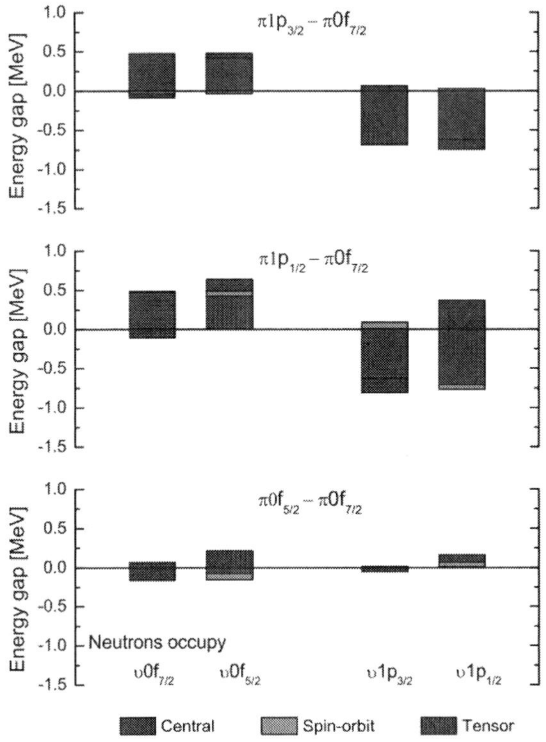

Figure 2. Contribution of proton-neutron central, spin-orbit and tensor force in the proton-single particle energy gaps. The gaps are represented as the difference of two matrix elements ($\bar{V}^{\pi\nu}_{j_1 j} - \bar{V}^{\pi\nu}_{j_2 j}$) at a time.

gaps by about 0.7 MeV, again, irrespective of j of neutron $1p$ orbits. Such a behaviour infers that the effective central force possesses strong-orbital node (nl) and weak-spin (j) dependency. The strong-orbital node and weak-spin dependency of central force has also been reported in Ref. [29], where it is proved using the spin-exchange zero range interaction. In the spin-orbit partners energy gap, the central force does not cause any effects as the relevant proton-neutron matrix elements have the same orbital nodes. The contribution of the spin-orbit force in this gap is also insignificant. Although the tensor force contributes relatively more when neutrons occupy $0f_{7/2}$ and $0f_{5/2}$ orbits, this is not large enough to cause a noteworthy change. Thus, the spin-orbital partner energy gap remains nearly constant in Sc isotopes.

3.2 The first $3/2^-$ state of ^{49,53,55}Sc

Above we have seen that the $1p_{3/2} - 0f_{7/2}$ single-particle energy gap reduces from $N = 28$ to $N = 34$, however, this is not compulsory that its effect will directly appear in the excited states [30], because the single-particle energy gaps are calculated using simple configurations of nucleons. To correlate the $1p_{3/2} - 0f_{7/2}$ proton gap with the first $3/2^-$ state of ^{49,53,55}Sc, we first need to take into account the effects of other configurations. Thus, we have performed full space shell model calculation using the shell model code NuShellX [31]. We have calculated the excitation energy of the $3/2^-$ state and also the proton transfer spectroscopic factor C^2S for the ground state $7/2^-$ and the $3/2^-$ state. These spectroscopic factors are determined with respect to $^{A-1}$Ca isotopes, which signifies how likely it is for the transferred proton to occupy $0f_{7/2}$ and $1p_{3/2}$ proton orbits for the $7/2^-$ and $3/2^-$ states, respectively. The theoretical results are given in Fig. 3 along with experimental data [23, 24, 32]. It can be seen that theory nicely reproduce the experimental data.

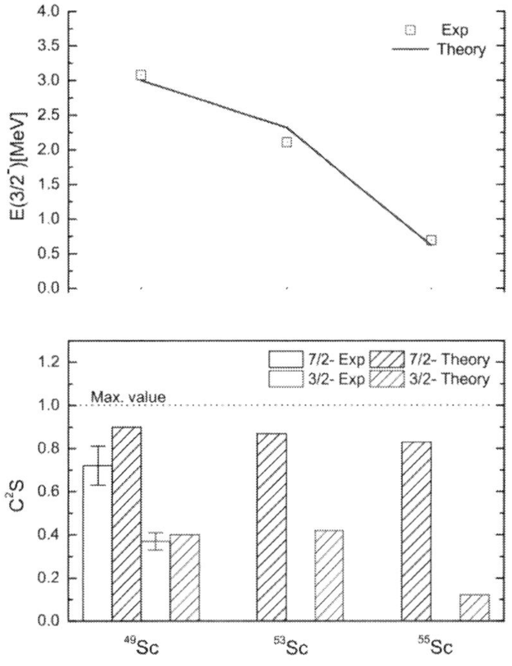

Figure 3. Upper part: The excitation energy of the first $3/2^-$ state of ^{49,53,55}Sc. Lower part: proton transfer spectroscopic factor for the ground state and the first $3/2^-$ state of ^{49,53,55}Sc.

As can be observed for ^{49,53}Sc that the proton single-particle strength is about 90 percent in the ground state, and about 40 percent in the $3/2^-$ state, which is a splendid value for excited states. Thus, the drop of the excitation energy of the $3/2^-$ state from ^{49}Sc to ^{53}Sc can be attributed to the narrowing of $1p_{3/2}$–$0f_{7/2}$ single-particle energy gap. In the $3/2^-$ state of ^{55}Sc, the single-particle strength is small, so , the further drop of the excitation energy cannot be accounted to $1p_{3/2} - 0f_{7/2}$ energy gap. The occupation numbers of neutron $1p_{1/2}$, and $0f_{5/2}$ orbits for this state are 1.06, and 1.07, respectively, which infer that it originates from the transition of a neutron across $N = 34$ semi-magic shell gap. However, as its excitation energy is small, this manifests that $N = 34$ semi-magic shell gap is fragile by nature. The third $3/2^-$ state of ^{55}Sc is found to own about 40 percent of total proton single-particle strength, but, it lies at 2.5 MeV. The high excitation energy of this state seems to be endowed by the higher order correlations.

4 Summary

In this work, we investigated the effects of proton-neutron force and its different components on the proton single-particle energy gaps. The calculations were performed for Sc isotopes within the nuclear shell model framework. It was found that the $0f_{5/2} - 0f_{7/2}$ energy gap remains nearly constant, while, $1p_{3/2} - 0f_{7/2}$ and $1p_{1/2} - 0f_{7/2}$ energy gaps enhance and reduce depending on the orbit occupied by neutrons. Mainly the central force was identified causing these changes in the gaps. Further, it was observed that the drop of the excitation energy of the first $3/2^-$ from ^{49}Sc to ^{53}Sc is due to the shrink of $1p_{3/2} - 0f_{7/2}$ gap. However, the subsequent drop of the excitation energy at ^{55}Sc is not associated with this gap. The $3/2^-$ state of ^{55}Sc gain its energy particularly from the transition of neutron across N = 34 semi-magic shell gap, but since the excitation energy is low, this exhibits that the $N = 34$ semi-magic shell gap is fragile by nature.

Acknowledgment

This research is funded by the Deutsche Forschungsgemeinschaft (DFG, German Research Foundation) under Project number 448588010.

References

[1] A. Ozawa, *et al.*, Phys. Rev. Lett. **84**, 5493 (2000).
[2] O. Sorlin *et al.*, Prog Part, Nucl. Phys. **61**, 602 (2008)
[3] Y. Kondo *et al.*, Phys. Rev. Lett. **116**, 102503 (2016).
[4] P. Doornenbal *et al.*, Phys. Rev. C **95**, 041301(R) (2017).
[5] S. M. Brown *et al.*, Phys. Rev. C **85**, 011302(R) (2012).
[6] H. N. Liu *et al.*, Phys. Lett. B **767**, 58 (2017).
[7] M. Petri *et al.*, Phys. Lett. B **748**, 173 (2015).
[8] B. V. Pritychenko *et al.*, Phys. Rev. C **63**, 011305(R) (2000).
[9] T. Motobayashi *et al.*, Phys. Lett. B **346**, 9 (1995).
[10] P. Doornenbal *et al.*, Phys. Rev. Lett. **111**, 212502 (2013).
[11] S. Takeuchi *et al.*, Phys. Rev. Lett. **109**, 182501 (2012).
[12] E. Becheva *et al.*, Phys. Rev. Lett. **96**, 012501 (2006).
[13] K. Tshoo *et al.*, Phys. Rev. Lett. **109**, 022501 (2012).
[14] F. Wienholtz *et al.*, Nature **498**, 346 (2013).
[15] D. Steppenbeck *et al.*, Nature **502**, 207 (2013).
[16] T. Otsuka *et al.*, Phys. Rev. Lett. **95**, 232502 (2005).
[17] T. Otsuka *et al.*, Phys. Rev. Lett. **104**, 012501 (2010).
[18] A. Umeya and K. Muto, Phys. Rev. C **69**, 024306 (2004).
[19] N. A. Smirnova *et al.*, Phys. Lett. B **686**, 109 (2010).
[20] N. A. Smirnova *et al.*, Phys. Rev. C **86**, 034314 (2012).
[21] M. W. Kirson, Phys. Lett. B **47**, 110 (1973).
[22] K. Klingenbeck *et al.*, Phys. Rev. C **15**, 1483 (1977).
[23] S. McDaniel *et al.*, Phys. Rev. C **81**, 024301 (2010).
[24] D. Steppenbeck *et al.*, Phys. Rev. C **96**, 064310 (2017)
[25] P. Kumar *et al.*, Phys. Rev. C **100**, 024328 (2019).
[26] E. Caurier *et al.*, Rev. Mod. Phys. **77**, 427 (2005).
[27] K. Heyde and J. L. Wood, Rev. Mod. Phys. **83**, 1467 (2011).
[28] Y. Utsuno, J. Phys. Conf. Ser. **445**, 012008(2013).
[29] N. A. Smirnova *et al.*, Phys. Rev. C **69**, 044306 (2004).
[30] P. Kumar *et al.*, Nucl. Phys. A **983**, 210 (2019).
[31] B. A. Brown *et al.*, Nucl. Data Sheets **120**, 115 (2014).
[32] J. W. Watson *et al.*, Phys. Rev. C **40**, 570 (1989)

 https://doi.org/10.1051/epjconf/202329002021

The β-decay of ^{71}Kr: precise measurement of the half-life

A. Vitéz-Sveiczer[1,2,3], *A.* Algora[1,2*], *G. G.* Kiss[2,4**], *B.* Rubio[1], *A. I.* Morales[1], *P.* Sarriguren[5], *G.* de Angelis[7], *F.* Recchia[8,9], *S.* Nishimura[4], *J.* Agramunt[1], *V.* Guadilla[1], *A.* Montaner-Pizá[1], *S. E. A.* Orrigo[1], *Á.* Horváth[10], *D.* Napoli[7], *S.* Lenzi[8,9], *A.* Boso[8,9], *V.* Phong[4,11], *J.* Wu[4], *P.-A.* Söderström[4], *T.* Sumikama[4], *H.* Suzuki[4], *H.* Takeda[4], *D. S.* Ahn[4], *H.* Baba[4], *P.* Doornenbal[4], *N.* Fukuda[4], *N.* Inabe[4], *T.* Isobe[4], *T.* Kubo[4], *S.* Kubono[4], *H.* Sakurai[4], *Y.* Shimizu[4], *C.* Sidong[4], *B.* Blank[12], *P.* Ascher[12], *M.* Gerbaux[12], *T.* Goigoux[12], *J.* Giovinazzo[12], *S.* Grévy[12], *T.* Kurtukián Nieto[12], *C.* Magron[12], *W.* Gelletly[1,13], *Zs.* Dombrádi[2], *Y.* Fujita[14,15], *M.* Tanaka[15], *P.* Aguilera[16], *F.* Molina[16,27], *J.* Eberth[17], *F.* Diel[17], *D.* Lubos[18], *C.* Borcea[19], *E.* Ganioglu[20], *D.* Nishimura[21], *H.* Oikawa[22], *Y.* Takei[22], *S.* Yagi[22], *W.* Korten[23], *G.* de France[6], *P.* Davies[24], *J.* Liu[25], *J.* Lee[25], *T.* Lokotko[25], *I.* Kojouharov[26], *N.* Kurz[26], *H.* Shaffner[26], and *A.* Petrovici[19]

[1]Instituto de Fisica Corpuscular, CSIC-Universitat de Valéncia, E-46071 Valéncia, Spain
[2]Institute for Nuclear Research (Atomki), H-4026 Debrecen, Hungary
[3]University of Debrecen, Doctoral School of Physics, H-4032 Debrecen, Egyetem tér 1, Hungary
[4]RIKEN – Nishina Center, Wako, Saitama 351-0198, Japan
[5]Instituto de Estructura de la Materia, IEM-CSIC, E-28006 Madrid, Spain
[6]Grand Accélérateur National d Ions Lourds, B.P. 55027, F-14076 Caen Cedex 05, France
[7]Istituto Nazionale di Fisica Nucleare, Laboratori Nazionali di Legnaro, Legnaro I-35020, Italy
[8]Istituto Nazionale di Fisica Nucleare, Sezione di Padova, Padova I-35131, Italy
[9]Dipartimento di Fisica dell Universitá degli Studi di Padova, Padova I-35131, Italy
[10]Eötvös Loránd University, Institute of Physics, H-1053 Budapest, Hungary
[11]Faculty of Physics, VNU Hanoi University of Science, 334 Nguyen Trai, Thanh Xuan, Hanoi, Vietnam
[12]Centre d Etudes Nucléaires de Bordeaux-Gradignan, UMR 5797 Université de Bordeaux CNRS/IN2P3, Gradignan Cedex, France
[13]Department of Physics, University of Surrey, Guildford GU2 7XH, United Kingdom
[14]Research Center for Nuclear Physics, Osaka University, Ibaraki, Osaka 567-0047, Japan
[15]Department of Physics, Osaka University, Toyonaka, Osaka 560-0043, Japan
[16]CEFNEN, Comisión Chilena de Energía Nuclear, Nueva Bilbao, 12501, Las Condes, Santiago, Chile
[17]Institute of Nuclear Physics, University of Cologne, D-50937 Cologne, Germany
[18]Physik Department E12, Technische Universität München, D-85748 Garching, Germany
[19]National Institute for Physics and Nuclear Engineering IFIN-HH, P.O. Box MG-6, Bucharest-Magurele, Romania
[20]Department of Physics, Istanbul University, Istanbul 34134, Turkey
[21]Department of Natural Sciences, Tokyo City University, Setagaya-ku, Tokyo 158-8557, Japan
[22]Department of Physics, Tokyo University of Science, Noda, Chiba 278-8510, Japan
[23]IRFU/DPhN, CEA, Université Paris-Saclay, F-91191 Gif-sur-Yvette, France
[24]Department of Physics, University of York, Heslington, York YO10 5DD, United Kingdom
[25]Department of Physics, University of Hong Kong, Pokfulam, Hong Kong
[26]GSI, Planckstrasse 1, D-64291 Darmstadt, Germany
[27]Millennium Institute for Subatomic Physics at High Energy Frontier - SAPHIR

Abstract. The very proton-rich ^{71}Kr isotope was produced through the in-flight fragmentation of ^{78}Kr on a beryllium target at RIKEN – Nishina Center in order to study its β-decay properties. A stack of double-sided silicon strip detectors, called WAS3ABi, was used as the decay station, where the detection of ion implants, β-decays and β-delayed protons took place. Beta-delayed γ-rays were measured using a system of 84 HPGe detectors, called EURICA, surrounding the decay station. The main goal of the present study was the precise measurement of the half-life of ^{71}Kr, as in the literature there is an almost 10 σ difference between the most precise independent results. Implant–β time correlations, implant–proton time correlations and implant–β–γ time correlations were all used to derive the half-life value, followed by a thorough investigation of systematic uncertainties for each method. As these values were found to be consistent, the weighted average $t_{1/2} = 94.40^{+19}_{-21}$ ms is reported as a new half-life value in this work. Furthermore a total of 26 previously unreported γ-transitions following the β-decay of ^{71}Kr were also identified in the analysis.

*e-mail: algora@ific.uv.es
**e-mail: ggkiss@atomki.hu

https://doi.org/10.1051/epjconf/202329002021

1 Introduction

The origin of *p*-nuclei is still not fully understood since it is likely that more than one process is responsible for the creation of these isotopes [1]. One of the conceivable candidates is the astrophysical *rp*-process taking place in X-ray bursts [2–4] as the result of hydrogen-rich material accreting from a low-mass star toward a neutron star in such binary systems. During these bursts thermonuclear ignition takes place and via a series of rapid proton capture reactions isotopes up to ^{105}Te can be formed [5]. Near the proton drip-line proton capture starts to compete with photodisintegration and eventually an equilibrium is reached until the relatively slow β-decay of these so called waiting-points takes place. Hence the modeling of the *rp*-process requires a precise knowledge of the β-decay half-lives.

Since it lies on the *rp*-process path, the half-life of ^{71}Kr has been measured in numerous radioactive ion beam experiments using implant-β time correlations, most recently by Sinclair *et al* in 2019 ($t_{1/2}^{\mathrm{Sinc.}}$ = 98.8 ± 0.3 ms) and Waniganeththi *et al* in 2022 ($t_{1/2}^{\mathrm{Wani.}}$ = 94.9 ± 0.4 ms) [6, 7]. As there was a 10 σ difference between these results, they derived the half-life based on the decay-curves of verified γ-transitions as well, leading to $t_{1/2}^{\mathrm{Wani.},\gamma}$ = 95.3 ± 2.2 ms. However the results were inconclusive since the statistical uncertainty was too high, as had been the case in the earlier results of $t_{1/2}^{\mathrm{Oin.}}$ = 100 ± 3 ms by Oinonen *et al* and $t_{1/2}^{\mathrm{Rog.}}$ = 92 ± 9 ms by Rogers *et al* [8, 9]. In this work the half-life value was measured using three different methods and better statistics than ever before in order to settle the disagreement between the reported results in literature.

2 Experimental method

The experiment was performed at RIKEN – Nishina Center [10], where a ^{78}Kr primary ion beam with an average intensity of 40 pnA was accelerated to 345 MeV/nucleon kinetic energy and directed on to a 5-mm-thick ^{9}Be target, resulting in the in-flight fragmentation of the ^{78}Kr nuclei. The fragments were separated and identified using the standard ΔE-$B\rho$-ToF method by the BigRIPS separator [11]. 9.8 million ^{71}Kr ions were implanted in the decay station WAS3ABi, consisting of a stack of double-sided silicon strip detectors (DSSSDs) [12], which was used for the detection of implanted ions, β-decays and β-delayed proton emission. The decay station was surrounded by EURICA, an array of twelve EUROBALL cluster germanium detectors, each containing seven crystals. The average distance between the center of the DSSSDs and the front face of the HPGe detectors is 22 cm, corresponding to a total absolute detection efficiency of about 8% at the energy of 1332 keV [13]. The energy spectrum of β-delayed γ-rays recorded by EURICA in coincidence with WAS3ABi signals, after the subtraction of the background and daughter contribution is shown in Figure 1.

Positrons and β-delayed protons were separated using an energy threshold and an exponential function was used to extrapolate and subtract the overlapping high-energy tail of positrons, as in our earlier work on the decay of ^{70}Kr [14], while the event selection criteria is detailed in [15, 16]. The resulting energy spectrum of β-delayed protons is shown in Figure 2.

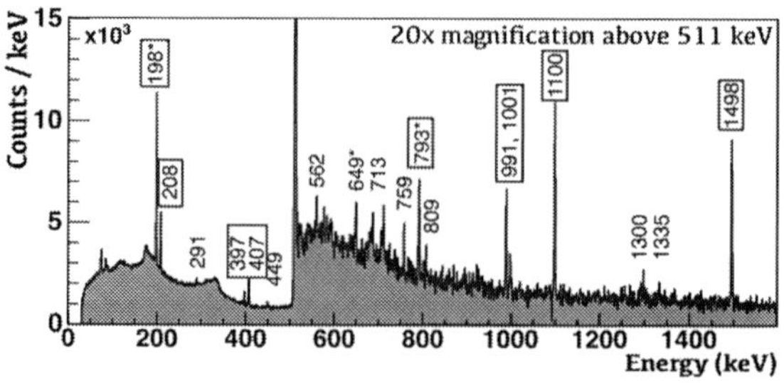

Figure 1. Energy spectrum of γ-rays emitted following the β-decay of ^{71}Kr. Doublets with energy differences within the order of the resolution of the HPGe detectors are marked with an asterisk. Transitions used for the half-life measurement are highlighted with black frames. The region above 511 keV is scaled up with a factor of 20 for the convenience of the reader.

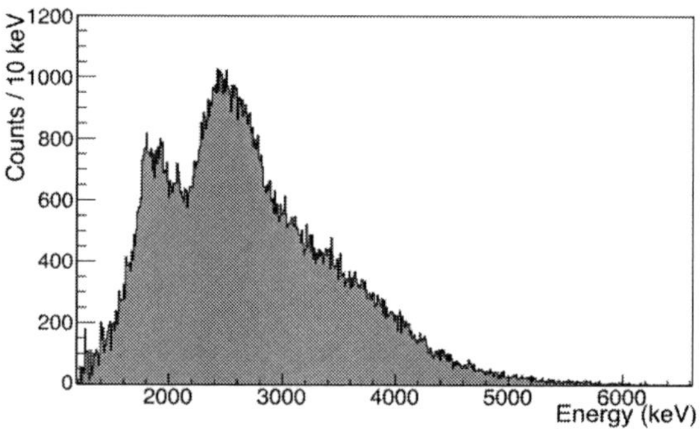

Figure 2. Intensity distribution of the β-delayed protons detected in the DSSSD as a function of the deposited energy. The high energy tail of β-particles was extrapolated by an exponential function, and then subtracted.

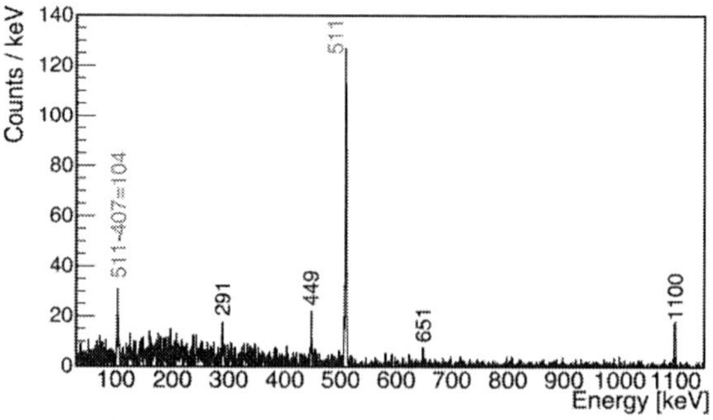

Figure 3. Energy spectrum of γ-transitions in coincidence with the E_γ = 407 keV transition. The peaks at E_γ = 511 keV and E_γ = 104 keV are coincidences originating from 511 keV annihilation photons and their Compton-scattering, respectively.

Coincidences between γ-rays during the same β-event were also analysed in order to look for cascades of γ-rays, used for verifying low intensity γ-rays and building the level-scheme. A fine example of such spectra is shown in

https://doi.org/10.1051/epjconf/202329002021

Figure 3, revealing four earlier unseen γ-transitions feeding the $E_x = 407$ keV level of ^{71}Br.

In this study we used three different methods to measure the half-life of the decay of this isotope and evaluated the systematic uncertainties of each method one-by-one, in order to derive a new solid half-life value:

- a fit of the Bateman equation to the the decay-curve of β-decays,
- exponential fit of the decay-curve of β-delayed protons,
- exponential fit of the decay-curve of β-delayed γ-rays.

2.1 Time distribution of β-events

The time distribution of implant–β correlations can be described by the Bateman-equation which is parametrised by the half-lives of the isotopes in the decay chain, the number of parent isotopes and the proton-emitting branching ratio (ε_p). In order to measure the half-life a binned maximum likelihood fit was used, with the two free parameters (the half-life and number of ^{71}Kr isotopes). All other parameters were constrained on a 95% confidence range obtained from literature. The time correlation histogram with the total fit function and its individual decay curves is shown in Figure 4. Systematic uncertainties arise from the Bateman-formula being based on the assumption, that β-detection efficiency is the same for every isotope in the chain and that literature data is accurate for the other isotopes in the decay chain. In this case these effects are expected to be small as all other isotopes in the decay chain have at least 2 orders of magnitude longer half-lives than ^{71}Kr. In order to evaluate any residual systematic uncertainties each *ad hoc* parameter of the fitting method – namely the endpoints of the fitting range, the width of bins and the lower energy threshold – were chosen randomly on a suitable range for every one of the 500 fits performed. Asymmetrical systematic uncertainties were calculated from the distribution of results of the fits.

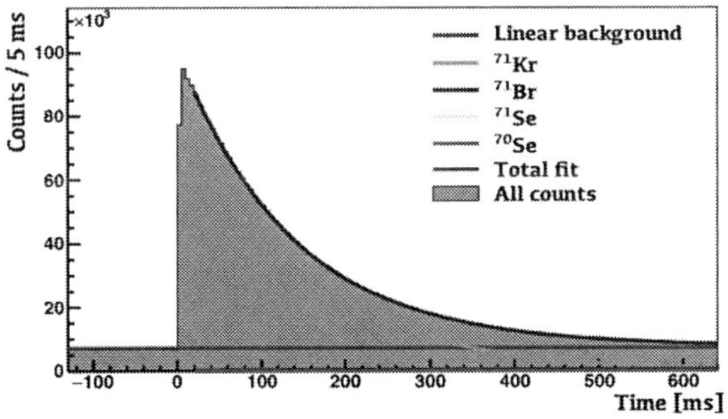

Figure 4. Time correlations of implant events and β-events of ^{71}Kr. The total fit is the sum of a Bateman formula and a linear background, and it is shown by a black line. The individual contributions are represented with colour coded lines. A half-life of $t^{i\beta}_{1/2} = 94.50 \pm 0.06$ (stat.) $^{+0.21}_{-0.23}$ (sys.) ms was derived.

2.2 Time distribution of β-delayed protons

The histogram was built from time correlations between implantation events and β-delayed proton emissions defined by every charged-particle event within the same pixel with deposited energy above a threshold energy (E_{cut}). Systematic uncertainties are effectively suppressed as the only proton emitting decay in this chain is the decay of interest. The time evolution of the overlapping high-energy tail of β-events was estimated through extrapolation and then subtracted from the implant–proton time correlations. The time correlation histogram with the exponential fit to the decay curve plus a linear background is shown in Figure 5 together with the subtracted contribution of high-energy β-events. As the most significant systematic uncertainty is expected to be from the subtraction of the estimated β-contribution, the same *ad hoc* parameters were varied as for the first method, but with the addition of the proton separation threshold being varied between $E_{cut} \in 1300 - 1500$ keV.

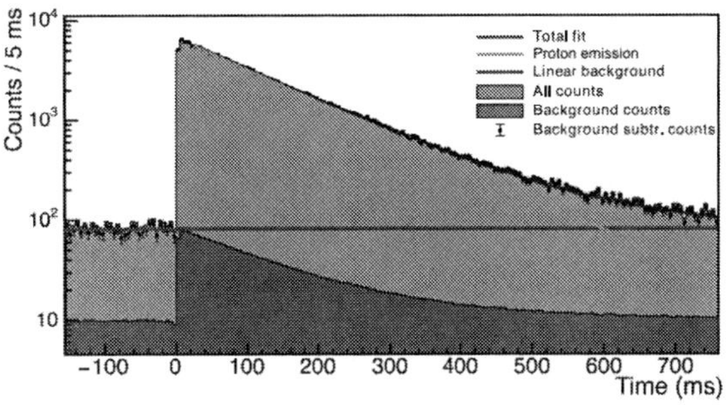

Figure 5. Time correlations of β-delayed proton events and ^{71}Kr implants. The time evolution of the subtracted β-contribution is shown in grey. The total fit is an exponential function with a linear background. A half-life of $t^{ip}_{1/2} = 94.08 \pm 0.22$ (stat.) $\pm$ 0.33 (sys.) ms was derived.

2.3 Time distribution of β-delayed γ-rays

Verified γ-transitions in the daughter nucleus, ^{71}Br, were also used to build decay-curves as well. However, as ^{71}Br has a ground state doublet with a first excited state at $E_\gamma = 9.8$ keV, we see a detailed decay-scheme with many transitions only a few keV from each other (as shown in Figure 1). Thus careful verification of a γ-transition required fulfilling any of the following criteria:

- It was reported in the high quality in-beam spectroscopic study of ^{71}Br by Fischer *et al* [17],
- It was confirmed to be in cascade with such a transition. For example $E_\gamma = 1100$ keV from Figure 3.
- It gave a consistent half-life value with other verified γ-transitions and it was in coincidence with 511 keV annihilation photons from the β^+-decay, but not with any other γ-transition, meaning it was a single transition straight to the ground state doublet.

The time correlation histogram was built using verified γ-transitions with a favourable signal-to-noise ratio, highlighted in Figure 1. Systematic uncertainties are effectively suppressed in principle, since only characteristic γ-rays are used resulting in a purely exponential decay curve. However the β-delayed γ-rays of the other isotopes in the decay chain are also present in the Compton-background. Thus this background was sampled at the proximity of each peak then subtracted after normalisation. The time correlation histogram with the exponential fit to the decay curve plus a linear background is shown in Figure 6. In order to evaluate any unaccounted systematic uncertainties, every *ad hoc* parameter of the fitting method was varied as in the first method, with the addition of the endpoints of the peak and Compton sampling energy gates.

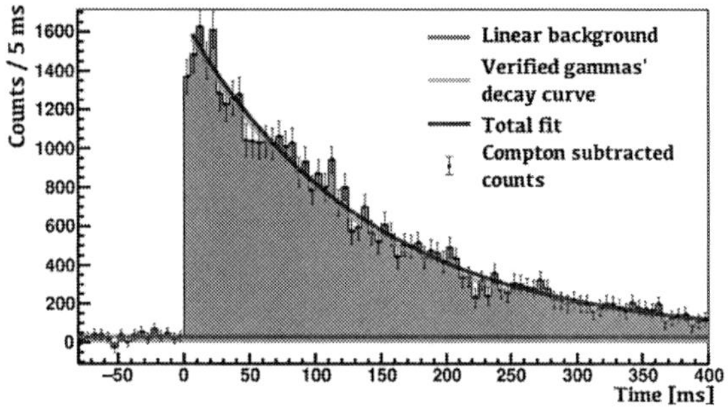

Figure 6. Time correlations of implant events and verified γ-transitions of the β-decay of ^{71}Kr. The total fit is the sum of an exponential function and a linear background, and it is shown by a black line. A half-life of $t_{1/2}^{i\beta\gamma} = 94.7 \pm 0.5$ (stat.) $^{+1.6}_{-1.5}$ (sys.) ms was derived.

3 Conclusion

In this work three half-life values were derived from different methods, $t_{1/2}^{i\beta} = 94.50 \pm 0.06$ (stat.) $^{+0.21}_{-0.23}$ (sys.) ms from the decay-curve of β-events, $t_{1/2}^{ip} = 94.08 \pm 0.22$ (stat.) $\pm$ 0.33 (sys.) ms from the decay-curve of β-delayed protons, and $t_{1/2}^{i\beta\gamma} = 94.7 \pm 0.5$ (stat.) $^{+1.6}_{-1.5}$ (sys.) ms from the decay-curve of β-delayed γ-rays, in consistency with each other. The weighted average is $t_{1/2} = 94.40^{+19}_{-21}$ ms, which is consistent with the result of Waniganeththi *et al*, but in strong disagreement, by more than 10 σ difference, with the result of Sinclair *et al.*

Acknowledgements

This work was carried out at the RIBF operated by RIKEN Nishina Center and CNS, University of Tokyo. We acknowledge the EUROBALL Owners Committee for the loan of germanium detectors and the PreSpec Collaboration for the readout electronics of the cluster detectors. This work was supported by the Spanish MICINN grants FPA2014-52823-C2-1-P, FPA2017-83946-C2-1-P (MCIU/AEI/FEDER); Ministerio de Ciencia e Innovacion grant PID2019-104714GB-C21; Centro de Excelencia Severo Ochoa del IFIC SEV - 2014 - 0398; *Junta para la Ampliación de Estudios* Programme (CSIC JAE-Doc contract) co-financed by FSE, by NKFIH (NN128072), the National Research, Development, and Innovation Fund of Hungary Project No. K 128729 the STFC (UK) through Grant No. ST/P005314/, the PROMETEO/2019/007 project and by the ÚNKP-20-5-DE-02 New National Excellence Program of the Ministry of Human Capacities of Hungary and by the JSPS KAKENHI of Japan (Grant No. 25247045). G. G. Kiss acknowledges support from the János Bolyai research fellowship of the Hungarian Academy of Sciences. A. A. acknowledges partial support of the JSPS Invitational Fellowships for Research in Japan (ID: L1955) P. S. acknowledges support from MCI/AEI/FEDER,UE (Spain) under grant PGC2018-093636-B-I00. F. M. acknowledges support from ANID FONDECYT Regular Project 1221364 and ANID - Millennium Science Initiative Program - ICN2019-044. Supported by the ÚNKP-22-3 New National Excellence Program of the Ministry for Culture and Innovation from the source of the National Research, Development and Innovation Fund (Project No. ÚNKP-22-3-II-DE-132).

References

[1] T. Rauscher *et al.*, Rep. Prog. Phys. **76** 066201 (2013).
[2] R. K. Wallace and S. E. Woosley, Astro. J. **45**, 389 (1981).
[3] H. Schatz *et al.*, Phys. Rep. **294**, 167 (1998).
[4] A. Parikh *et al.*, Prog. Part. Nucl. Phys. **69**, 225 (2013).
[5] H. Schatz *et al.*, Phys. Rev. Lett. **86**, 3471 (2011).
[6] L. Sinclair *et al.*, Phys. Rev. C **100**, 044311 (2019).
[7] S. Waniganeththi *et al.*, Phys. Rev. C **106**, 044317 (2022).
[8] M. Oinonen *et al.*, Phys. Rev. C **56**, 745 (1997).
[9] A. M. Rogers *et al.*, Nucl. Data Sheets **120**, 41 (2014).
[10] Y. Yano, Nucl. Inst. Meth. B **261**, 1009 (2007).
[11] N. Fukuda *et al.*, Nucl. Inst. Meth. B **317**, 323 (2013).
[12] S. Nishimura, Prog. Theor. Exp. Phys. 03C006 (2012).
[13] P.-A. Söderström *et al.*, N. I. M. B **317**, 649 (2013).
[14] A. Vitéz-Sveiczer *et al.*, Phys. Lett. B **830**, 137123 (2022).
[15] A. Morales *et al.*, Phys. Rev. C **95**, 064327 (2017).
[16] A. Vitéz-Sveiczer *et al.*, Acta Phys. Pol. **51(3)**, 587-594.
[17] S. M. Fischer *et al.*, Phys. Rev. C **72**, 024321

 https://doi.org/10.1051/epjconf/202329002022

Sr isotopes: the interplay between shape coexistence and quantum phase transitions

E. Maya-Barbecho[1,*] and *J.E.* García-Ramos[1,2,**]

[1]Departamento de Ciencias Integradas y Centro de Estudios Avanzados en Física, Matemática y Computación, Universidad de Huelva, 21071 Huelva, Spain
[2]Instituto Carlos I de Física Teórica y Computacional, Universidad de Granada, Fuentenueva s/n, 18071 Granada, Spain

Abstract. In this contribution we study the even-even Sr isotopes considering the influence of particle-hole intruder states. We find that the rapid onset of deformation at A = 98 and onwards can be explained by the crossing of regular and intruder states. We compare the systematics of Sr with the ones of Zr, Pt, and Hg nuclei.

1 Introduction

The understanding of the appearance of nuclear deformation is still of major interest in nuclear physics. The collective phenomena in the atomic nucleus is the origin of the onset of deformation, but it is also the responsible of the vibrational character of nuclei. The range of phenomena between both limits is modulated by a fine balance between the pairing, which tends to produce spherical shapes, and the quadrupole force, which makes the nucleus deformed. In others words, there is a competition between the stabilizing effect of the shell closure energy gap and the residual interaction between like and unlike nucleons.

Nuclei near to a shell closure tends to be spherical and excitations are relatively costly, specially the particle-hole ones, but, surprisingly, enough of the energy of such excitations can be considerably reduced due to the increase of the residual interaction because of the presence of a larger number of effective valence nucleons. This effect is particularly important when protons are near to a shell closure, while neutrons are at the middle of a shell, or viceversa. In this case, a particle-hole excitation will enhance the proton-neutron quadrupole interaction and it will considerable reduce the energy of such an excitation, being possible to appear at very low energy, even becoming the ground state of the system. Moreover, these configurations are expected to be more deformed than the normal ones. We will refer to them as intruder and regular configurations, respectively, and to the interplay between them as configuration mixing [1].

Along this contribution we will study the connection between shape coexistence and the so called Quantum Phase Transition (QPT) [2]. We will focus on the case of Sr, but a detailed comparison with Zr, Pt, and Hg nuclei is included.

*e-mail: esperanza.maya@alu.uhu.es
**e-mail: enrique.ramos@dfaie.uhu.es

2 Shape coexistence and quantum phase transitions: faces of a single phenomenon?

Shape coexistence in nuclear physics was first proposed by Morinaga in the 1950's and nowadays becomes a concept that appears throughout the entire nuclear landscape, especially in those nuclei at or near shell or sub-shell closures [1]. Shape coexistence presents certain key experimental features: an U shape pattern in the energy systematics of a set of states, the lowering of certain excited 0^+ states (intruder states), a rapid change in the value of the spectroscopic quadrupole moments, and the existence of strong E0 transitions. All of them show an almost symmetric trend with respect to the corresponding mid-shell and they are enhanced at this place.

Shape coexistence can be understood in terms of two well-known theoretical approaches: the spherical shell model and the mean field. According to the shell model approach, the intruder configurations correspond to the promotion of pairs of nucleons across the shell or subshell gap closures. The other major approach is the mean-field theory where, roughly speaking, one obtains an energy surface depending on certain deformation parameters. In this landscape, the ground band will correspond to the deepest minimum (regular states), but the other local minima can be interpreted as intruder configurations, each of them owning to different deformations.

The concept of QPT [2] implies the sudden change of the ground state structure of the system as a function of a control parameter. Such a control parameter can be in the case of the atomic nucleus, e.g., the neutron number. Therefore, a QPT can appear in an isotopic chain where the ground state deformation varies in an abrupt way when passing from isotope to isotope. Of course, the latter should be considered as an approximation because a control parameter should be, strictly speaking, continuous. A QPT is characterized by the discontinuity of some derivative of the system energy. Assuming that the two-neutron separation energy (S_{2n}) is somehow proportional to the derivative of the energy with respect to the neutron number, a first-order QPT involves a discontinuity in S_{2n}, while a second-order one a discontinuity in the derivative of S_{2n}. The even-even Sr isotope chain, as will be shown along the text, corresponds to the case of a first-order QPT.

An ideal framework to deal with QPTs is the interacting boson model (IBM) [3] which is a paradigmatic example of a nuclear algebraic model. In that framework, a QPT can be understood considering a Hamiltonian that is a combination of two given symmetries, combined through a control parameter, $H = x\,H_{\mathrm{sym1}} + (1-x)\,H_{\mathrm{sym2}}$ that allows to go from one to the other limit. The presence of a QPT is revealed by the existence of a critical value of the control parameter, x_c, for which the underlying structure of the system (phase) passes from *symmetry 1* to *symmetry 2*.

The IBM can be also used to deal with shape coexistence, allowing to define a framework where different particle-hole excitations are present in the spectrum. The formalism is known as IBM with configuration mixing, IBM-CM in short [4, 5].

Is there any connection between shape coexistence and QPT? First, one should note several similarities, namely, both phenomena involve a rapid change in the structure of certain set of states, either ground or excited states and, in both cases, at the mean-field level, several minima coexist. In the case of shape coexistence, the regular and the intruder configurations can be thought as having independent energy surfaces that interact among them, especially when their minima are close in energy, while under a QPT a single energy surface exists where two or more minima are present. The evolution of the relative position of the minima as a function of the neutron number is at the origin of the change of the nuclear deformation and it can be understood in terms of either shape coexistence or a QPT. On the other hand, from a quantum point of view clear differences should exist between both approaches. The most obvious one is the existence of a larger Hilbert space when two configurations are present, because of the extra nucleons that occupy the single particle levels. This extra set of levels presents a different energy systematics than the regular one because for them the effective nuclear interaction is different. Moreover, the intruder configuration is expected to be found at low energy when the number of valence nucleons is large, i.e. around mid shell, therefore a parabolic-like energy systematics for the intruder states appears frequently. However, the situation can be blurred due to the interaction between intruder and regular configurations or because their crossing [6]. When a QPT exist without the presence of multiple particle-hole excitations, a larger density of states can be also observed at low energy as for configuration mixing, but it never happens around mid-shell [7]. The remaining open question is whether a QPT can be induced by shape coexistence, which we will try to answer along the rest of the contribution.

3 Sr and Zr isotopes versus Pt and Hg nuclei

The regions around the sub-shell closure Z = 40 are, together with the area around Z = 82, two of the best examples of shape coexistence in nuclei. In Hg and Pt nuclei, two types of configurations show up with a clear presence of low lying 0^+ states, combined with a parabolic-like shape in the spectra of Hg, whereas this behavior is not so obvious in Pt. On the other hand, Zr and Sr are known to exhibit the faster change in deformation in an isotope chain at N = 60, with two particle-hole configuration coexisting and crossing at this point.

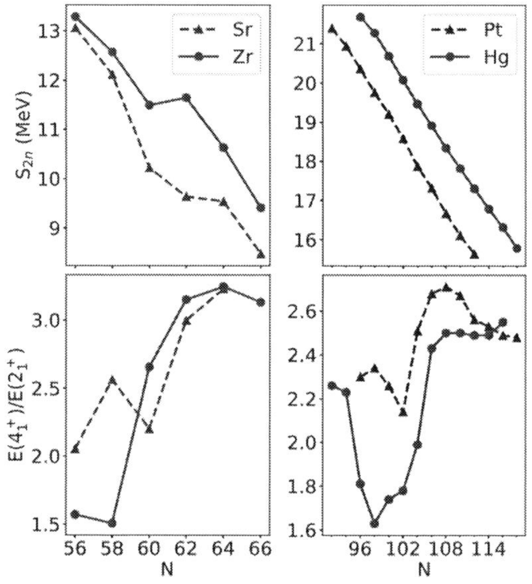

Figure 1. S_{2n} and $E(4_1^+)/E(2_1^+)$ as a function of the neutron number, *N*, for Sr, Zr, Pt and Hg isotopes.

From the point of view of a QPT, there are two key observables that can be used as indicators of its presence, namely, the two-neutron separation energy, S_{2n}, that will experience a discontinuity at a first-order transition point and the ratio $E(4_1^+)/E(2_1^+)$ which provides a hint for the appearance of deformation or, in a qualitative way, it is connected with the order parameter of the QPT. In Fig. 1, we present the systematics of S_{2n} and $E(4_1^+)/E(2_1^+)$, for Sr and Zr, on the one hand and, for Pt and Hg, on the other. Regarding the value of S_{2n}, one notices a clear change of slope in Sr and Zr, but a fully linear behavior in Hg and Pt. On the other hand, $E(4_1^+)/E(2_1^+)$ shows for Sr and Zr the typical rapid change that is expected in a QPT, passing from a value close to 2 (even 1.5 for $^{96-98}$Zr) to another close to 3.3. However, in Pt and Hg the interpretation of the systematics is not so evident and the physics involved in these cases seems to be quite different from the one of Sr and Zr.

To have a more clear view of the analogies and differences between both situations, we present in Fig. 2 some relevant IBM calculations based on [8] for Sr, [7] for Zr,

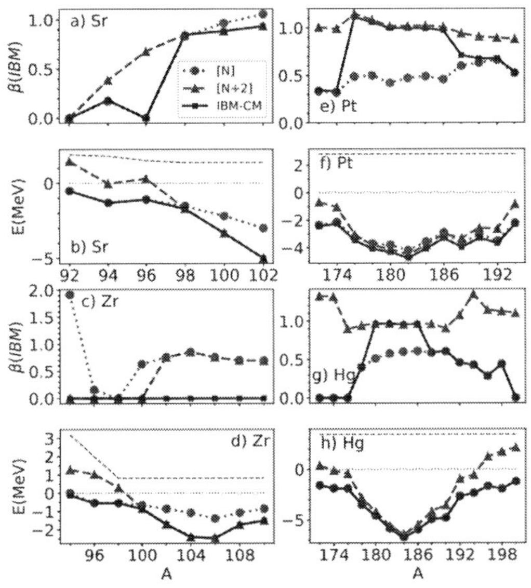

Figure 2. a), c), e), and g) panels: IBM value of β for the unperturbed regular, $[N]$, and intruder configurations, $[N+2]$, and full calculation, IBM-CM. b), d), f), and h) panels: ground state energy for the unperturbed regular and the intruder configurations and for the full calculation. The green thin dashed line stands for the energy needed to create a particle-hole excitation and the red thin dotted one stands for the reference level.

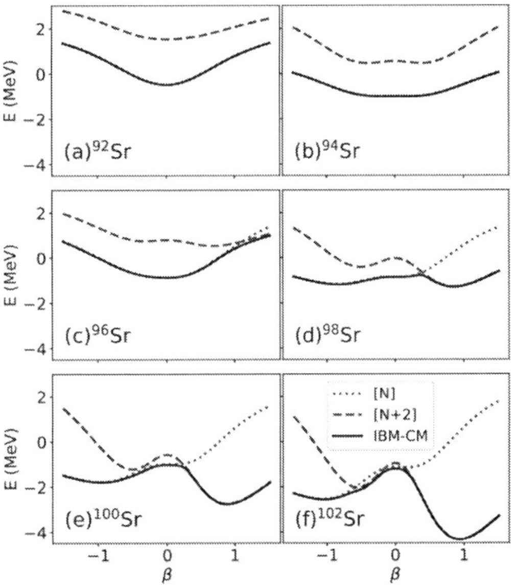

Figure 3. Axial IBM-CM mean-field energy surfaces for $^{92-102}$Sr isotopes. Red dotted line stands for the unperturbed regular configuration, [N], green dashed one for the unperturbed intruder configuration, [N+2], while the black continuous line for the full IBM-CM calculation.

[9] for Pt, and [10] for Hg. In particular, the results correspond to the value of the IBM deformation parameter β for the regular ([N]), the intruder configuration ([N+2]) and the complete system (IBM-CM) (panels a), c), e), and g)). Moreover, the unperturbed energy for the lowest regular, intruder, and ground state (IBM-CM) are depicted in panels b), d), f), and h). Concerning Sr and Zr, the distinctive feature is the crossing of the regular and the intruder configurations at A = 98 and 100, respectively, corresponding to N = 60, inducing a change in the slope of the ground state energy and also a rapid onset of deformation, which has a dramatic effect on the trend of the radii [8, 11]. In the case of Pt and Hg, both configurations remain quite close, without crossing in the case of Hg, and crossing before and after the mid-shell (N = 104) in the case of Pt. The ground state energy is hardly modified, but the deformation parameter β passed from the less deformed configuration (the regular one) to the most deformed one (the intruder one) around mid-shell. The crossing or the close approaching of regular and intruder configurations in Pt and Hg, respectively, have also a noticeable influence in the radii of these nuclei, as shown in [6, 10]. In summary, in Sr, Zr and Pt nuclei, a intruder state becomes the ground state when approaching the mid-shell, while it never happens in Hg. Note that in Pt, due to the strong mixing between both configurations [9], no dramatic effect are observed in the spectra, although the energy crossing exists.

The above features have direct consequences in the systematics of S_{2n} and $E(4_1^+)/E(2_1^+)$. In the case of Sr and Zr, what it is observed in Fig. 1 can be easily understood in terms of the crossing of configurations. Indeed, the change in the nature of the ground state produces a sudden change in the slope of the energy systematics and therefore a discontinuity in the value of S_{2n}. On the other hand, because the crossing is so fast, not only the ground state presents an intruder character, but also the rest of low-lying members of the yrast band involved in the energy ratio $E(4_1^+)/E(2_1^+)$, which makes that for $N < 60$ all the states present a vibrational character, while for $N \geq 60$ a rotational one appears. On the other hand, in the case of Hg, because the ground state presents all the way a regular character and, moreover, there is no interaction between intruder and regular sector, it can be understood the linear tendency of the S_{2n} even at the mid-shell. Besides, the crossing between 2_1^+ and 4_1^+ regular and intruder states can explain the drop of the ratio $E(4_1^+)/E(2_1^+)$ around the mid-shell. Finally, to explain the systematics of Pt in Fig. 1 is a challenge. First, the crossing of regular and intruder 0^+ energies would suggest a discontinuity in S_{2n} as the observed one in Sr and Zr. However, experimental values show a fully linear systematics. The reason for such a behavior is the strong mixing between the regular and the intruder sectors (see [9]). Note that according to [12] and figure Fig. 2, the deformation changes as it should be in a QPT, however the strong mixing somehow hides its effect in both the S_{2n} and $E(4_1^+)/E(2_1^+)$ observables.

A different way of seen the evolution of the nuclear deformation is to study the mean-field energy surfaces in a chain of isotopes, which can be calculated also using the IBM-CM formalism [13]. This is depicted in Fig. 3 for Sr isotopes, where the axial energy surfaces for the regular and the intruder configurations, as well as the one for the complete system are presented. Concerning the regular configuration (red dotted line), it presents a smooth evolution being spherical for A = 92 – 96, becoming very flat and finally axially oblate deformed for A = 98 – 102. Regarding the intruder configuration (green dashed line), it starts being also spherical for A = 92 and 94, but rapidly becomes axially prolate deformed from A = 96 and onwards. Finally, the IBM-CM full calculation (black full line), which essentially follows the lowest of the intruder and regular curves, starts also with a spherical minimum, A = 92 – 96 and also becomes an axially prolate deformed one for A = 98 – 102. A very relevant feature is that both configurations evolve separately from a spherical to a deformed shape, more rapidly in the case of the intruder configuration. Therefore, most probably in both configurations separately, regular and intruder, a QPT is present. Moreover, the fact that both configurations cross at A = 98 makes much more abrupt the change of deformation, also inducing a QPT in the whole system. The QPT present in the separated regular and the intruder configurations is called type-I QPT [14] and involves the existence of a single configuration energy surface where several minima are present. On the other hand, for the whole energy surface a type-II QPT [14] appears because two energy surfaces that cross are present. A type-II QPT is expected to induce much more abrupt changes than the type-I and it will own the typical phenomenology of a first order QPT. Note that the type-I QPT is also able to reproduce the S_{2n} trend of Zr isotopes, as was shown in [15], although the type-II QPT provides a more consistent framework [7] to explain the rapid onset of deformation. It is worthy to mention that this is not the only mechanism to induce a discontinuity in the S_{2n}, as happens, for instance, in the rare-earth region at $N \approx 90$.

4 Conclusions

In this contribution, we have studied the onset of deformation in Sr on the light of the coexistence of several particle-hole configurations. We have compared with Zr isotopes that is another example of $Z \approx 40$ nucleus and that presents a common trend regarding many observables and with Pt and Hg nuclei that are paradigmatic examples of nuclei where shape coexistence plays a major role in $Z \approx 82$. With the latter case notable differences exist.

The origin of the onset of deformation in $Z \approx 40$ region can be understood in terms of the residual proton-neutron interactions resulting in major modifications in the occupation of the $1g_{9/2}$ proton and $1g_{7/2}$ neutron orbitals. This approach suggested in Federman and Pittel [16, 17] emphasizes the importance of the simultaneous occupation of neutrons and protons spin-orbit partners. In this work, we have tried to define an effective approach for the Federman-Pittel mechanism using the IBM-CM.

In the framework of the IBM, the rapid changes observed in transitional regions, as is the case of Sr, can be understood in terms of a Quantum Phase Transition (QPT), however, the use of configuration mixing, IBM-CM, explains in a more consistent way the very abrupt changes in terms of the crossing of two configurations, calling to this situation type-II QPT. Note that the crossing does not always creates a QPT, as happens in the case of Pt, where the strong interaction between the involved particle-hole configurations hinder the observation of abrupt changes.

Sr, together with Zr, isotopes are very clear examples of QPTs that can be explained in terms of the crossing of two configurations, in other words, examples of Type-II QPTs.

Acknowledgements

This work was partially supported by grant PID2019-104002GB-C21 funded by MCIN/AEI/10.13039/501100001103 and "ERDF A way of making Europe". Resources supporting this work were provided by the CEAFMC and Universidad de Huelva High Performance Computer (HPC@UHU) funded by ERDF/MINECO project UNHU-15CE-2848.

References

[1] K. Heyde, J.L. Wood, Rev. Mod. Phys. **83**, 1467 (2011)
[2] P. Cejnar, J. Jolie, Progress in Particle and Nuclear Physics **62**, 210 (2009)
[3] F. Iachello, A. Arima, *The interacting boson model* (Cambridge University Press, Cambridge, 1987)
[4] P.D. Duval, B.R. Barrett, Phys. Lett. B **100**, 223 (1981)
[5] P.D. Duval, B.R. Barrett, Nucl. Phys. A **376**, 213 (1982)
[6] J.E. García-Ramos, V. Hellemans, K. Heyde, Phys. Rev. C **84**, 014331 (2011)
[7] J.E. García-Ramos, K. Heyde, Phys. Rev. C **102**, 054333 (2020)
[8] E. Maya-Barbecho, J.E. García-Ramos, Phys. Rev. C **105**, 034341 (2022)
[9] J. García-Ramos, K. Heyde, Nucl. Phys. A **825**, 39 (2009)
[10] J.E. García-Ramos, K. Heyde, Phys. Rev. C **89**, 014306 (2014)
[11] J.E. García-Ramos, K. Heyde, Phys. Rev. C **100**, 044315 (2019)
[12] J.E. García-Ramos, K. Heyde, L.M. Robledo, R. Rodríguez-Guzmán, Phys. Rev. C **89**, 034313 (2014)
[13] A. Frank, O. Castaños, P.V. Isacker, E. Padilla, AIP Conf. Proc. **638**, 23 (2002)
[14] N. Gavrielov, A. Leviatan, F. Iachello, Phys. Rev. C **99**, 064324 (2019)
[15] J.E. García-Ramos, K. Heyde, R. Fossion, V. Hellemans, S. De Baerdemacker, Eur. Phys. J. A **26**, 221 (2005)
[16] P. Federman, S. Pittel, Phys. Lett. B **69**, 385 (1977)
[17] P. Federman, S. Pittel, Phys. Rev. C **20**, 820 (1979)

Quarteting in deformed $N = Z$ nuclei

M. Sambataro[1,*] and *N.* Sandulescu[2,**]

[1]Istituto Nazionale di Fisica Nucleare - Sezione di Catania, Via S. Sofia 64, I-95123 Catania, Italy
[2]National Institute of Physics and Nuclear Engineering, P.O. Box MG-6, Magurele, Bucharest, Romania

Abstract. We describe deformed $N = Z$ nuclei in a formalism of α-like quartets. Quartets are constructed variationally by resorting to the use of proper intrinsic states. Various types of intrinsic states are introduced which generate different sets of quartets for a given nucleus. Energy spectra are generated via configuration-iteraction calculations in the spaces built with these quartets. The approach is applied to ^{24}Mg and ^{28}Si in the *sd* shell and to ^{48}Cr in the *pf* shell. In all cases a good description of the low-lying spectra is achieved. As a peculiarity of the approach, a close correspondence is observed between the various sets of quartets employed and the occurrence of well defined band-like structures in the spectra of the systems under study.

1 Introduction

The study of $N = Z$ nuclei is one of the key issues of contemporary nuclear structure physics. $N = Z$ nuclei are characterized by having an equal number of protons and neutrons sharing the same orbitals. Owing to this fact, the proton-neutron (*pn*) interaction is expected to play a relevant role in these nuclei.

Understanding the type of correlations induced by the *pn* pairing force in the wave function of $N = Z$ nuclei has proved to be a not trivial task [1–3]. An unambigous indication of the type of correlations that are generated by the *pn* pairing in $N = Z$ nuclei is provided by the exact analytical solutions of realistic Hamiltonians. In the case of the isovector pairing, finding such solutions has gone through various works spread over several decades [4–9], the first complete description of both eigenvalues and eigenstates being provided in 2006 by Dukelsky et al. [7]. However, the complicated mathematical structure of these exact solutions has hindered a simple understanding of the underlying correlations.

In a recent work [10], we have provided an alternative and more transparent description of the exact solutions of the isovector pairing which has evidenced a peculiar aspect of these solutions which had escaped the previous investigations. Indeed, we have shown on an analytic basis that what characterizes the isovector pairing Hamiltonian in $N = Z$ systems is the occurrence in its eigenstates of correlated four-body α-like structures made by two protons and two neutrons coupled to $T = 0$ ("quartets").

One needs to remark that, well before the exact treatment just discussed the essential role played by four-body correlations in $N = Z$ systems subject to an isovector pairing force, but limited to the special case of degenerate single-particle levels (the so-called SO(5) model), had already been emphasized in Ref. [11]. More generally, quartets have definitely a long history in nuclear structure physics [12–19], but their complexity has undoubtedly represented a hindrance to the development of quartet models.

In a recent past, the spectra of $N = Z$ nuclei have been constructed in terms of $T = 0$ quartets with various angular momenta [20–22]. In this approach, hereafter referred to as Quartet Model (QM), spectra of both positive and negative parity states have been generated by carrying out configuration-interaction calculations in a space of states formulated as products of these quartets. A crucial aspect of these calculations has consisted in the definition of the quartets to involve in the calculations. In these early works, we have adopted the criterion of assuming as $T = 0$ quartets those defining the low-lying eigenstates of the nearest $T = 0$ one-quartet systems [20–22]. While having the advantage of being straightforward, this "static" definition of the quartets is clearly not the most appropriate one since it fully neglects the effect of the Pauli principle on the amplitudes of the quartets when two or more of these quartets have to coexist in the same nucleus. Finding the most appropriate quartets to be employed in a QM calculation is a matter of primary importance for the validity of this approach and this has represented one of the goals of the present work.

The approach that we illustrate in this paper is based on the use of "dynamical" quartets, namely quartets which are fixed in correspondence with each nucleus [23]. Quartets result from the minimization of special intrinsic states. We present three different applications of our approach, two referring to the *sd* shell (^{24}Mg and ^{28}Si) and one to the *pf* shell (^{48}Cr). We show that the new criterion to fix the quartets provides not only a good description of the low-lying states of these nuclei but, as a peculiarity, it also allows the identification of band-like structures in their spectra.

*e-mail: michelangelo.sambataro@ct.infn.it
**e-mail: sandulescu@theory.nipne.ro

https://doi.org/10.1051/epjconf/202329002023

In Section 2, we describe our approach and present the applications. In Section 3, we summarize the results and draw the conclusions.

2 Formalism and applications

The quartets that we employ are formed by two neutrons and two protons coupled to total isospin $T = 0$ and angular momentum J. The associated creation operator is defined as

$$q^+_{JM} = \sum_{i_1 j_1 J_1} \sum_{i_2 j_2 J_2} \sum_{T'} q_{i_1 j_1 J_1, i_2 j_2 J_2, T'} [[a^+_{i_1} a^+_{j_1}]^{J_1 T'} [a^+_{i_2} a^+_{j_2}]^{J_2 T'}]^{JT=0}_M, \tag{1}$$

where $a^+_{i\tau}$ creates either a proton or a neutron (depending on the isospin projection τ) on the spherically-symmetric state $i \equiv \{n_i, l_i, j_i\}$. No restrictions on the intermediate couplings $J_1 T'$ and $J_2 T'$ are introduced and the amplitudes $q_{i_1 j_1 J_1, i_2 j_2 J_2, T'}$ are supposed to guarantee the normalization of the operator.

In the representation spanned by the quartets (1) we introduce a set of intrinsic states, the simplest of which has the form [23]

$$|\Theta_g\rangle = N_g (Q^+_g)^n |0\rangle, \tag{2}$$

where n denotes the number of quartets which can be formed with the valence nucleons outside the closed core, denoted by $|0\rangle$. $|\Theta_g\rangle$ is a condensate of the quartet Q^+_g defined by

$$Q^+_g = \sum_J \alpha^{(g)}_J (q^+_g)_{J0}, \tag{3}$$

where

$$(q^+_g)_{J0} = \sum_{i_1 j_1 J_1} \sum_{i_2 j_2 J_2} \sum_{T'} q^{(g)}_{i_1 j_1 J_1, i_2 j_2 J_2, T'} [[a^+_{i_1} a^+_{j_1}]^{J_1 T'} [a^+_{i_2} a^+_{j_2}]^{J_2 T'}]^{JT=0}_0 \tag{4}$$

In order to fix Q^+_g, we minimize the energy of the state $|\Theta_g\rangle$ with respect to the coefficients $q^{(g)}_{i_1 j_1 J_1, i_2 j_2 J_2, T'}$ and $\alpha_{g,J}$. The state (2) will be referred to as "ground" intrinsic state.

In addition to the state (2), we introduce a set of intrinsic states which are generated by promoting one of the quartets Q^+_g of $|\Theta_g\rangle$ to an excited $T = 0$ configuration. These states have the general form [23]

$$|\Theta_k\rangle = N_k Q^\dagger_k (Q^\dagger_g)^{(n-1)} |0\rangle, \tag{5}$$

with

$$Q^\dagger_k = \sum_J \alpha^{(k)}_J (q^\dagger_k)_{Jk}, \tag{6}$$

$$(q^+_k)_{Jk} = \sum_{i_1 j_1 J_1} \sum_{i_2 j_2 J_2} \sum_{T'} q^{(k)}_{i_1 j_1 J_1, i_2 j_2 J_2, T'} [[a^+_{i_1} a^+_{j_1}]^{J_1 T'} [a^+_{i_2} a^+_{j_2}]^{J_2 T'}]^{JT=0}_k \tag{7}$$

Assuming that the quartet Q^+_g has already been fixed, we construct the new quartet Q^+_k by minimizing the energy of $|\Theta_k\rangle$ with respect to the coefficients $q^{(k)}_{i_1 j_1 J_1, i_2 j_2 J_2, T'}$ and $\alpha^{(k)}_J$ (under the constraint of orthogonality when various states with the same k are involved). The states (5) will be identified with the value of the quantum number k. In particular, for $k = 0$ and 2 the states (5) will be referred to as "β" and "γ" intrinsic states, respectively. We will also indicate with the symbols q^+_β and q^+_γ the associated quartets (7).

Having defined the quartets, we are ready to carry out configuration-interaction calculations in a space spanned by these quartets. To this purpose we define the set of states (we work in the m-scheme)

$$|\Psi^{(n)}_{\overline{M}}, \{N_{JM}\}\rangle = \prod_{J \in (0, J_{max}); M \in (-J, J)} (q^+_{JM})^{N_{JM}} |0\rangle \tag{8}$$

with the conditions

$$\sum_{JM} N_{JM} = n, \qquad \sum_{JM} M N_{JM} = \overline{M}. \tag{9}$$

We then orthonormalize the states (8) and diagonalize the Hamiltonian in this new basis for the various $\overline{M}$. Calculations have been carried out in the sd and pf shells by adopting the USDB [24] and KB3G [25] interactions, respectively. The vacuum state $|0\rangle$ in Eq. (8) stands for the nucleus ^{16}O for sd shell nuclei and for ^{40}Ca for pf shell nuclei.

In Figs. 1,2 and 3, columns (A) and (B), we show what changes occur in the low-lying spectra of the nuclei under study when a set of dynamical quartets replace the analogous set of static quartets. These static quartets are those describing the lowest states of ^{20}Ne for the sd shell nuclei and those of ^{44}Ti for ^{48}Cr. In the columns (A) we plot all the positive parity states up to the 6_1 obtained using a set of static J=0,2,4 (plus J=6 in the case of ^{48}Cr) quartets. In spite of the quite good results for the ground state correlation energy (defined as the difference between the ground state energies with and in absence of interaction), in all cases the spectra show marked differences with respect to the shell model results. In the columns (B) of the same figures, the dynamical quartets with the same J's as above have been employed. In all three nuclei one observes a lowering of the yrast states $J = 0, 2, 4, 6$ forming the ground state bands while most of the remaining states are pushed up in energy. The new ground state bands are all closer in energy to the shell model ones and a considerable improvement is observed also in the accuracy of the ground state correlation energies. Thus adopting the quartets $(q^+_g)_{J0}$ associated with $|\Theta_g\rangle$ has had a positive effect only on the ground state bands of the nuclei under study.

Still in Figs. 1, 2 and 3 we observe how the use of the intrinsic states $|\Theta_k\rangle$ (5) can help to improve the spectra of the nuclei under study. In ^{24}Mg, Fig. 1(B), one notices that a band formed by the $J_k = 2_2, 3_1, 4_2, 5_1$ states has been shifted higher in energy when passing from static to dynamical quartets. By making use of the definition of the γ intrinsic state, we construct new quartets $(q^\dagger_\gamma)_J$ with $J = 2, 3, 4$ and perform a configuration-interaction calculation that includes these new quartets in addition to those already used for the calculation of Fig. 1(B). The new result is shown in Fig. 1(C). We observe a clear lowering of the energies of the states $J_k = 2_2, 3_1, 4_2, 5_1$ while the states of the ground state band have remained basically unmodified with respect to those of column (B). The inclusion of the quartets derived with the help of the γ intrinsic state

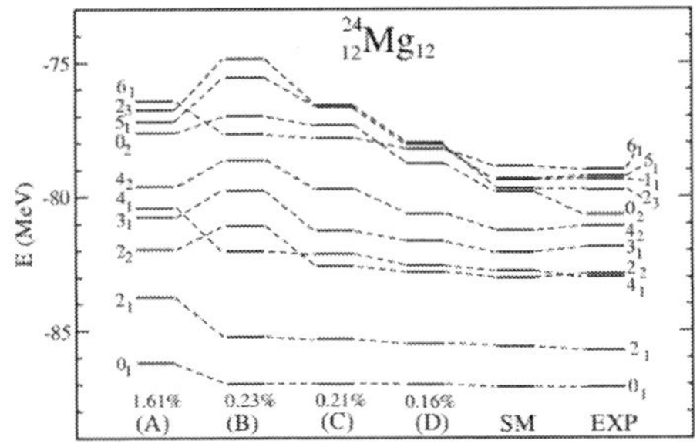

Figure 1. Spectra of ^{24}Mg obtained by performing configuration-interaction calculations in spaces built with various sets of $T = 0$ quartets (see text): (A), $J = 0, 2, 4$ static quartets from ^{20}Ne; (B), $J = 0, 2, 4$ dynamical quartets from the ground intrinsic state (2); (C), the same set as in (B) plus $J = 2, 3, 4$ quartets from the γ intrinsic state; (D), the same sets as in (C) plus $J = 0, 2, 4$ quartets from the β intrinsic state. SM, shell model results; EXP, experimental spectrum. The numbers above the symbols (A)-(D) are the relative errors in the ground state correlation energy with respect to the shell model value.

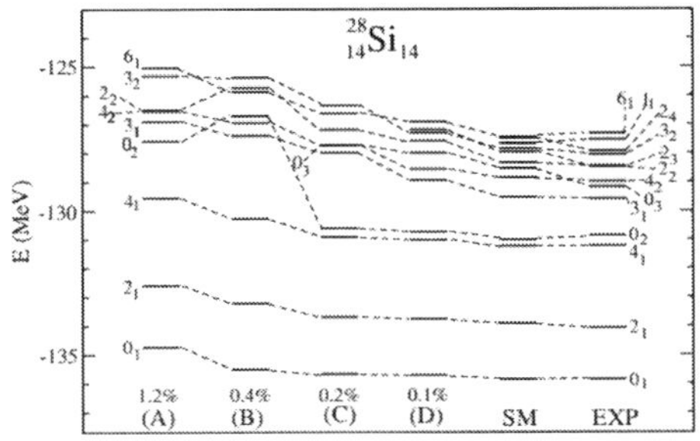

Figure 2. Spectra of ^{28}Si obtained by performing configuration-interaction calculations in spaces built with various sets of $T = 0$ quartets (see text): (A), $J = 0, 2, 4$ static quartets from ^{20}Ne; (B), $J = 0, 2, 4$ dynamical quartets from the ground intrinsic state (2); (C), the same set as in (B) plus $J = 0, 2, 4$ quartets from the β intrinsic state; (D), the same sets as in (C) plus $J = 3, 4$ quartets from the $k = 3$ intrinsic state (5). SM, shell model results; EXP, experimental spectrum. The numbers above the symbols (A)-(D) are the relative errors in the ground state correlation energy with respect to the shell model value.

has therefore essentially affected only those states which can be associated to a γ band of ^{24}Mg (and, in addition, the state 2_3). As a final step, we have explored the effect on this spectrum of the inclusion of a set of quartets $(q^\dagger_\beta)_J$ with $J = 0, 2, 4$. The basic effect which can be observed in Fig. 1(D) is a lowering the states $J = 0_2, 2_3$ which can be associated to a β band of ^{24}Mg. As a result of the

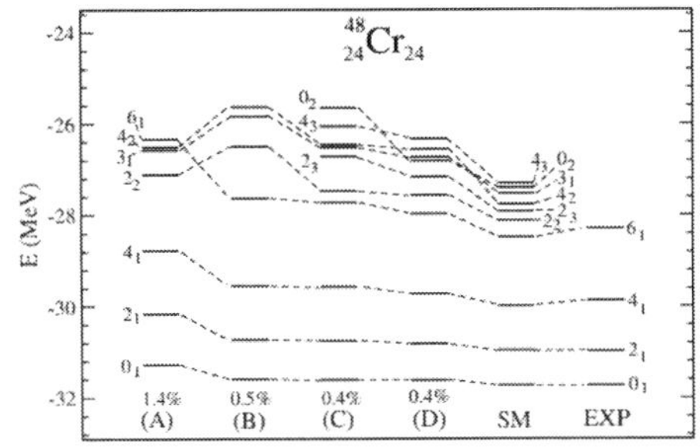

Figure 3. Spectra of ^{48}Cr obtained by performing configuration-interaction calculations in spaces built with various sets of $T = 0$ quartets (see text): (A), $J = 0, 2, 4, 6$ static quartets from ^{44}Ti; (B), $J = 0, 2, 4, 6$ dynamical quartets from the ground intrinsic state (2); (C), the same set as in (B) plus $J = 2, 3, 4$ quartets from the γ intrinsic state; (D), the same sets as in (C) plus $J = 0, 2, 4$ quartets from the β intrinsic state. SM, shell model results; EXP, experimental spectrum. The numbers above the symbols (A)-(D) are the relative errors in the ground state correlation energy with respect to the shell model value.

new diagonalization, also the 5_1 state is lowered in energy. The final spectrum shows a good agreement with the shell model one.

For what concerns ^{28}Si, what is most striking in the spectrum of Fig. 2(B) is the absence of a state 0_2 close to the state 4_1 as observed in both the shell model and the experimental spectrum. By interpreting this state 0_2 as the head of a β band, as a next step, we enlarge the model space by also including the quartets $(q^\dagger_\beta)_J$ with $J = 0, 2, 4$ associated with the β intrinsic state. The result of the new configuration calculation can be seen in Fig.2(C). The inclusion of the new quartets mostly affects the yrare $J = 0, 2$ states by giving rise, in particular, to a surprising lowering of the 0_2 state which positions itself immediately above the 4_1 state, where it is expected to be. In spite of the fact that a reasonably good agreement with the shell model spectrum has already been achieved, we perform an additional calculation which also includes the quartets $(q^\dagger_3)_{J3}$, with $J = 3, 4$, associated with the $k = 3$ intrinsic state (5). Such a new calculation is stimulated by an old analysis of ^{28}Si [26] in which the band head of a $K = 3$ band is positioned immediately above the 0_2 state. As it can be seen in Fig. 2(D), the new calculation essentially lowers the energy of the 3_1 and 4_2 states and, in addition, that of the 3_2 state. This new calculation further improves the quality of the QM spectrum which compares well with the shell model one.

The last case under investigation, ^{48}Cr, shares some analogies with the corresponding one of ^{24}Mg. Indeed one observes in Fig. 3(B) that, also in this case, the states $2_2, 3_1, 4_2$ have been shifted higher in energy when replacing the static quartets with the corresponding dynamical ones associated with the ground intrinsic state. These states reminding those of a γ band, we proceed as for ^{24}Mg by introducing the quartets $(q^{\dagger}_{\gamma})_J$, associated with the γ intrinsic state, with $J = 2, 3, 4$. The new calculation, Fig. 3(C), leaves unaffected the states of the ground rotational band (as for ^{24}Mg) while it lowers significantly the states $2_2, 3_1, 4_2$. The new calculation also leads to the appearance of new states $(2_3, 4_3, 0_2)$ in the highest part of the spectrum. These states have a correspondence with the shell model ones but with a 0_2 still too high in energy. By interpreting this state as a possible head of a β band and wishing to lower its energy, we perform a final calculation which includes also the quartets $(q^{\dagger}_{\beta})_J$ with $J = 0, 2, 4$. This calculation leads to a significant lowering of the 0_2 state (Fig. 3(D)) which improves the agreement between exact and approximate spectra.

3 Summary and conclusions

In this paper we have provided a description of deformed $N = Z$ nuclei in a formalism of α-like quartets. Quartets have been constructed variationally by resorting to the use of proper intrinsic states. Spectra have been obtained by carrying out configuration-interaction calculations in spaces built with these quartets. For each nucleus more sets of quartets have been used in correspondence with the various types of intrinsic states introduced. The intriguing aspect of these calculations has been the observation of band-like structures associated with the various sets of quartets. Also intriguing has been observing that these structures interfere little with each other in the sense that once one structure has been generated by means of a set of quartets, enlarging the configuration space with a new set of quartets moves down the states associated with the new structure without, however, strongly affecting the previous structure. The procedure has been applied to ^{24}Mg, ^{28}Si and ^{48}Cr nuclei and it has provided a good description of the low-lying spectra in all three cases. A merit of this description is that of relying on only a few degrees of freedom. For what concerns the ground states, in particular, we have shown that already the $T = 0$ quartets with $J = 0$ and $J = 2$ can guarantee an accurate approximation of these states. As a general conclusion, the results achieved in this work promote the new method proposed for the definition of the quartets as well as the use of the latter as basic structures for an effective description of the ground and excited states of deformed $N = Z$ nuclei.

Acknowledgements

This work was supported by a grant of the Romanian Ministry of Research and Innovation, CNCS - UEFISCDI, project number PCE 160/2021, within PNCDI III.

References

[1] B. Belyaev, V. Zacharev, V. Soloviev, Soviet Physics JETP **11**, 686 (1960)
[2] A. Romero, J. Dobaczewski, A. Pastore, Phys. Lett. B **795**, 177 (2019)
[3] E. Rrapaj, A. Machiavelli, A. Gezerlis, Phys. Rev. C **99**, 014321 (2019)
[4] R. Richardson, Phys. Rev. **144**, 874 (1966)
[5] F. Pan, J. Draayer, Phys. Rev. C **66**, 044314 (2002)
[6] J. Links, H. Zhou, M. Gould, R. McKenzie, J. Phys. A: Math. Gen. **35**, 6459 (2002)
[7] J. Dukelsky, V. Gueorguiev, P.V. Isacker, S. Dimitrova, B. Errea, S.H. Lerma, Phys. Rev. Lett. **96**, 072503 (2006)
[8] T. Skrypnyk, Nucl. Phys. B **863**, 435 (2012)
[9] T. Skrypnyk, Nucl. Phys. B **864**, 770 (2012)
[10] M. Sambataro, N. Sandulescu, J. Phys. G: Nucl. Part. Phys. **47**, 045112 (2020)
[11] J. Dobes, S. Pittel, Phys. Rev. C **57**, 688 (1998)
[12] V. Soloviev, Nucl. Phys. **18**, 161 (1960)
[13] B. Flowers, M. Vujicic, Nucl. Phys. **49**, 586 (1963)
[14] A. Arima, V. Gillet, J. Ginocchio, Phys. Rev. Lett. **25**, 1043 (1970)
[15] H. Faraggi, A. Jaffrin, M.C. Lemaire, M. Mermaz, J.C. Faivre, J. Gastebois, B. Harvey, J.M. Loiseaux, A. Papineau, Ann. of Phys. **66**, 905 (1971)
[16] J. Eichler, M. Yamamura, Nucl. Phys. A **182**, 33 (1972)
[17] F. Catara, J.M. Gomez, Nucl. Phys. A **215**, 85 (1973)
[18] M. Hasegawa, S. Tazaki, R. Okamoto, Nucl. Phys. A **592**, 45 (1995)
[19] R. Senkov, V. Zelevinsky, Phys. At. Nucl. **74**, 1267 (2011)
[20] M. Sambataro, N. Sandulescu, Phys. Rev. Lett. **115**, 112501 (2015)
[21] M. Sambataro, N. Sandulescu, Phys. Rev. C **91**, 064318 (2015)
[22] M. Sambataro, N. Sandulescu, Phys. Lett. B **763**, 151 (2016)
[23] M. Sambataro, N. Sandulescu, Phys. Lett. B **827**, 136987 (2022)
[24] B. Brown, W. Richter, Phys. Rev. C **74**, 034315 (2006)
[25] A. Poves, G. Martinez-Pinedo, Phys. Lett B **430**, 203 (1998)
[26] R. Sheline, S. Kubono, K. Morita, M. Tanaka, Phys. Lett. B **119**, 263 (1982)

Proton-neutron pairing, α-like quarteting and ground/excited states of N=Z nuclei

Nicolae Sandulescu[1,*]

[1]National Institute of Physics and Nuclear Engineering, Magurele - Bucharest, Romania

Abstract. Various studies have shown that the proton-neutron (pn) pairing correlations can be accurately described not by a condensate of Cooper pairs, as considered in the majority of mean-field calculations, but by a condensate of α-like quartets. After a short review of the quartet condensate model (QCM), it is discussed the effect of the pn pairing on the ground states of $N = Z$ nuclei, analysed recently in the framework of Skyrme-HF+QCM calculations. An interesting aspect pointed out by these calculations is the strong interdependence between all types of pairing correlations. In particular, when the isoscalar pn pairing channel is switched on, the pairing correlations are redistributed among all the pairing channels without changing significantly the total pairing energy. Due to this reason, for the majority of $N = Z$ nuclei, the binding energy is not affected much when the isoscalar pairing channel is switched on. Yet, in all calculations which include both the isovector and the isoscalar pairing forces, the isoscalar pairing correlations contribute significantly to the binding energies and coexist always with the isovector pn pairing. Finally it is presented a recent extension of the QCM approach to the excited stated of pn pairing Hamiltonians. It will be shown that the low-lying excited states of these Hamiltonians can be described by breaking a quartet from the ground state condensate and replacing it with an "excited" quartet, an approach which is analogous to the one-broken-pair approximation which is commonly used for like-particle systems.

1 Introduction

The role of proton-neutron (pn) pairing correlations in $N = Z$ nuclei is an issue debated for many years [1]. In the $N = Z$ nuclei are usually considered two types of pn pairing correlations, corresponding to spin-singlet isovector (S=0,T=1) and spin-triplet isoscalar (S=1,T=0) pn pairs. The isovector pn pairing is supposed to have similar properties as the neutron-neutron (nn) and proton-proton (pp) pairing. What is the role played by the isoscalar pn pairing in nuclei is not yet clear. In particular, the isoscalar pairing is commonly associated to a long standing question in nuclear structure, i.e., whether a condensate of deuteron-like proton-neutron pairs could exist in N = Z nuclei. In spite of many experimental and theoretical studies, up to now there is not a compelling evidence for a deuteron-like condensation in nuclei.

The majority of theoretical studies on pn pairing have been done in the Hartree-Fock-Bogoliubov (HFB) approach (see [2] and the references quoated therein). The great advantage of the HFB model is the unitary treatment of both isovector and isoscalar pairing in a simple approach based on the generalised Bogoliubov transformation. This advantage comes however with important drawbacks : the non-conservation of the particle number, the isospin and the angular momentum. To restore the conservation of these quantities employing projection techniques is a difficult task and some results on this line exist only for the trivial case of degenerate levels [3, 4]. Realistic beyond-HFB calculations with particle number and angular momentum projections have been done recently, but with the projection performed after the variation [5].

Many years ago it was pointed out that the isovector pairing generates, through the isospin conservation, α-like quartet structures formed by two neutrons and two protons coupled to the total isospin $T = 0$ [6–8]. In one of the first quartet formalisms, Flowers and Vujici supposed that the ground state of the isovector pairing Hamiltonian can be approximated by a BCS-type condensate expressed in term of quartets. This approach, which conserves exactly the isospin but not the particle number, is rather complicated and it was never applied in realistic calculations. An exact quartet solution was found later on for the particular case of pn pairing interactions acting on degenerate single-particle levels [3] . It was thus shown that, in this case, the ground state of $N = Z$ systems interacting by an isovector force or by isovector and isoscalar pairing forces of equal strength, can be expressed exactly by a quartet condensate. A realistic quartet condensation model (QCM) , for non-degenerate levels and general pairing forces, was proposed later on in Refs. [9–11]. These QCM models will be revised below and then applied for estimating the contribution of pn pairing to the binding energies of $N = Z$ nuclei [12].

Recently the QCM model was extended to describe the excited states of the pn Hamiltonians [13]. It was shown that the low-lying excites states of these Hamiltonians can

*e-mail: nsandulescu@theory.nipne.ro

be described accurately by by breaking a pair from the quartet condensate and replacing it with an excited quartet. This formalism will be presented in the second part of the article.

2 QCM approach for axially-deformed pairing Hamiltonians

In this section we consider the case of isovector-isoscalar pairing forces which scatter pairs of nucleons in time-reversed states. This is a typical situation for the mean field plus paiirng calculations done in the intrinsic system of axially-deformed nuclei. The QCM formalism corresponding to this case, proposed in Ref. [10], will be summarised below.

The Hamiltonian which describes the pairing correlations in the axially-deformed system has the expression [10],

$$\begin{aligned} H &= \sum_{i,\tau=\pm 1/2} \varepsilon_{i\tau} N_{i\tau} + \sum_{i,j} V_{i,j}^{(T=1)} \sum_{t=-1,0,1} P_{i,t}^{\dagger} P_{j,t} \\ &+ \sum_{i,j} V_{i,j}^{(T=0)} D_{i,0}^{\dagger} D_{j,0}, \end{aligned} \tag{1}$$

where $\varepsilon_{i,\tau}$ are the single-particle energies of neutrons (τ=1/2) and protons (τ=-1/2), while $N_{i,\tau}$ are the particle number operators. The second term is the isovector pairing interaction expressed by the isovector pair operators $P_{i,1}^{\dagger} = \nu_i^{\dagger}\nu_{\bar{i}}^{\dagger}$, $P_{i,-1}^{\dagger} = \pi_i^{\dagger}\pi_{\bar{i}}^{\dagger}$, $P_{i,0}^{\dagger} = (\nu_i^{\dagger}\pi_{\bar{i}}^{\dagger} + \pi_i^{\dagger}\nu_{\bar{i}}^{\dagger})/\sqrt{2}$. The third term is the isoscalar pairing interaction and $D_{i,0}^{\dagger} = (\nu_i^{\dagger}\pi_{\bar{i}}^{\dagger} - \pi_i^{\dagger}\nu_{\bar{i}}^{\dagger})/\sqrt{2}$ is the isoscalar pair operator. By $\nu_i^{\dagger}$ and $\pi_i^{\dagger}$ are denoted the creation operators for neutrons and protons in the state i, while $\bar{i}$ is the time conjugate of the state i. The states i, which correspond to the axially deformed mean-field, are characterised by the quantum numbers $i \equiv \{a_i, \Omega_i\}$, where Ω_i is the projection of the angular momentum on the symmetry axis.

By construction, in Eq. (1) the pairs operators have $J_z = 0$ but not a well-defined angular momentum J. In fact, when expressed in the laboratory frame, the isovector and the isoscalar intrinsic pairs can be written as a superposition of pairs with $J = 0, 2, 4, ..$ and, respectively, $J = 1, 3, 5, ...$ Therefore, the Hamiltonian (1) takes into account, in an effective way, pairing correlations which are not restricted only to the standard (J=0,T=1) and (J=1,T=0) channels.

In order to find the ground state energy of the Hamiltonian (1), we employ the quartet condensation model (QCM). Thus, according to QCM, the ground state of Hamiltonian (1) for even-even $N = Z$ systems is approximated by the trial state [10]

$$|QCM\rangle = (Q_{iv}^{\dagger} + \Delta_0^{\dagger 2})^{n_q}|0\rangle, \tag{2}$$

where $n_q = (N + Z)/2$, while $|0\rangle$ is the "vacuum" state represented by the nucleons which are supposed to be not affected by the pairing interaction. The operator $A^{\dagger}$ is the isovector quartet built by two isovector non-collective pairs coupled to the total isospin $T = 0$, i.e.,

$$Q_{iv}^{\dagger} = \sum_{i,j} x_{ij} [P_i^{\dagger} P_j^{\dagger}]^{T=0}. \tag{3}$$

Assuming that the mixing coefficients are separable, i.e., $x_{ij} = x_i x_j$, the isovector quartet takes the form

$$Q_{iv}^{\dagger} = 2\Gamma_1^{\dagger}\Gamma_{-1}^{\dagger} - (\Gamma_0^{\dagger})^2, \tag{4}$$

where

$$\Gamma_t^{\dagger} = \sum_i x_i P_{i,t}^{\dagger} \tag{5}$$

are collective pair operators for neutron-neutron pairs (t = 1), proton-proton pairs ($t = -1$) and proton-neutron pairs ($t = 0$). The isoscalar degrees of freedom are described by the collective isoscalar pair

$$\Delta_0^{\dagger} = \sum_i y_i D_{i,0}^{\dagger}. \tag{6}$$

The QCM states depend on the mixing amplitudes of the collective pair operators. They are determined variationally by minimizing the average of the Hamiltonian under the normalization condition imposed to the trial state. Details about these calculations are presented in Ref. [10].

3 The effect of pn pairing on the ground states of $N = Z$ nuclei

The QCM formalism presented above has been used recently to evaluate the contribution of the pn pairing to the ground state of nuclei close to the $N = Z$ line [12]. Below we present shortly the calculation scheme and the main results for the $N = Z$ nuclei.

To calculate the binding energies, we use a self-consistent mean-field plus pairing formalism. The calculations are done in the intrinsic system defined by an axially deformed mean-field generated by a Skyrme functional. The QCM calculations for the Hamiltonian (1) are performed iteratively with the Skyrme-HF calculations in a similar way as in the axially-deformed Skyrme+BCS calculations. Thus, at a given iteration, the QCM equations are solved for the single-particle states generated by the Skyrme functional. Then, the occupation probabilities of the single-particle states provided by QCM are employed to get new densities and a new Skyrme functional which, in turn, is generating new single-particle states. At the convergence, the binding energy is obtained by adding to the mean-field energy the contribution of the pairing energy. The latter is calculated as the average of the pairing force from which it is extracted out the contribution of self-energy terms. For the like-particle pairing these terms are

$$E_{n(p)}^{mf} = \sum_i V^{T=1}(i,i) v_{i,n(p)}^4, \tag{7}$$

while for the pn pairing the expresions are

$$E_{pn}^{mf}(T) = \sum_i V^T(i,i) v_{i,p}^2 v_{i,n}^2. \tag{8}$$

In the expressions above, $v^2_{i,n(p)}$ are the occupation probabilities for neutrons (protons) corresponding to the states included in the pairing calculations.

In the present calculations, for the isovector-isoscalar pairing interaction we employ a zero range force of the form:

$$V^T(r_1, r_2) = V_0^T \delta(r_1 - r_2) \hat{P}^T_{S,S_z} \tag{9}$$

where $\hat{P}^T_{S,S_z}$ is the projection operator on the spin of the pairs, namely, $S = 0$ for the isovector force and $S = 1, S_z = 0$ for the isoscalar force. The matrix elements of the pairing interaction (11) for the single-particle states provided by the Skyrme functional are calculated as shown in the Appendix of Ref. [14].

The most representative results for the binding energies of $N = Z$ nuclei are presented in Fig. 1. The figure shows the binding energies residuals, i.e., the difference between the theoretical and experimental binding energies. The parameters employed in the calculations are indicated in the figure. In what follows, we shall focus on the results corresponding to the pairing force of strength V_0=350. First of all, it can be seen that the Skyrme-HF results, obtained by using the parametrisation UNE1 [15], underestimate the binding energies by about 3-4 MeV in the middle mass region, while for the nuclei with $A > 90$ the calculated binding energies are larger than the experimental ones. As expected, the Skyrme-HF+BCS calculations, which take into account only the neutron-neutron (nn) and proton-proton (pp) pairing, is smoothing out the fluctuations of the HF results caused by the shell effects. For the N=Z nuclei with $60 < A < 80$, where the HF fluctuations are small, in BCS approximation the residuals are decreasing by about 1 MeV compared to the HF values.

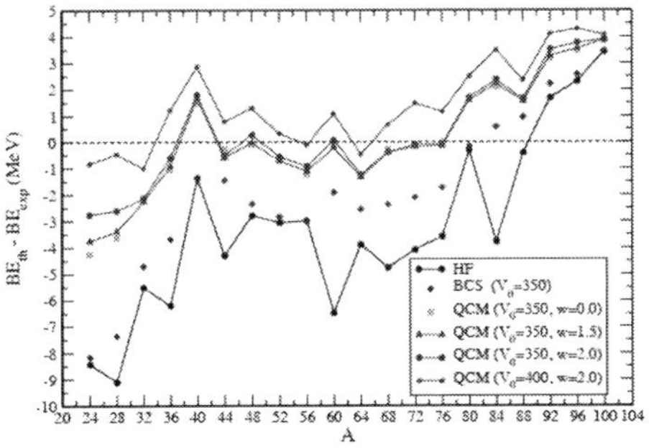

Figure 1. Binding energies residuals, in MeV, for even-even N=Z nuclei as a function of A=N+Z. The results correspond to the pairing forces indicated in the figure.

From Fig. 1 it can be seen that the binding energies are increasing significantly when are taken into account the isovector pn pairing correlations, treated in the QCM approach. On the other hand, except for A=24 and A=28, the effect of the isoscalar pn pairing on the binding energies is surprisingly small. This fact is caused by the competition between various pairing channels and between pairing and mean field. Thus, when the isoscalar pn channel is switched on, the like-particle and isovector pn pairing energies are decreasing. This decrease of the isovector pairing is not compensated by the pairing energy gained by opening the isoscalar channel. This is due to the fact that the off-diagonal matrix elements of the isoscalar interaction, which have the dominant contribution to the pairing energy, are small. One of the important conclusions of these calculations is that the isovector pn pairing has the most significative contribution to the binding energies of $N = Z$ nuclei. The isoscalar pairing correlations contribute less to the binding energies but they are present in all the calculated nuclei.

4 QCM approach for the spherical-symmetric pairing Hamiltonians

In this section we present briefly the the QCM approach for the isovector-isoscalar pairing Hamiltonians with spherical symmetry, introduced in Ref. [11].

The isovector-isoscalar pairing Hamiltonian consider here has the expression

$$\begin{aligned} H &= \sum_i \epsilon_i N_i + \sum_{i,j} V^{T=1}_{J=0}(i,j) \sum_{T_z} P^+_{i,T_z} P_{j,T_z} \\ &+ \sum_{i \le j, k \le l} V^{T=0}_{J=1}(ij,kl) \sum_{J_z} D^+_{ij,J_z} D_{kl,J_z}. \end{aligned} \tag{10}$$

The first two terms are the same as in Eq. (1) while the last term is the isoscalar pairing interaction written in term of the isoscalar pair operator where

$$D^+_{j_1 j_2 J_z} = \frac{1}{\sqrt{1+\delta_{j_1 j_2}}} [a^+_{j_1} a^+_{j_2}]^{J=1,T=0}_{J_z} \tag{11}$$

For even-even $N = Z$ systems the QCM ansatz for the ground state has formally the same expression as in the case of isovector pairing

$$|\Psi_{gs}\rangle = (Q^+_{ivs})^{n_q}|0\rangle. \tag{12}$$

The difference is that now the quartet operator Q^+_{ivs}, still having total isospin $T = 0$, is the sum of two quartets

$$Q^+_{ivs} = Q^+_{iv} + Q^+_{is}, \tag{13}$$

where

$$Q^+_{iv} = \sum_{ij} x_{ij} [P^\dagger_i P^\dagger_j]^{T=0} \tag{14}$$

is the collective quartet built by a linear combination of two non-collective isovector pairs coupled to the total isospin $T = 0$. The isoscalar quartet Q^+_{is} is formed by two isoscalar pairs coupled to total $J = 0$, i.e.,

$$Q^+_{is} = \sum_{j_1 j_2 j_3 j_4} y_{j_1 j_2 j_3 j_4} [D^+_{j_1 j_2} D^+_{j_3 j_4}]^{J=0}. \tag{15}$$

The parameters which define the collective quartets are determine from the minimisation of the average of the Hamiltonian on the QCM state (12). In Ref. [11] it was shown that the QCM approach provides accurate ground state energies for $N = Z$ systems interacting by realistic isovector-isoscalar pairing forces.

5 Excited states of pn pairing Hamiltonians in the QCM approach

Recently the QCM approach presented in the previous section was extended for describing the low-lying excited states of $N = Z$ systems interacting by isovector-isoscalar pairing forces [13]. Below we present briefly this extension and the most important findings.

In correspondence with the QCM ansatz (12) for the ground state, we construct a class of excited states by replacing a quartet of the condensate with an "excited" quartet. For the case of a spherically-symmetric mean field, these states take the form

$$|\Phi_{\nu,JJ_z}\rangle = \tilde{Q}_{\nu,JJ_z}(Q^+_{ivs})^{n_q-1}|-\rangle, \tag{16}$$

where

$$\tilde{Q}^+_{\nu,JM_\nu} = \sum_{ijkl} Y^{(\nu)}_{ijkl,J}[a_i^+a_j^+a_k^+a_l^+]^{T=0,J} \tag{17}$$

are the excited collective quartets. In order to define its coefficients $Y^{(\nu)}_{JJ_z}$, one has now to diagonalize the Hamiltonian (10) in the basis of non-orthogonal states

$$[P^+_{J_1,T'}(i_1,j_1)P^+_{J_2,T'}(i_2,j_2)]^{J,T=0}_{J_z}(Q^+_{ivs})^{n_q-1}|-\rangle. \tag{18}$$

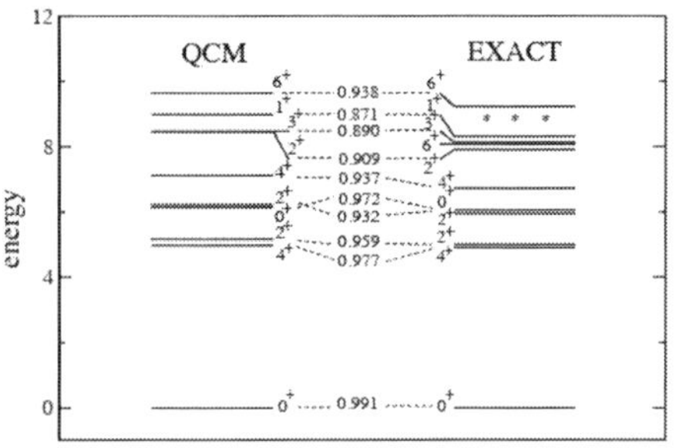

Figure 2. The low-lying spectrum provided by the QCM approximation (16) for the valence nucleons of ^{28}Si interacting by an isovector-isoscalar pairing force extracted from the USDB interaction. The numbers are the overlaps between the QCM and the exact wave functions. Energies are in MeV.

To illustrate the accuracy of the approximation (16) we take as example the valence nucleons in ^{28}Si and assume an isovector-isoscalar pairing force corresponding to the $(J = 0, T = 1)$ and $(J = 1, T = 0)$ channels of the the USDB interaction. Exact and approximate spectra are shown in Fig. 2. It can be seen that the overall agreement is good. For more details see Ref. [13].

6 Summary and Conclusions

We have shown how the iosvector and isoscalar proton-neutron (pn) pairing is treated in the quartet condensation model (QCM). In the QCM approach the pairing correlations are taken into account in terms of collective quartets formed by two neutrons and two protons coupled to total isospin $T = 0$ and total angular momentum $J = 0$. In this approach the ground state of $N = Z$ systems is described by a condensate of quartets while the excited states are obtained by breaking a quartet from the ground state condensate and replacing it with an "excited " quartet. As an application we have analysed, in the framework of Skyrme+QCM, the effect of pn pairing on the ground states of $N = Z$ nuclei. It is concluded that the pn pairing, in particular the isovector pairing, has a significant contribution to the binding energies of these nuclei.

Acknowledgments In this conference paper it is summarized the work done together with Danilo Gambacurta, Daniel Negrea and Michelangelo Sambataro. This work was supported by a grant of Romanian Ministry of Research and Innovation, CNCS - UEFISCDI, project number PCE 160/2021, within PNCDI II.

References

[1] S. Frauendorf and A. O. Macchiavelli, Prog. Part. Nucl. Phys. **78**, 24-90 (2014).
[2] A. L. Goodman, Phys. Rev. C **63**, 044325 (2001).
[3] J. Dobes, and S. Pittel, Phys. Rev. C **57**, 688 (1998).
[4] A. M. Romero, J. Dobaczewski, A. Pastore, Phys. Lett. B **795**, 177 (2019).
[5] E. Rrapaj, A. O. Macchiavelli, A. Gezerlis, Phys. Rev. C **99**, 014321 (2019).
[6] B.H. Flowers and M. Vujicic, Nucl. Phys. **49**, 586 (1963).
[7] B. Bremond and J. G. Valatin, Nuclear Physics **41**, 640 (1963).
[8] J. Eichler and M. Yamamura, Nucl. Phys. A **182**, 33 (1972).
[9] N. Sandulescu, D. Negrea, J. Dukelsky, C. W. Johnson, Phys. Rev. C **85**, 061303(R) (2012).
[10] N. Sandulescu, D. Negrea, D. Gambacurta, Phys. Lett. B **751**, 348 (2015).
[11] M. Sambataro and N. Sandulescu, Phys. Rev. C **93**, 054320 (2016).
[12] D. Negrea, N. Sandulescu, D. Gambacurta, Phys. Rev. C **105**, 034325 (2022).
[13] M. Sambataro and N. Sandulescu, Phys. Lett. B **820**, 136476 (2021).
[14] D. Gambacurta, D. Lacroix, Phys. Rev. C **91**, 014308 (2015).
[15] K. Kortelainen et al, Phys. Rev.C **85**, 024304 (2012).

β-delayed neutron spectroscopy of ^{85,86}As with MONSTER

A. Pérez de Rada[1,*], *D.* Cano-Ott[1], *T.* Martínez[1], *V.* Alcayne[1], *E.* Mendoza[1], *J.* Plaza[1], *A.* Sanchez-Caballero[1], *D.* Villamarín[1], *J.* Äystö[2], *A.* Jokinen[2], *A.* Kankainen[2], *H.* Penttilä[2], *S.* Rinta-Antila[2], *J.* Agramunt[3], *A.* Algora[3], *C.* Domingo-Pardo[3], *J.* Lerendegui-Marco[3,8], *J. L.* Taín[3], *K.* Banerjee[4,5], *C.* Bhattacharya[4,5], *P.* Roy[4,5], *F.* Calviño[6], *G.* Cortés[6], *C.* Delafosse[2,7], *I.* Matea[7], *J.* Benito[9], *S.* Alhomaidhi[10,11], and *A. K.* Mistry[10,11]

[1]Centro de Investigaciones Energéticas, Medioambientales y Tecnológicas (CIEMAT), E-28040 Madrid, Spain
[2]JYFL, The Accelerator Laboratory of the Department of Physics, University of Jyväskylä, FI-40014 Jyväskylä, Finland
[3]Instituto de Física Corpuscular (IFIC), CSIC-Universidad de Valencia, E-46071 Valencia, Spain
[4]Variable Energy Cyclotron Centre (VECC), IN-700064 Kolkata, India
[5]Homi Bhabha National Institute (HBNI), IN-400094 Anushakti Nagar, Mumbai, India
[6]Universitat Politècnica de Catalunya (UPC), Departament de Física, E-08034 Barcelona, Spain
[7]IJCLab, Université Paris-Saclay, CNRS/IN2P3, FR-91405 Orsay Cedex, France
[8]Universidad de Sevilla (US), Facultad de Física, E-41012 Sevilla, Spain
[9]Universidad Complutense de Madrid (UCM), Grupo de Física Nuclear, E-28040 Madrid, Spain
[10]Institut für Kernphysik, Technische Universität Darmstadt, D-64289 Darmstadt, Germany
[11]GSI Helmholtzzentrum für Schwerionenforschung, D-64291 Darmstadt, Germany

Abstract. The β-delayed neutron emission in the ^{85,86}As β-decays has been measured at the IGISOL facility of the Accelerator Laboratory of the University of Jyväskylä (JYFL-ACCLAB). The complete β-decays have been studied with a complex setup that consists of a plastic scintillator, the **MO**dular **N**eutron time-of-flight **S**pectrome**TER** (MONSTER), and two types of γ-ray detectors—an HPGe clover and four $LaBr_3$ crystals. The β-delayed neutron energy distributions have been determined by unfolding the TOF spectra with an innovative methodology based on the iterative Bayesian unfolding method and accurate Monte Carlo simulations.

1 Introduction

The field of β-delayed neutrons has experienced an increased activity during the last decades [1] thanks to the advances in nuclear experimental techniques and radioactive ion beam (RIB) facilities producing increased yields of neutron-rich isotopes. Properties from individual precursors like the emission probability, β-feeding, and energy spectrum are being measured with advanced neutron detectors [2–4], digital data acquisition (DAQ) systems [5], and high intensity RIBs [6–9].

In this paper, the results obtained in the measurement at the IGISOL facility of the Accelerator Laboratory of the University of Jyväskylä (JYFL-ACCLAB) on the β-delayed neutron energy distributions following the ^{85,86}As β-decays are reported. Some relevant properties of the β-decays of interest are presented in Table 1 [10].

Table 1. Relevant properties of the β-decays.

Isotope	$T_{1/2}$ (s)	Q_β (keV)	$Q_{\beta n}$ (keV)	P_n (%)
^{85}As	2.021	9224.5	4687.2	59.4
^{86}As	0.945	11541.0	5380.2	35.5

*e-mail: alberto.rada@ciemat.es

2 Experimental setup

The ^{85,86}As isotopes were produced by proton-induced fission reactions in a ^{nat}U target. In each case, the isobars were separated from the bulk of fission products by the IGISOL dipole magnet with resolution $M/\Delta M = 500$ [9] and implanted on a movable tape inside of the β-detector.

The β-detector [11] consisted of a BC408 cup-like shaped plastic detector wrapped with reflector foil and coupled to an R5924-70 PMT from Hamamatsu. The detector was placed inside the vacuum pipe and had a 10 mm aperture on the side of the detector cup that allowed for the implantation on the tape to take place.

The emitted β-delayed neutrons were detected with the **MO**dular **N**eutron time-of-flight **S**pectrome**TER** (MONSTER) [4, 12]. MONSTER consists of cylindrical cells of 200 mm diameter and 50 mm height, filled with either BC501A or EJ301 scintillating liquid. Each cell was coupled through a light guide of 31 mm thickness to either a R4144 or R11833 PMT from Hamamatsu. For this experiment, 48 cells arranged in two different setups, with 30 and 18 cells at 2 and 1.5 m flight paths, respectively, were used.

The signals from the β-detector and MONSTER were used as the start and stop signals, respectively, to measure the neutrons' time-of-flight (TOF). A picture of the whole ensemble can be seen in Figure 1.

https://doi.org/10.1051/epjconf/202329002025

Figure 1. Experimental setup mounted at JYFL-ACCLAB.

All the detector signals were registered with a custom digital DAQ system developed at CIEMAT based on ADQ14DC digitizers from Teledyne SP Devices with 14 bits vertical resolution, 1 GSample/s sampling rate, and 4 channels [5]. The complete setup required 60 channels. The system integrates custom pulse shape analysis software also developed at CIEMAT to analyze the registered signals online. The used pulse shape analysis routines did not add any dead time to the measurement.

The data was collected in cycles of implantation and decay adapted to the half-lives of the isotopes involved for maximizing the activity of the neutron emitters with respect to the other implanted isobars. During the cycle, the ion beam was switched on and off by electrostatic deflection at the switchyard of the IGISOL beam line. The cycle also included an initial time interval for background measurement and a final time interval for removal of the long-lived activity of the decay products.

3 Efficiency characterization

The β-detector efficiency was determined experimentally at two different endpoint energy values from the ^{92}Sr into ^{92}Y β-decay through the β-γ coincidence method using the HPGe and $LaBr_3$ detectors. The Monte Carlo simulations were used for extending the detector efficiency as a function of the endpoint energy and for estimating the detection threshold of 260 keV. The results of the efficiency calculation and the comparison with the experimental data are shown in Figure 2.

The MONSTER neutron detection efficiency was determined through Monte Carlo simulations and validated with the data from a measurement with a 1.00 ± 0.15 GBq ^{252}Cf source. The γ-rays from the spontaneous fission of ^{252}Cf detected in the $LaBr_3$ detectors served for establishing a TOF between the emission and the detection of the spontaneous fission neutrons. Neutron signals in MONSTER were selected by performing pulse shape discrimination, allowing for a reduction of the background due to γ-rays of one order of magnitude. The response of MONSTER to different neutron energies was obtained by setting different cuts in the γ-neutron TOF spectrum. The Monte Carlo simulations were performed with the light output response obtained from previous calibrations with monochromatic energy neutron beams [12]. There is an

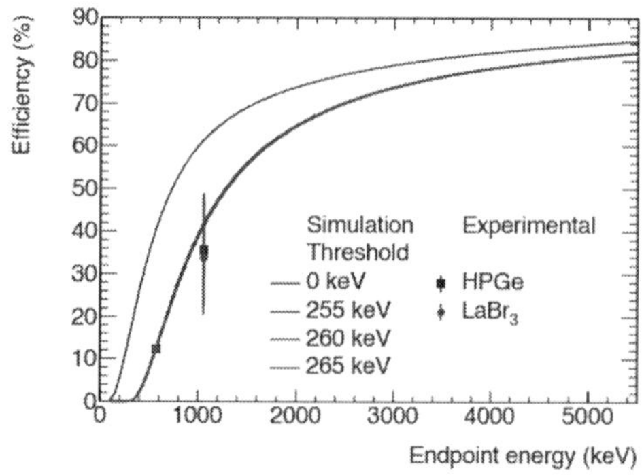

Figure 2. β-efficiency determined from the ^{92}Sr into ^{92}Y β-decay and comparison with Monte Carlo simulations.

excellent agreement between the experimental values and the Monte Carlo simulations of the MONSTER arrays for the two different flight paths, as shown in Figure 3.

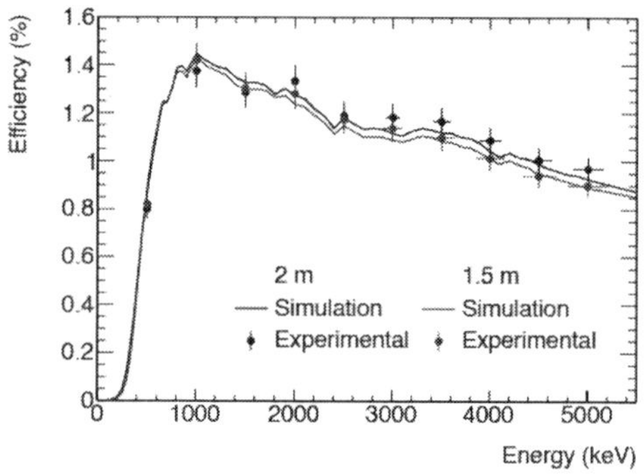

Figure 3. Total neutron detection efficiency for both MONSTER arrays determined from the ^{252}Cf source and comparison with Monte Carlo simulations.

4 Data analysis and results

4.1 β-activity distributions

The time evolution of the β-activity during each cycle due to all isotopes in the decay chain is governed by the Bateman equations. The experimental β-activity distribution is fitted with a function based on the generic solution given in [13] that describes the contribution of each of the isotopes involved in the decay chain:

$$A(t) = \sum_{i=1}^{n} \overline{\epsilon_i} \lambda_i N_i(t), \qquad (1)$$

where $\overline{\epsilon_i}$ is the average efficiency to detect a β-particle coming from the β-decay of the i-th member of the chain, λ_i is the decay constant of the i-th member of the chain, and $N_i(t)$ is the number of nuclei of the i-th member of the chain given by the solution of the Bateman equations, which depends on the respective implantation rate R_i.

The experimental β-activity distributions for both decays, together with the total fits and the individual contributions, can be seen in Figure 4. The measurement cycle in the case of ^{86}As was shorter in order to maximize its contribution with respect to the other implanted isobars.

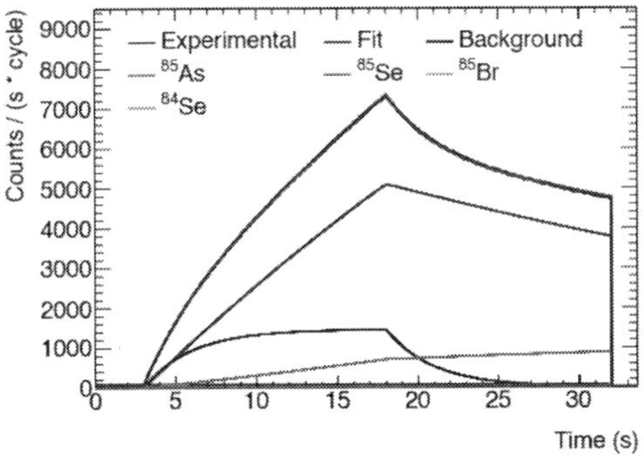

(a) ^{85}As

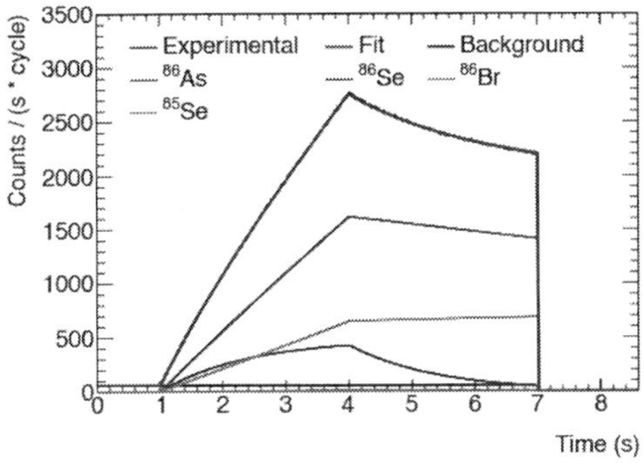

(b) ^{86}As

Figure 4. Time evolution of the β-activity distributions and the corresponding fits, including individual contributions.

The results of the fits are presented in Table 2. The other values that are in the table are the calculated average detection efficiencies as well as the total number of decays of each of the isotopes. The average detection efficiencies for each of the β-decays were calculated through Monte Carlo simulations using available data.

4.2 TOF spectra unfolding

An innovative analysis procedure based on the iterative Bayesian unfolding method [14] was developed to obtain the neutron energy distributions from the measured TOF spectra [15].

Assuming an initial probability distribution of independent causes (neutron energies) and the probability of each of the causes to produce specific effects (TOF spectrum), the conditional probability of a specific effect (E_j) being due to a certain cause (C_i) is given by:

Table 2. Results of the fit to the β-activity curve, along with the calculated average efficiencies and the total number of decays.

Isotope	$\overline{\epsilon}$ (%)	R (ions/s)	Total decays ($\times 10^6$)
^{85}As	80.8 ± 4.0	1800 ± 100	48 ± 3
^{85}Se	76.0 ± 3.8	23300 ± 900	239 ± 9
^{85}Br	69.7 ± 3.5	13900 ± 2500	42 ± 5
^{84}Se	55.1 ± 2.8	0 ± 0	1.8 ± 0.1
^{86}As	83.5 ± 4.2	570 ± 40	12.7 ± 0.8
^{86}Se	72.9 ± 3.6	16100 ± 400	74 ± 2
^{86}Br	77.5 ± 3.9	21500 ± 2100	30 ± 3
^{85}Se	76.0 ± 3.8	0 ± 0	0.32 ± 0.02

$$P(C_i|E_j) = \frac{P(E_j|C_i)P_0(C_i)}{\sum_{l=1}^{n_c} P(E_j|C_l)P_0(C_l)}. \quad (2)$$

The conditional probability $P(E_j|C_i)$ is the response matrix of the detection system and can be estimated through Monte Carlo simulations. The construction of the response matrix is crucial to achieve an accurate result. The built response matrices covered the whole energy range up to the $Q_{\beta n}$ values of the decays in intervals chosen according to the energy resolution of the system, and included an extra flat contribution to account for the γ-ray background.

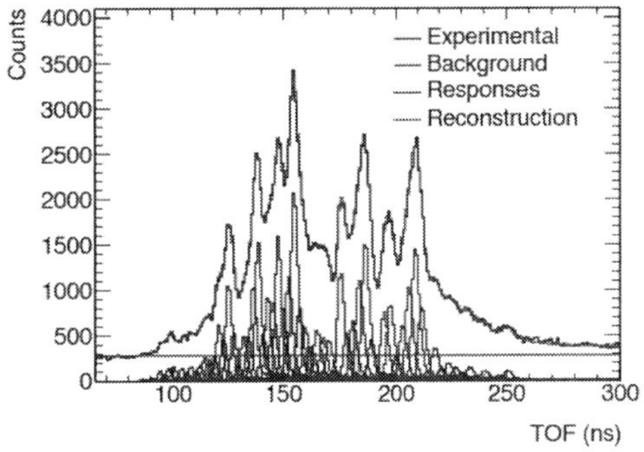

(a) ^{85}As

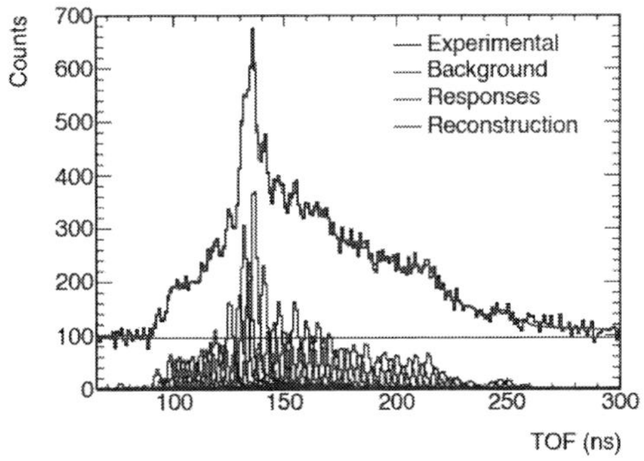

(b) ^{86}As

Figure 5. Experimental TOF spectra together with the reconstructions and individual causes' contributions.

The measured experimental TOF spectra for both decays and their reconstructions can be seen in Figure 5 for the MONSTER array at 2 m.

Finally, the neutron energy distributions obtained with both MONSTER arrays are presented in Figure 6. The agreement of the results obtained with both arrays is excellent within their energy resolution. The obtained results are also compared to existing data [10, 16]. The existing data has been rebinned to the binning of the results of this work. Also, the obtained energy distributions have been scaled so that the intensities over the experimental threshold of around 300 keV are equal to those of the existing data. The results obtained for ^{85}As are in excellent agreement with evaluations, which include experimental data obtained by the Mainz group and are supplemented with theoretical calculations. It's worth noting that no neutron intensity over around 3 MeV, which is predicted by calculations, is observed. No previous experimental data has been found for ^{86}As on the bibliography. The results of this work are in reasonable agreement with evaluations.

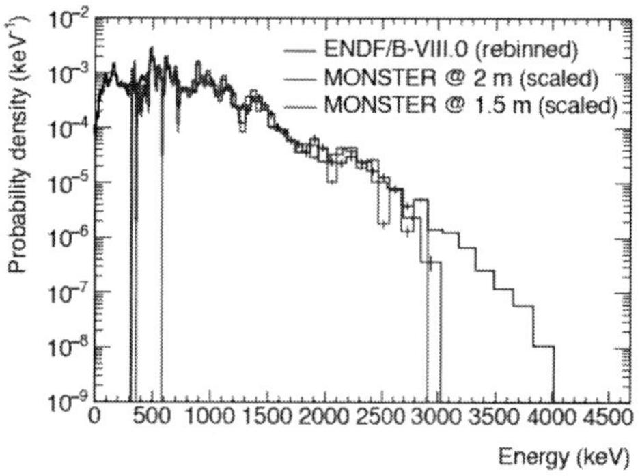

(a) ^{85}As

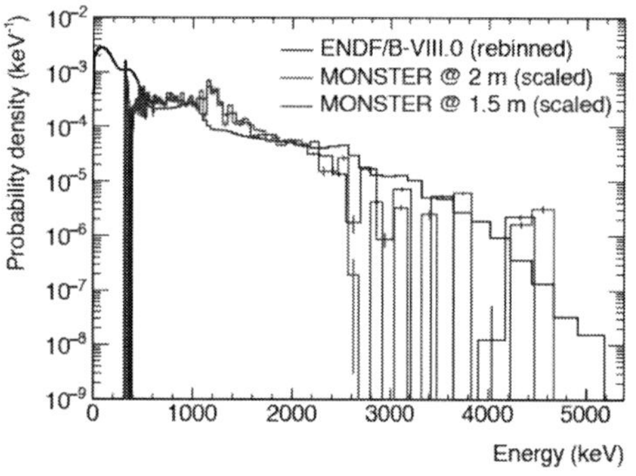

(b) ^{86}As

Figure 6. Neutron energy distributions obtained with both MONSTER arrays and compared to existing data.

5 Summary

This work is a successful commissioning and measurement with MONSTER, a neutron TOF spectrometer with excellent neutron/γ-ray discrimination capabilities and neutron energy resolution.

The efficiency of the different detectors has been obtained experimentally and reproduced with accurate Monte Carlo simulations.

An innovative analysis methodology has been developed. It is based on the fit of the β-activity curves with the Bateman equations and on the unfolding of the neutron TOF spectra with the iterative Bayesian unfolding method based on accurate Monte Carlo simulations.

The β-delayed neutron energy spectrum of ^{85}As has been procured. This result is in excellent agreement with previous experimental data and evaluations, and serves as the validation of the developed analysis methodology.

Lastly, the β-delayed neutron energy spectrum of ^{86}As has been procured for the first time. The obtained energy distribution is in reasonable agreement with evaluations.

References

[1] P. Dimitriou *et al.*, Nucl. Data Sheets, **173**, 144-238 (2021)
[2] A. Buţă *et al.*, Nucl. Instrum. and Methods A, **455**, 412-423 (2000)
[3] W. A. Peters *et al.* Nucl. Instrum. and Methods A, **836**, 122-133 (2016)
[4] A. R. García *et al.*, JINST, **7**, C05012 (2012)
[5] D. Villamarín *et al.*, Design of a high performance Digital data AcquIsition SYstem (DAISY) for innovative nuclear experiments. Submitted for publication
[6] P. Spiller *et al.*, Nucl. Instrum. and Methods A, **561**, 305-309 (2006)
[7] H. Okuno *et al.*, Prog. Theor. Exp. Phys., 03C002 (2012)
[8] R. Catherall *et al.*, Nucl. Instrum. and Methods B, **317**, 204-207 (2013)
[9] I. D. Moore *et al.*, Nucl. Instrum. and Methods B, **317**, 208-213 (2013)
[10] D. A. Brown *et al.*, Nucl. Data Sheets, **148**, 1-142 (2018)
[11] Z. Radivojevic, PhD Thesis (2001)
[12] T. Martínez *et al.*, Nucl. Data Sheets, **120**, 78-80 (2014)
[13] K. Skrable *et al.*, Health Physics, **27**, 155-157 (1974)
[14] G. D'Agostini, Nucl. Instrum. and Methods A, **362**, 487-498 (1995)
[15] A. Pérez de Rada, PhD Thesis (2022)
[16] M. C. Brady, PhD Thesis (1989)

The kink effect of the nuclear charge radii in some isotopic chains and the nucleon-nucleon tensor force within nonlinear relativistic models in the Hartree-Fock approximation.

Saturnino Marcos[1*], *Ramón* Niembro[1], and *Mercedes* López-Quelle[2]

[1]Universidad de Cantabria, Departamento de Física Moderna, 39005 Santander, Spain
[2]Universidad de Cantabria, Departamento de Física Aplicada, 39005 Santander, Spain

Abstract. Relativistic nonlinear models in the Hartree-Fock approach, with σ, ω, ρ and π mesons, are used to explore the influence of the nucleon-nucleon tensor force (*tf*) on the behaviour of the nuclear charge radii of the Pb isotopic chain. It is found that a part of its effect on the charge radii is channelled through its opposite effect to the spin-orbit interaction. The kink effect seems to be produced by the combination of the geometrical features of the $1i_{11/2}$ neutron orbital and its binding energy.

1 Introduction

The marked change in trend of the evolution of the charge radii of some isotopic families of nuclei versus the mass number A is known as kink effect (KE). This is a consequence of the shell structure of nuclei. The fact that the density-dependent Hartree-Fock model with standard Skyrme functionals [1] or Gogny forces [2] were not able to reproduce this effect for the Pb isotopic chain, for example, whereas the relativistic models in the simple mean-field approach did reasonably well [3, 4], has increased the interest in understanding the mechanism responsible for the KE [5-8]. However, it seems that the full theoretical understanding has not been reached yet.

The aim of this work is to consider a nuclear relativistic Lagrangian and use the Hartree-Fock approximation, without pairing correlations, to explore the influence of the NN tensor force (*tf*) on the behaviour of the nuclear charge radii of the Pb isotopic chain.

2 The model

The effective Lagrangian density L includes the nucleons and the σ, ω, ρ and π meson fields (φ, σ, ω_μ, ρ_μ and π, respectively), and the electromagnetic field A_μ [9]:

$$L = L_0(\varphi, \sigma, \omega_\mu, \rho_\mu, \pi, A_\mu) + L_{int.} - U_{NL}. \quad (1)$$

L_0 describes the free system and

$$L_{int.} = -g_\sigma \bar{\varphi}\sigma\varphi - g_\omega \bar{\varphi}\gamma^\mu \omega_\mu \varphi - g_\rho \bar{\varphi}\gamma^\mu \boldsymbol{\rho}_\mu \cdot \boldsymbol{\tau}\varphi - \frac{f_\pi}{m_\pi}\bar{\varphi}\gamma_5\gamma^\mu \partial_\mu \boldsymbol{\pi}\cdot\boldsymbol{\tau}\varphi - e\bar{\varphi}\gamma^\mu \frac{1+\tau_3}{2} A_\mu \varphi, \quad (2)$$

represents the interaction of nucleons with the boson fields. The nonlinear potential energy density

$$U_{NL} = \frac{1}{3}\bar{b}M(g_\sigma\sigma)^3 + \frac{1}{4}\bar{c}(g_\sigma\sigma)^4 \quad (3)$$

takes into account the σ-meson self-interactions. g_σ, g_ω, g_ρ and f_π are the σ, ω, ρ and π meson-nucleon coupling constants, respectively. $\boldsymbol{\tau}$ is the usual isovector operator, $\bar{b}$ and $\bar{c}$ are dimensionless parameters. $\frac{f_\pi^2}{2\pi} \cong 0.08$. The quantities m_σ, g_σ, g_ω, g_ρ, $\bar{b}$ and $\bar{c}$ are adjusted to nuclear ground state experimental data [9], see Table 1.

Table 1. Columns 3-8 give the fitted parameters of the model without (η=0) and with (η=1) tensor force contribution. m_σ is the σ meson mass, whereas g_σ, g_ω, g_ρ, $\bar{b}$ and $\bar{c}$ are the dimensionless parameters entering eqs. (2) and (3). The last 4 columns give the nuclear matter saturation values for the nuclear density (ρ_0), the binding energy per particle ($-E/A$), the symmetry energy coefficient (a_4) and the compressibility modulus (K).

Model	η	m_σ, MeV	g_σ	g_ω	g_ρ	$\bar{b}$ $\times 10^3$
HF0	0	441.7	5.015	9.510	0.67	–5.26
HF1	1	443.3	5.322	10.39	0.72	–4.36

Model	$\bar{c}$ $\times 10^3$	ρ_0, fm^{-3}	E/A, MeV	a_4, MeV	K, MeV
HF0	–8.95	0.146	–16.3	36.8	294
HF1	–7.26	0.149	–16.3	35.5	285

3 Results

3.1 Effect of the tensor force

Fig. 1 shows the neutron single-particle energies of ^{208}Pb. It can be seen that the *tf* reduce the spin-orbit splittings. Fig. 2 shows the charge radius isotope shift $\Delta\langle r_c^2(\mathrm{Pb})\rangle = \langle r_c^2(^A\mathrm{Pb})\rangle - \langle r_c^2(^{208}\mathrm{Pb})\rangle$. The *tf* reduces the difference between the radii of the two spin-orbit partners but it has little influence on $\Delta\langle r_c^2\rangle$ both for the i-conf. ($\ldots 1i_{11/2}^{N-126}$), where the KE is observable, and for the g-conf. ($\ldots 2g_{9/2}^{N-126}$), where it is not.

* Corresponding author: marcoss@unican.es

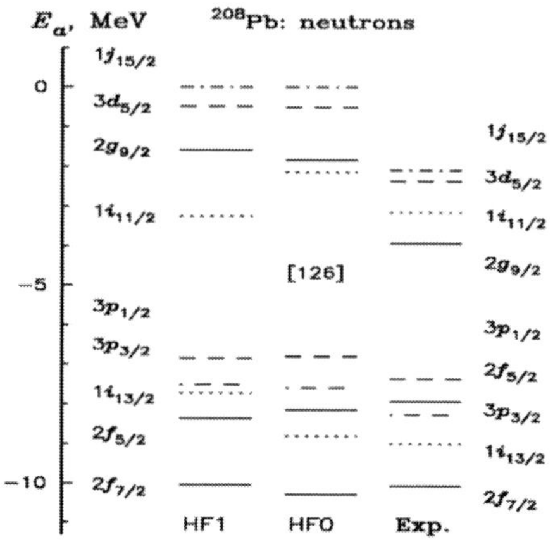

Figure 1: Neutron sp energies of ^{208}Pb for models HF0 (η=0) and HF1 (η=1), and Experiment data

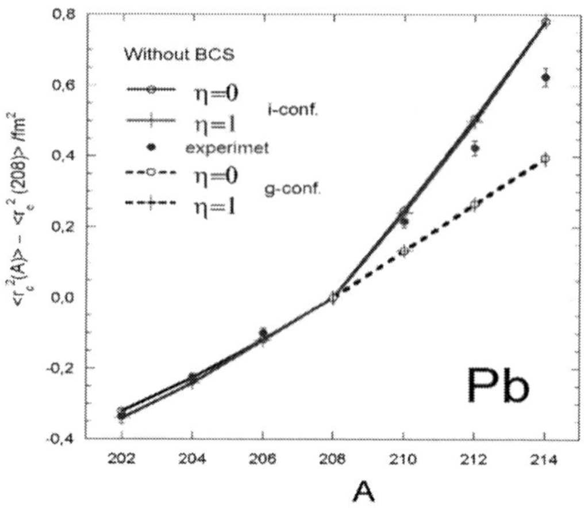

Figure 2: Charge radius isotope shift $\Delta\langle r_c^2\rangle$ for the Pb isotopic chain with respect to that of the ^{208}Pb for models HF0 (η=0) and HF1 (η =1) and for the neutron *i*- and *g*-confs.

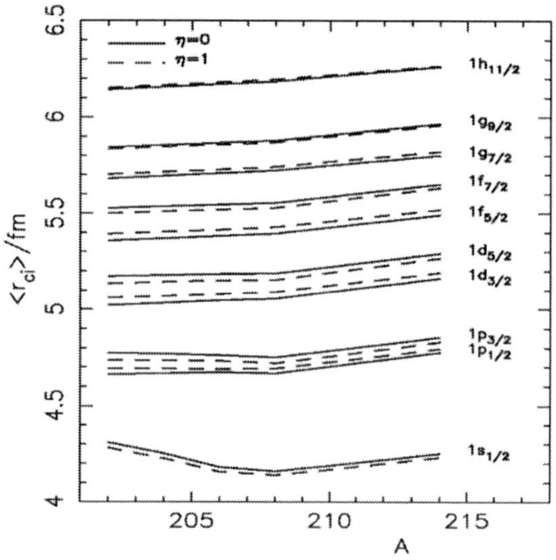

Figure 3: The sp r.m.s. charge radii of the proton nodeless orbitals of the Pb isotopic chain for the HF0 (η=0) and HF1 (η=1) models and *i*-conf.

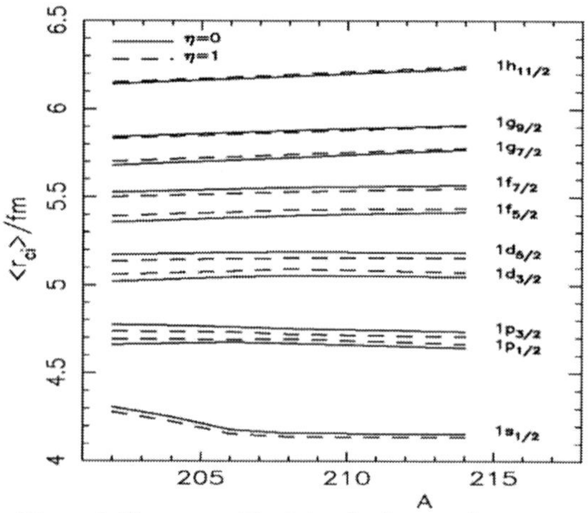

Figure 4: The same as Fig. 3, but for the *g*-conf.

Figs. 3 and 4 show the charge radii of the proton nodeless orbitals of the Pb isotopic chain. The effect of the *tf* on $\langle r_{ci}\rangle$ for the two orbitals of a spin-orbit doublet is opposite to each other and decreases as *l* increases. The effect of the *tf* is observable and it seems (only) qualitatively equivalent to that of reducing the spin-orbit interaction.

4 Conclusions

The effect of the tensor force (*tf*) on the nuclear charge radii of the nodeless proton orbitals of the Pb isotopic chain is opposite to that of the spin-orbit interaction and it decreases as *l* increases. For orbitals with nodes, the effect is more involved and less important. The global effect of the *tf* on the nuclei charge radii is very small.

References

1. N. Tajima, P. Bonche, H. Flocard, P.-H. Heenen, M. S. Weiss, Nucl. Phys. A **551**, 434 (1993)
2. T. Gonzalez-Llarena, J. L. Egido, G. A. Lalazissis, P. Ring, Phys. Lett. B **379**, 13 (1996)
3. M. M. Sharma, G. A. Lalazissis, P. Ring. Phys. Lett. B **317**, 9 (1993)
4. P. G. Reinhard, H. Flocard. Nucl. Phys. A **584**, 467 (1995)
5. S. Marcos, L. N. Savushkin, M. López-Quelle, R. Niembro, P. Bernardos, Phys. Lett. B **507**, 135 (2001)
6. R. Niembro, S. Marcos, M. López-Quelle, L. N. Savushkin, Phys. At. Nucl. **75**, 269 (2012)
7. P. M. Goddard, P. D. Stevenson, A. Rios, PRL **110**, 032503 (2013)
8. U. C. Perera, A. V. Afanasjev and P. Ring, Phys. Rev. C **104**, 064313 (2021)
9. S. Marcos, M. López-Quelle, R. Niembro and L. N. Savushkin, Phys. At. Nucl. **77**, 29 (2014)

Fission hindrances in transfermium nuclei

A. Lopez-Martens[1,*], *K.* Hauschild[1], *A.I.* Svirikhin[2], *Z.* Asfari[3], *M.L* Chelnokov[2], *V.I.* Chepigin[2], *O.* Dorvaux[3], *M.* Forge[3], *B.* Gall[3], *A.V.* Isaev[2], *I.N.* Izosimov[2], *K.* Kessaci[3], *A.A.* Kuznetsova[2], *O.N.* Malyshev[2], *R.S.* Mukhin[2], *A.G.* Popeko[2], *Yu.A.* Popov[2], *B.* Sailaubekov[2,4], *E.A.* Sokol[2], *M.S.* Tezekbayeva[2,5], and *A.V.* Yeremin[2]

[1]IJClab, Université Paris Saclay and CNRS, F-91405 Orsay, France
[2]FLNR, JINR, Dubna, Russia
[3]Université de Strasbourg, CNRS, IPHC UMR 7178, F-67000 Strasbourg
[4]L.N. Gumilyov Eurasian National University, Nur-Sultan, Kazakhstan
[5]The Institute of Nuclear Physics, 050032, Almaty, Kazakhstan

Abstract. Very heavy nuclei owe their stability against spontaneous fission to quantum shell effects, which depend on the local density of single-particle states. The height but also the width and the structure of the barrier in multi-dimensional deformation space determine the fission half-lives. Other effects come into play, such as the conservation of quantum numbers and superfluidity or stiffness of the system in the fission process. This is why odd nuclei have longer fission partial half-lives with respect to their even neighbours and also why multi-quasi-particle states, such as high-*K* states, are thought to be more stable against fission than the ground state. We will report here on two different fission studies carried out with the GABRIELA detector array at the focal plane of the recoil separator SHELS. The first study concerns the fission properties of ^{253}Rf, the most neutron deficient Rf isotope known to date. The second study focusses on a new measurement of the fission hindrance of the known 8^- isomer in ^{254}No.

1 Introduction

The existence of transfermium nuclei is solely due to quantum shell effects, which create a pocket at a given deformation in the potential energy surface of the nucleus and provide a barrier structure against spontaneous fission. These shell effects are clearly observed in the plot of fission partial half-lives of even-even very heavy and super heavy nuclei as a function of neutron number N (see Fig. 10 of ref.[1]), where fission rates are seen to increase drastically on either side of N=152 for the heaviest systems. This is especially visible for the No isotopic chain, where the lifetime drops by 10 orders of magnitude from ^{254}No to ^{250}No. For N≤152, the trend of lifetimes in Rf isotopes follows the one of No isotopes, accelerated by a factor of the order of $10^6 - 10^7$. This acceleration is understood as being due to a decline of the second barrier peak in Rf isotopes.

Fission half-lives in odd N/Z systems are known to be much longer than the ones of their even-even neighbours. This fact is theoretically understood, as detailed in Ref. [2]. The conservation of spin on the pathway to fission (and parity in the cases where axial symmetry is not broken) leads to a higher fission barrier by an amount called specialisation energy. The blocking of orbitals by the quasiparticle also lowers pairing energies. This in turn leads to larger collective masses, and therefore to longer fission times. These considerations also apply to multi-quasiparticle states, and in particular to high-K isomers, which are calculated to exhibit a larger stability against fission than the ground state of the same nucleus [3]. Until recently, however, there was no obvious experimental evidence for longer fission times from high-K states than from the corresponding ground state, except for ^{250}No (see Fig. 13 of Ref. [7]). New data on ^{254}Rf [6] and ^{250}No [4, 5] have shown that the fission hindrances ($T_{SF,iso}/T_{SF,gs}$) of high-K isomers in these nuclei are much larger than 1. These new findings cast a doubt on the 3 data points of Ref. [7], which are not lower limits, as pointed out by Clark [8]. In ^{256}Fm, the data comes from the observation of 2 beta-delayed fission events [9] and probably needs confirmation. The data point for ^{262}Rf no longer exists due to a mis-assigned fission activity [1]. Remains the 8^- isomer in ^{254}No, for which a 0.02% fission branch has been reported [10].

So the questions addressed in this talk are related to the limits of the nuclear chart for neutron-deficient Rf isotopes and the fission hindrance of the low-lying $K^\pi=8^-$ isomer in ^{254}No.

2 Experimental Setup

To address the questions of interest, experiments were performed at the FLNR in Dubna, using the GABRIELA de-

*e-mail: araceli.lopez-martens@ijclab.in2p3.fr

 https://doi.org/10.1051/epjconf/202329002027

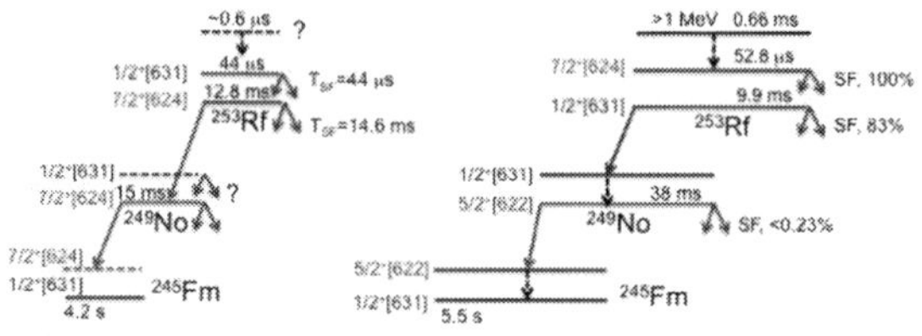

Figure 1. Left: Decay scheme of ^{253}Rf established from data taken at GSI (the figure is adapted from [13]). Right) Decay scheme obtained by compiling the data on ^{253}Rf and ^{249}No collected with GABRIELA and performing a Geant4-assisted analysis (see Ref. [16] for details).

tector setup [11] at the focal plane of the velocity filter SHELS [12]. Intense heavy-ion beams are provided by the U400 cyclotron and are used to produce transfermium nuclei in fusion-evaporation reactions on stable and actinide targets. GABRIELA is composed of an array of Double Sided Silicon-strip Detectors to detect the super heavy nuclei transported by the separator and their subsequent charged particle emissions (alpha particles, fission fragments, internal conversion electrons) and an array of Compton-suppressed Ge detectors to detect the emitted photons and X rays.

3 Fission stability of Rf isotopes

The most neutron-deficient Rf isotope known to date is ^{253}Rf. To produce it, we used the 2-neutron evaporation channel of the fusion reaction between ^{50}Ti ions and the atoms of an isotopically enriched ^{204}Pb target. A few months after our experiment, GSI published a paper [13], in which they confirmed the existence of 2 fission activities in this nucleus (it was not clear from Ref. [14] whether the longer-lived activity was from the production of ^{256}Rf on contaminants of the target). They also could observe a small alpha branch to the newly discovered ^{249}No [13, 15, 16]. On the basis that fission from a low-K state occurs faster than from a high-K state, the faster fission activity was assigned to the 7/2+[624] neutron configuration, while the longer-lived state was assigned the 1/2+[631] configuration, following the systematics of N=147 isotones. In our experiment, we collected a factor 10 more data. The lifetimes of the 2 fission activities were found to be in agreement with the GSI values. We also confirm the existence of an alpha decay branch to ^{249}No. What we also observed was the internal decay of an isomer, which has to be a high-K isomer given its measured excitation energy of more than 1 MeV. This internal decay is followed by the shorter-lived fission activity of ^{253}Rf. This finding leads us to change the assignments made in Ref. [13], and also the decay scheme of ^{253}Rf to ^{249}No, as shown in Fig. 1.

The relative excitation energies of the $7/2^+$ and $1/2^+$ states in ^{253}Rf could not be measured, but given the decreasing trend of the $1/2^+$[631] state in N=149 isotones (see Fig. 6 of Ref. [18]) , it is likely that the $1/2^+$ is the ground state in ^{253}Rf, in which case the $7/2^+$ state must not be more than $\approx$ 15 keV above the ground state in order to be isomeric. If, on the other hand, the $1/2^+$ is not the ground state, it must lie within less than 100 keV from the 7/2+ state. This leads to very a unique situation: the existence of 2 nearly degenerate states in the same nucleus, with very different fission properties.

The known fission half-lives of No and Rf isotopes are resumed in FIg. 2. The drop of 6-7 orders of magnitude in spontaneous fission half-lives in going from No to Rf isotones continues in ^{253}Rf and extrapolating further, one can conclude that the ground state of the next even Rf isotope (^{252}Rf) will probably be very close to the limit of stability of an atom and that if the ground state of ^{251}Rf has the same configuration as ^{249}No, it should be accessible to experiments. The difference in fission half-lives between the 2 close-lying neutron configurations in ^{253}Rf can be understood with the help of some preliminary axial calculations using the BSkG2 Skyrme parametrisation [19]. These show that the $7/2^+$ orbital follows closely the Fermi level all the way to scission, while the $1/2^+$ orbital, undergoes many avoided crossings and moves away from it. This should lead to a larger specialisation energy and could therefore explain the factor $\approx$250 difference in fission time. Of course, a full microscopic calculation, including the triaxial degree of freedom, with pairing and proper treatment of the avoided crossings is required to confirm the difference, but this is quite challenging and requires more computer time.

4 Fission from the 8^- isomer in ^{254}No

The ^{254}No nucleus is the most investigated nucleus in the transfermium region [10, 20–22]. Two isomers are known; a low-lying 264 ms 8^- state, which decays to an intermediate 3^+ band and to the ground state band and a higher-lying shorter-lived state, which decays to the lower-lying isomer and whose spin and parity are still under debate (in the level scheme established by the Berkeley group [22], a spin 16 has been assigned to the 184 μs isomer). In Ref. [10], one can find a value for the fission branch from the 8^- state, but it should be mentioned that the corresponding number of fission events was obtained by correcting for the contribution of the fission of the ground-states of ^{254}No and ^{252}No (produced on ^{206}Pb contaminants in the target). No fission has been observed from the short-lived isomer. In order to verify the fission hindrance of the low-lying isomer, we performed a long irradiation of an enriched 340 μg-thick ^{208}PbS target (99.57% ^{208}Pb, 0.01% ^{206}Pb) with an intense beam of ^{48}Ca. A total beam dose of 4.2 10^{18} was measured and close to 1 million full-energy alpha particles corresponding to the decay of ^{254,255}No were observed (more details on the experiment can be found in Refs. [23, 24]).

With such statistics, the fission events from the ground state as well as from the 8^- state can be isolated from all other possible sources of fission by selecting specific decay paths. In the case of the ground state fission, we demand that the implanted ions first undergo the characteristic decay of the long-lived isomer of ^{254}No. The time of the subsequent decays is shown as a function of decay energy

 https://doi.org/10.1051/epjconf/202329002027

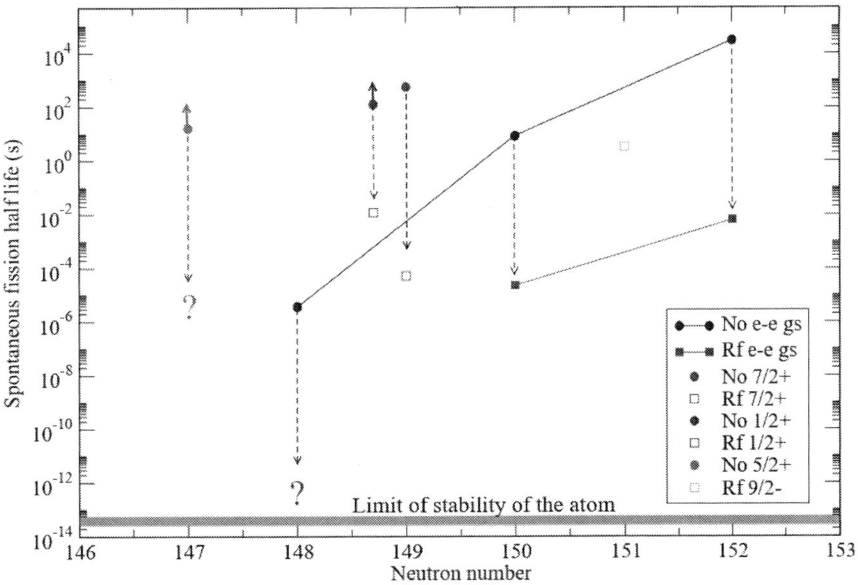

Figure 2. Fission half-lives of No (circles) and Rf (squares) isotopes as a function of neutron number. The spins and parities for the odd-N cases are given in the legend. In the case of ^{253}Rf (N=149), the data points are drawn at N=148.5 and 149 for clarity. The typical formation time of an electronic shell ($\approx 10^{-14}$ s) is shown and denotes the limit of stability of an atom.

in the top left panel of Fig. 3. The characteristic alpha decay of ^{254}No (highlighted in red) and the ground-state fission at large amplitude (surrounded by the blue box) are clearly visible. From the ratio of fission to alpha-decay events, a fission branch of 1.61(8) 10^{-3} could be extracted, which is in good agreement with previous measurements. The ground state half-life can be measured by fitting the time distribution of the alpha-decay events highlighted in red, as shown in the bottom left panel of Fig.3.
Now if we demand that the decay flux first go through the higher-lying isomer, the subsequent decay of the 8^- isomer is cleanly isolated in the correlation plot on the top right panel of Fig. 3. The signals arising from the isomeric cascade of internal conversion electrons and accompanying fluorescence and auger emissions are seen at low ADC channel values between 1 ms and 4 s (y-axis values of 10 and 22 respectively in the top right panel of Fig.3). In the similar time interval at large amplitudes, where one expects the isomeric fission events, no counts are observed in the corresponding blue box. The absence of fission events gives an upper limit of 2.26 10^{-5} for the fission branch of the 8^- isomer. The half-life of the isomeric decay is obtained by fitting the time distribution of the events highlighted in red with a 3-component fit, as shown in the bottom panel of Fig. 3. With the measured fission branches and half-lives, the fission hindrance of the isomer is found to be greater than 0.37, which is ~9 times larger than the previously established value.

5 Conclusion and perspectives

To conclude, the fission properties of 2 states in ^{253}Rf have been revisited, giving insights into the properties of even more neutron-deficient Rf isotopes and providing a new tool to probe the spectrum of single-particle states as a function of elongation. The fission of the 8^- state in ^{254}No has been searched for in a dedicated high-statistics run but no evidence for it could be found. Fission from a high-K state has therefore yet to be confirmed. The continued development of high-sensitivity and high-spectroscoppic-quality set-ups at various facilities around the world should allow to extend these studies to heavier systems.

References

[1] F. P. Hessberger, Eur. Phys. J. A **53**, 75 (2017)
[2] R. Rodrguez-Guzman and L.M. Robledo , Eur. Phys. J. A **52**, 348 (2016)
[3] F.R. Xu et al., Phys. Rev. Lett. **92**, 252501 (2004)
[4] J. Kallunkathariyil et al. Phys. Rev. C **101**, 011301(R) (2020)

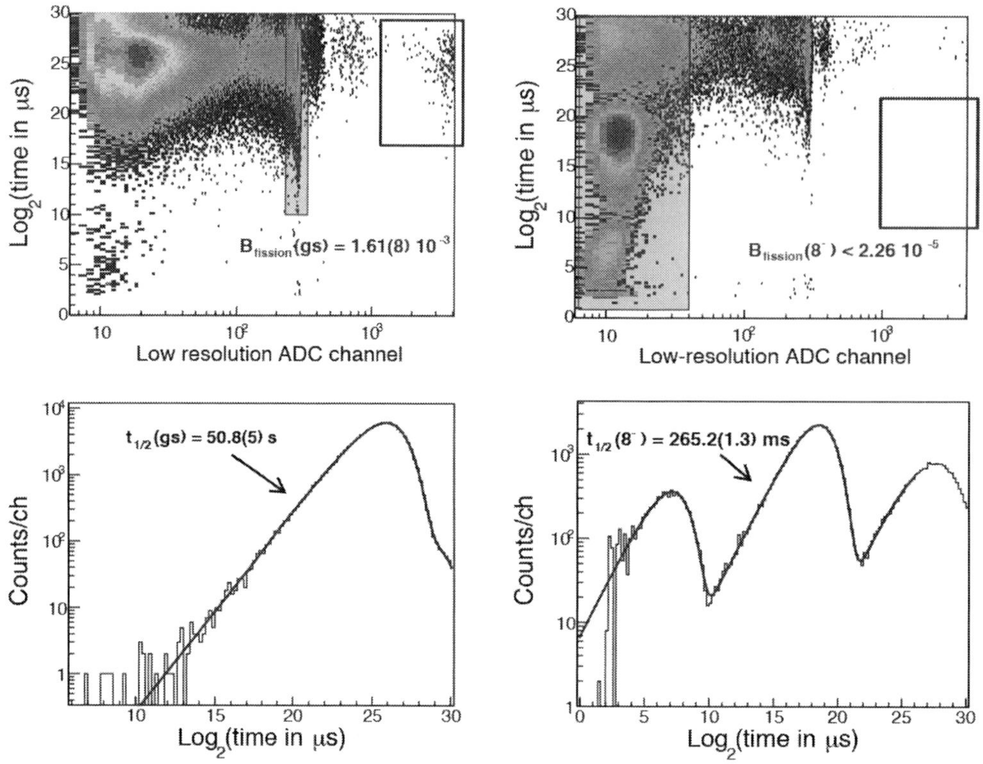

Figure 3. Top left) Time distribution of decay events following the decay of the 8$^-$ isomer in ^{254}No as a function of low-resolution ADC channel. In this particular instance, the low-resolution amplification range of the DSSD ADCs allowed to measure energies up to ~100 MeV. Top right) Same as on the left, but for decay events which follow the detection of the short-lived isomer of ^{254}No. Bottom panels) Projection onto the time axis of the shaded areas in the corresponding top panels. The solid lines represent multi-component fits to the resulting time distributions. In particular, due to the existence of short-lived isomers in cascade in ^{255}No [24, 25], the characteristic decay of the 11/2$^-$ state of ^{255}No appears in the right panel between x-axis values of 3 and 10 (8 μs-1 ms) and needs to be accounted for in the fit of the time distribution. The extracted fission branches and half-lives of the ground-state (left) and 8$^-$ isomeric state (right) are indicated in the corresponding panels. See text for more details.

[5] J. Khuyagbaatar et al., Phys. Rev. C **106**, 024309 (2022)
[6] H. David et al., Phys. Rev. Lett. **115**, 132502 (2015)
[7] F. Kondev, G. Dracoulis and T. Kibedi, At. Dat. And Nucl. Dat. Tables **103-104**, 50 (2015)
[8] R.M. Clark, EPJ Web of Conferences **131**, 02002 (2016)
[9] H.L. Hall et al., Phys.Rev C **39**, 1866 (1989)
[10] F.P. Hessberger et al., Eur. Phys. J. Q **43**, 55 (2010)
[11] R. Chakma et al., EUr. Phys. J. A **56**, 245 (2020)
[12] A. Popeko et al., Nucl. Instr. Meth. B **376**, 140 (2016)
[13] J. Khuyagbaatar et al., Phys. Rev. C 104 (2021) L031303
[14] F.P. Hessberger et al., Z. Phys. A **359**, 415 (1997)
[15] A. Svirikhin et al., Phys. of Part. and Nucl. Lett. **18**, 445 (2021)
[16] M. S. Tezekbayeva et al., Eur. Phys. J. A **58**, 52 (2022)
[17] A. Lopez-Martens et al., Phys. Rev. C 105, L021306 (2022)
[18] A. Lopez, K. Hauschild, Eur. Phys. J. A **58**, 134 (2022)
[19] W. Ryssens et al., Eur. Phys. J. A **58**, 246 (2022) and W. Ryssens, priv. comm.
[20] S.K. Tandel et al., Phys. Rev. Lett. **97**, 082502 (2006)
[21] R.-D. Herzberg et al., Nature **442**, 896 (2006)
[22] R.M. Clark et al., Phys. Lett. B **690**, 19 (2010)
[23] M. Forge et al., in preparation
[24] K. Kessaci et al., submitted to Phys. Rev. C
[25] A. Bronis et al., Phys. Rev. C **106**, 014602 (2022)

s-process Nuclear Reaction Rates

David Rapagnani[1,*]
for the LUNA collaboration

[1]Physics Department of the University of Naples "Federico II" and National Institute for Nuclear Physics (INFN) - Naples division, Complesso Universitario di Monte Sant'Angelo, viale Cintia 80126 Naples, Italy

Abstract. In stars the $^{13}C(\alpha, n)^{16}O$ and $^{22}Ne(\alpha, n)^{25}Mg$ reactions are the two main sources of neutrons for the so-called slow neutron capture process (s-process), which is one of the main mechanisms for the stellar synthesis of heavy elements. About $^{13}C(\alpha, n)^{16}O$, in despite of many efforts in measuring its cross section at the lower energies, only high uncertainty data above the s-process Gamow window (150 keV < E_{cm} < 230 keV) were available, due mostly to the difficulties on suppress the natural background. Indeed, only recently the LUNA collaboration performed high precision underground measurements of the reaction cross section inside the Gamow window, improving the accuracy of its extrapolation at the lower energies. Again due to natural background, only upper limits for the $^{22}Ne(\alpha, n)^{25}Mg$ reaction cross section are currently known in the s-process Gamow window (450 keV < E_{cm} < 750 keV). For this, the ERC founded project SHADES (Unina/INFN) aims to perform high precision and high sensitivity measurements of the $^{22}Ne(\alpha, n)^{25}Mg$ reaction cross section down to the neutron threshold. A sensitivity improvement of at least two orders of magnitude over the state of the art is expected thanks to the low natural background environment of the INFN-LNGS laboratory in Italy, the high beam current of the new LUNAMV accelerator and the Beam Induced Background events suppression performed by SHADES hybrid detectors array. In this paper I will present the LUNA efforts to estimate nuclear reaction rates for $^{13}C(\alpha, n)^{16}O$, with a focus on the R-Matrix analysis performed with the code AZURE2 to extrapolate nuclear reaction rates at stellar energies and the estimate of their uncertainty through Monte Carlo methods. I will also present an overview of the SHADES project to measure $^{22}Ne(\alpha, n)^{25}Mg$ in the Gamow window and the first results on the setup commissioning.

1 Introduction

Most of the heavy elements in the Universe are produced in low-mass Asymptotic Giant Branch (AGB) stars [1] where $^{13}C(\alpha, n)^{16}O$ burns the ^{13}C-pocket formed in the He layer [2, 3]. The neutron flux associated to this scenario is low and one usually refers to it as s-process (slow neutron capture). It is responsible for the synthesis of elements heavier than A = 90. Although a smaller contribution is also given by the $^{22}Ne(\alpha, n)^{25}Mg$ reaction, it is in more massive stars (M > 8 $M_\odot$, solar masses) that this last becomes the main neutron source for the s-process and it is the main contributor for the synthesis of A = 60 ~ 90 elements [4–6].

A precise knowledge of these reactions cross section at stellar energies is required by stellar models to better constraint s-process nucleosynthesis [7, and references therein]. Being at sub-Coulomb energies, direct measurement in the Gamow window are extremely challenging and particular care must be used to suppress environmental background. In the following sections are presented the LUNA collaboration efforts to measure and extrapolate $^{13}C(\alpha, n)^{16}O$ reaction cross section at stellar energies (Sec. 2), as well as the Scintillator-He3 Array for Deep-underground Experiments on the S-process (SHADES) project founded by the European Research Council (ERC), which aims at the measurements of $^{22}Ne(\alpha, n)^{25}Mg$ down to neutron threshold energy (Sec. 3).

2 $^{13}C(\alpha, n)^{16}O$

The s-process energy range (i.e. the Gamow window) for $^{13}C(\alpha, n)^{16}O$ is located at E_0= 150-230 keV[1], far below the coulomb barrier of ~ 3.5 MeV. At these energies, precise measurements of the $^{13}C(\alpha, n)^{16}O$ reaction cross section are very challenging being its value extremely small ($\ll$ pb). Indeed, despite of the many attempts during the years (e.g. [8–12]) measurements could never be performed below 280 keV, with an associate uncertainty of 40% at the lower energies. To improve the status of the art, the LUNA collaboration [13, and references therein] decided to perform a measurement campaign in the low-background environment of the Gran Sasso Laboratories (LNGS) of the Italian Institute of Nuclear Physics (INFN). Reaction neutrons were measured in the LUNA400 facility [14] with 18 ^{3}He counters in 4π configuration placed in a High Density Polyethylene Moderator, which acted both as detector support and as neutron moderator. Due

*e-mail: david.rapagnani@unina.it

[1]Energies are always in the center of mass framework, unless differently stated.

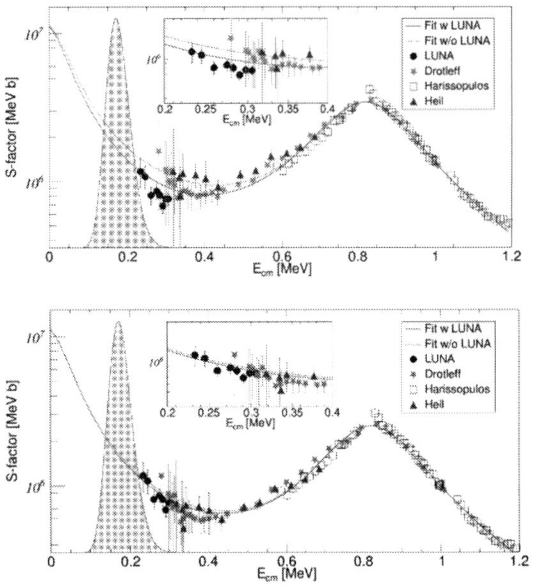

Figure 1. Top: Astrophysical S-factor of $^{13}C(\alpha,n)^{16}O$ using [20] and [21] normalization reference. Red lines are results of two the R-matrix analyses, with and without the LUNA data [19]. The blue curve is the "Gamow peak" at 90 MK. Bottom: Similar chart but with [22] as normalization reference.

to the very low background environment, intrinsic detectors background caused by trace presence of actinides in the aluminum detectors walls had a non negligible contribution, which was reduced by the employ of custom stainless wall detectors. Also the use of the Pulse Shape Discrimination (PSD) technique to identify acquired particles [15–18] allowed for a reduction of the background events to 1.2 $\pm$ 0.1 counts/hour which, in addition to the use of 99% enriched ^{13}C targets and the intense α beam current of 150 μA from LUNA400, allowed to perform measurements at beam energies of E_{lab} = 305 - 400 keV with a final statistical uncertainty of 2-18 % [19]. New LUNA data pushed the present low energy limit down to 230 keV entering for the first time the Gamow window.

An R-Matrix extrapolation with the AZURE2 code [23] was then performed including LUNA results to a selection of data from literature [20–22], being representative for the different normalization factors present in data published so far and extending at the lowest energies (see [19] for more details). In Figure 1 are reported the results of the R-Matrix evaluation for the two normalizations. It is worth to notice that the ANC value of the α-bounded $\frac{3}{2}^+$ near threshold state at 7215 keV was fixed with the value proposed [24] and confirmed [12] by indirect measurements. A Monte Carlo analysis sampling resonance parameters in their uncertainty was then performed to extrapolate S-Factor uncertainty. A 13% precision was obtained at the relevant stellar energy for $^{13}C(\alpha, n)^{16}O$, whereas the main contributor is the uncertainty on the ANC value. Considering that lower energies will be hardly accessible for this reaction in the near future [2] an improvement of $^{13}C(\alpha, n)^{16}O$ knowledge could be achieved with a more precise measurement of the 7215 keV state ANC. Furthermore, an investigation at higher energy is required to try solving the normalization discrepancy between data set available in literature.

3 $^{22}Ne(\alpha, n)^{25}Mg$

The s-process Gamow window for $^{22}Ne(\alpha, n)^{25}Mg$ extends down to about 400 keV, whereas the Coulomb barrier is located at a much more high energy of about 5 MeV. This and the neutron background of surface laboratories allowed so far to measure only upper limit for the reaction cross section at the lower relevant energies (see e.g. [20, 26, 27]). To increase our knowledge on s-process and better constraint heavy element abundance in the Universe, the ERC founded SHADES project aims at the measurement of $^{22}Ne(\alpha, n)^{25}Mg$ down to neutron threshold. The measurement will take place at the new LUNAMV facility located at INFN-LNGS where the 1.4 km of overburden rock can already grant a suppression of 2 order of magnitude of the neutron background respect the status of the art. This reduction will be further enhanced by the employ of a new detection array composed by liquid scintillators and neutron proportional counters: neutrons reaching thermal energies inside the scintillators will be measured by the counters. The array will then be able to quantify detected neutron energy to discriminate environmental and beam induced events from reaction ones. Also the use of an enriched ^{22}Ne gas target, recirculated and purified, together with aluminum sealing of the vacuum elements and Tantalum coating of the parts possibly exposed to the beam, will ensure a reduction of the beam induced events, mostly from $^{13}C(\alpha, n)^{16}O$, $^{11}B(\alpha, n)^{13}N$ and $^{11}B(\alpha, n)^{13}N$. Their cross sections are up to 6 order of magnitude more intense respect the one of $^{22}Ne(\alpha, n)^{25}Mg$ and even a small presence of carbon and boron must be avoided.

A cut-view of the SHADES setup is presented in Figure 2. The reaction chamber is surrounded by 18 ^3He counters and 12 EJ-309 scintillators kept in position by a polyethylene holder which serves at the same time as a shield against environmental neutrons. Several gauges and temperature sensors will monitor gas target performances while a Rutherford Back-Scattering Spectroscopy port will allow to measure ion beam current during the reaction yield measurement, for normalization purposes. Collimators at the entrance of the reaction chamber and of the different pumping stages will allow to operate the target in windowless mode.

SHADES detectors are currently under characterization. Preliminary background measurement performed at LNGS showed that the intrinsic background from the α decay of ^{238}U and ^{232}Th (possibly present as trace elements in the detector walls) is clearly identifiable in the scintillators. A forthcoming paper will discuss the PSD analysis

[2]Recently also the JUNA collaboration measured the same reaction reaching the same low energy limit [25]. Results are in agreement with the one of LUNA.

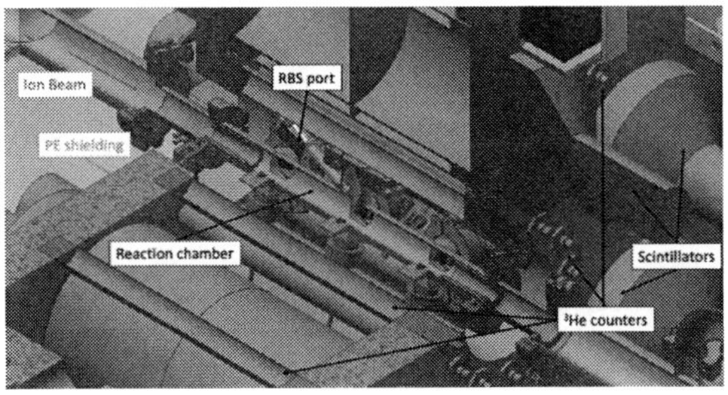

Figure 2. SHADES setup cut-view.

performed to identify particles and the evaluated detector walls activity. Energy calibration measurements were also performed at the Goethe University in Frankfurt through the ^{7}Li(p, n)^{7}Be reaction in the proton energy range E_p of 1.9 - 2.5 MeV, corresponding to neutron energies E_n ranging from 50 to 700 keV. Data analysis is presently ongoing. Similarly, the gas target setup is presently installed at the Tandem accelerator laboratory of the University of Campania "Luigi Vanvitelli" in Caserta (Italy): originally built for radiocarbon measurements [28], this facility was later expanded to perform AMS [29], nuclear astrophysics [30–32] and ion beam analysis with ^{7}Be radioactive beams [33–35]. Ion beam analysis measurements to characterize SHADES ^{22}Ne gas target thickness and density profile are currently ongoing and outputs will be described in a forthcoming paper.

Acknowledgements

D.R. acknowledges the ERC-SHADES project (ERC-StG 2019 852016) for funding.

References

[1] C. Sneden, J.J. Cowan, R. Gallino, Annu. Rev. Astron. Astrophys. **46**, 241 (2008)

[2] O. Straniero, R. Gallino, S. Cristallo, Nucl. Phys A **777**, 311 (2006)

[3] S. Cristallo, M.L. Cognata, C. Massimi, A. Best, S. Palmerini, O. Straniero, O. Trippella, M. Busso, G.F. Ciani, F. Mingrone et al., **859**, 105 (2018)

[4] S. Woosley, T. Weaver, Astrophysical Journal Supplement **101**, 181 (1995)

[5] P. Adsley, U. Battino, A. Best, A. Caciolli, A. Guglielmetti, G. Imbriani, H. Jayatissa, M. La Cognata, L. Lamia, E. Masha et al., Phys. Rev. C **103**, 015805 (2021)

[6] F. Ferraro, G.F. Ciani, A. Boeltzig, F. Cavanna, S. Zavatarelli, Frontiers in Astronomy and Space Sciences **7** (2021)

[7] H. Schatz, A.D.B. Reyes, A. Best, E.F. Brown, K. Chatziioannou, K.A. Chipps, C.M. Deibel, R. Ezzeddine, D.K. Galloway, C.J. Hansen et al., Journal of Physics G: Nuclear and Particle Physics **49**, 110502 (2022)

[8] C.N. Davids, Nuclear Physics A **110**, 619 (1968)

[9] J.K. Bair, F.X. Haas, Phys. Rev. C **7**, 1356 (1973)

[10] S. Kellogg, R. Vogelaar, R. Kavanagh, Bulletin of the American Physical Society **34**, 1192 (1989)

[11] K.K. Sekharan, A.S. Divatia, M.K. Mehta, S.S. Kerekatte, K.B. Nambiar, Phys. Rev. **156**, 1187 (1967)

[12] O. Trippella, M.L. Cognata, The Astrophysical Journal **837**, 41 (2017)

[13] F.R. Pantaleo, A. Boeltzig, A. Best, R. Perrino, M. Aliotta, J. Balibrea-Correa, F. Barile, D. Bemmerer, C. Broggini, C.G. Bruno et al. (LUNA Collaboration), Phys. Rev. C **104**, 025802 (2021)

[14] A. Formicola, G. Imbriani, M. Junker, D. Bemmerer, R. Bonetti, C. Broggini, C. Casella, P. Corvisiero, H. Costantini, G. Gervino et al., Nucl. Instr. Meth. A **507**, 609 (2003)

[15] J. Balibrea-Correa, G. Ciani, R. Buompane, F. Cavanna, L. Csedreki, R. Depalo, F. Ferraro, A. Best, Nucl. Instr. Meth. A **906**, 103 (2018)

[16] Ciani, G. F., Csedreki, L., Balibrea-Correa, J., Best, A., Aliotta, M., Barile, F., Bemmerer, D., Boeltzig, A., Broggini, C., Bruno, C. G. et al., Eur. Phys. J. A **56**, 75 (2020)

[17] L. Csedreki, G. Ciani, J. Balibrea-Correa, A. Best, M. Aliotta, F. Barile, D. Bemmerer, A. Boeltzig, C. Broggini, C. Bruno et al., Nuclear Instruments and Methods in Physics Research Section A: Accelerators, Spectrometers, Detectors and Associated Equipment **994**, 165081 (2021)

[18] A. Badalà, M.L. Cognata, R. Nania, et al., Riv. Nuovo Cim. **45**, 189–276 (2022)

[19] G.F. Ciani, L. Csedreki, D. Rapagnani, M. Aliotta, J. Balibrea-Correa, F. Barile, D. Bemmerer, A. Best, A. Boeltzig, C. Broggini et al. (LUNA Collaboration), Phys. Rev. Lett. **127**, 152701 (2021)

[20] H.W. Drotleff, A. Denker, H. Knee, M. Soine, G. Wolf, J.W. Hammer, U. Greife, C. Rolfs, H.P. Trautvetter, **414**, 735 (1993)

[21] M. Heil, R. Detwiler, R.E. Azuma, A. Couture, J. Daly, J. Görres, F. Käppeler, R. Reifarth, P. Tischhauser, C. Ugalde et al., Phys. Rev. C **78**, 025803 (2008)

[22] S. Harissopulos, H.W. Becker, J.W. Hammer, A. Lagoyannis, C. Rolfs, F. Strieder, **72**, 062801 (2005), `nucl-ex/0509014`

[23] R. Azuma, E. Uberseder, E. Simpson, C. Brune, H. Costantini, R. de Boer, J. Görres, M. Heil, P. LeBlanc, C. Ugalde et al., Phys. Rev. C **81**, 045805 (2010)

[24] M.L. Avila, G.V. Rogachev, E. Koshchiy, L.T. Baby, J. Belarge, K.W. Kemper, A.N. Kuchera, D. Santiago-Gonzalez, Phys. Rev. C **91**, 048801 (2015)

[25] B. Gao, T.Y. Jiao, Y.T. Li, H. Chen, W.P. Lin, Z. An, L.H. Ru, Z.C. Zhang, X.D. Tang, X.Y. Wang et al. (JUNA Collaboration), Phys. Rev. Lett. **129**, 132701 (2022)

[26] M. Jaeger, R. Kunz, A. Mayer, J.W. Hammer, G. Staudt, K.L. Kratz, B. Pfeiffer, Phys. Rev. Lett. **87**, 202501 (2001)

[27] V. Harms, K.L. Kratz, M. Wiescher, Phys. Rev. C **43**, 2849 (1991)

[28] F. Terrasi, N. De Cesare, A. D'Onofrio, C. Lubritto, F. Marzaioli, I. Passariello, D. Rogalla, C. Sabbarese, G. Borriello, G. Casa et al., Nuclear Instruments and Methods in Physics Research Section B: Beam Interactions with Materials and Atoms **266**, 2221 (2008), accelerators in Applied Research and Technology

[29] Y.J. Guan, et al., Chin. Phys. C 34(11), 1729-1732 (2010)

[30] M. Romoli, L. Morales-Gallegos, M. Aliotta, C.G. Bruno, R. Buompane, A. D'Onofrio, T. Davinson, M. De Cesare, A. Di Leva, P. Di Meo et al., Phys. J. **A 54**, 1 (2018)

[31] F. Brandi, L. Labate, D. Rapagnani, R. Buompane, A. Di Leva, L. Gialanella, L.A. Gizzi, Scientific Reports **10**, 5087 (2022)

[32] R. Buompane, A. Di Leva, L. Gialanella, A. D'Onofrio, M. De Cesare, J. Duarte, Z. Fülöp, L. Gasques, G. Gyürky, L. Morales-Gallegos et al., Physics Letters B **824**, 136819 (2022)

[33] D. Rapagnani, M. De Cesare, D. Alfano, R. Buompane, S. Cantoni, M. De Stefano Fumo, A. Del Vecchio, A. D'Onofrio, G. Porzio, G. Rufolo et al., Nuclear Instruments and Methods in Physics Research Section B: Beam Interactions with Materials and Atoms **467**, 53 (2020)

[34] D. Rapagnani, M.D. Cesare, R. Buompane, A.D. Vecchio, A.D. Leva, A. D'Onofrio, G. Porzio, L. Gialanella, Journal of Physics D: Applied Physics **54**, 32LT01 (2021)

[35] M. De Cesare, L. Savino, A. Di Leva, D. Rapagnani, A. Del Vecchio, A. D'Onofrio, L. Gialanella, Nuclear Instruments and Methods in Physics Research Section B: Beam Interactions with Materials and Atoms **479**, 264 (2020)

Recent THM investigations on neutron induced reactions at astrophysical energies

Maria Letizia Sergi[1,2*], *Livio* Lamia[1,2], *Silvio* Cherubini[1,2], *Giuseppe* D'Agata[1,2], *Alessia* Di Pietro[2], *Juan Pablo* Fernandez-Garcia[2], *Pierpaolo* Figuera[2], *Giovanni Luca* Guardo[2], *Marisa* Gulino[2,3], *Seiya* Hayakawa[4], *Iolanda* Indelicato[2], *Marco* La Cognata[2], *Marco* La Commara[5,6], *Dario* Lattuada[2,3], *Marcello* Lattuada[1,2], *Marco* Mazzocco[7,8], *Alessandro Alberto* Oliva[1,2], *Sara* Palmerini[9,10], *Rosario Gianluca* Pizzone[2], *Giuseppe Gabriele* Rapisarda[1,2], *Stefano* Romano[1,2], *Roberta* Spartà[2], *Claudio* Spitaleri[1,2], *Domenico* Torresi[2], and *Aurora* Tumino[2,3]

[1]Dipartimento di Fisica e Astronomia, via S. Sofia 64, I-95123 Catania, Italy
[2]INFN Laboratori Nazionali del Sud, Catania, Italy
[3]Università degli Studi di Enna "Kore", Enna, Italy
[4]Center for Nuclear Study, University of Tokyo, Japan
[5]INFN-Sezione di Napoli, Napoli, Italy
[6]Dipartimento di Farmacia, Università di Napoli "Federico II", Italy
[7]Dipartimento di Fisica e Astronomia, Univ. di Padova, Padova, Italy
[8]INFN-Sezione di Padova, Padova, Italy
[9]Istituto Nazionale di Fisica Nucleare, Sezione di Perugia, Via A. Pascoli s/n 06123, Perugia
[10]Dipartimento di Fisica e Geologia, Università degli Studi di Perugia, via A. Pascoli s/n 06123, Perugia

Abstract. Neutron induced reactions on stable and unstable nuclei play a significant role in the nucleosynthesis of the elements in the cosmos. In the last years, several efforts have been made to investigate the possibility of applying the Trojan Horse Method (THM) to neutron induced reactions mostly by using deuteron as "TH-nucleus". Here, the main advantages of using THM will be given together with a more focused discussion on the recent $^{7}Be(n,\alpha)^{4}He$ and the $^{14}N(n,p)^{14}C$ reactions.

1 Introduction

Neutron induced reaction play a key role in several astrophysical domains, ranging from primordial nucleosynthesis [1] to high mass nuclei stellar nucleosynthesis via the so-called *r*-process and *s*-process [2].

In the context of primordial nucleosynthesis, the cosmological lithium problem is one of the most challenging open questions in astrophysics and cosmology, introducing inconsistencies in the framework given by the standard Big Bang nucleosynthesis (BBN) for which primordial abundances are described by means of the only free baryon-to-photon ratio $\eta = n_b/n_\gamma$ parameter [3].

Possible solutions have been proposed, such as a different number of neutrino families or a revision of neutron lifetime. Among these, nuclear physics processes involving the unstable ^{7}Be ($t_{1/2}$=53.22 ± 0.06 days) are of particular interest. In more detail, at $\eta_{CMB} = (6.07 \pm 0.07) \times 10^{-10}$ [4] (value of η obtained through independent study on the cosmic microwave background (CMB) anisotropy) ^{7}Li is mainly produced from ^{7}Be that undergoes the electron capture process $e^{-}+{}^{7}Be \rightarrow {}^{7}Li+\nu_e$ at late times (i.e., long after the ^{7}Be synthesis). For this reason, ^{7}Be destruction channels are currently a matter of study because of their influence on the ^{7}Li cosmological abundances. In particular, the (n,α) reaction channel has been the subject of recent works [5-8].

Regarding the *s*-process (where "*s*" stands for *slow*), when a neutron is captured, the resulting nucleus experiences β-decay to a more stable nucleus. The neutron capture time τ_n is higher than the half-life time τ_β of the unstable nucleus, thus ensuring the occurrence of β-decay before further neutron captures could occur. The $^{14}N(n,p)^{14}C$ reaction plays an important role in the s-process of nucleosynthesis: ^{14}N is very abundant since it is a dominant product of hydrogen-burning in the CNO cycle, the stage prior to the s-process. So, with its relatively high cross section, this reaction can act as a strong neutron poison in the reaction chain to heavier elements. Also, ^{14}N is of crucial importance in the nucleosynthetic origin of fluorine, whose only stable isotope is ^{19}F [9,10]. The He-burning shell in asymptotic giant branch stars is thought to be the most likely site for the synthesis of fluorine, mainly through the nuclear chain $^{14}N(\alpha,\gamma)^{18}F(\beta^{+})^{18}O(p,\alpha)^{15}N(\alpha,\gamma)^{19}F$.

In this sense, the $^{14}N(n,p)^{14}C$ reaction plays a key role since its double effect of *removing* neutrons and, at the same time, *producing* protons, captured by ^{18}O, that could trigger the $^{13}C(p,\gamma)^{14}N$ which represents the competing channel of the neutron source $^{13}C(\alpha,n)^{16}O$.

* Corresponding author: sergi@lns.infn.it

https://doi.org/10.1051/epjconf/202329003002

Here, we report first on the cross section measurement of the $^{7}Be(n,\alpha)^{4}He$ performed via THM applied to the quasi-free $^{2}H+^{7}Be$ reaction and then on the preliminary results of the $^{14}N(n,p)^{14}C$ measurement by selecting the Quasi Free (QF) contribution to the $^{2}H(^{14}N,p^{14}C)p$ reaction.

2 Basic features of THM

Nuclear reaction cross section measurements of interest for nuclear astrophysics are difficult to be performed in terrestrial laboratories mainly because of Coulomb barrier penetration and electron screening effects [11]. In order to overcome these difficulties, indirect methods, as THM and Asymptotic Normalization Coefficient (ANC) method [12], have been proved to be a valid complementary tool (see for example [13] and [14]) for experimental nuclear astrophysics since they allow the extraction of the astrophysically relevant cross section by selecting a precise reaction mechanism on a suitable reaction process thanks to devoted theoretical formalism.

The THM allows to obtain the cross section of a binary reaction A(x,c)C of astrophysical interest for which direct measurements require the use of the extrapolation procedure to reach the relevant Gamow peak energy [15-19 and references therein].

In the last decades, the THM gave a relevant contribution to solve several problems, varying from pure nuclear physics to nuclear astrophysics.

The direct reaction mechanism approach is the basic theory of the THM, having its background in the research of the Quasi Free (QF) reaction mechanism [20-22].

By referring to the pole diagram of Fig. 1, nucleus *a* is chosen because of its large $a=x\oplus s$ configuration, its relatively low $x-s$ binding energy, and its known radial wavefunction for the $x-s$ configuration. It represents the so-called “Trojan-horse nucleus.” The three-body reaction is induced at energies well above the Coulomb barrier of the $A+a$ interacting particles in order to induce the sub-process $A-x$ in the nuclear field.

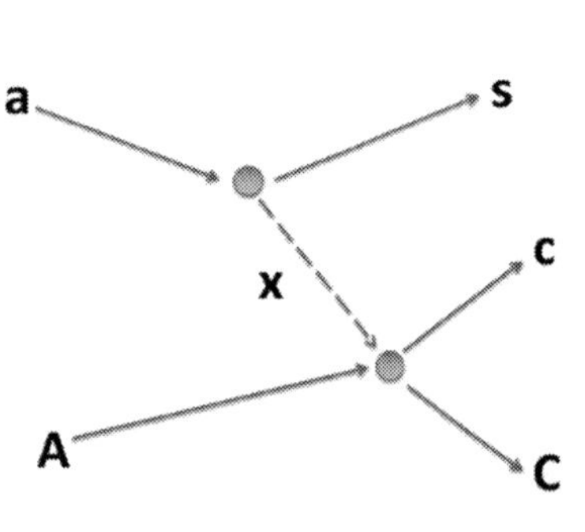

Fig. 1. Pole diagram for the quasi-free *a(A,cC)s* reaction, *s* being the so-called spectator, while *x* represents the participant to the astrophysically relevant *A(x, c)C* reaction.

Thus, the breakup of a is quasi-free because only cluster x takes part in the binary process, while the other counterpart s acts as spectator, i.e., it maintains in the exit channel the same momentum distribution it had inside a before its break-up. In addition, a specific role is played by the $x-s$ binding energy, as pointed out in [15-19]: it compensates for the projectile energy down to low, i.e., astrophysically relevant, energies, thus making it of immediate help for nuclear astrophysics purposes. In particular, since the A−x interaction occurs directly in the nuclear field, no Coulomb barrier penetration effects or screening phenomena [23] affect the THM data, in comparison with the direct impacts where these two effects cause the well known exponential decrease of the cross-section values and the enhancement of the cross-section values, respectively.

In the most simple theoretical description of THM by means of the plane wave impulse approximation, the cross section of the quasi-free a(A,cC)s reaction can be related to the one of the binary A(x,c)C processes via the formula [16,18]

$$\frac{d^3\sigma}{dE_c d\Omega_c d\Omega_C} \propto \mathrm{KF} \cdot |\Phi(\boldsymbol{p}_{xs})|^2 \cdot \frac{d\sigma}{d\Omega}\Big|_{cm}^{HOES},$$

where:

1. KF represents the kinematical factor, depending on the masses, momenta, and angles of the outgoing particles, that takes into account the final-state phase space factor.
2. $|\Phi(p_{xs})|^2$ is the square of the Fourier transform of the radial wavefunction describing the x−s intercluster motion, usually in terms of Hänkel, Eckart, or Hulthén functions, depending on the x−s system.
3. $\boldsymbol{d\sigma/d\Omega}|_{cm}^{HOES}$ is the half-off-energy-shell (HOES) differential cross section for the two-body reaction at the center-of-mass energy $E_{cm} = E_{cC}-Q$, where Q represents the Q-value of the HOES *A(x,c)C* reaction while E_{cC} represents the relative *c–C* energy measured in laboratory. The deduced cross section is HOES because in the entrance channel, the transferred particle x having mass m_x is *virtual*, thus its energy and momentum are not related by the mass-shell equation $\boldsymbol{E}_x = k_x^2/(2m_x)$.

Under QF conditions, the relative A−x energy is then determined by relation $E_{Ax} = \frac{p_{Ax}^2}{2\mu_{Ax}} - \epsilon_{sx}$, with ϵ_{sx} being the binding energy of the TH-nucleus. In the exit channel, the relation is restored because the emitted c–C particles are real (see [15-19] for details).

In recent past, an extension of the THM pointed out the ability of the method to also overcome the centrifugal barrier, measuring the bare nuclear cross section in neutron induced reactions [24-27].

In addition, a brand-new application of the method demonstrated the possibility to measure astrophysical relevant reaction cross section involving radioactive ion beams [28,29], with clear consequences in shedding light on research of neutron-induced reaction with unstable beams, usually tricky to perform with direct experiments [30]. In both the measurements described in the following sections, the deuteron was used as "trojan horse nucleus" because of its p⊕n cluster structure.

3 The ^{7}Be(n,α)^{4}He measurement

The ^{2}H(^{7}Be,αα)p experiment was performed at the EXOTIC facility of Laboratori Nazionali di Legnaro (INFN-LNL) using a 20.4 MeV ^{7}Be beam impinging on a CD_2 target with a thickness of 400 μg cm^{-2}. The detection setup was developed with the aim of detecting the two emerging alpha particles while the kinematical quantities of the undetected proton were reconstructed via momentum-energy conservation laws. In addition, since only quasi-free (QF) events were considered for the THM analysis, the setup covered the kinematical region corresponding to the QF-angular pairs, i.e., the angular pairs at which the spectator maintains the same momentum distribution it had inside the deuteron before its breakup [16,18].

The detailed description of the experimental set-up as well as the data analysis and the results have been already described step by step in [30], here only the resulting ^{7}Be(n,α) ^{4}He reaction cross section is shown in Fig. 2 as blue filled square with corresponding statistical uncertainties.

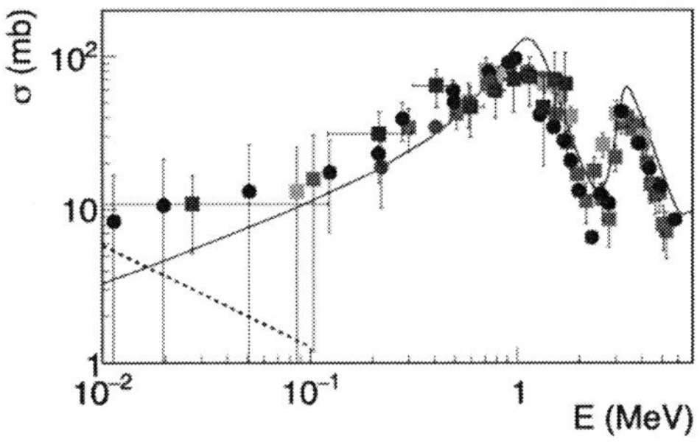

Fig. 2. THM ^{7}Be(n,α)^{4}He cross-section measurement (blue points with the statistical error) compared with the direct measurements [7] (red points) and [5] (black points). The magenta and green symbols refer to the reverse measurement obtained in [8]. The solid line is the ENDF/B-VII.1 evaluation [31] (https:// www.nndc.bnl.gov/sigma), while the dotted line gives the trend of the direct radiative capture (DRC) cross section given in [6].

The THM data span a broad energy region, i.e., from ~30 keV up to ~2 MeV, providing the cross section in the region of interest for big bang nucleosynthesis. The THM investigation nicely overlaps with the direct data of [7] (red points), and at lower energies with the ones derived via CSH in [5]; black points). Additionally, the present results agree with the ones derived in [8] reported as green and purple filled squares. In the same figure, the dashed line refers to the evaluation of the total s-wave component for the present reaction as derived in [6] while the solid line represents the ENDF/B-VII.1 evaluation by [31].

The binary cross section extracted here by using the THM allowed us, for the first time, to span a wide range in energy in a single experiment overlapping both the high-energy region and the one of interest for BBN. In addition to the abovementioned agreement, also note that the present THM investigation significantly reduces the uncertainty in the BBN energy region, thus further constraining the nuclear physics input for the cosmological lithium problem.

4 The ^{14}N(n,p)^{14}C measurement

The ^{2}H(^{14}N,p^{14}C)p experiment has been performed at Laboratori Nazionali del Sud of Catania (INFN-LNS). The SMP Tandem accelerator provided a 40 MeV ^{14}N beam with a spot size on target of about 1.5 mm ad intensities up to 2-3 pnA. Deuterated polyethylene target (CD_2) of about 150 μg/cm^2 was placed at 90° with respect to the beam axis. The experimental setup consisted of one 500 μm DSSSD (double sided silicon strips detectors) with an active area of 50 x 50 mm^2 with 16 strips per side orthogonally oriented and one 1000 μm position sensitive detector (PSD) devoted to proton detection. They were placed at distance d=28.4 cm and d= 25.5 cm from the target, respectively. The carbon particles were detected by means of a ΔE-E telescope (T) made of 20 μm strip detector as ΔE and 500 μm DSSSD as E-detector, placed at a distance d = 80 cm from the target and placed on the opposite side with respect to the beam direction. The detectors covered the laboratory angles between 19.4° and 30.6° (DSSSD), between 35.0° and 45.0° (PSD) and between -3.2° and -6.8° (ΔE-E telescope).

As a first stage of the experiment, energy and angular calibrations of the adopted experimental equipment have been performed by means of 10 MeV, 20 MeV and 25 MeV ^{14}N beam impinging onto ^{197}Au, CD_2, CH_2 thin targets as well as the use of a standard 8-peaks ^{228}Th alpha-source.

The reaction channel selection has been performed via the prior selection of the Z=6 locus in the ΔE-E telescope and the full kinematic reconstruction of the selected events with the hypothesis of a unitary mass for the third *undetected* particle. To validate the validity of the assumption, the experimental Q-value spectrum has been reconstructed and the result is shown as black points in Fig. 3. The agreement, within the experimental uncertainties, is a signature of a good detector calibration.

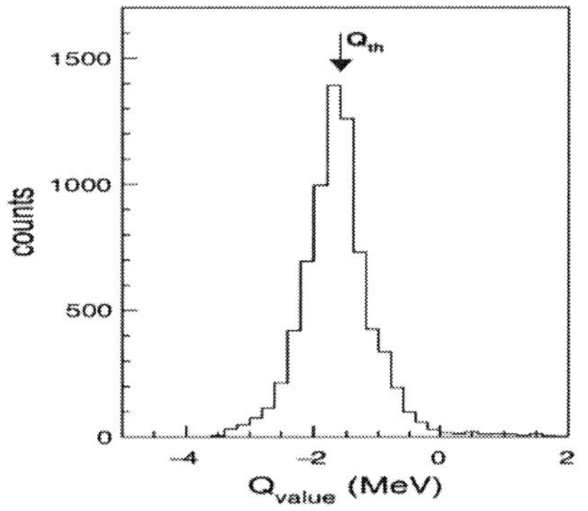

Fig. 3. Experimental Q-value spectrum. A peak centered at about −1.6 MeV, corresponding to the ^{2}H(^{14}N,p^{14}C)p channel. The black arrows correspond to the theoretical Q-value, Q_{th} =−1.599 MeV.

Because the $^{2}H(^{14}N, p^{14}C)p$ reaction has three particles in the exit channel, events from such reaction gather along a well-defined kinematic locus. By selecting the locus of interest, a kinematic identification of the reaction channel and, consequently, of the detected reaction products can be performed.

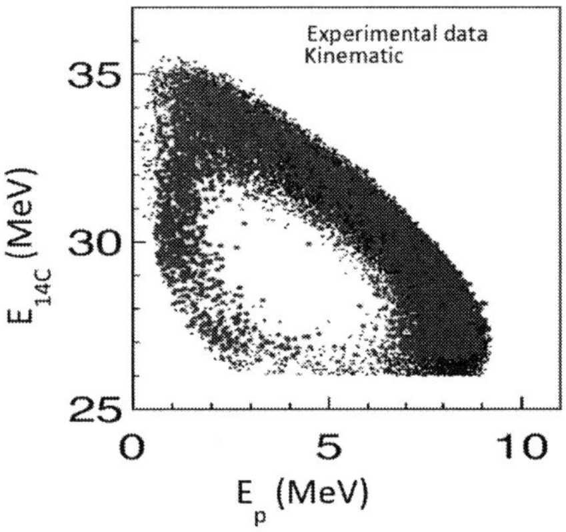

Fig. 4. Experimental kinematic locus (red circles) for the three-body $^{2}H(^{14}N,p^{14}C)p$ reaction induced at $E_{beam} = 40$ MeV superimposed onto the simulated kinematical locus (black circles).

For this purpose, the energy correlation between ^{14}C and p only for the coincidences T-DSSSD is shown in Fig 4.

By comparison with the corresponding three-body kinematic calculation (black circles) it turns out that a single reaction channel contributes to the E_{14C}– E^{p} correlation plot and the observed kinematic locus corresponds to the one for the $^{2}H(^{14}N,p^{14}C)p$ three-body reaction.

As a further stage of the analysis, the identification of the QF-reaction mechanism is mandatory by studying, for instance, the shape of the spectator experimental momentum distribution (see [8] for details). Preliminary studies suggest the strong presence of such a reaction mechanism in the energy window of our interest. Further progress in data analysis is still required in order to obtain the cross section for the $^{14}N(n,p)^{14}C$ reaction.

This research was funded by "Programma ricerca di ateneo UNICT 2020-22 linea2 (PIA.CE.RI 2020-22)" and by "Starting grant 2020" of the University of Catania.

References

1. R.G. Pizzone, et al., ApJ, **786**, 112, (2014)
2. C.A. Bertulani, *Nuclei in the Cosmos*, World Scientific, (2013), ISBN: 981441766
3. C.A. Bertulani, T. Kajino, PrPNP, **89**, 56, (2016)
4. P.A.R. Ade, N. Aghanim, M. Arnaud, et al. Planck Collaboration XIII, A&A, **594**, A13, (2016)
5. S.Q. Hou, J.J. He, S. Kubono, Y.S. Chen, Phys. Rev. C, **91**, 055802, (2015)
6. M. Barbagallo, A. Musumarra, L. Cosentino, et al., PhRvL, **117**, 152701 (2016)
7. T. Kawabata, Y. Fujikawa, T. Furuno, et al., PhRvL, **118**, 052701 (2017)
8. L. Lamia, C. Spitaleri, C.A. Bertulani, et al., ApJ, **850**, 175, (2017)
9. M. Forestini, S. Goriely, A. Jorissen, M. Arnould, A&A **261**, 157 (1992)
10. G. D'Agata, et al., ApJ **860**, aac207, (2018)
11. C.E. Rolfs, W.S. Rodney, *Cauldrons in the Cosmos*, Eds: Chicago, Illinois, USA.: Univ of Chicago Pr, 1988
12. G.G. Kiss, et al., Phys. Lett. B **807**, 135606 (2020)
13. A. Caciolli et al., EPJA **52**, 136 (2016)
14. C. Spitaleri, et al., Phys. Rev. C **90**, 035801 (2014)
15. C. Spitaleri, et al., Phys. Atom. Nucl. **74**, 1725, (2011)
16. C. Spitaleri, et al., Eur. Phys. J. A **52**, 77, (2016)
17. C. Spitaleri, et al., Eur. Phys. J. A **55**, 161 (2019)
18. R.E. Tribble, C.A. Bertulani, M. La Cognata, A.M. Mukhamedzhanov, C. Spitaleri, RPPh, **77**, 106901, (2014)
19. M. La Cognata, et al., Journ Phys. C -Nuclear and Part. Phys. **35**, 014014 (2008)
20. G. Baur, PhLB **178**, 135, (1986)
21. M. Zadro, D. Miljanić, C. Spitaleri, et al. 1989, Phys. Rev. C, **40**, 181
22. G. Calvi, et al., Phys. Rev. C **41**, 1848, (1990)
23. H. Assenbaum, K.L. Langanke, C. Rolfs, ZPhyA, **327**, 461 (1987)
24. M. Gulino, et al., J. Phys. G Nucl. Phys. **37**, 125105 (2010)
25. M. Gulino, et al., Phys. Rev. C **87**, 012801, (2013)
26. G.L. Guardo, et al., Phys. Rev. C **95**, 025807, (2017)
27. G.L. Guardo, et al., Eur. Phys. J. A **55**, 211, (2019)
28. S. Cherubini, S. et al., Phys. Rev. C **92**, 015805, (2015)
29. R.G. Pizzone, et al., Eur. Phys. J. A **52**, 24, (2016);
30. L. Lamia, et al., ApJ **879**, 23, (2019)
31. M.B. Chadwick, et al., NDS, **112**, 2887 (2011)
32. S.Q. Hou, J.J. He, S. Kubono, Y.S. Chen, Phys. Rev. C, **91**, 055802, (2015)

Populating α-unbound states in ^{16}O via $^{19}F(p,\alpha\gamma)^{16}O$ reaction

V.G. Távora[1,*], *O.* Tengblad[1], *M.J.G.* Borge[1], *J.A.* Briz[1,2], *D.* Fernandez Ruiz[1,2], *E.A.M.* Jensen[3], *A.N.* Nerio[1,2], and *A.* Perea[1]

[1]Instituto de Estructura de la Materia, CISC, Serrano 113bis, E-28006 Madrid, Spain
[2]Grupo de Física Nuclear, Facultad de CC. Físicas, Universidad Complutense de Madrid, CEI Moncloa, 28040 Madrid, Spain
[3]Department of Physics and Astronomy, University of Aarhus, DK-8000, Aarhus, Denmark

Abstract. The $^{12}C(\alpha,\gamma)^{16}O$ reaction is important in the production of universal ^{16}O, but its cross-section at the relevant energies of static helium burning is complex and uncertain. The total cross-section originates from a sum of resonance tails and direct captures, making the contributions of sub-threshold states difficult to estimate. One proposed method to estimate these contributions involves determining relevant reduced α-widths of the sub-threshold states through indirect measurements. Therefore, $^{19}F(p,\alpha\gamma)^{16}O$ reaction was used to populate α-unbound states in ^{16}O using a 2.6 MeV proton beam on a CaF_2 target. A detection system consisting of single and telescope configurations of Si detectors was used to detect 2α-particles and a ^{12}C particle in coincidence. Scintillator detectors were included in the setup to study the de-excitation of the states populated in ^{16}O to the ground state. Under good event conditions a preliminar indentification of the particles detected has been conducted.

1 Introduction: $^{12}C(\alpha,\gamma)^{16}O$ key reaction

One of the reactions in Nuclear Astrophysics that requires a better understanding is $^{12}C(\alpha,\gamma)^{16}O$. The reason for this is both, the unmitigated importance of the reaction as it is involved in the production of universal ^{16}O, and the complexity of its cross section at the relevant energies of static helium burning (300 keV).

As far as the cross-section is concerned the uncertainty is still undesirably large. As a consequence of the fact that, in the energy region of interest, there is no state of natural parity in ^{16}O near the α-threshold to serve as resonance for radiative capture, the total cross section originates from a sum of resonance tails and direct captures, both, to the ground and excited bound states of ^{16}O. Among the resonance tails contributing are two bound sub-threshold states, i.e., the 1^- state of -45 keV and the 2^+ state of -200 keV below the $\alpha+^{12}C$ threshold [1]. Fig. 1 provides a schematic representation of the $^{12}C(\alpha,\gamma)^{16}O$ reaction.

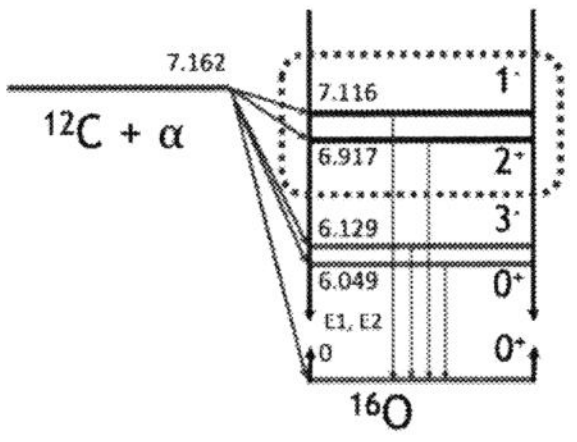

Figure 1. Schematic representation of $^{12}C(\alpha,\gamma)^{16}O$ reaction.

*e-mail: vicente.garcia@csic.es

One of the methods to estimate these contributions consist in determining the reduced α-widths of the sub-threshold states by indirect measurements, that are more sensitive to this width than the direct radiative capture measurement. Therefore, the $^{19}F(p,\alpha\gamma)^{16}O$ reaction has been used for the purpose of populating α-unbound states in ^{16}O (Fig. 2) [2].

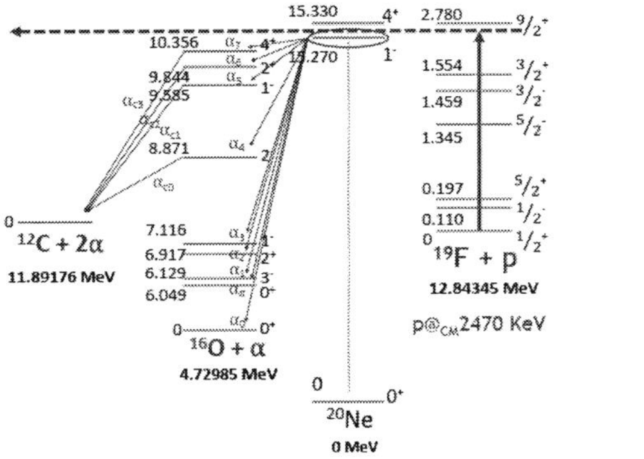

Figure 2. Population of unbound states on ^{16}O by $^{19}F(p,\alpha\gamma)^{16}O$ and their possible decay.

2 Experimental Methodology

In order to properly explore the excitation scheme of ^{16}O, the detection of 2α-particles and a ^{12}C particle in coincidence is required for above threshold levels as well as the de-excitation γ-rays are relevant for those sub-threshold levels.

A study of the $^{19}F(p,\alpha\gamma)^{16}O$ reaction has been performed at CMAM (Centre for Micro Analysis of Materials) facility (Madrid, Spain), using a 2.6 MeV proton beam, that impacts on a CaF_2 target with a thickness of 0.4 μm on 0.08 μm of Carbon backing.

The detection system (Fig. 3) consist of 14 5x5 cm^2 divided in 2x2 silicon detectors (Si-Ball)[3] and 4 telescopes (ΔE-E) consisting of a 5x5 cm^2 Double-Sided Stripped Silicon Detector (DSSD) as ΔE detector and non-segmented silicon detectors (PAD) as E detector. DSSD detector has 16x16 strips, which means they worked as a set of 256 pixels.

The Si-Ball was used in forward angles with respect to the target since it offers a good angular coverage and the possibility of avoiding angles around 0°-30° where the Rutherford elastic scattering dominates. Also at forward angles, two of the telescopes were placed following a configuration that allow the scattered protons from the beam pass through ΔE and stop in E detector while all the α-particles, ^{16}O and ^{12}C ions are stopped in ΔE detector. At backwards angles, two telescopes were placed, in this case the protons and α_0 traverse ΔE and stops in E detector while the rest of the particles emitted in the reaction reaction stops in ΔE detector. Thickness of the detectors used is given in Fig. 3.

In addition, 8 GAGG scintillator detectors and CEPA4 phoswich [4] detector were included in the setup in order to study the de-excitation of the excited states populated in ^{16}O to the ground state. Fig. 3 shows the setup used.

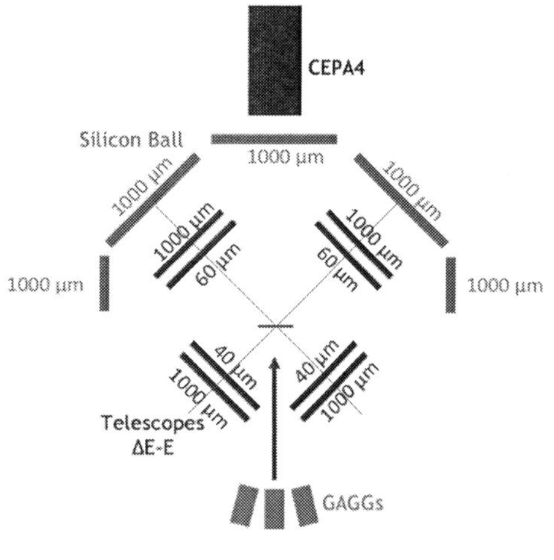

Figure 3. Scheme of the setup used used to measure the ejectiles produced in $^{19}F(p,\alpha\gamma)^{16}O$ reaction.

3 Analysis procedure

For a single particle detection in DSSD detectors, the junction (P-side) and the ohmic (N-side) is readout separately. For good event these signals correspond to the same deposited energy. Experimentally a certain deviation between both sides is expected so $|E_p - E_n| \leq 2\sigma$ condition has been included in the data processing, where σ value is obtained from the gaussian fit applied to $|E_p - E_n|$ distribution for each detector.

Under those conditions Fig. 4 shows a preliminary identification of the α-particles detected during the experiment as labeled in Fig. 2. The following step will be to remove proton background by anti-coincidence conditions with E detector and study γ emissions in coincidence with α-particles.

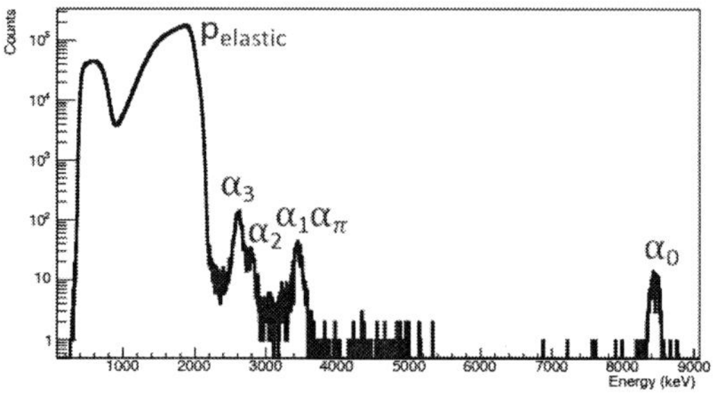

Figure 4. Preliminary particle identification as identified for a strip placed at $\theta = 60°$ in forward telescopes where we can see α-particles produced due to $^{19}F(p,\alpha\gamma)^{16}O$ reaction when populating different levels in ^{16}O, as well as elastic scattered protons that come from the beam with an energy of $\sim$ 2.46 MeV after being scattered and pass through ΔE, depositing $\sim$ 1.97 MeV.

Acknowledgements

We acknowledge the funding through the grant MCIN/AEI PID2019-104390GB-I00.

References

[1] L. Buchmann, The Astrophysical Journal **468**, L127 (1996)

[2] R.J. deBoer, O. Clarkson, A.J. Couture, J. Görres, F. Herwig, I. Lombardo, P. Scholz, M. Wiescher, Phys. Rev. C **103**, 055815 (2021)

[3] L. Fraile, J. Äystö, Nuclear Instruments and Methods in Physics Research Section A: Accelerators, Spectrometers, Detectors and Associated Equipment **513**, 287 (2003)

[4] E. Nácher, M. Mårtensson, O. Tengblad, et al., Nuclear Instruments and Methods in Physics Research Section A: Accelerators, Spectrometers, Detectors and Associated Equipment **769**, 105 (2015)

https://doi.org/10.1051/epjconf/202329003004

Phase Transitions under Strong Magnetic Fields in Neutron Star

Ishfaq A. Rather[1,*], *Veronica* Dexheimer[2,**], and *Ilídio* Lopes[1,***]

[1]Centro de Astrofísica e Gravitação-CENTRA, Instituto Superior Técnico-IST, Universidade de Lisboa-UL, Av. Rovisco Pais, 1049-001 Lisboa, Portugal
[2]Department of Physics, Kent State University, Kent OH 44242 USA

Abstract. A comprehensive study is carried out on the impact of strong magnetic fields on the deconfinement phase transition inside massive neutron stars. The matter equation of state and the general relativity solutions, which also fulfill Maxwell's equations, are modified when taking magnetic-field effects into account. We observe that the maximum mass and canonical-mass radius of stars computed using spherically-symmetric TOV equations and axisymmetric solutions obtained through the LORENE library differ significantly for large values of magnetic dipole moment. The discrepancies depend on the stellar mass being studied, as well as the stiffness of the equation of state. This indicates that the matter composition and interactions determine the magnetic field thresholds for the acceptable approximation of isotropic stars and the appropriate application of TOV equations.

1 Introduction

One of the most prominent and intriguing areas of research at the moment is the behavior of matter under extreme densities, temperatures, and magnetic fields. For studying physics in extreme conditions, neutron stars (NSs) are the ideal environment. Intriguing conclusions about NSs maximum mass, canonical mass, radius, tidal deformability, and other characteristics have been drawn as a result of recent advancements in observation. The equation of state (EoS) of the dense matter is, nevertheless, still widespread with uncertainty. Given that the density inside NSs is several times that of nuclear saturation ($\rho_0 \approx 10^{14}$ g cm^{-3}), it is challenging to ascertain their internal structure. Theoretical models describing cold and dense matter, calibrated around ρ_0 for isospin-symmetric nuclear matter (SNM), must be extrapolated in both density and isospin asymmetry as stellar inner densities approach 10^{15} g cm^{-3}. The structure of NSs is computed starting with the energy-momentum tensor, solving equations for hydrostatic equilibrium, and comparing them to astrophysical observations in order to evaluate these theories. The ability to explain the maximum measured NS mass, which according to the latest astrophysical data is above $2M_\odot$, is one criterion that all NS models must achieve [1, 2].

NSs have been known to possess immensely strong magnetic fields in addition to being extremely dense objects. Surface magnetic fields of$\sim 10^{12}-10^{13}$ G are already generated by combining a straightforward magnetic dipole model with known pulsar spin-down data [3]. The virial theorem is widely used to estimate the maximum magnetic field in the core of magnetars because it is not possible to measure it directly. According to most estimations, the field will have a theoretical maximum of 10^{18} G [4]. Strong magnetic fields can affect NSs in a number of ways, including changing the energy-momentum tensor, changing the EoS due to Landau quantization of the constituent charged particles, and breaking stellar spherical symmetry. It should be emphasized that the deviations from spherical symmetry for intense magnetic fields, however, can already be significantly below the threshold at which Landau quantization effects on the EoS become non-negligible [5]. In this case, Spherically symmetric TOV equations can no longer be used to describe the macroscopic structure of NS configurations. In the following, we discuss what this limit is in terms of the magnetic field strength and EoS.

*e-mail: ishfaq.rather@tecnico.ulisboa.pt
**e-mail: vdexheim@kent.edu
***e-mail: ilidio.lopes@tecnico.ulisboa.pt

2 Formalism

2.1 Nuclear Matter

For the description of hadronic matter, the DD-RMF model is used with the Lagrangian density given by

$$\begin{aligned}\mathcal{L}_m = \sum_b \bar{\psi}_b \Big\{ \gamma_\mu \Big(iD^\mu - g_\omega(\rho_b)\omega_\mu - \frac{1}{2} g_\rho(\rho_b)\rho_\mu \tau \Big) \\ - \Big(M_b - g_\sigma(\rho_b)\sigma \Big) \Big\} \psi_b + \frac{1}{2}\Big(\partial^\mu \sigma \partial_\mu \sigma - m_\sigma^2 \sigma^2 \Big) \\ - \frac{1}{4} W^{\mu\nu} W_{\mu\nu} + \frac{1}{2} m_\omega^2 \omega_\mu \omega^\mu - \frac{1}{4} R^{\mu\nu} R_{\mu\nu} + \frac{1}{2} m_\rho^2 \rho_\mu \rho^\mu \\ + \sum_l \bar{\psi}_l (i\gamma_\mu D^\mu - m_l)\psi_l \,, \qquad (1)\end{aligned}$$

where b sums over the baryon octet and l over non-interacting leptons. ψ_b and ψ_l represent the baryon and lepton Dirac fields and M_b and m_l the baryon and lepton masses, respectively. The density-dependent coupling

 https://doi.org/10.1051/epjconf/202329003004

constants for the DD-RMF parameter set are written as a function of baryon density ρ_B

$$g_i(\rho_B) = g_i(\rho_0)f_i(x) , \quad (2)$$

where the function $f_i(x)$ is given by

$$f_i(x) = a_i \frac{1 + b_i(x + d_i)^2}{1 + c_i(x + d_i)^2}, \; i = \sigma, \omega . \quad (3)$$

The coupling of the hyperons to the σ meson coupling constants is determined at saturation from fitting the Λ hyperon optical potential for SNM to results obtained from lattice calculations [6], reproducing the well-known potentials.

2.2 Quark Matter

The vector-enhanced bag (vBag) model is used to describe the quark matter while allowing hybrid stars to fulfill the $2M_\odot$ maximum mass limit [8–10].

The Lagrangian density for the vBag model is given by [7]

$$\mathcal{L} = \sum_{f=u,d,s} [\bar{\psi}_f (i\gamma_\mu \partial_\mu - m_f - B)\psi_f]\Theta_H - G_V \sum_f (\bar{\psi}_f \gamma_\mu \psi_f)^2 + \sum_l \bar{\psi}_l \gamma_\mu (i\partial_\mu - m_l)\psi_l , \quad (4)$$

where B denotes the bag constant and Θ_H is the Heaviside step function that allows for a confinement/deconfinement of the bag [11]. The vector interaction is introduced via the coupling of vector-isoscalar meson to the quarks with coupling constant G_V.

The expressions for the energy density and pressure for a single quark flavor are defined as

$$\mathcal{E}_{\rm vBag,f} = \mathcal{E}_f(\mu_f^*) + \frac{1}{2}K_\nu n_f^2(\mu_f^*) + B_{\chi,f} , \quad (5)$$

$$P_{\rm vBag,f} = P_f(\mu_f^*) + \frac{1}{2}K_\nu n_f^2(\mu_f^*) - B_{\chi,f} . \quad (6)$$

Here $B_{\chi,f}$ is the bag constant for a single quark flavor. The vector interaction parameter K_ν controls the stiffness of matter [12].

2.3 Mixed Phase

Charge neutrality is achieved globally using a Gibbs construction [13]. The extension of the mixed phase in density depends on the stiffness of the phases undergoing chemical equilibrium. Within the mixed-phase region, the expressions for the chemical potential and pressure are defined as

$$\mu_{B,H} = \mu_{B,Q}; \; \mu_{e,H} = \mu_{e,Q} , \quad (7)$$

and

$$P_H(\mu_B, \mu_e) = P_Q(\mu_B, \mu_e) = P_{MP} , \quad (8)$$

with the subscripts H, Q, and MP representing the hadronic phase, quark phase, and mixed-phase, respectively.

3 Magnetic Field

To investigate the effects of magnetic fields on our microscopic description of matter with a phase transition, we use a chemical-potential dependent magnetic field that is fitted from the solutions of the Einstein-Maxwell equations. The magnetic field-chemical potential relationship is dependent on the magnetic dipole moment and is provided by the relation [14]

$$B^*(\mu_B) = \frac{(a + b\mu_B + c\mu_B^2)}{B_c^2}\mu , \quad (9)$$

where μ_B is the baryon chemical potential in MeV and μ is the dipole magnetic moment in units of Am^2, so as to produce B^* in units of the electron critical field $B_c = 4.414 \times 10^{13}$ G. The coefficients a, b, and c taken as $a = -0.786$ G^2/(Am^2), $b = 1.24 \times 10^{-3}$ G^2/(Am^2 MeV) and $c = -3.51 \times 10^{-7}$ G^2/(Am^2 MeV^2) are obtained from a fit for the magnetic field in the polar direction of a star with a baryon mass of 2.2 M$_\odot$.

For the effective bag constant $B_{\rm eff}^{1/4} = 130$ and 150 MeV, the magnetic field at the stellar center is 3×10^{17} G and 1×10^{18} G, respectively for $\mu = 5 \times 10^{31}$ Am^2. With $\mu = 2{\times}10^{32}$ Am^2, the magnetic field goes upto $3.2{\times}10^{17}$ G and $1.2{\times}10^{18}$ G for $B_{\rm eff}^{1/4} = 130$ and 150 MeV, respectively.

The expressions for the free quark matter energy density and pressure in the presence of the magnetic field are shown in [15]. In presence of the magnetic field, the total energy density is

$$\mathcal{E} = \mathcal{E}_m + \frac{B^2}{8\pi} , \quad (10)$$

and the total pressure in transverse and polar directions to the local magnetic field is

$$P_\perp = P_m - \mathcal{M}B + \frac{B^2}{8\pi}, \quad P_\parallel = P_m - \frac{B^2}{8\pi} , \quad (11)$$

where the magnetization is calculated as

$$\mathcal{M} = \partial P_m / \partial B . \quad (12)$$

To ascertain the stellar properties of magnetic NSs, we employ the LORENE library [16, 17], which solves the Einstein-field Maxwell's equations with an axisymmetric deformation. When instead pressure in the local perpendicular direction to the magnetic field is applied in all directions, the spherically symmetric solutions obtained by solving the TOV equations lead to an overestimation of the mass and an underestimation of the equatorial radius and therefore cannot be used to determine stellar properties [5, 16]. In this work, we use both approaches for comparison.

4 Results and Discussion

To study the effect of a strong magnetic field on the hadron-quark phase transition, we employed the DD-MEX parameter set for the hadronic matter as it produces a very stiff EoS. For quark matter, two values of the effective bag

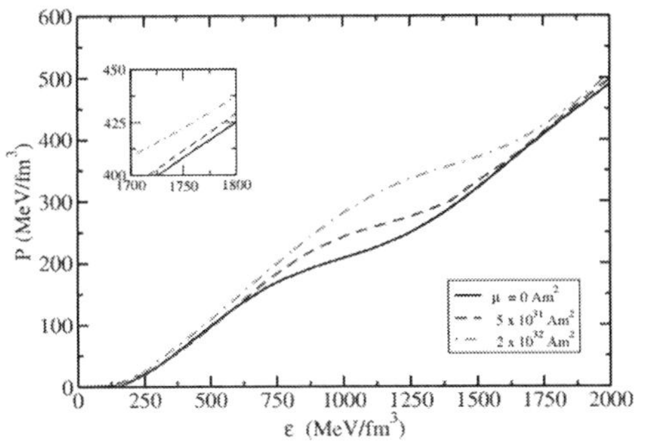

Figure 1. Energy density vs. transverse pressure at different values of dipole magnetic moment for the EoS with effective bag constant $B_{\text{eff}}^{1/4} = 130$ MeV.

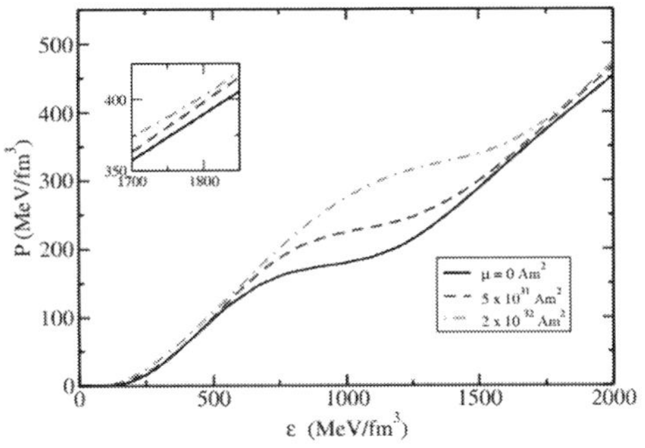

Figure 2. Same as Fig. 1, but for $B_{\text{eff}}^{1/4} = 150$ MeV.

constant, $B_{\text{eff}}^{1/4} = 130$ and 150 MeV are used, and the coupling constant parameter K_v is fixed at 4 GeV^{-2}.

Fig. 1 displays the NS EoS with quark deconfinement with effective bag constant $B_{\text{eff}}^{1/4} = 130$ MeV in the presence of the magnetic field profile for different values of the magnetic dipole moment. For the lower effective bag constant, the hadronic phase in the low-density region is largely unchanged by the magnetic field, whereas the pure quark phase in the high density is slightly stiffer than the EoS without the magnetic field for the magnetic dipole moment of $\mu = 5 \times 10^{31}$ Am^2, which corresponds to a high-density magnetic field of $\sim 10^{17}$ G. Both the hadronic phase and the pure quark phases stiffen up for the magnetic dipole moment of $\mu = 2 \times 10^{32}$ Am^2, which is equivalent to a $\sim 10^{18}$ G high-density magnetic field.

Fig. 2 displays the EoS for effective bag constant $B_{\text{eff}}^{1/4} = 150$ MeV. The pure quark phase only comprises a minimal portion (if any) of NSs since it only appears at very high densities. The beginning of the mixed phase and the pure quark phase are both significantly impacted by the increased effective bag constant. This is because the magnetic field is significantly greater at the densities where it occurs. Consequently, it is clear that the inclusion of the magnetic field and its strength have an impact on the transitions to the mixed phase and the pure quark phase. The magnetic field also suppresses the hyperons that emerge around the onset of mixed-phase due to an increased proton density.

The mass-radius diagram for hybrid stars computed for effective bag constant $B_{\text{eff}}^{1/4} = 130$ MeV without and with magnetic field estimated for various values of the magnetic dipole moment is shown in Fig. 3. The left panel displays the results obtained from solving Einstein and Maxwell's equations with an axisymmetric deformation (LORENE library), and the right panel displays the results using answers from spherically symmetric TOV equations. For $\mu = B = 0$, they are exactly equivalent, which should be the case. The maximum mass star has a radius of 12.6 km is $2.13 M_{\odot}$ without a magnetic field. The canonical mass has a radius of about 13 km, which satisfies the recent NICER radius constraints. With increasing magnetic dipole moment, the mass and radius of entire stellar sequences increase. For $\mu = 5 \times 10^{31}$ Am^2, which corresponds to a magnetic field of 3×10^{17} G in the center of the maximum-mass star, the radius of low/intermediate-mass stars obtained using the LORENE library differs significantly from the radius obtained from TOV, 0.6 km for the canonical mass when compared with 0.1 km for the maximum mass. In the left panel of Fig. 3, the radius shown is the equatorial radius, and under the influence of strong poloidal magnetic fields, stars become oblate.

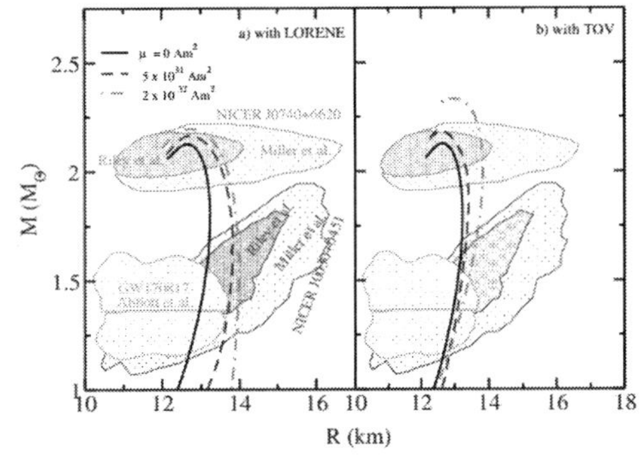

Figure 3. Mass-radius profile for hybrid stars with effective bag constant $B_{\text{eff}}^{1/4} = 130$ MeV shown at different values of magnetic dipole moment. The left panel shows results obtained using the LORENE library, while the right panel shows solutions from TOV. Recent constraints on mass and radius are also shown [1, 18–23].

Fig. 4 also displays the mass-radius profiles for hybrid stars but with effective bag constant $B_{\text{eff}}^{1/4} = 150$ MeV. For $\mu = 0$, the maximum mass obtained lowers to 2.08 $M_{\odot}$ with corresponding radius of 12.76 km. The radius at the canonical mass is 12.85 km. Our results are still in agreement with NICER, LIGO/VIRGO, and mass constraints. Using LORENE, as the magnetic dipole moment increases, the maximum mass increases more than for the previous effective bag constant considered. The radius of the whole sequence also increases more. The solutions from TOV, do not depend as much on the effective bag constant, and in this case, are farther from reproducing low/intermediate mass results from LORENE. A difference of around 0.5 km is observed in the radius at

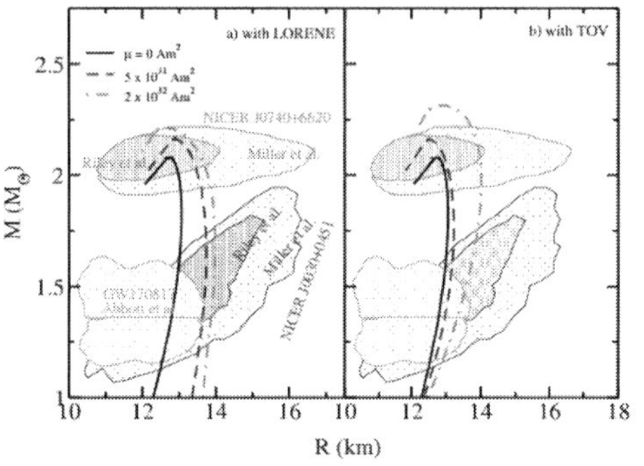

Figure 4. Same as Fig. 3, but for $B_{\rm eff}^{1/4} = 150$ MeV.

the canonical mass and −0.3 for the maximum mass with measurements from LORENE and TOV. This implies that the difference between the two approaches depends on the EoS (particle composition and interactions) and the mass of the star we are calculating the deformation for.

Considering the rotation along with the magnetic field, the ratio between the polar and the equatorial radii would increase and the resulting star would be deformed more. The difference between these two radii would depend on the star's rotation rate, the strength of the magnetic field, and the properties of the stellar matter.

5 Summary and Conclusion

We investigated how strong magnetic fields influence the transition from baryons to quarks in NS cores. A magnetic field profile with a quadratic relationship with the chemical potential is used to investigate the effects of magnetic fields on the EoS. In this setup, the magnetic field only really depends on the magnetic dipole moment at high densities, reaching 10^{18} G for the largest magnetic dipole moment studied. Magnetic fields cause the EoS to stiffen, although the effects are more pronounced in the mixed phase, which is wider in terms of energy density and occurs at higher densities. Larger magnetic dipole moments and effective bag constants have stronger impacts.

The poloidal magnetic field makes NSs oblate using the LORENE library, enlarging the equatorial radius. While the magnetic field affects the mass and radius of all stars in a family, this is not the case for TOV, which results in the rise of only massive stars and an unreasonably huge increase in their masses. The effective bag constant has no bearing on this qualitative behavior. The mass and radii of the entire family of stars defined by a lower effective bag constant (softer EoS) utilizing LORENE are quantitatively impacted by the strong magnetic field.

The difference in radius between LORENE and TOV for particular constant stellar masses is interestingly higher with lower values of μ and B. The canonical radius is consistently smaller for TOV overall, and the disparity grows with the effective bag constant. TOV can have a smaller or larger radius than LORENE for the radius of the maximum-mass star, depending upon whether the μ is small or large. The sequence's maximum mass is always too large for TOV, and this disparity gets more as μ and the effective bag constant increases.

Acknowledgement

I. A. R. and I. L. acknowledge the Fundação para a Ciência e Tecnologia (FCT), Portugal, for the financial support to the Center for Astrophysics and Gravitation (CENTRA/IST/ULisboa) through the grant Project No. UIDB/00099/2020 and grant No. PTDC/FIS-AST/28920/2017. V. D. acknowledges support from the National Science Foundation under grants PHY1748621, MUSES OAC-2103680, and NP3M PHY-2116686.

References

[1] J. Antoniadis, P.C.C. Freire *et al.*, Science **340** (2013)
[2] E. Fonseca et al., Astrophys. Jour. L. **915**, L12 (2021)
[3] S.B. Popov, R. Turolla, A. Possenti, MNRAS **369**, L23 (2006)
[4] D. Lai, S.L. Shapiro, Astrophys. Jour. **383**, 745 (1991)
[5] R.O. Gomes, H. Pais, V. Dexheimer, C. Providência, S. Schramm, Astron. Astrophys. **627**, A61 (2019)
[6] T. Inoue (HAL QCD), JPS Conf. Proc. **26**, 023018 (2019), AIP Conf. Proc. **2130**, 020002 (2019)
[7] T. Klähn, T. Fischer, Astrophys. J. **810**, 134 (2015)
[8] M. Cierniak, T. Klähn, T. Fischer, N. Bastian, Univ. **4**, 30 (2018)
[9] I.A. Rather, U. Rahaman, V. Dexheimer, A.A. Usmani, S.K. Patra, Astrophys. J. **917**, 46 (2021)
[10] I.A. Rather, A.A. Usmani, S.K. Patra, Jour. Phys. G: Nucl. and Part. Phys. **48**, 085201 (2021)
[11] E. Farhi, R.L. Jaffe, Phys. Rev. D **30**, 2379 (1984)
[12] W. Wei, B. Irving, M. Salinas, T. Klähn, P. Jaikumar, Astrophys. J. **887**, 151 (2019)
[13] N.K. Glendenning, Phys. Rev. D **46**, 1274 (1992)
[14] V. Dexheimer, B. Franzon, R. Gomes, R. Farias, S. Avancini, S. Schramm, Phys. Lett. B **773**, 487 (2017)
[15] I.A. Rather, A.A. Rather, I. Lopes, V. Dexheimer, A. Usmani, S. Patra, Astrophys. J. **943**, 52 (2023)
[16] D. Chatterjee, T. Elghozi, J. Novak, M. Oertel, MNRAS **447**, 3785 (2015)
[17] LORENE, *Langage objet pour la relativité numérique*, http://www.lorene.obspm.fr (-)
[18] R. Abbott, T.D. Abbott et al., Astrophys. J. **896**, L44 (2020)
[19] P.B. Demorest, T. Pennucci, S.M. Ransom, M.S.E. Roberts, J.W.T. Hessels, Nature **467**, 1081 (2010)
[20] H.T. Cromartie, E. Fonseca *et al.*, Nature Astron. **4**, 72 (2019)
[21] M.C. Miller *et al.*, Astrophys. J. **887**, L24 (2019), Astrophys. J. Lett. **918**, L28 (2021)
[22] T.E. Riley *et al.*, Astrophys. J. **887**, L21 (2019), Astrophys. J. Lett. **918**, L27 (2021)
[23] G. Raaijmakers et al., Astrophys. J. **887**, L22 (2019)

Recent results from Borexino on solar neutrinos

D. Basilico[1,*], *S.* Appel[2], *Z.* Bagdasarian[3], *G.* Bellini[1], *J.* Benziger[4], *R.* Biondi[5], *B.* Caccianiga[1], *F.* Calaprice[6], *A.* Caminata[7], *A.* Chepurnov[8], *D.* D'Angelo[1], *A.* Derbin[9], *A.* Di Giacinto[5], *V.* Di Marcello[5], *X.F.* Ding[6], *A.* Di Ludovico[6], *L.* Di Noto[7], *I.* Drachnev[9], *D.* Franco[10], *C.* Galbiati[6,11], *C.* Ghiano[5], *M.* Giammarchi[1], *A.* Goretti[6], *A.S.* Göttel[3,12], *M.* Gromov[8,13], *D.* Guffanti[14], *Aldo* Ianni[5], *Andrea* Ianni[6], *A.* Jany[15], *V.* Kobychev[16], *G.* Korga[17,18], *S.* Kumaran[3,12], *M.* Laubenstein[5], *E.* Litvinovich[19,20], *P.* Lombardi[1], *I.* Lomskaya[9], *L.* Ludhova[3,12], *G.* Lukyanchenko[19], *I.* Machulin[19,20], *J.* Martyn[14], *E.* Meroni[1], *L.* Miramonti[1], *M.* Misiaszek[15], *V.* Muratova[9], *R.* Nugmanov[19,20], *L.* Oberauer[2], *V.* Orekhov[14], *F.* Ortica[21], *M.* Pallavicini[7], *L.* Pelicci[3,12], *Ö.* Penek[3], *L.* Pietrofaccia[6], *N.* Pilipenko[9], *A.* Pocar[22], *G.* Raikov[19], *M.T.* Ranalli[5], *G.* Ranucci[1], *A.* Razeto[5], *A.* Re[1], *M.* Redchuk[3,12], *N.* Rossi[5], *S.* Schönert[2], *D.* Semenov[9], *G.* Settanta[3], *M.* Skorokhvatov[19,20], *A.* Singhal[3,12], *O.* Smirnov[13], *A.* Sotnikov[13], *R.* Tartaglia[5], *G.* Testera[7], *E.* Unzhakov[9], *F.* Villante[5,23], *A.* Vishneva[13], *R.B.* Vogelaar[24], *F.* von Feilitzsch[2], *M.* Wojcik[15], *M.* Wurm[14], *S.* Zavatarelli[7], *K.* Zuber[25], and *G.* Zuzel[15]

[1]Dipartimento di Fisica, Università degli Studi and INFN, 20133 Milano, Italy.
[2]Physik-Department, Technische Universität München, 85748 Garching, Germany
[3]Institut für Kernphysik, Forschungszentrum Jülich, 52425 Jülich, Germany
[4]Chemical Engineering Department, Princeton University, Princeton, NJ 08544, USA
[5]INFN Laboratori Nazionali del Gran Sasso, 67010 Assergi (AQ), Italy
[6]Physics Department, Princeton University, Princeton, NJ 08544, USA
[7]Dipartimento di Fisica, Università degli Studi e INFN, 16146 Genova, Italy
[8]Lomonosov Moscow State University Skobeltsyn Institute of Nuclear Physics, 119234 Moscow, Russia
[9]St. Petersburg Nuclear Physics Institute NRC Kurchatov Institute, 188350 Gatchina, Russia
[10]AstroParticule et Cosmologie, Université Paris Diderot, CNRS/IN2P3, CEA/IRFU, Observatoire de Paris, Sorbonne Paris Cité, 75205 Paris Cedex 13, France
[11]Gran Sasso Science Institute, 67100 L'Aquila, Italy
[12]III. Physikalisches Institut B, RWTH Aachen University, 52062 Aachen, Germany
[13]Joint Institute for Nuclear Research, 141980 Dubna, Russia
[14]Institute of Physics and Excellence Cluster PRISMA+, Johannes Gutenberg-Universität Mainz, 55099 Mainz, Germany
[15]M. Smoluchowski Institute of Physics, Jagiellonian University, 30348 Krakow, Poland
[16]Institute for Nuclear Research of NAS Ukraine, 03028 Kyiv, Ukraine
[17]Department of Physics, Royal Holloway University of London, Egham, Surrey,TW20 0EX, UK
[18]Institute of Nuclear Research (Atomki), Debrecen, Hungary
[19]National Research Centre Kurchatov Institute, 123182 Moscow, Russia
[20]National Research Nuclear University MEPhI (Moscow Engineering Physics Institute), 115409 Moscow, Russia
[21]Dipartimento di Chimica, Biologia e Biotecnologie, Università degli Studi e INFN, 06123 Perugia, Italy
[22]Amherst Center for Fundamental Interactions and Physics Department, University of Massachusetts, Amherst, MA 01003, USA
[23]Dipartimento di Scienze Fisiche e Chimiche, Università dell'Aquila, 67100 L'Aquila, Italy
[24]Physics Department, Virginia Polytechnic Institute and State University, Blacksburg, VA 24061, USA
[25]Department of Physics, Technische Universität Dresden, 01062 Dresden, Germany

Abstract. Borexino has been a neutrino detector based on ultrapure liquid scintillator, located at the Laboratori Nazionali del Gran Sasso, Italy. Its main scientific goal was the real-time measurement of solar neutrino fluxes, which play an irreplaceable role for the comprehension of the mechanisms powering our star. Over the past two years, the Borexino collaboration has pursued the improvement of the CNO flux measurement, obtaining further indications about the solar metallicity. In a parallel way, Borexino has demonstrated for the first time the possibility of exploiting the directional Cherenkov information, in a liquid scintillator detector, for the detection of sub-MeV solar neutrinos.

1 Introduction

Solar neutrinos are generated by the thermonuclear reactions taking place in the innermost layers of the Sun. The main contribution to the energy production in the Sun is given by the *pp* chain sequence, which make up the $\approx 99\%$ of solar luminosity. The CNO cycle reactions, which involves employs Carbon, Nitrogen and Oxygen as catalysts nuclei, contribute marginally in the Sun, but they are expected to play a dominant role in more massive and older stars. The measurements of solar neutrino fluxes can be

*e-mail: davide.basilico@mi.infn.it

used either to test the MSW-LMA paradigm assuming the Standard Solar Model (SSM) flux predictions [1] or, in a complementary way, to probe our understanding of solar physics assuming the validity of the neutrino oscillation mechanism [2]. Indeed, they represent a unique and helpful tool in settling the solar metallicity problem.

In the following, the most recent results obtained by Borexino on solar neutrinos will be reviewed. Firstly, the improvement of CNO solar neutrino spectroscopy will be described. Then, a novel technique to exploit the scarce Cherenkov light for directional measurements will be introduced.

2 Borexino detector

Borexino [3] has been a large-volume liquid scintillator detector, whose primary scientific goal consisted in the real-time measurement of solar neutrino fluxes. It was located deep underground (3800 m m.w.e.) in the Hall C of the Laboratori Nazionali del Gran Sasso (Italy), where the atmospheric muon flux is suppressed by a factor of 10^{-6} with respect to the Earth surface. The detector design, as shown in Fig. 1, is driven by the principle of graded shielding: the inner scintillating core is located at the center of a set of spherical layers of increasing radiopurity from outside to inside. This configuration suppresses as much as possible the external background due to the surrounding rocks and to the cosmic muons. The detector is instrumented with 2112 PMTs, mounted inwardly on a mechanical support sphere: they measure the intensity and the arrival time of this light, allowing the reconstruction of the energy and position of the events.

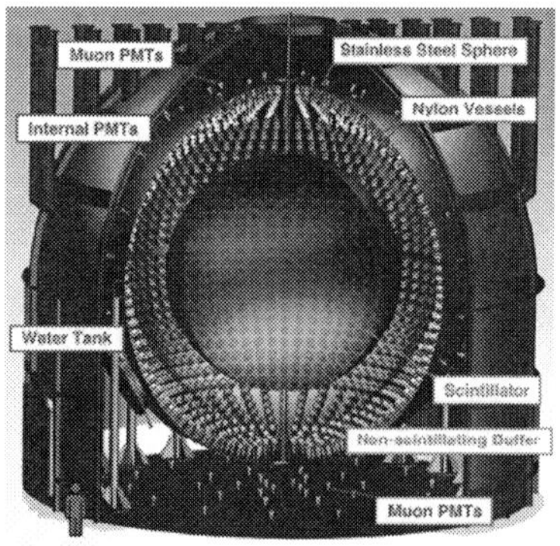

Figure 1. Schematic view of the Borexino detector.

Thanks to its excellent radiopurity levels, Borexino completed the spectroscopy of the solar neutrinos emitted from the primary *pp* chain reactions [4–9]. By analyzing the Phase-III dataset (2016-2020), it also claimed the first detection of the CNO neutrinos flux, paving the way for a solution to the long-standing solar metallicity problem [10].

3 Improved measurement of solar neutrinos flux from the CNO cycle

The key requirements to perform solar neutrino spectroscopy with Borexino consist in the excellent radiopurity levels, the low cosmogenic ^{11}C rate and the application of techniques to reduce it furtherly [11], the precise understanding of the detector backgrounds, and a comprehensive modeling of the detector response. In addition, the spectral shapes of recoiled electrons by CNO neutrinos and of ^{210}Bi background electrons show a strong similarity; for this reason, a constraint of this annoying background, obtained independently from the spectral fit, is required to proceed with the CNO analysis [12]. This is achieved by exploiting the secular equilibrium of ^{210}Bi background with its daughter, the α-decaying ^{210}Po, whose rate can be easily measured thanks to particle identification techniques [13]. This strategy can be adopted only in those scintillator regions which are thermally stable: in such a way the scintillator convective currents, which could move out-of-equilibrium ^{210}Po, are strongly suppressed.

With respect to the previous data release, the CNO analysis has been improved on multiple sides. The exposure of the *final Phase-III* dataset (Jan 2017 - Oct 2021), which is considered for this analysis, is 33.5% larger with respect to the previous analysis one. Most importantly, the final Phase-III dataset is characterized by a better thermal stability: as a consequence, the ^{210}Po convective motions within the fiducial volume are furtherly attenuated. These features led to the extraction of an enhanced ^{210}Bi constraint, directly impacting on the CNO neutrino rate precision. In addition the treatment of the time evolution of the PMTs effective quantum efficiencies, based on the ^{14}C data, has been improved.

The analysis ROI extends from 320 keV to 2640 keV, while the reconstructed events are selected within a software-cut defined 71.3 tonne fiducial volume. The analysis relies on a Monte Carlo based multivariate fit, performed on two energy spectra and the radial distribution of events passing the selection cuts. The Three-Fold Coincidence (TFC), an algorithm for the cosmogenic event reduction, is applied to create the two complementary datasets on which the fit is performed: one depleted (TFC-subtracted) and one enriched (TFC-tagged) in ^{11}C. The spectral fit of the final Phase-III TFC-subtracted dataset is reported in Fig. 2.

By including the Borexino results in the global analysis of solar neutrinos data, along with the KamLAND reactor data, the resulting solar neutrino fluxes are found in good agreement with the high metallicity SSM B16-GS98 [14]. Considering a binary hypothesis test based on the Borexino ^{7}Be, ^{8}B and CNO results, the SSM B16-AGSS09met [15, 16] at 3.1σ C.L. is disfavored at 3.1σ C.L. as an alternative to SSM B16-GS98. Moreover, one can exploit the direct dependence of the CNO cycle flux from the C and N abundances in the solar core, to break the opacity and metallicity degeneracy; therefore, the C and N abundances in the Sun ($N_{\rm CN}$) can be measured. The results are displayed in Fig. 3 and show a good agreement

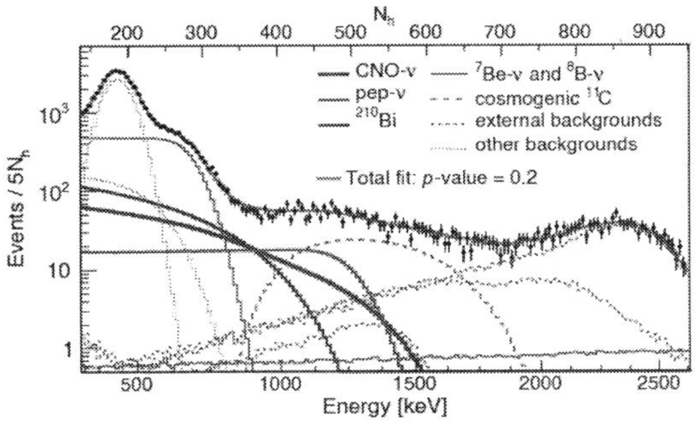

Figure 2. Spectral fit of the final Borexino Phase-III data (black points) after the TFC algorithm application (TFC-subtracted dataset). The sum of the individual components from the fit (magenta) is superimposed on the data (black points). The CNO neutrinos, ^{210}Bi and *pep* neutrinos contributions are reported in solid red, dashed blue and dotted green lines; the other neutrino and background components are shown in grey. The energy estimator N_h, shown in upper scale, is the number of detected photoelectrons normalized to 2,000 live channels.

with the high-metallicity sets (as MB22 [17], GS98), confirming instead a 2σ tension with the low-metallicity ones (AGSS09met, C11 [18], AAG21 [19])

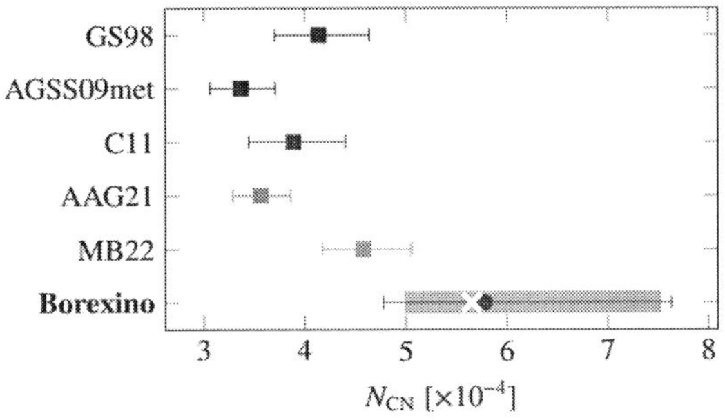

Figure 3. Abundance of Carbon and Nitrogen in the solar photosphere (N_{CN}) obtained from spectroscopy (squares) and from Borexino analysis (circle). The uncertainty due to the precision of CNO flux measurement is shown as the grey area. The white cross refers to the same study repeated after considering the B16-AGSS09met as reference SSM instead of B16-GS98.

4 First directional measurement of sub-MeV solar neutrinos with Borexino

In liquid scintillator detectors as Borexino, the emitted photons following an electron recoil are mainly due to scintillation processes. The Cherenkov photons play a sub-dominant role, accounting for few percent of the total emitted light: this scarcity prevents the possibility to perform an event-by-event direction reconstruction.

Nevertheless, Borexino has been able to perform a measurement of sub-MeV solar neutrinos, by exploiting this feeble Cherenkov light [20, 21]. We employed a novel analysis technique named *Correlated and Integrated Directionality* (CID): the detected PMT-hit pattern of an event is correlated to the position of the Sun, which is well-known for each event. This process is repeated for a large number of events, leading to an angular distribution between the hit PMTs and the solar direction. The directional angle α is defined for each PMT hit as the angle between the known solar direction and the photon direction, given by the reconstructed event vertex of the recoil electron and the hit PMT position. An example of the angular correlation of photon hits is shown in Fig. 4. The dominant scintillation light is emitted isotropically and is uncorrelated with the Sun position. For this reason, the analysis is restricted to the first and second hits as they are characterized by the highest Cherenkov over scintillation ratio. In this analysis, as shown in Fig. 5, the cos α first hit distribution of solar neutrinos is expected to have a signature Cherenkov peak at ~ 0.7, while the one for radioactive background should be almost flat.

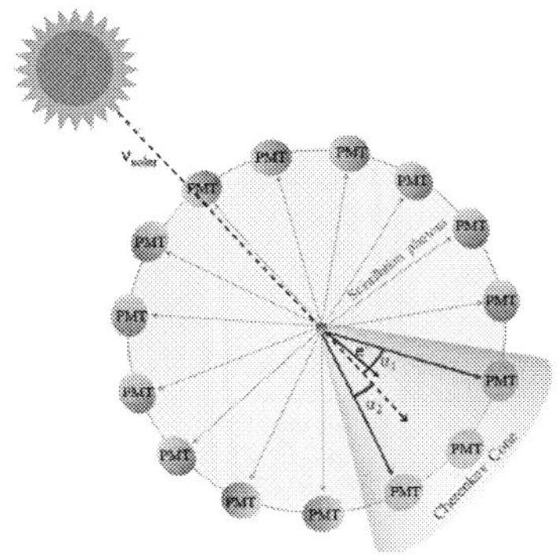

Figure 4. Angular correlation of photon hits expressed in terms of the directional angle α, given by the reconstructed vertex of the solar neutrino event and the position of the Sun. In this case, the first two event hits, with directional angles α_1 and α_2, are both induced by Cherenkov photons.

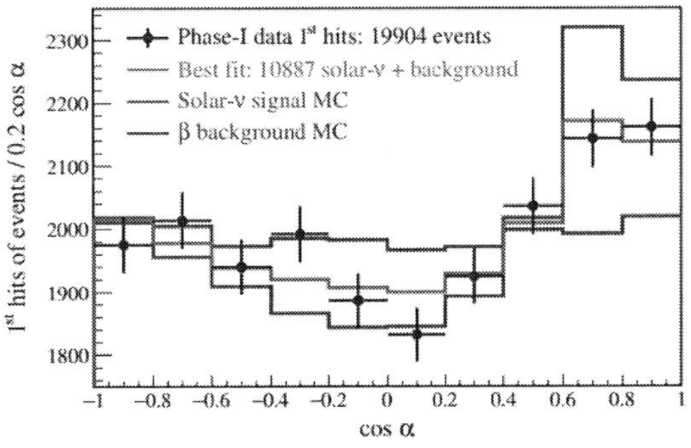

Figure 5. The cos α distributions of the 1st hits of all the selected events (black points) compared with the best fit curve (magenta) of the resulting number of solar neutrinos plus background. The Monte Carlo based probability distributions of neutrino signal and β background used in the fit are reported in red and blue respectively.

The considered dataset is Phase-I one (2007-2010) with a 0.54-0.74 MeV ROI, which includes ^{7}Be, *pep*, and CNO neutrinos. The best fit for the number of solar neutrino events returns $N_{\text{solar}-\nu} = 10887^{+2386}_{-2103}$ (stat.) $\pm$ 947 (syst.) (68% C.I.) out of the 19904 selected events. By fixing the *pep* and CNO rates to the SSM prediction, the resulting ^{7}Be rate is found in agreement with the Borexino spectroscopy results. The no-neutrino hypothesis is excluded with a $> 5\sigma$ confidence level, resulting in the first observation of sub-MeV solar neutrinos using the CID method.

In conclusion, this analysis provides experimental proof that the directional information of Cherenkov light is accessible even for sub-MeV neutrinos in a liquid scintillator detector.

References

[1] N. Vinyoles, A.M. Serenelli, F.L. Villante, S. Basu, J. Bergström, M. Gonzalez-Garcia, M. Maltoni, C. Peña-Garay, N. Song, Astrophys. J. **835**, 202 (2017), 1611.09867

[2] A.M. Serenelli, W.C. Haxton, C. Pena-Garay, Astrophys. J. **743**, 24 (2011), 1104.1639

[3] G. Alimonti et al. (Borexino), Nucl. Instrum. Meth. A **600**, 568 (2009)

[4] G. Bellini et al. (Borexino), Phys. Rev. D **89**, 112007 (2014), 1308.0443

[5] M. Agostini et al. (Borexino), Nature **562**, 505 (2018)

[6] M. Agostini et al. (Borexino Collaboration), Phys. Rev. D **100**, 082004 (2019)

[7] G. Bellini et al. (Borexino), Nature **512**, 384 (2014)

[8] M. Agostini et al. (Borexino), Phys. Rev. D **101**, 062001 (2020), 1709.00756

[9] M. Agostini et al. (BOREXINO), Astropart. Phys. **92**, 21 (2017), 1701.07970

[10] M. Agostini et al. (Borexino), Nature **587**, 577 (2020), 2006.15115

[11] M. Agostini et al. (Borexino), Eur. Phys. J. C **81**, 1075 (2021), 2106.10973

[12] M. Agostini et al. (BOREXINO), Eur. Phys. J. C **80**, 1091 (2020), 2005.12829

[13] F. Villante et al., Phys. Lett. B **701**, 336 (2011)

[14] N. Grevesse, A.J. Sauval, Space Sci. Rev. **85**, 161 (1998)

[15] P. Scott, N. Grevesse, M. Asplund, A.J. Sauval, K. Lind, Y. Takeda, R. Collet, R. Trampedach, W. Hayek, Astron. Astrophys. **573** (2015)

[16] P. Scott, M. Asplund, N. Grevesse, M. Bergemann, A.J. Sauval, Astron. Astrophys. **573** (2015)

[17] E. Magg, M. Bergemann, A. Serenelli, M. Bautista, B. Plez, U. Heiter, J. Gerber, H.G. Ludwig, S. Basu, J. Ferguson et al., A&A **661**, A140 (2022), 2203.02255

[18] E. Caffau et al., Sol. Phys. **268**, 255 (2011)

[19] M. Asplund, A.M. Amarsi, N. Grevesse, A&A **653**, A141 (2021), 2105.01661

[20] M. Agostini et al. (Borexino), Phys. Rev. Lett. **128**, 091803 (2022)

[21] M. Agostini et al. (Borexino), Phys. Rev. D **105**, 052002 (2022)

BINGO: background reduction techniques for $0\nu2\beta$ bolometric experiments

Hawraa Khalife[1,*] *On behalf of the BINGO collaboration*[**]

[1]IRFU, CEA, Université Paris-Saclay, F-91191 Gif-sur-Yvette, France

Abstract. BINGO is a project dedicated to explore new methods for background reduction in experiments searching for $0\nu2\beta$ decay. It is based on bolometers, one of the most promising techniques to search for $0\nu2\beta$. CUORE and CUPID-Mo/0 are the main bolometric experiments that have illustrated the most relevant limiting factors on sensitivity the bolometers are facing. Surface αs are the main source of background in CUORE, and this contribution has been mitigated in CUPID-Mo/0 using dual heat-light channels, i.e., a main absorber embedding the 2β decay isotope facing a light detector. In this case, surface α rejection is achieved thanks to the lower α light output compared to β/γ. However, there are still other background components that limit the sensitivity of the experiments, such as pile-ups due to random coincidences of physical events, external γ background and β surface radioactivity. BINGO's proposed technology aims at reducing the background index down to 10^{-5} counts/(keV kg yr) in the region of interest, thus boosting the sensitivity on the effective Majorana neutrino mass. This can be achieved by: (i) having a revolutionary detector assembly with a reduction in the passive materials facing the detector; (ii) increasing the light detector sensitivity thanks to Neganov-Luke amplification; (iii) using an active shield, based on BGO scintillators with bolometric light detector readout to surround the experimental volume. The main text describes the listed approaches in detail, as well as the recent results of prototype tests.

1 Introduction

Neutrinoless double-beta decay ($0\nu2\beta$) is a hypothetical lepton-number violating process that was proposed by W. H. Furry in 1939 [1]. It can occur when an even-even nucleus (A,Z) decays into its isobar (A, Z+2) with the emission of two electrons. Searching for this decay is currently the only way to prove experimentally that neutrinos are Majorana particles ($\nu = \bar{\nu}$). The signature of $0\nu2\beta$ decay is a monochromatic peak at the Q-value ($Q_{\beta\beta}$) of the transition corresponding to the energy sum of the two emitted electrons. There are many experiments digging their way to discover this decay, however, there are many challenges on the way that needs to be overcome to boost their sensitivity. One of the most limiting factors to search for this decay are the backgrounds from various components. In fact, we expect very few events in the region of interest (ROI) coming from this decay because of the very long half-life (the best current limits are $> 10^{24-26}$ yr, depending on the 2β isotope [2]), which can be easily buried in the background. BINGO (Bi-Isotope $0\nu2\beta$ Next Generation Observatory), an ERC funded project, focuses on introducing new methods to mitigate the backgrounds. It aims to set the ground for large-scale next generation bolometric experiments to be capable of exploring the inverted hierarchy region and a part of the normal region thanks to the very low expected background index of about 10^{-5} counts/(keV kg yr). The three main background components that will be fought against are: pile-ups (especially those due to random coincidences of events from the relatively fast $2\nu2\beta$ decay of ^{100}Mo), external γ radioactivity and near surface radioactivity. BINGO will study two 2β decay isotopes: ^{100}Mo ($Q_{\beta\beta}$= 3034 keV) and ^{130}Te ($Q_{\beta\beta}$= 2527 keV), they will be embedded respectively in Li_2MoO_4 and TeO_2 crystals, which will be operated as bolometric detectors. The project will bring original improvements to the well-established technology of the dual heat-light readout, demonstrated by CUPID-Mo [3, 4] and CUPID-0 [5]: (i) a revolutionary detector assembly that will reduce the total surface radioactivity contribution by at least one order of magnitude; (ii) the light-detector sensitivity will be increased by one order of magnitude thanks to Neganov-Luke amplification that can help achieve pile-ups rejection; (iii) for the first time in an array of macrobolometers, an internal active shield will be used, based on BGO scintillators with scintillation light readout, that will suppress the external γ background and surface contamination from crystals facing it using coincidence cuts. This paper will describe the foreseen BINGO demonstrator "mini-BINGO" that will be operated in a pulse-tube cryostat at Modane Underground Laboratory (LSM), in addition to some of the prototype tests results.

2 Mini-BINGO demonstrator

The mini-BINGO demonstrator will be composed of two towers: one tower will consist of 12 Li_2MoO_4 crystals ($45 \times 45 \times 45$ cm^3) and the other of 12 TeO_2 crystals ($50 \times 50 \times 50$ cm^3). The tower is arranged in 6 floors,

*e-mail: hawraa.khalife@cea.fr
**http://www.bingo-neutrino.eu/

https://doi.org/10.1051/epjconf/202329004002

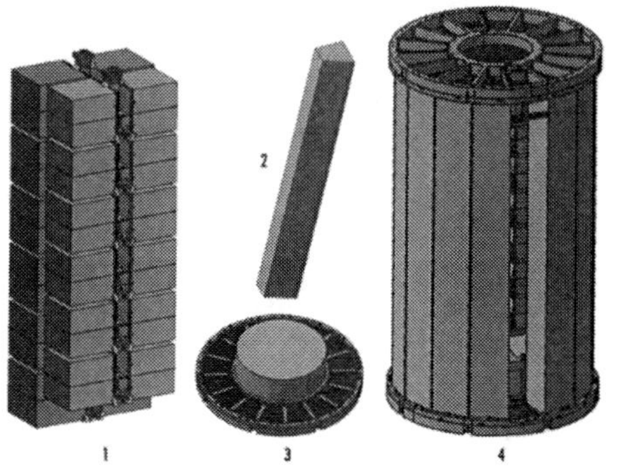

Figure 1. All the components of mini-BINGO. (1) The Li_2MoO_4 and TeO_2 towers coupled to LDs. (2 & 3) The active veto, consisting of 16 BGO bars on the lateral and 2 BGO cylinders on top and bottom. Each scintillator will be coupled to a NL LD. (4) The full mini-BINGO design, the $0\nu2\beta$ detectors (towers) surrounded by the active veto (one BGO scintillator is not shown for the internal set-up structure visibility).

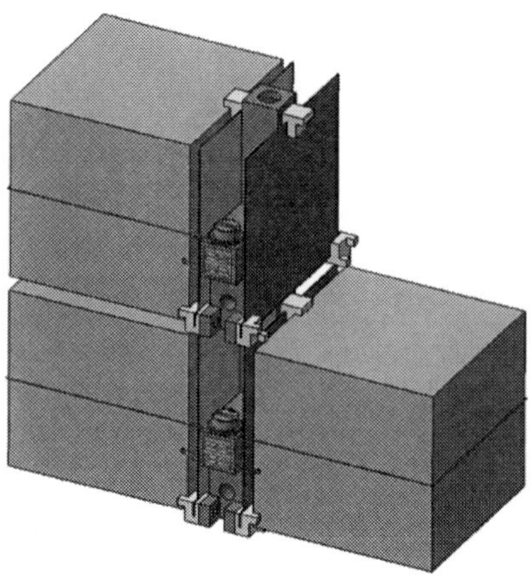

Figure 2. A scheme showing two detectors module. Each module consists of two crystals and two LDs that are fixed using a nylon wire.

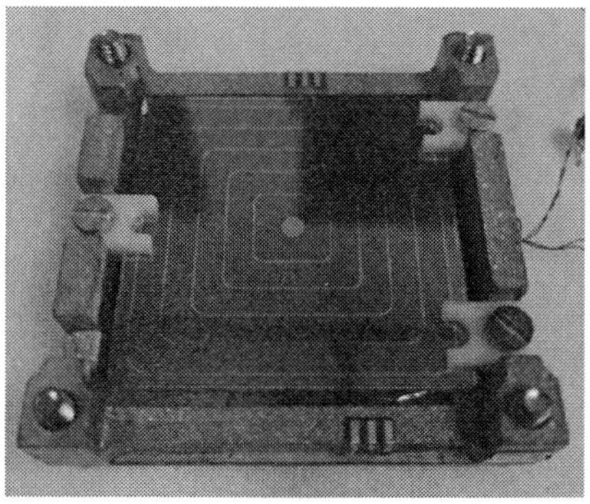

Figure 3. A NL square LD with concentric rounded corners square electrodes.

each floor will host two crystals, each crystal will be coupled to a Ge light detector with the same dimensions as the crystal face. The light detectors (LD) will be based on the Neganov-Luke (NL) effect in order to improve their performance [6, 7]. In addition, the towers will be surrounded by an active veto which will consist of BGO scintillators coupled to NL LDs to register the scintillation signal after particle interaction. Fig. 1 summarizes the design.

2.1 Assembly design

The design that BINGO proposes for the detector's assembly is based on three concepts: less passive material in the detectors array volume, open structure to exploit coincidence cuts and easiness of assembly. Each module consists of a copper piece that acts as the main support for the module. Three PTFE pieces are fixed on each side of the copper piece to keep the LDs and the crystals. A 0.35 mm slot is made in all PTFE pieces to sandwich the 0.3 mm LDs (Fig. 2). Against the same PTFE pieces the crystal is fixed. A nylon wire with a diameter of 0.45 mm presses the crystal to the copper holder, its ends are locked with a screw to the copper support. Like this, the crystals and the LDs are fixed from moving. The tension exerted by the nylon wire on the crystals is calibrated to 4 kg per wire side using a torque screwdriver. All the single modules are connected together using a rode to get the full tower. The detectors are equipped with a Ge-NTD (neutron transmutation doped) thermistors to register the temperature variation after particles interaction. The Ge-NTD ($3\times3\times1$ cm^3 on crystals and $3\times1\times1$ cm^3 on LDs) is glued to the crystals using Araldite® rapid epoxy glue. The Ge-NTDs have gold pads which are bonded to kapton with also gold pads that are placed on the copper piece to have the electrical and thermal contact.

This compact design allows us to have a reduction by more than one order of magnitude in passive materials surrounding the 2β decay detectors compared to the standard way of keeping bolometers [3–5]. Furthermore, this design allows us to have surface background rejection through anti-coincidence between crystals. Another feature for this design is exploiting the LDs to shield the crystals from the main copper pieces, the only passive material along with the nylon wire. The latter composes only a tiny fraction of passive material surface, around 2 cm^2.

2.2 Neganov-Luke light detectors

The second technological innovation in BINGO is the development of a bolometric light detector enhanced by the Neganov-Luke effect. It will consist of a thin Ge wafer (0.3 mm) with Al electrodes deposited on it. Biasing the electrodes will result in an electric field in the wafer which will drift the electron-hole pairs produced by the absorbed light. The work done by the field on the charge carriers contributes with an additional heat (it is a manifestation of the Joule effect), magnifying the signal-to-noise ratio in the thermal signal by a significant factor (10-20

Figure 4. Two modules of the nylon wire assembly hosting four Li_2MoO_4 (1) each coupled to a standard LD (2) tested at LSC.

Figure 5. A section of the active veto, composed of two trapezoidal BGO bars (1), each bar is coupled to two standard Ge LDs (2). The BGOs are facing TeO_2 crystal (3) that has a U source deposited on its surface to mimic surface contamination.

in gain depending on the maximum applied bias possible, i.e. the strength of electric field). The target of BINGO is to be able to detect the light signal corresponding to a $0\nu2\beta$ event, which amounts to ~3 keV in the case of the scintillation signal in Li_2MoO_4 and to only ~100 eV in the case of the prevalent Cherenkov signal in TeO_2. An α particle releasing the same energy as a $0\nu2\beta$ event would produce only ~15% of scintillation light in Li_2MoO_4 [3] and no Cherenkov light at all in TeO_2 [8], allowing for its safe rejection by the comparison of the heat and light signals. BINGO light detectors aim at a baseline width of only ~3 eV RMS. This low value is essential to achieve the desired α rejection in the challenging TeO_2 case, but it is also crucial for Li_2MoO_4, where a high signal-to-noise ratio in the light signal is required to reject the background induced by the random coincidences of the $2\nu2\beta$ events, which is very challenging in the ^{100}Mo case because of the relatively fast $2\nu2\beta$ process ($T_{1/2} = 7 \times 10^{18}$ yr). Fig. 3 shows an example of a NL LD with concentric electrodes.

2.3 Active veto

For the first time in a large array of macro-bolometers, an almost-hermetic active inner shield will be developed, surrounding completely the BINGO detectors. The first function of this shield is to reduce the contribution coming from external γ radioactivity, especially from the 2615 keV line of ^{208}Tl, but also some low-intensity characteristic γ's of ^{214}Bi (belonging to the ^{238}U chain and related to ^{226}Ra and radon progeny) above 2615 keV. The second function is to reject surface radioactivity in the external crystals of the arrays, which face directly the active shield. It is very important that this shield is sensitive to surface events, otherwise these peripheral elements will not take advantage of a complete β surface rejection. We remark that this shield will operate also as a very efficient muon veto. The active shield will consist of BGO scintillator bars on the lateral and also two BGO cylinders on top and bottom. Each scintillator will be coupled to NL LD to register the light signal after particle interaction. It is important for the scintillator to have a low energy threshold (around 50 keV) to reject the 2615 keV γ that is specially dangerous in the case of TeO_2 ($Q_{\beta\beta}$ = 2527 keV). 50 keV energy threshold in the scintillator translates to < 1 keV energy threshold in the LD, depending on the light yield (LY) which is given in keV/MeV, meaning the energy in keV detected in the LD for a 1 MeV energy deposition in the scintillator. Of course, such an energy threshold in the LD needs to be improved, and this can be realised thanks to the NL LD which will lead to an amplification of the signal to achieve few keV energy threshold (depends on LY and gain induced by the NL effect).

3 Prototype tests results

Emerging to mini-BINGO requires some prototype tests validation on all mini-BINGO ingredients: starting with the nylon wire assembly, the BGO active veto and the NL LDs. The first prototype tests on the nylon wire assembly was performed on two small 2×2×2 cm^3 Li_2MoO_4 crystals each coupled to a 2×2 cm^2 standard LD. The performances allowed us to validate the first prototype test of the design, the baseline resolutions FWHM of the crystals were on average around 0.9 keV [9]. Then we moved to test mini-BINGO size crystals and LDs (Fig. 4). The assembly was tested first above-ground at the IJCLab (results

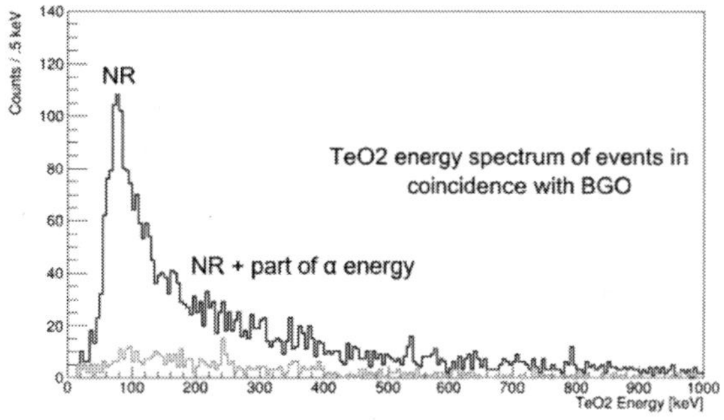

Figure 6. *Blue*: The energy spectrum of TeO_2 events in coincidence with BGO. The peak at low energy (~80 keV) represents the nuclear recoil energy and the tail to the right represents the nuclear recoil plus part of the α energy deposited in TeO_2 before escaping to BGO. *Green*: The accidental coincidences between TeO_2 and BGO.

can be found in paper [10]) before moving to Canfranc underground laboratory (LSC) in the CROSS pulse-tube cryostat. At LSC, we have measured baseline resolutions 1.9–5.9 keV FWHM for the four crystals with an average of 3.3 keV. The LDs baseline resolutions RMS were ranging between 70–111 eV. These values are achieved without exploiting the NL effect in which we expect more than 10 times better baseline resolution. The assembly with the nylon wires undergone two thermal cycles during the run ranging from room temperature to mK stages, which validates that the nylon wire preserves its properties and doesn't stretch out.

In addition, we performed a to study a small section of the BGO active veto, in particular the anti-coincidence between a crystal with surface contamination and the LD reading the light signal from the BGO. For this purpose, we have assembled in a copper holder a section of the active veto consisting of two BGO scintillators (height 12 cm) with a trapezoidal cross-section (Fig. 5). The crystals are in direct contact with the PTFE piece, that sandwich also the trapezoidal shaped LD, making the required thermal decoupling from the copper holder and protecting the LDs from vibrations. Facing the BGOs, a TeO_2 crystal is placed with a solution of U α source deposited on the surface facing the BGOs. The source has two main α lines at ~4.2 and ~4.7 MeV from the ^{238}U and ^{234}U decays, respectively, and is used to mimic surface radioactivity. An α decay will result in two scenarios: (i) the α escapes the TeO_2 surface and deposits all of its energy in BGO and the recoil nucleus is completely absorbed in TeO_2; (ii) the α deposits a part of its energy in TeO_2 before escaping to BGO and the recoil nucleus is absorbed in TeO_2 also. In both cases, when the α interacts in BGO, a scintillation light will be emitted and detected by the LDs. The nuclear recoil energy is around 80 keV, so the TeO_2 energy spectrum will show a peak at this energy with a tail extending to the higher energy region representing the degraded energy of the α escaping to BGO. Fig. 6 shows the energy spectrum of TeO_2 events in coincidence with BGO.

The second important study of the active veto is to have a rough estimation of the efficiency at the energy threshold of the LD, keeping in mind that we are not exploiting the NL effect, yet in this set-up. The LY was estimated to be around 7 keV/MeV, so for a 50 keV energy threshold in BGO and an expected gain of at least 10 when having a NL LD, the energy threshold in the LD is foreseen to be around 3-4 keV. To estimate the efficiency at this value, fake pulse with such an energy were injected into the streaming data. After the data processing, the efficiency at 3 keV was found to be >92%. The result is quite encouraging knowing also that we can improve even better by having a NL LD giving a higher gain in amplification.

Currently the group is doing some R&D on NL LD to study various electrode geometries and their method of deposition (evaporation and lithography) to achieve the best performance.

4 Conclusions

In this paper we have summarized the BINGO project innovative ideas. The proposals will allow next-next generation bolometric experiments to boost their sensitivity on $0\nu2\beta$ decay by working on reducing the background from various components to achieve a background index of the order of 10^{-5} counts/(keV kg yr). The first demonstrator of mini-BINGO that will take place at LSM is currently under development. Various prototype tests have been carried out successfully, and some studies are still ongoing, especially for the NL LD. In addition, performing screening on different samples of BGO for γ-ray spectroscopy is needed to reduce the scintillation material radioactivity, particularly on ^{207}Bi contamination that can affect the active veto purpose, if it is present at high values.

5 Acknowledgments

The BINGO project has received funding from the European Research Council (ERC) under the European Union's Horizon 2020 research and innovation program (grant agreement No 865844). The INR NASU group was supported in part by the National Academy of Sciences of Ukraine in the framework of the project "Development of bolometric experiments for the search for double beta decay".

References

[1] W.H. Furry, Phys. Rev. **56**, 118 (1939)
[2] M. Agostini et al., arXiv:2202.01787 (2022)
[3] E. Armengaud et al., Eur. Phys. J. C **80**, 4 (2020)
[4] C. Augier et al., Eur. Phys. J. C **82**, 1033 (2022)
[5] O. Azzolini et al., Phys. Rev. Lett. **129**, 111801 (2022)
[6] B. Neganov and V. Trofimov, Otkrytia i izobretenia **146**, 215 (1985)
[7] P. N. Luke, J. Appl. Phys. **64**, 6858 (1988).
[8] L. Bergé et al., Phys. Rev. C **97**, 032501(R) (2018)
[9] A. Armatol et al., arXiv:2204.14161 (2021)
[10] A. Armatol, arXiv:2301.06946 (2023)

https://doi.org/10.1051/epjconf/202329005001

Constraining Neutron-Star Matter — Combination of heavy-ion experiments and multi-messenger astronomy

Arnaud A. Le Fèvre[1,*]

[1]GSI Helmholtzzentrum für Schwerionenforschung GmbH, Darmstadt, Germany

Abstract. Describing supernova explosions or neutron-star collisions requires a deep understanding of properties of nuclear matter at supra-saturation densities, and extreme neutron over proton asymmetries. So far, our knowledge about dense matter provided by astrophysical observations in the cores of neutron stars remains limited. However, dense nuclear matter is also probed in terrestrial heavy-ion collision (HIC) experiments. We demonstrate how, within a novel approach, using Bayesian inference, combining data from astrophysical multi-messenger observations of neutron stars and from HICs at relativistic energies, one can improve our understanding of dense nuclear matter. The inclusion of HIC data probing the nuclear matter equation-of-state (EoS) at supra-saturation density has the effect of increasing the predicted pressure in the core of neutron stars relative to previous analyses, and shifts the neutron-star radii expectation towards larger values, in accordance with recent observations by the Neutron Star Interior Composition Explorer mission. More remarkable is that, though the sources and methods of observation are orthogonal, the constraints from HIC experiments and multi-messenger observations are consistent with each other. It shows that both methods can be complementary at intermediate densities, and strengthen each other. Another conclusion is that in order to be even more constraining, the constraint of the EoS of asymmetric nuclear matter by HIC methods should be improved above twice saturation density, which should be feasible with future experiments with enhanced precision and higher bombarding energy.

1 Introduction

The nuclear equation of state (EoS) can be probed and constrained by terrestrial experiments with atomic nuclei and astrophysical observations of neutron stars and supernovae. At densities well above nuclear saturation density, $n_{sat}= 0.16 fm^{-3}$, experimental and theoretical descriptions of the EoS are less accurate, and other open questions remain, such as whether a possible phase transition to exotic phases of nuclear matter is taking place [1]. At densities below 1-$2n_{sat}$, the nuclear EoS is better constrained, from both experimental observations [2, 3] and nuclear theory of microscopic calculations based e.g. on chiral effective field theory (EFT) of QCD [4–9]. Beyond these densities, further experimental and observational data are necessary. In the last years, a very promising tool has been developed [10, 11], combining, within a Bayesian framework, data from the multi-messenger astrophysics analysis of neutron stars and their collisions, as they provide access to dense neutron-rich matter which has not been so far explored in terrestrial experiments. Multi-messenger astronomy is a flourishing source of data since recent years, with the advent of gravitational-wave observations (GW) [12] and electromagnetic measurements of neutron stars [13–15]. This new analysis method led to much improved constraints on the EoS at high density [10, 16–24]. However, for densities below $2n_{sat}$, it shows still considerable uncertainties, mainly due to the range of (higher) densities that probe neutron-star (NS) radii. HICs have the potential to provide a precise and complementary information in this lower density range. HIC experiments performed with heavy-ion beam energies of up to few GeV per nucleon, can probe the nuclear EoS mainly in a density range of around 0.5-$3n_{sat}$ [2, 3, 25], and can provide new input for the global knowledge of the nuclear EoS [26–28].

2 Bayesian analysis method

To illustrate quantitatively the power of HIC studies to complement multimessenger astronomy for constraining the EoS of neutron-star matter, we will report in the following the promising new method recently introduced in [29], that has been followed very recently by a rather similar approach [30]. This pioneering work is based on a Bayesian multimessenger framework [10, 11], where numerous EoS's are probed with respect to their agreement with a large variety of observational data. There, the novelty has consisted in combining informations from nuclear theory, astrophysical observations of neutron stars and results from HIC experiments. In order to quantify the influence of HIC informations, authors have selected HIC experiments where the EoS has been constrained with the so far highest accuracy, in the density range of interest for the multi-messenger astronomy, i.e. above saturation density.

*e-mail: A.LeFevre@gsi.de

2.1 Building the prior

A Bayesian framework is first built with a so-called 'prior', which consists of a large amount of EoS's (here 15,000), each having a well-defined weight. In this work, the chiral effective field theory (ChEFT) [6, 31–33] has been chosen for its capability to provide theoretical uncertainties [8, 34] for each sampled EoS. In addition, the local chiral two- and three-nucleon interaction Monte-Carlo sampling of the chiral-EFT, as used here, is among the most precise many-body methods to solve the nuclear many-body problem [35]. But for the being, ChEFT has a limit of validity in density, around max. $1.5n_{sat}$, corresponding to the breakdown scale of the expansion that is estimated to be about 500-600 MeV/c [8]. Therefore, above $1.5n_{sat}$, each EOS must be extrapolated within limits defined by known constraints, such as the speed of sound (c_s) in neutron-star matter [36], constrained only by causality ($c_s \leq c$) and stability of neutron-star matter ($c_s \geq 0$). Since very high densities are not addressed in this work, no information from perturbative QCD calculations have been taken into account. The second constraint for extrapolations is the mass of heaviest neutron stars (pulsars) so far observed [37–39], of at least 1.9 solar masses ($1.9M_{\odot}$) that must be supported by EoSs. All sampled extrapolated EoSs that cannot fulfil both constraints are discarded from the prior. As can be seen in figure 1 panel a, extrapolations within both limits, which are minimalist assumptions, induce a broad uncertainty for the prior set of EoSs at high densities, therefore too for neutron-star masses and radii (figure 2a of [29]) which are mostly influenced by high densities. Other assumptions taken in recent reference [30] to build the prior show that it may bias the final result of a Bayesian analysis (by shifting final probability distributions of e.g. pressure and mass in/of neutrons stars), but within reasonable limits, as long as the chosen prior is broadly enough distributed not to artificially cut-off experimental constraints taken as inputs of the analysis.

2.2 Input from multi-messenger astronomy

Various constraints from multi-messenger astronomy observations have been applied to the Bayesian analysis to obtain the final posterior distribution of EoSs, with the following sequence of constraints: (1) The measured range of the maximum mass of neutron stars, constrained by massive pulsars (here PSR J0348+0432 [37] and PSR J1614-2230 [38]) and by the binary neutron-star collision GW170817 [40, 41]. (2) Constraints from X-ray pulse-profile modelling of PSR J0030+0451 and PSR J0740+6620 using data from NICER and the X-ray Multi-Mirror Mission (XMM-Newton) [14, 16, 42]. (3) Gravitational wave (GW) information from the neutron-star mergers GW170817 [12] and GW190425 [43]. (4) Constraint from the kilonova AT2017gfo [13] associated with a GW signal. Figure 1 panel b shows that the convolution of these astrophysical inputs to the prior lead to large constraints on the neutron-star EoS, strongest above $1.5n_{sat}$. As illustrated by figure 2b of Ref. [44], the high-density

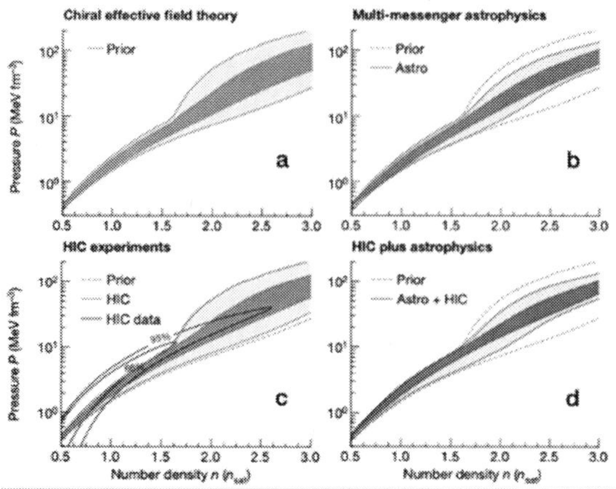

Figure 1. Constraints on the EoS of neutron-star matter. a-d, Evolution of the pressure as a function of baryon number density for the EoS prior (a, grey), when including only data from multi-messenger neutron-star observations (b, green), when including only HIC data (c, orange), and when combining both (d, blue). The shading corresponds to the 95% and 68% credible intervals (lightest to darkest). The impact of the HIC experimental constraint (HIC data, purple lines at 95% and 68%) on the EoS is shown in c. In b-d, the 95% prior bound is shown for comparison (grey dashed lines). (Figure taken from [29])

astrophysical constraints affect mostly the high-mass region in the NS mass-radius relation and exclude the stiffest EoSs that lead to the largest NS radii.

2.3 Input from HIC experiments

Chosen experimental constraints of the nuclear EoS have been taken from FOPI [2] (see figure 2, red area) and ASY-EOS [3] (see figure 3, red area) experiments performed at the GSI Helmholtz Centre for Heavy Ion Research of Darmstadt, with Au+Au collisions from 0.4 to 1.5 A.GeV incident energy. FOPI describes the symmetric-nuclear-matter (SNM) component of the EoS, and probes densities in the 0.7-3 n_{sat} range. ASY-EOS on its side describes the (neutron-to-proton) asymmetric nuclear matter (ANM) contribution of the EoS (called '(a)symmetry' energy) within the range of densities around 0.7-2 n_{sat}. This lower range of sensitivity in density limits that of the combination of both SNM and ANM contributions necessary to reconstruct the EoS of NS matter. However, the density cut-off of the HIC input in the Bayesian analysis has been set at $3n_{sat}$, although the sensitivity of ASY-EOS results is very low at this density. Between 2 and 3 n_{sat}, where FOPI results are still very accurate, another SNM EoS constraint, with broader uncertainties, has been taken instead as a test, coming from other HIC experiments at the Bevalac accelerator at Lawrence Berkeley National Laboratory and the Alternating Gradient Synchrotron at Brookhaven National Laboratory [25]. This latter contribution has shown very little influence on final results of the Bayesian analysis, due to the above mentioned very low weight of the symmetry energy constraint at these densities. Figure 1 panel c shows that the impact of the combined HIC experimental SNM-ANM constraints for the neutron-star EoS. HICs tends to favor even more stiffer EoSs than astrophysical observations do. That is, as seen in Figure 4 panel a, EoSs that have higher pressures around $1.5n_{sat}$ density.

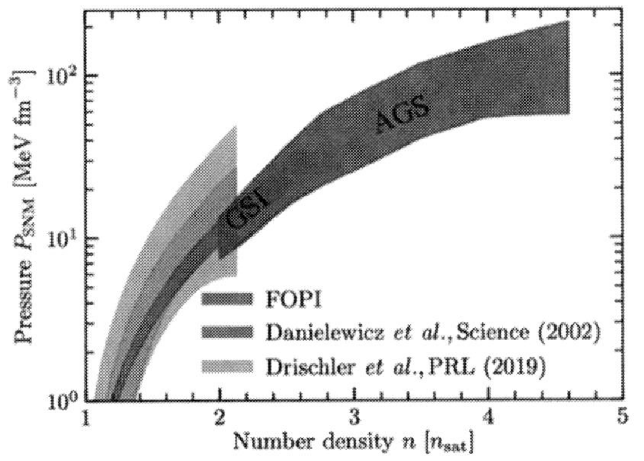

Figure 2. Constraints on the EoS of symmetric nuclear matter obtained from various HIC experiments: FOPI (GSI) in red, AGS in green, compared with nuclear theory (ChEFT) in blue. (Figure taken from [29])

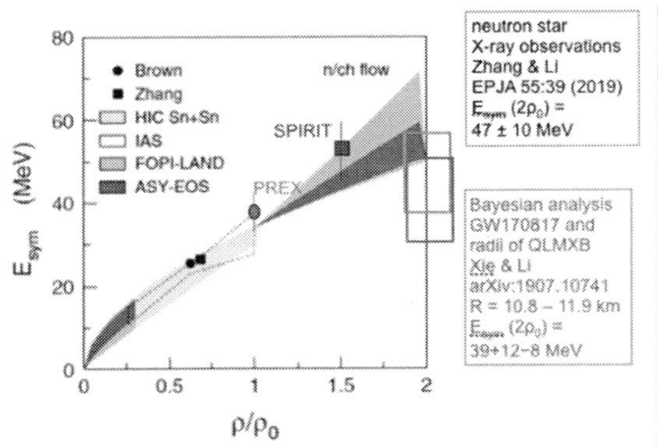

Figure 3. Compilation of nuclear symmetry energy constraints obtained with HIC experiments, compared with recent astrophysical constraints. Taken from [3] and complemented with results from [44]. Recent PREX [45] and SπRIT [46] data have been added for comparison, illustrating the better accuracy of the ASY-EOS constraint.

2.4 Combining astrophysical and HIC inputs

Figure 1 panel d shows the final EoS constraints obtained by combining both astrophysical multi-messenger observations and HIC information. Whereas the astrophysical data rule out the most extreme EoS's, HIC inputs indicate larger pressures around 1-$1.5n_{sat}$, the density range where the HIC experimental sensitivity is highest. This goes in the direction of recent NICER observations on the

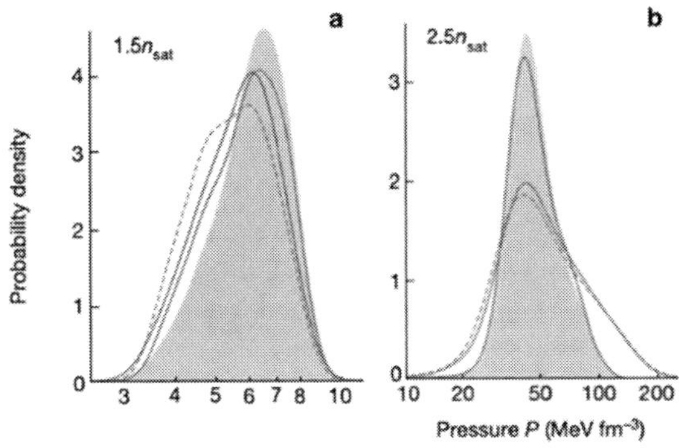

Figure 4. Posterior distributions for the pressure in neutron-star matter at $1.5n_{sat}$ (a) and $2.5n_{sat}$ (b) at different stages of our analysis, with the result of combined Astro + HIC shaded in light blue; lines: see legend of figure 1 (Figure taken from [29])

EoS [16, 17]. Most remarkable, as illustrated by Figure 4 panel a, HIC experiments and astrophysical observations, show an almost perfect agreement. At low densities, HICs exhibit a clear impact on the posterior EoS distribution, whereas the EoS at higher density ($\geq 2n_{sat}$) is mostly determined by astrophysical observations (Figure 4 panel b). As astrophysical observations mainly probe neutron stars with M $\geq 1.4M_{\odot}$, for which the relevant densities are higher than those probed by ASY-EOS, used HIC inputs influence the radii of these neutron stars to a smaller degree. HIC data start to have some influence on radii of low-mass stars with M$\approx 1.0\ M_{\odot}$.

3 Perspectives

As we have seen, constraints from collisions performed at GSI with FOPI and ASY-EOS experiments show a remarkable consistency with astrophysical observations even though they are obtained with completely orthogonal methods. Including data of heavy-ion collisions in the analyses has enabled additional constraints near 1.5 n_{sat} where both nuclear theory and astrophysical observations are less sensitive. Including HIC inputs provides indeed a more complete understanding of dense nuclear matter than using the sole astrophysical information. However, the so-far main limitation of the impact of HIC inputs on the knowledge of the EoS of NS matter is coming from the still too low maximum densities probed by experiments measuring the symmetry energy like ASY-EOS. Concerning this latter experiment, the first reason was that the beam energy was limited to 0.4A GeV. The second reason was that flow of protons could not be discriminated from other hydrogen isotopes, as detailed in [27]. In the future, the priority should be given to experiments that probe higher densities, for both SNM (above 3 n_{sat}) and ANM (above 1.5 n_{sat}) components of the EoS, while reducing the experimental uncertainties. Concerning the symmetry energy, several projects are on-going, such as (1) at RIKEN in Japan with the Sπrit Collaboration [46] by measuring pion yields with an improved precision, (2) at GSI Darmstadt with the R3B project of a second generation of ASY-EOS experiments — benefiting in particular from the high performance of the neutron detector NeuLAND — having great potential to provide new constraints of neutron star properties at around 2.5 n_{sat} where results are currently driven by astrophysical observations. With the upocoming FAIR facility, R3B will provide unique opportunities to produce and study neutron-rich matter at densities comparable to those in the interior of neutron stars or in neutron star mergers. In parallel, still at GSI, the HADES experiment has the potential to probe the symmetry at equally high densities using e.g. kaon (K^+/K^0) yield ratios with high accuracy, produced in HIC central collisions close to the production threshold. Similar methods maybe used afterwards by CBM at FAIR in Au-Au collisions ($\sqrt{s_{NN}}$ = 2.9-4.9 GeV). In addition, CBM might detect new phases of QCD matter, possibly involving hyperons and, ultimately, the transition to a deconfined quark matter phase at the highest densities. All these perspectives, recently detailed in [27, 28], have the potential to revolutionise our understanding of the nuclear EoS and neutron stars.

References

[1] E. Annala, T. Gorda, A. Kurkela, J. Nättilä and A. Vuorinen, Nat. Phys. **16**, 907-910 (2020).

[2] A. Le Fèvre, Y. Leifels, W. Reisdorf, J. Aichelin and C. Hartnack, Nucl. Phys. A **945**, 112-133 (2016).

[3] P. Russotto, et al., Phys. Rev. C **94**, 034608 (2016).

[4] K. Hebeler, J.M. Lattimer, C.J. Pethick and A. Schwenk, Astrophys. J. **773**, 11 (2013).

[5] I. Tews, T. Krüger, K. Hebeler, and A. Schwenk, Phys. Rev. Lett. **110**, 032504 (2013).

[6] J.E. Lynn, et al., Phys. Rev. Lett. **116**, 062501 (2016).

[7] C. Drischler, K. Hebeler, and A. Schwenk, Phys. Rev. Lett. **122**, 042501 (2019).

[8] C. Drischler, R.J. Furnstahl, J.A. Melendez, and D.R. Phillips, Phys. Rev. Lett. **125**, 202702 (2020).

[9] S. Huth, C. Wellenhofer an A. Schwenk, Phys. Rev. C **103**, 025803 (2021).

[10] T. Dietrich, et al., Science **370**, 1450-1453 (2020).

[11] P.T.H. Pang, et al., Astrophys. J. **922**, 14 (2021).

[12] B.P. Abbott, et al., Phys. Rev. Lett. **119**, 161101 (2017).

[13] B.P. Abbott, et al., Astrophys. J. **848**, L13 (2017).

[14] M. C. Miller et al., Astrophys. J. Lett. **887**, L24 (2019).

[15] T.E. Riley, et al., Astrophys. J. Lett. **887**, L21 (2019).

[16] M.C. Miller, et al., Astrophys. J. Lett. **918**, L28 (2021).

[17] G. Raaijmakers, et al., Astrophys. J. Lett. **918**, L29 (2021).

[18] A. Bauswein, O. Just, H.-T. Janka and N. Stergioulas, Astrophys. J. **850**, L34 (2017).

[19] E. Annala, T. Gorda, A. Kurkela and A. Vuorinen, Phys. Rev. Lett. **120**, 172703 (2018).

[20] E.R. Most, L.R. Weih, L. Rezzolla and J. Schaffner-Bielich, Phys. Rev. Lett. **120**, 261103 (2018).

[21] B.P. Abbott, et al., Phys. Rev. Lett. **121**, 161101 (2018).

[22] D. Radice and L. Dai, Eur. Phys. J. A**55**, 50 (2019).

[23] C.D. Capano, et al., Nat. Astron. **4**, 625-632 (2020).

[24] I. Legred, K. Chatziioannou, R. Essick, S. Han and P. Landry, Phys. Rev. D **104**, 063003 (2021).

[25] P. Danielewicz, R. Lacey, and W.G. Lynch, Science **298**, 1592-1596 (2002).

[26] M.B. Tsang, W.G. Lynch, P. Danielewicz, and C.Y. Tsang, Phys. Lett. B **795**, 533-536 (2019).

[27] P. Russotto, M.D. Cozma, E. De Filippo, A. Le Fèvre, Y. Leifels, J. Lukasik, La Rivista del Nuovo Cimento (2023).

[28] A. Sorensen, et al., arXiv:2301.13253 [nucl-th] (2023).

[29] S. Huth, P.T.H. Pang, et al., Nature **606** (2022).

[30] C.Y Tsang, M.B. Tsang, W. Lynch, R. Kumar, C. Horowitz, pre-print March 2023 - https://doi.org/10.21203/rs.3.rs-2494603/v1

[31] I. Tews, J. Carlson, S. Gandolfi and S. Reddy, Astrophys. J. **860**, 149 (2018).

[32] E. Epelbaum, H.-W. Hammer and U.-G. Meissner, Rev. Mod. Phys. **81**, 1773-1825 (2009).

[33] R. Machleidt, and D.R. Entem, Phys. Rep. **503**, 1-75 (2011).

[34] E. Epelbaum, H. Krebs and U.-G. Meißner, Eur. Phys. J. A **51**, 53 (2015).

[35] J. Carlson, et al., Rev. Mod. Phys. **87**, 1067 (2015).

[36] I. Tews, J. Margueron, and S. Reddy, Phys. Rev. C**98**, 045804 (2018).

[37] J. Antoniadis, et al., Science **340**, 1233232 (2013).

[38] Z. Arzoumanian, al., Astrophys. J. Suppl. Ser. **235**, 37 (2018).

[39] E. Fonseca, et al., Astrophys. J. Lett. **915**, L12 (2021).

[40] B. Margalit, and B.D. Metzger, Astrophys. J. **850**, L19 (2017).

[41] L. Rezzolla, E.R. Most, and L.R. Weih, Astrophys. J. **852**, L25 (2018).

[42] T.E. Riley, et al., Astrophys. J. Lett. **918**, L27 (2021).

[43] B. Abbott, et al., J. Lett. **892**, L3 (2020).

[44] Horowitz et al., J. Phys. G: Nucl. Part. Phys. **41** 093001 (2014)

[45] D. Adhikari et al. (PREX Collaboration), Phys. Rev. Lett. **126**, 172502 (2021)

[46] J.Estee et al., Phys. Rev. Lett. **126**(16), 162701 (2021)

EPJ Web of Conferences **290**, 05002 (2023) https://doi.org/10.1051/epjconf/202329005002
EuNPC 2022

Constraining the QCD phase diagram

Owe Philipsen[1]

[1]Institute for Theoretical Physics, Goethe University Frankfurt,
Max-von-Laue-Str. 1, D-60438 Frankfurt am Main, Germany

Abstract. Lattice QCD and functional methods are making significant progress in constraining the QCD phase diagram. As an important milestone, the chiral phase transition with massless u, d-quarks at zero density is now understood to be of second order for all strange quark masses, and a smooth crossover as soon as $m_{u,d} \neq 0$. Together with information on fluctuations and refined reweighted simulations, this bounds a possible critical point to be at $\mu_B/T \gtrsim 3$. On the other hand, an approximately chiral-spin symmetric temperature window has been discovered above the chiral crossover, $T_{\rm ch} \lesssim T \lesssim 3T_{\rm ch}$, with distinct correlator multiplet patterns and a pion spectral function suggesting resonance-like degrees of freedom, which dissolve graduallly with temperature.

1 Introduction

Knowledge of the QCD phase diagram is of great importance for the physics of heavy ion collisions and neutron stars. Theoretical predictions from lattice QCD are severely hampered by a fermion sign problem, which prohibits straightforward Monte Carlo simulations at finite baryon chemical potential. Nevertheless, methods working for $\mu_B \lesssim 3T$ as well as studies of the general parameter dependence of the chiral phase transition have reached a new level of maturity, providing phenomenologically relevant constraints on the location of a possible critical point.

The expected scenario for physical QCD is intimately connected to the situation in the chiral limit of massless u, d-quarks [1–3], as sketched in Fig. 1. For $m_{u,d} = 0$ there is an exact chiral symmetry, whose breaking/restoration across $T_c(\mu_B)$ must proceed by a non-analytic phase transition. If this transition is second-order at $\mu_B = 0$ and first-order at $T = 0$, as in several low-energy models, there must be a tricritical point where the order of the transition changes. On quite general grounds, a Z(2)-critical line emanates from a tricritical point in in the direction of the symmetry breaking field (i.e. $m_{u,d}$), with a known tricritical exponent [4], and represents the critical endpoint at physical mass values. However, if nothing is known about the chiral limit, other possibilities are transition lines that are entirely first or entirely second order for $m_{u,d} = 0$, and the situation at physical masses would be different. This illustrates the importance of the chiral limit to both constrain and understand the physical phase diagram.

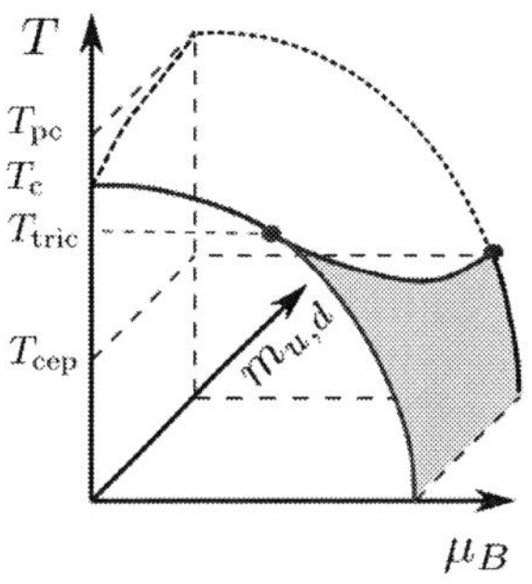

Figure 1. Connection of the putative QCD phase diagram for physical light quark masses to the chiral limit in the front plane.

2 The chiral transition at $\mu_B = 0$

In the chiral limit the quark determinant in the partition function diverges, prohibiting direct simulations and making its approach computationally expensive. For decades expectations have thus been based on an analysis of 3d sigma models as low energy effective theories, augmented by a 't Hooft term for the $U(1)_A$ anomaly, whose renormalisation group flow based on the epsilon expansion [5] predicts the chiral phase transition to be first-order for $N_f \geq 3$. The case of $N_f = 2$ was found to crucially depend on the anomalous $U(1)_A$ symmetry: If it remains broken at T_c, the chiral transition should be second order in the $O(4)$-universality class, whereas its effective restoration would enlarge the symmetry and push the transition to first order.

Early QCD simulations on coarse lattices were consistent with the scenario shown in Fig. 2 (left): A first-order region could be seen for $N_f = 3$, whereas the smallest available masses showed a continuous crossover for $N_f = 2$. However, the location of the Z(2)-boundary varies widely between different discretisations, indicating large cutoff effects. The general pattern is for the first-order region to shrink when the lattice is made finer, while improved staggered actions see no trace of a first-order transition at all. For a more detailed discussion and list of references see [7].

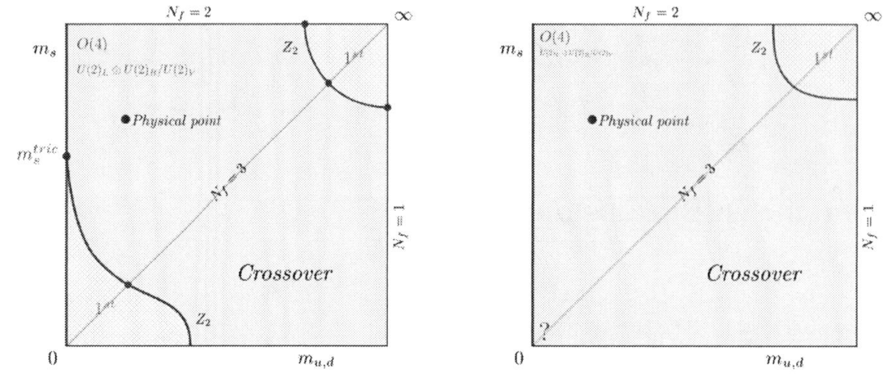

Figure 2. The order of the QCD thermal transition as a function of the quark masses. Left: Scenario proposed in [5] and observed on coarse lattices. Right: Emerging continuum limit [6]. The universality class in the three-flavour chiral limit is not yet known. From [6].

Recent investigations using the Highly Improved Staggered Quark (HISQ) action start at the $N_f = 2 + 1$ physical point and then gradually reduce the light quark mass until $m_{PS} \approx 55$ MeV [8]. Fig. 3 (left) shows the chiral susceptibility normalised to its peak value at the physical pion mass. The peak location defines the pseudo-critical temperature T_{pc}. However, the peak height stays finite and no sign of a first-order transition is detectable. Analytic fRG calculations in infinite volume [9] are compatible with this behaviour, Fig. 3 (left). Similar results are found using twisted mass Wilson fermions at quark masses at and above the physical point [10]. Thus either a $Z(2)$-critical point bounding a very narrow first-order region is approached, or a second-order transition in the chiral limit.

Eventually one would like to calculate the critical exponents characterising the approach to criticality, and thus determine the universality class from first principles. However, this requires exponential accuracy in the numerically expensive light mass regime and is not feasible. In practice the data are fitted to scaling formulae with exponents fixed to known values for the universality classes in question, such as the approach of the crossover temperature to the critical temperature T_c in the chiral limit,

$$T_{pc}(m_l) = T_c\left(1 + K\, m_l^{1/\beta\delta}\right) + \text{sub-leading}\,. \quad (1)$$

The extrapolated chiral critical temperatures are

$$\begin{aligned} T_c &= 132^{+3}_{-6}\ \text{MeV} \quad \text{HISQ fermions [8]}\,, \\ T_c &= 134^{+6}_{-4}\ \text{MeV} \quad \text{twisted mass Wilson [10]}\,, \\ T_c &= 142^{+0.5}_{-0.5}\ \text{MeV} \quad \text{fRG methods [9]}\,. \end{aligned}$$

The variation between the possible sets of critical exponents is covered by the errors, so that an extrapolation makes sense even without definite knowledge of the true universality class.

A first-order chiral transition for $N_f \in [2, 7]$ could recently be practically ruled out by investigating the effects of the number of flavours, their masses and the lattice spacing [6] on the transition. The study exploits the fact that a change from a first-order to a second-order chiral transition necessarily passes through a tricritical point, such as in the scenario Fig. 2 (left). This implies tricritical scaling with known exponents for the Z(2) boundary line as it approaches the chiral limit, allowing for a controlled extrapolation. Rather than varying quark masses independently, ref. [6] continuously varies the number N_f of degenerate quarks [12]. A tricritical m_s^{tric} as in Fig. 2 (left) then translates into a tricritical $2 < N_f^{\text{tric}} < 3$. On the lattice, such a tricritical point additionally depends on the lattice spacing, $N_f^{\text{tric}}(a)$, which can be inverted to $a^{\text{tric}}(N_f)$.

Fig. 3 (middle) shows the $Z(2)$-critical boundary line separating the light bare quark mass region featuring first-order transitions from the crossover region as a function of lattice spacing for the unimproved staggered fermion action. In agreement with previous studies, the first-order region grows as more flavours are added, and it shrinks as the lattice is made finer. Those N_f-theories with three available lattice spacings show tricritical scaling and a tricritical point in the lattice chiral limit, which is moving to the left as N_f is increased. Note that the continuum limit is in the origin at $(am, aT) = (0, 0)$. This implies a tricritical point in the continuum limit to be beyond $N_f^{\text{tric}} > 7$. Conversely, the first-order regions observed in these simulations are *not* connected to the continuum limit and must be regarded as lattice artefacts. A powerful check on this finding is provided by $N_f = 3$ $O(a)$-improved Wilson fermions [11]. Plotting those data in terms of the appropriate scaling variable, perfect tricritical scaling is observed, Fig. 3 (right), so that the continuum transition is of second order as in the staggered case. Two further discretisations with $N_f = 3$, HISQ fermions [13] and domain wall fermions [14], do not see any phase transition at the smallest available quark masses, and thus are fully consistetnt with those findings. A recent fRG study of 3d sigma models including a ϕ^6 term [15] and numerical bootstrap methods applied to $U(m)\times U(n)$ [16] models are also compatible with a second-order chiral transition. Thus, the Columbia plot in the continuum looks as in Fig. 2 (right), with a second-order chiral transition for $m_{u,d} = 0$ and any value of m_s, and crossover as soon as $m_{u,d} \neq 0$.

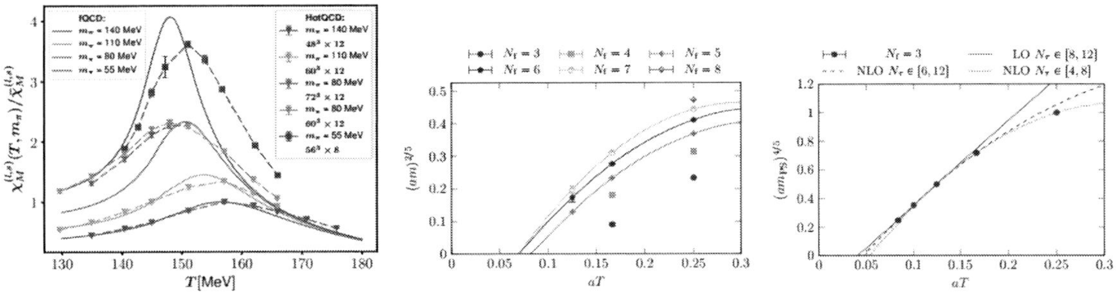

Figure 3. Left: Chiral susceptibility at physical m_s for a range of decreasing $m_{u,d}$ with the HISQ lattice action [8] and from fRG calculations [9]. From [9]. The $Z(2)$-critical line separating first-order transitions (below) from crossover (above), for unimproved staggered fermions [6] (middle) and $O(a)$-improved Wilson fermions [11] (right), with tricritical scaling fits to both. From [6].

3 The search for a critical point

Lattice searches for a critical point are based on indirect methods to extract information about the phase structure for small baryon density, $\mu_B/T \lesssim 3$: (i) Reweighting [17], (ii) Taylor expansion in μ/T [18] and (iii) analytic continuation from imaginary chemical potential, for which there is no sign problem [19, 20]. When the QCD pressure is expressed as a series in baryon chemical potential,

$$\frac{p(T,\mu_B)}{T^4} = \frac{p(T,0)}{T^4} + \sum_{n=1}^{\infty} \frac{1}{2n!}\chi_{2n}^B(T)\left(\frac{\mu_B}{T}\right)^{2n}, \quad (2)$$

$$\chi_{2n}^B(T) = \left.\frac{\partial^{2n}(\frac{p}{T^4})}{\partial(\frac{\mu_B}{T})^{2n}}\right|_{\mu_B=0},$$

the Taylor coefficients are the baryon number fluctuations evaluated at zero density, which can also be computed by fitting to untruncated results at imaginary μ_B. This permits full control of the systematics between (ii) and (iii). These coefficients are presently known up to $2n = 8$.

Quite generally power series are limited by their radius of convergence, corresponding to the nearest singularity of the full function (relative to the expansion point) in the complex variable. If such a singularity is located on the real axis, it signals a phase transition.

In [21] a cluster expansion model (CEM) was developed, with all expansion coefficients recursively related to the lower ones, and the first two are fixed by lattice data. As in the virial expansion, such a recursion is possible as long as the system is sufficiently dilute and dominated by two-body interactions. The model is quantitatively consistent with all known lattice pressure coefficients, as e.g. in Fig. 4 (left), and since its coefficients are known to all orders, it allows a controlled extraction of the radius of convergence. The only phase transition predicted in this way is the known Roberge-Weiss transition in the direction of imaginary chemical potential $\mu_B = i\pi T$ [22], which implies that any phase transition in the real direction is further away than that. While this is just a model, it tells us that nothing in the available lattice fluctuation data enforces singular behaviour.

Another option is to use the coefficients calculated directly from the lattice. From these one can construct Padé-approximants, which are rational functions (i.e. infinite order in the expansion variable) whose first coefficients agree with the explicitly computed ones. These approximants can be further constrained by simulation results for the full pressure at imaginary chemical potential. Their singularities are then taken as estimates for the singularities of the full function. Locations of the Lee-Yang edge singularities (indicating a branch point in the pressure) extracted in this way are shown in Fig. 4 (middle), based on HISQ fermion data [23]. So far all singularities are at complex values of the chemical potential, but note the closing in toward the real axis as temprerature is decreased. This bounds a true phase transition to lower temperatures than those for which coefficients are currently available.

Table 1. Lattice bounds on the location of a critical point

Criterion	$(\mu_B/T)^{cep} \gtrsim$	Ref.
$T_{cep} < T_c$ as in Fig. 1	3.1	[7]
CEM of lattice fluctuations	π	[21]
Padé-appr. LY-zeroes	2.5	[23]
Reweighted simulation	2.5	[24]

Taylor expansions are avoided by using reweighting techniques in Monte Carlo simulations. Recent new calculations with stout smeared staggered fermions [24], using techniques considerably refined compared to earlier ones, evaluate the renormalised chiral condensate,

$$\langle\bar{\psi}\psi\rangle_R(T,\mu) = -\frac{m_{ud}}{f_\pi^4}\left[\langle\bar{\psi}\psi\rangle_{T,\mu} - \langle\bar{\psi}\psi\rangle_{0,0}\right], \quad (3)$$

as shown in Fig. 4 (right). The reweighted real μ_B simulations are fully compatible with the analytic continuation from imaginary μ_B simulations, but with smaller errors. Neither method shows a sign of non-analyticity so far.

In summary, lattice information on physical QCD in equilibrium at $\mu_B/T \leq 3$ is increasing and increasingly controlled. Table 1 collects the current bounds on the location of a critical endpoint resulting from these analyses. These are also consistent with the critical endpoint candidates found in the most recent truncations of Dyson-Schwinger equations [26] and their combination with functional renormalisation group methods [27], which predict $(\mu_B/T)^{cep} \approx 5.6$.

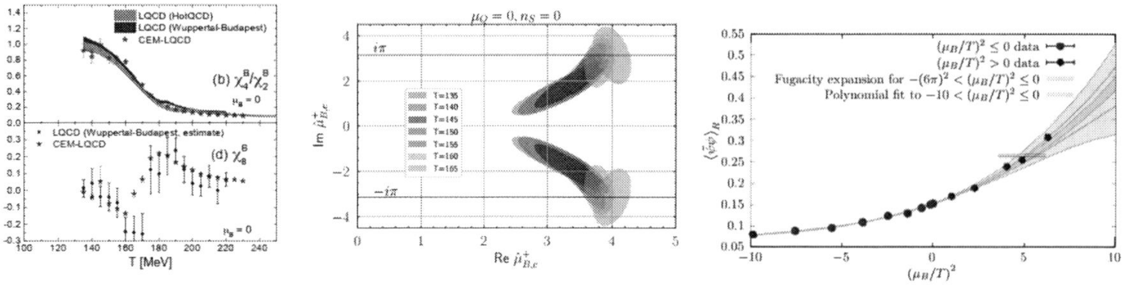

Figure 4. Left: Baryon number fluctuations from the lattice in comparison with the CEM model. From [25]. Middle: Lee-Yang zeros indicating the radius of convergence of the pressure series in complex chemical potential. From [23]. Right: Renormalised chiral condensate as a function of real and imaginary chemical potential. From [24].

4 Emergent chiral spin symmetry

While a critical point remains elusive on the lattice so far, an emergent approximate chiral spin symmetry has been discovered above the chiral crossover at $\mu_B = 0$. A $SU(2)_{CS}$ chiral spin transformation acts on Dirac fields as

$$\psi(x) \to \exp\left(i\vec{\Sigma}\cdot\vec{\epsilon}\right)\psi(x)\,, \quad \vec{\Sigma} = (\gamma_k, -i\gamma_5\gamma_k, \gamma_5)\,. \tag{4}$$

Here $k = 1, \ldots 4$ can be any of the euclidean gamma matrices. From the generators it is apparent that $SU(2)_{CS} \supset U(1)_A$. When combined with ordinary vectorial isospin transformations, $SU(2)_{CS} \otimes SU(2)_V$ can be embedded into the larger $SU(4)$, which contains the usual chiral symmetry of the massless QCD Lagrangian, $SU(4) \supset SU(2)_L \times SU(2)_R \times U(1)_A$.

The QCD Lagrangian is not invariant under chiral spin transformations. A thermal medium implies a preferred Lorentz frame, and the massless quark action can be decomposed as

$$\bar{\psi}\gamma_\mu D_\mu\psi = \bar{\psi}\gamma_0 D_0\psi + \bar{\psi}\gamma_i D_i\psi\,. \tag{5}$$

Explicit calculation shows the colour-electric part of the quark-gluon interaction to be CS- and $SU(4)$-invariant, while the colour-magnetic interaction and kinetic terms (and thus the free Dirac action) are not. Hence, chiral spin symmetry is never exact in physical QCD, but its approximate realisation is possible if the colour-electric quark-gluon interaction dominates the quantum effective action in some dynamical range, which would then be strongly coupled.

On the lattice, symmetries become apparent in degeneracy patterns of euclidean correlation functions,

$$C_\Gamma(\tau, \mathbf{x}) = \langle O_\Gamma(\tau, \mathbf{x})\, O_\Gamma^\dagger(0, \mathbf{0})\rangle\,, \tag{6}$$

with Γ some appropriate Dirac matrix. Information about all excitations in each quantum number channel is carried by the spectral functions $\rho_\Gamma(\omega, \mathbf{p})$,

$$\begin{aligned} C_\Gamma(\tau, \mathbf{p}) &= \int_0^\infty \frac{d\omega}{2\pi}\, K(\tau, \omega)\rho_\Gamma(\omega, \mathbf{p})\,, \\ K(\tau, \omega) &= \frac{\cosh(\omega(\tau - 1/2T))}{\sinh(\omega/2T)}\,. \end{aligned} \tag{7}$$

For an isotropic system in equilibrium, it is sufficient to probe the spatial and temporal correlators averaged over the orthogonal directions,

$$C_\Gamma^s(z) = \sum_{x,y,\tau} C_\Gamma(\tau, \mathbf{x})\,, \quad C_\Gamma^\tau(\tau) = \sum_{x,y,z} C_\Gamma(\tau, \mathbf{x})\,. \tag{8}$$

Numerical results for spatial $J = 0, 1$ meson correlators from $N_f = 2$ JLQCD domain wall fermions with physical quark masses, good chiral symmetry and lattice spacings < 0.1 fm [28] are shown in Fig. 5. Three multiplets of spatial correlators, $E_{1,2,3}$, at different temperatures are seen. Of these, E_1 is due to $U(1)_A$ restoration whereas E_3 requires the full chiral symmetry. Both multiplets are expected above the chiral crossover. Not expected is the multiplet E_2, which does *not* correspond to a representation of chiral symmetry, but to one of the larger $SU(4)$. For the representations and associated meson states, see [29, 30]. Appearance of E_2 demonstrates the dynamical emergence of chiral spin symmetry in this regime. As temperature is further increased, E_2 gradually disappears as a separate multiplet and only those belonging to the expected chiral symmetry survive. Similar findings are reported from temporal correlators at $T = 220$ MeV and the same lattice spacing [31]. Recently, the same patterns were observed with the quark content increased to $N_f = 2 + 1 + 1$ QCD, including physical strange and charm quarks [32].

A related observable which is sensitive to the entry and exit of the CS-symmetric temperature range are screening masses [33]. These correspond to the exponential decay of the large-separation spatial correlators in (8), $C_\Gamma^s(z) \sim \exp(-m_{scr}z)$ for $z \to \infty$. While not directly accessible experimentally, they can be readily evaluated non-perturbatively and perturbatively. Around the chiral crossover temperature, the screening masses show the expected degeneracy due to chiral symmetry. However, the temperature dependence predicted by pertubation theory is only attained for $T \gtrsim 500$ MeV. This can be understood by the approximate chiral spin symmetry in between, which perturbation theory about free quarks cannot reproduce.

If there is a chiral-spin symmetric band at zero density, it must necessarily continue to finite baryon density because the μ_B-term respects that symmetry. From the known behaviour of screening masses with μ_B one

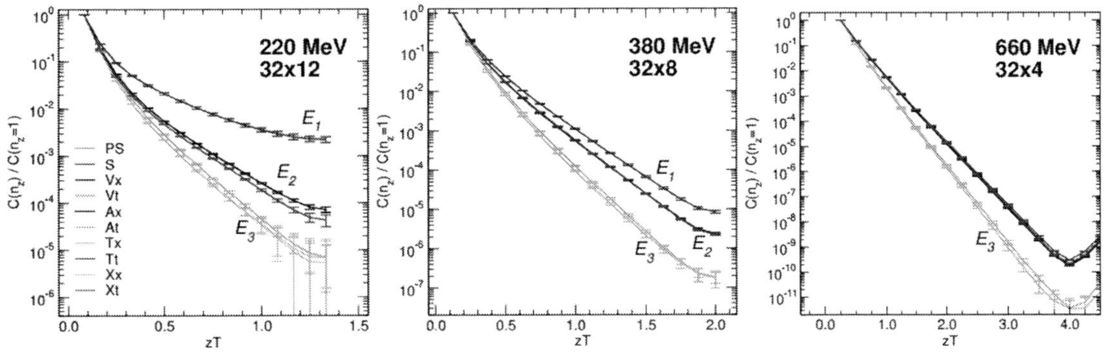

Figure 5. Spatial correlation functions with domain wall fermions show distinct E_1, E_2, E_3 multiplets of the approximate $SU(4)$ chiral spin symmetry, at temperatures above the crossover. At large temperatures, these reduce to the multiplets of the ordinary chiral symmetry. From [28].

infers that the chiral-spin symmetric band curves downwards, as in the possible phase diagram Fig. 6 (left). In the absence of lattice data the continuation to higher densities and lower temperatures is speculative, of course, but baryon parity doublets and quarkyonic matter both are CS-symmetric candidates to populate such a band [33].

Information about the nature of the effective degrees of freedom in different regimes is encoded in the spectral functions, Eq. (8). Unfortunately, their extraction from discrete sets of lattice correlator data represents an ill-posed inversion problem. Recently, a new method was attempted [34], which applies to stable scalar particles in a heat bath, i.e. to the pion in the case of QCD, which allows to circumvent the integral inversion.

The method exploits locality of quantum field theories to ensure a representation of the spectral function [35, 36],

$$\rho_{PS}(\omega, \mathbf{p}) = \int_0^\infty ds \int \frac{d^3u}{(2\pi)^2}\, \epsilon(p_0)\, \delta\big(\omega^2 - (\mathbf{p}-\mathbf{u})^2 - s\big)\, \widetilde{D}_\beta(\mathbf{u}, s)\,, \tag{9}$$

with $\beta = 1/T$, the thermal spectral density $\widetilde{D}_\beta(\mathbf{u}, s)$, and the standard Källen-Lehmann vacuum representation arising as $T \to 0$. For stable massive particles the analytic vacuum structure of the spectral density is preserved in the absence of a true phase transition. The authors therefore propose an ansatz with separable particle and scattering contributions,

$$\widetilde{D}_\beta(\mathbf{u}, s) = \widetilde{D}_{m,\beta}(\mathbf{u})\, \delta(s - m^2) + \widetilde{D}_{c,\beta}(\mathbf{u}, s)\,. \tag{10}$$

In an isotropic medium the spatial correlators and the spectral density are then related by [34]

$$C^s_{PS}(z) = \frac{1}{2}\int_0^\infty ds \int_{|z|}^\infty dR\, e^{-R\sqrt{s}} D_\beta(R, s). \tag{11}$$

For temperatures below the threshold to the scattering states the first term in Eq. (10) should dominate. Once the continuum part is neglected, the calculation is straightforward. First, spatial pion correlators from [28] are fitted by the sum of two exponentials representing the π, π^*. This provides the $D_{m,\beta}(|\mathbf{x}|) = \alpha_{\pi,\pi^*} \exp(-\gamma_{\pi,\pi^*}|\mathbf{x}|)$, from which the spectral function can be reconstructed using Eqs. (9,10) and the known vacuum masses m_π, m_{π^*}. The result is shown in Fig. 6 (middle) and, as a non-trivial check, correctly predicts the temporal lattice correlator [31] for $\tau > m_{\pi^*}^{-1}$, Fig. 6 (right). The spectral function shows resonance-like peaks for both the pion and its first excitation. As the temperature increases, the peaks widen and gradually disappear, consistent with sequential hadron melting, but at temperatures significantly above the chiral crossover. This suggests non-perturbative, hadron-like excitations within the approximately chiral-spin symmetric temperature range.

5 Conclusions

The last few years have seen remarkable progress in the determination of the QCD phase structure. A major milestone is the understanding of the chiral transition at zero density in the massless limit, which is nearly completed. The transition temperature is known fairly accurately and there is strong evidence for the transition to be of second order for all $N_f \in [2, 7]$. Together with data on baryon number fluctuations, this constrains the location of a critical point to $(\mu_B/T)^{\rm cep} > 3$ and $T_{\rm cep} < 132$ MeV.

A new development is the discovery of an approximate chiral spin symmetry, which emerges dynamically in a temperature band above the chiral crossover, $T_{pc} \lesssim T \lesssim 3T_{pc}$. It can be identified in the multiplet structure of correlation functions and affects associated observables like screening masses and spectral functions. Together these suggest a regime with chiral symmetry restored but hadron-like degrees of freedom. It would be most interesting to investigate if and how this affects experimental observables in heavy ion collisions.

Acknowledgments: The author acknowledges support by the Deutsche Forschungsgemeinschaft (DFG) through the grant CRC-TR 211 "Strong-interaction matter under extreme conditions" and by the State of Hesse within the Research Cluster ELEMENTS (Project ID 500/10.006).

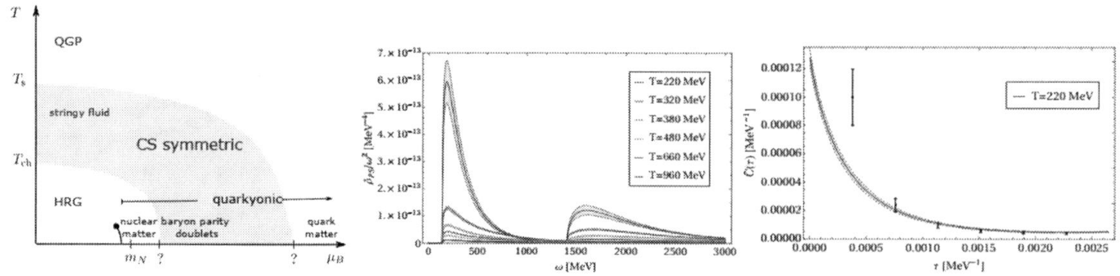

Figure 6. Left: a possible QCD phase diagram with a chiral-spin symmetric band. From [33]. Middle: Pion spectral function extracted from spatial lattice correlators. Right: Temporal correlator predicted by that spectral function, compared to lattice data. From [34].

References

[1] A.M. Halasz, A.D. Jackson, R.E. Shrock, M.A. Stephanov, J.J.M. Verbaarschot, Phys. Rev. D **58**, 096007 (1998), hep-ph/9804290

[2] K. Rajagopal, F. Wilczek, *The Condensed matter physics of QCD* (2000), pp. 2061–2151, hep-ph/0011333

[3] Y. Hatta, T. Ikeda, Phys. Rev. D **67**, 014028 (2003), hep-ph/0210284

[4] I. Lawrie, S. Sarbach, in *Phase transitions and critical phenomena*, edited by C. Domb, J. Lebowitz (1984), Vol. 9, p. 1

[5] R.D. Pisarski, F. Wilczek, Phys. Rev. D **29**, 338 (1984)

[6] F. Cuteri, O. Philipsen, A. Sciarra, JHEP **11**, 141 (2021), 2107.12739

[7] O. Philipsen, Symmetry **13**, 2079 (2021), 2111.03590

[8] H.T. Ding et al. (HotQCD), Phys. Rev. Lett. **123**, 062002 (2019), 1903.04801

[9] J. Braun, W.j. Fu, J.M. Pawlowski, F. Rennecke, D. Rosenblüh, S. Yin, Phys. Rev. D **102**, 056010 (2020), 2003.13112

[10] A.Y. Kotov, M.P. Lombardo, A. Trunin, Phys. Lett. B **823**, 136749 (2021), 2105.09842

[11] Y. Kuramashi, Y. Nakamura, H. Ohno, S. Takeda, Phys. Rev. D **101**, 054509 (2020), 2001.04398

[12] F. Cuteri, O. Philipsen, A. Sciarra, Phys. Rev. D **97**, 114511 (2018), 1711.05658

[13] L. Dini, P. Hegde, F. Karsch, A. Lahiri, C. Schmidt, S. Sharma, Phys. Rev. D **105**, 034510 (2022), 2111.12599

[14] Y. Zhang, Y. Aoki, S. Hashimoto, I. Kanamori, T. Kaneko, Y. Nakamura, PoS **LATTICE2022**, 197 (2023), 2212.10021

[15] G. Fejos, Phys. Rev. D **105**, L071506 (2022), 2201.07909

[16] S.R. Kousvos, A. Stergiou (2022), 2209.02837

[17] Z. Fodor, S.D. Katz, Phys. Lett. B **534**, 87 (2002), hep-lat/0104001

[18] C.R. Allton, S. Ejiri, S.J. Hands, O. Kaczmarek, F. Karsch, E. Laermann, C. Schmidt, L. Scorzato, Phys. Rev. D **66**, 074507 (2002), hep-lat/0204010

[19] P. de Forcrand, O. Philipsen, Nucl. Phys. B **642**, 290 (2002), hep-lat/0205016

[20] M. D'Elia, M.P. Lombardo, Phys. Rev. D **67**, 014505 (2003), hep-lat/0209146

[21] V. Vovchenko, J. Steinheimer, O. Philipsen, H. Stoecker, Phys. Rev. D **97**, 114030 (2018), 1711.01261

[22] A. Roberge, N. Weiss, Nucl. Phys. B **275**, 734 (1986)

[23] D. Bollweg, J. Goswami, O. Kaczmarek, F. Karsch, S. Mukherjee, P. Petreczky, C. Schmidt, P. Scior (HotQCD), Phys. Rev. D **105**, 074511 (2022), 2202.09184

[24] S. Borsanyi, Z. Fodor, M. Giordano, S.D. Katz, D. Nogradi, A. Pasztor, C.H. Wong, Phys. Rev. D **105**, L051506 (2022), 2108.09213

[25] V. Vovchenko, J. Steinheimer, O. Philipsen, A. Pasztor, Z. Fodor, S.D. Katz, H. Stoecker, Nucl. Phys. A **982**, 859 (2019), 1807.06472

[26] P.J. Gunkel, C.S. Fischer, Phys. Rev. D **104**, 054022 (2021), 2106.08356

[27] F. Gao, J.M. Pawlowski, Phys. Lett. B **820**, 136584 (2021), 2010.13705

[28] C. Rohrhofer, Y. Aoki, G. Cossu, H. Fukaya, C. Gattringer, L.Y. Glozman, S. Hashimoto, C.B. Lang, S. Prelovsek, Phys. Rev. D **100**, 014502 (2019), 1902.03191

[29] L.Y. Glozman, Eur. Phys. J. A **51**, 27 (2015), 1407.2798

[30] L.Y. Glozman, M. Pak, Phys. Rev. D **92**, 016001 (2015), 1504.02323

[31] C. Rohrhofer, Y. Aoki, L.Y. Glozman, S. Hashimoto, Phys. Lett. B **802**, 135245 (2020), 1909.00927

[32] T.W. Chiu (2023), 2302.06073

[33] L.Y. Glozman, O. Philipsen, R.D. Pisarski, Eur. Phys. J. A **58**, 247 (2022), 2204.05083

[34] P. Lowdon, O. Philipsen, JHEP **10**, 161 (2022), 2207.14718

[35] J. Bros, D. Buchholz, Nucl. Phys. B **429**, 291 (1994), hep-th/9807099

[36] J. Bros, D. Buchholz, Ann. Inst. H. Poincare Phys. Theor. **64**, 495 (1996), hep-th/9606046

EPJ Web of Conferences **290**, 05003 (2023) https://doi.org/10.1051/epjconf/202329005003
EuNPC 2022

Study of strange particle p_T spectra in heavy-ion collisions at relativistic energies

Oana Ristea[1], *Catalin* Ristea[1,2,*], *Alexandru* Jipa[1], *Marius* Calin[1], and *Tiberiu* Esanu[1,3]

[1]University of Bucharest, Faculty of Physics, Bucharest, Romania
[2]Institute of Space Science (ISS), Bucharest, Romania
[3]Horia Hulubei National Institute of Physics and Nuclear Engineering (IFIN-HH), Bucharest, Romania

Abstract. The transverse momentum distributions of strange and multi-strange hadrons produced in most central Au+Au collisions at RHIC-BES energies were studied using the Boltzmann-Gibbs blast-wave model. The energy dependence of the kinetic freeze-out parameters is presented and discussed.

1 Introduction

One of the main goals of the relativistic nuclear collisions studies is to investigate the behavior of nuclear matter under extreme conditions of temperature and energy density [1, 2]. In a relativistic nuclear collision a large number of secondary particles are produced, and based on their analysis information about the evolution of the system produced in the collision can be obtained. An important phenomenon that can be studied at kinetic freeze-out stage is the collective transverse flow of nuclear matter as it is entirely generated in the collision. This collective flow of the system occurs due to large pressure gradients generated by the intense interactions between the produced particles. As the particles interact within the system, they acquire a collective flow velocity, having the effect of increasing their momentum. Thus, the shape of the transverse momentum spectra is influenced by the presence of the transverse collective flow and, using different parameterizations, this collective flow can be quantified.

One of the models used to extract the kinetic freeze-out parameters is the Boltzmann-Gibbs blast-wave (BW) model [3, 4]. The shape of each spectrum is determined by the kinetic (thermal) freeze-out temperature, the collective transverse expansion velocity, the particle mass, and the choice of flow profile (the dependence of the transverse flow velocity on the radius of the expanding source). The p_T spectrum of produced particles can be described by the following formula:

$$\frac{dN}{p_T dp_T} \sim \int_0^R r\,dr\, m_T\, I_0\left(\frac{p_T \sinh\rho}{T}\right) K_1\left(\frac{m_T \cosh\rho}{T}\right) \quad (1)$$

where T is the kinetic freeze-out temperature, $m_T = \sqrt{p_T^2 + m^2}$ is the transverse mass, m is the particle mass, $\rho_T = \tanh^{-1}\beta_T(r)$ and I_0 and K_1 are the modified Bessel functions. The radial transverse flow velocity profile can be parametrized as:

$$\beta_T(r) = \beta_s \left(\frac{r}{R}\right)^n \quad (2)$$

where β_s is the surface flow velocity, the n is the exponent which describes the flow profile and r/R is the relative radial position in the thermal source. Assuming a uniform particle density, the average transverse collective flow velocity can be calculated as:

$$\langle\beta_T\rangle = \frac{2}{2+n}\beta_s \quad (3)$$

2 Results

The p_T spectra of strange and multi-strange hadrons such as K^0_S, Λ, $\overline{\Lambda}$, Ξ^-, $\overline{\Xi}^+$, ϕ, Ω^-, and $\overline{\Omega}^+$ produced in most central Au+Au collisions at RHIC-BES energies ($\sqrt{s_{NN}}$ = 7.7, 11.5, 19.6, 27 and 39 GeV) were fitted with Eq. 1. Data are taken from [5, 6]. Figure 1 shows the BW fits to the STAR p_T spectra, fitted up to $p_T < 3$ GeV/c. The fit parameters are the kinetic freeze-out temperature, T, the average transverse flow velocity, $\langle\beta_T\rangle$, and the exponent of the flow profile, n.

The kinetic freeze-out temperature as a function of average transverse flow velocity is presented in Figure 2. The strange BW fit parameters are compared to the corresponding parameters extracted from the bulk ($\pi^\pm$, $K^\pm$, p and $\overline{p}$) p_T spectra BW fits. The bulk T and $\langle\beta_T\rangle$ values are taken from [7]. For all energies, the values of strange particles T are larger than the corresponding T values extracted from BW fits on bulk particles p_T spectra [7].

The average flow velocities of strange hadrons are smaller than the corresponding values of bulk particles [7]. These results could indicate that, with incresing energy, the strange particles decouple earlier in time in the system's evolution, at a higher kinetic freeze-out and having a smaller flow velocity because collective flow builds up

*e-mail: catalin.ristea@spacescience.ro

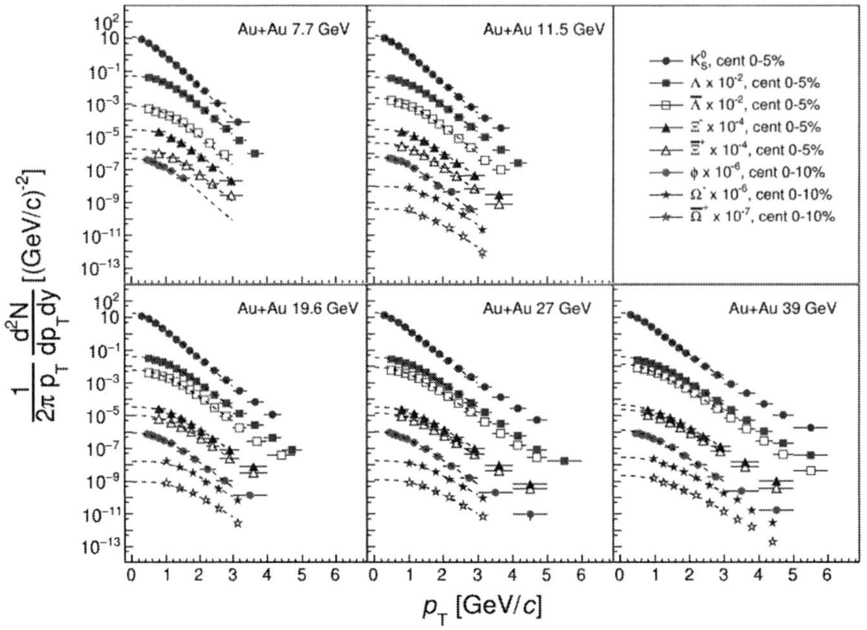

Figure 1. The transverse momentum spectra of K^0_S, Λ, $\overline{\Lambda}$, Ξ^-, $\overline{\Xi}^+$, ϕ, Ω^-, and $\overline{\Omega}^+$ produced at midrapidity ($|y| < 0.5$) in most central Au+Au collisions at $\sqrt{s_{NN}}$ = 7.7, 11.5, 19.6, 27, and 39 GeV as obtained by the STAR Collaboration [5, 6]. Dotted lines are the simultaneous fits to data using BGBW model.

over the entire evolution of the collision system. Therefore, if the strange particles are emitted when the system is hotter, their collective flow is smaller. These results are consistent with previous SPS and RHIC results [8–13]. Due to different hadronic interaction cross sections, the produced particles freeze-out from the fireball at different times and therefore the system can have sequential freeze-out stages: particles with smaller interaction cross-section can decouple sooner from the system, followed by the bulk particles at a later FO stage [15, 16].

Because it is known that the strange particles may exhibit different production mechanisms and interaction cross-sections [17–19], the particles have been divided into two groups — those which share valence quarks with the nucleons (K^0_S, Λ, and Ξ^-) and those which do not ($\overline{\Lambda}$, $\overline{\Xi}^+$, ϕ, Ω^-, and $\overline{\Omega}^+$). The fit parameters obtained from fits on p_T spectra of the two groups of strange particles in most central Au+Au collisions at $\sqrt{s_{NN}}$ = 7.7–39 GeV are presented in the Figs. 3-5.

The n exponent extracted from the $\overline{\Lambda}$, $\overline{\Xi}^+$, ϕ, Ω^-, and $\overline{\Omega}^+$ spectra fits is very close to 0 for all energies suggesting a constant flow profile for these particles in the analyzed energy range. The values of n parameter of the two groups of particles are similar for $\sqrt{s_{NN}}$ = 7.7–11.5 GeV. For $\sqrt{s_{NN}}$ = 19.6–39 GeV, the values of n parameter of K^0_S, Λ, and Ξ^- group are larger compared to $\overline{\Lambda}$, $\overline{\Xi}^+$, ϕ, Ω^-, and $\overline{\Omega}^+$ group and increase with energy. This behaviour could

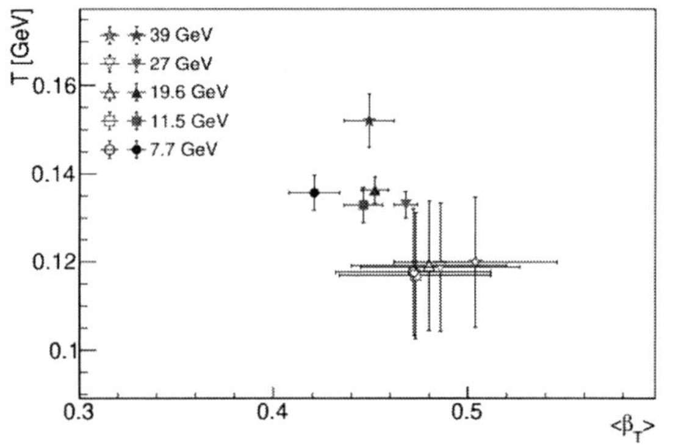

Figure 2. The dependence of kinetic freeze-out temperature on average transverse flow velocity. Full symbols are from BGBW fits on K^0_S, Λ, $\overline{\Lambda}$, Ξ^-, $\overline{\Xi}^+$, ϕ, Ω^-, and $\overline{\Omega}^+$ p_T spectra. The open symbols are the $\pi^\pm$, $K^\pm$, p and $\overline{p}$ BW fits results taken from Ref. [7].

indicate that the K^0_S, Λ, and Ξ^- group gives a stronger constraint to n parameter.

The two groups of strange particles have different p_T spectra shapes. The K^0_S p_T spectrum is steeper than Λ, and Ξ^- spectra and due to higher statistics, fit results seems to be driven by the K^0_S p_T spectrum rather than the other two spectra. For steeper spectra, the n parameter has a

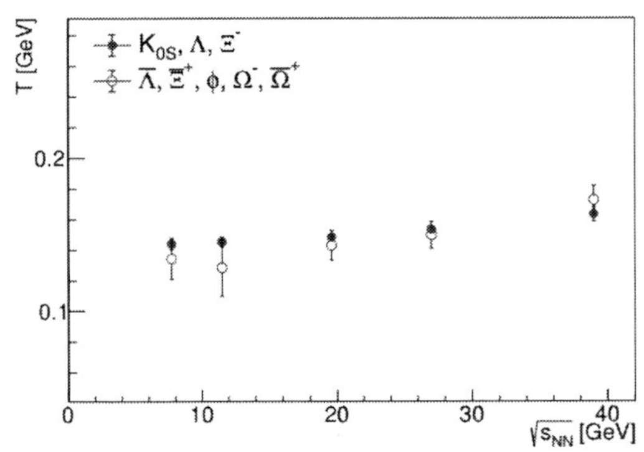

Figure 3. The kinetic freeze-out temperature from BW fits on the K^0_S, Λ, and $Ξ^-$ spectra (full symbols) and on the $\overline{Λ}$, $\overline{Ξ}^+$, φ, $Ω^-$, and $\overline{Ω}^+$ spectra (open symbols) in most central Au+Au collisions at $\sqrt{s_{NN}}$ = 7.7–39 GeV as a function of energy.

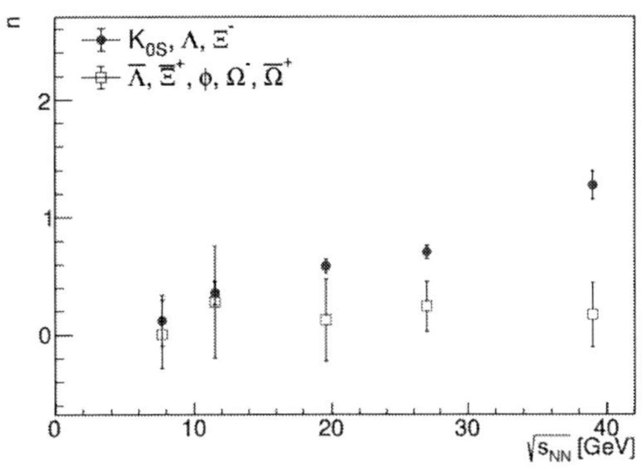

Figure 5. The n parameter extracted from BW fits on the K^0_S, Λ, and $Ξ^-$ spectra (full symbols) and on the $\overline{Λ}$, $\overline{Ξ}^+$, φ, $Ω^-$, and $\overline{Ω}^+$ spectra (open symbols) in most central Au+Au collisions at $\sqrt{s_{NN}}$ = 7.7–39 GeV as a function of energy.

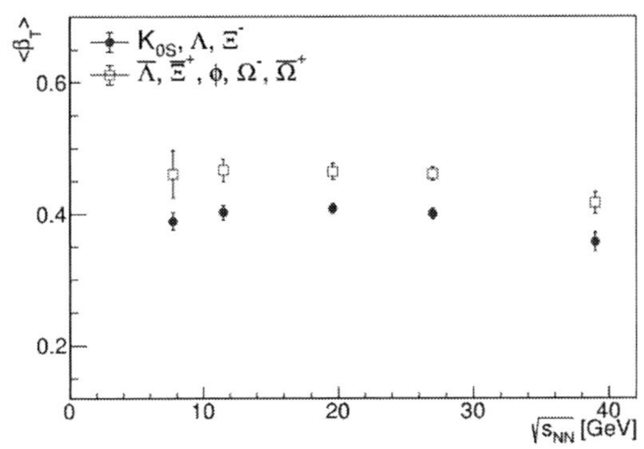

Figure 4. The average transverse flow velocity from BW fits on the K^0_S, Λ, and $Ξ^-$ spectra (full symbols) and on the $\overline{Λ}$, $\overline{Ξ}^+$, φ, $Ω^-$, and $\overline{Ω}^+$ spectra (open symbols) in most central Au+Au collisions at $\sqrt{s_{NN}}$ = 7.7–39 GeV as a function of energy.

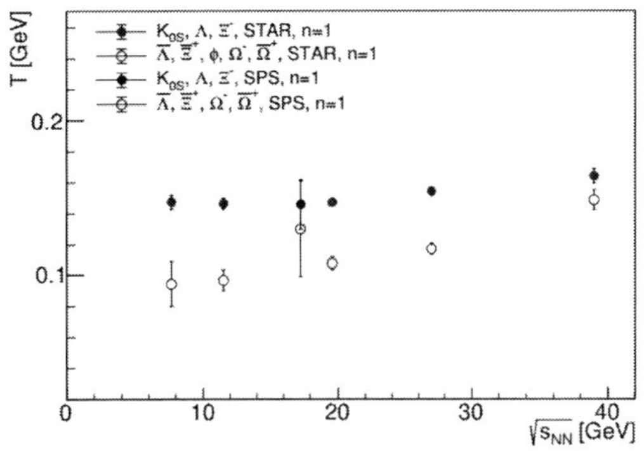

Figure 6. The T parameter obtained from BW fits on different strange particle p_T spectra using a linear flow profile (n = 1, fixed parameter). The NA57 results obtained considering the same value of n in most central Pb+Pb collisions at $\sqrt{s_{NN}}$ = 17.3 GeV are also shown.

larger value and corresponds to a higher T and a smaller $\langle β_T \rangle$. For the other group, their p_T spectra are more curved at low p_T compared to K^0_S p_T spectrum, therefore are described by a smaller value of n parameter. For smaller values of n parameter, $\langle β_T \rangle$ parameter is larger and the corresponding T is smaller. The average transverse flow velocity of $\overline{Λ}$, $\overline{Ξ}^+$, φ, $Ω^-$, and $\overline{Ω}^+$ group is larger by ~ 15% than that of K^0_S, Λ, and $Ξ^-$ group.

We compared our results with previous results obtained in Pb+Pb collisions at $\sqrt{s_{NN}}$ = 17.3 GeV by the NA57 Collaboration [8, 9]. The NA57 results using the BW model on p_T spectra of strange particles were obtained assuming a linear transverse flow velocity profile (n = 1). Therefore, we fixed the n parameter describing the flow profile at the value n = 1.

The T and $\langle β_T \rangle$ obtained considering a linear dependence of the flow velocity on the radial position in the thermal source (n = 1) are shown in Fig. 6 and Fig. 7. For a fixed value of n parameter, the more curved p_T spectra in the low p_T region are described by a larger $\langle β_T \rangle$. The average transverse flow velocity is larger for $\overline{Λ}$, $\overline{Ξ}^+$, φ, $Ω^-$, and $\overline{Ω}^+$ group. The kinetic temperature is larger for K^0_S, Λ, and $Ξ^-$ group and increases very slowly with energy for both strange particle groups. Our results are consistent within errors with SPS results.

3 Conclusions

The p_T spectra of strange particles produced in most central Au+Au collisions at RHIC-BES energies were studied with BW model and the radial transverse flow and the kinetic freeze-out temperature of the system were extracted. In the RHIC-BES energy range, the strange hadrons have a

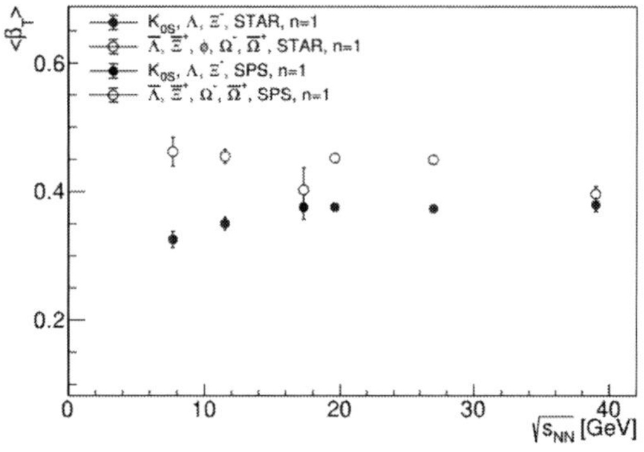

Figure 7. The $\langle\beta_T\rangle$ parameter obtained from BW fits on different strange particle p_T spectra using a linear flow profile (n = 1, fixed parameter). The NA57 results obtained considering the same value of n in most central Pb+Pb collisions at $\sqrt{s_{NN}} = 17.3$ GeV are also shown.

larger freeze-out temperature than that of bulk hadrons and a smaller average flow velocity. These results can indicate a double kinetic freeze-out scenario due to the separate decoupling of bulk and strange particles. In addition, the particles containing strangeness were examined separately to study their radial flow and freeze-out. These results are consistent with previous SPS results.

Acknowledgements

This work was supported by IFA, FAIR-RO program, project number FAIR-08/2020. CR was partially supported by Romanian Ministry of Research, Innovation and Digitalization under Romanian National Core Program LAPLAS VII – contract no. 30N/2023.

References

[1] Y. Akiba *et al.*, The Hot QCD White Paper: Exploring the Phases of QCD at RHIC and the LHC, arXiv:1502.02730 [nucl-ex], (2015).

[2] W. Busza, K. Rajagopal, and W. van der Schee, Heavy Ion Collisions: The Big Picture, and the Big Questions, Ann. Rev. Nucl. Part. Sci. **68**, 339 (2018)

[3] E. Schnedermann, J. Sollfrank, and U. Heinz, Thermal phenomenology of hadrons from 200A GeV S+S collisions, Phys. Rev. C **48**, 2462 (1993)

[4] E. Schnedermann and U. Heinz, Hydrodynamical assessment of 200A GeV collisions, Phys. Rev. C **50**, 1675 (1994)

[5] J. Adams *et al.* (STAR Collaboration), Strange hadron production in Au+Au collisions at $\sqrt{s_{NN}} = 7.7, 11.5, 19.6, 27$, and 39 GeV, Phys. Rev. C **102**, 34909 (2020)

[6] L. Adamczyk *et al.* (STAR Collaboration), Probing Parton Dynamics of QCD Matter with Omega and Phi Production, Phys. Rev. C **93**, 21903 (2016)

[7] L. Adamczyk *et al.* (STAR Collaboration), Bulk properties of the medium produced in relativistic heavy-ion collisions from the beam energy scan program, Phys. Rev. C **96**, 044904 (2017)

[8] G. E. Bruno (for the NA57 Collaboration), Blast-wave analysis of strange particle m_T spectra in Pb-Pb collisions at the SPS, J. Phys. G **31**, S127 (2005)

[9] F. Antinori *et al.* (NA57 Collaboration), Study of the transverse mass spectra of strange particles in Pb-Pb collisions at 158 A GeV/c, J.Phys.G **30**, 823 (2004)

[10] J. Adams *et al.* (STAR Collaboration), Experimental and Theoretical Challenges in the Search for the Quark Gluon Plasma: The STAR Collaboration's Critical Assessment of the Evidence from RHIC Collisions, Nucl. Phys. A **757**, 102 (2005)

[11] J. Adams *et al.* (STAR Collaboration), Multistrange Baryon Production in Au-Au Collisions at $\sqrt{s_{NN}} = 130$ GeV, Phys. Rev. Lett. **92**, 182301 (2004)

[12] O. Barannikova (for the STAR Collaboration), Freeze-out dynamics at RHIC, Acta Phys.Hung. A **24**, 39 (2005)

[13] O. Barannikova (for the STAR Collaboration), Probing collision dynamics at RHIC, arXiv:nucl-ex/0403014

[14] O. Ristea, C. Ristea, A. Jipa, Study of strange particle p_T spectra in relativistic heavy-ion collisions with blast-wave model, Int. J. Mod. Phys. E **31** (2022) 09, 2250090

[15] S. Chatterjee, R. M. Godbole, S. Gupta, Strange freeze-out, Phys. Lett. B **727**, 554 (2013)

[16] M. Waqas, G. X. Peng, Fu-Hu Liu, An evidence of triple kinetic freeze-out scenario observed in all centrality intervals in Cu-Cu, Au-Au and Pb-Pb collisions at high energies, J. Phys. G **48**, 075108 (2021)

[17] H. van Hecke, H. Sorge, and N. Xu, Evidence of early multi-strange hadron freeze-out in high energy nuclear collisions, Phys. Rev. Lett. **81**, 5764 (1998)

[18] S. Margetis, K. Safarik and O. Villalobos Baillie, Strangeness Production in Heavy-Ion Collisions, Annu. Rev. Nucl. Part. Sci **50**, 299 (2000)

[19] C. Blume, C. Markert, Strange hadron production in heavy ion collisions from SPS to RHIC, Prog. Part. Nucl. Phys. **66**, 834 (2011)

Centrality Dependent Lévy Analysis of two-pion BEC Functions at PHENIX

Tamás Novák[1,2]* *for the PHENIX Collaboration*

[1] MATE KRC, H-3200 Gyöngyös, Mátrai út 36, Hungary
[2] BGE, H-1054 Budapest, Alkotmány út 9-11, Hungary

Abstract. We present most recent PHENIX preliminary data on centrality dependence of two-pion Bose-Einstein correlation functions measured in $\sqrt{s_{NN}} = 200$ GeV Au+Au collisions at the Relativistic Heavy Ion Collider (RHIC). The data are well described by assuming the source to be a Lévy-stable distribution. The Lévy parameters, R, α, λ are measured in 18 bins of transverse mass (m_T) for 4 centrality intervals. The Lévy scale parameter $R(m_T)$ decreases with m_T and exhibits a clear centrality ordering which supports its geometrical interpretation. The Lévy exponent $\alpha(m_T)$ is independent of m_T in every centrality bin but shows some centrality dependence. At all centralities α is significantly different from that of a Gaussian ($\alpha = 2$) or Cauchy ($\alpha = 1$) source distribution. The Lévy strength parameter λ was also investigated. We observed that $\lambda(m_T)$ decreases at low m_T and saturates around $m_T = 0.6$ GeV/c^2. In the already published 0-30% centrality class, Monte Carlo simulations are found to be inconsistent with the measurements, unless a significant reduction of the in-medium mass of the η' meson is included.

1 Introduction

In particle and nuclear physics intensity interferometry provides a direct experimental method for the determination of sizes, shapes and lifetimes of particle-emitting sources (for recent reviews see [1, 2]). In particular, boson interferometry provides a powerful tool for the investigation of the space-time structure of particle production processes, since Bose-Einstein correlations (BEC) of two identical bosons reflect both geometrical and dynamical properties of the particle radiating source.

The technique was discovered in astrophysics by R. Hanbury Brown and Q. R. Twiss in correlation measurements, that were performed in radio and optical astronomy to measure the angular diameters of stars [3]. Hanbury Brown and Twiss are considered the experimental founders of the HBT effect. Independently, the intensity correlations of identical pions were observed in proton-antiproton annihilation. These correlations were explained by G. Goldhaber, S. Goldhaber, W-Y. Lee and A. Pais [4] on the basis of the Bose-Einstein symmetrization of the wave-function of identical pion pairs.

If interactions between the created hadrons, higher order correlations, decays and all other dynamical two-particle correlations may be neglected, then the two-particle Bose-Einstein correlation function is simply related to the source function $S(x, k)$ (which describes the probability density of particle creation at the space-time point x and with four-momentum p).

2 The PHENIX Detector

The PHENIX experiment at RHIC was designed to study various type of particles produced in heavy ion collisions, including photons, electrons, muons and charged hadrons.

PHENIX has two central arm spectrometers (East and West), each covering $|\eta| < 0.35$ in pseudorapidity and $\Delta\varphi = \pi/2$ in azimuth. In each central arm, charged particle tracks are reconstructed using hit information from the drift chamber (DC), the first layer of pad chambers (PC1) and the collision z-vertex position measured by the beam-beam counters (BBC).

The charged pion identification was done by lead scintillator Electromagnetic Calorimeter (PbSc) as well as the high resolution time-of-flight detectors (TOF East and TOF West). We identified charged pions in 0.2 GeV/$c \leq p_T \leq 0.85$ GeV/c transverse momentum range.

The detailed description of the basic experimental configuration (without the upgrades made after early 2000s) can be found here [5].

3 Parametrizations of BEC

The Bose-Einstein correlation function, C_2, is measured by $C_2(Q) = \rho(Q)/\rho_0(Q)$, where $\rho(Q)$ is the density of identical boson pairs with invariant four-momentum difference $Q = \sqrt{-(p_1 - p_2)^2}$ and $\rho_0(Q)$ is the similar density in an artificially constructed reference sample, which should differ from the data only in that it does not

* Corresponding author: novak.tamas@uni-bge.hu

https://doi.org/10.1051/epjconf/202329005004

contain the effects of Bose symmetrization of identical bosons. It is often parametrized as

$$C_2 = \gamma[1 + \lambda G](1 + \varepsilon Q) \tag{1}$$

with

$$G = \exp(-(RQ)^2). \tag{2}$$

The source is frequently assumed to be Gaussian which leads to Gaussian correlation function. So, the corresponding distribution of boson emission points in space-time is a spherically symmetric Gaussian with standard deviation R. However, there is no good reason to assume Gaussian source function – except simplicity. Recently many measurements suggested phenomena beyond Gaussian distribution in heavy ion collisions.

Past measurements of two-pion Bose-Einstein correlation functions went beyond the Gaussian approximation show that the precise shape of BEC is indeed not Gaussian. For example, L3 experiment applied Lévy stable distributions first time [6]. The PHENIX experiment also measured correlation functions based on Lévy [7]. So, the best way to depart from the Gaussian is the generalization to a symmetric Lévy stable distribution. It yields to

$$G = \exp(-(RQ)^\alpha), \tag{3}$$

where $0 < \alpha \leq 2$ is the so-called index of stability which was introduced to BEC in [8]. In the case of $\alpha = 2$ the Gaussian case is restored, the case of $\alpha = 1$ corresponds to the Cauchy distribution.

There can be many reasons for the appereance of Lévy distributions: criticality, QCD jets, anomalous diffusion and other phenomena. Lévy-type of analyses were discussed before in Refs. [9 – 11].

It is also important to mention the final state Coulomb-correction which is a crucial part of any correlation measurement dealing with charged particles. One cannot fit the above functional form to the measured correlation functions before properly taking the final state Coulomb repulsion of the identically charged pions into account. For an exhaustive description of this, in particular the functions used in fitting our data, see Refs. [12 – 14].

The Lévy parametrization (incorporating the effect of the final-state Coulomb interactions) is found to give a statistically acceptable description of the data with Lévy R, α and λ as it can be seen on Fig. 2.

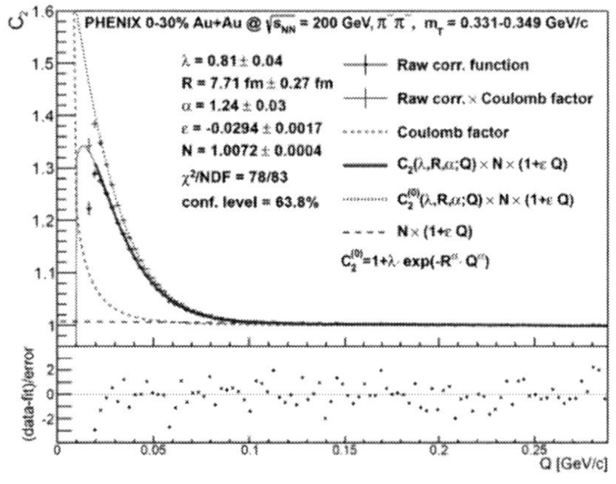

Fig. 2. Example fit of Bose-Einstein correlation function.

3.1 BEC at the second order QCD phase transition

According to recent lattice QCD calculations the line of the first order phase transition ends at the critical end-point (CEP), where the transition from hadron gas to quark-gluon-plasma (QGP) becomes a second order phase transition.

In case of second order phase transition the η exponent describes the power-law behavior of the spatial correlations at the critical point: $\sim r^{-(d-2+\eta)}$, where d is the dimension [15, 16]. For the three-dimensional Lévy stable sources the correlation between the initial and actual positions decays also as a power-law, where the exponent is given by the Lévy index of stability: $\sim r^{-(1+\alpha)}$ [8, 16]. From comparing these exponents, one can easily see that the Lévy exponent is identical to η at the critical point.

Rieger determined the correlation exponent using 3d Ising model [15]: $\alpha(\text{Lévy}) = \eta(\text{3d Ising}) = 0.50 \pm 0.05$. If one extracts the value of the Lévy exponent α at different center-of-mass collision energies, the data may yield information on the nature of the quark-hadron phase transition, particularly it may shed light on the location of the critical endpoint on the phase-diagram.

3.2 The shape parameters of BEC

The most important physical parameters (R, α, λ) are measured in this work as a function of pair transverse mass (m_T). The transverse mass is defined with $m_T = \sqrt{m^2 + p_T{}^2}$.The R is the length of homogeneity (measures the size), α is the index of stability and λ is the so-called impact parameter (responsible for instance particle creation mechanism, like core-halo model [17, 18]).

The effect of the parameters can be seen in Fig. 3.

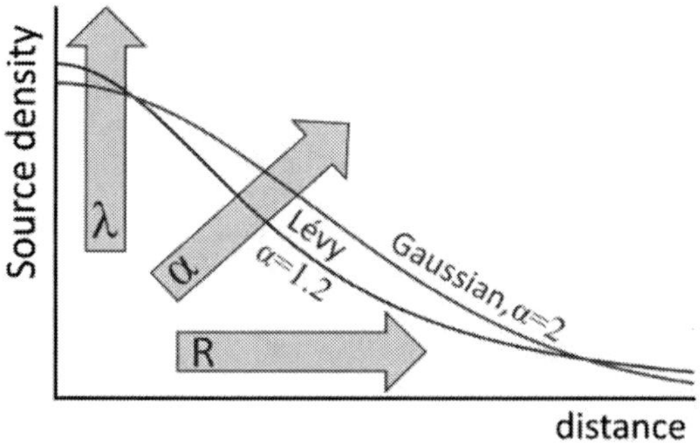

Fig. 3. Comparison of the Gaussian and the Lévy distribution with a given R, α and λ. It can be seen that the Lévy distribution has a power-law-like tail.

If there is no halo part and the source is chaotic and fully thermal then $\lambda = 1$. If $\lambda < 1$ it is the sign of the presence of the decay mesons. The α shape parameter or Lévy-index indicates the deviation from the Gaussian case. Moreover, it could be associated – as it was mentioned before – to one of the critical exponents namely to the critical exponent of the spatial correlation in 3-dimension [15]. The R is frequently referred as the Lévy scale parameter.

4 Results

We studied the Lévy-parameters in 10% wide centrality bins in 0–40% range and in 18 transverse mass (m_T) bins at $\sqrt{s_{NN}} = 200$ GeV collision energy in Au+Au system. We determine the m_T dependencies of the Lévy parameters.

4.1 The Lévy scale parameter *R*

First, we measured the $R(m_T)$ dependencies in the above mentioned centrality ranges. The results can be seen in Fig. 4.

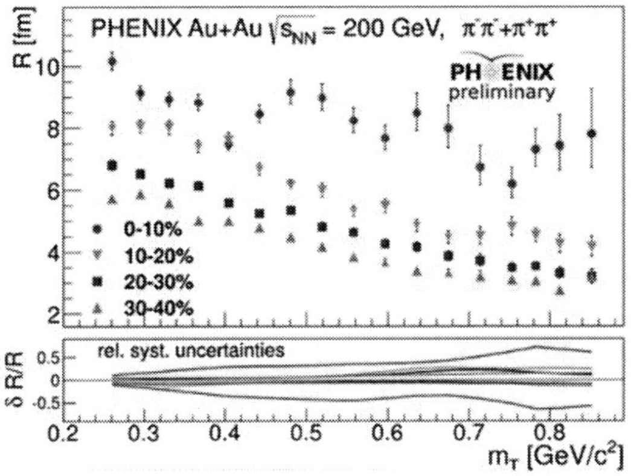

Fig. 4. The Lévy scale parameter R as a function of m_T and centrality.

The Lévy scale exhibits decreasing trends as a function of m_T, and one can clearly observe a geometrical centrality dependence. In more central collisions, the R parameter (which is related to the physical size of the system) takes larger values.

4.2 The Lévy shape parameter α

In case of the Lévy exponent α we can see that for all centrality and m_T ranges the values are between 1 and 2. The measured values are far from the Gaussian $\alpha = 2$ case and also far from the conjectured $\alpha \leq 0.5$ at the critical point.

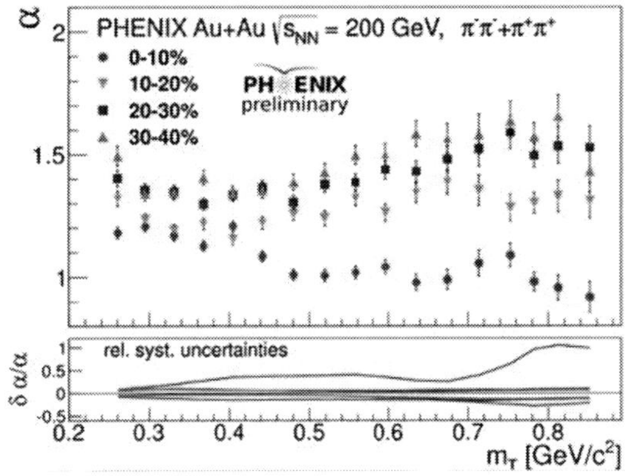

Fig. 5. The Lévy shape parameter α as a function of m_T and centrality.

4.3 The Lévy strength parameter λ

As it was mentioned above the λ parameter could be related to the core-halo ratio. $\lambda(m_T)$ is shown in Fig. 6. As one can see from the figure, the strength of the correlation function is not equal to unity, and not even constant as a function of m_T, the reason for which may be the fact that a large fraction of low m_T pions are produced from decay of long-lived resonances.

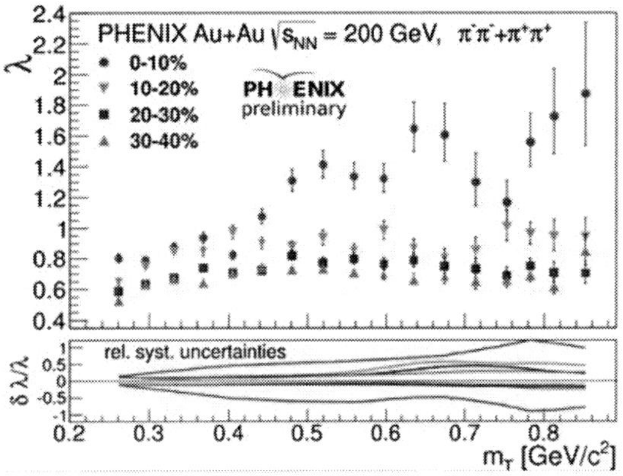

Fig. 6. The Lévy strength parameter λ as a function of m_T and centrality.

As seen in Fig. 6. λ appears to increase with m_T until it saturates around $m_T = 0.6$ GeV/c^2. To further study the dependence of λ on m_T it is advantageous to use the ratio λ/λ_{max} where λ_{max} is the saturated value of λ which we determine in the region $m_T > 0.55$ GeV/c^2. Fig. 7. shows the resulting λ/λ_{max} dependence on m_T.

https://doi.org/10.1051/epjconf/202329005004

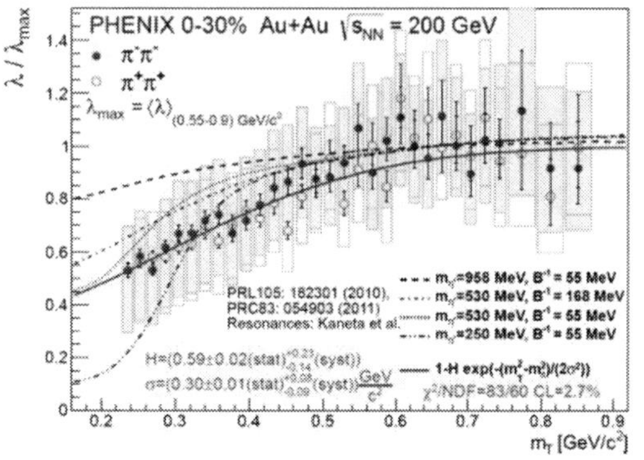

Fig. 7. Normalized correlation strength parameter $\lambda/\lambda_{\max}$ versus average m_{T}.

The decreasing trend at lower m_{T} values can clearly be observed in the case of the $\lambda(m_{\mathrm{T}})$, but it is clearer in the case of $\lambda(m_{\mathrm{T}})/\lambda_{\max}$. It has been suggested [19] that $U_A(1)$ symmetry restoration and its related in-medium mass reduction of the η' meson in hot, dense hadronic matter would cause a reduction in the value of λ at low m_{T}. In Fig. 7, our data are compared with parameter scans from Refs. [20, 21], using different values for the in-medium η' mass $m^*_{\eta'}$, and the η' condensate temperature (slope parameter) $B^{-1}_{\eta'}$. Our data are seen to be suppressed compared to the prediction with no in-medium η' mass modification. Within systematics, our data are not inconsistent with selected parameter scan results of Refs. [20, 21] using a modified in-medium η' mass. This phenomenon can be explained in more detail in Refs. [2, 7]. Our data allows for a scrutiny of several theoretical predictions for the in-medium mass modification of η' [22 – 27].

It is worth noting that this effect was not observed in the S+Pb data on $\lambda(m_{\mathrm{T}})$ at E_{LAB} = 200 AGeV CERN SPS energies [28]. Similarly, a result consistent with lack of suppression was reported recently by the NA61 collaboration for Be+Be collisions at E_{LAB} = 150 AGeV CERN SPS energies [29].

5 Conclusions

We measured correlation functions in Au+Au collisions at $\sqrt{s_{NN}} = 200$ GeV colliding energies in the PHENIX experiment in 18 m_{T} and 4 centrality range. We fit the measured data with Lévy distribution, yield the Lévy parameters and determine their m_{T} and centrality dependencies.

The Lévy scale parameter R has decreasing trends in all observed centrality bin as a function of m_{T} and clearly shows geometrical centrality dependence. We also observed that the α is far from the Gaussian case corresponds to $\alpha = 2$, and also it was found to be inconsistent with $\alpha \leq 0.5$, the conjectured value at the QCD critical point. We found a statistically significant decrease of the strength of the correlation function $\lambda(m_{\mathrm{T}})$ at low values of the transverse mass. We also observed that the normalized $\lambda/\lambda_{\max}$ function has similar behavior as $\lambda(m_{\mathrm{T}})$. However, this result is not consistent with predictions without in-medium η' mass modification.

The author is thankful for the support of NKFIH grant K138136 and also K133046.

References

1. W. Kittel, Acta Phys. Pol. B**32** (2001)
2. T. Csörgő, Heavy Ion Physics **15**, (2002)
3. R.H. Brown, R.Q. Twiss, Nature **177** (1956)
4. G. Goldhaber, S. Goldhaber, W. Lee, A. Pais, Phys. Rev. Lett. **3**: 181 (1959)
5. K. Adcox *et al.* [PHENIX Collaboration], Nucl. Inst. Methods Phys. Res., Sec. A **499** (2003)
6. P. Achard *et al.* [L3 Collaboration], Eur Phys. J. C **71** (2011)
7. A. Adare *et al.* [PHENIX Collaboration], Phys. Rev. C **97** (2018)
8. T. Csörgő, S. Hegyi, W. Zajc, Eur. Phys. J. C **36** (2004)
9. M. Csanád, Nucl. Phys. A. **774** (2006)
10. D. Kincses, Universe 4 **31** (2018)
11. S. Lökös, Universe 4 **11** (2018)
12. M. Csanád, S. Lökös, M. Nagy, Phys. Part. Nuclei **51** (2020)
13. D. Kincses, M. Nagy, M. Csanád, Phys. Rev. C **102** (2020)
14. B. Kurgyis, D. Kincses, M. Nagy, M. Csanád, https://arxiv.org/abs/2007.10173 (2023)
15. H. Rieger, Phys. Rev. B. **52** (1995)
16. T. Csörgő, S. Hegyi, T. Novák, W. Zajc, AIP. Conf. Proc. **828** (2006)
17. T. Csörgő, B. Lorstad, J. Zimányi, Z. Phys. C Part Fields **71** (1996)
18. J. Bolz, U. Ornik, M. Plumer, B.R. Schlei, Phys. Rev. D **47** (1993)
19. S.E. Vance, T. Csörgő, Phys. Rev. Lett. **81** (1998)
20. T. Csörgő, R. Vértesi, J. Sziklai, Phys. Rev. Lett. **105** (2010)
21. R. Vértesi, T. Csörgő, J. Sziklai, Phys. Rev. C **83** (2011)
22. R.D. Pisarski, F. Wilczek, Phys. Rev. D **29** (1984)
23. J.I. Kapusta, D. Kharzeev, Larry D. McLerran, Phys. Rev. D **53** (1996)
24. S. Weinberg, Phys. Rev. D **11** (1975)
25. Z. Huang, X.N. Wang, Phys. Rev. D **53** (1996)
26. Y. Kwon, S.H. Lee, K. Morita, Gy. Wolf, Phys. Rev. D **86** (2012)
27. D. Horvatic, D. Kekez, D. Klabučar, Phys. Rev. D **99** (2019)
28. H. Bekker *et al.* [NA44 Collaboration], Phys. Rev. Lett. **74** (1995)
29. H. Adhikary *et al.* [NA61/SHINE Collaboration] https://arxiv.org/abs/2302.04593 (2023)

Multiplicity-dependent production of heavy mesons with strangeness in small systems at LHCb

Clara Landesa Gómez[1,*]

[1]On behalf of the LHCb collaboration
Instituto Galego de Física de Altas Enerxías, Universidade de Santiago de Compostela, Santiago de Compostela, A Coruña, Spain

Abstract. The phenomenon of strangeness enhancement, originally proposed as a signature of quark-gluon plasma formation, has received considerable new interest following recent observations in small collision systems. LHCb is uniquely well suited to study such effects in the heavy quark sector, down to very low transverse momentum. Here we will present new LHCb results on the production rates of B_s^0 relative to B^0 mesons versus multiplicity in *pp* collisions. Potential implications for the hadronization mechanism of heavy quarks and our understanding of the factorization of fragmentation functions will be discussed.

1 Introduction

The Quark Gluon Plasma (QGP) is a state of matter in which quarks and gluons are not confined within hadrons, but evolve as quasi-free particles. The QGP is of great interest to the scientific community since it constitutes a somewhat unexplored region in the Quantum Chromo Dynamics phase diagram. Besides, it is predicted to have been formed microseconds after the Big Bang and to constitute the inner core of neutron stars [1].

The production of the QGP requires achieving high energy densities. Theoretical [2] and experimental evidence [3, 4] agree that it is possible to reach this colour deconfinement in the laboratory when colliding two heavy nuclei. In these kinds of collisions, there are several indicators for the formation of the QGP. An enhancement in the production of hadrons with strange valence quarks with respect to light hadrons in heavy-ion collisions has long been considered a probe of the QGP [5]. Only in collisions with a high number of participating nucleons (participants) is the density high enough to produce the QGP. In a collision of two heavy ions, the charged particle multiplicity is highly correlated to the number of participants [6], so a strangeness enhancement with multiplicity can be associated with plasma production.

Over recent years there has been an increase in interest in the search for QGP traits in smaller collision systems, such as *p*Pb and *pp*. For instance, the ALICE collaboration has measured a strangeness enhancement in high multiplicity *pp* collisions [7]. The LHCb measurements of the ratio of B_s^0 to B^0 cross sections, $\sigma_{B_s^0}/\sigma_{B^0}$, as a function of charged particle multiplicity and p_T [8] constitutes a new test for the presence of QGP properties in small collision systems.

The production mechanism of hadrons (hadronisation) in hadronic collisions is substantially affected by the nature of the colliding matter. *B* mesons are exceptional probes for hadronisation for two main reasons: There are no *b* quarks in the beam protons, thus their production is dominated by hard parton scatterings in the initial stages of the collisions [9] and, due to their large mass, their production is well described by pQCD [10–12].

The usual hadronisation mechanism for *B* mesons in hadronic collisions is called fragmentation, where showers of partons produced by outgoing quarks form into hadrons [13, 14]. However, in the case of a very dense medium, readily available quarks with overlapping wavefunctions can combine into colour singlets, hadronising through coalescence. Coalescence is especially important in high-energy heavy ion collisions where the QGP is formed [15]. In *pp* collisions, if hadronisation via coalescence emerges as a mechanism for forming final state *B* hadrons, the production of B_s^0 (*s*,*b*) relative to B^0 (*d*,*b*) mesons could increase with particle multiplicity.

2 B_s^0/B^0 production ratio with multiplicity

2.1 Data Sample and Selection

The *pp* collisions data used for these measurements were collected by the LHCb experiment at a centre of mass energy $\sqrt{s}$ = 13 TeV, corresponding to a total integrated luminosity of 5.4 fb^{-1}. The LHCb detector is a single-arm forward spectrometer covering the pseudorapidity range $2 < \eta < 5$, designed to study heavy flavour physics at the LHC [16]. LHCb is composed of several subdetectors. The LHCb tracking system consists of the VErtex LOcator system (VELO) and four planar tracking stations. The VELO is a silicon strip tracking detector that encloses the beams' interaction point (primary vertex). The main

*email: *clara.landesa.gomez@cern.ch*

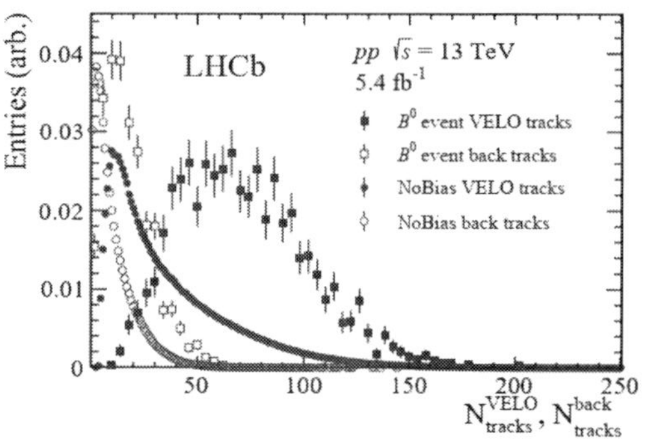

Figure 1. Distribution of N_{tracks}^{VELO} and N_{tracks}^{back} for NoBias events (blue squares) and B^0 signal events (red circles). The vertical scale is arbitrary. From [8].

purpose of the VELO is to reconstruct the primary vertices. Particle identification in LHCb is provided by four different detectors: the calorimeter system, the two RICH stations (that use Cherenkov radiation) and the muon stations.

A simulation sample was used in order to account for the effects of the reconstruction and the selection. The *pp* events are generated using Pythia [17, 18] with a specific LHCb configuration [19]. Decays of unstable particles are described by EvtGen [20] and the detector response is mimicked by Geant4 [21, 22] as described in Ref. [23].

The yields of B_s^0 and B^0 are both measured through the same decay channel,

$$B_s^0 \text{ or } B^0 \rightarrow (J/\psi \rightarrow \mu^-\mu^+)\pi^+\pi^-, \qquad (1)$$

that gives similar yields for the two mesons.

Only events with one primary vertex are considered. The events are further selected through triggers that select the decay $J/\psi \rightarrow \mu^-\mu^+$. In order to identify the B meson candidates within these events, certain kinematic and particle-identification requirements are implemented to the tracks of the daughter particles.

The sample is divided into multiplicity classes. The multiplicity is characterised by the number of tracks reconstructed by the VELO (N_{tracks}^{VELO}), and a subset of these that point backwards with respect to LHCb (N_{tracks}^{back}), in the pseudorapidity interval $-3.5 < \eta < -1.5$. The distributions of N_{tracks}^{VELO} and N_{tracks}^{back} are shown in Figure 1, both for NoBias events and for B^0 signal events. NoBias events do not have any selection, aside from requiring a bunch crossing and a primary vertex. The B^0 events distribution was achieved by applying the *sPlot* method [24] to remove the background from the selected sample.

The NoBias events are used to normalise the multiplicity. Precisely, for each of the $N_{tracks}^{VELO}/N_{tracks}^{back}$ intervals, the measurement is quoted in terms of the number of tracks at the centre of the interval, divided by the mean number of tracks in NoBias events. These take the values $\langle N_{tracks}^{VELO}\rangle_{NoBias} = 37.7$ and $\langle N_{tracks}^{back}\rangle_{NoBias} = 11.1$, with negligible uncertainties.

2.2 Determination of the B_s^0/B^0 production ratio

The ratio of cross-sections $\sigma_{B_s^0}/\sigma_{B^0}$ is calculated as

$$\frac{\sigma_{B_s^0}}{\sigma_{B^0}} = \frac{N_{B_s^0}}{N_{B^0}} \times \frac{\mathcal{B}_{B^0}}{\mathcal{B}_{B_s^0}} \times \frac{\varepsilon_{B^0}^{acc}}{\varepsilon_{B_s^0}^{acc}} \times \frac{\varepsilon_{B^0}^{trig}}{\varepsilon_{B_s^0}^{trig}} \times \frac{\varepsilon_{B^0}^{PID}}{\varepsilon_{B_s^0}^{PID}} \times \frac{\varepsilon_{B^0}^{reco}}{\varepsilon_{B_s^0}^{reco}}. \qquad (2)$$

Here, $N_{B_s^0}$ and N_{B^0} are the measured yields for B_s^0 and B^0, respectively, and $\frac{\mathcal{B}_{B^0}}{\mathcal{B}_{B_s^0}}$ is the ratio of the branching ratios to the decay in Eq. 1. The four remaining factors are the ratios of LHCb acceptance and the trigger, particle identification and reconstruction efficiencies, in order of appearance.

For each of the multiplicity (and p_T) intervals that the data sample is divided into, the yield ratios are determined through a likelihood fit on the $J/\psi\pi^+\pi^-$ invariant mass spectrum. The PDFs[1] that model the two signals are Crystal Ball functions, which have tail parameters constrained to values determined by the simulation. The background is modelled by an exponential function. Figure 2 shows examples of the fit for two multiplicity ranges.

The ratios of the acceptances and the efficiencies are found to be consistent with unity and with uncertainties of $\sim$ 1%, due to the similarities of the B_s^0 and B^0 decays. However, for the case of the reconstruction efficiencies, the ratio is significantly different from one, $\varepsilon_{B^0}^{reco}/\varepsilon_{B_s^0}^{reco} = 0.86 \pm 0.04$, for the full data sample. The reason for this is that, due to its larger mass, the $\pi^+\pi^-$ spectrum for the B_s^0 is dominated by the intermediate-mass resonances $f_0(980)$ and $f_0(1500)$ [25], while for the B^0 the main contribution is that of the lower-mass $\rho^0(770)$ [26], the reconstruction of which is less efficient.

2.3 Results

The B_s^0 to B^0 ratio was measured for the full integrated sample and found to be $\sigma_{B_s^0}/\sigma_{B^0} = 0.30 \pm 0.01 \pm 0.03$, being the first uncertainty from statistical and the second from systematic sources.

The ratio was studied for several normalised multiplicity intervals, defined for both N_{tracks}^{VELO} and N_{tracks}^{back}. These results are shown in Figure 3. The ratio consistently increases with N_{tracks}^{VELO}, which is not the case for N_{tracks}^{back}. Previous measurements from e^+e^- collisions at the $\Upsilon(5S)$ and Z^0 resonances [27] are represented and show good agreement with the data at low multiplicity.

In order to understand to a greater extent what is the mechanism responsible for the increase in the ratio with N_{tracks}^{VELO}, the data sample is divided into p_T intervals. Figure 4 shows the results for three p_T regimes. For the lowest p_T interval, $0 < p_T < 6$ GeV/c, where the B mesons have p_T smaller or similar to their mass, the ratio shows an increasing behaviour with N_{tracks}^{VELO}. If the data is fit to a linear model, the slope differs from zero by 3.4 standard deviations. At the higher p_T intervals, $6 < p_T < 12$ GeV/c and $12 < p_T < 20$ GeV/c, no such multiplicity dependence is found, and the results are consistent with the measurements from e^+e^- collisions.

[1] Probability Distribution Functions

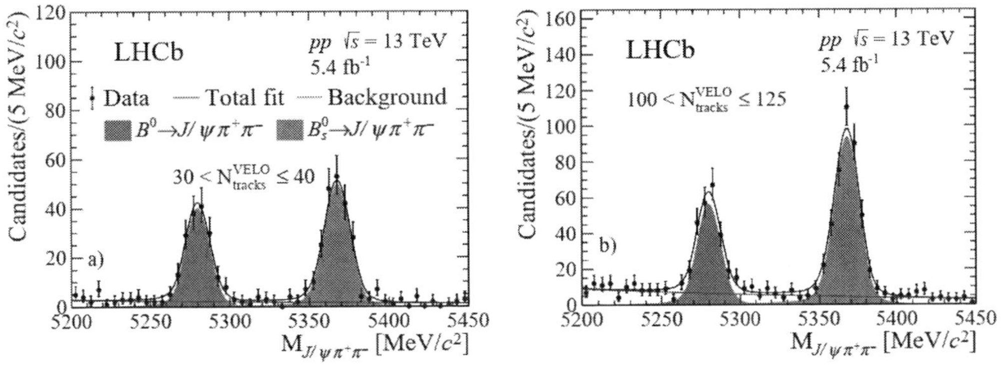

Figure 2. $J/\psi\pi^+\pi^-$ invariant mass spectrum and fit, with signal and background contributions, in the multiplicity ranges $30 < N_{\text{tracks}}^{\text{VELO}} \leq 40$ (left) and $100 < N_{\text{tracks}}^{\text{VELO}} \leq 125$ (right). From [8].

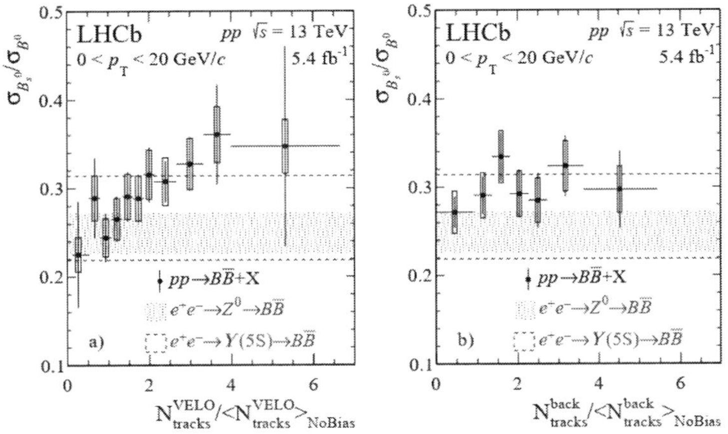

Figure 3. Ratio of cross sections $\sigma_{B_s^0}/\sigma_{B^0}$ measured against normalised multiplicity: in the left, all VELO tracks are used as multiplicity proxy and in the right, backward VELO tracks. The vertical error bars (boxes) represent point-to-point uncorrelated (fully correlated) uncertainties. The horizontal bands show the values measured in e^+e^- collisions. From [8].

3 Conclusion and Future Prospects

The cross-section ratio $\sigma_{B_s^0}/\sigma_{B^0}$ is found to be consistent with previous measurements from e^+e^- collisions for high p_T and for low p_T at relatively low multiplicity.

There is an enhancement of the B_s^0 meson production with respect to the B^0 production with global event multiplicity at low p_T, but not with backward multiplicity. The lack of dependence on backward multiplicity could indicate that the mechanism responsible for the increase in the ratio is related to the local particle density in a similar rapidity interval as the B mesons. These measurements are qualitatively consistent with the emergence of quark coalescence as an additional hadronisation mechanism: the increase of the ratio at high multiplicity only happens for B_s^0 mesons with low p_T and relatively low velocity, that have more overlap with the low-p_T bulk of the s quarks produced in the collision, creating $b-s$ pairs through coalescence; while at high-p_T they hadronise via fragmentation, as in e^+e^- collisions.

The LHCb Collaboration is progressing on a similar analysis for the ratio of D_s^+ to D^+ mesons with multiplicity in pPb collisions. Other related work includes the study of the production of Ξ_c^+ in pPb collisions and of light strange hadrons and their dependence with multiplicity for pp and pPb collisions.

https://doi.org/10.1051/epjconf/202329005005

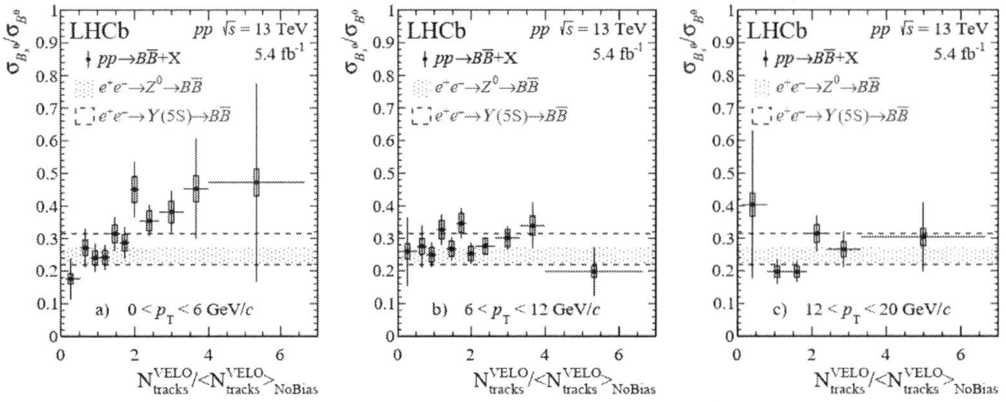

Figure 4. Ratio of cross sections $\sigma_{B_s^0}/\sigma_{B^0}$ measured against normalised N^{VELO}_{tracks}, for three p_T ranges: $0 < p_T < 6$ GeV/c (left), $6 < p_T < 12$ GeV/c (middle), and $12 < p_T < 20$ GeV/c (right). The vertical error bars (boxes) represent point-to-point uncorrelated (fully correlated) uncertainties. The horizontal bands show the values measured in e^+e^- collisions. From [8].

References

[1] K. Rajagopal, F. Wilczek, in *At The Frontier of Particle Physics* (World Scientific, 2001), pp. 2061–2151, μhttps://doi.org/10.1142%2F9789812810458_0043
[2] L.D. McLerran, Rev. Mod. Phys. **58**, 1021 (1986)
[3] D. Teaney, J. Lauret, E.V. Shuryak, Phys. Rev. Lett. **86**, 4783 (2001), nucl-th/0011058
[4] E.V. Shuryak, Nucl. Phys. A **750**, 64 (2005), hep-ph/0405066
[5] P. Koch, B. Muller, J. Rafelski, Phys. Rept. **142**, 167 (1986)
[6] M.L. Miller, K. Reygers, S.J. Sanders, P. Steinberg, Ann. Rev. Nucl. Part. Sci. **57**, 205 (2007), nucl-ex/0701025
[7] J. Adam et al. (ALICE Collaboration), Nature Physics **13**, 535 (2017)
[8] R. Aaij et al. (LHCb Collaboration), *Modification of b–quark hadronization in high-multiplicity pp collisions at $\sqrt{s}$ = 13 TeV* (2022), submitted to PRL, 2204.13042
[9] E. Norrbin, T. Sjostrand, Eur. Phys. J. **C17**, 137 (2000), hep-ph/0005110
[10] R. Aaij et al. (LHCb Collaboration), Phys. Rev. Lett. **118**, 052002 (2017), 1612.05140
[11] R. Aaij et al. (LHCb Collaboration), JHEP **12**, 026 (2017), 1710.04921
[12] S. Acharya et al. (ALICE), JHEP **05**, 220 (2021), 2102.13601
[13] B. Andersson, G. Gustafson, G. Ingelman, T. Sjostrand, Phys. Rept. **97**, 31 (1983)
[14] B.R. Webber, Nucl. Phys. **B238**, 492 (1984)
[15] R.J. Fries, B. Muller, C. Nonaka, S.A. Bass, Phys. Rev. Lett. **90**, 202303 (2003), nucl-th/0301087
[16] A.A. Alves, Jr. et al. (LHCb Collaboration), JINST **3**, S08005 (2008)
[17] T. Sjöstrand, S. Mrenna, P. Skands, Comput. Phys. Commun. **178**, 852 (2008), 0710.3820
[18] T. Sjöstrand, S. Mrenna, P. Skands, JHEP **05**, 026 (2006), hep-ph/0603175
[19] I. Belyaev et al., J. Phys. Conf. Ser. **331**, 032047 (2011)
[20] D.J. Lange, Nucl. Instrum. Meth. **A462**, 152 (2001)
[21] J. Allison, K. Amako, J. Apostolakis, H. Araujo, P. Sjostrand:2006zaDubois et al. (Geant4 collaboration), IEEE Trans.Nucl.Sci. **53**, 270 (2006)
[22] S. Agostinelli et al. (Geant4 collaboration), Nucl. Instrum. Meth. **A506**, 250 (2003)
[23] M. Clemencic et al., J. Phys. Conf. Ser. **331**, 032023 (2011)
[24] M. Pivk, F.R. Le Diberder, Nucl. Instrum. Meth. **A555**, 356 (2005), physics/0402083
[25] R. Aaij et al. (LHCb Collaboration), Phys. Rev. D **89**, 092006 (2014)
[26] R. Aaij et al. (LHCb Collaboration), Phys. Rev. D **90**, 012003 (2014)
[27] Y.S. Amhis et al. (HFLAV), Eur. Phys. J. **C81**, 226 (2021), 1909.12524

Constraining Neutron-Star Matter — Combination of heavy-ion experiments, astronomy and theory.

Arnaud Le Fèvre[8*], *Sabrina* Huth[1,2,13], *Peter T. H.* Pang[3,4,13], *Ingo* Tews[5], *Tim* Dietrich[6,7], *Achim* Schwenk[1,2,9], *Wolfgang* Trautmann[8], *Kshitij* Agarwal[10], *Mattia* Bulla[11], *Michael W.* Coughlin[12] and *Chris* Van Den Broeck[3,4]

[1]Department of Physics, Technische Universität Darmstadt, Darmstadt, Germany.
[2]ExtreMe Matter Institute EMMI, GSI Helmholtzzentrum für Schwerionenforschung GmbH, Darmstadt, Germany.
[3]Nikhef, Amsterdam, The Netherlands.
[4]Institute for Gravitational and Subatomic Physics (GRASP), Utrecht University, Utrecht, The Netherlands.
[5]Theoretical Division, Los Alamos National Laboratory, Los Alamos, NM, USA.
[6]Institut für Physik und Astronomie, Universität Potsdam, Potsdam, Germany.
[7]Max Planck Institute for Gravitational Physics (Albert Einstein Institute), Potsdam, Germany.
[8]GSI Helmholtzzentrum für Schwerionenforschung GmbH, Darmstadt, Germany.
[9]Max-Planck-Institut für Kernphysik, Heidelberg, Germany.
[10]Physikalisches Institut, Eberhard Karls Universität Tübingen, Tübingen, Germany.
[11]The Oskar Klein Centre, Department of Astronomy, Stockholm University, AlbaNova, Stockholm, Sweden.
[12]School of Physics and Astronomy, University of Minnesota, Minneapolis, MN, USA.

Abstract. Interpreting high-energy, astrophysical phenomena, such as supernova explosions or neutron-star collisions, requires a robust understanding of matter at supranuclear densities. However, our knowledge about dense matter explored in the cores of neutron stars remains limited. Fortunately, dense matter is not probed only in astrophysical observations, but also in terrestrial heavy-ion collision experiments. In a novel approach, using Bayesian inference, we combine data from astrophysical multi-messenger observations of neutron stars [1-9] and from heavy-ion collisions at relativistic energies [10,11] with microscopic nuclear theory calculations [12-17] to improve our understanding of dense matter. We find that the inclusion of heavy-ion collision data indicates an increase in the pressure in dense matter relative to previous analyses, shifting neutron-star radii towards larger values, consistent with recent observations by the Neutron Star Interior Composition Explorer mission [5-8,18]. It is found in addition that constraints from heavy-ion collision experiments show a remarkable consistency with multi-messenger observations and provide complementary information on nuclear matter at intermediate densities. This work combines nuclear theory, nuclear experiment and astrophysical observations, and shows how joint analyses can shed light on the properties of neutron-rich supranuclear matter over the density range probed in neutron stars. (Work primarily published in Ref. [44]: S. Huth, et al., Nature 606 (2022)).

1 Introduction

The nuclear equation of state (EOS) describes dense matter probed in terrestrial experiments with atomic nuclei as well as in astrophysical observations of neutron stars. The nuclear EOS has to be determined through approximate theoretical calculations or from experimental or observational data. As a result, at densities well above nuclear saturation density, $n_{sat} = 0.16\ fm^{-3}$, for which experimental and theoretical information are less robust, the nuclear EOS is still highly uncertain and many open questions remain, such as whether a possible phase transition to exotic phases of matter exists in nature [19].

At densities below 1–2n_{sat}, the EOS and its theoretical uncertainty can be obtained from microscopic calculations based on chiral effective field theory (EFT) of QCD [12–17]. To probe dense matter beyond these densities, further approaches, based on experimental and observational data, are necessary. A very promising tool is the multi-messenger astrophysics analysis of neutron stars and their collisions, which provides access to dense neutron-rich matter not accessible in terrestrial experiments at present. In recent years, the advent of gravitational-wave (GW) astronomy [1] and new electromagnetic observations of neutron stars [3,5,6], including the Neutron Star Interior Composition Explorer (NICER) mission of the National Aeronautics and Space Administration (NASA) [5,6], led to new constraints on the EOS [7,9,18,20–26]. However, these observations mainly probe the EOS at densities $\gtrsim 2n_{sat}$ and still carry considerable uncertainties, reflected in the ranges for predictions of neutron-star radii. More precise or new complementary information is required to reduce the uncertainties further.

The gap between our current knowledge of the EOS stemming from nuclear theory and experiment at low densities and astrophysical observations of neutron stars at higher densities can be bridged by heavy-ion collision (HIC) experiments. These experiments, performed with heavy-ion beam energies of up to 2 GeV per nucleon, probe the nuclear EOS mainly in a

* Corresponding author: a.lefevre@gsi.de

density range of 1–2n_{sat} at present [10,11,27], representing a new source of information [28].

2 Method.

In this work, within a Bayesian multimessenger framework [9,39], which analyses each EOS with respect to its agreement with a variety of observational data, we have combined information from nuclear theory (Fig. 1a), astrophysical observations of neutron stars (Fig. 1b) and results from HIC experiments (Four-Pi (FOPI) [10] and the Asymmetric-Matter EOS (ASY-EOS) experimental campaigns [11]) that were performed at the GSI Helmholtz Centre for Heavy Ion Research (Fig. 1c). The combination of these experiments provides new constraints for neutron-rich matter in the range around 1–2n_{sat}. The EOS constraint from ref. 27 for symmetric nuclear matter obtained from HIC experiments at the Bevalac accelerator at Lawrence Berkeley National Laboratory and the Alternating Gradient Synchrotron at Brookhaven National Laboratory has been included as well. In all experiments, gold nuclei were collided. The information from this series of HIC experiments allows us to further constrain the EOS in a density range for which theoretical calculations become less reliable.

3 Nuclear theory input

15,000 EOS's have been sampled for the local chiral effective field theory (EFT) interactions [14,29-31] to build the prior set of the Bayesian analysis. The EFT expansion enables estimates of theoretical uncertainties [16,32] needed by the Bayes method. The local chiral two- and three-nucleon interaction Monte-Carlo sampling of the chiral-EFT are among the most precise many-body methods to solve the nuclear many-body problem [33]. The breakdown scale of the expansion was estimated to be about 500–600 MeV/c [16]. Therefore, the EOS set using chiral EFT input has been limited only up to 1.5n_{sat} (corresponding to Fermi momenta of the order of 400 MeV/c), but a variation in the range 1–2n_{sat} has shown no substantial impact on our final results for neutron-star radii [34]. Above the 1.5n_{sat} limit, each EOS is extrapolated according to the speed of sound (c_s) in neutron-star matter [35], constrained only by causality ($c_s \leq c$) and stability of neutron-star matter ($cs \geq 0$). No information at asymptotically high densities from perturbative QCD calculations is taken into account. In addition, selected EOSs support neutron stars with masses of at least 1.9 solar masses (1.9$M_{\odot}$). EOSs that support only neutron stars with maximum masses well below the lower limit from the combined observations of heavy pulsars [36–38] are removed from the sample.

These general assumptions lead to a broad uncertainty for the EOS at higher densities (Fig. 1a), as well as for neutron-star masses and radii (Fig. 2a of Ref. [44]). The EOS prior is then used to analyse astrophysical observations and HIC experiments.

4 Multi-messenger astrophysics information

Various constraints from multi-messenger astronomy observations have been applied to the Bayesian analysis, with following sequence of constraints: (1) The lower and upper bounds of the maximum mass of neutron stars, respectively constrained by measurements of massive neutron stars (pulsars) PSR J0348+0432 [36] and PSR J1614-2230 [37] on the one side, and from the observation of the binary neutron-star collision GW170817 [40,41] on the other side. (2) Constraints from X-ray pulse-profile modelling of PSR J0030+0451 and PSR J0740+6620 using data from NICER and the X-ray Multi-Mirror Mission (XMM-Newton) [5,7,8]. (3) GW information from the two neutron-star mergers GW170817 [1] and GW190425 [2] by matching the observed GW data with theoretical GW models that depend on neutron-star properties [42], using an improved version of the main waveform model used by the Laser Interferometer Gravitational-Wave Observatory/Virgo Collaboration for the study of GW170817 [43] and GW190425 [2]. (4) Information from the kilonova AT2017gfo [3] associated with the GW signal. Fig. 1b shows that the above astrophysical informations leads to important constraints on the neutron-star EOS. The effect is strongest above 1.5n_{sat}, where the extrapolation in the speed of sound is used for the EOSs of the prior. The high-density astrophysical constraints affect mostly the high-mass region in the mass–radius plane and exclude the stiffest EOSs that lead to the largest radii (Fig. 2b of Ref. [44]).

5 Data from HIC experiments

As a last layer of constraints in the Bayesian analysis, the EOS constraint from HIC experiments has been applied. The FOPI [10] and ASY-EOS [11] experiments performed at GSI provide information respectively on the symmetric nuclear matter EOS (that is, matter with the same amount of protons and neutrons) and on the symmetry energy, which describes the energy cost of changing protons into neutrons in nuclear matter. For both experiments,

[197]Au nuclei were collided at relativistic energies (0.4 to 1.5 GeV per nucleon), forming an expanding fireball in the collision region. This expansion is dictated by the achieved compression and therefore depends on the EOS of hot and dense matter. Owing to the initial neutron-to-proton asymmetry of the Au–Au system, the expansion of the emitted nucleons is sensitive to the nuclear symmetry energy. Constraints on the symmetry energy (from ASY-EOS) can be translated into a constraint on the pressure of neutron-star matter as a function of the baryon density when empirical information on symmetric nuclear matter from experiments (FOPI) with atomic nuclei is used. Whereas the FOPI experiment delivers an EOS constraint for symmetric nuclear matter at densities in the range 1–3n_{sat}, the ASY-EOS experiment probes the symmetry energy roughly between 1 and 2n_{sat}. Additional constraints on the pressure of symmetric nuclear matter have been applied at larger densities, obtained from experimental data from Lawrence Berkeley National Laboratory and Brookhaven National Laboratory [27] in which [197]Au nuclei were collided at energies up to 10 GeV per nucleon. These are sensitive to higher densities, 2–4.5n_{sat}, but we include their constraints only up to 3n_{sat}, where the sensitivity of the ASY-EOS experiment ends. We find that the inclusion of this further constraint has only minimal impact. Fig. 1c shows that the combined HIC experimental constraints for the neutron-star EOS tend to prefer EOSs stiffer than the ones favored by astrophysical observations (that is, EOSs that have higher pressures at densities up to 2n_{sat}; Figs. 1c, 2a).

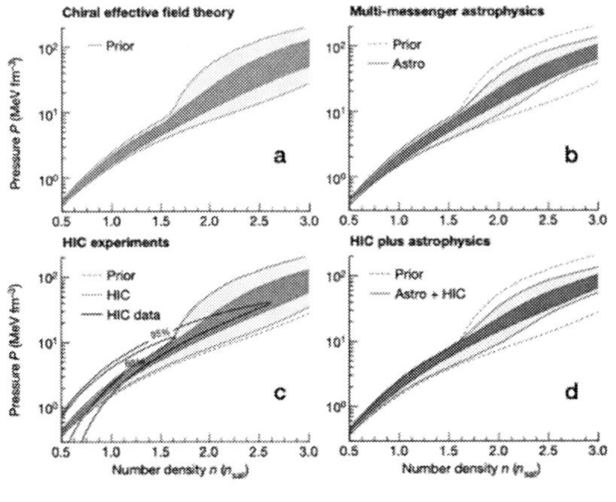

Fig. 1. Constraints on the EOS of neutron-star matter. a–d, Evolution of the pressure as a function of baryon number density for the EOS prior (a, grey), when including only data from multi-messenger neutron-star observations (b, green), when including only HIC data (c, orange), and when combining both (d, blue). The shading corresponds to the 95% and 68% credible intervals (lightest to darkest). The impact of the HIC experimental constraint (HIC data, purple lines at 95% and 68%) on the EOS is shown in c. In b–d, the 95% prior bound is shown for comparison (grey dashed lines).

6 Combining microscopic and macroscopic collisions

The final EOS constraints are obtained through the combination of both the HIC information and astrophysical multi-messenger observations (Fig. 1d). The multi-messenger data rule out the most extreme EOS behavior, and the HIC data favor larger pressures around 1–1.5n_{sat}, the density range where the HIC experimental sensitivity is highest. This is similar to the effect of recent NICER observations on the EOS [7,18]. Hence, the two complementary approaches, HIC experiments and astrophysical observations, show a remarkable agreement (Fig. 2a). At low densities, HIC results have a clear impact on the total posterior for the EOS, whereas the EOS at higher densities ($\gtrsim 2n_{sat}$) is mostly determined by astrophysical observations. At these densities, HIC results deviate only mildly from the prior (Fig. 2b). This is also reflected in the radii of neutron stars shown in Fig. 2e, f of Ref. [44]. As astrophysical observations mainly probe neutron stars with $M \gtrsim 1.4 M_\odot$, for which the relevant densities are higher, HIC information influences the radii of these neutron stars to a smaller degree. On the other hand, the radius of low-mass stars with $M \approx 1.0\ M_\odot$ is also constrained by HIC information. Our final result for a typical $1.4 M_\odot$ neutron star is $12.01^{+0.37}_{-0.38}$ km at 68% uncertainty ($12.01^{+0.78}_{-0.77}$ km at 95% uncertainty). Comparing this value to the result without any HIC information, $11.93^{+0.39}_{-0.41}$ km at 68% confidence, highlights the benefit of combining these various sources of information in a statistically robust framework.

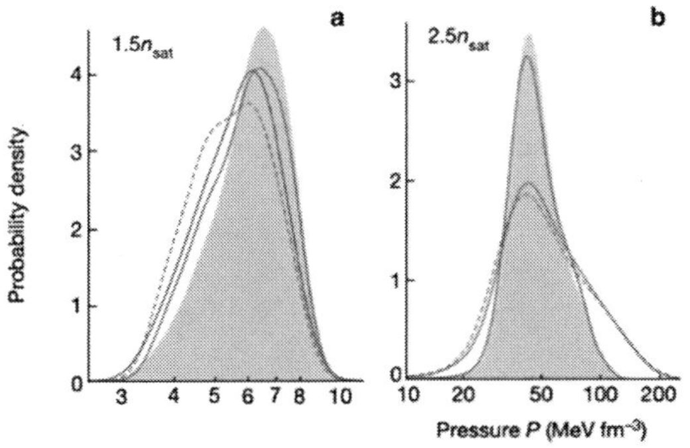

Fig. 2. Posterior distributions for the pressure at 1.5n_{sat} (a) and 2.5n_{sat} (b) at different stages of our analysis, with the combined Astro + HIC region shaded in light blue.

7 Perspectives.

Advancing HIC experiments to probe higher densities, above 2–3n_{sat}, will be key. Combining the latter with a reduction of experimental uncertainties, data from HICs have great potential to provide complementary EOS information, bridging nuclear theory and astrophysical observations. In the next few years, the ASY-EOS-II and Compressed Baryonic Matter experiments at the upcoming Facility for Antiproton and Ion Research at GSI will provide a unique opportunity to study nuclear matter at densities probed in the core of neutron stars and their mergers, and might detect new phases of QCD matter, possibly involving hyperons and, ultimately, the transition to a deconfined quark matter phase at the highest densities. Together with experiments at the Rare Isotope Beam Facility at RIKEN in Japan and the Nuclotron-Based Ion Collider Facility in Russia, the robust combination of experimental

HIC constraints and astrophysical observations has the potential to revolutionize our understanding of the EOS.

References

1. B.P. Abbott, et al., Phys. Rev. Lett. **119**, 161101 (2017).
2. B. Abbott, et al., J. Lett. **892**, L3 (2020).
3. B.P. Abbott, et al., Astrophys. J. **848**, L13 (2017).
4. M. Coughlin, et al., Astrophys. J. **849**, 12 (2017).
5. M. C. Miller et al., Astrophys. J. Lett. **887**, L24 (2019).
6. T.E. Riley, et al., Astrophys. J. Lett. **887**, L21 (2019).
7. M.C. Miller, et al., Astrophys. J. Lett. **918**, L28 (2021).
8. T.E. Riley, et al., Astrophys. J. Lett. **918**, L27 (2021).
9. T. Dietrich, et al., Science **370**, 1450–1453 (2020).
10. A. Le Fèvre, Y. Leifels, W. Reisdorf, J. Aichelin and C. Hartnack, Nucl. Phys. A **945**, 112–133 (2016).
11. P. Russotto, et al., Phys. Rev. C **94**, 034608 (2016).
12. K. Hebeler, J.M. Lattimer, C.J. Pethick and A. Schwenk, Astrophys. J. **773**, 11 (2013).
13. I. Tews, T. Krüger, K. Hebeler, and A. Schwenk, Phys. Rev. Lett. **110**, 032504 (2013).
14. J.E. Lynn, et al., Phys. Rev. Lett. **116**, 062501 (2016).
15. C. Drischler, K. Hebeler, and A. Schwenk, Phys. Rev. Lett. **122**, 042501 (2019).
16. C. Drischler, R.J. Furnstahl, J.A. Melendez, and D.R. Phillips, Phys. Rev. Lett. **125**, 202702 (2020).
17. S. Huth, C. Wellenhofer an A. Schwenk, Phys. Rev. C **103**, 025803 (2021).
18. G. Raaijmakers, et al., Astrophys. J. Lett. **918**, L29 (2021).
19. E. Annala, T. Gorda, A. Kurkela, J. Nättilä and A. Vuorinen, Nat. Phys. **16**, 907–910 (2020).
20. A. Bauswein, O. Just, H.-T. Janka and N. Stergioulas, Astrophys. *J.* **850**, L34 (2017).
21. E. Annala, T. Gorda, A. Kurkela and A. Vuorinen, Phys. Rev. Lett. **120**, 172703 (2018).
22. E.R. Most, L.R. Weih, L. Rezzolla and J. Schaffner-Bielich, Phys. Rev. Lett. **120**, 261103 (2018).
23. B.P. Abbott, et al., Phys. Rev. Lett. **121**, 161101 (2018).
24. D. Radice and L. Dai, Eur. Phys. J. **A55**, 50 (2019).
25. C.D. Capano, et al., Nat. Astron. **4**, 625–632 (2020).
26. I. Legred, K. Chatziioannou, R. Essick, S. Han and P. Landry, Phys. Rev. D **104**, 063003 (2021).
27. P. Danielewicz, R. Lacey, and W.G. Lynch, Science **298**, 1592–1596 (2002).
28. M.B. Tsang, W.G. Lynch, P. Danielewicz, and C.Y. Tsang, Phys. Lett. B **795**, 533–536 (2019).
29. I. Tews, J. Carlson, S. Gandolfi and S. Reddy, Astrophys. J. **860**, 149 (2018).
30. E. Epelbaum, H.-W. Hammer and U.-G. Meissner, Rev. Mod. Phys. **81**, 1773–1825 (2009).
31. R. Machleidt, and D.R. Entem, Phys. Rep. **503**, 1–75 (2011).
32. E. Epelbaum, H. Krebs and U.-G. Meißner, Eur. Phys. J. A **51**, 53 (2015).
33. J. Carlson, et al., Rev. Mod. Phys. **87**, 1067 (2015).
34. R. Essick, I. Tews, P. Landry, S. Reddy, and D.E. Holz, Phys. Rev. C **102**, 055803 (2020).
35. I. Tews, J. Margueron, and S. Reddy, Phys. Rev. **C98**, 045804 (2018).
36. J. Antoniadis, et al., Science **340**, 1233232 (2013).
37. Z. Arzoumanian, al., Astrophys. J. Suppl. Ser. **235**, 37 (2018).
38. E. Fonseca, et al., Astrophys. J. Lett. 915, L12 (2021).
39. P.T.H. Pang, et al., Astrophys. J. 922, 14 (2021).
40. B. Margalit, and B.D. Metzger, Astrophys. J. 850, L19 (2017).
41. L. Rezzolla, E.R. Most, and L.R. Weih, Astrophys. J. 852, L25 (2018).
42. T. Dietrich, et al., Phys. Rev. D 100, 044003 (2019).
43. B.P. Abbott, et al., Phys. Rev. X 9, 011001 (2019).
44. S. Huth, P.T.H. Pang, et al., Nature 606 (2022).

EPJ Web of Conferences **290**, 06001 (2023)
EuNPC 2022
https://doi.org/10.1051/epjconf/202329006001

Deeply Virtual Compton Scattering with CLAS12 at Jefferson Lab

Adam Hobart[1,*] for the CLAS collaboration

[1]Université Paris-Saclay, CNRS/IN2P3, IJCLab, 91405 Orsay, France

Abstract. A key step toward a better understanding of the nucleon structure is the study of Generalized Parton Distributions (GPDs). GPDs are the object of an intense effort of research since they convey an image of the nucleon structure where the longitudinal momentum and the transverse spatial position of the partons inside the nucleon are correlated. Moreover, GPDs give access, via Ji's sum rule, to the contribution of the orbital angular momentum of the quarks to the nucleon spin, which is important to the understanding of the origins of the nucleon spin. Deeply Virtual Compton scattering (DVCS), the electroproduction of a real photon off the nucleon at the quark level, is the golden process directly interpretable in terms of GPDs of the nucleon. The GPDs are accessed in DVCS mainly through the measurements of single- or double- spin asymmetries. Combining measurements of asymmetries from DVCS experiments on both the neutron and the proton will allow us to perform the flavor separation of the u and d quarks GPDs via linear combinations of proton and neutron GPDs. This paper introduces recent DVCS measurements from the CLAS12 experiment at Jefferson Lab with the upgraded 11 GeV polarized electron beam. Details on the data analysis along with results on Beam Spin Asymmetries are presented.

1 Introduction

Generalized Parton Distributions (GPDs) [1–8] are structure functions that factorize the electroproduction of a photon or a meson on the nucleon. They correspond to the Fourier transform of the QCD non-local and non-diagonal operators and can provide key information in understanding the nucleon structure. The GPDs describe the correlations between the longitudinal momentum and transverse spatial position of the partons inside the nucleon [1, 4, 7, 8], and they give access to the contribution of the orbital momentum of the quarks to the nucleon spin [2, 3]. The nucleon GPDs are accessed in experiments of exclusive lepto-production of a photon (e.g. Deeply Virtual Compton Scattering - DVCS process), or of a meson (e.g. Deeply Virtual Meson Production) off the nucleon at sufficiently large Q^2, where Q^2 is the virtuality of the exchanged photon. This is defined as $-(k-k')^2$, where k and k' are the momenta of the initial and final electrons respectively.

Considering only helicity-conserving quantities and the quark sector, there are four GPDs, $H, \tilde{H}, E, \tilde{E}$, which depend, in leading-order and leading-twist QCD, upon three variables: x, ξ and t (see [1] for more details). The quantities $x-\xi$ and $x+\xi$ are the longitudinal momentum fractions of the quarks coming out and going back into the nucleon, respectively and t is the squared four-momentum transfer between the final and initial nucleon and can equally be defined as $(q-q')^2$, where q and q' are the momenta of the virtual and final real photons respectively. Figure 1 illustrates the leading-order Feynman diagram for DVCS and the definitions of the relevant kinematic variables mainly x_B defined as $x_B = Q^2/2qp$ with p being the initial momentum of the struck nucleon.

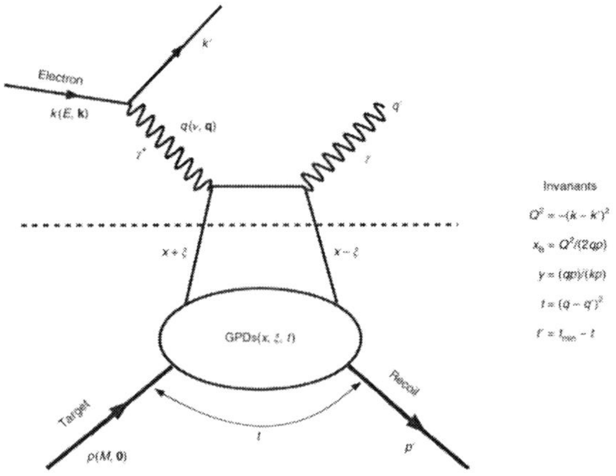

Figure 1. The handbag diagram for DVCS. Kinematical variables are defined in the figure and in the text.

The DVCS process, which naturally interferes with the Bethe-Heitler (BH) process, where the final state photon is emitted by the lepton, allows access to combinations of GPD-related quantities called the Compton Form Factors $CFF(\xi, t)$. In an experimental configuration, where the beam is polarized and the target is unpolarized, one of the accessible observables is the Beam Spin Asymmetry

*e-mail: ajhobart@jlab.org

(BSA), which relates to the *CFF* as

$$A_{LU} \propto \sin(\Phi)\Im(F_1\mathcal{H}+\xi(F_1+F_2)\mathcal{H}+-kF_2\mathcal{E}+...)\,[9], \quad (1)$$

where Φ is the angle between the lepton scattering and photon production planes, F_1 and F_2 are the Dirac and Pauli form factors, k defined as $k = -t/4M^2$ with M being the nucleon's mass, and $\mathcal{H}$ and $\mathcal{E}$ are the *CFF* related to the GPDs H and E, respectively. The BSA observable will be mainly sensitive to the *CFF* $\mathcal{H}$ of the proton, if the target is a proton, and to the *CFF* $\mathcal{E}$ of the neutron, if the target is a neutron. This is mainly due to the different contributions that arise from the Dirac and Pauli form factors for each nucleon type. Therefore, in order to separate the GPDs of a given nucleon, all the possible experimental configurations need to be considered for the DVCS process with the given nucleon as a target. The quark-flavor separation of the GPDs becomes possible when experimental measurements are performed on both nucleons.

In this paper, we present preliminary results and analysis of the BSA of DVCS from a deuterium target (neutron-DVCS: nDVCS, $ed \rightarrow e'n'\gamma(p)$ and proton-DVCS: pDVCS, $ed \rightarrow e'p'\gamma(n)$; initial electron: e, scattered electron: e', deuterium: d, scattered neutron: n', spectator proton: p, scattered proton: p', spectator neutron: n, photon: γ) using the CLAS12 detector [10] at Jefferson Lab. The GPD E is one of the least known GPDs, and only one previous measurement exists from pioneering experiments performed in Hall-A at Jefferson Lab [11]. However, the experiment reported here is the first measurement with tagging of the active neutron and hence, allowing for a precise determination of the BSA. E is one of the two GPDs that enter into the Ji's sum rule [2, 3], which links the total angular momentum (J_q) carried by each quark q to the sum of the GPDs H and E. The measurement of the GPD H of the bound proton is a first measurement of its kind and is of importance in quantifying nuclear medium effects on GPDs.

2 Experiment and Event Selection

The experiment has taken place in the Hall-B of JLab utilizing the large acceptance spectrometer CLAS12 [10]. The CEBAF accelerator produces a high-intensity, high-duty-factor, polarized electron beam. An average beam polarization of ~85% was measured throughout the experiment in dedicated Møeller scattering measurement runs. The target was an unpolarized liquid deuterium cell measuring 5 cm long. The experiment ran during three different periods, collecting roughly 40×10^9 triggers. A quarter of the data were taken with a 10.6 GeV beam energy, another quarter at 10.2 GeV energy and the other half at 10.4 GeV energy.

To guide the data analysis, in particular for the determination of the process selection, a Monte-Carlo simulation was used. An event generator for incoherent electroproduction of photons on deuterium was adopted, which produces either $ed \rightarrow e'n'\gamma(p)$ or $ed \rightarrow e'p'\gamma(n)$ events, proportionally to their relative cross-sections, coming from the nDVCS/pDVCS and BH reactions [12]. The

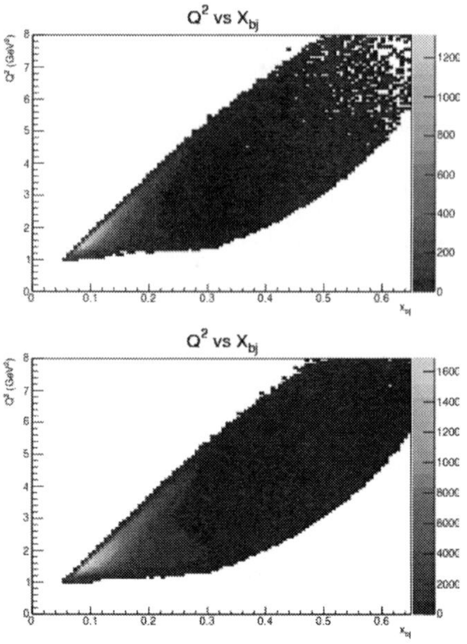

Figure 2. Preliminary. Q^2 versus x_B for the nDVCS data sample (top) and pDVCS data sample (bottom) with all selection cuts applied. The large kinematic reach of CLAS12 is clearly demonstrated in this distribution.

DVCS amplitude is calculated according to the BMK formalism [13]. The Fermi-motion distribution is implemented in the simulations using the Paris potential [14].

Events with at least one identified electron, one photon with energy above 2 GeV, and one nucleon (proton for pDVCS and neutron for nDVCS) with momentum above 0.35 GeV/c were selected. The particle identification of CLAS12 has been used to identify all particles within each event [15]. In the case where multiple final-state particles of the same type are present in the event, all possible combinations of the exclusive final state were considered for treatment. The best combination was taken to be the one minimizing a multi-dimensional χ^2-like quantity calculated using variables related to the exclusive final state.

Several cuts were applied in order to select the relevant kinematic region for the DVCS reactions. The electron momentum was selected to be above 1 GeV/c. We selected nucleon momentum to be above 0.35 GeV/c in order to remove spectator-nucleon events. A cut on the cone angle formed by the electron and the neutron ($\theta_{e,n} < 5°$) was applied to remove radiative photons reconstructed as neutrons. A cut on the cone angle formed by the electron and the photon ($\theta_{e,\gamma} < 5°$) was applied to remove radiative photons that were likely emitted close to the electron. We cut on $Q^2 > 1$ GeV^2/c^2 and $W > 2$ GeV/c to ensure the applicability of the leading-twist GPD formalism and remove contributions from the resonance region. We also require $E_\gamma > 2$ GeV as photons from a DVCS process are expected at high energies. Moreover, fiducial cuts were applied to remove regions where the detector acceptance varies rapidly.

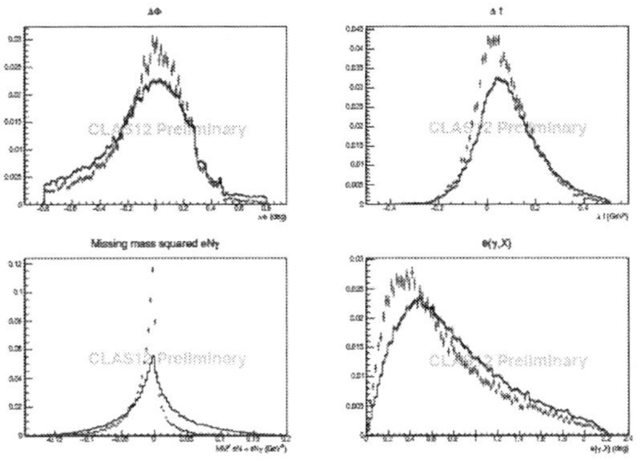

Figure 3. Preliminary. nDVCS data. Top left: $\Delta\Phi$. Top right:Δt. Bottom left: the missing mass in the $en \to enX$ reaction. Bottom right: the cone angle between the detected and the reconstructed photon $\theta_{\gamma,X}$. Data (in blue) contains background from partially reconstructed π^0 decays. Simulations (in red) of DVCS/BH on deuterium containing both the neutron and proton DVCS/BH channels.

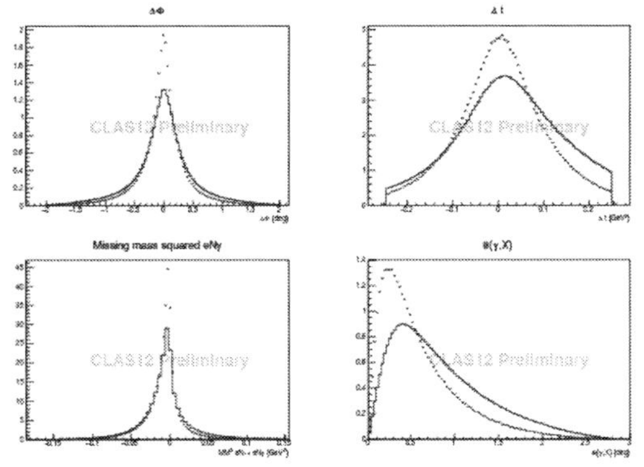

Figure 4. Preliminary. pDVCS data. Top left: $\Delta\Phi$. Top right:Δt. Bottom left: the missing mass in the $en \to enX$ reaction. Bottom right: the cone angle between the detected and the reconstructed photon $\theta_{\gamma,X}$. Data (in blue) contains background from partially reconstructed π^0 decays. Simulations (in red) of DVCS/BH on deuterium containing both the neutron and proton DVCS/BH channels.

Exclusivity cuts were applied to select the $e'n'\gamma(p)$ and $e'p'\gamma(n)$ final states while minimizing the background. For a nucleon N, cuts on the missing mass in the $ed \to e'N'\gamma X$, $eN \to e'N'\gamma X$ and $eN \to e'N'X$ reactions and the missing momentum in the $ed \to e'N'\gamma X$ reaction were considered. Cuts were also applied on the difference between the angle Φ between the lepton scattering and photon production planes, computed using the nucleon and the virtual photon and using the virtual and the real photon ($\Delta\Phi$). Similarly, a cut was applied on the difference between two ways to compute t, either using the scattered nucleon or using the virtual and real photon (Δt). Finally, a cut on the cone angle between the detected and the reconstructed photon from the missing particle selection, $\theta_{\gamma,X}$ was applied to reduce the contamination of photons from the partially reconstructed π^0 decay.

Due to inefficiencies in the tracking system of the CLAS12 detector and to the existence of dead regions in the tracker acceptance, some reconstructed neutrons were actually misidentified protons that were cut out using a multivariate analysis technique (Boosted Decision Trees [16]) and utilizing final state exclusivity variables.

Figure 2 shows the two dimensional plane in Q^2 and $x_B = \dfrac{Q^2}{2M\nu}$ after all selection cuts are applied. The distribution clearly shows the large kinematic reach that the large acceptance CLAS12 detector can provide for the measurement of asymmetry observables in the DVCS processes.

Figures 3 and 4 shows some exclusivity variables for the selected $e'n'\gamma$ and $e'p'\gamma$ events in data (in blue) compared to simulations (in red). In order to understand the discrepancies between data and simulations, it is important to know that the data sample contains the background originating from the partially reconstructed π^0 decay produced in the process $ed \to e'n'\gamma(\gamma(\pi^0), p)$ and $ed \to e'p'\gamma(\gamma(\pi^0), n)$ where only one of the two photons from the π^0 decay is reconstructed and considered as that of the DVCS process.

The procedure adopted in this analysis to compute the π^0 contamination to the DVCS events sample was done separately for each bin in Q^2, x_B, t and ϕ. The procedure consists of estimating the ratio, in simulations, of partially reconstructed $e'N'\pi^0$(1 photon) decay, selected as if it was true DVCS process, to fully reconstructed, with dedicated exclusive selection, $e'N'\pi^0$ decays. The fact that the ratio is evaluated separately in each bin is beneficial since acceptance and simulation-data discrepancies cancel in the ratio. This ratio, which is obtained from simulations, is expected to be equal to that from data. It is hence applied in data by multiplying it with the number of reconstructed $e'N'\pi^0$ events and therefore evaluating the number of $e'N'\pi^0$(1 photon) in data. The obtained number of $e'N'\pi^0$(1 photon) is subtracted from DVCS reconstructed decays in data for each kinematics bin, and this contamination is evaluated to range between 10-45% depending on the bin under consideration.

3 Extraction of the Beam Spin Asymmetry Observable

The BSA, A_{LU}, is defined as $A_{LU} = \dfrac{1}{P}\dfrac{N^+ - N^-}{N^+ + N^-}$ where P is the average beam polarisation and $N^{+(-)}$ stands for the number of DVCS events for positive (negative) beam helicity. Figures 5 and 6 shows the distribution of A_{LU} for nDVCS and pDVCS respectively, integrated over the measured kinematic domain.

Given the available statistics for nDVCS, we performed an extraction of the BSA in two-dimensional bins in Q^2, x_B or t versus Φ. We considered three bins in Q^2 [1, 1.9], [1.9, 2.9], and [2.9, 6] GeV^2, three bins in X_B [0.05, 0.14], [0.14, 0.2], and [0.2, 0.6] and three bins in $-t$ [0, 0.3], [0.3, 0.5], and [0.5, 1.1] GeV^2. For pDVCS,

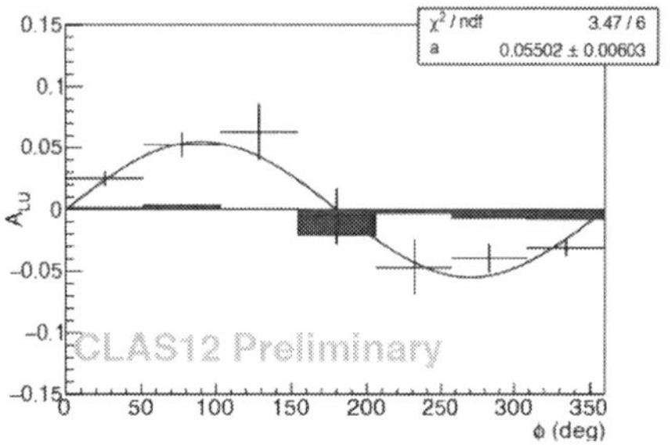

Figure 5. Preliminary. nDVCS BSA integrated over the measured kinematic domain. The distribution is modeled with the function (a sin(Φ)). The red histogram demonstrates the total systematic uncertainty.

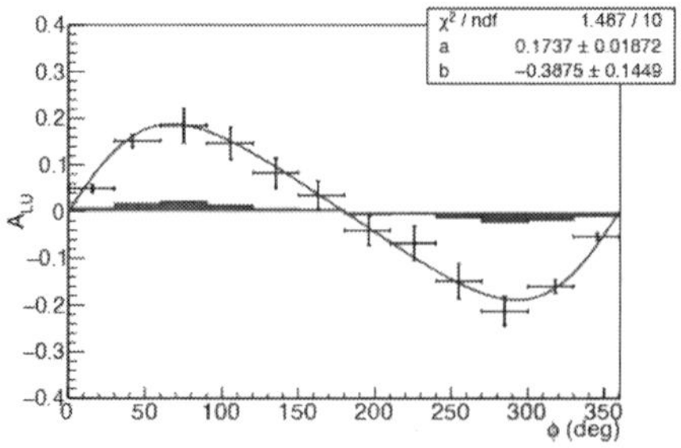

Figure 8. Preliminary. pDVCS BSA of bin in $X_B \in [0.2, 0.6]$, $Q^2 \in [2.9,3.8]$ GeV^2 and $-t \in [0.54, 1.1]$ GeV^2. The distribution is modeled with the function (a sin(Φ)/(1 + b cos(Φ))). The red histogram demonstrates the total systematic uncertainty.

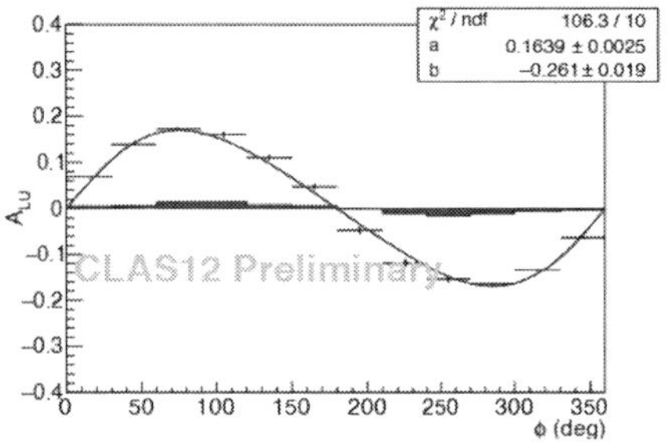

Figure 6. Preliminary. pDVCS BSA integrated over the measured kinematic domain. The distribution is modeled with the function (a sin(Φ)/(1 + b cos(Φ))). The red histogram demonstrates the total systematic uncertainty.

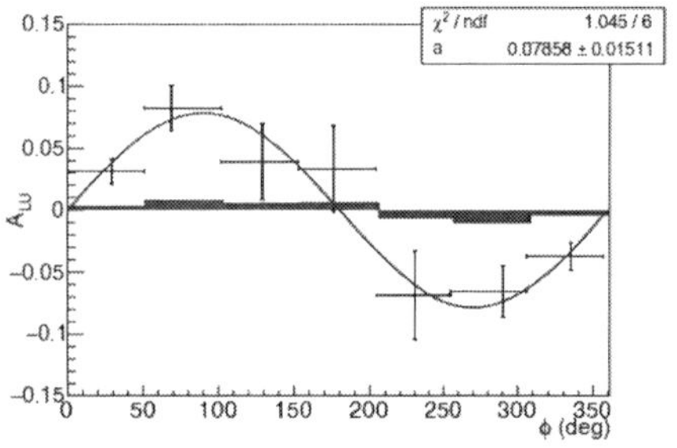

Figure 7. Preliminary. nDVCS BSA of bin in $X_B \in [0.14, 0.2]$. The distribution is modeled with the function (a sin(Φ)). The red histogram demonstrates the total systematic uncertainty.

we performed a 3 dimensional binning of the data in Q^2, x_B and t versus Φ. We considered 5 bins in Q^2 [1,1.6], [1.6,2], [2,2.9], [2.9,3.8] and [3.8,8] GeV^2, 4 bins in X_B [0.04,0.13], [0.13,0.16], [0.16,0.2] and [0.2,0.6] and 3 bins in $-t$ [0, 0.28], [0.28, 0.54], and [0.54, 1.1] GeV^2. Figure 7 shows the BSA of nDVCS in the kinematic bin defined by $X_B \in$[0.14, 0.2] and figure 8 show the BSA of pDVCS in the kinematic bin defined by $X_B \in [0.2, 0.6]$, $Q^2 \in [2.9,3.8]$ GeV^2 and $-t \in [0.54, 1.1]$ GeV^2. The fit function shown in the figures is approximated from physics considerations, however, no physics conclusion is intended from it. A low χ^2 per number of degrees of freedom should not be interpreted as overfitting of the data while a large value might hint to the necessity to include a cos(2ϕ) term in the denominator. Nevertheless, this large χ^2 per number of degrees of freedom is observed on the integrated BSA of pDVCS and can be an effect of integrating over the whole measured kinematic domain. The systematic uncertainties include the uncertainty related to proton contamination cut (for nDVCS BSA only), the uncertainty related to the applied exclusivity selection and the uncertainty on the beam polarization which subsumes the differences between the integrated luminosity for the beam polarization states. The uncertainty related to π^0 subtraction is still under evaluation. Radiative corrections are applied.

4 Conclusions

We have presented the analysis status of nDVCS and pDVCS experimental data obtained from polarized electron beam scattering off a deuterium target. We have observed a positive BSA for nDVCS. The observed BSA for pDVCS is twice as large as that of nDVCS. The integrated BSA of pDVCS is consistent with the integrated BSA observed for a free proton hinting for weak medium effects on the proton GPDs. A more thorough analysis in kinematics bins is to be performed in order to pin any underlying medium effects. The analysis exploits the full kinematic coverage of the CLAS12 detector. Therefore, we performed an extraction of the BSA observable in two-dimensional bins in Q^2, x_B or t versus Φ for the nDVCS BSA and in three-dimensional bins in Q^2, x_B and t versus Φ. Further results, including BSA of DVCS from free proton target can be consulted in [17].

References

[1] D. Müller, D. Robaschik, B. Geyer, F.-M. Dittes, and J. Horejsi. Fortschr. Phys. 42 (1994) 101.
[2] X. Ji, Phys. Rev. Lett. 78 (1997) 610.

https://doi.org/10.1051/epjconf/202329006001

[3] X. Ji, Phys. Rev. D 55 (1997) 7114.
[4] A.V. Radyushkin, Phys. Lett. B 380 (1996) 417; Phys. Rev. D 56 (1997) 5524.
[5] J.C. Collins, L. Frankfurt and M. Strikman, Phys. Rev. D 56 (1997) 2982.
[6] K. Goeke, M. V. Polyakov and M. Vanderhaeghen, Prog. Part. Nucl. Phys. 47 (2001) 401.
[7] M. Diehl, Phys. Rept. 388 (2003) 41.
[8] A.V. Belitsky, A.V. Radyushkin, Phys. Rept. 418 (2005) 1.
[9] M. Guidal *et al.* Rep. Prog. Phys. 76 (2013) 066202
[10] V.D.Burkert *et al.*, NIM Physics A. 959 (2020) 163419.
[11] M. Benali *et al.*, Nature Phys. 16 (2020) 191.
[12] A. El Aloui and E. Voutier, private communication.
[13] A.V. Belitsky, D. Müller, A. Kirchner, Nucl. Phys. B **629** (2002) 323-392.
[14] M. Lacombe *et al.*, Phys. Rev C **21** (1980) 861.
[15] V. Ziegler *et al.*, NIM Physics A. 959 (2020) 163472.
[16] Y. Coadou, Artificial Intelligence for High Energy Physics, pp. 9-58 (2022) arXiv 2206.09645
[17] G. Christiaens, M. Defurne, D. Sokhan V.Ziegler *et al.*, arXiv (2022) 221111274.

https://doi.org/10.1051/epjconf/202329006002

J/ψ Near-Threshold Photoproduction off the Proton and Neutron with CLAS12

R. Tyson[1,*], *B.* McKinnon[1], *D. G.* Ireland[1], and the CLAS Collaboration

[1]SUPA, School of Physics and Astronomy, University of Glasgow, Glasgow G12 8QQ, United Kingdom

Abstract. Near-threshold J/ψ photoproduction is a key aspect of the physics program at the Thomas Jefferson National Accelerator Facility (JLab) 12 GeV upgrade due to the wealth of information it has to offer. J/ψ photoproduction proceeds through the exchange of gluons in the t-channel and is expected to provide unique insight about the nucleon gravitational form factors and the nucleon mass radius. The JLab based CLAS Collaboration, which uses the CEBAF Large Acceptance Spectrometer (CLAS12), aims to measure the J/ψ near threshold photoproduction cross section using both a proton and a deuteron target, from threshold up to 10.6 GeV. The deuteron target further offers the possibility of comparing the proton and neutron gluonic form factors and mass radii in a first measurement of the cross sections off a proton or neutron within the deuteron target. The analysis towards these measurements is ongoing and well advanced, with machine learning based techniques for particle identification already designed and validated with CLAS12 data.

1 Introduction

J/ψ, a $c\bar{c}$ meson, has played an important role in modern day particle and nuclear physics since its discovery in 1974 [1, 2]. As such it has been extensively studied in collider experiments with centre of mass energies well beyond the threshold energy required to produce J/ψ. However, theoretical developments at the start of the century has prompted renewed interest in J/ψ photoproduction near the 8.2 GeV threshold, which is close to the energy at which the CEBAF Large Acceptance Spectrometer (CLAS12) operates.

Currently-available measurements of the J/ψ photoproduction cross section off the proton in the near threshold region (SLAC [3], Cornell [4], GlueX [5]) span the energy range from threshold up to about 20 GeV [5]. Several existing theoretical calculations of the exclusive J/ψ photoproduction cross section off the proton [6–8] aim to describe the J/ψ photoproduction production mechanism. In particular, [7] implies a large gluonic contribution to the nucleon mass and the predicted cross section is in a very good agreement with the GlueX results [5] that have a similar energy range to that covered by CLAS12 data. This model is very interesting as it relates to other models within the framework of which the real part of the J/ψ-nucleon scattering amplitude contains the anomalous trace of the energy-momentum tensor [9]. These models are based on the assumption of a two gluon exchange as being the dominant production mechanism. Further evidence that the two gluon exchange is dominant over other production mechanisms comes from the fact that a two gluon form factor is shown to adequately describe the J/ψ differential cross section as a function of t [10]. The relation between the J/ψ photoproduction differential cross section and the trace anomaly of Quantum Chromodynamics (QCD) was also studied by means of a holographic QCD approach [11]. Furthermore, estimates of the b parameter that quantifies the magnitude of the contribution to the nucleon mass due to the trace anomaly can be extracted from VMD based models of J/ψ photoproduction [12, 13].

*e-mail: r.tyson.1@research.gla.ac.uk

The matrix elements of the trace anomaly known as gravitational form factors (GFFs), which encode mechanical properties of the nucleon [14, 15], can then be directly related to J/ψ photoproduction [14, 15]. A scalar GFF is estimated from the J/ψ differential cross section as a function of t [14] by assuming a simple dipole form:

$$G(t) = M(1 - \frac{t}{m_s^2})^{-2}, \tag{1}$$

where m_s is a free parameter which relates to the rms mass radius of the nucleon as [14]:

$$< R_m^2 >= \frac{12}{m_s^2}. \tag{2}$$

The first estimate of the mass radius of the proton, shown in Figure 1 of [14], based on GlueX data [5], yields $m_s = 1.24 \pm 0.07$ GeV and $R_m = \sqrt{< R_m^2 >} = 0.55 \pm 0.03$ fm. Other estimates by the J/ψ-007 experiment carried out in JLab's Hall C are in good agreement [15]. The fact that the charge radius of the proton $R_C = 0.8409 \pm 0.0004$ fm is statistically significantly larger than the mass radius of the proton suggests that the quark radius of the proton is significantly larger than its gluon radius [14].

Before moving on, it is worth emphasising that the assumption of a dominant two gluon exchange for J/ψ photoproduction sees two spin-1 gluons exchanged, mimicking a spin-2 system such as a graviton [11, 16]. Due to the extreme weakness of the gravitational field of the proton and the fact that graviton-proton scattering is currently off limits [14], the theoretical framework described above offers new opportunities in hadron physics whereby the mass and mechanical properties of the nucleon can be measured from J/ψ photoproduction. Although the mass radius of the proton has already been quantified, CLAS12 will provide the first estimates of the mass radius of the bound neutron and proton in deuteron.

2 Methodology

CLAS12 [17] is a large acceptance spectrometer located in Hall B at JLab where the Continuous Electron Beam Accelerator Facility (CEBAF) [18] was recently upgraded to produce a 12 GeV electron beam, with beam energies up to 11 GeV delivered to CLAS12. CLAS12, shown in Figure 1, has a forward detector covering polar angles of 5 to 35 degrees and a central detector covering the range 35 to 125 degrees. Both have full azimuthal coverage. Only the forward detector is used in the measurement of J/ψ photoproduction off the free and bound nucleon, motivating a short description of its main components. The High Threshold Cherenkov Counter (HTCC) is used to identify electrons [17]. The Drift Chambers (DC), in combination with the solenoid magnet, allow to determine the charge and momentum of final-state particles [17]. The Forward Time Of Flight (FTOF) counters were designed to identify charged hadrons by measuring their speed [17]. The Electromagnetic Calorimeters (PCAL and EC) are used to detect neutrals and identify electrons and muons by measuring their energy deposition [17].

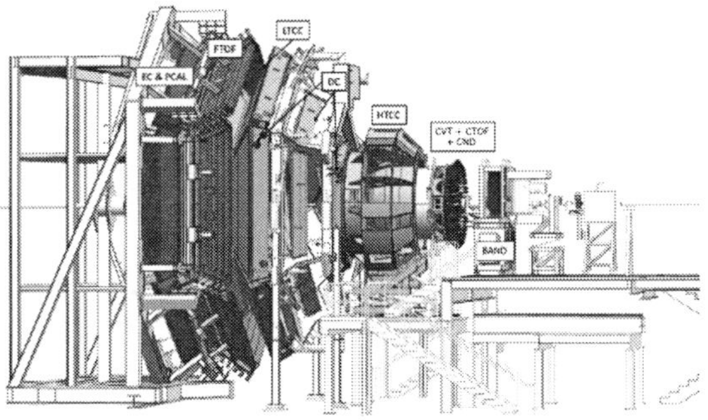

Figure 1. The CEBAF Large Acceptance Spectrometer (CLAS12) pictured with its main subsystems highlighted. In the forward detector, magnetic spectrometry is carried out by means of a superconducting solenoid magnet. CLAS12 is comprised of six identical sectors covering the full azimuthal-angle range, with three of these visible here.

In recent years, CLAS12 took data with both proton and deuteron targets, which means that data on several J/ψ reaction channels is available. On a hydrogen target, J/ψ is produced in the interaction of the electron beam with a free proton. The interaction is mediated by a quasi-real photon. With a deuteron target, J/ψ is produced off either the bound proton or neutron, and the selection of the reaction channel requires the detection of the final-state proton or neutron. The scattered electron is undetected by CLAS12 and J/ψ is identified by detecting the lepton pair (electron-positron or di-muon) to which it decays. The recoil nucleon on which J/ψ was produced is also detected to select exclusive photoproduction events.

The final state particles are identified in the first step of this analysis. Electrons and positrons are required to produce a signal in the HTCC and have a ratio of the energy deposition to momentum around 0.25. Muons are minimum ionising particles selected with cuts on their energy deposition in the calorimeters. The lepton particle identification (PID) is refined by training a machine learning classifier on variables from several CLAS12 detector subsystems, such as the energy deposition and cluster information in the calorimeters and the number of photoelectrons produced in the HTCC. The type of classifier used here is a boosted decision tree [19]. The training sample contains simulated pions, the main background in di-lepton identification in CLAS12 data, and simulated positrons or muons. The classifier learns to distinguish between these particle types. The PID process is then reduced to a cut on the response of the classifier. The J/ψ and background yields as a function of the cut on the response are shown in Figure 2, for J/ψ produced on a hydrogen target and decaying to a di-muon pair. For protons (and charged hadrons in general) a cut is made on the β versus momentum parametrization. For neutrons, only a neutral charge is required. No further cuts are applied as there is not any strong evidence of photon contamination. Some corrections are then applied to the detected particles. When a photon has also been detected within a small angular range from the final-state electron or positron, the photon energy is added to the lepton energy. This method recovers electron (positron) energy loss due to bremsstrahlung. Neutrons produce secondary clusters that are reconstructed as independent neutral tracks and give rise to a non-physical background in the neutron sample. These are removed by taking the earliest neutral in a given sector. The reconstructed path length for neutrons is corrected by comparing the reconstructed to expected path length in exclusive reactions producing a neutron. This leads to a more accurate estimate of the neutron momentum. Fiducial cuts are applied to remove electron or positron hits close to the edges of the PCAL where the shower is not fully contained within the calorimeter. Fiducial cuts in the drift chambers are applied to electrons, positrons, protons, and muons by removing hits at the edge of the layers.

The next step of the analysis is to select J/ψ photoproduction events. First, to ensure quasi-real photoproduction events, the invariant mass Q^2 of the virtual photon mediating the interaction between the electron beam and the target must be close to zero. An upper limit of 0.3 GeV^2 on Q^2 restricts the virtuality of the photon and ensures the selection of quasi-real photoproduction events. Next, the invariant mass of the state X in the reaction $eN \rightarrow l^+l^-NX$ should be consistent with the electron rest mass as the scattered electron is not detected, as mentioned previously. For all reactions of interest discussed here, the assumption is made that the target nucleon N is at rest. At GeV energies, m_X is essentially zero, with the Fermi motion of the bound proton and bound neutron in deuteron decreas-

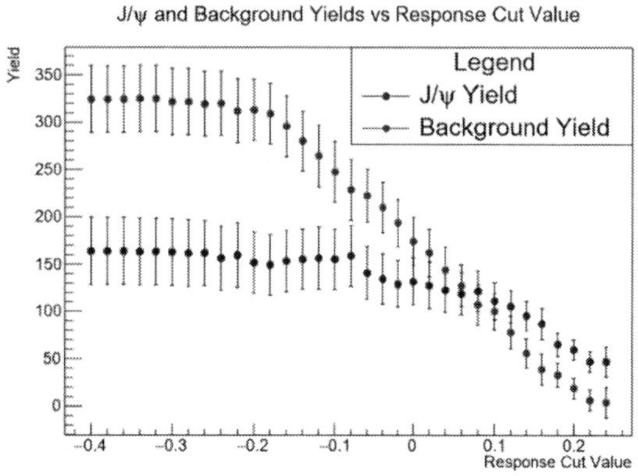

Figure 2. The J/ψ and background yields as a function of the cut on the muon machine-learning classifier response, for J/ψ produced on a hydrogen target and decaying to a di-muon pair. The background yield drops off faster than the J/ψ yield. This allows for an improved identification of muon pairs with a cut at 0.1, for example.

ing the resolution on the invariant mass of X. The invariant mass squared m_x^2 is referred to as the missing mass squared of the event, and is typically restricted to a range from −0.3 to 0.3 GeV^2. Cuts on the missing mass squared and Q^2 allow to select quasi-real photoproduction of di-lepton pairs. Such pairs can be produced in the decay of J/ψ, ρ, ω or ϕ, Bethe Heitler photoproduction, or Time-like Compton Scattering [20]. The number of J/ψ events in the data sample is then measured by fitting the di-lepton invariant mass.

3 Results and Outlook

Once the final state particles have been identified and corrected and the final states of interest have been selected, the total and differential cross sections can be extracted from the experimental yields. The total cross section for the $ep \rightarrow (e')J/\psi p$ reaction, where J/ψ is reconstructed by means of its di-muon decay, as for the $ep_{bound} \rightarrow (e')J/\psi p$ and $en_{bound} \rightarrow (e')J/\psi n$ reactions, where J/ψ is reconstructed by means of its electron-positron decay, is shown in Figure 3. Here, p_{bound} and n_{bound} denote the bound proton and neutron in deuteron, respectively. The absolute normalisation is yet to be evaluated and, thus, the cross section is given in arbitrary units. As these results are still preliminary, they are not compared to previous data. Work is ongoing to better understand the normalisation. However, some interesting observations can be derived from Figure 3. First, the cross section off the free and bound proton agree within the statistical uncertainties, suggesting that any contributions due to final-state interactions in the quasi-free sample are smaller than the statistical precision of the measurement. The cross section off the bound proton and neutron also agree with one another within their uncertainties, which suggests that if there are any isospin violation effects, they are smaller than the precision on these cross sections. Two- and three-

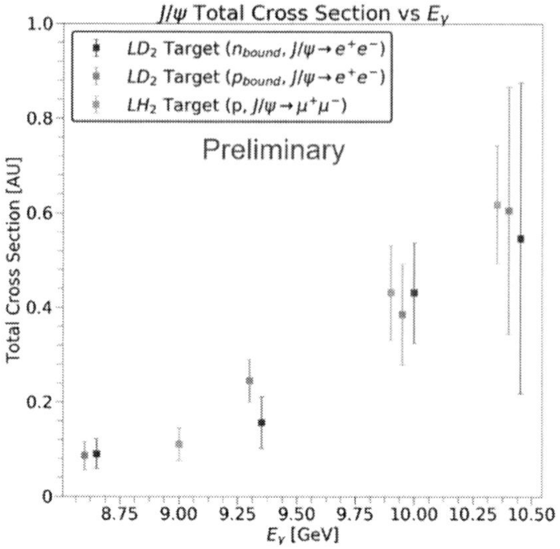

Figure 3. The J/ψ total cross section for J/ψ produced exclusively off a hydrogen (LH_2) target decaying to a di-muon pair (orange), J/ψ produced exclusively off the bound proton (green) or bound neutron (red) in the deuteron target (LD_2) decaying to an electron positron pair (blue). The cross section is given in arbitrary units as these results are still preliminary. Only statistical uncertainties are shown here, with the data points reported at the centre of the E_γ bin. The uncertainty increases as a function of E_γ due to the decreasing photon flux (taken as the sum of the real and quasi-real fluxes).

gluon exchange mechanisms are isospin conserving and if they dominate the reaction dynamics, one would expect the same cross sections off the proton and neutron. Finally the estimate of the total cross section in the di-muon decay channel is the result of the first CLAS12 analysis using muons. This demonstrates the viability of carrying out muon physics with CLAS12 when machine learning techniques are employed. While not shown here, very preliminary estimates of the differential cross sections of the three final states of interest are consistent with each other within their uncertainties. These estimates will enable a quantitative comparison of the mass radii of the free and the bound proton, and the bound neutron.

The near-threshold exclusive J/ψ photoproduction total and differential cross sections are sensitive to the J/ψ production mechanism and the nucleon gravitational form factors. The J/ψ differential cross section as a function of t can be directly connected to the nucleon mass radius. CLAS12 is aiming for a first measurement directly comparing the J/ψ photoproduction cross section off the free proton, bound proton, and bound neutron. This will constitute a first measurement of the bound proton and bound neutron mass radii. AI based improvements in the tracking reconstruction at CLAS12 show an average 40% increase in the efficiency for 3 charged particles. The data already taken at CLAS12 is about to be reprocessed with the new tracking improvements. The experiment aiming for the measurements of J/ψ photoproduction on deuterium still has about 40% of beam time left to run. Both of these increases in statistics will enable higher precision estimates of the exclusive total and differential cross sections. Future luminosity upgrades at JLab and CLAS12 will enable a further increase in statistics of J/ψ photoproduction at CLAS12. An energy upgrade of the accelerator at JLab would provide complementary studies of higher mass charmonium states.

Acknowledgements

This work has been funded by the U.K. Science and Technology Facilities Council under grants ST/P004458/1 and ST/V00106X/1. I would also like to thank Adam Hobart who presented this talk in my absence.

References

[1] J.J. Aubert, *et al.*, E598 Collaboration, Experimental Observation of a Heavy Particle J, Phys. Rev. Lett. **33** 1404 (1974)

[2] J.E. Augustin, *et al.*, SLAC-SP-017 Collaboration, Discovery of a Narrow Resonance in e+e− Annihilation, Phys. Rev. Lett. **33** 1406 (1974)

[3] U. Camerini, *et al.*, Photoproduction of the ψ Particles, Phys. Rev. Lett. **35** 483 (1975)

[4] B. Gittelman, K. M. Hanson, D. Larson, E. Loh, A. Silverman, and G. Theodosiou, Photoproduction of the ψ(3100) Meson at 11 GeV, Phys. Rev. Lett. **35** 1616 (1975)

[5] A. Ali *et al.* (GlueX Collaboration), First Measurement of Near-Threshold J/ψ Exclusive Photoproduction off the Proton, Phys.Rev.Lett. **123** 072001 (2019).

[6] S.J. Brodsky, E. Chudakov, P. Hoyer, J.M. Laget, Photoproduction of charm near threshold, Phys.Lett. B **498** 23 (2001).

[7] D. Kharzeev, H. Satz, A. Syamtomov, and G. Zinovev, J/ψ-Photoproduction and the Gluon Structure of the Nucleon, Nucl.Phys. A **661** 568 (1999).

[8] A.N. Hiller Blin, C. Fernandez-Ramirez, A. Jackura, V. Mathieu, V.I. Mokeev, A. Pilloni, A. P. Szczepaniak, Studying the P_c (4450) resonance in J/ψ photoproduction off protons, Phys. Rev. D **94** 034002 (2016).

[9] D. Kharzaeev, Quarkonium Interactions in QCD, Proc. Int. Sch. Phys. Fermi **130** 105 (1996) Preprint available at arXiv:nucl-th/9601029 (1996).

[10] L. Frankfurt, M. Strikman, Two-gluon form factor of the nucleon and J/ψ photoproduction, Phys. Rev. D. **66**, 031502 (2002)

[11] Y. Hatta and D.-L. Yang, Holographic J/ψ production near threshold and the proton mass problem, Phys. Rev. D **98** 074003 (2018)

[12] R. Wang, X. Chen, J. Evslin, The origin of proton mass from J/ψ photo-production data, Eur. Phys. J. C **80** 507 (2020)

[13] W. Kou, R. Wang, X. Chen, Extraction of Proton Trace Anomaly Energy from Near-Threshold ϕ and J/ψ photo-productions, Eur. Phys. J. A **58** 155 (2022)

[14] D. Kharzeev, Mass radius of the proton, Phys. Rev. D **104** 054015 (2021)

[15] D. Duran, *et al.* (J/ψ-007 Collaboration), When Color meets Gravity; Near-Threshold Exclusive J/ψ Photoproduction on the Proton, (2022). Preprint available at arXiv:2207.05212

[16] Y.Guo, X.Ji, Y.Liu, QCD analysis of near-threshold photon-proton production of heavy quarkonium, Phys. Rev. D **103** 096010 (2021)

[17] V. Burkert, *et al.*, The CLAS12 Spectrometer at Jefferson Laboratory, *Nucl. Inst. Methods A* **959** 163419 (2020)

[18] H.A. Grunder, J.J. Bisognano, W.I. Diamond, B.K. Hartline, C.W. Leemann, J. Mougey, R.M. Sundelin, R.C. York, The continuous electron beam accelerator facility, *Nuclear Physics A* **478** (1988)

[19] Y. Freund, R. E. Schapire, Experiments with a new boosting algorithm, *Proceedings of the 13th international conference on machine learning* 148–156 (1996)

[20] P. Chatagnon, *et al.* (CLAS Collaboration), First Measurement of Timelike Compton Scattering, Phys. Rev. Lett. **127**262501 (2021)

Investigating the E2 Nuclear Resonance Effects in Kaonic Atoms: The KAMEO Proposal

L. De Paolis[1,*], *D.* Bosnar[2], *M.* Bragadireanu[3], *M.* Cargnelli[4], *M.* Carminati[5], *A.* Clozza[1], *G.* Deda[5], *R.* Del Grande[6,1], *K.* Dulski[1,7,8], *C.* Fiorini[5], *I.* Friščić[2], *C.* Guaraldo[1], *M.* Iliescu[1], *M.* Iwasaki[9], *A.* Khreptak[1,7], *S.* Manti[1], *J.* Marton[4], *M.* Miliucci[1], *P.* Moskal[7,8], *F.* Napolitano[1], *S.* Niedźwiecki[7,8], *O.* Ohnishi[10], *K.* Piscicchia[11,1], *Y.* Sada[10], *A.* Scordo[1], *F.* Sgaramella[1], *M.* Silarski[7,8], *D. L.* Sirghi[11,1,3], *F.* Sirghi[1], *M.* Skurzok[7,8], *S.* Wycech[12], *A.* Spallone[1], *K.* Toho[9], *M.* Tüchler[4,13], *C.* Yoshida[10], *J.* Zmeskal[4], and *C.* Curceanu[1]

[1]INFN, Laboratori Nazionali di Frascati, Via E. Fermi 54, I-00044 Frascati(RM), Italy
[2]Department of Physics, Faculty of Science, University of Zagreb, 10000 Zagreb, Croatia
[3]Horia Hulubei National Institute of Physics and Nuclear Engineering, Str. Atomistilor No. 407, P.O. Box MG-6 Bucharest-Măgurele, Romania
[4]Stefan Meyer Institute for Subatomic Physics, Kegelgasse 27, 1030 Wien, Austria
[5]Politecnico di Milano, Dipartimento di Elettronica, Informazione e Bioingegneria and INFN Sezione di Milano, I-20133 Milano, Italy
[6]Physik Department E62, Technische Universiät München, James-Franck-straße 1, 85748 Garching, Germany
[7]Faculty of Physics, Astronomy, and Applied Computer Science, Jagiellonian University, Lojasiewicza 11, Krakow, 30-348, Poland
[8]Center for Theranostics, Jagiellonian University, Kopernika 40, Krakow, 31-501, Poland
[9]RIKEN, Institute of Physical and Chemical Research, Wako, Tokyo 351-0198, Japan
[10]Research Center for Electron Photon Science (ELPH), Tohoku University, Sendai, Japan
[11]Centro Ricerche Enrico Fermi—Museo Storico della Fisica e Centro Studi e Ricerche "Enrico Fermi", Via Panisperna 89a, I-00184 Roma, Italy
[12]National Centre for Nuclear Research, 02-093, Pasteura, Warsaw
[13]University of Vienna, Vienna Doctoral School in Physics, Wien, Austria

Abstract. The E2 nuclear resonance effect in kaonic atoms occurs when the energy of atomic de-excitation closely matches the energy of nuclear excitation, leading to the attenuation of some X-ray lines in the resonant isotope target. This phenomenon provides crucial information on the strong interaction between kaons and nuclei. The only nuclear E2 resonance effect observed so far was in the $K^- - {}^{98}_{42}Mo$ isotope, measured by G. L. Goldfrey, G- K. Lum, and C. E. Wiegand at Lawrence Berkeley Laboratory in 1975. However, the 25 hours of data taking were not sufficient to yield conclusive results. In four kaonic Molybdenum isotopes (${}^{94}_{42}Mo$, ${}^{96}_{42}Mo$, ${}^{98}_{42}Mo$, and ${}^{100}_{42}Mo$), the nuclear E2 resonance effect is expected to occur at the same transition with similar energy values. To investigate this, the KAMEO (Kaonic Atoms Measuring Nuclear Resonance Effects Observables) experiment plans to conduct research on kaonic Molybdenum isotopes at the DAΦNE e^+e^- collider during the SIDDHARTA-2 experiment. The experimental strategy involves exposing four solid strip targets, each enriched with one Molybdenum isotope, to negatively charged kaons and using a germanium detector to measure X-ray transitions. In addition, a non-resonant ${}^{92}_{42}Mo$ isotope solid strip target will be used as a reference for standard non-resonant transitions.

1 Introduction

Kaonic atoms are formed when a negatively charged kaon (K^-) is captured by an atomic system due to the electromagnetic interaction with the nucleus, replacing an electron in an atomic shell. After capture, the K^- starts an electromagnetic cascade process, which ends with its nuclear absorption. As the kaon approaches the nucleus and fills the innermost levels of the kaonic atom, the strong kaon-nucleus interaction alters the atomic structure by shifting and broadening the levels which can be detected through dedicated X-ray spectroscopy techniques [1–3].The investigation of kaonic atoms began in the 1970s [4–6] and continues to this day [7–13], providing a crucial tool for understanding the strong kaon-nucleon interaction in the low-energy regime. In 2023, the SIDDHARTA-2 experiment will perform the first measurement of kaonic deuterium at the DAΦNE e^+e^- collider at the National Laboratories of Frascati (LNF) in Italy [14]. Combined with the kaonic hydrogen measurement performed by the SIDDHARTA experiment in 2011 [10], this measurement will enable the extraction of antikaon-nucleon scattering lengths with isospin dependence. Whenever a nuclear excitation energy is very close to an atomic de-excitation energy, a resonance occurs in the atomic system, known as E2 nuclear resonance [15]. Several kaonic atoms are predicted to be resonant, including four isotopes of kaonic

*e-mail: luca.depaolis@lnf.infn.it

Molybdenum ($^{94}_{42}$Mo, $^{96}_{42}$Mo, $^{98}_{42}$Mo, and $^{100}_{42}$Mo) [16, 17]. These isotopes may provide insight into the properties of deeply bound kaonic atomic levels, which are otherwise difficult to access through K^- cascades due to nuclear absorption. By comparing the measurements of these isotopes, information about the strong nuclear potential can be obtained by investigating variations in the resonance's parameters with increasing neutron number along the Molybdenum isotopes. The first measurement of the E2 nuclear resonance effect in kaonic atoms was performed by G. L. Goldfrey, G. K. Lum, and C. E. Wiegand in $^{98}_{42}$Mo at the Lawrence and Berkeley Laboratory (LBL) in California, in 1975 [18, 19]. Unfortunately, only 25 hours of data were collected which was not sufficient for a definitive result. The E2 nuclear resonance effect has been measured in other exotic atoms such as pionic atoms and anti-protonic atoms. In particular, the measurement of the E2 nuclear resonance effect in even-A anti-protonic Tellurium isotopes has been used to determine the properties of the neutron density in the nuclear periphery of Te isotopes [20]. The same investigation can be performed on kaonic molybdenum isotopes, revealing the properties and parameters of the neutron density in the nuclear periphery. This may have an important role for the observation of the neutrinoless double beta decay of $^{98}_{42}$Mo which could demonstrate that neutrinos are Majorana particles [21]. The nuclear matrix elements for this decay are calculated using models that depend on the relative distance between the two neutrons involved in the decay. A more precise estimation of the root mean square (rms) of the neutron radius, obtainable through the study of the E2 resonance in $K^- - ^{98}_{42}$Mo, could provide further constraints to define the relative distance among neutrons in the isotope. Finally, this paper briefly presents the possibility and advantages of investigating kaonic Molybdenum isotopes at the DAΦNE collider with the KAMEO setup, running in parallel with the SIDDHARTA-2 experiment.

2 The KAMEO proposal: the measurement of the E2 nuclear resonance effects in kaonic molybdenum isotopes

The E2 nuclear resonance effect occurs when the energy required for a nuclear excitation closely matches the de-excitation energy of an atomic transition, resulting in a mixing of atomic states due to electrical quadrupole excitations of nuclear rotational states. This mixing produces a wave function ϕ that includes a small mixture of excited nuclear and de-excited atomic wavefunctions, described by the equation:

$$\psi = \sqrt{1-|\alpha|^2}\phi(n,l,0^+) + \alpha\phi(n,l-2,2^+) \qquad (1)$$

where $\alpha = \pm\frac{\langle n',l-2,2^+|H_q|n,l,0^+\rangle}{E_{(n',l-2,2^+)}-E_{(n,l,0^+)}}$ is the admixture coefficient determined by the electric quadrupole interaction between the kaon and the nucleus.

The rate of nuclear absorption increases significantly in kaonic atoms for each unit of decreasing in orbital angular momentum, and the induced width is determined by the equation:

$$\Gamma^{Ind}_{n,l} = |\alpha|^2\Gamma^0_{n',l-2} \qquad (2)$$

In this way, the E2 nuclear resonance can lead to a significant attenuation of the involved kaonic X-ray line and of any lower lines, in resonant kaonic atoms. In kaonic Molybdenum isotopes 94, 96, 98, and 100, the E2 nuclear resonance effect occurs by mixing (6h,0^+) and (4f,2^+) states, described by the wave function:

$$\psi = \sqrt{1-|\alpha|^2}\phi(6h,0^+) + \alpha\phi(4f,2^+) \qquad (3)$$

where the admixture coefficient $\alpha = \pm\frac{\langle 4f,2^+|H_q|6h,0^+\rangle}{E_{(4f,2^+)}-E_{(6h,0^+)}}$. Parameters and energies of the E2 nuclear resonance effects in kaonic Mo isotopes are reported in Tab. 1. A schematic representation of the E2 nuclear resonance effect in $^{98}_{42}$Mo is presented in Fig. 1. The investigation of the E2 nuclear resonance effects in kaonic Mo isotopes represents a unique and interesting opportunity to extract fundamental information for a better understanding of the properties of kaon-nucleon strong interaction. Moreover, the comparison among the measured spectra of kaonic Mo isotopes could allow the investigation of the distribution of the neutrons in the nuclear periphery [20].

The E2 nuclear resonance effect in kaonic atoms was first measured in 1975 at Lawrence and Berkeley Laboratory in California, using a negative kaon beam and solid targets of $^{98}_{42}$Mo, with non-resonant $^{92}_{42}$Mo as reference [16, 18, 19]. Germanium detectors feeding a pulse height analyzer were used to collect the spectra, and the E2 nuclear resonance effect was observed as attenuation of X-ray lines. However, the experiment only yielded 25 hours of data, which was not enough for a conclusive result. Subsequently, no further investigations into the E2 nuclear resonance effect in kaonic atoms were conducted to today. The KAMEO experiment plans to measure the E2 nuclear resonance effects in kaonic Mo isotopes at the DAΦNE e^+e^- collider of the LNF, in Italy. The DAΦNE collider is a ϕ factory producing low energetic K^- (p=127 MeV/c), suitable for the investigation of kaonic atoms [5, 14]. The KAMEO apparatus consists of several solid target strips of enriched Mo isotopes (>99%): $^{94}_{42}$Mo, $^{96}_{42}$Mo, $^{98}_{42}$Mo, $^{100}_{42}$Mo and $^{92}_{42}$Mo (used as the reference for standard transition not affected by E2 nuclear resonance); and a High Purity Germanium (HPGe) detector. The surface of the solid target strips will be 4 cm x 8 cm. The thickness will be estimated with a dedicated Monte Carlo simulation to maximize the efficiency in kaonic Mo isotope production and measurement of the X-ray transitions. In 2023, the SIDDHARTA-2 experiment will perform the first measurement of the 2p→1s X-ray transition in kaonic deuterium, using a cylindrical target placed above the Interaction Point (IP) of the DAΦNE collider. The KAMEO experiment is able to run in parallel with SIDDHARTA-2, exploiting the horizontal plane, as shown in Figure 2. The SIDDHARTA-2 luminometer[23], consisting of plastic scintillators (80 mm × 40 mm × 2 mm each), read by pairs of Photo Multiplier Tubes (PMTs) [24, 25], will be used as trigger for the HPGe detector X-ray spectroscopy. The luminometer is placed at a distance of 110 mm from

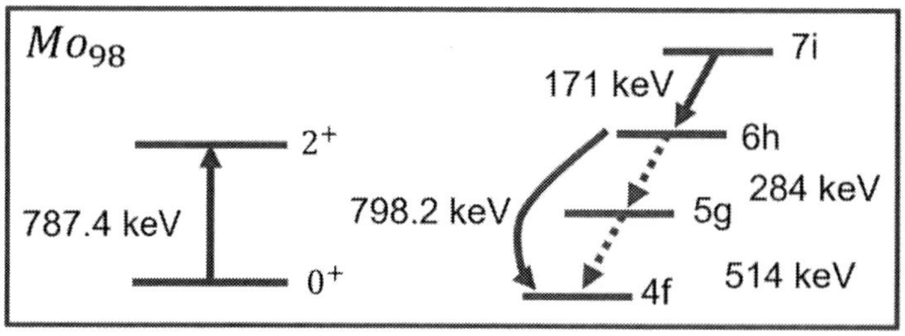

Figure 1. Schematic views of the E2 nuclear resonance effect in kaonic Molybdenum 98. The resonance mixes (6h,0$^+$) and (4f,2$^+$) states, attenuating the 6h→4f transition line and any lower lines with respect to a non-resonant Molybdenum isotope.

Table 1. Table with energies and parameters of the E2 resonant Molybdenum isotopes. [15]

Atom	$E_{0^+\to 2^+}[keV]$	Mixing(n,l)	$E_{(n,l)\to(n',l-2)}[keV]$	Line	E[keV]	Atten.
${}^{94}_{42}Mo$	871	(6,5) ÷ (4,3)	799	6→5	284	0.18
${}^{96}_{42}Mo$	778	(6,5) ÷ (4,3)	799	6→5	284	0.71
${}^{98}_{42}Mo$	787	(6,5) ÷ (4,3)	798	6→5	284	0.81
${}^{100}_{42}Mo$	536	(6,5) ÷ (4,3)	798	6→5	284	0.04

the Interaction Point (IP) of the DAΦNE collider. The Mo target strips will be positioned 5 mm outside the luminometer. The p-type HPGe detector is designed by Baltic Scientific Instruments to work under high-rate conditions (to 150 kHz). It has a cylindrical active volume having 59.3 mm of height and 59.8 mm of base diameter. The energy resolution of the HPGe detector, tested with ^{133}Ba and ^{60}Co sources (activity < 1 μCi), are 0.87 keV at 81 keV, 1.06 keV at 302.9 keV. These parameters make the HPGe detector ideal to perform the measurement of the 7i→6h and 6h→5g atomic transitions in kaonic Mo isotopes (having energies of ∼170 keV and ∼280 keV, respectively) in the high-rate environment of the DAΦNE collider. The HPGe detector will be positioned 115 mm far from the Mo target strips. The detection efficiency will be estimated with a dedicated Monte Carlo simulation.

The KAMEO experimental proposal aims the measurement of the mixing coefficient α in the ${}^{94}_{42}$Mo, ${}^{96}_{42}$Mo, ${}^{98}_{42}$Mo, and ${}^{100}_{42}$Mo resonant isotopes, caused by the E2 nuclear resonance effect, with a precision better than 10%. The mixing coefficient is obtained from the attenuation of the 6h→5g transition in the spectra of resonant Mo isotopes, compared to non-resonant ${}^{92}_{42}$Mo, used as a reference. Additionally, KAMEO will provide the first measurement of the shift ($\epsilon_{6h,0^+}$) and width ($\Gamma_{6h,0^+}$) of the 6h atomic level in kaonic resonant Mo isotopes due to the E2 nuclear resonance effect, and the first extraction of the shift ($\epsilon_{4f,2^+}$) and width ($\Gamma_{4f,2^+}$) of the 4f atomic level in excited-nucleus isotopes of kaonic Mo due to the kaon-nucleus strong interaction. The 4f state could be a Coulomb-assisted nuclear quasi-bound state of K$^-$ meson [20]. To extract the shifts and broadenings of atomic levels, the X-ray transitions in kaonic Molybdenum E2-resonant isotopes are compared to purely electromagnetic values determined using QED. This procedure is similar to that used for antiprotonic Tellurium isotopes [20], with an expected precision of some keV. Finally, the high-precision measurements performed by KAMEO enable the study of the neutron density in the nuclear periphery of kaonic Mo isotopes, whose number of nucleons in the nucleus growth of pairs of neutrons from the lightest resonant isotope. The difference between neutron and proton *rms* radii can be determined with a precision of 0.1 fm, and the neutron *rms* radius can be precisely estimated (following a similar procedure used in [20]). The extraction of the neutron *rms* radius in ${}^{98}_{42}$Mo would be crucial for research on the neutrinoless double beta decay of ${}^{98}_{42}$Mo, which violates lepton number conservation. Observing this decay would demonstrate that the neutrino is a Majorana particle [21].

3 Conclusion and outooks

The experimental investigation of the E2 nuclear resonance effect in the resonant isotopes of kaonic Molybdenum (94, 96, 98, and 100) presents a novel perspective into the strangeness sector's strong interaction at low energies. KAMEO could perform the best measurement of the E2 nuclear resonance effects in kaonic Mo isotopes. Moreover, this measurement could enhance the understanding of the neutron distribution in the nuclear periphery, providing also important information for the investigation of the neutrinoless double beta decay in ${}^{98}_{42}$Mo. Finally, the first measurement of shifts and widths of the 4f energy level in kaonic Mo E2-resonant isotopes, which could be a Coulomb-assisted nuclear quasi-bound state of K$^-$ meson, could reveal fundamental details of the strong kaon-nucleus interaction, providing parameters which allow updates of the theoretical models. The KAMEO experiment plans precise measurements of the E2 nuclear resonance effects in kaonic molybdenum isotopes 94, 96, 98 and 100

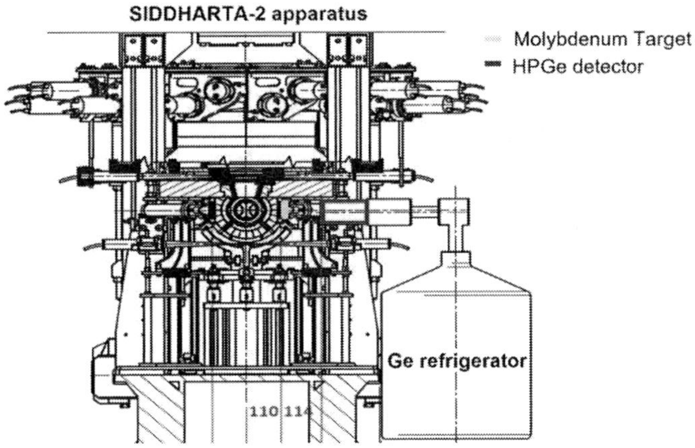

Figure 2. Schematic view of the KAMEO installed on the horizontal line close to the interaction point of the DAΦNE collider, running in parallel with the SIDDHARTA-2 experiment.

at LNF's DAΦNE collider. The measurement will be performed simultaneously with the SIDDHARTA-2 experiment and with an optimized experimental setup developed through a dedicated Monte Carlo simulation.

We thank C. Capoccia from LNF-INFN and H. Schneider, L. Stohwasser, and D. Pristauz-Telsnigg from Stefan Meyer-Institut for their fundamental contribution in designing and building the SIDDHARTA-2 setup. We thank as well the DAΦNE staff for the excellent working conditions and permanent support. Part of this work was supported by the Austrian Science Fund (FWF): [P24756-N20 and P33037-N]; the EXOTICA project of the Minstero degli Affari Esteri e della Cooperazione Internazionale, PO21MO03; the Croatian Science Foundation under the project IP-2018-01-8570; the EU STRONG-2020 project (Grant Agreement No. 824093); the EU Horizon 2020 project under the MSCA (Grant Agreement 754496); the Japan Society for the Promotion of Science JSPS KAKENHI Grant No. JP18H05402; the Polish Ministry of Science and Higher Education grant No. 7150/E-338/M/2018 and the Polish National Agency for Academic Exchange(grant no PPN/BIT/2021/1/00037); the EU Horizon 2020 research and innovation programme under project OPSVIO (Grant Agreement No. 101038099). The authors acknowledge support from the SciMat and qLife Priority Research Areas budget under the program Excellence Initiative—Research University at the Jagiellonian University.

References

[1] Lee C H, Jung H, Min D P and Rho M, Phys. Lett. B **326**, 14-20 (1994)
[2] Bernard V, Prog. Part. Nucl. Phys. **60**, 82-160 (2008)
[3] Cieplý A, Mai M, Meißner U G and Smejkal J, Nucl. Phys. A **954**, 17-40 (2016)
[4] Davies J D *et al.*, Phys. Lett. B **83**, 55 (1979)
[5] Izycki M *et al.*, Z. Phys. A **297**, 11 (1980)
[6] Baird S *et al.*, Nucl. Phys. A **392**, 297 (1983)
[7] Iwasaki M *et al.*, Phys. Rev. Lett. **78**, 3067 (1997)
[8] Beer G *et al.*, Phys. Rev. Lett. **94**, 212302 (2005)
[9] Okada S *et al.*, Phys. Lett. B **653**, 387 (2007)
[10] Bazzi M *et al.*, Phys. Lett. B **681**, 310 (2009)
[11] Bazzi M *et al.*, Phys. Lett. B **704**, 113 (2011)
[12] Bazzi M *et al.*, Phys. Lett. B **697**, 199 (2011)
[13] Bazzi M *et al.*, Phys. Lett. B **714**, 40 (2012)
[14] Curceanu C *et al.* 2021 *Few-Body System* **62** 83 (2021)
[15] Leon M, Nucl. Phys. A **260**, 461-473 (1976)
[16] Leon M, Informal Report United States: N. p. (1975)
[17] Leon M, Phys. Lett. B **50**, 425 (1974)
[18] Goldfrey G L, Lum G K and Wiegand C E, Phys. Lett. B **65**, 43-45 (1976)
[19] Wiegand C E and Goldfrey G L, Phys. Rev. A **9**, 2282 (1974)
[20] Klos B, Wycech S *et al.*, Phys. Rev. C **69**, 044311 (2004)
[21] Nesterenko D A, Jokiniemi L, Kotila J *et al.*, Eur. Phys. J. A **58**, 44 (2022)
[22] Bradbury J N, Leon M, Daniel H and Reidy J J, Phys. Rev. Lett. **34**, 303 (1975)
[23] Skurzok M *et al.*, JINST **15**, P10010 (2020)
[24] Moskal P *et al.*, Nucl. Inst. and Meth. A **764**, 317 (2014)
[25] Niedźwiecki S *et al.*, Acta Phys. Polon. B **48**, 1567 (2017)

NP effects in $\Lambda_b \to \Lambda_c^{(*)}$ semileptonic decays.

Neus Penalva[1,*], *Eliecer* Hernández[2], and *Juan* Nieves[1]

[1]Instituto de Física Corpuscular (centro mixto CSIC-UV), Institutos de Investigación de Paternam C/Catedrático Beltrán 2, E-46980 Paterna, València, Spain
[2]Departamento de Física Fundamental e IUFFyM, Universidad de Salamanca , Plaza de la Merced s/n, E-37008 Salamanca, Spain

Abstract. In the context of lepton flavor universality violation (LFUV) studies, we study different observables related to the $b \to c\tau\bar{\nu}_\tau$ semileptonic decays. These observables are expected to help in distinguishing between different New Physics (NP) scenarios. Since the τ lepton is very short-lived, we consider three subsequent τ-decay modes, two hadronic $\pi\nu_\tau$ and $\rho\nu_\tau$ and one leptonic $\mu\bar{\nu}_\mu\nu_\tau$. This approach enables the differential decay width to be written in terms of visible (experimentally accessible) variables of the massive particle created in the τ decay. We present numerical results for the observables that can be accessed through the visible kinematics for the $\Lambda_b \to \Lambda_c$ and the $\Lambda_b \to \Lambda_c^*(2595)$ transitions. This work is based on Refs.[1–3].

1 Motivation

This work is motivated by the latest $\mathcal{R}_{\Lambda_c}$ result from LHCb [4] and other observables measured, such as $\mathcal{R}_{H_c} = \dfrac{\Gamma(H_b \to H_c\tau\bar{\nu}_\tau)}{\Gamma(H_b \to H_c\ell\bar{\nu}_\ell)}$, the averaged tau-polarization asymmetry $P_\tau(D^*)$ and the longitudinal D^* polarization $F_L^{D^*}$. Although some of them are compatible with the Standard model (SM) within errors, all together point to possible new physics (NP) effects [5]. As these discrepancies have been seen in decays where the out-going lepton is a τ^- particle, it is believed that these NP effects only affect the 3th quark and lepton generations.

For studying the NP effects, we use the most general Hamiltonian including effects from light right-handed neutrinos eq. (1) [6].

$$H_{\rm eff} = \frac{4G_F V_{cb}}{\sqrt{2}}\Big[\underbrace{(1+C_{LL}^V)O_{LL}^V + C_{RL}^V O_{RL}^V}_{\rm (axial-)vector} + \underbrace{C_{LL}^T O_{LL}^T}_{\rm tensor}$$
$$+ \underbrace{C_{LL}^S O_{LL}^S + C_{RL}^S O_{RL}^S}_{\rm (pseudo-)scalar} + \underbrace{C_{LR}^V O_{LR}^V + C_{RR}^V O_{RR}^V}_{\rm (axial-)vector\ RH\ \nu's}$$
$$+ \underbrace{C_{LR}^S O_{LR}^S + C_{RR}^S O_{RR}^S + C_{RR}^T O_{RR}^T}_{\rm (pseudo-)scalar/tensor\ RH\ \nu's}\Big] + h.c., \quad (1)$$

The Wilson coefficients ($C_{LR}^{V,S,T}$) parameterize the deviations from SM and are fitted to experimental anomalies in $\bar{B}$ meson decays. Hence all different NP models give the same results for $R_{D^{(*)}}$ and in this work we look for other observables that could distinguish among NP models/fits. For the predictions showed here we will use Fits 6 and 7 from ref. [7] and scenario 7a from ref. [6]. In Fits 6 and 7 there are effects coming only from left-handed neutrinos while scenario 7a includes right-handed ν's effects. Here, we will show a summary of the results for $\Lambda_b \to \Lambda_c, \Lambda_b \to \Lambda_c^*(2595)$ and $\Lambda_b \to \Lambda_c^*(2625)$ published in refs. [2, 3].

All the physics in the decays is encoded in the Wilson coefficients and the 10 independent functions of ω that can be seen in table 1 [1]. In the first row, we can see the observables accessible through the decay involving unpolarized τ^-'s that correspond with the angular asymmetries. The measuring of the spin and spin-angular asymmetries of the second and third rows requires the knowledge of the τ^- polarization. Moreover, the two observables in the last row are only different from zero when some of the Wilson coefficients are complex. The ones that have been measured in some of the decays are the forward-backward asymmetry ($A_{\rm FB}$), the integrated longitudinal polarization ($\langle P_L^{\rm CM}\rangle$) and the unpolarized decay width $\left(\dfrac{d\Gamma_{\rm SL}}{d\omega} \propto n_0\right)$.

2 Final state visible kinematics

One of the problems for measuring these observables is that τ^- particles decay very fast and must be reconstructed from one of its many decay channels. All of them involve at least one neutrino that cannot be measured. Therefore, to avoid having to reconstruct the τ energy and direc-

	observables
unpolarized τ^-	n_0, $A_{\rm FB}$, A_Q
polarized τ^-	$\langle P_L^{\rm CM}\rangle$, $\langle P_T^{\rm CM}\rangle$, Z_L, Z_Q, $Z_\perp$
complex WC's	$\langle P_{TT}\rangle$, Z_T

Table 1: Observables that contain all physics of the decay

*e-mail: neuspe@ific.uv.es

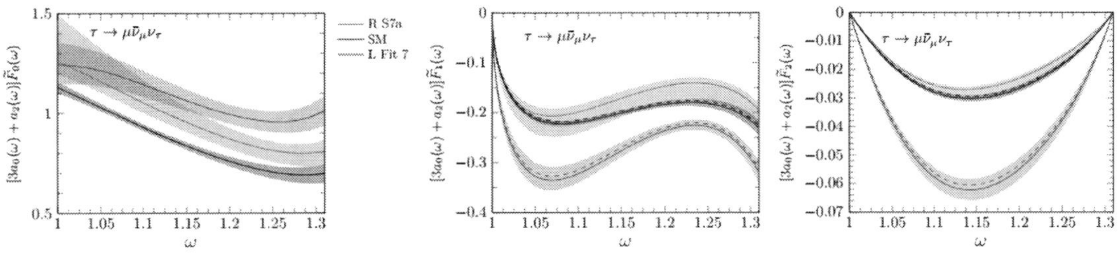

Figure 1: $n_0\tilde{F}^{\mu\bar{\nu}_\mu}_{0,1,2}(\omega)$ for $\Lambda_b \to \Lambda_c$ decay in SM and different NP fits.

tion, we propose using variables related to its visible decay products.

The starting point for our work is the triple differential decay rate of eq. (2), where ξ_d is related to the energy of the charged particle coming from the τ^- decay and θ_d is the angle that is formed by the same particle and the out-going hadron directions. To obtain the expression in eq. (2), we have used the zero width approximation for the τ decay [1, 8].

$$\frac{d^3\Gamma_d}{d\omega d\xi_d d\cos\theta_d} = \mathcal{B}_d \frac{d\Gamma_{\rm SL}}{d\omega}\Big\{F_0^d(\omega,\xi_d)+ F_1^d(\omega,\xi_d)\cos\theta_d + F_2^d(\omega,\xi_d)P_2(\cos\theta_d)\Big\}, \qquad (2)$$

where

$$\begin{aligned} F_0(\omega,\xi_d) &= C_n(\omega,\xi_d) + C_{P_L}(\omega,\xi_d)\langle P_L^{\rm CM}\rangle \\ F_1(\omega,\xi_d) &= C_{A_{FB}}(\omega,\xi_d)A_{FB} + C_{Z_L}(\omega,\xi_d)Z_L \\ &\quad + C_{P_T}(\omega,\xi_d)\langle P_T^{\rm CM}\rangle \\ F_2(\omega,\xi_d) &= C_{A_Q}(\omega,\xi_d)A_Q + C_{Z_Q}(\omega,\xi_d)Z_Q \\ &\quad + C_{Z_\perp}(\omega,\xi_d)Z_\perp. \end{aligned} \qquad (3)$$

All the asymmetries, except the CP-violating ones, can be obtained from this differential decay rate taking into account that the $C_i(\omega,\xi_d)$ functions that enter in $F_{0,1,2}$ are kinematical factors that can be computed analytically and depend on the τ decay mode ($d = \pi, \rho, \mu$) (see appendix G of [1]).

This triple differential decay rate can be difficult to measure and we could be interested in increasing statistics by integrating over some of the variables. Integrating over the energy and angle of the τ−decay massive product gives us back the unpolarized differential decay width. In the following, we will study how much information is kept in the angular and the energy distributions.

3 Angular distributions

We start by integrating the energy of the τ-decay massive product eq. (6). The new $\widetilde{F}_i$ functions are obtained from the previous F_i after integrating over the energy. As the only part that depends on the energy are the $C_i(\omega,\xi_d)$ analytic kinematical factors, we still have access to some observables [2].

$$\frac{d^2\Gamma_d}{d\omega d\cos\theta_d} = \mathcal{B}_d \frac{d\Gamma_{\rm SL}}{d\omega}\Big[\widetilde{F}_0^d(\omega) + \widetilde{F}_1^d(\omega)\cos\theta_d + \widetilde{F}_2^d(\omega)P_2(\cos\theta_d)\Big], \qquad (4)$$

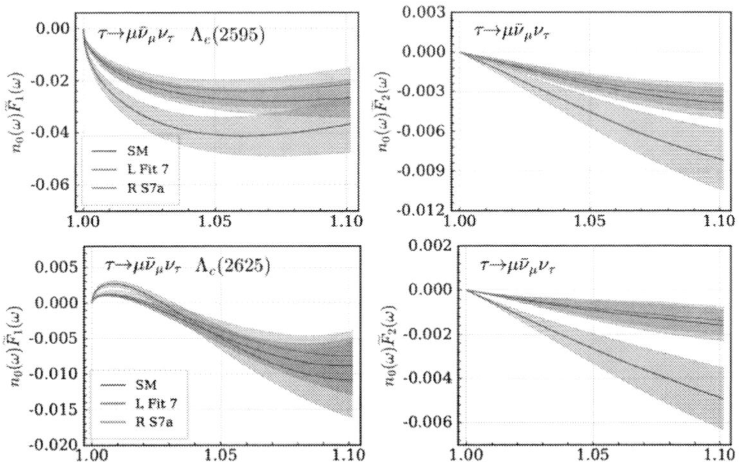

Figure 2: $n_0\tilde{F}^{\mu\bar{\nu}_\mu}_{1,2}(\omega)$ for $\Lambda_b \to \Lambda_c^*(2595)$ and $\Lambda_b \to \Lambda_c^*(2625)$ decays in SM and different NP fits.

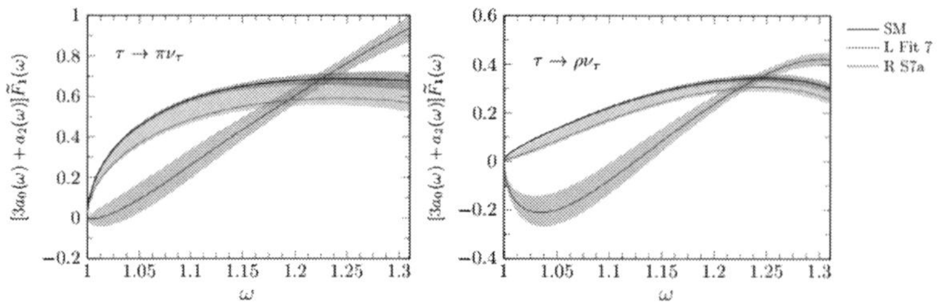

Figure 3: $n_0 \widetilde{F}_1^{\pi(\rho)}(\omega)$ for $\Lambda_b \to \Lambda_c$ decays.

For all τ-decay modes, $\widetilde{F}_0 = \frac{1}{2}$ and thus the information in $\langle P_L^{\rm CM}\rangle$ is lost [2]. In figs. 1 and 2 we show these $\widetilde{F}_i$ functions for the leptonic decay mode of the τ particle. In all cases, Fit 7 can be easily distinguished from the other NP models, except where the form factor errors for $\Lambda_b \to \Lambda_c^*$ decays are still too big. We also can see in the plot of fig. 1 that the limit for a massless μ (dashed line) works fine for all cases. The previous results can be compared also with the hadronic modes for the τ decay. In fig. 3, we show $\widetilde{F}_1$ for the hadron modes of the tau decay and it can be seen that the shape is different from that of the leptonic mode. That can be explained when comparing analytically the $C_i(\omega, \xi_d)$ in the π-massless and μ-massless limit. By doing this, we find an extra factor 3 in the spin and spin-angular asymmetries, as it is seen at eq. (5), that give us more discriminating power in the hadronic decay [2].

$$C^{\pi}_{A_{FB},A_Q}(\omega) = C^{\mu\bar{\nu}_\mu}_{A_{FB},A_Q}(\omega) + O(y^2),$$
$$C^{\pi}_{P_T,Z_L,Z_Q,Z_\perp}(\omega) = -3\,C^{\mu\bar{\nu}_\mu}_{P_T,Z_i}(\omega) + O(y^2) \qquad (5)$$

The next step is integrating over ω to get the angular distribution (eq. (6)). The predictions are shown in fig. 4 for the three τ-decay modes discussed here. Because of that extra factor 3, it is easier to distinguish among models in the hadronic τ-decay.

$$\frac{d\Gamma_d}{d\cos\theta_d} = \mathcal{B}_d\Gamma_{\rm SL}\Big[\frac{1}{2} + \widehat{F}_1^d\cos\theta_d + \widehat{F}_2^d\,P_2(\cos\theta_d)\Big]. \quad (6)$$

4 Energy distributions

To obtain the energy distributions, we come back to eq. (2) and integrate over the angle this time. Now, all the information, except the one from F_0, disappear [2]. That allows us to extract $\langle P_L^{\rm CM}\rangle(\omega)$. That observable, integrated over ω, has been measured by the Belle collaboration in the context of $B \to D^*$ decays [9]. We show in fig. 5 its dependence on ω for the $\Lambda_b \to \Lambda_c^{(*)}$. As before, we can see that Fit 7 can be easily distinguished in all Λ_b decays.

Finally, we integrate over ω, to get[1]:

$$\widehat{F}_0^d(E_d) = \frac{1}{\Gamma_{\rm SL}}\int_1^{\omega_{\rm sup}(E_d)} \frac{1}{\gamma}\frac{d\Gamma_{\rm SL}}{d\omega}\Big\{C_n^d(\omega, E_d) + C^d_{P_L}(\omega, E_d)\,\langle P_L^{\rm CM}\rangle(\omega)\Big\}\,d\omega, \qquad (7)$$

where $\hat{F}_0$ is the normalized energy distribution. $\langle P_L^{\rm CM}\rangle$ does not contribute to the normalization of $\hat{F}_0$ but it affects its shape as seen in fig. 6. Again, we see more discriminating power for the hadronic τ-decay modes.

5 Conclusions

In conclusion, we have seen that using visible final-state kinematics helps to avoid using τ variables that are difficult to reconstruct. With this approach, we can obtain all the observables except the CP-violating ones. Moreover, one can increase statistics by integrating in some of the variables. In the energy distribution we keep the $\langle P_L^{\rm CM}\rangle$ information and the angular distribution includes all other angular-spin asymmetries. $\frac{d\Gamma}{d\cos\theta_d}$, $\frac{d\Gamma}{dE_d}$ and $\frac{d\Gamma}{d\omega}$ are also useful and give complementary information. Finally, we have seen (through eq. (5) and figs. 3, 4 and 6) that the hadronic τ-decay modes have more discriminating power.

[1] We make the change $E_d = \xi_d$ and exchange the integration order of ω and ξ_d. The details of this computation are in ref. [2])

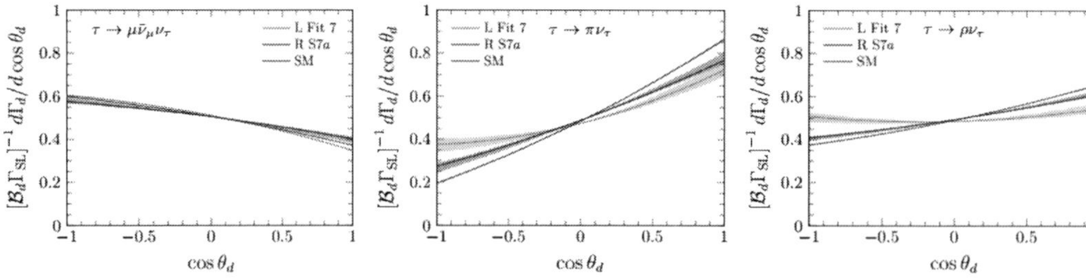

Figure 4: Angular $d\Gamma/d\cos\theta_d$ distribution for the $\Lambda_b \to \Lambda_c$ decays. Same NP scenarios as before.

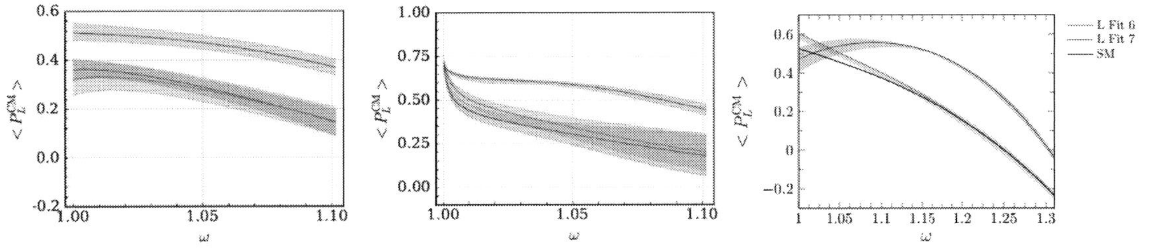

Figure 5: $\langle P_L^{\mathrm{CM}} \rangle$ for the $\Lambda_b \to \Lambda_c^*(2595)$, $\Lambda_b \to \Lambda_c^*(2625)$ and $\Lambda_b \to \Lambda_c$ decays.

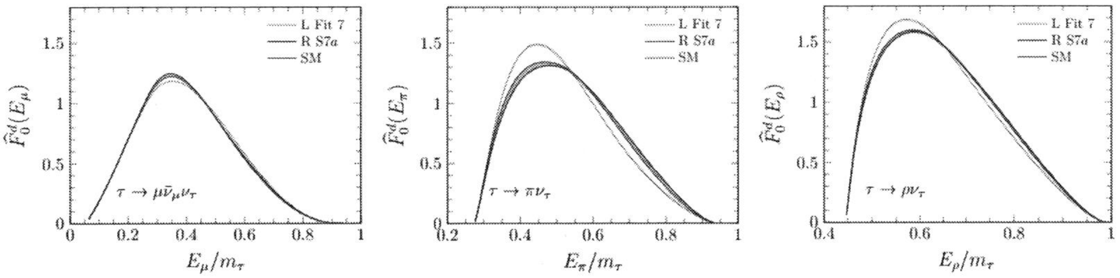

Figure 6: $\widehat{F}_0^d(E_d)$ for the $\Lambda_b \to \Lambda_c$ decays.

Acknowledgments

This research has been supported by the Spanish Ministerio de Ciencia e Innovación (MICINN) and the European Regional Development Fund (ERDF) under contracts PID2020-112777GB-I00 and PID2019-105439GB-C22, the EU STRONG-2020 project under the program H2020-INFRAIA-2018-1, grant agreement no. 824093 and by Generalitat Valenciana under contract PROMETEO/2020/023.

References

[1] N. Penalva, E. Hernández, J. Nieves, JHEP **10**, 122 (2021), 2107.13406
[2] N. Penalva, E. Hernández, J. Nieves, JHEP **04**, 026 (2022), 2201.05537
[3] M.L. Du, N. Penalva, E. Hernández, J. Nieves, Phys. Rev. D **106**, 055039 (2022), 2207.10529
[4] R. Aaij et al. (LHCb), Phys. Rev. Lett. **128**, 191803 (2022), 2201.03497
[5] Y.S. Amhis et al. (HFLAV), Eur. Phys. J. C **81**, 226 (2021), 1909.12524
[6] R. Mandal, C. Murgui, A. Peñuelas, A. Pich, JHEP **08**, 022 (2020), 2004.06726
[7] C. Murgui, A. Peñuelas, M. Jung, A. Pich, JHEP **09**, 103 (2019)
[8] P. Asadi, A. Hallin, J. Martin Camalich, D. Shih, S. Westhoff, Phys. Rev. D **102**, 095028 (2020), 2006.16416
[9] S. Hirose et al. (Belle), Phys. Rev. Lett. **118**, 211801 (2017), 1612.00529

SIDDHARTA-2 veto system design and performance for kaonic atoms studies at DAΦNE

F. Sgaramella[1*], M. Bazzi[1], D. Bosnar[2], M. Bragadireanu[3], M. Cargnelli[4], M. Carminati[5,6], A. Clozza[1], G. Deda[5,6], R. Del Grande[7,1], K. Dulski[1,9,10], L. De Paolis[1], L. Fabbietti[7], C. Fiorini[5,6], I. Friščić[2], C. Guaraldo[1], M. Iliescu[1], M. Iwasaki[8], A. Khreptak[1,9], S. Manti[1], J. Marton[4], M. Miliucci[1,**], P. Moskal[9,10], F. Napolitano[1], S. Niedźwiecki[9,10], H. Ohnishi[11], K. Piscicchia[12,1], Y. Sada[11], A. Scordo[1], M. Silarski[9], D. Sirghi[1,3,12], F. Sirghi[1,3], M. Skurzok[9,10], A. Spallone[1], K. Toho[11], M. Tüchler[4,13], C. Yoshida[11], J. Zmeskal[4], and C. Curceanu[1]

[1]INFN-LNF, Istituto Nazionale di Fisica Nucleare-Laboratori Nazionali di Frascati, Frascati, 00044 Roma, Italy
[2]Department of Physics, Faculty of Science, University of Zagreb, 10000 Zagreb, Croatia
[3]Horia Hulubei National Institute of Physics and Nuclear Engineering IFIN-HH Măgurele, Romania
[4]Stefan-Meyer-Institut für Subatomare Physik, Vienna, 1030, Austria
[5]Politecnico di Milano, Dipartimento di Elettronica, Informazione e Bioingegneria, Milano, 20133, Italy
[6]INFN Sezione di Milano, 20133, Italy
[7]Physik Department E62, Technische Universität München, Garching, 85748, Germany
[8]RIKEN, Tokyo 351-0198, Japan
[9]Faculty of Physics, Astronomy, and Applied Computer Science, Jagiellonian University, Krakow, 30-348, Poland
[10]Center for Theranostics, Jagiellonian University, Krakow, Poland
[11]Research Center for Electron Photon Science (ELPH), Tohoku University, Sendai, 982-0826, Japan
[12]Centro Ricerche Enrico Fermi – Museo Storico della Fisica e Centro Studi e Ricerche "Enrico Fermi", Roma, 00184, Italy
[13]University of Vienna, Vienna Doctoral School in Physics, Vienna, 1090, Austria

Abstract. Light kaonic atoms spectroscopy provides a unique approach to study the low-energy strong interaction in the strangeness sector. Precise measurements of X-ray emission from light kaonic atoms provide valuable information on kaon-nucleus interaction at threshold without the need for extrapolation as required in scattering experiments. The SIDDHARTA-2 experiment at the DAΦNE collider of INFN-LNF is now poised to perform the challenging measurements of the K^-- d 2p→1s transition to extract the isospin-dependent antikaon-nucleon scattering lengths. To achieve this goal, the background reduction is a crucial factor. This paper provides an overview of the SIDDHARTA-2 Veto-1 system, which uses scintillators outside the vacuum chamber to detect charged particles produced by K^- absorption by the nucleus. The arrival time of these particles is correlated with the position where the kaonic atom has been created inside the setup, allowing for the rejection of kaons stopped outside the target cell, which is a critical component for reducing the background and improve the accuracy of the measurement.

1 Introduction

SIDDHARTA-2 is a cutting-edge experiment that aims to measure the properties of kaonic atoms, which are exotic atomic systems consisting of a negatively charged kaon and a nucleus. The study of kaonic atoms is a unique tool to understand the strong interaction between particles in the non perturbative regime with strangeness, with implications from nuclear and particle physics to astrophysics [1–4]. In 2009, the SIDDHARTA experiment performed precise measurements of atomic transitions in kaonic hydrogen. However, to determine the isospin dependent antikaon-nucleon scattering lengths, the crucial measurement of kaonic deuterium is still missing due to the expected low yield of the kaonic deuterium transition to the fundamental level. To perform this challenging measurement, the SIDDHARTA-2 collaboration developed a completely new apparatus. The SIDDHARTA-2 setup (Fig. 1) is currently installed at the interaction region of the DAΦNE [5] collider of Istituto Nazionale di Fisica Nucleare - Laboratori Nazionali di Frascati (INFN-LNF). DAΦNE delivers low momentum (127 MeV/c) charged K^+K^- pairs coming from the ϕ meson decay. The trigger of the experiment consists of two plastic scintillators placed in the vertical plane above and below the beam pipe to detect the two charged kaons. Once the kaons are triggered, they pass through a thin degrader made of few hundred microns of mylar sheets and then enter in a cylindrical target cell filled with gas. The degrader is tuned to maximize the kaons stopped in the gas target [6]. Upon stopping, a K^- is captured into a kaonic atom that emits X-rays in the range of several keV before being absorbed by the nucleus. The gaseous cryogenic target is surrounded by 384 state-of-the-art Silicon Drift Detectors (SDDs) devel-

*e-mail: francesco.sgaramella@lnf.infn.it
**current address: ASI, Agenzia Spaziale Italiana, Roma, Italy

 https://doi.org/10.1051/epjconf/202329006005

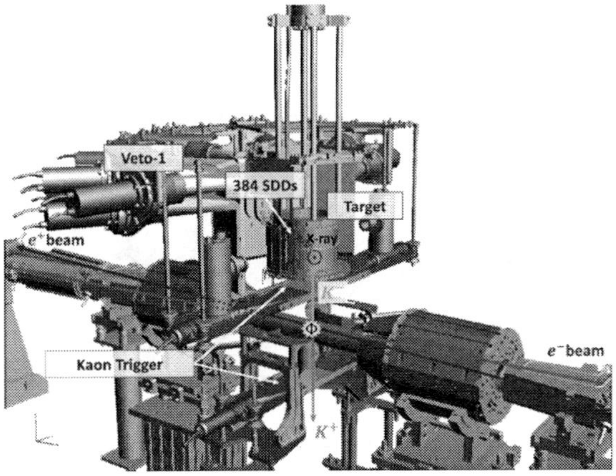

Figure 1. Schematic layout of the SIDDHARTA-2 experimental apparatus

oped specifically for high precision X-ray spectroscopy. The good energy and time resolution ($\Delta E/E < 10^{-3}$) as well as the excellent linear response of the SDD system are key features for performing the kaonic deuterium measurement with high precision [7–9]. In addition, the setup is equipped with several veto systems crucial to increase the signal to background ratio. In this work we present the Veto-1 system of the SIDDHARTA-2 experiment, based on the detection of the charged pions produced by kaons absorption on various materials, including the gas target.

2 The Veto-1 system

2.1 Working principle

To successfully measure kaonic deuterium transitions while dealing with low signal and significant background, it is crucial to select events that occur solely within the gas target and eliminate the hadronic background due to kaons stopped elsewhere. The Veto-1 working principle involves identifying events within the gas, based on their characteristic timing, related to kaon moderation. The Veto-1 system for the SIDDHARTA-2 experiment aims to provide a signal from particles, mainly pions, as result of K^- nuclear absorption in the gas target. The signal is easily distinguishable from events related to K^- stopped in the solid elements of the setup, due to their longer moderation time. The slowing-down time elapsed between the K^- entrance in high density gas and its subsequent capture by an atom has a typical range from 3 to 7 ns and is correlated to the number of collisions needed to transfer the majority of its kinetic energy to the gas atoms.
The system takes advantage of the high probability (around 90%) for the production of charged pions following the final kaon absorption by the atomic nucleus [10–13], making the Veto-1 an efficient tool to provide information on where the kaonic atom has been created.

2.2 Design

Considering that pions coming from hadronic interaction easily pass through SDDs and vacuum chamber, the design for the Veto-1 system consists of a barrel-shape scintillator, surrounding the cylindrical vacuum chamber. A dedicated Monte Carlo Geant4 based simulation has been performed, to determine the optimal set of geometric parameters that result in a high count of the detected photons, while minimizing the spread of signal time.
The barrel is made of 12 distinct (260×110×10) mm^3 plastic scintillator (ELJEN TECHNOLOGY, EJ-200) segmented into 3 equal strips, read by two Hamamatsu R10533 photomultipliers (PMT) through a complex multi-reflection light guides, in order to fit the available space around the vacuum chamber, limited by the external lead shielding (Fig. 2). To increase the covered solid angle, two additional scintillators have been placed under the vacuum chamber at the sides of the kaon trigger.

2.3 Performance

A Veto-1 prototype was tested at the PSI using a 170 MeV/c momentum pion beam. The measured mean time resolution was (746±53) ps FWHM, while the efficiency was (96±2)% [14], in agreement with the requirements for the optimal operation of the system. In 2022 the full Veto-1 system was installed around the SIDDHARTA-2 vacuum chamber and was tested for the first time with the rest of the apparatus during the commissioning run of the experiment. The target cell was filled with Helium at a temperature of 27 K and pressure of 1 bar, which corresponds to 1.0% LHeD (Liquid Helium Density). The high yield of the kaonic helium allowed to verify the operation of the full SIDDHARTA-2 setup with an integrated luminosity of 5 pb^{-1}.
The timing information given by the difference between the between a pion-induced signal in the Veto-1 scintillators and the kaon trigger signal is related to the moderation time of kaons in the gaseous target, allowing for the separation of the events corresponding to kaons stopped in the gas from background events caused by the decay of K^+ or K^- stopped in the setup materials. This discrimination can be clearly seen in Fig. 3. The plot shows the time distribution, given in arbitrary units of the Time to Digital Converter (TDC), obtained during the kaonic helium run. The first peak corresponds to particles produced by the K^- absorption in the setup materials, such as the target entrance window, the degrader or the several mechanical support frames. Instead, the second peak corresponds to the K^- stopped inside the gas target, since the kaon moderation time in the gas is longer with respect to the moderation time in solid elements. Finally, the right tail is related to the K^+ decay and no peak is seen because there is no nuclear absorption.
The sensitivity of the Veto-1 to the kaon moderation time can be used also to optimize the degrader, since the amplitude and the position of the peaks shown in Fig. 3 change depending on the thickness and density of the materials through which the kaon passes. The degrader is a fundamental element of the apparatus and the optimization of its

 https://doi.org/10.1051/epjconf/202329006005

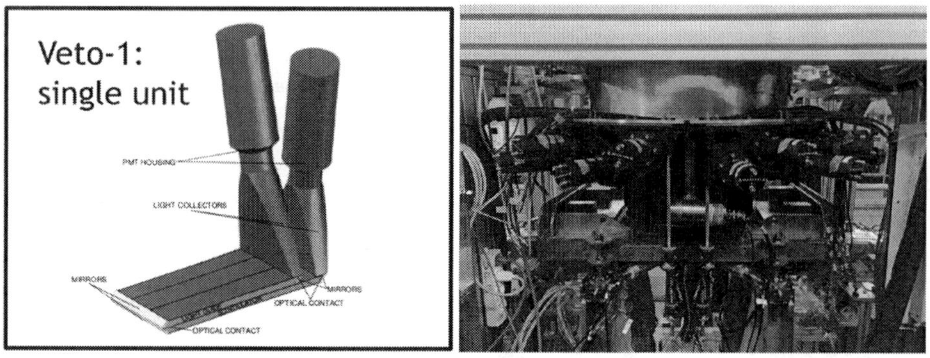

Figure 2. Left: Veto-1 single unit layout. Right: Photo of the Veto-1 system installed around the SIDDHARTA-2 vacuum chamber.

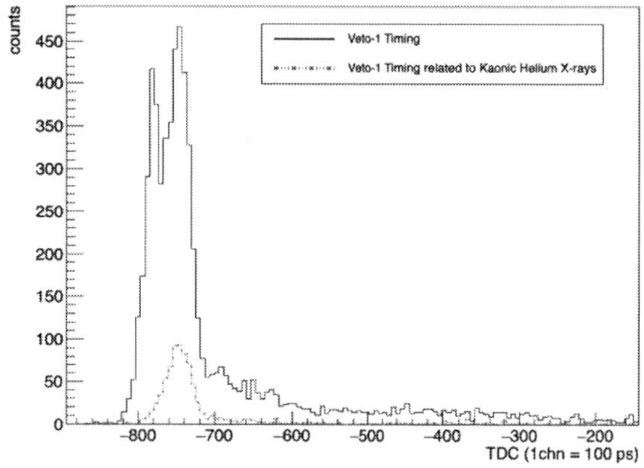

Figure 3. Veto-1 time distribution, in arbitrary unit of the Time to Digital Converter (TDC), measured during the SIDDHARTA-2 Kaonic Helium run. The red dotted line represents the time distribution of events related to kaonic Helium atoms.

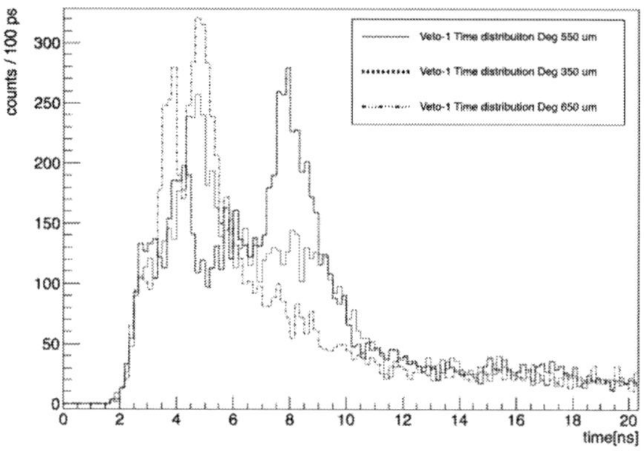

Figure 4. Veto-1 timing simulations for different degraders and the target filled with helium at 1.4% LHeD.

thickness is essential to maximize the kaons stopped inside the gas [6]. As an examples Fig. 4 shows the simulated time distributions for different degrader thicknesses and a target cell filled with helium at 1.4% LHeD. The difference between the plots is clearly visible and comparing the simulations with the real data will help to cross-check the Monte Carlo simulation and optimize the degrader thickness.

Moreover, the time distribution also changes if there is a variation in gas pressure or temperature, since the kaon stopping distribution strongly depends on the gas density. Therefore, the Veto-1 can be used to monitor the experiment's parameters and helps to ensure the stability of the data taking conditions.

3 Conclusions

In conclusion, the Veto-1 system is a critical component of the SIDDHARTA-2 experiment. The system's design and performance were carefully evaluated, and the results showed that the Veto-1 system can identify the pions produced by kaons absorption with high efficiency and good time resolution. The Veto-1 system's ability to provide timing information is essential for the success of the experiment improving the signal to background ratio, by rejecting the events related to K^+ decay or K^- stopped outside the gas target. Furthermore, the Monte Carlo simulations demonstrate that the Veto-1 system also provides valuable information for monitoring the experiment's condition and ensuring its stability.

Acknowledgments

We thank C. Capoccia from LNF-INFN and H. Schneider, L. Stohwasser, and D. Pristauz-Telsnigg from Stefan Meyer-Institut for their fundamental contribution in designing and building the SIDDHARTA-2 setup. We thank as well the DAΦNE staff for the excellent working conditions and permanent support. We acknowledge support from the SciMat and qLife Priority Research Areas budget under the program Excellence Initiative—Research University at the Jagiellonian University. Part of this work was supported by the Austrian Science Fund (FWF): [P24756-N20 and P33037-N]; the EXOTICA project of the Minstero degli Affari Esteri e della Cooperazione Internazionale, PO21MO03; the Croatian Science Foundation under the project IP-2018-01-8570; the EU STRONG-2020 project (Grant Agreement No. 824093; the EU Horizon 2020 project under the MSCA (Grant Agreement 754496); the Japan Society for the Promotion of Science JSPS KAKENHI Grant No. JP18H05402; the Polish Ministry of Science and Higher Education grant No. 7150/E-338/M/2018 and the Polish National Agency for Academic Exchange (grant no PPN/BIT/2021/1/00037).

References

[1] C. Curceanu, C. Guaraldo, M. Iliescu, M. Cargnelli, R. Hayano, J. Marton, J. Zmeskal, T. Ishiwatari, M. Iwasaki, S. Okada et al., Rev. Mod. Phys. **91**, 025006 (2019)

[2] C. Curceanu et al., Symmetry **12**, 547 (2020)

[3] M. Merafina, F.G. Saturni, C. Curceanu, R. Del Grande, K. Piscicchia, Phys. Rev. D **102**, 083015 (2020), 2007.03024

[4] A. Drago, M. Moretti, G. Pagliara, Astron. Nachr. **340**, 189 (2019)

[5] C. Milardi et al., *Preparation Activity for the Siddharta-2 Run at DAΦNE*, in *9th International Particle Accelerator Conference, IPAC2018, Vancouver BC Canada* (2018)

[6] D. Sirghi et al. (SIDDHARTA-2), J. Phys. G: Nucl. Part. Phys. **49**, 055106 (2022)

[7] M. Miliucci et al., Measur. Sci. Tech. **32**, 095501 (2021)

[8] M. Miliucci et al., Measur. Sci. Tech. **33**, 095502 (2022)

[9] F. Sgaramella et al., Physica Scripta **97**, 114002 (2022)

[10] J.e.a. Beringer (Particle Data Group), Phys. Rev. D **86**, 010001 (2012)

[11] C. Vander Velde-Wilquet, J. Sacton, J.H. Wickens, D.N. Tovee, D.H. Davis, Nuovo Cim. A **39**, 538 (1977)

[12] C. Vander Velde-Wilquet, J. Sacton, J.H. Wickens, C. Comber, D.H. Davis, D.N. Tovee, Nuovo Cim. A **38**, 178 (1977)

[13] C. Vander Velde-Wilquet, G.S. Keyes, J. Sacton, J.H. Wickens, D.H. Davis, D.N. Tovee, Nucl. Phys. A **241**, 511 (1975)

[14] M. Bazzi et al., JINST **8**, T11003 (2013)

EPJ Web of Conferences **290**, 07001 (2023) https://doi.org/10.1051/epjconf/202329007001
EuNPC 2022

Pulsed Production of Antihydrogen in AEgIS

N. Zurlo[1,2,*], *M.* Auzins[3], *B.* Bergmann[4], *G.* Bonomi[2,5], *R.S.* Brusa[6,7], *P.* Burian[4], *A.* Camper[8], *F.* Castelli[9], *R.* Ciury[10], *G.* Consolati[11,12], *M.* Doser[13], *A.* Farricker[14], *L.* Glöggler[13], *Ł.* Graczykowski[15], *M.* Grosbart[13], *F.* Guatieri[7], *N.* Gusakova[13,16], *S.* Haider[13], *S.* Huck[13,17], *M.* Janik[15], *G.* Kasprowicz[15], *G.* Khatri[13], *Ł.* Kłosowski[10], *G.* Kornakov[15], *V.* Krumins[3,13], *L.* Lappo[15], *A.* Linek[10], *J.* Malamant[8,13], *C.* Malbrunot[13], *S.* Mariazzi[6,7], *L.* Nowak[13], *D.* Nowicka[15], *E.* Oswald[13], *D.* Pagano[2,5], *L.* Penasa[6,7], *M.* Piwiński[10], *S.* Pospisil[4], *L.* Povolo[6,7], *F.* Prelz[12], *S.* Rangwala[18], *B.* Rienäcker[13,14], *O.M.* Røhne[8], *A.* Rotondi[2,19], *H.* Sandaker[8], *P.* Smolyanskiy[4], *T.* Sowiński[20], *D.* Tefelski[15], *G.* Testera[21], *M.* Volponi[4,13], *C.P.* Welsch[14], *T.* Wolz[13], *M.* Zawada[10], and *J.* Zielinski[15]
(AEgIS collaboration)

[1]Department of Civil, Environmental, Architectural Engineering and Mathematics, University of Brescia, via Branze 43, 25123 Brescia, Italy
[2]INFN Pavia, via Bassi 6, 27100 Pavia, Italy
[3]University of Latvia, Department of Physics Raina boulevard 19, LV-1586, Riga, Latvia
[4]Institute of Experimental and Applied Physics, Czech Technical University in Prague, Husova 240/5, 11000 Prague 1, Czech Republic
[5]Department of Mechanical and Industrial Engineering, University of Brescia, via Branze 38, 25123 Brescia, Italy
[6]TIFPA/INFN Trento, via Sommarive 14, 38123 Povo, Trento, Italy
[7]Department of Physics, University of Trento, via Sommarive 14, 38123 Povo, Trento, Italy
[8]Department of Physics, University of Oslo, Sem Sælandsvei 24, 0371 Oslo, Norway
[9]Department of Physics, University of Milano, via Celoria 16, 20133 Milano, Italy
[10]Institute of Physics, Faculty of Physics, Astronomy and Informatics, Nicolaus Copernicus University in Toruń, Grudziadzka 5, 87-100 Torun, Poland
[11]Politecnico of Milano, Piazza Leonardo da Vinci 32, 20133 Milano, Italy
[12]INFN Milano, via Celoria 16, 20133, Milano, Italy
[13]Physics Department, CERN, 1211 Geneva 23, Switzerland
[14]University of Liverpool, UK and The Cockcroft Institute, Daresbury, UK
[15]Warsaw University of Technology, Faculty of Physics ul. Koszykowa 75, 00-662, Warsaw, Poland
[16]Department of Physics, NTNU, Norwegian University of Science and Technology, Trondheim, Norway
[17]Department of Physics, University of Hamburg, Jungiusstraße 9, 20355 Hamburg, Germany
[18]Raman Research Institute, C. V. Raman Avenue, Sadashivanagar, Bangalore 560080, India
[19]Department of Physics, University of Pavia, via Bassi 6, 27100 Pavia, Italy
[20]Institute of Physics, Polish Academy of Sciences, Aleja Lotnikow 32/46, PL-02668 Warsaw, Poland
[21]INFN Genova, via Dodecaneso 33, 16146 Genova, Italy

Abstract. Low-temperature antihydrogen atoms are an effective tool to probe the validity of the fundamental laws of Physics, for example the Weak Equivalence Principle (WEP) for antimatter, and -generally speaking- it is obvious that colder atoms will increase the level of precision.

After the first production of cold antihydrogen in 2002 [1], experimental efforts have substantially progressed, with really competitive results already reached by adapting to cold antiatoms some well-known techniques previously developed for ordinary atoms. Unfortunately, the number of antihydrogen atoms that can be produced in dedicated experiments is many orders of magnitude smaller than of hydrogen atoms, so the development of novel techniques to enhance the production of antihydrogen with well defined (and possibly controlled) conditions is essential to improve the sensitivity.

We present here some experimental results achieved by the AEgIS Collaboration, based at the CERN AD (Antiproton Decelerator) on the production of antihydrogen in a pulsed mode where the production time of 90% of atoms is known with an uncertainty of $\sim$ 250 ns [2]. The pulsed antihydrogen source is generated by the charge-exchange reaction between Rydberg positronium (Ps^*) and an antiproton ($\bar{p}$): $\bar{p} + Ps^* \rightarrow \bar{H}^* + e^-$, where Ps^* is produced via the implantation of a pulsed positron beam into a mesoporous silica target, and excited by two consecutive laser pulses, and antiprotons are trapped, cooled and manipulated in Penning-Malmberg traps. The pulsed production (which is a major milestone for AEgIS) makes it possible to select the antihydrogen axial temperature and opens the door for the tuning of the antihydrogen Rydberg states, their de-excitation by pulsed lasers and the manipulation through electric field gradients.

In this paper, we present the results achieved by AEgIS in 2018, just before the Long Shutdown 2 (LS2), as well as some of the ongoing improvements to the system, aimed at exploiting the lower energy antiproton beam from ELENA [3].

 https://doi.org/10.1051/epjconf/202329007001

1 Introduction

Experimental evidence has demonstrated that things fall in the Earth's gravitational field, at the same place, with the same acceleration regardless of their mass and composition. In fact, already in the seventeenth century Newton [4] realised that inertial and gravitational masses must be equal, establishing what is now called the Weak Equivalence Principle (WEP).

More than two centuries later, the Einstein Equivalence Principle (EEP), a cornerstone of the General Relativity [5], was proposed by Einstein in 1916 as an extension of the WEP. Einstein stipulated the WEP's validity as a prerequisite in his formulation of the equivalence principle.

Today, the WEP has undergone considerable experimental testing, and very strict constraints on its potential violation with ordinary matter have been established ($< 10^{-15}$), see the review [6] as well as the most recent results [7]. Moreover, the WEP should hold for antimatter, according to various theoretical and experimental considerations although these arguments are indirect and rely on certain theoretical presumptions. Conversely, the majority of attempts at a quantum theory of gravity predict novel interactions that could violate the WEP for antimatter (see e.g. [8]).

The goal of the AEgIS (Antimatter Experiment: gravity, Spectroscopy, Interferometry) experiment is to perform a direct test of the WEP on antimatter by measuring the acceleration of a cold antihydrogen beam in Earth's gravitational field [9], and its development is based on the idea of determining the vertical displacement due to gravity of an antihydrogen beam passing through a moiré deflectometer coupled to a position sensitive detector, akin to the one presented in [10]. The next section will provide an overview of the AEgIS experiment and procedure, which finds its foundations in a resonant charge-exchange process ([11, 12]).

2 Experimental apparatus and method

Differently from other antihydrogen experiments also based at the CERN AD (ATHENA [1], ATRAP [13], ALPHA [14] and ASACUSA [15]), that require the adjacent confinement of clouds of oppositely charged cold particles (antiprotons and positrons), our experiment necessitate to set up a sample of trapped antiprotons as well as a pulsed source of Ps^*. In the past, another experiment exploited the resonant charge exchange reaction between antiprotons and Ps^*, mediated by laser-excited cesium atoms (ATRAP [16]), but -as the experiments mentioned before- it led to a continuous antihydrogen production, while AEgIS method has the advantage to produce antihydrogen in a very short (~ 250 ns) temporal window.

Since the charge exchange cross section depends only on the relative velocity v_r between $\bar{p}$ and Ps^*, absolute velocities are in principle irrelevant, but in our conditions ($T_{\bar{p}} \simeq 400$ K, $v_{Ps^*,avg} \simeq 10^5$ m/s [17]) we can say that the relative velocity is mainly given by the Ps^* velocity itself. As a consequence, $\bar{p}$ cooling would have no effect on the production rate for us, although we need that antiprotons are cold enough to have the full efficiency in $\bar{H}$ detection (since the velocity of the resulting $\bar{H}$ is dominated by that of the $\bar{p}$, and the detector efficiency is reduced right after Ps^* is produced). On the other hand, it is clear that also cold Ps^* would be beneficial, because the reaction cross section drops suddenly for v_r higher than the e^+ velocity in the classical Ps orbit [18, 19]: in our case, $n_{Ps} \simeq 17$ and the cross section is negligible for $v_r \gtrsim 1.3 \cdot 10^5$ m/s.

Figure 1a shows the main elements of the AEgIS apparatus where $\bar{p}$ and Ps^* are prepared. All the charged particles, namely e^+, $\bar{p}$ and e^- (used to cool antiprotons by Coulomb collisions), are confined and manipulated inside Malmberg-Penning traps [20, 21] built with a sequence of cylindrical electrodes of different lengths, disposed in Ultra High Vacuum at cryogenic temperatures, aligned with the magnetic field produced by superconducting solenoids. Each electrode has a slice replaced by a metallic grid to get Ps^* into the trap.

Summarising, we catch and stack multiple antiprotons bunches (typically 8), delivered by the AD, in a trap region called "5 T trap" and we cool them by collisions with electrons which, in turn, loose their radial energy in the high magnetic field (precisely 4.46 T). The mixed $\bar{p}/e^-$ non-neutral plasma is then radially compressed [22–24] and then antiprotons are ballistically transferred along the expanding magnetic field lines into the region called "1 T trap" because of its lower magnetic field, where further cooling with electrons is performed.

At the end of this preparation, that usually takes ~ 15 minutes, we obtain up to 10^6 $\bar{p}$ trapped with about 10^6 e^- in a quasi-harmonic potential ready for possible $\bar{H}$ formation. We infer the radial shape of the plasma with an imaging system made of a downstream on-axis Micro Channel Plate (MCP) coupled to a phosphor screen read out by a CMOS camera [22], while the number of e^- is measured by means of signals collected on a Faraday cup and the number of $\bar{p}$ is inferred from signals in external scintillators. The size of the $\bar{p}$ plasma is typically around 1 mm. We keep $\bar{p}$ plasma confined for some thousands of seconds with some expansion and negligible antiproton losses.

While antiprotons are prepared in this way, e^+ are stacked in a Surko-type accumulator based on a ~ 1 GBq ^{22}Na source. Once the antiprotons are ready, a bunch of typically $2 \cdot 10^6$ positrons (with a temporal width shorter than $\sim$10 ns) is released every ~ 100 s, is accelerated in flight to 4.6 keV and is magnetically transported along the B-field. This bunch is eventually smashed into the $e^+ - Ps$ converter, a cryogenic nanoporous target [25] which is situated close to the 1 T $\bar{p}$ trap (see Figure 1b).

The formed ortho-positronium (the para-positronium being too short lived in this regard) is emitted by the converter with kinetic energy $\sim$3 eV [26] and it is slowed down by collisions with the nanochannel walls until it is emitted in vacuum. After leaving the converter, it is ex-

*e-mail: zurlo@cern.ch

https://doi.org/10.1051/epjconf/202329007001

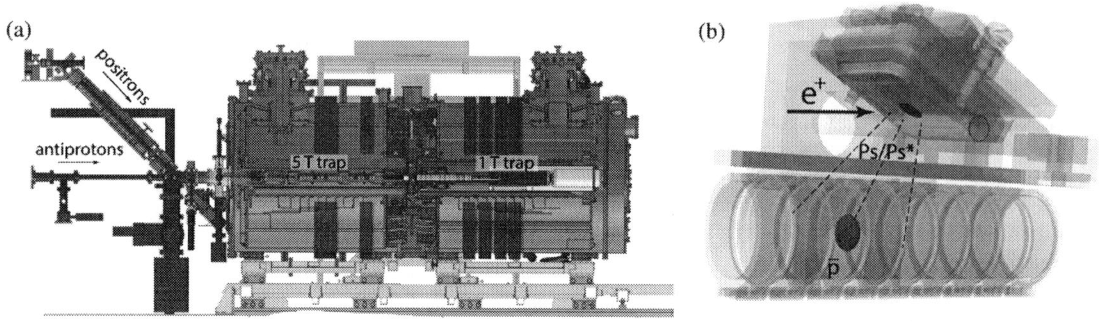

Figure 1. Drawing of the experimental apparatus. In violet, the external scintillator detector array, mentioned in the text (b): Enlargement of the $\bar{H}$ production region, located in the centre of the 1 T trap. In red: the antiproton plasma, inside a classical multi-ring trap hosting a mesh in the upper part. In blue: the nanoporous silica converter region where positrons impinge producing *Ps*, which is subsequently excited by two pulsed laser beams (sketched in yellow).

cited to a Rydberg state with principal quantum number $n \simeq 17$ through two subsequent laser pulses [27, 28] (the first from $n = 0$ to $n = 3$; the second to the targeted n). The time when the first laser is fired is the reference time for our pulsed scheme: this time is known with few-ns-accuracy.

Given the typical time spread of the target-hitting e^+ pulse of < 10 ns, and the *Ps* cooling time in the converter of the order of ten ns, the major uncertainty on the $\bar{H}$ production time is due to the Ps^* velocity spread that affects the time of flight needed to reach the $\bar{p}$ cloud. With the *Ps* velocity distribution measured in dedicated experiment [17], this time is a few hundreds of ns.

3 Results

First, we should recall that any antihydrogen produced, differently from trapped antiprotons, would be unconfined by the Penning trap fields and would eventually annihilate on the trap walls after a short drift time. The annihilation products from the antiproton annihilation (manly charged pions) can be then detected by the external scintillator detector array. In practice, there are also small, continuous losses of antiprotons in the Penning trap, due to collisions on the residual gas and to trap imperfections that make the antiproton plasma slowly expand and the peripherals antiprotons collide with the trap wall. These "lost antiprotons" produce a signal that in our detectors is totally similar to the one generated by antihydrogen: this is the reason why it so important to estimate accurately this "background signal", as we will explain later.

The detection of the events discussed here was performed with the so-called "External scintillator detector array" (see Figure 1a), situated outside the superconducting solenoids. To be precise, it consisted of 12 plastic (EJ200) arc-shaped scintillating slabs, 1 cm thick and 10 or 20 cm wide, coupled to two photomultipliers (PMTs), one at each end. Each scintillator was designed to cover a 120° angle centred around the vertical plane containing the solenoid axis and was situated either above or below the main cryostat.

Differently from "standard" data acquisition, relying on the signal produced by CAEN N413, LeCroy 623A and LeCroy 622 discriminators and essentially taken via the coincidence between the two photomultipliers connected to the same slab (that was not sensitive enough to effectively count the few antihydrogen annihilations over the antiproton annihilations/cosmic rays/environmental background), we developed a different, more accurate approach for the record and the analysis of the scintillator data in a convenient time window around the time of the positron injection, where antihydrogen production had to be investigated. In fact, through CAEN V1720 modules we recorded the full digitised photomultiplier signals with a sampling frequency of 250 MHz in a time window that can be as wide as a ~5 ms (actually, 1/8 of this duration was sufficient in our case). These signals were subsequently exploited offline only after averaging the signal between the two photomultipliers reading the same scintillating slab, to compensate for light attenuation inside each slab. After an accurate calibration of each PMT, the signal amplitude could be considered an excellent proxy for the energy deposit inside the scintillator by the charged particles passing through it, as reported in [29], so different efficiencies are expected for antiproton/antihydrogen annihilations, *Ps* annihilations and cosmic/environmental background, depending on the energy loss distribution and on the selected threshold. The mentioned digital acquisition of the full PMT signals, together with an accurate threshold optimisation, guided by a detailed Monte Carlo simulation of the full system based on Geant4 (using cosmic muons energy loss for reference, see [30–32] for further details on the cosmic muon generator), allowed us to improve the signal over background ratio for $\bar{p}$ annihilation excess, which in turn is a clear signature for $\bar{H}$ annihilation.

To demonstrate antihydrogen production, we have performed experiments in 3 different conditions, always with

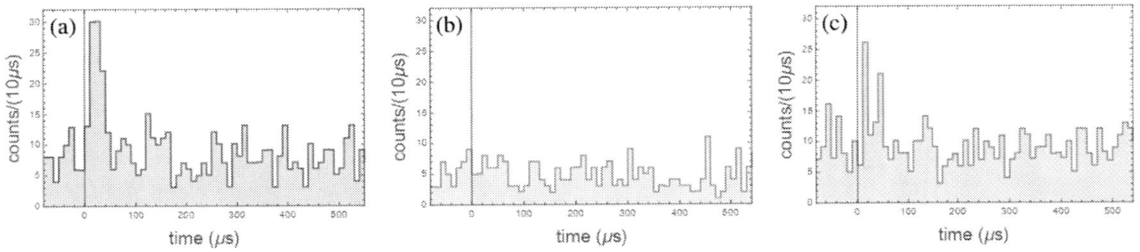

Figure 2. Time distribution scintillator pulses. (a) Experiment with $\bar{p}$, e^+ and laser (2206 cycles, $1.08 \times 10^9 \bar{p}$). (b) Experiment with $\bar{p}$, e^+ and without laser (1211 cycles, $6.08 \times 10^8 \bar{p}$). (c) Experiment with $\bar{p}$, laser and without e^+ (3498 cycles, $1.58 \times 10^9 \bar{p}$). Notice that the overall number of cycles is different for each typology of data; in order to make a direct comparison, histogram renormalization should be applied.

antiprotons stored in the Penning-Malmberg trap. First, we took sequences of runs with e^+ injection, subsequent *Ps* production and laser excitation in nominal conditions (Figure 2a): these are all the ingredients needed to form $\bar{H}$. Then we took cycles without extracting e^+ from the accumulator (Figure 2b) and finally cycles without firing the laser (Figure 2c). In the last two kind of runs, $\bar{H}$ formation is inhibited by the absence of some of the essential ingredients to produce it.

The full signal was acquired for a 650 μs-long time window, starting 50 μs before the laser shot ($t = 0$), in order to have a convenient control time window. Unfortunately, the first 1 μs after the positron injection the detector was blind because it needed time to recover from the saturation experienced right after the positron bunch injection, which produced a large number of simultaneous e^+ annihilations.

From Figure 2a we can infer that there is a signal just after $t = 0$ when we had antiprotons, e^+ injection and laser. This signal occurs between t=1 μs and t=26 μs, consistently with the mentioned saturation of the detector in the first μs. On the other hand, without firing the laser (Figure 2b), the signal was reduced to the temporally-uniform background, due to cosmic rays and $\bar{p}$ losses, as expected. It must be also mentioned (see Figure 2c) that some (but fewer) excess events are present also in the runs with $\bar{p}$ and the laser but without e^+ injection, that we interpret as a possible background of $\bar{p}$ annihilations due to the laser-induced outgassing from the cryogenic surfaces. Summarising, we observed 79 events in the signal region while we expect to detect 33.4 ± 4.6 events under the hypothesis of absence of $\bar{H}$ formation (see [2] for a more detailed discussion of the statistical significance).

4 Improvements after 2018

During and after the CERN LS2, an appreciable number of improvements has been implemented on the system to enhance the overall performance (and antihydrogen yield in particular).

First, the focus was on positrons and positronium. The experiment design has been revised: since the old setup (Figure 1b) was based on a *Ps*-producing converter situated on top of the $\bar{p}$ trap, the motional Stark effect prevented *Ps* atoms moving perpendicularly to the magnetic field from being excited to higher Rydberg state [17]. During the LS2, it was modified into a fully collinear setup, where the converter is aligned with axis of the $\bar{p}$ trap whose electrodes in turn can have larger size since their radius is not limited by the distance to the converter. In principle, this should lead approximately to a 4-fold improvement in antihydrogen yield.

At the same time, the preparation of the e^+-*Ps* converter was improved [33] ($\sim$ 3-fold improvement in antihydrogen yield) and new diagnostic tools were developed [34, 35].

On the antiproton side, the commissioning of the ELENA ring allows now the capture of a much larger number of antiprotons, assuming equal catching potentials and optimal degrader foil thicknesses (besides the fact that in the new system the high voltage for catching antiprotons has been slightly increased with respect to 2018). In fact, the AD delivered a bunch of $\sim 3 \cdot 10^7 \bar{p}$ with 5.3 MeV kinetic energy (100 MeV/c momentum) every 100 s that where reduced by a factor of $\sim$ 100 after catching, while ELENA releases $\sim 6 \cdot 10^6 \bar{p}$ with 100 keV kinetic energy (13.7 MeV/c momentum) with comparable repetition rate, but with much higher catching efficiency for the related experiments (first measurements suggest that for AEgIS it could be better than $\sim$ 60%). In principle, this should reflect a $\sim$ 4-fold improvement in antihydrogen yield.

Other upgrades concern the laser system (in particular a new alexandrite laser system for a more efficient *Ps* excitation), the electronics for the nanosecond control system to set the voltages on the trap electrodes (which has been moved to the Sinara hardware family driven by the ARTIQ library system, which in turn is based on Python) as well as the overall software (where a new distributed, modular control system arranged in micro-services and programmed in Labview™ has been implemented to run the experiment). The new laser system is meant essentially to improve the long term stability of the cavity and, because of the wider laser bandwidth, also the antihydrogen yield should benefit. The control system upgrades were developed mainly for an improvement in ease of operation of the experiment.

In principle, if everything works as per design, the antihydrogen yield should increased by one or two orders of magnitude.

5 Conclusions

We have presented the first AEgIS results about $\bar{H}$ pulsed production, based the 'resonant" charge exchange reaction between $\bar{p}$ and laser-excited Ps^*. Till now, this is the $\bar{H}$ production scheme with the most precise time tagging ever achieved.

Although we have confirmed that the the $\bar{H}$ signal is not consistent with the background, the production rate is still extremely low: on average, less than 0.1 $\bar{H}$ for a typical cycle operated in 2018. In fact, substantial upgrades are being implemented to improve the overall experiment.

References

[1] M. Amoretti et al., *Production and detection of cold antihydrogen atoms*, Nature 419, 456 (2002).

[2] C. Amsler et al., *Pulsed production of antihydrogen*, Commun. Phys. 4, 19 (2021). DOI:10.1038/s42005-020-00494-z

[3] W. Bartmann et al. on behalf of the ELENA and AD teams, *The ELENA facility*, Phil. Trans. R. Soc. A 376: 20170266 (2018). http://dx.doi.org/10.1098/rsta.2017.0266

[4] I. Newton, *Philosophiae Naturalis Principia Mathematica*, Londini: Jussu Societatis Regae ac Typis Josephi Streater: Prostant Venales apud plures Bibliopolas (1687).

[5] A. Einstein, *Die Grundlage der allgemeinen Relativitätstheorie*, Annalen der Physik, Vol. 49, p.769 (1916).

[6] C. M. Will, *The Confrontation between General Relativity and Experiment*, Living Reviews in Relativity, Vol. 17, 4 (2014).

[7] P. Touboul et al., *MICROSCOPE Mission: Final Results of the Test of the Equivalence Principle*, Phys. Rev. Lett. 129, 121102 (2022).

[8] V. A. Kostelecký and J. D. Tasson, *Matter-gravity couplings and Lorentz violation*, Phys. Rev. D 83, 016013 (2011).

[9] M. Doser et al. (AEgIS Collaboration), *Exploring the WEP with a pulsed beam of cold anihydrogen*. Class. Quantum Grav. Vol. 29, p. 213 184009 (2012).

[10] S. Aghion et al. (AEgIS Collaboration), *A moiré deflectometer for antimatter*, Nat. Commun., Vol 5, p.4538, (2014). DOI: http://dx.doi.org/10.1038/ncomms5538

[11] B. I. Deutch, A. S. Jensen, A. Miranda and G. C. Oades, Proceedings of The First Workshop on Antimatter Physics at Low Energy, 371, FNAL (1986).

[12] M. Charlton, *Antihydrogen production in collisions of antiprotons with excited states of positronium*, Physics Letters A, Vol. 143 (3) 143 (1990).

[13] G. Gabrielse et al., *Driven production of cold antihydrogen and the first measured distribution of antihydrogen states*. Phys. Rev. Lett. 89, 213401 (2002). DOI:10.1103/PhysRevLett.89.213401

[14] G. B. Andresen et al. *Trapped antihydrogen*, Nature Vol. 468, p. 673 (2010).

[15] N. Kuroda et al. *A source of antihydrogen for in-flight hyperfine spectroscopy*, Nat. Commun., Vol. 5, p.3089 (2014).

[16] C. H. Storry et al., *First Laser-Controlled Antihydrogen Production* , Phys. Rev. Lett. 93, 263401 (2004).

[17] M. Antonello et al., *Rydberg-positronium velocity and self-ionization studies in a 1T magnetic field and cryogenic environment*, Phys. Rev. A 102, 013101 (2020).

[18] D. Krasnicky, R. Caravita, C. Canali and G.Testera, *Cross section for Rydberg antihydrogen production via charge exchange between Rydberg positroniums and antiprotons in a magnetic field*, Phys. Rev. A 94, 022714(14) (2016).

[19] D. Krasnicky, G. Testera and N. Zurlo, *Comparison of classical and quantum models of anti-hydrogen formation through charge exchange*, J. Phys. B: At. Mol. Opt. Phys. 52, 115202 (2019).

[20] F. G. Major, V. N. Gheorghe and G. Werth, *Charged Particle Traps Springer Series on Atomic, Optical, and Plasma Physics*, Springer-Verlag Berlin Heidelberg (2005).

[21] D. H. E. Dubin and T. M. O'Neil, *Trapped non-neutral plasmas, liquids, and crystals (the thermal equilibrium states)*. Rev. Mod. Phys. 71, 87-172 (1999).

[22] S. Aghion et al., (AEgIS Collaboration), *Compression of a mixed antiproton and electron non-neutral plasma to high densities*. Eur. Phys. J. D 72, 76-86 (2018).

[23] G. B. Andresen et al., (ALPHA Collaboration), *Compression of antiproton clouds for antihydrogen trapping*. Phys. Rev. Lett. 100, 203401 (2008).

[24] N. Kuroda et al., *Radial compression of an antiproton cloud for production of intense antiproton beams*. Phys. Rev. Lett. 100, 203402 (2008).

[25] S. Mariazzi, P. Bettotti and R. S. Brusa, *Positronium Cooling and Emission in Vacuum from Nanochannels at Cryogenic Temperature*, Phys. Rev. Lett. Vol. 104 p. 243401 (2010).

[26] Y. Nagashima et al., *Origin of positronium emitted from SiO_2*, Phys. Rev. B 58, 12676-12679 (1998).

[27] S. Aghion et al. (AEgIS Collaboration), *Laser excitation of the $n = 3$ level of positronium for antihydrogen production*, Phys. Rev. A 94, 012507 (2016).

[28] M. Antonello et al. (AEgIS Collaboration), *Rydberg-positronium velocity and self-ionization studies in 1T magnetic field and cryogenic environment*, Phys. Rev. A 102, 013101 (2020).

[29] N. Zurlo et al. (AEgIS Collaboration), *Calibration and equalisation of plastic scintillator detectors for antiproton annihilation identification over positron/positronium background*, Acta Physica

Polonica B, Vol. 51, No 1, p. 213 (2020).

[30] D. Pagano, G. Bonomi, A. Donzella, A. Zenoni, G. Zumerle and N. Zurlo, *EcoMug: An Efficient COsmic MUon Generator for cosmic-ray muon applications*, Nucl. Instrum. Meth. Phys. Res. A 1014, 165732 (2021). DOI: https://doi.org/10.1016/j.nima.2021.165732

[31] G. Bonomi, A. Donzella, D. Pagano, A. Zenoni, G. Zumerle and N. Zurlo, *A Monte Carlo Muon Generator for Cosmic-Ray Muon Applications* Journal of Advanced Instrumentation in Science, Vol. 2022 (2022). DOI: https://doi.org/10.31526/jais.2022.290.

[32] N. Zurlo, G. Bonomi, A. Donzella, D. Pagano, A. Zenoni and G. Zumerle, *A new Monte Carlo muon generator for cosmic-ray muon applications*, Proceedings of Computational Tools for High Energy Physics and Cosmology — PoS(CompTools2021), Vol.409, p.019 (2022). DOI: https://doi.org/10.22323/1.409.0019

[33] S. Mariazzi et al. (AEgIS Collaboration), *High-yield thermalized positronium at room temperature emitted by morphologically tuned nanochanneled silicon targets* J. Phys. B: At. Mol. Opt. Phys. 54 085004 (2021).

[34] C. Amsler et al. (AEgIS Collaboration), *A ~100 μm-resolution position-sensitive detector for slow positronium* Nucl. Instrum. Meth. Phys. Res. B 457, 44-48 (2019).

[35] L. Glöggler et al. (AEgIS Collaboration), *High-resolution MCP-TimePix3 imaging/timing detector for antimatter physics* Meas. Sci. Technol. 33 115105 (2022).

EPJ Web of Conferences **290**, 07002 (2023) https://doi.org/10.1051/epjconf/202329007002
EuNPC 2022

Electric dipole moment of charged particles using storage rings

Vera Shmakova[1,*] for the JEDI collaboration

[1]GSI Helmholtzzentrum für Schwerionenforschung GmbH, 64291 Darmstadt, Germany

Abstract. The Standard Model (SM) of Particle Physics cannot explain the matter-antimatter asymmetry in the Universe. Especially, new sources of CP-violation are necessary, and this is why physics beyond the SM is being explored. The search for permanent Electric Dipole Moments (EDMs) of elementary particles has the potential to provide a powerful tool to discover new sources of CP-violation. Finding an EDM would be a convincing indication for physics beyond the SM. Moreover, a rotating polarized beam used in EDM searches is sensitive to oscillating EDM caused by axions or axion-like particles (ALPs), the candidates for the dark matter. Storage rings are promising devices to measure EDMs of charged particles by observing the effect of the EDM on the spin motion in the ring. The Cooler Synchrotron COSY at the Forschungszentrum Jülich provides polarized protons and deuterons with momenta up to 3.7 GeV/c and it constitutes an suitable testing ground and starting point for an experimental program to search for EDMs of charged particles. This report provides an overview of the current status of the JEDI project for the first direct measurement of the deuteron EDM at COSY. The analysis is currently ongoing. Due to the complexity of storage rings, this study combines high precision measurements and thorough understanding of the systematic uncertainties involved.

1 Introduction

The Standard Cosmological Model (SCM) does not explain the baryon asymmetry in the Universe, which is quantified by the ratio of baryon to photon number densities, $\eta = \frac{n_b - n_{\bar{b}}}{n_\gamma}$, where n_b and $n_{\bar{b}}$ are, respectively, the numbers of baryons and antibaryons and n_γ is the number of relic photons. The SCM predicts that $\eta \sim 10^{-18}$ [3], but experimental observations reveal a much larger value of η of the order of 10^{-10} [4, 5]. In 1967, Andrei Sakharov proposed a set of requirements that are necessary for baryon-antibaryon asymmetry to occur in the Universe, one of which is the violation of both C and CP invariance [6]. Electric dipole moments (EDMs) of particles violate both parity (P) and time reversal symmetry (T). This implies that, if CPT symmetry is conserved, CP symmetry is also violated. Thus, detecting the EDM of a fundamental particle can provide valuable insights into the matter-antimatter asymmetry in the Universe.

The JEDI collaboration is working towards measuring the EDMs of charged hadrons using a storage ring. An overview of the existing limits of electric dipole moments of various particles is shown in Fig. 1. Currently, there is no direct measurement of the proton EDM. The present limit for the proton EDM was derived based on the EDM measurement of the mercury atom, and there exist no limits at all for the deuteron EDM.

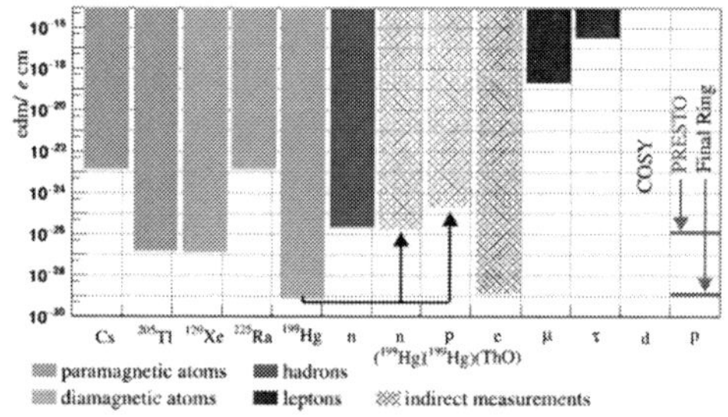

Figure 1. Existing limits for electric dipole moment for various particles (figure taken from [1]). The last two columns are showing the goals of the JEDI collaboration. The initial stage with a magnetic ring COSY is aiming at the first measurement of the deuteron EDM. The goal limits for the second stage "PRESTO" and for the third stage with the final ring are shown in the last column. The last two stages are described in more detail in Section 4.

2 Measurement of the EDM using storage rings

The motion of the spin vector $\vec{S}$ in a storage ring with respect to the cyclotron motion, subject to a radial electric field $\vec{E}$ and a vertical magnetic field $\vec{B}$, is described by the Thomas-BMT equation [7–9]

$$\frac{d\vec{S}}{dt} = [\vec{\Omega}_{\rm MDM} - \vec{\Omega}_{\rm cyc} + \vec{\Omega}_{\rm EDM}] \times \vec{S}. \quad (1)$$

The spin precession depends on the angular velocities due to the magnetic dipole moment (MDM) and the electric

*e-mail: v.shmakova@fz-juelich.de

dipole moment (EDM), $\vec{\Omega}_{\rm MDM} - \vec{\Omega}_{\rm cyc}$ and $\vec{\Omega}_{\rm EDM}$, given by

$$\vec{\Omega}_{\rm MDM} - \vec{\Omega}_{\rm cyc} = -\frac{q}{m}\left(G\vec{B} - (G - \frac{1}{\gamma^2 - 1})\frac{\vec{\beta}\times\vec{E}}{c}\right),$$
$$\vec{\Omega}_{\rm EDM} = -\frac{\eta_{\rm EDM}q}{2mc}\left(\vec{E} + c\vec{\beta}\times\vec{B}\right). \quad (2)$$

Here G is the anomalous gyromagnetic g-factor and $\eta_{\rm EDM}$ denotes a dimensionless electric dipole moment factor, while the electric dipole moment $\vec{d} = \eta_{\rm EDM}\frac{q}{2mc}\vec{S}$. When the spin of a particle remains aligned with its momentum in the absence of an EDM, this is referred to as the "frozen spin" condition. This condition arises when the term $\vec{\Omega}_{\rm MDM} - \vec{\Omega}_{\rm cyc}$ vanishes. For particles with $G > 0$, the frozen spin condition can be achieved in a purely electric ring if the momentum of the particle satisfies the condition $(G - \frac{1}{\gamma^2-1}) = 0$. In this case the radial electric field $\vec{E}$ will exert a torque $\vec{d}\times\vec{E}$, and the spin will precess around the radial axis in the ring leading to a vertical spin component in the beam. This is the idea behind the final stage of the JEDI project, making use of a dedicated purely electric ring to achieve the necessary precision for the proton EDM measurement of up to 10^{-29} e · cm (see last column of Fig. 1) [1]. Meanwhile, as the first stage, a proof-of-principle experiment is currently underway with the existing magnetic storage ring COSY in Jülich, Germany [1], depicted in Fig. 2.

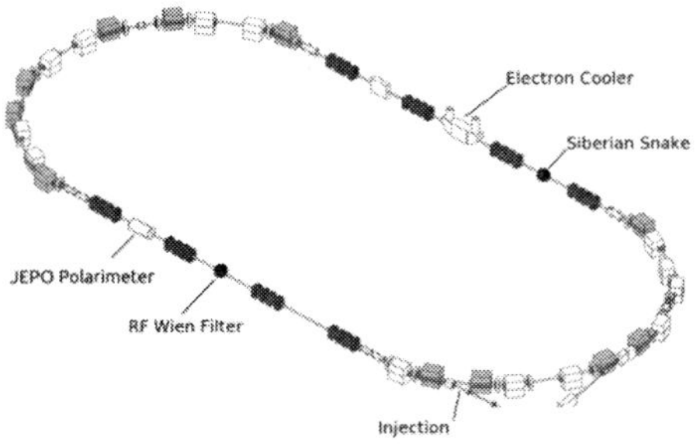

Figure 2. The schematic of Cooler Synchrotron COSY showing the positions of the principal elements.

2.1 EDM experiment at the Cooler Synchrotron COSY

The magnetic storage ring COSY has been previously used for hadron physics experiments and is capable of providing polarized protons and deuterons at momenta up to 3.7 GeV/c. It is an ideal starting point for the search of the EDMs of charged hadrons. The schematic layout of COSY and principal elements are shown in Fig. 2. The experiment was carried out with vertically polarized deuterons injected into the COSY ring and accelerated to a momentum of 970 MeV/c. The principal values for the deuteron EDM experiment at COSY were $G_{\rm d} = -0.143$, $\beta = 0.4594$, $\eta_{\rm EDM}(d = 10^{-26}$ e · cm$) = 1.9 \cdot 10^{-12}$ (see Eq. 2). After bunching (two bunches were stored in the machine, more details are provided further in the text) and electron cooling, the beam was extracted on a carbon block target and the elastically scattered deuterons were detected with the JEPO polarimeter. Subsequently, the spins of the particles were rotated in the horizontal plane by means of an radio-frequency (RF) solenoid. The spin coherence time of the in-plane polarization time was optimized using sextupole magnets [12, 13]. The coherence time of about 1000 s was reached for the purposes of the experiment. In

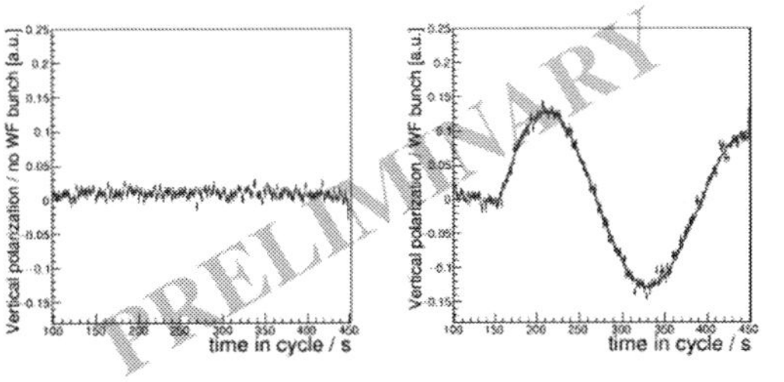

Figure 3. Example of a dataset obtained for one cycle for $\phi^{\rm WF} = -2°$, $\chi^{\rm sol} = -4°$. Left: vertical polarization for the bunch shielded from the Wien filter field, so the polarization stay intact during the cycle. Right: oscillations of the vertical polarization of the second bunch, being exposed to the full field of the Wien filter. Combined fit for the vertical and in-plane polarization is displayed.

a purely magnetic ring, according to Eq. (2), the spin rotation caused by the EDM receives a contribution in the vertical direction from the motional electric field $\vec{E}_r = c\vec{\beta}\times\vec{B}_y$. Due to the spin precession caused by the term $G\vec{B}$, in the absence of a static electrical $\vec{E}$ field, there will be only tiny oscillations of the vertical polarization. In order to accumulate the EDM signal, an RF Wien filter was used [2, 11]. It provides exactly orthogonal magnetic and electric fields and was operated on one of the harmonics K of the spin precession frequency $f_s = f_{\rm rev} \cdot \gamma G$, so that the Wien filter operated at $f_{\rm WF} = f_{\rm rev}\,|K + \gamma G|$, i.e., for deuterons at $K = -1$ and $f_{\rm rev} \sim 750$ kHz, $f_{\rm WF} \sim 871$ kHz.

A real-time feedback system was used to monitor the spin tune ν_s [14] and to keep the Wien filter frequency at $f_{\rm WF} = f_{\rm rev}(K + \nu_s)$ by maintaining the relative phase between spin precession and Wien filter RF oscillations constant throughout the measurements [15, 16]. With such settings a particle receives a spin kick in the same direction for each turn it passes through the RF Wien filter, resulting in a out-of-plane accumulation of the vertical polarization P_y.

In addition, the Wien filter was equipped with fast switches that are capable of turning the RF of the Wien filter on and off for each of several bunches in the ring, allowing us to manipulate the polarization of one chosen bunch in the ring. This way, having two bunches in the machine, one of the bunches can be shielded from the Wien filter field (Fig. 3, left) and be used to lock the phase and fulfill $f_{\rm WF} = f_{\rm rev}K + f_s$ with the feedback system, while the other bunch experiences the full field of the Wien filter, and its effect on P_y and P_x can be measured (Fig. 3, right).

https://doi.org/10.1051/epjconf/202329007002

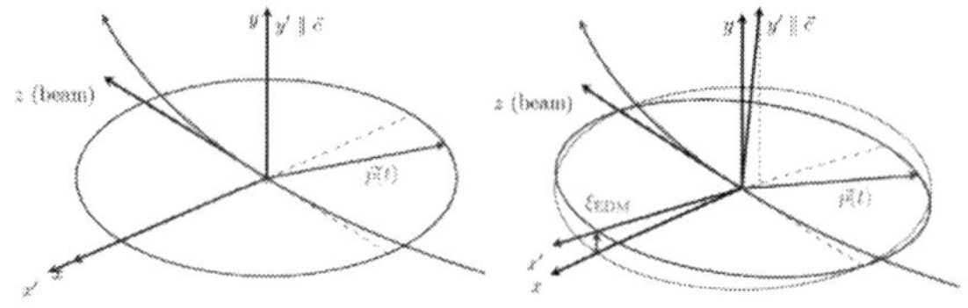

Figure 4. Invariant spin axis is vertical in case of an ideal ring in the absence of the EDM (left), tilted in radial direction due to EDM effect (right). Figure is taken from [2].

2.2 Determination of the invariant spin axis orientation

The observable that we are aiming to measure is the so-called invariant spin axis, an axis about which the spin of the particle is precessing at a given point of the orbit. In the absence of the EDM, this axis coincides with the direction of the magnetic bending field of the ring dipoles $\vec{B_y}$ in an ideal ring, i.e., it is precisely vertical, as shown in Fig. 4, left. As soon as the EDM or magnetic imperfections enter the picture, this will cause a tilt of the invariant spin axis away from the vertical direction. An EDM in an ideal ring will lead to a radial tilt of the invariant spin axis (see Fig. 4, right). In reality, magnetic misalignments will tilt the axis in both directions away from the vertical axis. Once the impact of magnetic imperfections is known, the direction of the invariant spin axis provides an experimental access to the EDM.

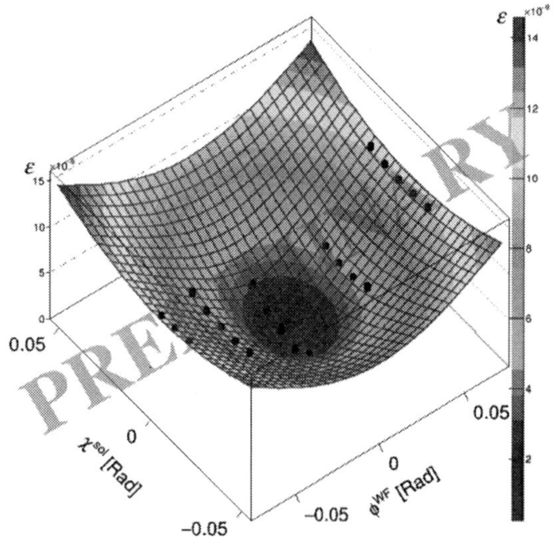

Figure 5. Map of the resonance strength ε^{EDM} (see Eq. (3)) for various Wien filter angle ϕ^{WF} and solenoid spin rotation angle χ^{sol}.

As soon as the magnetic field axis of the Wien filter is parallel to the invariant spin axis, there would be no polarization build-up observed. Therefore, the way to measure the orientation of the invariant spin axis in the experiment is to tilt to various angles either the $\vec{B}$ field of the RF Wien filter, or the invariant spin axis itself by making use of a Syberian snake on opposite side of the ring (see Fig. 2). This way one can search for the combination of ϕ^{WF} and χ^{sol} for which no out-of-plane rotation of the polarization is detected. For this purpose the Wien filter was rotated around the beam direction, providing a radial tilt of ϕ^{WF} in the range $\pm 4°$ and the Syberian snake was used to incline the invariant spin axis at the location of the Wien filter by χ^{sol} in the longitudinal direction.

The resonance strength ε can be defined as a ratio between spin angular frequency Ω^{p_y} and the revolution frequency Ω^{rev}:

$$\varepsilon = \frac{\Omega^{p_y}}{\Omega^{rev}}. \tag{3}$$

During the experiment, data for the vertical asymmetries were collected and fits of the datasets (Fig. 3) enable an accurate determination of the oscillation frequency. A preliminary map of the resulting resonance strengths ε for various Wien filter rotations ϕ^{WF} and the solenoid settings χ^{sol} is shown in Fig. 5. The minimum of the surface indicates the orientation of the invariant spin axis at the location of the Wien filter given by

$$\phi_0^{WF} = 2.53 \pm 0.02 \text{ mrad},$$

$$\chi_0^{sol} = -3.39 \pm 0.03 \text{ mrad}.$$

This large tilt of the axis suggests that our result is mainly dominated by systematic effects that arise from magnetic imperfections in the ring. Currently, realistic simulations of the COSY ring and further investigations regarding magnetic imperfections leading to systematic effects is ongoing.

3 Search for axions and axion like particles

Another experiment that was conducted at COSY is the first search for the axions and axion-like particles (ALPs) [17]. An axion is a hypothetical pseudoscalar particle that is postulated in Peccei-Quinn theory to resolve strong CP-problem [18]. Both axions and ALPs [20] might be candidates for cold dark matter in the universe [20, 21]. Due to the coupling between the axion field and the spin of a particle, a storage ring can be utilized for the search of ALPs. This coupling is based on the induction of an oscillating component d_{AC} to the total EDM of the particle by oscillating axion field:

$$d(t) = d_{DC} + d_{AC}cos(\omega_a(t) + \phi_a), \tag{4}$$

where d_{DC} is the permanent (static) component of the EDM. If the axion frequency matches the spin resonance, $\omega_a = \Omega_{MDM}$, there would be a build-up of vertical polarization out of the horizontal plane. With that in mind, the frequency scan in search for this resonance was done in range from 4.95 to 5.02×10^{-9} eV, looking for a change in the vertical polarization. No signal was observed.

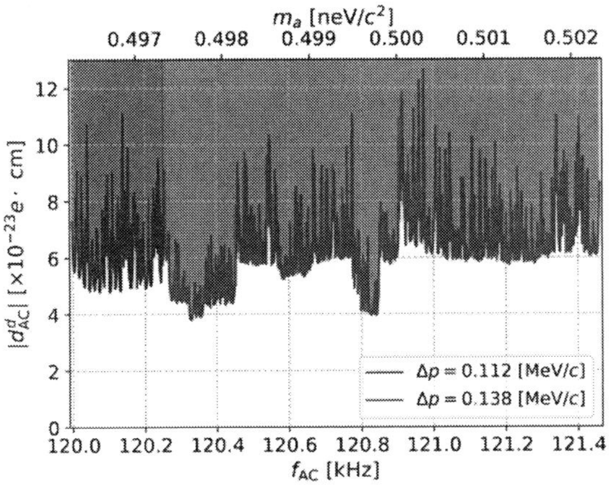

Figure 6. Preliminary 90% upper confidence level sensitivity for excluding the ALPs induced oscillating deuteron EDM in the frequency range from 120.0 to 121.4 kHz and the corresponding axion mass range of 4.95 to 5.02×10^{-9} eV. The green and blue colors show two scanning ramp rates in momentum change. The darker lines denote the upper bounds on the oscillating EDM, the shaded area is the excluded region. Figure taken from [17].

The results at the 90% confidence level sensitivity for excluding the ALPs induced oscillating deuteron EDM, d^d_{AC}, are shown on Fig. 6. The resulting upper bound of d^d_{AC}, based on the average of the individual limit points in Fig. 4, is $d^d_{AC} < 10 \cdot 10^{-23}$ $e \cdot$cm. For more details, see [17].

4 Conclusion and Outlook

Searches for Electric Dipole Moments are crucial for probing possible sources of CP-violation beyond the SM. Storage rings could be used for the search of the EDMs of charged particles. To this end, the first measurements of the deuteron EDM were performed at the Cooler Synchrotron COSY at Forschungszentrum Jülich, Germany.

The observed result is predominantly influenced by systematic effects that stem from magnetic imperfections present in the ring. Thus, extensive efforts are underway to further investigate these systematic effects, with the aim of differentiating them from the EDM signal. This would conclude the first stage of the JEDI project.

The second stage addresses a prototype ring that will employ deflector elements providing purely electric fields or combined electric and magnetic bending. This ring would allow for the operation of clockwise and counter-clockwise circulating beams in what is known as the frozen spin condition. This intermediate stage is necessary in order to master all the challenges concerning all electric ring operation, and the clockwise-counter clockwise beam operation with orbit differences up to few picometers and other aspects. For the second stage, the proposal "PRESTO" has been submitted as a Design Study in the framework of Horizon Europe (see last column of Fig. 1 for the expected level of precision at this stage).

Finally, the third stage aims to build a dedicated all electric ring for highest precision measurements of the proton EDM, aiming at a sensitivity of 10^{-29} e · cm (see the last column on Fig. 1).

5 Acknowledgements

I am very grateful to Frank Rathmann, Volker Hejny and Andro Kacharava for attentive reading of this document and providing many valuable suggestions.

References

[1] F. Abusaif *et al.*, Storage Ring Search for Electric Dipole Moments of Charged Particles - Feasibility Study, *CERN Yellow Report* **257** (2021), https://doi.org/10.23731/CYRM-2021-003.
[2] F. Rathmann, N.N. Nikolaev, J. Slim, Phys. Rev. Accel. Beams **23** 024601 (2020).
[3] W. Bernreuther, Lect. Notes Phys. **591** 237 (2002).
[4] WMAP collaboration, Astrophys. J. Suppl. **148** 1 (2003).
[5] V. Barger, J. P. Kneller, H.-S. Lee, D. Marfatia and G. Steigman, Phys. Lett. **B566** 8 (2003).
[6] A. D. Sakharov, Pisma Zh. Eksp. Teor. Fiz. **5** 32 (1967).
[7] V. Bargmann, L. Michel, and V. L. Telegdi, Phys. Rev. Lett. **2** 435, 1959.
[8] D. F. Nelson, A. A. Schupp, R. W. Pidd, and H. R. Crane, Phys. Rev. Lett. **2** 492, 1959.
[9] T. Fukuyama and A. J. Silenko, Int. J. Mod. Phys. **A28** 1350147, 2013.
[10] A. Saleev *et al.*, Phys. Rev. ST Accel. Beams **20** 072801 (2017).
[11] J. Slim *et al.*, Nucl. Instrum. Methods Phys. Res. A **828** 166 (2016).
[12] Z. Bagdasarian *et al.*, Phys. Rev. ST Accel. Beams **17** 052803 (2014).
[13] G. Guidoboni *et al.*, Phys. Rev. Lett. **117** 054801 (2016).
[14] D. Eversmann *et al.*, Phys. Rev. Lett. **115** 094801 (2015).
[15] N. Hempelmann *et al.*, Phys. Rev. Lett. **119** 014801 (2017).
[16] N. Hempelmann *et al.*, Phys. Rev. Accel. Beams **21** 042002 (2018).
[17] S. Karanth, E.J. Stephenson, S.P. Chang *et al*, https://arxiv.org/abs/2208.07293v3
[18] R. D. Peccei and H. R. Quinn, Phys. Rev. Lett. **38** 1440 (1977).
[19] R. D. Peccei and H. R. Quinn, Phys. Rev. D **16** 1791 (1977).
[20] P. Zylaet al, "Review of Particle Physics", PTEP **2020(8)** 083C01 (2020).
[21] P. Sikivie, Rev. Mod. Phys. **93(1)** 015004 (2021).

https://doi.org/10.1051/epjconf/202329008001

Radiobiology experiments with a laser driven x-ray source: exploring the UHDR regime

A. Reija[1,2,*], *R.* Contreras-Martínez[1,4], *A.* Coathup[1], *F.* Barranco[2], *J.* Juan[2], *A.* Alejo[1], *A.* Bembibre[1], *J.* Benlliure[1], *J.* Peñas[1], *M.E.* Aguado-Barrera[3], *A.* Vega[3], *C.* Ruiz[4], and *M.* Seimetz[2]

[1]Instituto Galego de Física de Altas Enerxías (IGFAE), Universidade de Santiago de Compostela, Spain.
[2]Instituto de Instrumentación para Imagen Molecular (I3M), CSIC - Universitat Politècnica de València, Spain.
[3]Fundación Pública Galega de Medicina Xenómica (FPGMX), Instituto de Investigación Sanitaria de Santiago de Compostela, Spain.
[4]Instituto de Física Fundamental y Matemáticas, Universidad de Salamanca, Spain.

Abstract. Laser Driven Plasma Accelerators (LDPA) can generate both characteristic and Bremsstrahlung x-rays in ultrashort bunches. Because of the high intensity of the laser pulses, these laser-induced x-ray sources can achieve peak dose rates compatible with the FLASH regime. At the Laser Laboratory for Acceleration and Applications (L2A2/USC), a setup for the irradiation of *in vitro* cell cultures with a laser-driven x-ray source has been developed. The suitability of this source for studying the influence of Ultra High Dose Rates (UHDR) on DNA damage has been proven through the irradiation of *in vitro* cell cultures.

1 Laser-driven acceleration and FLASH radiotherapy

Since the development of the Chirped Pulse Amplification (CPA) technique, the generation of ultra-short and ultra-intense laser pulses and its applications in medical physics have become a very active field of research. These laser pulses have a temporal component previous to the main pulse which is produced due to the amplification of the spontaneous emission, generated during the process of the stimulated emission in the laser's gain medium. If the laser pulse impinges focused on a solid target [1], this component can be intense enough to strongly ionize the surface of the target, creating a plasma. When the electromagnetic fields of the main pulse interact with the free electrons of the plasma, these will experience different acceleration mechanisms [2–4] depending on several parameters such as the laser intensity on focus, its angle of incidence or the target thickness. Consequently, the accelerated electrons will interact with the non-ionized atoms of the target producing a Bremsstrahlung continuum as they decelerate, and characteristic x-rays.

These specific particle acceleration mechanisms result in the capability of generating pulsed x-ray sources for preclinical research. In addition, since the acceleration occurs due to the generation of electric fields of the order of TV/m within a few microns, laser-driven acceleration gives rise to more compact sources if compared to conventional radiofrequency accelerators. Laser-induced radiation sources can produce relevant doses in short periods of time, which makes them suitable for radiobiological research related to Ultra High Dose Rate (UHDR/FLASH) therapy techniques.

The interest of delivering doses at very high rates relies on the so-called FLASH effect: while in conventional radiotherapy doses are fractionated and delivered at rates < 1 Gy/s, FLASH radiotherapy (FLASH-RT) proposes to shorten irradiation times while increasing the dose per fraction, so that the average dose rate becomes greater than 40 Gy/s; and often achieving peak dose rates of $> 10^4$ Gy/s. The main observed benefit of this new radiotherapy technique is a sparing of normal tissue without lowering the antitumoral efficacy achieved in a conventional treatment. This protection of healthy tissues has been experimentally observed in numerous pre-clinical *in vivo* studies in the last decade [5, 6] and the safety and feasibility of the procedure has been tested in human studies [7].

The underlying biochemical mechanism that results in the FLASH effect remains unclear, although there are three main hypothesis that could explain such reduced toxicity of FLASH-RT to normal tissue. The first one proposes the possibility of a greater oxygen depletion in healthy cells during the irradiation time if high dose rates are employed: a transient state of hypoxia (low oxygen concentration) in a cell makes it significantly radiorresistant [8]. The second hypothesis relies on the assumption that if the dose is delivered at a very short irradiation time, the number of lymphocytes damaged by the effects of ionizing radiation is significantly reduced, so the immune response that repares healthy cells within a FLASH irradiation may be higher if compared to

*email: alicia.reija@rai.usc.es

conventional treatments. Finally, the last hypothesis states that the irradiation at very high instantaneous dose rates enhances high concentrations of radicals in the process of water radiolysis, favoring its recombination instead of its reaction with molecular oxygen, which would increase the probability of producing DNA permanent damage [9][10].

Biological effects of laser-based x-ray sources have been investigated previously, addressing the dose and dose rate-dependent survival of certain mammalian cell lines [11, 12]. The setup presented here is characterized by dose concentration in ultra-short pulses. In addition, it allows for studying DNA damage induced by low-energy x-rays (<30 keV) which can differ significantly from those at 100-200 keV and beyond [13, 14].

2 Experimental setup of a laser-driven x-ray source at L2A2

In the Laser Laboratory for Acceleration and Applications (L2A2) at the University of Santiago de Compostela we have set up a table-top, laser-driven x-ray source for radiobiology studies in the FLASH regime. It is induced by Ti:Sa laser pulses of 1 mJ of energy and 35 fs of temporal FWHM operating at a 1 kHz repetition rate. These pulses are tightly focused with a microscope objective (Mitutoyo 20×) onto the surface of a copper plate. These laser pulses are p-polarized, and its angle of incidence with respect to the normal direction of the target is 45°. The FWHM of the laser focal spot in the horizontal direction is 1.5 μm, so at the focus the reached intensities are > 10^{18} W/cm^2.

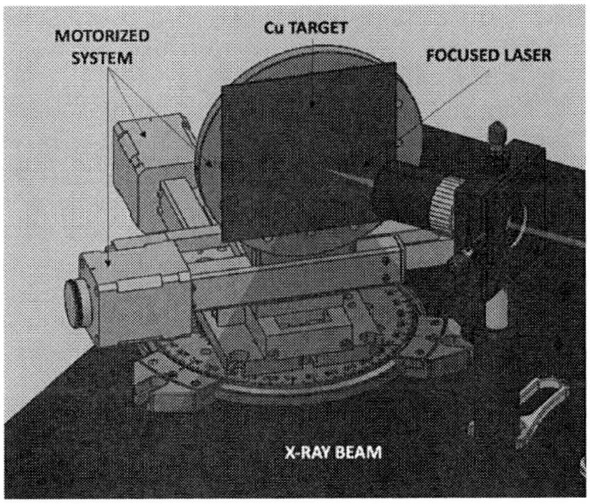

Figure 1. Scheme of the x-ray source setup.

At every laser shot, the copper surface becomes damaged, so the target is mounted on a rotatory motor, which combined with a linear motor that moves within the focal plane of the laser, refreshes the target surface at every laser shot, thus x-ray spectrum stabilisation is ensured. In addition, these two motors are mounted on another linear motor that corrects the wobble movement of the rotatory motor, so that during the entire irradiation the copper plate is placed within the focal plane of the laser with a precision comparable to the Rayleigh length. This entire motorized system was previously described in [15], and it is schemed in Figure 1. The induced x-rays are emitted in front of the copper target, with the direction of the maximum flux centered around the target's normal axis.

3 Energy distribution and dose evaluation

The energy spectrum of the x-ray source has been measured with a CdTe detector (Amptek XR-100T-CdTe) at 25 cm from the target-normal direction. A 1 mm diameter lead collimator was placed in front of its berylium window. In order to reduce pile-up effects, the spectrum was acquired with an aluminum filter of 165 μm width, and the de-piled spectrum was retrieved applying an algorithm described in [16]. In Figure 2, the energy distribution shows a maximum x-ray flux around the superimposed K_α and K_β characteristic Cu peaks (8.03-8.91 keV). However, the de-pile algorithm was not capable of correcting the characteristic Cu peak repetition, and recent measurements in our laboratory have confirmed that the maximum cutoff energy is still overestimated.

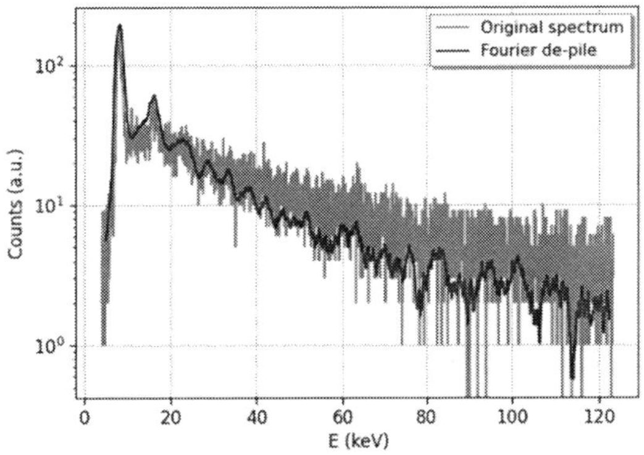

Figure 2. Raw vs depiled x-ray energy spectrum, measured at 25 cm and at 0° with respect to the target normal with a 165 μm Al filter.

On the other hand, the delivered dose was measured with unlaminated EBT3 radiochromic films (RCF). Since this type of films have a low energy dependency when irradiated in air [17], the batch was calibrated with a 6 MV Linac. The RCF were scanned with an Epson Perfection V750 Pro at 300 dpp and in tiff format, and the dose was evaluated in the red channel.

Several measurements of the average dose rate - defined as the relation between the dose read at the RCF and the total irradiation time- are displayed in Figure 3. The average value at 10 cm is 28.14 mGy/s and the peak dose rate within a pulse is 8.04 · 10^8 Gy/s, which was estimated assuming that the x-ray pulses are emitted at the same repetition rate and duration as the laser pulses,

$$\dot{D}_{peak} = \frac{\bar{\dot{D}}\ [\text{Gy/s}]}{1000\ [\text{Hz}] \cdot 35 \cdot 10^{-15}\ [\text{s}]} \tag{1}$$

The relative error of the average dose rate shown in Figure 3 states that the source is not totally stabilized. This instability comes from the difficulty in the online correction of the wobble movement of the rotatory motor: since the maximum laser intensity on target is only achieved within a micrometric spatial interval (Rayleigh length), possible imprecisions during this correction significantly lower the x-ray generation efficiency. This issue has currently been addressed through the addition of an ionization chamber, which coupled to the laser shutter ensures a better precision in the total delivered dose.

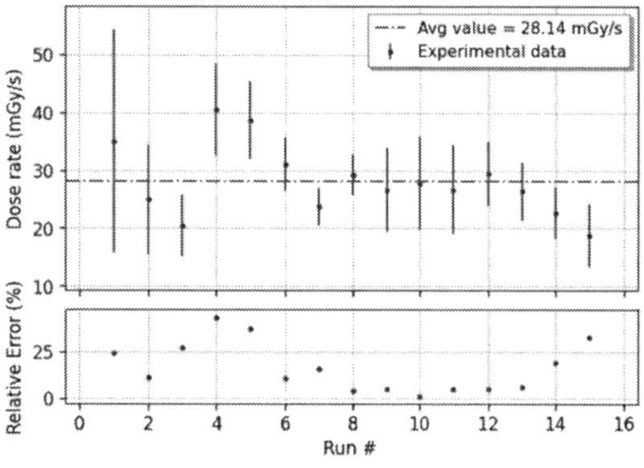

Figure 3. Dose rate values and relative error.

4 Irradiation and analysis of *in vitro* cell cultures

To demonstrate the suitability of the source to perform radiobiology experiments at ultra-high dose rates, we tested the antitumoral response of the laser-induced x-rays through the irradiation of *in vitro* cell cultures. We chose the A549 line (human basal alveolar epithelial adenocarcinoma cells), and the cultures were prepared, standardized, handled and analysed in the Fundación Pública Galega de Medicina Xenómica (FPGMX). A monolayer of cells was grown on a 70 μm thick polyethylene (PE) foil sealing a sterile culture flask. Cell cultures were safely transported between FPGMX and L2A2. The samples were placed at 10 cm from the source, with irradiation times that ranged from 60 s to 150 s.

The total radiation dose was measured on radiochromic films placed outside of the cell culture flask. This films were covered with 11 μm of Al to filter the possible reflected laser energy that could saturate the film.

To determine the dose on the cells we performed separate measurements with radiochromic films placed inside one of the flasks, after the PE window and without the Al filter. The following correlation between the inside and the outside film was revealed:

$$D_{\text{in}} = k_D\, D_{\text{out}} \tag{2}$$

The correction factor $k_D = 0.702$ was applied to the dose values recorded in the RCF placed after the Al foil.

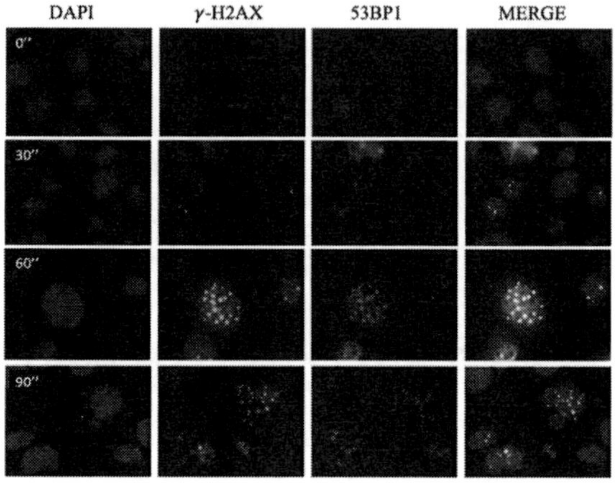

Figure 4. Analysis of the immunofluorescence staining of foci γ-H2AX and 53BP1 in A549 cells in response to irradiation.

The effects of radiation toxicity were evaluated 24 h after the irradiation with the quantification of DNA repair foci assessed by immunofluorescence analysis of γ-H2AX and 53BP1 foci (see Figure 4). For each of the samples, 60 cell nuclei have been identified and the corresponding damage foci have been counted using confocal microscopy. In Figure 5, the number of colocalized γ-H2AX and 53BP1 foci versus the dose absorbed by the cultures is represented: while the effects of radiation were comparable to those shown by control samples for doses up to 3 Gy, from 3 to 5 Gy the number of foci quantified per cell showed a linear response with the absorbed dose (represented in red in Figure 5).

In addition, as a proof-of-principle, a series of identical cell samples were irradiated with a conventional radiotherapy source (6 MV Varian Linac) at FPGMX, in a similar range of total dose in order to establish the optimum culture time and analysis protocols.

5 Conclusions and future prospects

We have developed a laser-induced x-ray source for the irradiation of *in vitro* cell cultures. In a commissioning experiment, DNA damage has been quantified through γ-H2AX and 53BP1 immunofluorescence, and the average foci per cells versus the absorbed dose showed a linear response for doses greater than 3 Gy. Details of

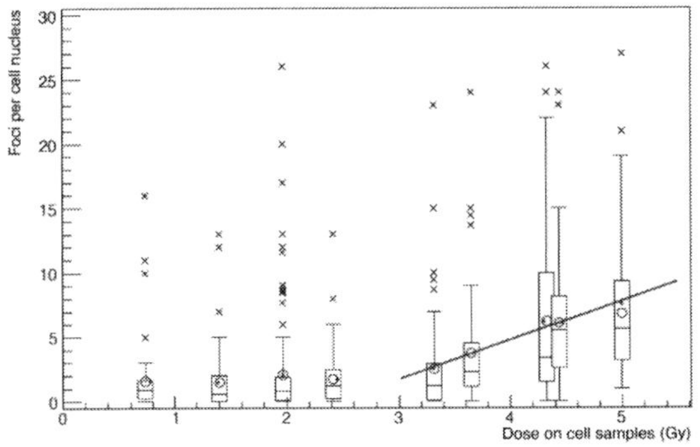

Figure 5. Combined γ-H2AX and 53BP1 foci per cell versus absorbed dose. Absorption on PE foil was taken into account in the dose evaluation. The red line is a linear fit to the mean values for doses > 3 Gy.

the total dose applied to the cells and the manipulation of biological samples have been addressed.

Further stabilisation of the dose rate is still ongoing. For real-time control of the total radiation dose an ionization chamber has recently been implemented. Our aim is to perform systematic studies of DNA damage by low-energy x-rays and to compare them to other radiation sources such as a clinical Linac or laser-accelerated protons with an experimental setup described previously [18].

Acknowledgements

Project RTI-2018-101578-B-C22 financed by MCIN/AEI/10.13039/501100011033 and FEDER "Una manera de hacer Europa". Project AICO/2020/207 financed by Generalitat Valenciana. Research co-financed by the European Union through Programa Operativo del Fondo Europeo de Desarrollo Regional (FEDER) of the Comunitat Valenciana, IDIFEDER/2021/004; the "María de Maeztu" Units of Excellence program MDM-2016-0692; and by "la Caixa" Foundation [LCF/BQ/PI20/11760027]. The authors also want to thank Prof. Jose Manuel Udias and his team from Universidad Complutense de Madrid for useful discussions regarding dosimetry with radiochromic films.

References

[1] A. Macchi, A Superintense Laser-Plasma Interaction Theory Primer (Springer Dordrecht, 2013)
[2] J.P. Freidberg, R.W. Mitchell, R.L. Morse, L.I. Rudsinski, Phys. Rev. Lett. **28**, 795 (1972)
[3] F. Brunel, Phys. Rev. Lett. **59**, 52 (1987)
[4] W. Kruer, K. Estabrook, The Physics of fluids **28**, 430 (1985)
[5] V. Favaudon, L. Caplier, V. Monceau, F. Pouzoulet, M. Sayarath, C. Fouillade, M.F. Poupon, I. Brito, P. Hupé, J. Bourhis et al., Science translational medicine **6**, 245ra93 (2014)
[6] P. Montay-Gruel, A. Bouchet, M. Jaccard, D. Patin, R. Serduc, W. Aim, K. Petersson, B. Petit, C. Bailat, J. Bourhis et al., Radiotherapy and Oncology **129**, 582 (2018)
[7] J. Bourhis, W.J. Sozzi, P.G. Jorge, O. Gaide, C. Bailat, F. Duclos, D. Patin, M. Ozsahin, F. Bochud, J.F. Germond et al., Radiotherapy and Oncology **139**, 18 (2019)
[8] M. Durante, E. Bräuer-Krisch, M. Hill, The British Journal of Radiology **91**, 20170628 (2018), pMID: 29172684, `https://doi.org/10.1259/bjr.20170628`
[9] A. Espinosa-Rodriguez, D. Sánchez-Parcerisa, P. Ibanez, J. Vera-Sánchez, A. Mazal, L. Fraile, J. Udías, International Journal of Molecular Sciences **23**, 13484 (2022)
[10] P. Montay-Gruel, M.M. Acharya, K. Petersson, L. Alikhani, C. Yakkala, B.D. Allen, J. Ollivier, B. Petit, P.G. Jorge, A.R. Syage et al., Proceedings of the National Academy of Sciences **116**, 10943 (2019)
[11] C. Tillman, G. Grafström, A.C. Jonsson, B.A. Jönsson, I. Mercer, S. Mattsson, S.E. Strand, S. Svanberg, Radiology **213**, 860 (1999), pMID: 10580967
[12] K. Shinohara, H. Nakano, N. Miyazaki, M. Tago, R. Kodama, Journal of Radiation Research **45**, 509 (2004)
[13] M. Mestres, M.R. Caballín, L. Barrios, M. Ribas, J.F. Barquinero, Radiation Research **170**, 93–100 (2008)
[14] E. Beyreuther, W. Dörr, A. Lehnert, E. Leßmann, J. Pawelke, Radiation and Environmental Biophysics **51**, 33 (2012)
[15] L. Martín, J. Benlliure, D. Cortina, J.J. Llerena, D. González, C. Ruiz, Journal of Physics: Conference Series **1079**, 012008 (2018)
[16] G. Hernández, F. Fernández, Applied Physics B **124**, 1 (2016)
[17] G. Massillon-Jl, A. Cabrera-Santiago, N. Xicohténcatl Hernández, Physica Medica **61**, 8 (2019)
[18] A. Torralba, L. Palenciano, A. Reija, J.P. Rigla, J. Peñas, J.J. Llerena, R. Contreras-Martínez, J. Benlliure, A. Vega, M.E. Aguado-Barrera et al., Quantum Beam Science **6** (2022)

Scientific programme at the Laser Laboratory for Acceleration and Applications

Aarón Alejo[1,*] and *José* Benlliure[1]

[1]Instituto Galego de Física de Altas Enerxías, Universidade de Santiago de Compostela, Santiago de Compostela 15782, Spain

Abstract. In recent years, laser-driven accelerators have emerged as a promising technology for the development of compact and cost-effective sources of high-energy particles and radiation. Here, we introduce our research programme at the Laser Laboratory for Acceleration and Applications, which focusses on high-power laser systems, development of high-repetition rate targetry, and biomedical applications of laser-based accelerators. Our recent results include the generation of high-energy proton and ion beams through laser-plasma acceleration, which have potential applications in materials activation for medical imaging. We have also investigated the feasibility of using laser-driven X-rays and ion beams for radiobiological research, including FLASH therapy and hadron therapy.

1 Introduction

Laser-plasma accelerators have attracted significant interest as a promising technology for generating high-energy particle beams with unprecedented characteristics. These compact accelerators benefit from the stronger accelerating electric fields that can be withstood by a plasma, orders of magnitude higher than those achieved in state-of-the-art conventional RF accelerators, enabling the acceleration of charged particles to very high energies over short distances. Thanks to this, laser-plasma accelerators have the potential to revolutionize the field of particle acceleration, offering a cost-effective and compact alternative to traditional accelerators,

Significant progress has been made in recent years in laser-plasma acceleration, with several key milestones achieved. These include the generation of high-quality electron beams with energies up to 8 GeV [1], acceleration of protons to energies in excess of 100 MeV [2], as well as demonstration of high-quality secondary radiation sources, such as X-rays, positrons, or neutrons [3, 4]. Furthermore, the ongoing progress in laser-driven accelerators and the unique characteristics of the produced radiation –e.g. ultra-short duration or very low emittance– have opened up the possibility of using these beams in proof-of-principle applications in a wide range of fields. including FLASH radiation therapy for cancer treatment, nuclear activation of materials for medical imaging, or sample characterisation in material science.

In this paper, we describe our research programme at the Laser Laboratory for Acceleration and Applications (L2A2), focused on high-power laser systems, development of high-repetition rate targetry for laser-plasma experiments, and biomedical applications of laser-based accelerators, with a specific focus on the acceleration of ion beams, particularly protons, for medical imaging applications. We highlight our efforts to develop stable laser-driven radiation sources operating at high repetition rates, by means of both the stabilisation of the in-house high-power laser STELA, and the development of new target systems. We also describe our ongoing work to explore the potential of laser-plasma accelerators for real-life applications, including the production of isotopes for Positron Emission Tomography medical imaging and phase-contrast imaging.

2 STELA laser system

The Laser Laboratory for Acceleration and Applications (L2A2) is a recently-built infrastructure at the *Universidade de Santiago de Compostela* (USC) for the investigation of laser-plasma interaction, which hosts a clean room, laboratory for development of acceleration targets and beam diagnostics, and a radio-protected experimental hall (Fig. 1) [5]. The core facility of L2A2 is STELA (Santiago TErawatt LAser), a compact Ti:Sapphire laser system built by Thales (Alpha 10/XS). The STELA laser offers two different outputs, each of which powers an independent beamline.

The main laser beamline, or *high-energy beamline*, delivers ultra-short pulses (25 fs) with energy per pulse up to 1 J at a 10 Hz repetition rate and peak power around 50 TW. As this beamline is mainly devoted to ion acceleration using solid targets, the laser chain has been modified to improve its operation. For instance, a key requirement for efficient ion acceleration is a high contrast ratio. In our case, the contrast is enhanced, reaching a ratio of $> 10^{10}$ at 5 ps, thanks to the implementation the Crossed-Polarised Wave (XPW) technique early in the laser chain. In addition to this, the interaction of the laser with a solid target

*e-mail: aaron.alejo@usc.es

 https://doi.org/10.1051/epjconf/202329008002

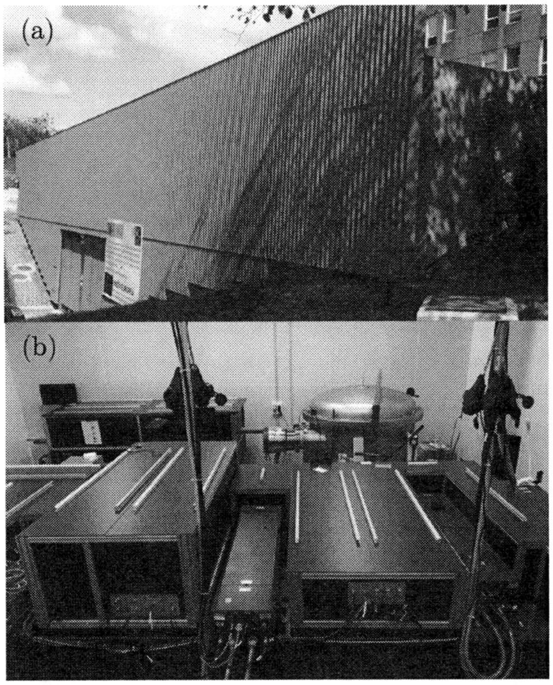

Figure 1. (a) Picture of the Laser Laboratory for Acceleration and Applications building. (b) Picture of the STELA laser system as installed in the clean room.

can lead to the back-propagation of a significant amount of energy back into the laser chain, which can result in damage of key components of the system. To mitigate this, a pulse picker has been installed at the end of the final amplification stage, which prevents the propagation of these retro-reflections into the system. Currently, there are significant efforts to further improve the stability of the laser system, a key requirement to achieve a stable laser-based accelerator, particularly in the context of laser chain evolution due to heating of the different components in the case of high-power, high-repetition-rate laser systems [6].

The second beamline, or *low-energy beamline*, which takes advantage of the pulses generated at the system front-end and extracted before the final amplification stages, is a versatile tool for a range of laser-based experiments, such as X-ray generation or material processing. In this case, the laser operates at a repetition rate of 1 kHz and an energy output of 1 mJ, producing pulses with 30 fs pulse duration.

3 High-repetition rate targetry and diagnostics

In the past, the production of stable, laser-driven radiation sources has been limited by the reduced repetition rate of the laser systems capable of producing such sources. However, as the ongoing developments in laser-technology remove this bottleneck, improvements on the targetry have become crucial to advancing the field of laser-plasma accelerators, particularly relevant in the case of solid targets. As these targets are destroyed after their irradiation, they need to be replaced for a fresh one and re-aligned with few-microns-precision before a new irradiation takes place.

Several solutions are currently being actively studied to address this limitation [7]. In our case, we are particularly focussed on the development of targetry systems for solid targets, specifically rotating wheels and tape drives. In the first case, targets are regularly distributed in the radial and angular dimensions of a precisely-machined wheel, attached to a motorised system allowing for three-dimensional movement. Albeit this target design has advantages, such as allowing for a large number of irradiations or the broad variety of target material and thickness that can be used, some limitations need to be addressed in order to be able to shoot continuously at multi-Hertz rates. In our case, the solution is based on pre-characterising the surface of each target using an optical position sensor, which allows to rapidly change between targets and precisely position them on the focal plane of the focussing optics, with micron-level precisions. Using this system, continuous stable ion acceleration at rates up to 10 Hz have been demonstrated, with a standard deviation of the proton cut-off energy $< 10\%$.

Despite the flexibility and robustness of the rotating wheel, this solution presents intrinsic shortcomings in the context of state-of-the-art laser systems, such as the maximum number of targets, typically constrained to a few thousand, or the maximum repetition rate of operation, due to the required separation between targets and maximum velocity for the wheel movements involved. For these reasons, we are currently working on the development of a tape-drive system [7]. In this case, a thin tape is used as target, ensuring a fresh surface interacts with the laser as the tape is wound onto a motorised spool [8]. By ensuring the tape remains at a constant tension as it slides between the rods defining the interaction point, the tape surface is ensured to remain stable on the focal plane as the target is replaced. Our tape drive system is currently being commissioned at the Laser Laboratory for Acceleration and Applications, with preliminary stability measurements already indicating standard deviations of the tape position $<10\,\mu$m [7].

4 Biomedical applications

Laser-driven radiation sources have attracted significant attention for their potential use in different biomedical applications. Among these applications, at L2A2 we particularly focus on the generation of radio-isotopes of relevance for positron emission tomography (PET), and next-generation radio-therapy cancer treatments.

Medical imaging based on radiation emitted by unstable nuclei is used on a daily basis as a fundamental tool to diagnose a number of diseases, including cancer. Arguably, one of the most powerful imaging techniques is positron emission tomography (PET). In PET, a positron emitting isotope is injected into the patient, accumulating in the target tissue. As the radionuclides decay, low-energy positrons are emitted and promptly annihilate from the combination with a nearby electron, resulting in the emission of two counter-propagating 511 keV photons, which can be detected and back-traced, enabling the characterisation of the distribution of the radioisotope in the tissue, and hence the tissue of interest.

However, the bottleneck in the use of certain radioisotopes currently lies on their short lifetimes. The production of these nuclides relies on the use of radio-frequency acceleration cavities, commonly cyclotrons, facilities characterised for their sizeable footprints (~10 m^3), the shielding requirements (~180 m^3 of concrete) and large costs (~ 10M€). As a result, a single production facility must provide nuclides to different hospitals in the area, in distances that range up to 100s of km. In order to ensure that sufficient activation levels are still present upon arrival to the hospital, only a narrow range of radioisotopes with longer lifetimes are being used for imaging, commonly ^{18}F with a half-life of ~ 110 minutes. However, there are radioisotopes with shorter lifetimes, such as ^{11}C with a half-life of ~ 20 minutes, that are highly demanded for particular diagnostics, such as neurodegenerative diseases, but cannot be provided by the available facilities as the nuclides will have decayed before being injected into the patient. In this context, there has been a growing interest in the development of cost-effective and compact accelerator systems capable of generating such nuclides on-demand, such as laser-driven ion accelerators. At L2A2, we are actively investigating the production of ^{11}C radio-isotopes using laser-driven ions, potentially capable of generating activation levels up to tens of MBq using the in-house laser, with the experimental setup shown in Fig. 2.

Figure 2. Experimental setup for the production of ^{11}C radioisotopes.

The other biomedical application being actively studied is novel radiation therapy techniques, particularly those related to Ultra-High Dose Rate (UHDR) effects, also known as FLASH effects. Conventional radiotherapy relies on the delivery of the required radiation dose at rates <1 Gy/s. Instead, in FLASH radiotherapy the dose is fractioned and delivered in the form of short bunches, allowing for an increase in the average dose rate to >40 Gy/s. Laser-driven radiation sources are ideally suited for such an irradiation, typically being delivered as separate ultra-short bunches, with durations ranging from picoseconds up to nanoseconds, leading to dose rates as high as 10^9Gy/s. Such ultra-high peak dose rates, alongside the ongoing developments on laser technology enabling Hertz-repetition rates to reach the required average dose rates, have opened up the possibility of characterising the radiobiological effects of pulsed ionising radiation at ultra-high dose rates.

UHDR techniques have been shown to present beneficial effects in pre-clinical in-vivo studies, which have been explained in terms of a lowering of the radiation effects over healthy tissue, while maintaining the antitumoral efficacy with respect to conventional treatment. Different hypotheses have been proposed to explain the underlying biochemical mechanism responsible of FLASH effects, such as the healthy cells undergoing a transient hypoxic state that enhances their radiorresistance, or an increase in the immune response repairing healthy cells due to the reduced number of lymphocytes damaged when higher dose rates are employed. There is, therefore, significant scope to explore the functioning of living tissues when exposed to UHDR of radiation.

In this context, L2A2 offers the unique possibility of studying FLASH effects of radiation on living cells using ultra-short bunches of protons or X-rays. Following initial developments to ensure the stability of the laser-driven sources, a key requirement for radiobiological studies, we are currently performing the first commissioning experiments using in-vitro cell cultures [9].

5 Phase contrast X-ray imaging

Tomographic imaging is a powerful technique used in various fields, including medical imaging, material science, and biology. It allows for the reconstruction of three-dimensional images of the internal structure of an object, which is crucial in many applications. X-ray imaging is one of the most widely used techniques for tomography due to its high penetration depth and spatial resolution. However, conventional X-ray imaging relies on the absorption of X-rays, which limits its sensitivity to soft tissues and low-density materials.

Phase contrast imaging is an emerging technique that overcomes the limitations of absorption-based imaging by detecting the phase shift of X-rays passing through an object. This shift can be induced by changes in the refractive index of the object, which is related to the object's internal structure. Phase contrast imaging can provide high-contrast images of soft tissues and low-density materials that are difficult to see using conventional X-ray imaging.

Despite the appeal of phase contrast imaging, the use of this technique has been constrained due to the limitations of conventional X-ray sources, such as their incoherence, large source size and low brightness. Laser-driven X-ray sources have recently appeared as a promising alternative for phase contrast imaging, thanks to the extremely small source size (few micron). They can produce X-rays with high brightness and coherence, making

https://doi.org/10.1051/epjconf/202329008002

them ideal for phase contrast imaging applications, with additional benefits such as compactness, portability, and cost-effectiveness.

At L2A2, we have developed a stable, laser-driven kHz X-ray source operating in air capable of producing copious amounts of X-rays with energies up to tens of keV, narrow divergence and few micron source size. This source has recently been shown to be capable of producing radiography images with phase-contrast enhancement, and is currently being employed for imaging of samples of interest in biomedicine.

6 Conclusions

Laser-driven sources of radiation have attracted significant attention due to their unparalleled intrinsic properties, such as ultra-short duration, high brightness and low emittance, alongside the compactness and cost-effectiveness with respect to conventional accelerators. The ongoing progress in laser technology is allowing for an increase in the number of installations that can reach the required laser intensities to produce these beams, including the laser at L2A2, a 50 TW commercial system operating at 10 Hz.

Here, we have described our research program, which is heavily based on the production of stable X-ray and ion beams for a variety of applications in fields such as medical imaging and cancer therapy. Our work on high-repetition rate targetry has enabled us to produce stable beams that can operate for the extended periods of time required. We have presented some of the main applications currently being studied, including using laser-driven ion beams for the production of radio-isotopes for medical imaging, studies of ultra-high dose rate effects on cells, and improvements on X-ray imaging thanks to phase-contrast techniques.

References

[1] A. Gonsalves, K. Nakamura, J. Daniels, C. Benedetti, C. Pieronek, T. De Raadt, S. Steinke, J. Bin, S. Bulanov, J. Van Tilborg et al., Physical review letters **122**, 084801 (2019)

[2] A. Higginson, R. Gray, M. King, R. Dance, S. Williamson, N. Butler, R. Wilson, R. Capdessus, C. Armstrong, J. Green et al., Nature communications **9**, 724 (2018)

[3] A. Alejo, G.M. Samarin, J.R. Warwick, G. Sarri, Frontiers in Physics **7**, 49 (2019)

[4] A. Alejo, H. Ahmed, A. Green, S. Mirfayzi, M. Borghesi, S. Kar, Il nuovo cimento C **38**, 1 (2015)

[5] J. Benlliure, D. Cortina-Gil, J. Llerena, C. Ruiz, Nuclear Instruments and Methods in Physics Research Section A: Accelerators, Spectrometers, Detectors and Associated Equipment **916**, 158 (2019)

[6] A. Alejo, A. Bembibre, J. Peñas, J. Benlliure, L. Martín, M.T. Flores-Arias, *Characterisation of the stability and long-term evolution of the properties of a 45TW laser operating at 10Hz*, in *European Physical Journal Web of Conferences* (2022), Vol. 266, p. 13001

[7] A. Bembibre, A. Alejo, J. Peñas, J. Benlliure, *Long-lasting laser-driven proton source*, in *European Physical Journal Web of Conferences* (2023), Vol. XXX, p. XXXXX

[8] M. Noaman-ul Haq, H. Ahmed, T. Sokollik, L. Yu, Z. Liu, X. Yuan, F. Yuan, M. Mirzaie, X. Ge, L. Chen et al., Physical Review Accelerators and Beams **20**, 041301 (2017)

[9] A. Reija et al., *Radiobiology experiments with a laser-induced x-ray source: exploring the UHDR regime*, in *European Physical Journal Web of Conferences* (2023), Vol. XXX, p. XXXXX

https://doi.org/10.1051/epjconf/202329008003

Long-lasting laser-driven proton source

A. Bembibre[1,*], *A.* Alejo[1], *J.* Peñas[1], and *J.* Benlliure[1]

[1]Instituto Galego de Física de Altas Enerxías (IGFAE), Universidade de Santiago de Compostela, Spain.

Abstract. Laser-driven ion accelerators are a promising alternative to conventional accelerators currently in use. Following the laser-plasma interaction, the target is destroyed, making it necessary to replenish it, a task particularly challenging in the case of state-of-the-art laser systems operating at multi-hertz repetition rates. Here, we will introduce some of the different target systems currently being developed for continuous operation at such rates. In particular, a target system based on a tape-drive that is currently being developed at the Laser Laboratory for Acceleration and Applications (L2A2, USC) will be presented, which addresses the need of a target system capable of operating continuously for thousands of shots at several hertz. Preliminary measurements show tape stabilities below 10 μm of standard deviation.

1 Introduction

The lower shielding requirements, compactness and reduced operational costs of laser-driven ion accelerators make them an interesting alternative to conventional accelerators, opening the possibility of increasing the number of facilities capable of offering ion beams. Characteristics such as ultra-short ion bunches, broad spectrum, simultaneous electron acceleration or gamma production, make these accelerators suitable for, among others, replicating outer space conditions, electron-driven fast ignition (FI) scheme in inertial confinement fusion (ICF), FLASH therapy for cancer treatment, or radioisotope and neutron production in combination with a secondary target [1].

To produce such ion bunches, an ultra-short, high-energy laser pulse is focused down to a few-micron focal spot onto a thin solid target, reaching intensities above $10^{18}\,\mathrm{W\,cm^{-2}}$. Under these conditions, ions are accelerated via the Target Normal Sheath Acceleration (TNSA) mechanism. In this regime, the pedestal and the prepulses typically preceding the main pulse create a plasma on the front side of the target. Subsequently, when the main pulse reaches the target, the ponderomotive force accelerates the electrons in the plasma. Those electrons travel through the target before escaping from the rear side along the direction normal to the surface. As a result, a quasi-static charge-separation sheath, with fields up to TV/m, is established, capable of accelerating protons and ions from the rear surface along the target normal direction [1].

Following the interaction with the laser, the target is damaged, making it necessary to replace it. In the past, lasers were typically operated in a single-shot mode, which meant that target replacement could be done individually. However, with the emergence of high repetition rate (HRR) lasers, it has become necessary to investigate alternative target designs compatible with these new systems. In the present paper, we will outline some of the recent developments on target systems for ion acceleration, particularly focusing on the tape-drive system designed at the Laser Laboratory for Acceleration and Applications (L2A2, Universidade de Santiago de Compostela).

2 High repetition rate targets

The acceleration of ions using ultra-intense lasers requires the use of overdense targets [1]. Such densities can be achieved through several alternatives such as solid targets, liquid jets or high density gas targets. Of these, solid targets are arguably the most commonly employed solution, given the simplicity of the setup. However, compared to the other options, the use of solid foils is also the most complicated in terms of replenishing after the irradiation. Several solutions to address this issue have been proposed, such as the use of pre-characterised wheel systems [2]. It consists of a wheel that can hold up to a few thousand targets and is attached to a three motor system that can precisely position the targets at focal plane. The alignment positions of each target are pre-characterised and then programmed to automatically replenish the target after the laser interaction. Such systems have been successfully used to accelerate ions at rates up to the hertz level. Nevertheless, their use at higher repetition rates is limited due to the velocity of movement required to reach the necessary separation among shots.

To further extend the number of shots, alternatives such as liquid targets have been explored [3]. They consist of one or two colliding jets, forming a thin sheet where the laser impinges. However, improvements are being developed as their operation is limited due to loss of vacuum caused by imperfections in liquid extraction, before they can be reliably used in the context of HRR PW lasers. Following the same principle, high pressure gas jets with overcritical densities can be used [4]. Originally, this kind

*email: *adrian.bembibre.fernandez@usc.es*

of systems were used with CO_2 (λ = 10.6 µm) lasers exploiting their lower critical density ($n_c \approx 9.8 \times 10^{18}$ cm^{-3}), enabling the study of novel radiation-pressure-based ion acceleration mechanisms. The development of gas jets reaching even higher densities, over the critical density of the Ti:Sapphire systems (λ = 800 nm, $n_c \approx 1.7 \times 10^{21}$ cm^{-3}) is an active field of research.

Despite the potential of these target systems, another possibility with relatively easy implementation is a tape-drive system [5]. In this system, the target is a tape that is continuously refreshed, and once it is placed at the desired plane, it is mechanically ensured to keep the position stable. In the following section, the tape-drive system developed at the L2A2 will be presented.

3 Tape-drive design and characteristics

The tape-drive system developed (Fig. 1) has two spools that can allocate tens of metres of few-micron thick tape, allowing for thousands of consecutive shots, and several rods to guide the tape though and place it on the shooting position. Those spools are both attached to rotary motors, giving the possibility of moving the tape forward and backwards, which, for tapes sufficiently wide, would significantly increase the shots available. For this reason, to change the shooting position, the main body is mounted over two motorised stages that can be remotely controlled and automatised. Moreover, these stages are used to precisely place the tape at the focal plane. Furthermore, the angle of incidence of the laser can also be precisely controlled using the micrometric screws (see *Angle screws* in Fig. 1a). Additionally, transverse probing of the interaction is enabled by a side window on the body of the tape-drive.

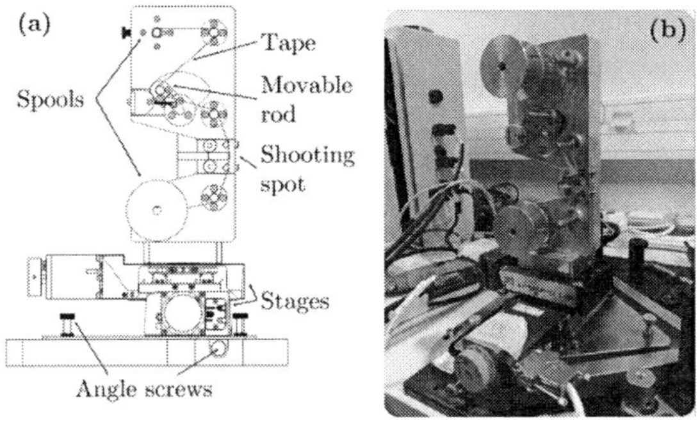

Figure 1: (a) Schematic and (b) picture of the tape-drive developed at L2A2.

Following the initial alignment of the system, the tape must remain stable on the focal plane as the target is replaced. In our case, this is achieved by using a closed-loop system which uses feedback from one of the rods. Such rod is movable and attached to a spring to ensure it stays in contact with the tape and give a certain margin to prevent tape breakage in case of excess tension. The position of the rod depends on the tension applied to the tape and is monitored with an angular encoder. The angle read by this encoder is used in the closed-loop to change the rotating speed of the motors and maintain the optimum tension in the tape to keep it stable at shooting spot.

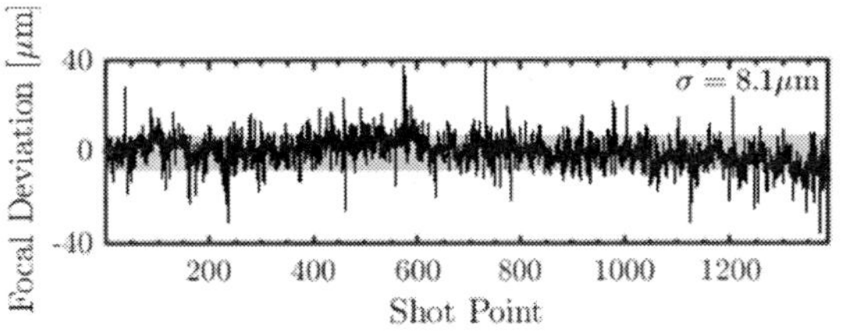

Figure 2: Stability of the tape position over almost 1400 consecutive shots. The standard deviation is σ = 8.1 µm.

Preliminary stability measurements were done using a Kapton tape of 13 µm. The tape surface position was measured using an optical position sensor (optoNCDT 1320). The results show a standard deviation of 8.1 µm (Fig. 2), a displacement below the Rayleigh length of the focusing optics typically employed to accelerate ions using lasers.

4 Conclusions and future prospects

We have developed a tape-drive target for laser-driven ion acceleration that can be used continuously for several thousands of shots. It has shown good stability using a 13 µm Kapton tape. We expect to use aluminium tapes in the near future and further improve the stability, specially peak to peak, improving the closed-loop algorithm and the tape winding.

Acknowledgements

This work has been supported by Spanish Ministerio de Ciencia e Innovación within the framework of the FPI predoctoral grant (PRE2019-090730), RETOS project (RTI2018-101578-B-C21) and by Xunta de Galicia and EU, LASERPET project (0213-AD009.01); the "María de Maeztu" Units of Excellence program MDM-2016-0692; and by "la Caixa" Foundation [LCF/BQ/PI20/11760027].

References

[1] A. Macchi, *A Superintense Laser-Plasma Interaction Theory Primer* (Springer Dordrecht, 2013)

[2] P. McKenna, K.W.D. Ledingham, I. Spencer, T. McCany, R.P. Singhal, C. Ziener, P.S. Foster, E.J. Divall, C.J. Hooker, D. Neely et al., Review of Scientific Instruments **73**, 4176 (2002), https://doi.org/10.1063/1.1516855

[3] P. Puyuelo-Valdes, D. de Luis, J. Hernandez, J.I. Apiñaniz, A. Curcio, J.L. Henares, M. Huault, J.A. Pérez-Hernández, L. Roso, G. Gatti et al., Plasma Physics and Controlled Fusion **64**, 054003 (2022)

[4] S.N. Chen, M. Vranic, T. Gangolf, E. Boella, P. Antici, M. Bailly-Grandvaux, P. Loiseau, H. Pépin, G. Revet, J.J. Santos et al., Scientific Reports **7**, 13505 (2017)

[5] M. Noaman-ul Haq, H. Ahmed, T. Sokollik, L. Yu, Z. Liu, X. Yuan, F. Yuan, M. Mirzaie, X. Ge, L. Chen et al., Phys. Rev. Accel. Beams **20**, 041301 (2017)

Evaluation of water equivalent thicknesses using the IEM-CSIC scanner prototype

Amanda Nathali Nerio[1,*], *José Antonio* Briz[1,2], *María José* García Borge[1], *Enrique* Nácher[3], *Olof* Tengblad[1], *Ángel* Perea[1], and *Vicente* García Távora[1]

[1]Instituto de Estructura de la Materia, CSIC, Serrano 113-bis, E-28006, Madrid, Spain
[2]Grupo de Física Nuclear (GFN) and IPARCOS, Universidad Complutense de Madrid (UCM), E-28040, Madrid, Spain
[3]Instituto de Física Corpuscular, CSIC-Universidad de Valencia, E-46980, Paterna, Valencia, Spain

Abstract. Proton imaging has been proposed as an alternative low-dose imaging technique for acquiring relative stopping power (RSP) maps and for patient positioning. In proton therapy and hadrontherapy, the water equivalent thickness (WET) of a material represents its radiological thickness, however, its calculation requires computationally intensive methods or approximate solutions. At IEM-CSIC, we have developed a scanner prototype for imaging with protons composed of two double-sided silicon strip detectors (DSSSD) and fast scintillators with high energy resolution, detectors that are commonly used in Experimental Nuclear Physics. Two custom-made samples of aluminum, air, and polymethyl methacrylate (PMMA) were imaged and their proton radiography images (pRads) are presented in terms of WET. Pattern reproduction, spatial resolution, and sensitivity to different materials were studied. We present pRads of our samples obtained using a rather traditional approach to compute WET values. All three materials of the studied samples, aluminum, PMMA, and air, were well distinguished; the resulting WET values were in good agreement between samples.

1 Introduction

Proton therapy is a modality in cancer treatment that improves the tumor-to-normal tissue dose ratio in comparison to conventional radiotherapy with X-rays by achieving conformal dose distributions. The proton therapy principle is based on the way protons slow down while traversing matter [1]. Protons are interesting for tumor treatment as they allow us to control the deposit of energy at a given depth in the patient by tuning the proton beam energy so that therapeutic doses are applied mostly at a given depth and vanish beyond the tumor. However, uncertainty in the proton range prediction compromises the quality of the treatment plans, as an important fraction of energy might be wrongly deposited in healthy tissue [2, 3]. The proton range depends on the proton energy and on the stopping power of the materials traversed. In proton therapy, it is very important to achieve precise proton range control, which can be improved with an accurate knowledge of the relative stopping power (RSP) map of the patient in the treatment position [4].

Currently, X-ray computed tomography (xCT) images are used to calculate the RSP maps that are needed for proton therapy treatment planning. This procedure can result in up to 3.5% errors in the determination of RSP, making proton therapy less beneficial as compared to conventional radiotherapy with X-rays [4]. Proton computed tomography (pCT) represents a more accurate imaging technique for providing RSP distributions without conversion from Hounsfield units, a measurement of the linear attenuation coefficient of the tissue; while also constituting an advantageous technique in terms of dose delivery when compared to xCT [5].

In proton therapy, the beam penetration range is commonly prescribed in terms of the water equivalent thickness (WET), or the beam penetration depth in liquid water. There are several suggested techniques for calculating WET values, however, they can be easily determined by the range shift related to placing a test object in the proton path [2].

The IEM-CSIC scanner prototype includes two double-sided silicon strip detectors (DSSSDs) as proton trackers. They provide both, spatial and energy-deposited information. Hence, the first DSSSD allows the determination of the initial proton beam energy, whilst the second DSSSD also helps to determine the residual energy of the protons. The residual energy detector of the prototype provides high energy resolution and fast response, which allows us to test it under realistic conditions of proton rates and energy. In this contribution, we present the results obtained from proton radiographs (pRads) using 100 MeV proton beams in terms of WET. The approach we used to determine the WET values is that described by Zhang *et al.* [6], where we compute proton ranges in liquid water taken from PSTAR database [7] for the initial and residual proton energies.

2 Materials and methods

2.1 Proton imaging scanner

The pRads presented in this work were taken at the Cyclotron Centre Bronowice (CCB) facility of the Institute of Nuclear Physics-Polish Academy of Sciences, Krakow, Poland. The proton imaging scanner is described in detail by J.A. Briz *et al.* [3]. It is composed of two tracking

*e-mail: amanda.nerio@csic.es

https://doi.org/10.1051/epjconf/202329008004

planes and a residual energy detector. The experimental setup is shown in Fig. 1.

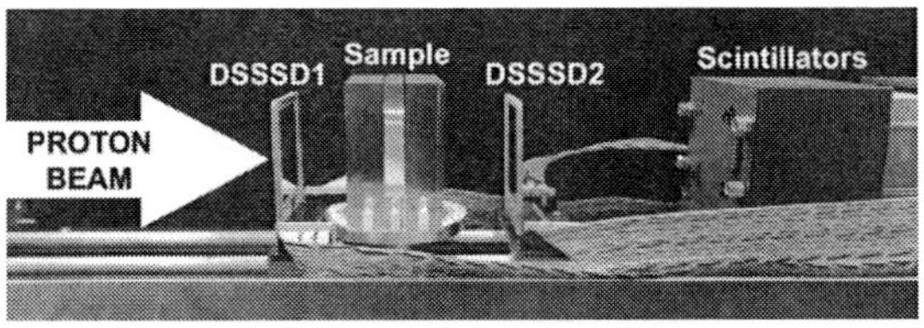

Figure 1. Experimental setup used to obtain the proton radiographs at the CCB facility. The proton-scanner prototype is formed by two DSSSDs, for proton tracking, and the CEPA4 scintillation detector, as residual energy detector. The sample under study is placed in between the DSSSDs.

The front and rear tracking planes incorporate a 50×50 mm^2 double-sided silicon strip detector (DSSSD) with a thickness of 1000 μm each. The design of our prototype allows us to obtain the entry and exit position of each proton.

The residual energy detector, the **C**ALIFA **E**ndcap **P**hoswich **A**rray (CEPA4 detector), is an array of four scintillation units, each of them comprised of two scintillator crystals of $LaBr_3$(Ce) and $LaCl_3$(Ce) in phoswich configuration with a common photomultiplier tube. Due to its high performance, in terms of energy resolution and fast time response, this detector has proven to be a good tool in proton spectroscopy, making it an ideal instrument for designing our pCT scanner prototype. A full description of the CEPA4 detector can be found in [8].

2.2 Known samples

The two custom-made samples containing aluminum, air, and polymethyl methacrylate (PMMA) shown in Fig. 2 were studied. Both samples are custom-made plates of PMMA of 10 mm thickness and 76 × 80 mm^2 surface which were placed in between two uniform PMMA plates of equal surface size and 20 mm thickness each. The first sample (A) is a cross pattern sample of aluminum with a uniform thickness of 10 mm, its dimensions are 40 × 40 mm^2, 54.5 mm diagonal, and 9 mm arm width. The second sample (B) is a regular spatial pattern of 39 × 40 mm^2 surface made of aluminum. On the left side of the pattern, there are holes of 1, 2, 3, and 5 mm diameters and equal separation. While on the right side of it, there are slits of 2, 3, and 5 mm width and separations. This sample was used to evaluate spatial resolution values ranging from 0.1 to 0.5 line pairs per mm.

2.3 Proton radiography generation

The position, generated from the Cartesian combination of the vertical and horizontal strips of each DSSSD, and the energy loss of all protons were registered into two-dimensional plots of hits and energy loss. These plots represent the number of protons traversing each pixel that deposit energy in each detector and the amount of energy they deposit. Although protons lose energy in a continuous way within the samples, the pRads were generated at

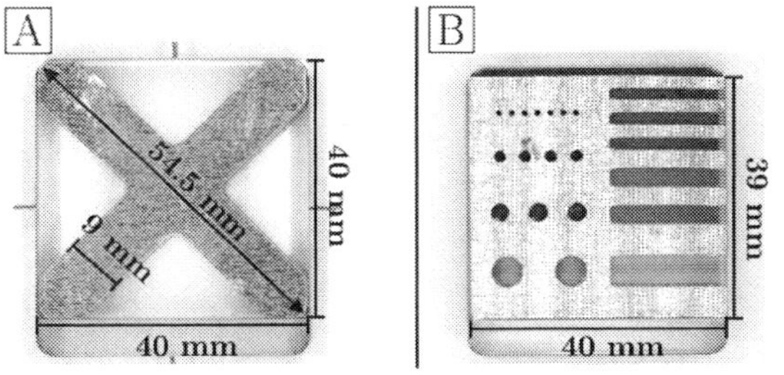

Figure 2. Pictures of the studied samples with the aluminum patterns and their dimensions. (A) Cross pattern, (B) Regular spatial pattern, from top to bottom: diameter and spacing size of the regions with circles of 1 mm, 2 mm, 3 mm, 5 mm; and bar width and separation width of 2 mm, 3 mm, 5 mm.

the *sample plane*, considered to be the central plane of the sample perpendicular to the beam.

To obtain the pRads we assumed a uniform distribution of the hits within the pixel area of each detector to describe the possible trajectories a proton can follow as a straight-line path through the tracking system (Fig. 3) and its corresponding residual energy. The modeling of the uniform distribution of the hits within each detector pixel was done using a random generator between 0 and 1 for the hit position in each DSSSD. This made possible the uniform distribution of the hits within the full area of a detector pixel, using Eq. 1.

$$\begin{pmatrix} x_P \\ y_P \end{pmatrix} = \frac{1}{2}\begin{pmatrix} (x_1 + R_1(0,1)) + (x_2 + R_2(0,1)) \\ (y_1 + R_3(0,1)) + (y_2 + R_4(0,1)) \end{pmatrix}, \quad (1)$$

where (x_P, y_P) represents the Cartesian coordinates of the proton hit at the image plane considered to be located at the central position of the sample; (x_1, y_1) and (x_2, y_2) are the coordinates of the proton hit on the front and rear DSSSD, respectively, and $R_i(0, 1)$ are uniform random values between 0 and 1. The uniform distribution of protons within a detector pixel allowed for an increase in the image granularity beyond the 32 × 32 pixels achievable with the experimental setup, where we used two tracking planes with 16 × 16 pixels each.

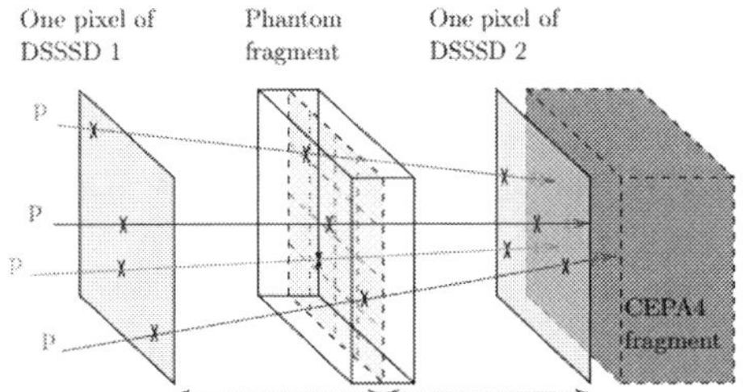

Figure 3. Possible proton trajectories through two pixels of the detection system.

With the IEM-CSIC scanner prototype, we are capable of determining the residual energy of the proton (E_f) defined as the sum of the energy measured on the second tracker and on the scintillators. The range of protons in water depends on the proton energy (see Fig. 4), hence, Eq. 2 is used to determine the WET of a proton given the initial and residual proton energies (E_0 and E_f) [6].

$$\mathbf{WET} = t_w = R(E_0) - R(E_f), \quad (2)$$

where $R(E)$ is the range in water of protons with energy E.

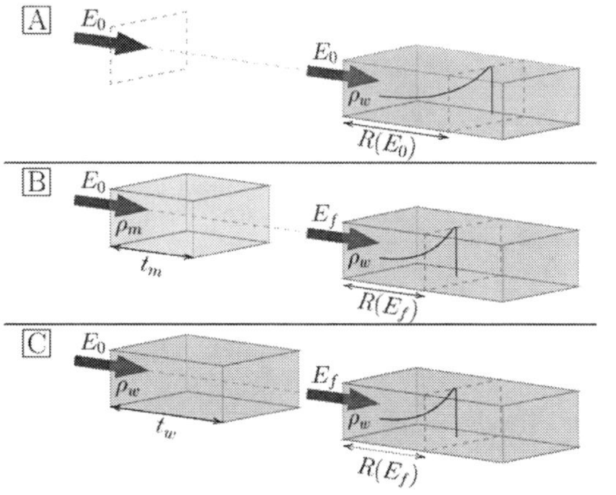

Figure 4. Scheme of the range in water for protons with initial energy E_0, after traversing (A) no material, (B) material with thickness t_m and density ρ_m, and (C) equivalent water thickness t_w, reproducing the same exit energy (E_f), therefore the same range in water $R(E_f)$.

Proton tracks were stored, and their WET were determined using Eq. 2. Proton ranges were directly obtained from the PSTAR database for protons in liquid water [7]. As expected from the Rutherford scattering suffered by the protons on the titanium scattering foil, regions on the right-side of the image showed higher particle counts because they were located closer to the incident beam direction (0° of polar scattering angle). This effect was compensated by normalizing the images by the number of protons traversing, thus, obtaining the average energy loss per traversing proton. Finally, the pRads were presented as a normalization of the WET values of the imaged objects.

3 Results

The pRads were obtained following the procedure described in Section 2. Let us remember that the samples were placed in between two PMMA pieces of 20 mm thickness each. Thus, protons passing by the air region traverse 40 mm of PMMA in addition to the 10 mm of air, a similar situation is observed for those protons traversing aluminum or PMMA.

3.1 Cross pattern sample

The image shown in Fig. 5 was obtained when using the sample shown in Fig. 2(A) in between two PMMA uniform slices 20 mm thick each. Three main regions are clearly distinguished, each one being consistent with the materials of the cross pattern sample. WET values are presented in millimeters. In Fig. 5, the black zones (WET≈ 58 mm) correspond to the air regions of the sample; the grey zone (WET≈ 66 mm) is the region where the PMMA is located, *i.e.* the square frame around the cross pattern; and the zone of the cross pattern made of aluminum appears in white (WET≈ 73 mm). Notice that the grey gradient (WET≈ 61 mm) observed between the black and white zones corresponds to the air-aluminum interface.

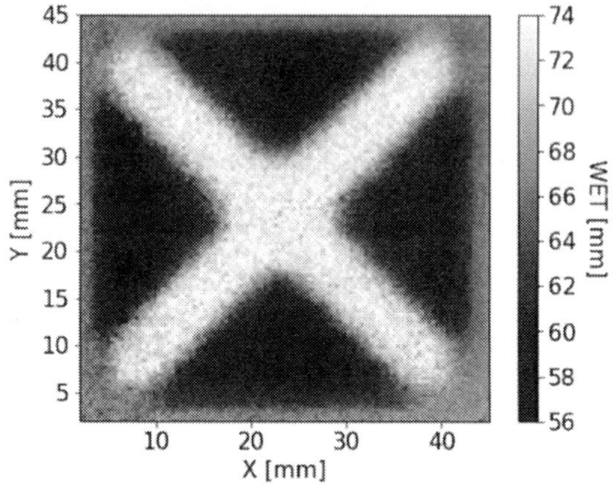

Figure 5. Experimental pRad obtained with the cross-shaped pattern sample. The image is a 2D map of the WET values in mm. The position on the image plane was obtained using the hit positions on DSSSD1 and DSSSD2, as described in Section 2.3.

The pixelation of pRads was set to 128 × 128. Each of the 16384 image pixels displays the averaged WET value of all protons passing through it. A histogram of the averaged WET was generated and it is shown in Fig. 6.

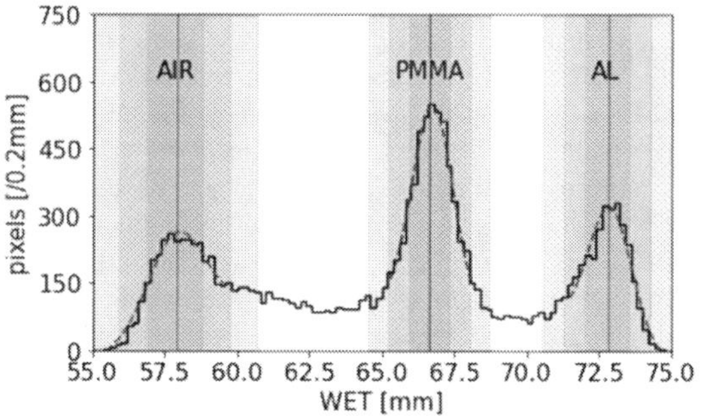

Figure 6. Histogram of the WET values obtained from the image pixels of the pRad of the cross pattern sample. The vertical axis represents the number of pixels shown in Fig. 5 with the same WET value (±0.1 mm). Three main peaks corresponding to the air (red), PMMA (blue), and aluminum (pink) regions are observed. The mean WET values and the regions corresponding to σ, 2σ, and 3σ are displayed in degraded colors.

The histogram shows three resolved structures corresponding to the different materials present in our sample: aluminum, PMMA, and air, all of them with 10 mm thickness approximately. A Gaussian fit was done for each of the three main peaks observed, resulting in WET values of 57.8(9) mm for the peak corresponding to air, 66.6(5) mm for the PMMA peak, and 72.7(7) mm for the aluminum region. In the same figure, the mean WET values and the regions corresponding to σ, 2σ, and 3σ for each material are shown in degraded colors. Looking back into the sample, it must be said that it was placed between two slices 20-mm thick of PMMA each. Hence, for protons of 100 MeV, a water thickness of 57.8 mm is equivalent to 40 mm of PMMA, 66.6 mm of water is equivalent to 50 mm of PMMA, and 72.7 mm of water is equivalent to a combination of 40 mm of PMMA and 10 mm of aluminum.

3.2 Regular spatial pattern sample

The image shown in Fig. 7 was obtained for the sample pattern shown in Fig. 2(B). The statistical study of the averaged WET values showed that, for protons of 100 MeV, WET values of the PMMA and aluminum regions were 66.8(6) mm and 72.8(6) mm, respectively. The WET values of PMMA and aluminum regions were consistent with the ones obtained for the cross-pattern sample. The larger air regions (3- and 5-mm) provided WET values in agreement with those obtained in Section 3.1.

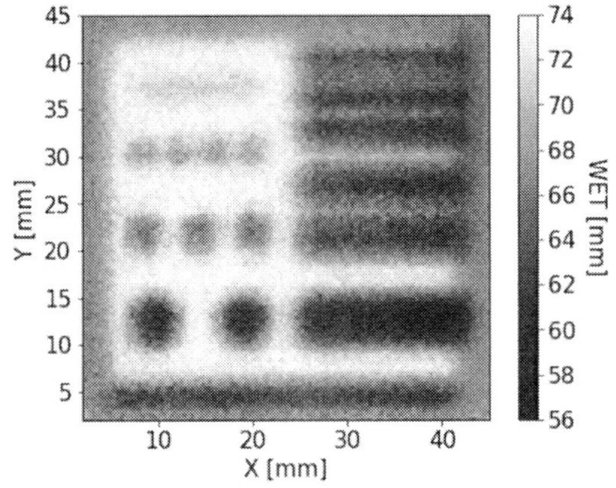

Figure 7. Experimental pRad obtained with the regular spatial pattern sample. The image is a 2D map of the WET values in mm. The position on the image plane was obtained using the hit positions on DSSSD1 and DSSSD2, as described in Section 2.3. The spatial resolution of the IEM-CSIC scanner prototype was determined to be better than 2 mm in [3].

4 Conclusion

The first steps in the performance tests of our proton imaging device have been completed by obtaining proton radiographs (pRads) of simple samples with known geometries and known materials. We presented a fast and functional, but at the same time useful, process to generate a pRad in terms of water equivalent thickness (WET) values. An important step in the WET calibration as a function of the initial and residual beam energy was introduced. The WET histogram of the cross-pattern sample indicated that the device has enough sensitivity to distinguish bulk structures of aluminum, PMMA, and air. As for the regular spatial pattern sample, the WET value results PMMA, aluminum, and the larger air regions were compatible between samples. We have made significant progress in the WET value determination, however, there is work to be done for the implementation of more sophisticated analysis methods, a more realistic description of the proton paths, and the improvement of the data acquisition rate to reduce scanning times, so that it would be possible to get closer to a preclinical stage with the IEM-CSIC prototype used as a proton radiography scanner.

Acknowledgements

This work was partially supported by the project Proton Therapy and Nuclear Techniques for Oncology (PRONTO) funded by the local government of Madrid, with reference B2017/BMD-3888; by the Horizon 2020 Research and Innovation Programme of the European Union under Grant 654002 (ENSAR2) and Grant 730983 (INSPIRE). This publication is part of the R&D grants PID2019-104714GB-C21, and PID2019-104390GB-I00, funded by the Spanish Ministry of Science MCIN/AEI/10.13039/501100011033, and grant CIPROM/2021/064 from Generalitat Valenciana. The work of A. N. Nerio was partially supported by the Erasmus Mundus Joint Master Degree in Nuclear Physics co-funded by the Erasmus+ Programme of the European Union.

References

[1] G. Dedes, J. Dickmann, K. Niepel, P. Wesp, R.P. Johnson, M. Pankuch, V. Bashkirov, S. Rit, L. Volz, R.W. Schulte et al., Physics in Medicine & Biology **64**, 165002 (2019)

[2] R. Zhang, W.D. Newhauser, Physics in Medicine & Biology **54**, 1383 (2009)

[3] J.A. Briz, A.N. Nerio, C. Ballesteros, M.J.G. Borge, P. Martínez, A. Perea, V.G. Távora, O. Tengblad, M. Ciemala, A. Maj et al., IEEE Transactions on Nuclear Science **69**, 696 (2022)

[4] H. Paganetti, Physics in Medicine & Biology **57**, R99 (2012)

[5] R. Hurley, R. Schulte, V. Bashkirov, A. Wroe, A. Ghebremedhin, H.W. Sadrozinski, V. Rykalin, G. Coutrakon, P. Koss, B. Patyal, Medical physics **39**, 2438 (2012)

[6] R. Zhang, P.J. Taddei, M.M. Fitzek, W.D. Newhauser, Physics in Medicine & Biology **55**, 2481 (2010)

[7] M.J. Berger, J.S. Coursey, M.A. Zucker, *AESTAR, PSTAR, and ASTAR: computer programs for calculating stopping-power and range tables for electrons, protons, and helium ions (version 2.0.1)* (2017), 08.03.2023, `http://physics.nist.gov/Star`

[8] E. Nácher, M. Mårtensson, O. Tengblad, H. Álvarez-Pol, M. Bendel, D. Cortina-Gil, R. Gernhäuser, T. Le Bleis, A. Maj, T. Nilsson et al., Nuclear Instruments and Methods in Physics Research Section A: Accelerators, Spectrometers, Detectors and Associated Equipment **769**, 105 (2015)

Search for D-D Fusion initiated by Nuclear Hotspots in Boranes

Hans Georg Priesmeyer[1*] and *Edward* Cazalas[2],

[1]IEAP, University of Kiel, 24118 Kiel, Germany
[2]Nuclear Engineering, University of Utah, Salt Lake City, USA

Abstract. In the group of boron hydrogen compounds, decaborane and aminoborane belong to those with a particularly high hydrogen density in the molecular volume. Almost 100% of the protons in the molecule can be replaced by deuterons. In the element boron, the isotope ^{10}B is 19.9% naturally present. By capturing thermal neutrons in this isotope, the high-energy reaction products ^{4}He and ^{7}Li are formed. It thus forms a hotspot in a spatially stable cloud of deuterium nuclei, which can be triggered by external action and whose energy could be transferred to deuterium in a similar way to neutral beam injection, a successful method to heat a fusion plasma.

1 Introduction

The molecules of certain boranes have a very high hydrogen density. These densities exist without externally acting forces and are stable. The hydrogen nuclei may be replaced by deuterium. Decaborane molecules contain two atoms of ^{10}B on the average. The reaction $^{10}_{5}B\ (n,\alpha)^{7}_{3}Li$ produces two energetic ions and might serve as a "hotspot" within a dense cloud of deuterons, triggered by thermal neutron capture. The effect of such "hotspot" energy release is comparable to neutral gas injection, an effectively used method of injecting deuterons or tritons with energies of 50 keV to 130 keV into a fusion plasma that is held together magnetically. The fast ions lose energy through coulomb collisions with electrons and ions in the plasma and thus act as a heat source [1]. In our case, just Coulomb interactions must be considered.

According to classical collision laws, the supplied energy could be sufficient to initiate D-D fusion. An experiment to show whether this energy can be transferred to deuterium nuclei, is proposed in this paper.

2 Decaborane

Decaborane has a history as an admixture to solid rocket fuel and recently for hydrogen storage. Deuterated decaborane has also found application in fusion research [2]. This shows that the use of decaborane is also possible for producing nuclear environments.

2.1 Properties of Decaborane

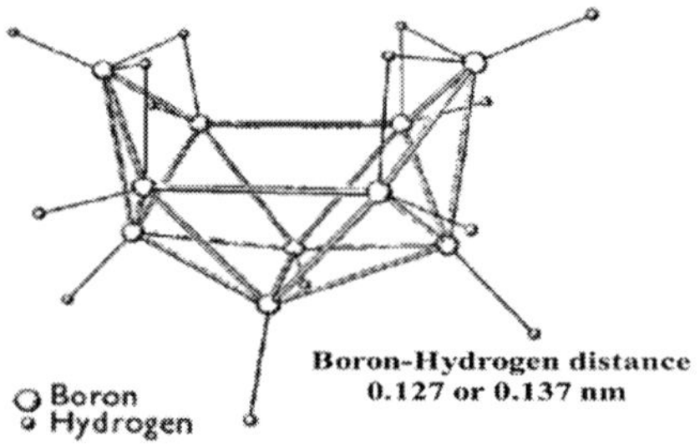

Fig. 1. Form of the molecule of $B_{10}H_{14}$ [3].

Kasper et al. [3] have determined the unitcell parameters of decaborane to: a0 = 1.445 nm, b0 = 2.088 nm, c0 = 0.568 nm. Therefore the unitcell volume is 1.714 x 10^{-27} m^3. Since there are 8 molecules per unitcell and 14 deuterons per molecule, the deuteron density in decaborane is n_D = 6.54 x 10^{28} /m^3. For Aminoborane the deuteron density amounts to n_D = 9.15 x 10^{28} /m^3. Typically about 98% deuteration of boranes can be achieved.

2.2 Uncertainties in the experimental procedure

A number of open questions exists: How does the time of molecular disintegration compare to the velocity of the ions created? Will the deuteron density remain high for the duration needed for an interaction? What is the probability of electric charges coming into the required proximity to each other? Can the same probability be assumed for the required energy transfer to initiate fusion or will it be lower?

* Corresponding author: priesmeyer@physik.uni-kiel.de

These questions led us to approach the task empirically. It has to be expected that the planned experiment has particularly high requirements with respect to a high flux of incident thermal neutrons, a very low background of fast neutrons, efficient fast neutron detectors with a high solid angle range. The detection of fast neutrons (both 2.45 MeV and 14 MeV) in time coincidence with the 480 keV γ – ray from the ^{7}Li* decay proves the initiation D-D fusion.

As an estimate for the conversion factor of a nuclear "hotspot" – reaction we consider the early developments to produce 14.1 MeV neutrons from thermal neutrons using converters based on lithium deuteride compounds and the reaction ^{6}LiD(n,α)T, which showed that about one fast neutron is produced per 10^4 thermal neutrons [4].

If we assume a factor of 10^{-4} for the probability to transfer energy from a primary deuteron to the next, it becomes clear that the energy transfer to initiate fusion has low probability. An estimate for the ratio of the number of ^{10}B(n,α)Li reactions to the number of deuterons accelerated by α particles or Lithium ions could be of the same order of magnitude. This would yield a reaction rate of one D-D fusion per 10^8 thermal neutrons captured. Such a probability is, however, subject to uncertainties in both directions, but it shows clearly that very great effort will be experimentally required to demonstrate this fusion. In any case, the experiment must be sensitive enough to detect the 2.45 MeV neutrons that are produced.

3 The Experiment

3.1 The reactions

$$\begin{aligned} {}^{10}_{5}\mathrm{B} + {}^{1}_{0}\mathrm{n} &\rightarrow {}^{7}_{3}\mathrm{Li}(1.01\ \mathrm{MeV}) + {}^{4}_{2}\mathrm{He}(1.78\ \mathrm{MeV}) \quad 6\% \\ &\rightarrow {}^{7}_{3}\mathrm{Li}^{*}(0.84\ \mathrm{MeV}) + {}^{4}_{2}\mathrm{He}\ (1.47\ \mathrm{MeV}) \quad 94\% \\ &\qquad + 480\ \mathrm{keV}\ \ \gamma\text{-ray} \end{aligned} \tag{1}$$

If the energy transfer from fast ^{7}Li and ^{4}He ions allow the initiation of D-D fusion, this will occur in two branches with an almost equal likelihood:

$${}^{2}_{1}D + {}^{2}_{1}D \rightarrow {}^{3}_{1}T(1.01MeV) + {}^{1}_{1}p(3.02MeV) \tag{2}$$

$${}^{2}_{1}D + {}^{2}_{1}D \rightarrow {}^{3}_{2}He(0.82MeV) + {}^{1}_{0}n(2.45MeV) \tag{3}$$

The detection of the 480 keV gamma ray from the excited ^{7}Li* nucleus defines the time of the energy release within the borane molecule. Contrasting to this, D-D fusion in about a half of the cases will produce a 2.45 MeV neutron. The detection of these neutrons in coincidence with the 480 keV gamma rays proves D-D fusion. The other half of the reactions will produce 1.01 MeV tritons. Due to the higher cross-section of D-T fusion, 14 MeV neutrons might be generated by this fusion type. Therefore, both branches of D-D fusion may serve as evidence of its initiation.

Besides the fusion reactions two other exothermal light ion reactions are possible (Table 1).

Table 1. List of light-nuclide exothermal reactions [5].

$D + T \Rightarrow n(14.4MeV) + \alpha(3.5MeV)$
$D + D \Rightarrow T(1.01MeV) + p(3.02MeV)$
$D + D \Rightarrow {}^{3}He(0.82MeV) + n(2.45MeV)$
${}^{3}He + D \Rightarrow \alpha(3.6MeV) + p(14.7MeV)$
$p + {}^{11}B \Rightarrow 3 \cdot \alpha(8.7MeV)$

These reactions could in principle contribute to the energy inventory within the reaction volume.

3.2 The experimental setup

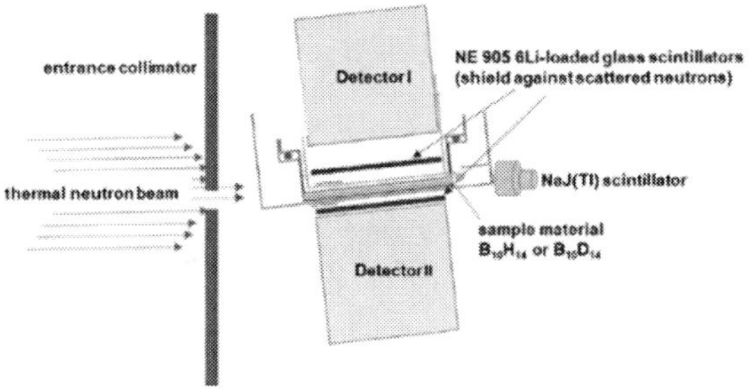

Fig. 2. Sample and detector arrangement.

References

1. J. Freidberg Plasma Physics and Fusion Energy, Cambridge University Press 2007
2. T. Nakano, S. Higashijima, H. Kubo, j. Yagyu, T. Arai, N. Asakura, K. Izami, Journal of Nuclear Materials **313-316**, 149 (2003)
3. J. S. Kasper, C. M. Lucht, D. Harker, Acta Cryst. **3**, 436 (1950).
4. W. H. Miller, W. S. Law and R. M. Brugger, Nucl. Instr. Meth. **216**, 219 (1983)
5. M. Kikuchi, K. Lackner, M. Q. Tran, Editors, Fusion Physics, STI/PUB/**1562**, IAEA Vienna, (2012)

Nuclear fragmentation cross section measurements with the FOOT experiment

Riccardo Ridolfi[1,2,*]

[1]INFN, Sezione di Bologna - Bologna, Italy
[2]Dipartimento di Fisica e Astronomia "Augusto Righi", Università di Bologna - Bologna, Italy

Abstract. Different fields can profit by nuclear fragmentation cross section measurements: among them hadrontherapy and space radioprotection are of particular interest. Hadrontherapy employs high-energy beams of charged particles (protons and heavier ions) to treat deep-seated tumours. In these treatments nuclear interactions have to be considered: beam particles can fragment in the human body releasing a non-zero dose beyond the tumour while fragments of human body nuclei can modify the dose released in healthy tissues.
On the radioprotection side, the interest in long-term manned space missions beyond Low Earth Orbit is growing in these years but it has to cope with significant health concerns from radiation in space, necessitating an accurate cross section data description. The FOOT (FragmentatiOn Of Target) experiment was proposed to cover these gaps in data. It was designed to detect, track and identify nuclear fragments and aims to measure double differential cross sections both in angle and kinetic energy which is the most complete information to address existing questions. The FOOT experimental setups, the experimental program and a first cross section analysis of 400 MeV/u ^{16}O beam on Carbon target data acquired in July 2021 at GSI (Darmstadt, Germany) will be presented.

1 Introduction

Hadrontherapy is an external radiation therapy technique in which protons and heavier ions are used to treat deep-seated solid tumours. The main advantage to use charged particles to treat tumours is their favourable depth-dose profile which is very different from that of photons used in conventional radiation therapy. Indeed, while a photon beam reduces its intensity exponentially as it enters into a material, charged particles are characterised by a low energy release in the beginning followed by a sharp rise after which the particle stops (Bragg peak). This behaviour makes charged particles particularly suitable to treat tumours near critical organs that must be spared by the radiation, especially in younger patients. Moreover, the electric charge of hadrons allows to actively move the beam to cover all the tumour volume. Furthermore, heavier ions such as Carbon and Oxygen can play an important role in treating radioresistant tumours thanks to their enhanced biological effectiveness. However, nuclear interactions have to be accounted for: beam particles can fragment in the human body producing lower mass nuclei able to release a non-zero dose beyond the Bragg peak. This contribution has to be properly described. On the other hand, nuclear interactions providing the fragmentation of nuclei of the human body give rise to fragments with low energy. These nuclear fragments can modify the dose released in healthy tissues and their effects are still in question given the lack of accurate cross sections data. The study of such nuclear interactions is of strong interest also in the space radioprotection field. Indeed, the interest in long-term manned space missions beyond the Low Earth Orbit is growing in these years. However, the health risks linked to space radiation are a major hazard which can potentially prevent any mission due to unacceptable risks for the astronauts since huge gaps in fragmentation cross sections data used to benchmark both deterministic and Monte Carlo models are present. Among all the possible measurements, double differential cross sections, both in angle and kinetic energy, would be the most significant contribution of experimental nuclear physics to these fields providing the most complete information to develop a new generation of treatment planning systems for patients and reliable transport codes and risk models for space radioprotection.

2 The FOOT experiment

To address all these questions, the FOOT (FragmentatiOn Of Target) experiment was proposed. It is composed by two independent and complementary setups, an Emulsion Cloud Chamber and an electronic setup composed by several subdetectors providing redundant measurements of kinematic properties of fragments produced in nuclear interactions between a beam and a target. FOOT was designed to detect, track and identify nuclear fragments and aims to measure double differential cross section both in angle and kinetic energy relevant for hadrontherapy and radioprotection in space. Thanks to its tabletop setup, the FOOT experiment can be mounted at research cen-

*e-mail: Riccardo.Ridolfi@bo.infn.it

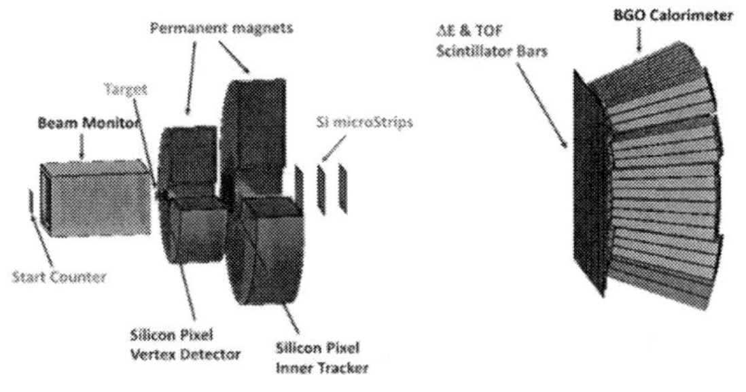

Figure 1. Schematic view of the FOOT electronic setup [1].

tres and at clinical facilities to harness the available variety of beams and energies. Indeed, the core program of the experiment foresees the use of ^{4}He, ^{12}C and ^{16}O beams with C, C_2H_4 and PMMA targets to measure fragmentation cross sections in different energy ranges both for hadrontherapy (200 – 400 MeV/u) and for space radioprotection (700 – 800 MeV/u). Thanks to its flexibility, the experiment will be able to extend its physics program to other beam-target settings to possibly cover other data gaps.

FOOT will be able to measure both target fragmentation in proton treatments by means of an inverse kinematic approach and projectile fragmentation in hadrontherapy from direct kinematics approach. Moreover, measurements performed with FOOT will be also interesting for radioprotection in space, especially for light fragments produced by ions up to Oxygen. To achieve its goal FOOT will measure differential cross sections with respect to the kinetic energy $d\sigma/dE_{\text{kin}}$ for the target fragmentation process with an accuracy better than 10% and double differential cross sections $d^2\sigma/d\Omega dE_{\text{kin}}$ for the projectile fragmentation process with an accuracy better than 5% on the determination of the fragment yields in angle and in kinetic energy.

2.1 The electronic setup

The electronic setup was designed to detect and identify fragments heavier than ^{4}He with an angular acceptance up to a polar angle of 10°.

To fit the strict requirements of measurement precision and ease of movement, the FOOT setup includes a ΔE-TOF system and a calorimeter to measure the velocity, the charge and the kinetic energy of the fragment together with a magnetic spectrometer equipped with three tracking stations and two permanent magnets to provide a measure of the momentum of the fragments. A schematic view of the electronic setup is shown in Fig. 1.

2.1.1 Start Counter and Beam Monitor

The Start Counter (SC), placed at the very beginning of the setup, provides the minimum bias trigger of the experiment and the start time for the Time Of Flight (TOF) measurement. The SC is made by a thin foil of EJ-228 plastic scintillator 250 μm thick with an active surface of 5×5 cm^2. The light produced in the scintillator is collected laterally by 48 3 × 3 mm^2 SiPMs. The readout and the power supply of the SiPMs is handled by the WaveDAQ system [2], which is able to sample signals at rates up to 5 GS/s in a dynamic range of 1 V.

The Beam Monitor (BM) is a drift chamber consisting of twelve wire layers able to reconstruct the beam direction in both views. The BM efficiency is higher than 90% for a wide range of beams and energies while a lower limit on the spatial resolution of 100 μm, in the central part of the BM cell, has been achieved [3].

The BM detector will be placed between the SC and the target and will be used to measure the direction and interacting point of the beam ions on the target while rejecting beam fragmentation before the target. The BM high spatial resolution is of great importance to measure the direction of the fragments with respect to the beam with an accuracy of few mrad which is necessary to measure the kinetic energy of the fragments in inverse kinematic with the required resolution.

2.1.2 Vertex detector, Inner Tracker and magnetic system

The Vertex detector (VTX) is organized in 4 pixel sensor layers covering an area of 2 × 2 cm^2, placed along the z axis, respectively at 0.6 – 0.9 – 2.1 – 2.4 cm from the target center ensuring a geometrical acceptance of about 40° for the produced fragments. The Inner Tracker (ITR) foresees two planes of the same pixel sensors to track the fragments between the two magnets at a distance of about 15 cm from the target covering a sensitive area of about 8 × 8 cm^2. In order to fulfill the required performances and constraints, the technology of the MIMOSA-28 (M28) Monolithic Active Pixel Sensors (MAPS) has been chosen for these detectors [1]. The magnetic system of the experiment is composed by two permanent magnets in Halbach configuration providing respectively a maximum intensity of 1.4 T and 0.9 T along the y axis in the internal cylindrical hole.

2.1.3 Micro Strip Detector

After the magnets, a third tracking station is present: this is essential both to measure the momentum of fragments and to match them with the hits in the ΔE-TOF detector and in the calorimeter. For such a tracking station a microstrip silicon detector (MSD), which can also provide a redundant measurement of dE/dx thanks to its analog readout, was chosen. The detector is composed by three x-y planes with an active area of 9.6 × 9.3 cm^2 , separated by a 2 cm gap along the beam direction and positioned right after the second magnet. In order to reduce the amount of material and to provide the x-y coordinate readout, a setup with two perpendicular 150 μm thick Single-Sided Silicon Detector (SSSD) sensors has been adopted for each MSD x-y plane. A strip pitch size of 50 μm has been chosen in order to minimize the fragment pile-up in the same strip. For each SSSD, one every three strips is read for a total of 640 channels.

2.1.4 ToF Wall

The ToF Wall detector (TW) is composed of two layers of 20 plastic scintillator bars (EJ-200 by Eljen Technology), arranged orthogonally and wrapped with reflective aluminum and darkening black tape. Each bar is 0.3 cm thick, 2 cm wide and 44 cm long. The two orthogonal x-y layers form a $40 \times 40\,\mathrm{cm}^2$ active area detector that provides the measurements of the energy deposited ΔE, the stop time to compute the TOF (the start time is provided by the SC), and the hit position. The TW transverse dimensions have been chosen to match the angular aperture of the heavy fragments at the distance of the detector from the target ($1-2$ m) depending on experimental room conditions. The simultaneous measurement of ΔE and TOF allows to identify the charge Z of the ions crossing the detectors. The TW performances meet the FOOT requirements of a TOF resolution better than 100 ps and an energy loss resolution $\sigma(\Delta E)/\Delta E \approx 5\%$, for the heavier fragments.

2.1.5 Calorimeter

The FOOT calorimeter is the most downstream detector and it is designed to measure the kinetic energy of the fragments, providing a necessary information to compute their mass A. $Bi_4Ge_3O_{12}$ (BGO) was chosen as the best material: the FOOT calorimeter will be composed by 320 BGO crystals covering an almost circular area of $\approx$ 20 cm radius. The crystals have a truncated pyramid shape with a front (back) area of about $2\times2\,\mathrm{cm}^2$ ($3\times3\,\mathrm{cm}^2$) and a length of 24 cm, chosen to minimize the energy leakage mainly due to neutrons escaping the calorimeter. The calorimeter will be readout by the WaveDAQ system as the SC and TW detectors at a rate of 1 GS/s in order to sample the full waveform. Several beam tests with different ions and energies were performed showing a very good linearity response and a energy resolution $\sigma(E_{\mathrm{kin}})/E_{\mathrm{kin}}$ below 2% for heavier fragments.

3 First analysis of GSI2021

In July 2021 the FOOT collaboration had the possibility to perform measurements on beam at GSI in Darmstadt using ^{16}O beam. The data taking involved a reduced apparatus of the FOOT electronic setup. Indeed, since the magnets, the Inner Tracker (IT) and the full Calorimeter (CALO) will be available within 2023, the setup was composed by the Start Counter (SC), the Beam Monitor (BM), the Vertex Detector (VTX), the Micro Strip Detector (MSD), the ToF Wall (TW) and one module (9 crystals) of the CALO. A preliminary analysis using the SC, BM and TW on a subset of data (400 MeV/u ^{16}O + C) is presented in the following.

3.1 MC simulations

The main information to extract from FLUKA Monte Carlo (MC) simulations are detection efficiencies for cross section calculations. In this context the efficiency was measured as follows:

$$\varepsilon(Z) = \frac{N_{\mathrm{TW}}(Z)}{N_{\mathrm{track}}(Z)} \tag{1}$$

where $N_{\mathrm{TW}}(Z)$ are the fragments with charge Z correctly reconstructed by the TW produced by primary particle in the target and $N_{\mathrm{track}}(Z)$ are the fragments generated by MC emitted from the target within the TW solid angle.

To not depend on fragmentation models in MC for the efficiency evaluation, for all MC analysis only fragments produced in the target were considered excluding secondary fragmentation outside the target. To extract N_{TW}, the number of TW points satisfying the same conditions of N_{track} were counted for each Z. Moreover, to not depend on the charge identification algorithm efficiency (even though it results to be high) N_{TW} were considered taking the true charge. This means that reconstruction efficiency should account for acceptance (both angular and in kinetic energy), intrinsic detector efficiency and clustering algorithm reconstruction efficiency. The same efficiency is then calculated for every angle for each charge for the differential cross section in angle.

3.2 GSI data

In the GSI2021 data taking the FOOT collaboration collected more than 40 million events using Carbon (C) and Polyethylene (C_2H_4) targets with ^{16}O beam at 200 and 400 MeV/u.

For this analysis, only runs with 400 MeV/u Oxygen beams on a Carbon target were selected. The goal of the preliminary analysis is to evaluate both angular and integrated elemental cross sections since there were no detectors providing the momentum or the kinetic energy of produced fragments. To this end, the elemental cross section for each Z can be written as follows:

$$\Delta\sigma(Z) = \int_{\beta_{\mathrm{min}}}^{\beta_{\mathrm{max}}} \int_0^{\theta_{\mathrm{max}}} \left(\frac{\partial^2\sigma}{\partial\theta\partial\beta}\right) \mathrm{d}\theta \mathrm{d}\beta = \frac{Y(Z)}{N_{\mathrm{prim}} \cdot N_{\mathrm{TG}} \cdot \varepsilon(Z)} \tag{2}$$

where θ_{max} is the maximum TW detector acceptance angle (4.85°), β_{min} and β_{max} are the minimum and maximum velocity, respectively, corresponding to the T_{tof} range in which charge identification algorithm works properly ($0.3-0.9$), $Y(Z)$ is the number of fragments of a given charge measured by TW, N_{prim} is the number of primaries impinging on the target, $\varepsilon(Z)$ is the total efficiency for a given charge as calculated in the previous section and N_{TG} is the number of interaction centres in the target per unit surface which can be written as

$$N_{\mathrm{TG}} = \frac{\rho \cdot \Delta x \cdot N_{\mathrm{A}}}{A} \tag{3}$$

where $\rho = 1.83\,\mathrm{g/cm}^3$ is the graphite target density, $\Delta x = 0.5$ cm is the target thickness, N_{A} is the Avogadro number and $A = 12.0107$ is the graphite mass number.

Regarding angular elemental differential cross section, in Eq. 2 only integration in velocity is performed leading to the following formula:

$$\frac{\mathrm{d}\sigma}{\mathrm{d}\theta}(Z) = \frac{Y(Z,\theta)}{N_{\mathrm{prim}} \cdot N_{\mathrm{TG}} \cdot \Delta\theta \cdot \varepsilon(Z,\theta)} \tag{4}$$

where $Y(Z,\theta)$ is the number of fragments of a given charge measured by TW within a given angle, $\varepsilon(Z,\theta)$ is the efficiency for a given charge in a given angle and $\Delta\theta$ is the

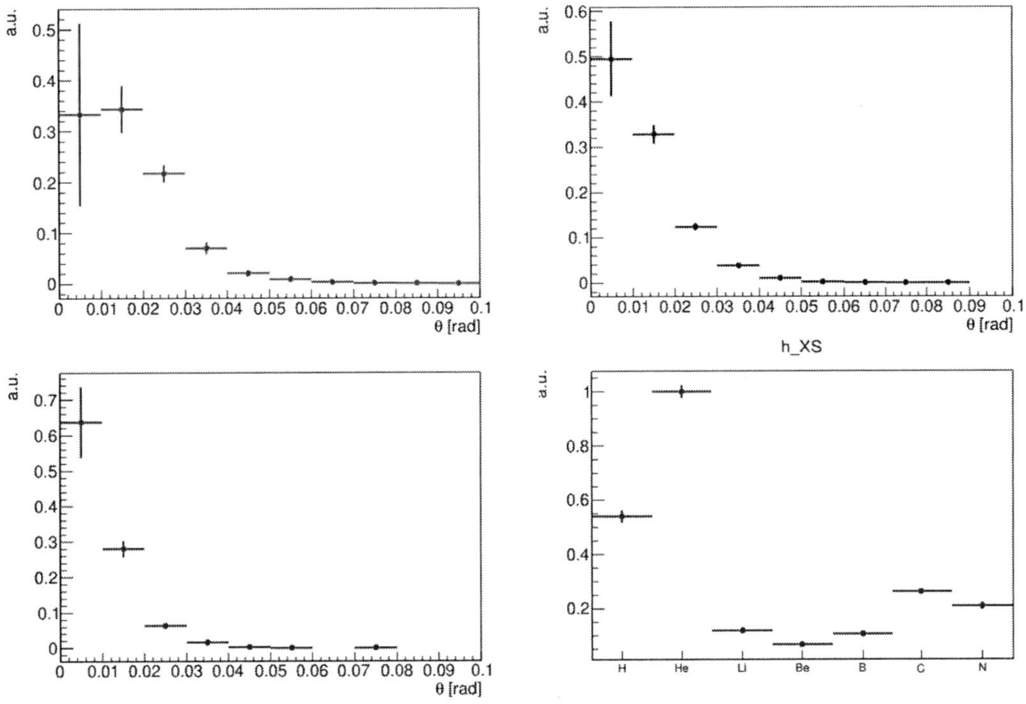

Figure 2. Preliminary angular differential cross sections for $Z = 5$, $Z = 6$, $Z = 7$ and total integrated cross sections.

bin width. The bin width was set to 10 mrad ($\approx 0.57°$) since it represents the angle covered by a TW unit at the target-TW distance.

Since tracking detectors are not included in this analysis a strategy to remove the background composed by primary beam fragmentation outside the target had to be developed. Indeed, the out-of target fragmentation turns to be not negligible given the path length of fragments travelling in air and in detectors accounting for almost 30% of the total number of fragmentations. For this reason, a run without target was included in the analysis to subtract the background contribution. A preliminary result of differential cross sections accounting for statistical errors only is reported in Fig. 2. Since the systematic uncertainties are still being evaluated, angular differential cross sections are normalized to the unit area to highlight their shape while total integrated cross section is normalized to the highest value.

The preliminary results show that the FOOT setup is able to properly address cross section measurements which will be presented in a future work.

4 Acknowledgments

The FOOT Collaboration acknowledges the INFN for its support in building and running the detector, the Open Physics Hub (OPH) for the support and the GSI laboratory. The GSI data were taken in the IBER_009 experiment, supported by European Space Agency (ESA)-IBER19 project, in the frame of FAIR Phase-0.

References

[1] G. Battistoni, M. Toppi, V. Patera, T.F. Collaboration, Frontiers in Physics **8**, 555 (2021)

[2] L. Galli, A. Baldini, F. Cei, M. Chiappini, M. Francesconi, M. Grassi, U. Hartmann, M. Meucci, F. Morsani, D. Nicolò et al., Nuclear Instruments and Methods in Physics Research Section A: Accelerators, Spectrometers, Detectors and Associated Equipment **936**, 399 (2019)

[3] Y. Dong, S. Gianluigi, C. Sofia, A. Andrey, A. Behcet, A. Giovanni, A. Stefano, R. Arteche Diaz, B. Mattia, B. Nazar et al., Nuclear Instruments and Methods in Physics Research Section A: Accelerators, Spectrometers, Detectors and Associated Equipment **986**, 164756 (2021)

https://doi.org/10.1051/epjconf/202329008007

Radiotracers for the study of Marine and Oceanic Ecosystems

E. Nácher[1,*], *B.* Rubio[1,**], *J.* Balibrea[1], *I.* Ladarescu[1], *J.* Lerendegui-Marco[1], *M.* Martínez[1], *L.* Sánchez[1], *D.* García-Párraga[2], *J. L.* Crespo-Picazo[2], *M.* Roche[2], *C.* Tomás[2], and *G.* de Angelis[3]

[1]IFIC (CSIC-Uni. Valencia), Spain
[2]Oceanogràfic Fundation, Valencia, Spain
[3]LNL-INFN, Legnaro, Italy

Abstract. Acidification of the oceans influences the physiology of aquatic organisms and, in general, the ecology of marine ecosystems. Here we report on the first stage of the REMO project, focused on the study of marine species especially affected by acidification of waters, using radiotracers and nuclear instrumentation techniques.

1 Motivation

Since the beginning of the industrial revolution, humans have produced CO_2 levels well above the natural level which stayed between 170 ppm and 280 ppm (parts per million) for the last 800000 years. The change started at the end of the XIX century when the levels started to increase reaching a value of more than 418 ppm in 2023 [1]. The oceans have adsorbed about one third of the carbon dioxide released to the atmosphere in the form of carbonic acid H_2CO_3 resulting from the combination of H_2O and CO_2. This has led to a continuous increase in acidification of the oceans, often called the other CO_2 problem, that adds to the well known global warming problem. It is estimated that if CO_2 continues to be released at the same rate as today, ocean acidity will increase by 170% compared to pre-industrial levels. Water acidification influences the physiology of aquatic organisms and, in general, the ecology of marine ecosystems. This anthropogenic effect affects ecosystems at remote places such as the coral reefs due to the oceanic currents. While it is of utmost importance to stop and if possible decrease the anthropogenic production of CO_2, it is also important to make a realistic diagnosis of how these changes affect marine organisms. This is important not only as part of the general health and integrity of the global ecosystem but also because it can affect food security. It is in the diagnosis of the problem that nuclear techniques can help and is the main purpose of the REMO (Radiotrazadores para el estudio de Ecosistemas Marinos y Oceánicos) project. The main idea is to monitor the influence of water acidity in the growth of marine and oceanic ecosystems, in a controlled and stable laboratory environment.

*e-mail: enrique.nacher@csic.es
**e-mail: berta.rubio@ific.uv.es

2 Radiotracers for the study of Marine and Oceanic Ecosystems, the REMO project

In order to follow the growth of the marine species under different conditions of acidity in water, we want to measure the absorption of calcium using as a radiotracer the radioactive isotope ^{45}Ca. Similar activity has been carried out at the IAEA in Monaco since the beginning of this century [2]. However, their methodology includes the destruction of the individuals (dissolving them in scintillator material) in order to measure the amount of ^{45}Ca that has been incorporated in the organism. This makes impossible to study the same animal along its entire life cycle, furthermore, it sometimes requires the culture of a huge number of individuals. A complete review of this kind of studies, with emphasis on the need for non-destructive studies can be found in Ref. [3]. In fact, we plan to use a non-destructive method, putting the individuals inside a set-up and measuring the amount of ^{45}Ca using ad-hoc nuclear instrumentation. We plan to start with two different water conditions. One with the present PH level of 8.1 and the other with the PH level expected at the end of the present century, 7.8, if the increase in CO_2 production continues at the rate of recent years, see figure 1.

The starting phase of the project will concentrate on two key ecosystems, a Mediterranean aquarium to grow bivalves, and a Tropical aquarium with corals. The first one is very important for the Spanish economy, food supply and export. The second is very important at a global level since the coral reefs contain more than 25% of marine biodiversity, and it is a key place for the reproduction of about a quarter of marine species.

The aquaria are now being installed at the research laboratories of the Oceanogràfic Foundation in Valencia. The team there has ample experience in the growing and maintenance of the most important marine echo-systems that are also presented to the general public in the form of the largest aquarium in Europe. They have an extensive labo-

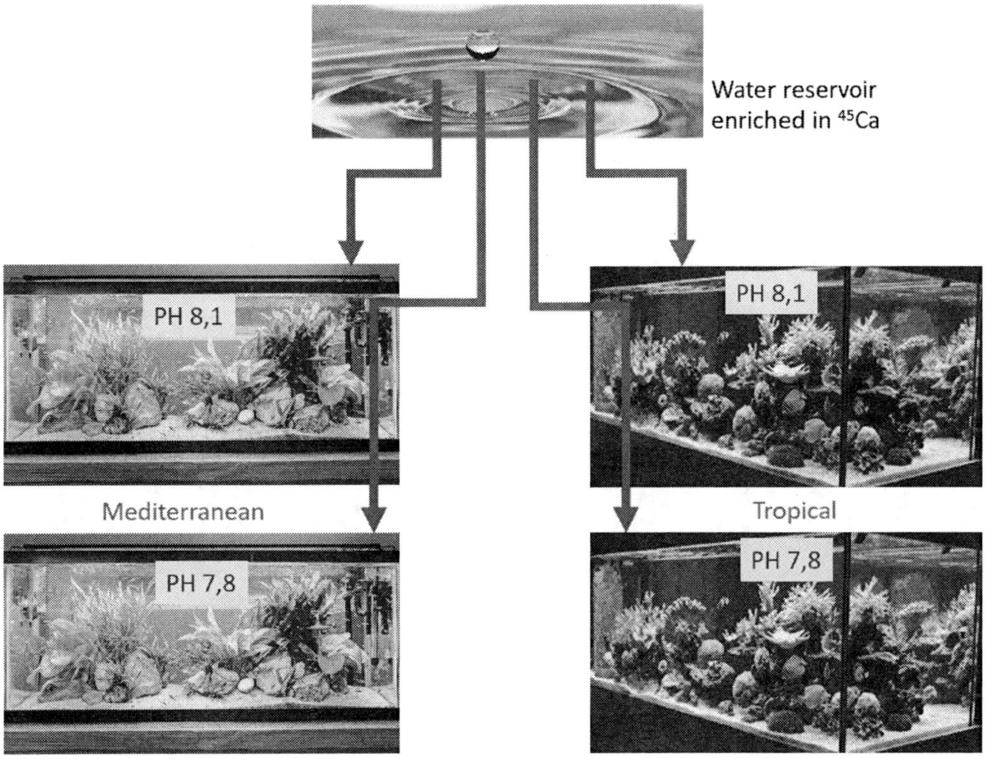

Figure 1. Schematic idea of the distribution of aquaria at the Oceanogràfic. On the left, the two Mediterranean aquaria with the present and end-of-century acidity conditions. On the right, the same acidity conditions, now for two tropical aquaria. For the ^{45}Ca enrichment, a common reservoir will be used.

ratory with the capacity to host the planned aquaria in the best controlled conditions. They also have the best quality natural sea water in the Valencia region. At the moment of writing this article the components of the set-up have already arrived at the Oceanogràfic Foundation and a first mounting with the illumination and filtration plant is ready for a first test with fresh water (see Fig. 2 upper panel).

As mentioned above, our goal is to measure the amount of Ca incorporated by the individual organism in the growing process. This will be done by dissolving a fixed amount of the radioactive isotope of ^{45}Ca into the water, which is chemically identical to the stable "normal" Ca existing in nature and consequently will be incorporated in the body in the same way.

^{45}Ca has a very simple decay scheme, as can be seen in figure 3. Almost 100% of the decays proceed via β^- decay to the ^{45}Sc ground state emitting an electron with an energy that covers a continuous spectrum with a maximum energy of 256 keV. It is this radiation that we want to measure. The half life of this isotope is 163 days.

Several circumstances have to be considered. a) The amount of radioactivity dissolved in the water, of the order of a few kBq, will not affect the physiology of the specimen. There are two reasons for this, firstly the activity is not strong and secondly the range of electrons of this energy in water is only a few mms. b) The growing process in molluscs as well as in corals occurs mainly in the most external part of the animal, hence minimising the self absorption. c) Both kinds of specimen are static. This facilitates the operation of taking them out of the aquatic medium and locating them inside the instrumental setup. d) The half life of ^{45}Ca is of the same order of magnitude as the growing life of the animals. All these considerations are positive and helpful for the proposed methodology. Less advantageous is the fact that corals are very sensitive to external stress and cannot be outside the water for more than 5 or 10 minutes without being affected. In consequence, this is the time scale for the measurements. Molluscs on the contrary are more resilient.

The instrumentation is currently under development at the Gamma and Neutron laboratories of IFIC in Valencia and at the LNL in Legnaro. Two main ideas are proposed, the first one is based on thin scintillating fibres that can bend and adapt to the shell of molluscs. We have already acquired three types of fibre of different manufacturer, thickness, scintillating material and cladding, and we are performing a comparative study to decide which one is best for measuring the low-energy betas emitted by ^{45}Ca. The lower panel of Fig. 2 shows a small sample of these fibres at our test bench at IFIC. The second detector

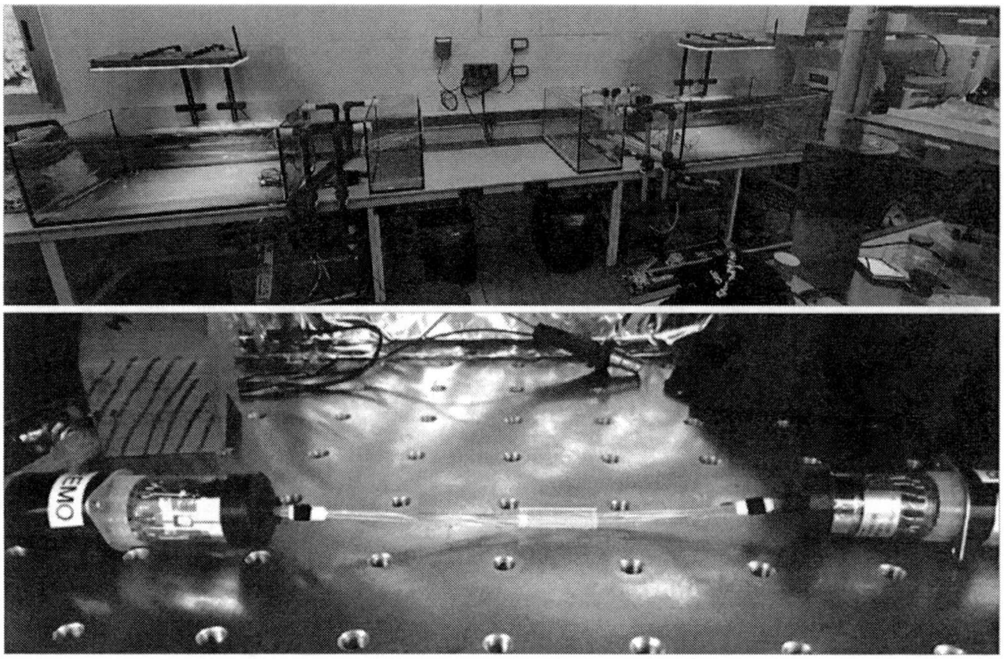

Figure 2. Upper panel: the aquarium system, still dry, as mounted at the Oceanogràfic. Lower Panel, a bundle of green scintillation fibers coupled to two small photomultiplier tubes in our test bench at IFIC.

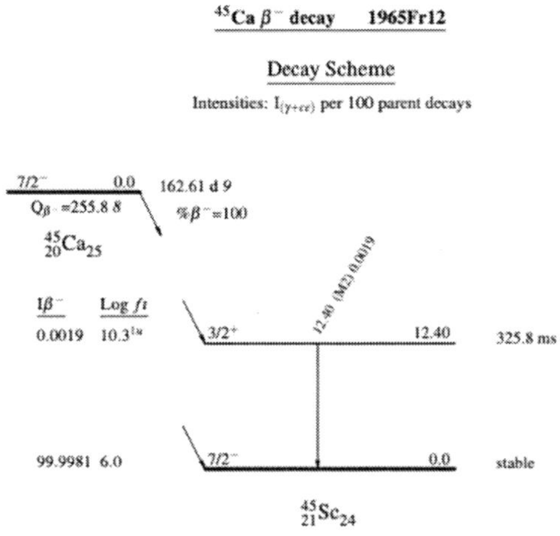

Figure 3. Decay scheme of ^{45}Ca. This radioactive Ca isotope decays by emitting beta particles of up to 256 keV in energy with a half life of 163 days. ^{45}Ca will be used as a radiotracer in the REMO project. Data taken from https://www.nndc.bnl.gov/.

approach is a large volume vessel filled with a commercial liquid scintillator with a special design that protects the mollusc or coral piece from the liquid. The vessel is coupled to a large area photomultiplier tube through an optical Quartz window. This second design is under construction at the time of writing.

3 Acknowledgments

This study forms part of the ThinkInAzul programme and was supported by MCIN with funding from European Union NextGenerationEU (PRTR-C17.I1) and by Generalitat Valenciana under project GVA-THINKINAZUL/2021/036.

References

[1] https://gml.noaa.gov/ccgg/
[2] https://www.iaea.org/newscenter/multimedia/videos/ocean-acidification-helping-to-tackle-this-global-threat-with-nuclear-techniques
[3] Cresswell T, Metian M, Fisher NS, et al., (2020). Frontiers in Marine Science 7, 406.

Interfering reaction channels observed in the $^{2}H(^{8}He,^{4}He)^{6}H$ reaction studies

I.A. Muzalevskii[1,2,*], *A.A.* Bezbakh[1,2], *E.Yu.* Nikolskii[2,3], *V.* Chudoba[1,2], *A.M.* Abakumov[2], *S.A.* Krupko[2], *S.G.* Belogurov[2,4], *D.* Biare[2], *A.S.* Fomichev[2,5], *E.M.* Gazeeva[2], *A.V.* Gorshkov[2], *L.V.* Grigorenko[2,3,4], *G.* Kaminski[2], *O.* Kiselev[6], *D.A.* Kostyleva[6], *B.* Mauyey[2,7], *I.* Mukha[6], *A.M.* Quynh[2,8], *S.I.* Sidorchuk[2,9], *N.B.* Shulgina[3,10], *R.S.* Slepnev[2], *A.* Swiercz[2], *G.M.* Ter-Akopian[2,5], *R.* Wolski[2], and *M.V.* Zhukov[11]

[1]Institute of Physics, Silesian University in Opava, 74601 Opava, Czech Republic
[2]Flerov Laboratory of Nuclear Reactions, JINR, 141980 Dubna, Russia
[3]National Research Centre "Kurchatov Institute", Kurchatov sq. 1, 123182 Moscow, Russia
[4]National Research Nuclear University "MEPhI", 115409 Moscow, Russia
[5]Dubna State University, 141982 Dubna, Russia
[6]GSI Helmholtzzentrum für Schwerionenforschung GmbH, 64291 Darmstadt, Germany
[7]L.N. Gumilyov Eurasian National University, 010008 Nur-Sultan, Kazakhstan
[8]Nuclear Research Institute, 670000 Dalat, Vietnam
[9]Voronezh state university, Universitetskaya sq. 1, 394018 Voronezh
[10]Bogoliubov Laboratory of Theoretical Physics, JINR, 141980 Dubna, Russia
[11]Department of Physics, Chalmers University of Technology, S-41296 Göteborg, Sweden

Abstract. In the recent work [Nikolskii et al., Phys. Rev. C **105**, 064605 (2022)] the $^{2}H(^{8}He,^{4}He)^{6}H$ reaction was used for the study of the extreme neutron-rich ^{6}H isotope. A broad bump was observed in the measured ^{6}H spectrum interpreted as the broad overlapping ground and some low-lying states of this nuclide. There could be certain doubts in the interpretation of this work: in conditions of the limited phase space it is not impossible that the structure in the missing mass spectrum of ^{6}H is actually induced by the resonant states populated by some other channels opened in the $^{8}He+^{2}H$ interaction. This work provides a body of the evidence for the correct channel identification and for the absence of the ^{6}H resonances at energy $E_T = 0 - 3.5$ MeV above the $^{3}H+3n$ decay threshold. In addition the first strong experimental evidence is given that the $^{6}H \rightarrow {}^{5}H^{*}+n \rightarrow {}^{3}H+3n$ sequential decay is the dominating ^{6}H decay channel.

1 Introduction

So far there is no answer to a fundamental question about the limits of the nuclear structure defined by the border between the unbound nuclei, still showing their resonance states, and the nuclear systems exhibiting only the continuous spectra. The search for the new isotopes beyond the driplines is a main experimental tool to find out where such limit is located. Not long ago a series of experiments were performed in order to obtain the superheavy hydrogen isotopes ^{7}H and ^{6}H [1–5]. The recently reported results [6, 7] shed light on the spectroscopy of these nuclei and decay mechanisms of their ground and excited states. However, the missing mass (MM) spectrum of ^{6}H reported in work [7] is characterized by the absence of any distinct narrow peak, and the observed resonance state is relatively wide and implicit. The latter leads to the question of weather the obtained structure could be formed by the final-state interaction appearing in some other reaction channels which are not so exotic and may occur with higher probabilities.

This work is dedicated to the analysis of the diverse reaction channels occurring in the $^{8}He+^{2}H$ interaction characterized by the detection of the same products as those which were used in Ref. [7] to identify the ^{6}H population. Such analysis allows one to elucidate the ^{6}H decay channels and its spectrum nature reported in [7].

2 Experiment

The experiment was performed at the ACCULINNA-2 facility, FLNR, JINR, producing 26 AMeV ^{8}He beam and focusing it on the cryogenic deuterium target. The detection system was intended to identify the products of the $^{2}H(^{8}He,^{4}He)^{6}H$ reaction and the further $^{6}H \rightarrow {}^{3}H+3n$ decay. The employed detector system is described in Ref. [7].

For the identification of beam nuclei two plastic scintillators were used, which allowed to measure the energy of the projectile from its time-of-flight (ToF) and identify the isotope by the dE-ToF method. The trajectories of the beam projectiles were tracked by the two pairs of multi-wire proportional chambers. The cryogenic target was filled with deuterium gas at atmospheric pressure and cooled to 27 K. For the detection of the charged reaction products two types of ΔE-E telescopes were used: the side assembly of the three (20 μm, 1 mm and 1 mm thick) silicon strip detector (SSD) telescopes, and the front tele-

*e-mail: muzalevsky@jinr.ru

https://doi.org/10.1051/epjconf/202329009001

scope made of the 1.5 mm double side SSD coupled to the CsI(Tl) scintillator array. The thin, 20 μm detectors in the side telescopes allowed one to reliably identify and reconstruct the low-energy particles (the recoil ^{4}He nuclei with energy ≥5 MeV) emitted from the target in the laboratory angular range between 8° and 26°, see Ref. [8]. The front telescope covered the lab angles ≤ 9°. It was used to measure the high-energy particles (tritons with energy up to 160 MeV) stopping in the CsI(Tl) crystal. The neutron detection was realized by the ToF stilbene modules [9].

3 "Background" reaction channels for ^{6}H

The first question to be solved is whether the low-energy part of the ^{6}H MM spectrum, see the E_T spectrum in Fig. 1 (a), can be caused by some other reaction mechanism. The ^{6}H events were identified by the detection of the ^{4}He recoil, emerging from the ^{2}H(^{8}He,^{4}He)^{6}H reaction, and the ^{3}H decay product of the ^{6}H decay (also the neutron coincidence information can be used on demand). How can we be sure that these nuclei were produced solely in the channel populating the ^{6}H states, but not in some different reaction channels, where they appear in the ^{3}H+n+n+n+^{4}He group? In reality the final state products of the of the ^{8}He+^{2}H interaction can be partitioned also as {^{3}H+n+n} + {^{4}He+n}, {^{3}H+n} + {^{4}He+n+n} or {^{3}H} + {^{4}He+n+n+n}. These outcomes correspond to the reactions ^{2}H(^{8}He, ^{5}He)^{5}H, ^{2}H(^{8}He,^{6} He*)^{4}H, and ^{2}H(^{8}He,^{7} He*)^{3}H. In this section, we provide data for the two of these reaction channels and study the correlations of these reaction channels with the obtained low-energy part of the ^{6}H spectrum.

The ^{7}He system was reconstructed assuming that ^{3}H is the nucleus formed immediately in the ^{2}H(^{8}He,^{7} He*)^{3}H reaction. The ^{4}He nucleus, detected in the side telescope, was used as the selection gate allowing one to reduce significantly the MM background conditions. Effective method was the so-called "kinematical triangle" selection used in works [6, 7, 10]. This allowed us to localize the background events in the correlation pattern of the kinetic energy of the emitted particles, taken in the center-of-mass frame (CMS) of the decaying system, with the MM energy of the decaying system (in this case it is the correlation of the ^{4}He kinetic energy in the ^{7}He CMS with the ^{7}He MM energy).

To study the correlations with the ^{7}He* MM spectrum we selected the low-energy part of the ^{6}H MM spectrum derived from the ^{2}H(^{8}He,^{4}He)^{6}H reaction data, see the green colored part in Fig. 1 (a). The ^{7}He* MM spectrum is presented in Fig. 1 (b) by the black histogram, while the green-line histogram corresponds to the low energy events in the ^{6}H MM spectrum. One may see that the low-energy events of ^{6}H are spread in the high-energy region of ^{7}He and do not form any pronounced peak. Thus, we conclude that the analyzed range of ^{6}H spectrum is very weakly affected by the final state interactions connected with the possible population of ^{7}He states in the ^{2}H(^{8}He,^{7} He*)^{3}H reaction.

The other possible "background" reaction ^{2}H(^{8}He, ^{5}He)^{5}H is the result of a triton transfer from

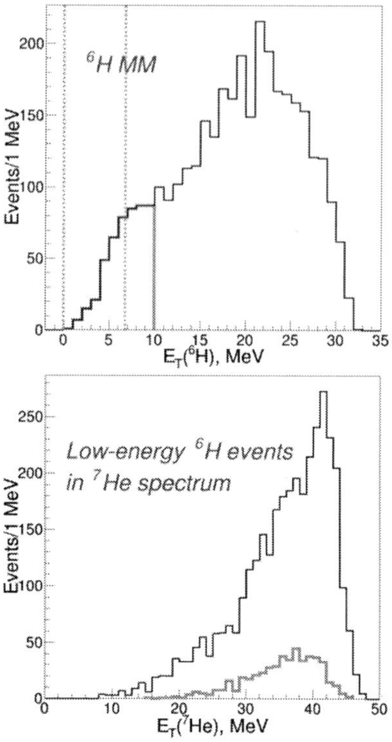

Figure 1. (a) The ^{6}H MM spectrum; the part of this spectrum at E_T < 10 MeV is indicated by green color. The vertical gray dotted lines show the ^{3}H+3n threshold and E_T(^{6}H)=6.8 MeV. (b) The ^{7}He MM spectrum built with respect to the ^{4}He+3n threshold (black histogram). The green histogram in (b) corresponds to the low-lying events with E_T < 10 MeV in the spectrum of ^{6}H.

^{8}He. This channel leads to the production of the target-like ^{5}He and beam-like ^{5}H. The latter moves forward and decays into ^{3}H+n+n, which does not allow to measure it directly by our setup. For that reason the ^{2}H(^{8}He, ^{5}He)^{5}H reaction examination can be performed only based on the triple ^{4}He-^{3}H-n coincidence events. The ^{6}H spectrum, stemming from the spectrum of Fig. 1 (a) by this coincidence requirement, is shown in Fig. 2 (a). For those events, the recoil ^{5}He invariant mass (IM) spectrum can be reconstructed from the measured ^{4}He and neutron, see the black histogram in Fig. 2 (b). The low-energy ^{6}H events with E_T < 10 MeV in the ^{5}He IM distribution are presented with the green histogram in Fig. 2 (b). One may find here the following:

- (i) The ^{5}He g.s. resonance is expected to be found with E_T(^{5}He) ~ 0.9 MeV. We really see a strong indication of this resonant state in the derived spectrum of ^{5}He.
- (ii) The ^{6}H low energy events are practically not correlated with the energy range of ^{5}He where the resonant states are known to exist. Thus, the ^{6}H spectrum can not be governed by the final state interactions connected with possible population of ^{5}He in the ^{2}H(^{8}He,^{5} He)^{5}H reaction.

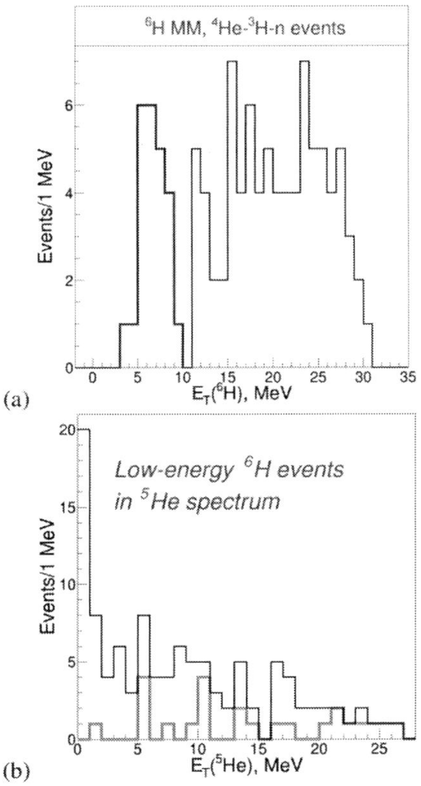

Figure 2. (a) The ^{6}H MM spectrum based on the ^{4}He-^{3}H-n coincidence events; the part of this spectrum at E_T < 10 MeV is given in green color. (b) The ^{5}He invariant mass spectrum constructed from the ^{4}He and neutron momentum vectors in the triple ^{4}He-^{3}H-n coincidence events.

The reconstruction of the ^{2}H(^{8}He,^{6} He*)^{4}H reaction requires the detection of 2n coincidence events, which is not possible because of the low efficiency of our neutron-wall setup. However, it is quite unexpected that such reaction is important, while the populations of ^{5}He and ^{7}He are found to be negligible.

4 ^{5}H populated in the ^{6}H decay

In work [7], the authors showed that the assumption of the ^{6}H→^{5}H(g.s.)+n →^{3}H+3n sequential decay leads to the evidence of an extremely strong "dineutron-type" correlation in the decay of the ^{5}H ground state. Here we apply the reaction-channel test for the decay-mechanism analysis of the ^{6}H spectrum obtained in this work.

Let us assume that in the process of the ^{8}He+^{2}H interaction the ^{5}H resonance is somehow populated. And moreover, we presume that all these events are characterized by the detection of the low-energy ^{4}He appearing in the side telescope in coincidence with the high-energy ^{3}H and neutron hitting the front telescope and stilbene wall, correspondingly. The only two channels satisfying these conditions are:

- The triton transfer reaction: ^{2}H(^{8}He,^{5}He)^{5}H.

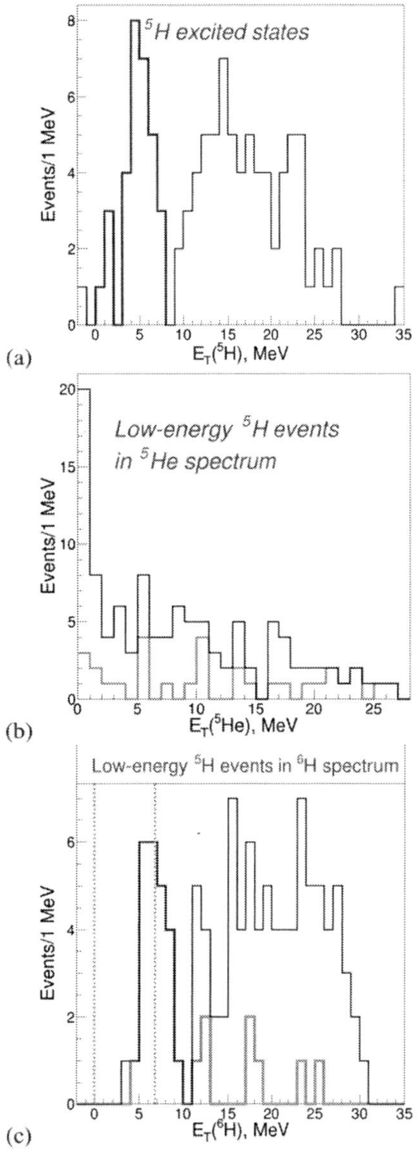

Figure 3. (a) The ^{5}H MM spectrum reconstructed via the IM spectrum of ^{5}He. The green histogram corresponds to the ^{5}H events with $0 < E_T(^5\mathrm{H}) < 9$ MeV. (b) The ^{5}He IM spectrum. (c) The ^{6}H MM spectrum formed by the triple ^{4}He-^{3}H-n coincidence events. The vertical lines show the ^{3}H+3n threshold and the resonance observed at $E_T(^6\mathrm{H})$ = 6.8 MeV.

- The deuteron transfer reaction ^{2}H(^{8}He,^{4}He)^{6}H in the assumption of the sequential ^{6}H→^{5}H(g.s.)+n →^{3}H+3n decay.

Both options imply that the studied ^{5}H can be reconstructed as a missing component from the measured ^{4}He and the neutron. The ^{5}H MM spectrum presented in Fig. 3 (a) was obtained from the analysis of the ^{2}H(^{8}He,^{4}He)^{6}H reaction data [7]. An important feature of this spectrum is the strong event concentration obtained around

https://doi.org/10.1051/epjconf/202329009001

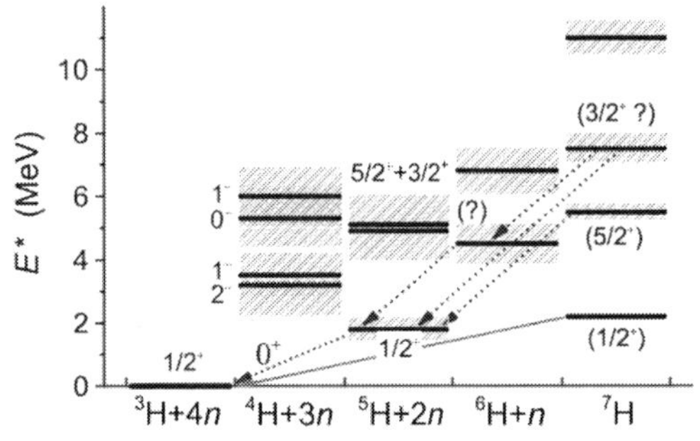

Figure 4. The level schemes of ^{6}H and the known neighboring ^{4}H, ^{5}H [11–13] and ^{7}H [6, 9] systems. The solid red arrow illustrates the decay mechanism of the ground state of ^{7}H which is expected to be the "true" $4n$ emission. The dotted blue arrows illustrate the decay mechanism of the higher excitations in ^{7}H, and the suggested in this work sequential decay of $^6\text{H}\rightarrow{}^5\text{H(g.s.)}+n\rightarrow{}^3\text{H}+3n$.

$E_T \sim 5$ MeV — the energy attributed in Ref. [13] to the $3/2^+ - 5/2^+$ doublet of the first excited states in ^{5}H. The expected $1/2^+$ ^{5}H g.s. with $E_T \sim 1.8$ MeV is practically not populated. This situation is quite consistent with the simplistic expectation of the ^{6}H structure as a proton hole in ^{7}He — the excited $^6\text{He}(2^+)$ configuration is known to dominate the ^{7}He g.s. structure.

The green line histograms in all three panels of Fig. 3 are formed by the low-energy ^{5}H events with $E_T(^5\text{H}) < 9$ MeV. It can be found from Fig. 3 (b) that the ^{5}H peak population do not show strong correlation with the population of the low-lying resonant states of ^{5}He. So, the hypothesis about the $^2\text{H}(^8\text{He},^5\text{He})^5\text{H}$ triton transfer reaction contribution can be rejected. In contrast, it can be found from Fig. 3 (c) that the correlation of the low-lying ^{5}H peak with the low-lying ^{6}H peak is unexpectedly strong: the higher excitations of ^{6}H practically do not populate the low-lying ^{5}H states. Thus, we conclude that the events of low-energy part of the ^{5}H MM spectrum are connected with the deuteron transfer reaction $^2\text{H}(^8\text{He},^4\text{He})^6\text{H}$ and the subsequent $^6\text{H}\rightarrow{}^5\text{H(g.s.)}+n\rightarrow{}^3\text{H}+3n$ decay.

5 Results

In this work we investigated the ancillary reaction channels which may impede the interpretation of the $^2\text{H}(^8\text{He},^4\text{He})^6\text{H}$ reaction data. The obtained results confirm those reported in Ref. [7] and provide more evidence for the suggested structure and decay mechanism of both ^{6}H and ^{7}H systems.

The analysis of the possible background channels confirms the nature of the low-energy ^{6}H MM spectrum presented in Ref. [7]. For the first time the spectrum of ^{5}H produced in the ^{6}H decay was reconstructed. The obtained strong correlation between the low-energy spectra of ^{6}H and ^{5}H makes the first experimental evidence that the notion about the two-step $^6\text{H}\rightarrow{}^5\text{H(g.s.)}+n\rightarrow{}^3\text{H}+3n$ decay is correct. Although, the obtained ^{5}H MM energy resolution did not allow us to resolve it into the known ^{5}H structure with the ground state at 1.8 MeV [11–13]. The presented strong correlation of the ^{5}H and ^{6}H spectra reproduces the conclusion of the absence of the ^{6}H resonances for $E_T < 3.5$ MeV. The latter, in conjunction with the observation of the ^{7}H g.s. at 2.2(5) MeV, increases the reliability of the level data presented in work [7] for the heavy hydrogen isotopes, see Fig. 4.

The reliable determination of the decay schemes for the superheavy ^{6}H and ^{7}H isotopes is a long-standing experimental challenge, and this work represents an important step towards the resolution of this problem.

6 Acknowledgement

We acknowledge the principal support of this work by the Russian Science Foundation grant No. 22-12-00054. The authors are grateful to Profs. Yu.Ts. Oganessian and S.N. Dmitriev for the long-term support and development of this activity. We acknowledge important contribution of Prof. M.S. Golovkov to the development of the experimental methods and useful discussions. Also, authors express their gratitude to the acceleration team for the stable work of U-400M cyclotron during all runs.

References

[1] M. Caamaño, D. Cortina-Gil, W. Mittig *et al.*, Phys. Rev. Lett. **99**, 062502 (2007).

[2] M. Caamaño, D. Cortina-Gil, W. Mittig *et al.*, Phys. Rev. C **78**, 044001 (2008).

[3] S. Fortier, E. Tryggestad, E. Rich *et al.*, AIP Conf. Proc. **912** (AIP, New York, 2006).

[4] E.Y. Nikolskii, A.A. Korsheninnikov, H. Otsu *et al.*, Phys. Rev. C **81**, 064606 (2010).

[5] Y. Gurov, B. Chernyshev, S. Isakov *et al.*, Eur. Phys. J. A **32**, 261 (2007).

[6] I.A. Muzalevskii, A.A. Bezbakh, E. Yu. Nikolskii *et al.*, Phys. Rev. C **103**, 044313 (2021).

[7] E.Yu. Nikolskii, I.A. Muzalevskii, A.A. Bezbakh *et al.*, Phys. Rev. C **105**, 064605 (2022).

[8] I.A. Muzalevskii, V. Chudoba, S.G. Belogurov *et al.*, Bulletin of the Russian Academy of Sciences: Physics, **84**:500–504 2020.

[9] A.A. Bezbakh, S.G. Belogurov, R. Wolski *et al.*, Instrum. Exp. Tech. **61**, 631 (2018).

[10] A.A. Bezbakh, V. Chudoba, S.A. Krupko *et al.*, Phys. Rev. Lett. **124**, 022502 (2020).

[11] A.A. Korsheninnikov, M.S. Golovkov, I. Tanihata *et al.*, Phys. Rev. Lett. **87**, 092501 (2001).

[12] M.S. Golovkov, L.V. Grigorenko, A.S. Fomichev *et al.*, Phys. Rev. Lett. **93**, 262501 (2004).

[13] M.S. Golovkov, L.V. Grigorenko, A.S. Fomichev *et al.*, Phys. Rev. C **72**, 064612 (2005).

EPJ Web of Conferences **290**, 09002 (2023) https://doi.org/10.1051/epjconf/202329009002
EuNPC 2022

Carbon related background estimation in the deuteron breakup reaction

Bogusław Włoch[1,*], *Izabela* Ciepał[1], *Paweł* Kulessa[2], *Adam* Kozela[1], *Jerzy* Łukasik[1], *Wiktor* Parol[1], *Piotr* Pawłowski[1], *Aleksandra* Podwysocka[3], *Michał* Prędota[4], *Izabela* Skwira-Chalot[3], *Bogdan* Sowicki[1], and *Mirosław* Ziębliński[1]

[1]Institute of Nuclear Physics Polish Academy of Sciences, Krakow, Poland
[2]Forschungszentrum Jülich IKP, Jülich, Germany
[3]Faculty of Physics, University of Warsaw, Poland
[4]Faculty of Physics, Warsaw University of Technology, Warsaw, Poland

Abstract. An experiment devoted to studies of relativistic effects in the proton-deuteron breakup reaction has been performed at Cyclotron Center Bronowice in Kraków, Poland, with the use of the Kratta detectors and a deuterated polyethylene target (CD_2). One of the main sources of background are reactions on carbon. They need to be subtracted from the measured spectra to obtain pure proton-deuteron breakup cross section. In order to determine this component, an additional measurement was carried out using a target made of natural graphite. A procedure of the background subtraction is discussed.

1 Introduction

The deuteron breakup reaction is often used as a tool to test theoretical approaches which describe the interaction between nucleons [1] in few-nucleon systems. The three-nucleon force (3NF) plays a fundamental role in the description of such processes. 3NF is an irreducible force which arises from the exchange of mesons between at least tree nucleons [2].

In recent years, there has been growing interest in the role of 3NF in the deuteron breakup reaction at higher energies, as it is believed that 3NFs may play an increasingly important role in this regime. As the energy increases, the relative contribution of the D-state to the deuteron wave function becomes more important, and the effects of 3NFs become more pronounced [3].

At higher energies the relativistic effects become also important, and it is crucial to include them in usually non-relativistic theoretical calculations. This has been demonstrated by a number of theoretical studies focused on the deuteron breakup recation [4, 5]. Already in much simpler process, a proton-deuteron elastic scattering, the relativistic effects are important. However, the calculations that account for such effects [6] do not reproduce the experimental data in a proper way [7]. As it was also demonstrated in [8], the situation is even worse in the case of the deuteron breakup. This raises the question of whether the relativistic treatment of the few-nucleon interaction is correct or the 3NFs is underestimated, or we are still missing some important ingredients of the system dynamics.

To test the theoretical predictions [4], a dedicated measurement of the deuteron breakup using a 200 MeV proton beam was conducted in February and March 2022 at Cyclotron Center Bronowice in Krakow, Poland. 30 Kratta detectors [9] were used to measure deutron-proton (*dp*) coincidences for specific *dp* angular configurations for which the calculations predict extremely large relativistic effects and practically negligible effects of 3NF. The main idea of this experiment is to answer the question whether the relativistic calculations performed by Skibiński et al. [4] are correct.

*e-mail: μbwloch@ifj.edu.pl

2 Target setup

The primary target used in this experiment was solid deuterated poliethylene (CD_2) of the diameter of 9.5 mm and surface density of 62 mg/cm^2. This target has been mounted on an aluminum frame using thin carbon threads, as shown in the Fig. 1. The second target was of the same diameter but made of pure natural graphite with the density of 50 mg/cm^2. It was used mainly to estimate the contribution of reactions on carbon contained in the polyethylene target. In addition to the previously mentioned targets, a set of auxiliary targets made of polyethylene, thick carbon-13 and lead was used for the detector calibration purposes, and a luminescent target based on ZnS was used for beam spot observation. All these targets were placed on a remotely controlled target ladder.

3 Background estimation

When analyzing the data measured with the CD_2 target, it is necessary to perform background subtraction to isolate the signal (proton-deuteron coincidences) from the reactions on carbon. Therefore, a dedicated measurement using the graphite target has been performed to establish background contribution originating from the carbon component in the CD_2 target. By applying the same analysis conditions as for the CD_2 data, one can determine

Figure 1. The deuterated polyethylene (CD_2) target suspended on the aluminum frame.

how much of the original signal came from carbon related events. However, due to the fact that the surface density of both targets was different and the overall dead time of the acquisition system was different, it was necessary to determine the relative scaling factor for both data sets. In this measurement proton-deuteron coincidences have been registered. The particle identification (PID) is based on the ΔE-E technique and the particles were selected by imposing the graphical cuts, see Fig. 2 as example. To determine the relative scalings and to fix the carbon component deuterons were selected in both data sets collected with CD_2 and carbon targets.

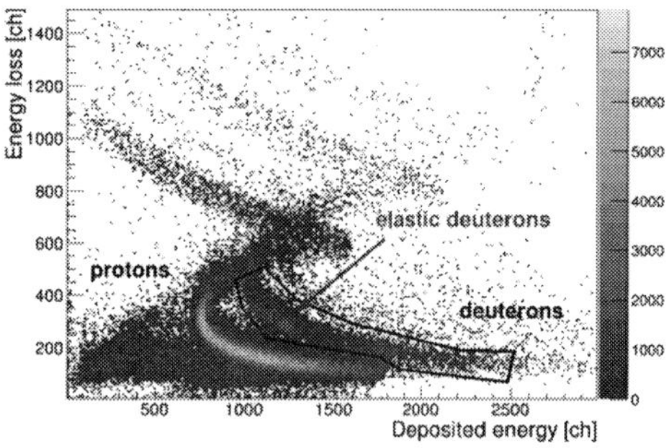

Figure 2. The E-ΔE events distribution for one of the Kratta detector, placed at 55°. Branches of protons and deuterons are clearly visible. The identification cut for deuterons is presented as a solid black line.

Our strategy was to investigate energy spectrum of the deuterons for carbon and CD_2 targets. In the case of CD_2 target we expect to have elastically scattered deuterons from proton-deuteron reactions as well as deuterons originating from the reactions on carbon inside the CD_2 target. Elastically scattered deuterons were mostly registered at backward angles while the corresponding protons were registered more in the forward detectors. Similarly, proton-deuteron coincidences are observed as a result of the reaction on carbon target. Deuterons originating from reaction on carbon form a continuous spectrum over the entire energy range as one can see in Fig. 3. The elastic proton-deuteron scattering contribution is visible as a peak in the CD_2 target spectrum. The scaling factor can be easily found by superimposing both data sets. The region to the left (lower energy) of the elastic deuteron peak may be contaminated with the breakup protons. Hence, only the region to the right of the elastic peak was chosen to find the scaling factor. It was determined by integrating the number of events obtained in the entire area of interest and making ratio of the result obtained for both targets.

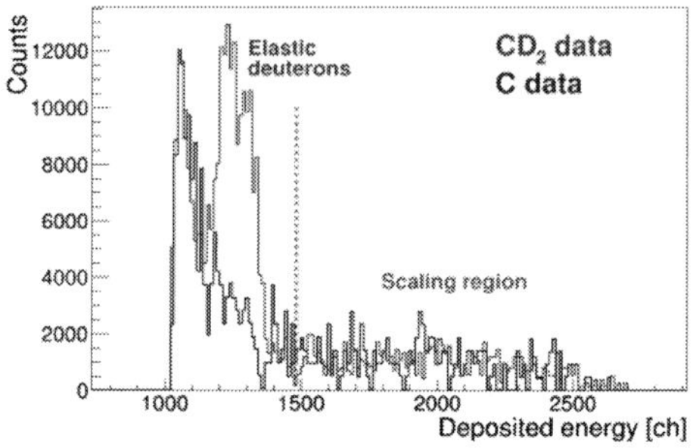

Figure 3. The energy spectra of the deuterons for the detector placed at 55° after applying PID cut shown in Fig. 2. The red line represents the data collected with the CD_2 target while the blue one, already scaled, with the graphite target. Vertical magenta line indicates the edge of the scaling region.

4 Summary

In conclusion, the relativistic treatment of the calculations is essential for a complete understanding of the 3NF effects in the deuteron breakup reaction at higher energies and can significantly affect the predictions of theoretical models. Therefore, it is very important to precisely investigate these effects for phase-space regions where the impact of the other dynamical components (Coulomb force, 3NF) of the nucleon interaction is negligible. The key element needed for determination of the cross sections in this experiment is to estimate the background contribution originating from protons colliding with carbon nuclei in a solid CD_2 target. This was done by measuring the particles on a separate target made of natural carbon. It was shown how to determine the relative scaling factor of these two datasets using registered deuterons.

Acknowledgements. This work was supported by the Polish National Science Center (NCN) from the 2019/35/N/ST2/03321 grant.

References

[1] I. Ciepał et al., Few Body Syst. **56**, 665 (2015)
[2] H. Miyazawa J. Fujita, Prog. Theor. Phys. **17**, 360 (1957)
[3] W. Glockle et al., Phys. Rep. **274**, 107-285 (1996)
[4] R. Skibiński et al., Eur. Phys. J. A **30**, 369 (2006)
[5] H. Witała et al., Phys. Rev. C **83**, 044001 (2011)
[6] H. Witała et al., Phys. Rev. C **88**, 069904 (2013)
[7] K. Hatanaka et al., Phys. Rev. C **66**, 044002 (2002)
[8] P. Adlarson et al., Phys. Rev. C **101**, 044001 (2020)
[9] J. Łukasik et al. Nucl. Instrum. Methods Phys. Res. A **709**, 120 (2013)

The role of deformation in the ^{17}C structure and its influence in transfer and breakup reactions

P. Punta[1,*], *J. A.* Lay[1,2,**], and *A. M.* Moro[1,2,***]

[1]Departamento de FAMN, Facultad de Física, Universidad de Sevilla, Apartado 1065, E-41080 Sevilla, Spain
[2]Instituto Interuniversitario Carlos I de Física Teórica y Computacional (iC1), Apdo. 1065, E-41080 Sevilla, Spain

Abstract. ^{17}C structure is studied within a two-body model, a weakly bound neutron moving in a deformed potential generated by the *core*. A semi-microscopic method has been used to generate the deformed valence-core potential. The method consists of the convolution of a realistic nucleon-nucleon (NN) interaction with the core transition densities, which are obtained by antisymetrized molecular dynamics (AMD). The results highlight the important role of deformation for this nucleus and can be easily applied to reaction calculations.

1 Introduction

The study of nuclei far from the stability line is one of the main topics in current nuclear physics research. Due to its excess of neutrons, ^{17}C is an example of these so-called exotic nuclei. It is a specially interesting case, because it is a weakly bound nucleus whose first exited sate is a strong candidate for halo nature.

In this contribution, we present a semimicroscopic model for ^{17}C nucleus, described as a neutron+*core* two-body system. The purpose is to have a model sufficiently manageable to be applied to calculations of transfer and breakup reactions but yet providing a realistic description of the ^{17}C low-lying spectrum. Weakly bound nuclei are conveniently described within few-body models, which usually ignore possible deformation of the fragments. However, *core deformations* are known to significantly affect both the structure and dynamics of these systems. For this reason, the structure of different weakly bound nuclei has been studied using deformed two-body models such as the particle-rotor or the particle-vibrator. For example, the particle-rotor model has been successfully applied to the study of the structure and resonant breakup of ^{11}Be [1–4]. In the present contribution, the application of a more sophisticated semi-microscopic model to the ^{17}C system is detailed.

2 Semi-microscopic particle-plus-AMD model

In the weak-coupling limit, the Hamiltonian of our two-body system can be written as

$$\mathcal{H} = T(\vec{r}) + h_{core}(\xi) + V_{vc}(\vec{r},\xi) + V_{\ell s}(r)(\vec{\ell}\cdot\vec{s}). \quad (1)$$

where $T(\vec{r})$ is the kinetic energy operator for the relative motion between the valence particle and the core, $h_{core}(\xi)$ is the Hamiltonian of the core and $V_{vc}(\vec{r},\xi)$ is the effective valence-core interaction. A phenomenological spin-obit term with the usual radial dependence $V_{\ell s}(r)$ is added to the effective interaction. The variable ξ denotes the internal coordinates of the core, so the dependence of $V_{vc}(\vec{r},\xi)$ on it accounts for core excitation effects.

The eigenfunctions of the Hamiltonian, for a given energy ε, can be characterized by the parity π and the total angular momentum $\vec{J}$, resulting from the coupling of the angular momentum $\vec{j}$ of the valence particle to the core angular momentum $\vec{I}$. These functions can be generically expressed as

$$\Psi^{J^\pi}_{\varepsilon M}(\vec{r},\xi) = \sum_{\alpha} R^{J^\pi}_{\varepsilon\alpha}(r)\left[\mathcal{Y}^{j}_{\ell s}(\hat{r})\otimes\phi_I(\xi)\right]_{JM}, \quad (2)$$

where $\vec{\ell}$ is the orbital angular momentum between the valence particle and core, which couples to the spin of the valence particle $\vec{s}$ to give the particle total angular momentum $\vec{j}$. The label α denotes the set of quantum numbers $\{\ell, s, j, I\}$.

In our semi-microscopic model, the $V_{vc}(\vec{r},\xi)$ interaction is calculated convoluting an effective inmedium NN interaction with microscopic transition densities of the core nucleus. In the present case, the NN interaction from Jeukenne, Lejeune, and Mahaux (JLM) is used [5]. The required transition densities are obtained from the antisymmetrized molecular dynamics (AMD) calculation [6, 7]. The formalism of this model, that we will call PAMD, is explained in detail in [8], where it is applied to ^{11}Be and ^{19}C.

2.1 Diagonalization in the THO basis

The Hamiltonian presented above is diagonalized in a pseudo-state basis. Therefore, the eigenfunctions of

*e-mail: ppunta@us.es
**e-mail: lay@us.es
***e-mail: moro@us.es

 https://doi.org/10.1051/epjconf/202329009003

Eq. (2) can be expressed as a linear combination of the basis functions. In our case, the transformed harmonic oscillator basis (THO) is chosen. This basis applies a locale scale transformation (LST) to the harmonic oscillator wave functions,

$$R_{n\ell}^{THO}(r) = \frac{s}{r}\sqrt{\frac{ds}{dr}}R_{n\ell}^{HO}[s(r)]. \tag{3}$$

$R_{n\ell}^{HO}(s)$ is the radial part of the usual HO functions and we adopted a parametric form for the LST, $s(r)$, from Karataglidis *et al.* [9]

$$s(r) = \left[\frac{1}{\left(\frac{1}{r}\right)^m + \left(\frac{1}{\gamma\sqrt{r}}\right)^m}\right]^{\frac{1}{m}}. \tag{4}$$

This basis has been successfully applied to the discretization of the continuum of weakly bound nuclei to describe breakup and transfer direct reactions both for two-body and three-body systems [10, 11].

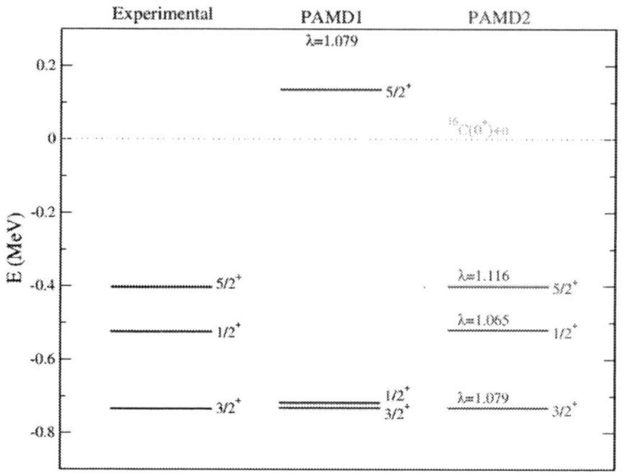

Figure 1. Experimental and calculated energy levels of ^{17}C. Starting from the left, the second column is the PAMD1 model, with a single renormalization factor, and the third the PAMD2, which uses different factors to reproduce the experimental values [14, 15].

3 Results of the application to ^{17}C

The method described above is applied to ^{17}C using the transition densities of ^{16}C from [7]. Some of the results of this model were first shown in [12] together with the discussion of the one-neutron transfer reaction ^{16}C(d,p). Here, a more detailed study of the ingredients of the model and its results is presented.

We assume that the only accessible core states are the ground state 0^+ and the first excited state 2^+. The latter is assigned the excitation energy of 1.766 MeV, which corresponds to the experimental energy of the first excited state of ^{16}C [13]. Furthermore, for the valence neutron, $\ell \leq 3$ is restricted.

In order to reproduce the experimental separation energy, $S_n = 0.734$ MeV [14], a global renormalization factor of $\lambda = 1.079$ needs to be applied to the effective potential $V_{vc}(\vec{r}, \xi)$. Although this factor is purely phenomenological, the change in interaction is less than 8%. This interaction is supplemented with a spin-orbit term with a standard strength of $V_{so} = 6$ MeV. It is parameterized, as usual, in terms of the derivative of a Wood-Saxon shape, and the values $R_{so} = 3$ fm, $a_{so} = 0.65$ fm are used to be consistent with the extension of the effective potential.

With these parameters, the Hamiltonian is diagonalized in the THO basis, giving rise to the energy levels shown in the central spectrum of figure 1. In the same figure, on the left, the experimental energies are shown [14, 15]. On the right are the renormalization λ factors that must be applied in each case to reproduce the experimental energies. Therefore, we distinguish between a PAMD1 model, which uses a single renormalization factor, and the PAMD2 model, which applies different factors according to the angular momentum of the ^{17}C state. The biggest difference with the PAMD1 model is that it obtains the second excited state $5/2_1^+$ as a resonance instead of a bound state. However, this resonance is located very close to the ^{17}C + n threshold, approximately 0.1 MeV above, with a difference of 0.5 MeV with respect to the experimental value. Similarly, the first excited state $1/2_1^+$ is only 0.2 MeV below the experimental level.

The functions $u_\alpha(r) = rR_{\varepsilon\alpha}^{J^\pi}(r)$, which describe the radial part of the wave function, are shown both for the ground state (figure 2) and for the two first excited states (figure 3). The ground state is the same for the PAMD1 and PAMD2 models, therefore its wave function is also identical. In the case of excited states, figure 3 compares the functions obtained for the PAMD1 and PAMD2 models. In the upper panel, they are compared for the first excited state $1/2_1^+$, and in the lower panel for the second excited state $5/2_1^+$. It can be verified that, in both cases, the wave functions are almost identical despite the energy differences of the states. For the $1/2_1^+$ case, our results support the halo nature of this state in view of the greater spatial extension of its associated wave function. This is quantitatively corroborated in table 1, where the mean square radii obtained for each state and model are shown. The values obtained for the first excited state are clearly larger than those obtained for the rest. As this state is less bound in the PAMD2 model, there is an appreciable increase in the mean square radius in this case. Note that, since the $5/2_1^+$ state appears as a resonance in the PAMD1 model, its root mean square radius is meaningless. In fact, in the lower panel of figure 3, it can be seen that the component $|d_{5/2} \otimes 0^+\rangle$ of this state does not vanish completely asymptotically for the PAMD1 model. Table 1 also shows the weights of the α components of the wave function for each state and each model. The differences between the values of the two models are not significant.

4 Conclusions and perspectives

Analyzing the results, the PAMD2 model does not have many advantages compared to PAMD1 model, aside from

Table 1. Weights of the components of the wave function and mean square radii for ^{17}C. The results obtained with the PAMD1 and PAMD2 models for the ground state and the two first excited states are shown

State	$3/2^+_{gs}$	$1/2^+_1$		$5/2^+_1$	
Model	PAMD1 - PAMD2	PAMD1	PAMD2	PAMD1	PAMD2
$\vert(\ell s)j\otimes 0^+\rangle$	0.028	0.512	0.530	0.326	0.333
$\vert s_{1/2}\otimes 2^+\rangle$	0.349	-	-	0.668	0.657
$\vert d_{3/2}\otimes 2^+\rangle$	0.131	0.040	0.032	0.000	0.000
$\vert d_{5/2}\otimes 2^+\rangle$	0.492	0.448	0.438	0.006	0.009
r_{ms} (fm)	4.030	5.246	5.748	-	4.480

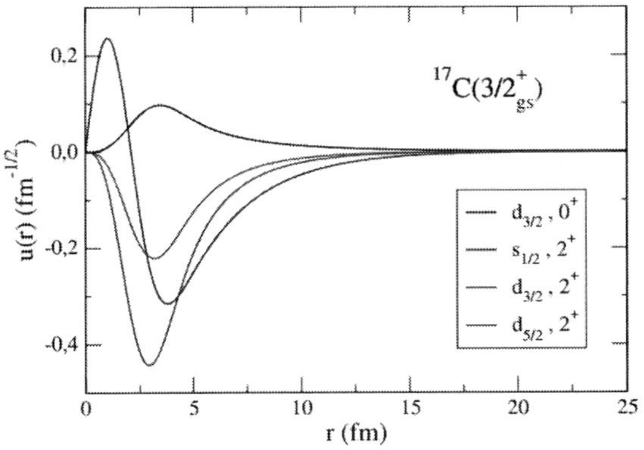

Figure 2. Radial part of the wave function obtained for the ground state of ^{17}C.

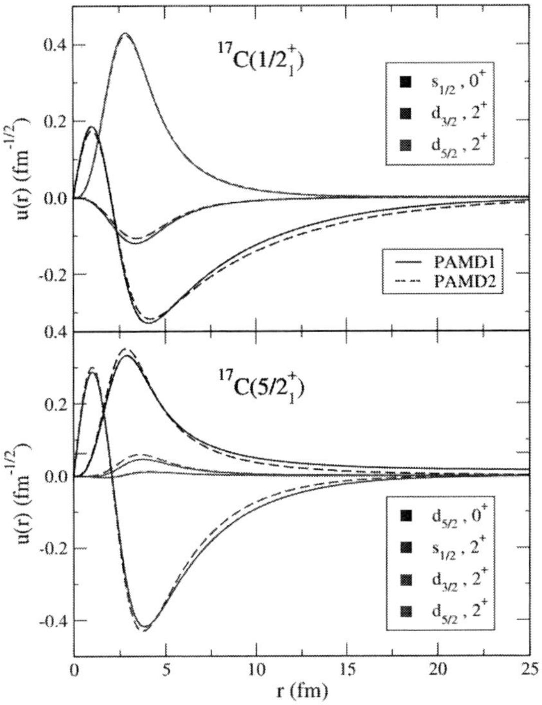

Figure 3. Radial part of the wave function obtained for the first excited state (upper panel) and second excited state of ^{17}C (lower panel). The results of the PAMD1 and PAMD2 models are compared

providing a better description of the energies. The inclusion of more phenomenological parameters in order to better reproduce the experimental energies of the bound states, does not cause significant changes in the wave functions of these states. Regarding the wave functions, there is a strong mixing of the components due to deformation, as shown by the weights in table 1. This is expected to produce strong effects on any reaction that depends on the structure of ^{17}C. These effects have already been proven for the case of ^{19}C, reaction calculations have been performed describing the ^{19}C structure with the same PAMD formalism [16].

Being a weakly bound nucleus, the study of one-neutron transfer reactions involving ^{17}C is especially interesting. Calculations for the reaction ^{16}C$(d,p)^{17}$C, for which there are recent experimental data from GANIL [12], are underway and the results will be presented soon. The referred work shows data of the angular distribution of the cross section for the transfer to the bound states of ^{17}C. The idea is that, to calculate this cross section theoretically, the overlap function $\langle^{17}\mathrm{C}|^{16}\mathrm{C}\rangle$ is required, which can be directly related to the functions $R^{J^\pi}_{\varepsilon\alpha}(r)$ obtained with the PAMD model.

In Ref. [12], spectroscopic factors (SF) for the $\langle^{17}\mathrm{C}|^{16}\mathrm{C}\rangle$ overlap were inferred from the comparison of the measured transfer data with ADWA calculations. In our assumed simplified two-body model, in which antisymmetrization between the valence neutron and the core is neglected, SF cannot be strictly obtained. However, as long as these antisymmetrization effects are not large, the weights listed in table 1 can be approximately regarded as SF and compared with the values from Ref. [12]. Therefore, as we start from ^{16}C in its ground state 0^+, the spectroscopic factors shown in [12] can be compared with the weights of the components $|(\ell s)j\otimes 0^+\rangle$ obtained with our model. For the $|d_{3/2}\otimes 0^+\rangle$ component of the ground state, a 0.3 factor is presented, perfectly compatible with our model. In the case of the $1/2^+_1$ state ($|s_{1/2}\otimes 0^+\rangle$), the range of values is between 0.5 and 1.0, so the values shown in table 1 fall in the lower part of the range. Nevertheless, the values we obtain for the component $|d_{5/2}\otimes 0^+\rangle$ of the $5/2^+_1$ state is below the experimental values $(0.5-0.7)$, what can be a limitation of our structure model.

The study of the low-lying continuum of ^{17}C is also in progress, with the goal of extending the application of the PAMD model to the transfer to the continuum for $^{16}C(d,p)^{17}C$ reaction. Furthermore, this model will be applied to the breakup reaction $^{17}C + p \rightarrow ^{16}C + n + p$, for which there are already published data [17].

Acknowledgments

The present research is funded from grant PID2020-114687GB-I00 by MCIN/AEI/10.13039/501100011033, the project PAIDI 2020 with Ref. P20_01247 by the Consejería de Economía, Conocimiento, Empresas y Universidad, Junta de Andalucía (Spain), and by ERDF A way of making Europe. P.P. acknowledges PhD grants from the Ministerio de Universidades and the Consejería de Transformación Económica, Industria, Conocimiento y Universidades, Junta de Andalucía

References

[1] J.A. Lay, A.M. Moro, J.M. Arias, J. Gómez-Camacho, Phys. Rev. C **85**, 054618 (2012)

[2] A.M. Moro, J.A. Lay, Phys. Rev. Lett. **109**, 232502 (2012)

[3] V. Pesudo, M.J.G. Borge, A.M. Moro, J.A. Lay, E. Nácher, J. Gómez-Camacho, O. Tengblad, L. Acosta, M. Alcorta, M.A.G. Alvarez et al., Phys. Rev. Lett. **118**, 152502 (2017)

[4] A.M. Moro, J.A. Lay, J. Gómez Camacho, Physics Letters B **811**, 135959 (2020)

[5] J.P. Jeukenne, A. Lejeune, C. Mahaux, Phys. Rev. C **16**, 80 (1977)

[6] Y. Kanada-En'yo, H. Horiuchi, A. Ono, Phys. Rev. C **52**, 628 (1995)

[7] Y. Kanada-En'yo, F. Kobayashi, T. Suhara, Journal of Physics: Conference Series **445**, 012037 (2013)

[8] J.A. Lay, A.M. Moro, J.M. Arias, Y. Kanada-En'yo, Phys. Rev. C **89**, 014333 (2014)

[9] S. Karataglidis, K. Amos, B.G. Giraud, Phys. Rev. C **71**, 064601 (2005)

[10] J.A. Lay, A.M. Moro, J.M. Arias, J. Gómez-Camacho, Phys. Rev. C **82**, 024605 (2010)

[11] J. Casal, E. Garrido, R. de Diego, J.M. Arias, M. Rodríguez-Gallardo, Phys. Rev. C **94**, 054622 (2016)

[12] X. Pereira-López et al., Phys. Lett. B **811**, 135939 (2020)

[13] D. Tilley, H. Weller, C. Cheves, Nuclear Physics A **564**, 1 (1993)

[14] W. Meng, G. Audi, F.G. Kondev, S. Naimi, Chinese Physics C **41**, 030003 (2017)

[15] Z. Elekes et al., Phys. Lett. B **614**, 174 (2005)

[16] J.A. Lay, R. de Diego, R. Crespo, A.M. Moro, J.M. Arias, R.C. Johnson, Phys. Rev. C **94**, 021602 (2016)

[17] Y. Satou et al., Phys. Lett. B **660**, 320 (2008)

EPJ Web of Conferences **290**, 09004 (2023) https://doi.org/10.1051/epjconf/202329009004
EuNPC 2022

Impact of core excitations in break-up reactions with halo nuclei

J. A. Lay[1,2,*], *P.* Punta[1,**], and *A. M.* Moro[1,2,***]

[1]Departamento de FAMN, Facultad de Física, Universidad de Sevilla, Apartado 1065, E-41080 Sevilla, Spain
[2]Instituto Interuniversitario Carlos I de Física Teórica y Computacional (iC1), Apdo. 1065, E-41080 Sevilla, Spain

Abstract. We revisit the resonant break-up of ^{19}C on protons at intermediate energies. In this reaction, it was found, for the first time in a halo nucleus to our knowledge, that the cross section was largely dominated by the excitation of the core. In this contribution, we study the robustness of this conclusion against the choice of different optical potentials for the core.

1 Introduction

Halo nuclei have been an important field of Nuclear Physics since its discovery, thanks in part to the dawn of radioactive beam facilities. Halo nuclei appear close to the neutron and proton driplines where one or two valence nucleons are so barely bound that they orbit far around the rest of nucleons that conforms a compact core. Valence nucleons wavefunction, therefore, extends far away from the core forming the matter halo that name these nuclei. Following this picture, halo nuclei are often treated within two- or three-body valence-core models, considering an inert core. Such picture breaks down in regions of the nuclear chart very far away from the stability valley which are nowadays being explored thanks to the advances in existing radioactive beam facilities. The cores of such halo nuclei will present very low excitation energies for their first excited states. Therefore, it will be also very easy to excite the core along any reaction.

Including internal degrees of freedom of the core in a few-body reaction formalism has been a tremendous challenge. Most few-body models have been recently extended to include the effect of core excitations, such as a non-recoil extension of the Distorted Waves Born Approximation (NR-DWBA) [1, 2], the Extended Continuum-Discretized Coupled-Channels (XCDCC) [3, 4], and the extension of the Faddeev/AGS equations [5]. Alongside these developments, there has been a revival of structure models which enable the incorporation of such excitations in a simple way to allow its inclusion in reaction calculations. That is the case of the particle-rotor and particle-vibrator models [6, 7]. In particular, the particle-rotor model has been successfully applied to ^{11}Be [8–10], a benchmark for the case of halo nuclei. These models incorporate certain free parameters which are adjusted to reproduce the available information of the halo nucleus or that of the nucleus corresponding to its core. However, especially with newly discovered halo nuclei, it is usual to have very little experimental information to constrain such parameters. It is also possible that the structure of the core goes beyond the picture of a rotor or a vibrator. For those cases, a semi-microscopic folding model based on core transition densities was introduced in [11]. The model succeeded to reproduce the main features of the low-lying spectrum of ^{11}Be and, more importantly, of the less known halo nucleus ^{19}C. In both cases, the transition densities for the respective cores were obtained with the Antisymmetrized Molecular Dynamics (AMD) method.

This model, named particle-AMD model, was also successful in the reproduction of the resonant breakup of ^{19}C [12] by analyzing the ^{19}C(p,p') experimental data by Satou *et al.* [13]. In [12], ^{19}C resonant breakup was found to be the first case of a halo nuclei where the break up is not governed by the excitation of the valence neutron but, on the contrary, by the excitation of the core.

In the present contribution, we revisit the case of ^{19}C(p,p') resonant breakup and discuss the robustness of such conclusion in detail with respect to the influence of the different optical potentials involved in the reaction calculation. The manuscript is structured as follows. In Sec. 2, we summarize the main features of the particle-AMD (PAMD) structure model. In Sec. 3, we discuss how the optical potentials enter into an XCDCC calculation. Finally, in Sec. 4 we show the different results obtain for different core-target potentials and we discuss on Sec. 5 the role of core excitations and its robustness against the core-target interaction.

2 Structure model

We discuss the structure of the halo nucleus in terms of a two-body core + valence-particle model where we will include the possibility of a non-inert core. Thus, a general wavefunction for the nucleus will include different eigenstates of the core, $\Phi_{IM_I}(\xi)$, being I the angular momentum, M_I its projection, and ξ the degrees of freedom of the core.

*e-mail: lay@us.es
**e-mail: ppunta@us.es
***e-mail: moro@us.es

If such wavefunction has total angular momentum J and projection M we can express it as:

$$\Psi_{JM}(\vec{r},\xi) = \sum_{\alpha} \left[\varphi_\alpha(\vec{r}) \otimes \Phi_I(\xi)\right]_{JM}, \quad (1)$$

where the functions $\varphi_\alpha(\vec{r})$ describe the relative motion between the valence particle and the core, being $\vec{r}$ the relative distance between the two. The index α, so-called channel, denotes the set of quantum numbers $\{l, s, j, I\}$, with l, s, and j being the orbital angular momentum, the intrinsic spin of the valence particle, and their sum ($\vec{j} = \vec{l} + \vec{s}$), respectively. Such a sum in α will cover all possible channels consistent with the total angular momentum ($\vec{J} = \vec{j} + \vec{I}$) and parity. Each channel will have a specific weight in each state of the nucleus. This weight can be regarded as a unit-normalized spectroscopic factor.

To obtain the eigenstates of the nucleus, we need a Hamiltonian for the system that includes a model for the internal Hamiltonian of the core $h_c(\xi)$ and the interaction between the core and the valence-particle $V_{vc}(\vec{r},\xi)$. In the weak-coupling limit we can assume:

$$\mathcal{H}_{\text{proj}} = T(\vec{r}) + h_{core}(\xi) + V_{vc}(\vec{r},\xi). \quad (2)$$

It is important to remark that the interaction $V_{vc}(\vec{r},\xi)$ is the only term that connects the inner degrees of freedom of the core and the motion of the valence particle, thus being the main responsible for the coupling of the different channels, determining the relative weights in the different eigenstates of the hamiltonian. We will follow two different models for such interaction: the particle-rotor model (PRM) [7] and the semi-microscopic particle-AMD model (PAMD) [11].

For the calculation of the eigenstates of the system, we employ the Pseudo-state method, in which the model hamiltonian is diagonalized in a THO basis [14]. This basis is constructed by applying an analytical Local Scale Transformation (LST) [15] to the Harmonic Oscillator (HO). Such LST is meant to change the Gaussian asymptotic behavior of HO into an exponential one, more appropriate to tackle potentials that vanish asymptotically like a Woods-Saxon.

3 Reaction model

The remaining ingredients are the optical potentials between the fragments and the target:

$$V_{\text{proj}-T} = V_{vT}(\vec{R}_{vT}) + V_{cT}(\vec{R}_{cT},\xi). \quad (3)$$

Both potentials are usually taken from global parameterizations where one evaluates the potential at the energy per nucleon of the incident projectile. Withing the XCDCC framework, the interaction enters into the reaction calculation through the coupling potentials:

$$\left\langle \Psi^f_{J'M'} \left| V_{vT}(\vec{R}_{vT}) + V_{cT}(\vec{R}_{cT},\xi) \right| \Psi^i_{JM} \right\rangle. \quad (4)$$

It can be seen that the only place in the interaction with the target that includes any information on the core excitation is V_{cT} that depends on the internal degrees of freedom of the core ξ. In other words, by neglecting this dependence one will exclude the effect of the excitation of the core along the reaction. These effects are usually called *dynamic* effects of the core. If *dynamic* effects are neglected, the effect of the core will still be present through the weights of the different components in the projectile wavefunction, $\Psi_{JM}(\vec{r},\xi)$. This other effect is usually called *static* effect.

4 Results

We analyze the resonant break of ^{19}C on protons at 70 MeV/nucleon measured at RIKEN by Satou *et al.* [13]. The resonance considered here appears as a prominent peak in the energy distribution of the breakup cross section at $E_x = 1.46 \pm 0.10$ MeV that was assigned spin and parity $5/2^+$ in [13]. In [12] it was found a good agreement of the angular distribution of such resonance with the first $5/2^+$ resonance in the PAMD model and an unexpected dominance of dynamic core excitations on that break-up. We will start by analyzing within the XCDCC framework if the latter conclusion is robust against a change in the core-target potential. To do so, the other optical potential, the valence-target potential, which corresponds to the *n-p* potential in this case, remains the same: a simple Gaussian potential from Refs. [1, 16], whose parameters were adjusted to reproduce a Faddeev calculation of the breakup of ^{11}Be+p with the more realistic *p-n* CD-Bonn potential.

We repeat the XCDCC calculations with two different core-target potentials. For the two choices we start from a spherical potential, namely, the Watson global parameterization [17] and the CH89 potential [18]. These two potentials are then deformed with the deformation length δ obtained from the analysis of the AMD transition densities of ^{18}C [11]. In Fig. 1, we compare the angular distributions of the resonant break-up obtained with both potentials with the original one published in [12], where a folding potential obtained from the JLM nucleon-nucleon interaction [19] and the AMD transition densities for ^{18}C was used [20]. It can be seen that the three potentials provide a very similar cross section, both in shape and magnitude. Recalling that the shape of the angular distribution is determined by the coherent sum of core and valence contribution, one can infer that the relative weight of both contributions is similar in the three cases.

5 Conclusions

We have analyzed the resonant break-up of ^{19}C on protons at 70 MeV/nucleon with emphasis in the relative importance of the core excitation mechanism and its dependence on the underlying core-target interaction. For that, have performed XCDCC calculations for the break-up to the first $5/2^+$ resonance described in the PAMD model. Three different prescriptions have been used for the core-target potential to assess the influence of this ingredient on the results, namely Watson, CH89 and a folding potential using the same AMD transition densities for ^{18}C as in the structure model.

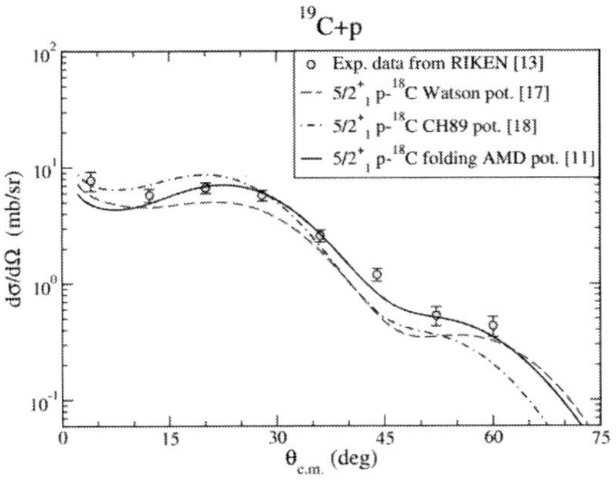

Figure 1. Angular distribution of the break-up cross section for the first $5/2^+$ resonance calculated in XCDCC with three different core-target potentials: Watson [17] (dashed red line), CH89 [18] (dot-dashed blue line) and a folding potential from AMD transition densities [11] (solid black line). The experimental data are taken from Satou *et al.* [13]. All calculations are convoluted with the experimental angular resolution.

All three calculations show a good agreement between themselves and are consistent with the experimental data, both in magnitude and shape. The best reproduction of the data is still found for the original folding potential used in [12]. The choice of the global parameterization by Watson is slightly below the experimental data but shows a very good agreement with the shape of the distribution.

Considering that the same valence-target potential is used for the three calculations and that its contribution was already found negligible in [12], in the three calculations the main mechanism that can explain the amount of break-up is the dynamic excitation of the core. This proves that such an extra break-up is not an effect of an inadequate choice of the optical potential. The only remaining free parameter is the deformation length of ^{18}C which is taken from the AMD transition densities. However, such deformation length is constrained by the quadrupole electric properties of ^{18}C which are well known and well reproduced by the AMD calculation. Therefore, it can be stated that the conclusion that ^{19}C resonant break-up is dominated by dynamic excitations of the core is robust against the choice of core-target optical potential.

Acknowledgments

The present research is funded from grant PID2020-114687GB-I00 by MCIN/AEI/10.13039/501100011033, the project PAIDI 2020 with Ref. P20_01247 by the Consejería de Economía, Conocimiento, Empresas y Universidad, Junta de Andalucía (Spain), and by ERDF A way of making Europe. P.P. acknowledges PhD grants from the Ministerio de Universidades and the Consejería de Transformación Económica, Industria, Conocimiento y Universidades, Junta de Andalucía

References

[1] A.M. Moro, R. Crespo, Phys. Rev. C **85**, 054613 (2012)

[2] A.M. Moro, J.A. Lay, Phys. Rev. Lett. **109**, 232502 (2012)

[3] N.C. Summers, F.M. Nunes, I.J. Thompson, Phys. Rev. C **74**, 014606 (2006)

[4] R. de Diego, J.M. Arias, J.A. Lay, A.M. Moro, Phys. Rev. C **89**, 064609 (2014)

[5] A. Deltuva, Phys. Rev. C **88**, 011601 (R) (2013)

[6] A. Bohr, B. Mottelson, *Nuclear Structure*, New York, W. A. Benjamin edn. (1969)

[7] J.A. Lay, A.M. Moro, J.M. Arias, J. Gómez-Camacho, Phys. Rev. C **85**, 054618 (2012)

[8] E. Kwan, C. Wu, N. Summers, G. Hackman, T. Drake, C. Andreoiu, R. Ashley, G. Ball, P. Bender, A. Boston et al., Physics Letters B **732**, 210 (2014)

[9] V. Pesudo, M.J.G. Borge, A.M. Moro, J.A. Lay, E. Nácher, J. Gómez-Camacho, O. Tengblad, L. Acosta, M. Alcorta, M.A.G. Alvarez et al., Phys. Rev. Lett. **118**, 152502 (2017)

[10] A.M. Moro, J.A. Lay, J. Gómez Camacho, Physics Letters B **811**, 135959 (2020)

[11] J.A. Lay, A.M. Moro, J.M. Arias, Y. Kanada-En'yo, Phys. Rev. C **89**, 014333 (2014)

[12] J.A. Lay, R. de Diego, R. Crespo, A.M. Moro, J.M. Arias, R.C. Johnson, Phys. Rev. C **94**, 021602 (R) (2016)

[13] Y. Satou et al., Phys. Lett. B **660**, 320 (2008)

[14] J.A. Lay, A.M. Moro, J.M. Arias, J. Gomez-Camacho, Phys. Rev. C **82**, 024605 (2010)

[15] S. Karataglidis, K. Amos, P. Fraser, L. Canton, J. Svenne, Nucl. Phys. **A813**, 235 (2008)

[16] R. Crespo, A. Deltuva, A.M. Moro, Phys. Rev. C **83**, 044622 (2011)

[17] B.A. Watson, P.P. Singh, R.E. Segel, Phys. Rev. **182**, 977 (1969)

[18] R.L. Varner, W.J. Thompson, T.L. McAbbe, E.J. Ludwig, T.B. Clegg, Phys. Rep. **201**, 57 (1991)

[19] J.P. Jeukenne, A. Lejeune, C. Mahaux, Phys. Rev. C **16**, 80 (1977)

[20] Y. Kanada-En'yo, F. Kobayashi, T. Suhara, Journal of Physics: Conference Series **445**, 012037 (2013)

 https://doi.org/10.1051/epjconf/202329009005

A new experiment on $^{19}F(p,\alpha)^{16}O$ reaction at low energies.

Luigi Redigolo[1,2*], *Ivano* Lombardo[1,2], and *Daniele* Dell'Aquila[3,4]

[1]Dipartimento di Fisica e Astronomia, Università degli Studi di Catania, via S. Sofia 64, 95123, Catania, Italy
[2]INFN - Sezione di Catania, via S. Sofia 64, 95123, Catania, Italy
[3]Dipartimento di Fisica "Ettore Pancini", Università degli Studi di Napoli "Federico II", via Cinthia 21, 80126, Napoli, Italy
[4] INFN - Sezione di Napoli, via Cinthia 21, 80126, Napoli, Italy

Abstract. $^{19}F(p,\alpha)$ reactions are important both in nuclear astrophysical field and for the study of the inner structure and the cluster behaviour in light nuclei. The available cross section data on such reaction, although being in reasonable agreement over a wide energy range, still show some discrepancies for both α_0 and α_π channels. In this work we discuss the preliminary results of a new experiment aimed to solve the disagreement around 1.6 MeV for the $^{19}F(p,\alpha_0)$ absolute cross section. The excellent resolution of the detection device allowed us also to resolve the α_π peak in the spectra obtained in the 1.0-1.5 MeV beam energy range, where very few data where previously reported in the literature.

1 Introduction

High excitation energy states in self-conjugate nuclei, as ^{16}O and ^{20}Ne, could bring up clues of their α-cluster structure when suitable spectroscopy studies are performed; this is especially important in energy regions close to the cluster decomposition thresholds indicated in the Ikeda diagram [1,2]. For the case of ^{20}Ne, this study can be profitably performed by analysing p+^{19}F induced reactions at energies close to the Coulomb barrier [3-5].

Furthermore, understanding the magnitude of the competition between the $^{19}F(p,\alpha)$ and $^{19}F(p,\gamma)$ reactions at sub-coulomb energies is crucial to determine possibly escape mechanisms and pathways from the ordinary CNOF cycle [6]. In this proceeding, we will show some preliminary results aiming at removing some disagreement, in both the shape and the magnitude, of different data sets. The detailed search for normalization factors between various data sets that would lead to a common behaviour of the S-factor for this reaction is discussed in [7], but while for all the other energy regions this common trend was found, a disagreement of a factor about 1.5 is still persisting for the $^{19}F(p,\alpha_0)$ reaction in the region $E_{cm} \simeq 1.5 - 1.7$ MeV between the data sets of Refs. [8,9]. Such disagreement prevents an accurate knowledge of the spectroscopical properties of 14.35 MeV (1^-) and 14.47 (0^+) excited states in ^{20}Ne (see Ref. [10] for further details). Similarly, very few and sparse data concerning the $^{19}F(p,\alpha_\pi)$ reaction channels are reported in the literature in the region around 1.1 and 1.3 MeV, preventing the accurate knowledge of the partial decay widths towards this channel of states at about 14 MeV [10,11]; this point is particularly important in the framework of the cluster structure in the nucleus, having a strongly pronounced $\alpha \oplus {}^{12}C$ structure.

2 Experimental setup

The experiment has been performed using a proton (^1H) beam accelerated at laboratory energies between 1.130-1.300 MeV and 1.620 -1.720 MeV, with 20 keV steps.

The beam was delivered by the Singletron accelerator at the Physics Department of the University of Catania (Italy). The beam energy stability was better than 1 keV; the typical beam intensity used in the experiment was $\simeq$ 100 nA. The calibration of the beam energy was carefully checked by the analysis of well-known proton elastic scattering resonances in ^{12}C. The target was made by a CaF_2 film of 30 μg/cm^2 thickness, deposited over a 10 μg/cm^2 carbon backing. Data from elastic scattering and alpha-particles emission have been collected via a movable high resolution silicon detector, similar to the ones used in [12], that was placed at laboratory angles of 115°, 125°, 135°, 145°, 155° and 165°, with angular opening of about 1°. The vacuum level of the scattering chamber was of the order of 10^{-7} mbar: this strongly prevents carbon build-up effects on the target.

2.1 Preliminary analysis and background

For every energy and every angle, we measured counts due to elastic scattering events on carbon, oxygen, fluorine, calcium and tantalum, and reaction particle counts (for α_0, α_π, and α_1 particle emissions). They have been extracted from the measured ejectile spectra by means of a multi-gaussian fit. As shown in Figure 1 for a typical spectrum taken at backward angles, the achieved energy resolution is several times better than the state of art one [13]. This feature is particularly

* Corresponding author: luigi.redigolo@ct.infn.it

important especially for the study of α_π channel, whose peak is in close vicinity with the α_1 peak: in this case, we got a peak to background ratio of ~12.5, with respect to the factor ~3 reported by Ref. [13].

For the elastic scattering peaks, we also included a polynomial function allowing for the reproduction of the small background in such energy region of the spectrum.

To avoid errors deriving from the variation of the Calcium vs Fluorine stoichiometry in the CaF_2 target under bombardment, we adopted an internal normalization technique by using, as reference, the counts under the elastic peak of fluorine. Similar normalization protocols were used also in [14] to derive absolute cross sections with the OSCAR hodoscope [15].

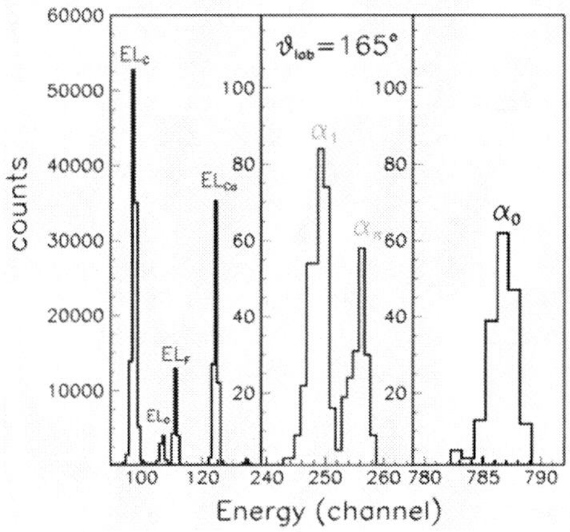

Fig. 1. Ejectile spectrum collected from the experiment at 165° detection angle in the laboratory system, at 1.3 MeV bombarding energy. Elastic scattering peaks on fluorine and other nuclides present in the target are labelled, together with peaks due to the emission of $\boldsymbol{\alpha}$ particles.

Absolute scattering cross section were deduced from *R*-matrix calculations at all the studied angles, by using the resonance parameters from [10], that describe very well elastic scattering data at backward angles. The number of detected particles for α_0 (and α_π) and elastic scattered protons ($N_{\alpha_{0,\pi}}$ and $N_{p,ela}$, respectively) are related to the corresponding cross sections as follows:

$$N_{\alpha_{0,\pi}} = N_i \mathcal{N} \frac{d\sigma^{\alpha_{0,\pi}}}{d\Omega}(\theta, E)\Delta\Omega \qquad (1)$$

$$N_{p,ela} = N_i \mathcal{N} \frac{d\sigma^{p,ela}}{d\Omega}(\theta, E)\Delta\Omega \qquad (2)$$

Here N_i is the number of incident particles, $\mathcal{N}$ the number of target atoms per unit area, θ the polar detection angle and ΔΩ the solid angle covered by the detector. By dividing eq. (1) to eq. (2) and knowing the proton scattering cross sections from the *R*-matrix calculations, we can simplify several factors and obtain the α_0 (and α_π, in the same way) differential cross section as follows:

$$\frac{d\sigma^{\alpha_0}}{d\Omega}(\theta, E) = \frac{N_{\alpha_0}}{N_{p,ela}} \cdot \frac{d\sigma^{p,ela}}{d\Omega}(\theta, E) \qquad (3)$$

To estimate the uncertainty level of the $\frac{d\sigma^{p,ela}}{d\Omega}$ *R*-matrix extrapolations, we compared published data of p+^{19}F elastic scattering at backward angles (>135°) with such extrapolations, and we find on the average an agreement better than 10%. We then assumed conservatively a 10% error on the estimate of the $\frac{d\sigma^{p,ela}}{d\Omega}(\theta, E)$ absolute differential cross section. The total uncertainty was then obtained by including also statistical uncertainties on counts in the error propagation.

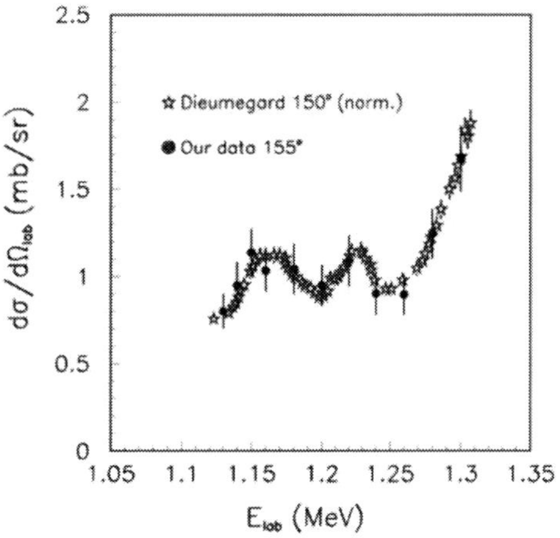

Fig. 2. Comparison of the excitation function at θ_{lab} = 155° between present work and data from Ref. [16] normalized by a factor 2.

3 Data analysis

As a first check on the data collected in the present experiment, we compared the excitation function for the α_0 channel at 155° with similar results reported by Ref. [16] at 150°. The shape of the two excitation functions is in excellent agreement, while the absolute cross section scale of data from Ref. [16] is a factor of 2 smaller than the present data. This scale factor is quite similar to the one used in Ref. [3] to compare their data and data from Ref. [16] in the 0.7-1.1 MeV energy region, pointing out a possible normalization problem in the data of [16]. At variance, the cross section scale reported by [17] at 150° in the 1.3 MeV region (not shown in Figure 2 for simplicity) is in close agreement with the one deduced in the present experiment.

To deduce the integrated cross section from the present data, that are measured only in the backward hemisphere, we need to know the trend of the angular distribution in the whole angular range. Fortunately, such trends are quite well known from the literature [8,18,19] in the energy ranges here investigated for the α_0 channel: 1.13-1.30 MeV and 1.62-1.72 MeV, and they are in good agreement. Similarly, the shapes of α_π angular distributions (and their expansion in Legendre polynomials) in the 1.13-1.30 MeV range have been reported in Ref. [20].

We performed spline fits of the Legendre polynomial coefficients (B_i) (in relative units and truncated to $\ell = 2$) as a function of energy, as taken from the literature. In this way, it was possible to interpolate the B_i values (in relative units) at the energy values of the present experiments. With such interpolations, we estimated, energy by energy, the trend of angular distributions; the absolute cross section scale was instead determined by fitting the backward angle cross section data collected in the present experiment with the estimated angular distributions. The integrated cross section and its uncertainty were then estimated analytically from the Legendre polynomial expansion of the angular distribution in absolute cross section units.

An example fit for the α_π angular distribution at the beam energy of 1.15 MeV can be seen in Figure 3. The presently measured data reproduce quite well the shape of the angular distribution interpolated from literature data at backward angles. Similar results have been deduced also for the other investigated energies and for the α_0 channel.

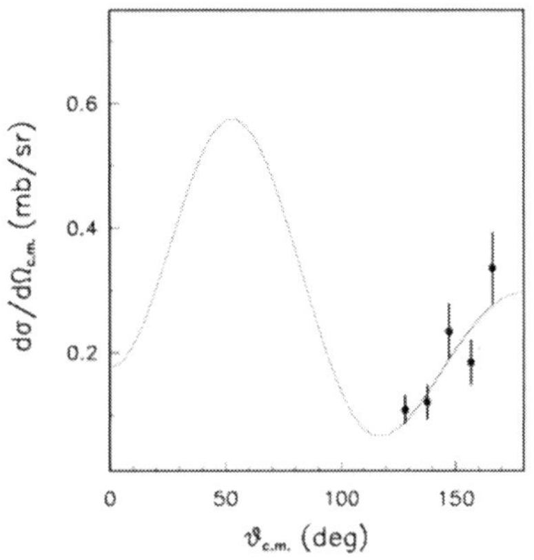

Fig. 3. Fit of the interpolated angular distribution for the ^{19}F(p,α_π)^{16}O reaction at E = 1.15 MeV with the experimental absolute cross section data coming from the present experiment. The interpolated curve is obtained by a spline fit of the coefficients of Legendre polynomial expansion reported at similar energies in Ref. [20].

The results of integrated cross sections data for both channels will be discussed in more details in forthcoming publications. As a preliminary observation concerning the ^{19}F(p,α_0) reaction in the region $E_{cm} \simeq 1.5-1.7$ MeV, the data obtained in the present work strongly agree, both in shape and in absolute values, with the [9] measurements. On the other hand, ^{19}F(p,α_π)^{16}O data in the region $E_{cm} \simeq 1.1-1.3$ MeV are in agreement, within the error bars, with the estimates reported in Ref. [10] based on unpublished results obtained during the '70s.

The impact of the new datasets here obtained on the spectroscopic properties of excited states in ^{20}Ne is currently under investigation and will be the subject of future publications.

4 Conclusions

The aim of this work was to improve our knowledge on the ^{19}F(p,α_0) and ^{19}F(p,α_π)^{16}O reactions at low energies, where the occurrence of contrasting data sets or energy regions with lacking or very sparse data have been reported. These reactions are important both for the spectroscopy of natural-parity states of the self-conjugate ^{20}Ne nucleus and to determine the behaviour of stellar nucleosynthesis beyond the CNOF cycle in various astrophysical contexts. In this proceeding we presented preliminary results on the new measured data, at various angles, in the 1.13-1.30 MeV and 1.5-1.7 MeV energy ranges for the α_0 and α_π channels coming from p + ^{19}F reactions. We described the internal normalization procedure that leads us to estimate the absolute cross sections for both reaction channels by avoiding any ambiguity due to target stoichiometry, beam current integration and solid angle measurements. The integrated cross section was then determined by coupling, through a fit procedure, the shape of angular distributions already reported in the literature in the full angular domain with the measured data obtained at backward angles. A further analysis is in progress to determine new spectroscopic parameters of ^{20}Ne states in the excitation energy region $\approx 14.0-14.4$ MeV from an R-matrix analysis of the present results and other data taken from the literature.

5 Acknowledgements

We are indebted with S. Mirabella (Univ. of Catania) and S. Tatì (Univ. of Catania) for delivering excellent quality beams from the Singletron accelerator. We acknowledge gratefully M. D'Andrea (INFN – Sez. Catania) for his help during the mounting phase of the experimental setup and the data taking.

References

1. W. von Oertzen, Martin Freer, Y. Kanada-En'yo, Phys. Rep., **432**, 43-113 (2006)
2. J. Bishop et al., Phys. Rev. C, **100,** 034320 (2019)
3. I. Lombardo, D. Dell'Aquila, L. Campajola, E. Rosato, G. Spadaccini, M. Vigilante, J. Phys. G: Nucl. Part. Phys., **40,** 125102 (2013)
4. I. Lombardo, L. Campajola, D. Dell'Aquila, M. La Commara, A. Ordine, E. Rosato, G. Spadaccini, M. Vigilante, J. Phys. Conf. Ser., **569,** 012068 (2014)
5. I. Lombardo, D. Dell'Aquila, L. Campajola, E. Rosato, G. Spadaccini, M. Vigilante, Bull. Russ. Acad. Sc., **78,** 1093-1096 (2014)
6. R. J. deBoer, O. Clarkson, A. J. Couture, J. Görres, F. Herwig, I. Lombardo, P. Scholz, M. Wiescher, Phys. Rev. C, **103,** 055815 (2021)
7. J. He, I. Lombardo, D. Dell'Aquila, Y. Xu, L. Zhang, W. Liu, Chin. Phys. G, **42,** 015001 (2018)
8. R. L. Clarke, E. B. Paul, Can. J. Phys., **35,** 155-164 (1957)

9. P. Cuzzocrea, A. De Rosa, G. Inglima, E. Perillo, E. Rosato, M. Sandoli, G. Spadaccini, Lett. Nuov. Cim., **28**, 515-522 (1980)
10. I. Lombardo, D. Dell'Aquila, J. He, G. Spadaccini, M. Vigilante, Phys. Review C, **100**, 044307 (2019)
11. D. R. Tilley, C. M. Cheves, J. H. Kelley, S. Raman, H. R. Weller, Nucl. Phys. A, **636**, 249-364 (1998)
12. I. Lombardo, L. Campajola, E. Rosato, G. Spadaccini, M. Vigilante, Nucl. Instr. Meth. Phys. Res. B, **302**, 19-23 (2013)
13. R. Caracciolo, P. Cuzzocrea, A. De Rosa, G. Inglima, E. Perillo, M. Sandoli, G. Spadaccini, Lett. Nuov. Cim., **11,** 33-38 (1974)
14. I. Lombardo et al., J. Phys. G: Nucl. Part. Phys., **48**, 065101 (2021)
15. D. Dell'Aquila et al., Nucl. Instr. Meth. Phys. Res. A, **877,** 227-237 (2018)
16. D. Dieumegard, B. Maurel, G. Amsel, Nucl. Inst. Meth., **168 (1-3),** 93-103 (1980)
17. I. Golicheff, M. Loeuillet, C. Engelmann, J. Rad. Chem., **22**, 113-129 (1974)
18. A. Isoya, H. Ohmura, T. Momota, Nucl. Phys., **7**, 116-125 (1958)
19. A. De Rosa, E. Perillo, P. Cuzzocrea, G. Inglima, E. Rosato, M. Sandoli, G. Spadaccini, Nuov. Cim., **44A,** 433-440 (1978)
20. A. Isoya, K. Goto, T. Momota, J. Phys. Soc. Jap., **11**, 899-906 (1956)

Dineutron correlations in knockout reactions with Borromean halo nuclei

Jesús Casal[1,*], *Mario* Gómez-Ramos[1], and *Antonio M.* Moro[1,2]

[1]Dpto. de Física Atómica, Molecular y Nuclear, Facultad de Física, Universidad de Sevilla, Apdo. 1065, E-41080 Sevilla, Spain
[2]Instituto Interuniversitario Carlos I de Física Teórica y Computacional (iC1), Apdo. 1065, E-41080 Sevilla, Spain

Abstract. We study dineutron correlations in proton-target knockout reactions induced by Borromean two-neutron halo nuclei. Using a core + n + n three-body model for the projectile and a quasifree sudden reaction framework, we focus on the correlation angle as a function of the intrinsic neutron momentum. Our results indicate that the correlations are strong in a range of neutron momenta associated to the nuclear surface. We also discuss on the role of core excitations for such correlations.

1 Introduction

Borromean halo nuclei are loosely bound systems formed by a compact core and two valence neutrons, such that the core + n subsystem is unbound [1, 2]. The diffuse matter distribution arising from this particular arrangement has deep implications for their dynamics [3, 4] and provides information about the limits of nuclear stability [5]. A two-neutron halo wast first observed in the seminal work by Tanihata *et al.*, where a relatively large interaction cross section for ^{11}Li was reported [6]. Other cases have been studied over the years, including ^{6}He, ^{14}Be, ^{17,19}B, ^{22}C or ^{29}F [7–14]. The Borromean nature of these systems implies that the correlations between the valence neutrons are a key aspect to understand their properties [15]. These correlations tend to favour a compact neutron-neutron structure typically referred to as dineutron configuration [16], the extent of which is somewhat related to the amount of mixing between different-parity orbitals [17]. Coulomb dissociation and knockout reactions have been employed to access these spatial correlations, for instance in ^{11}Li, typically linked to a small angle between the neutrons in coordinate space and a large opening angle in momentum space [8, 18, 19]. Recently, a similar approach has been introduced via quasifree knockout of the halo neutrons with proton targets, (p, pn), which provides spectroscopic information of the projectile ground state while probing the unbound binary subsystems [10, 13, 20]. Interestingly, this enables the exploration of such correlations as a function of the neutron density by analyzing the average correlation angle as a function of the intrinsic neutron momentum [21, 22]. For ^{11}Li, it was reported that the dineutron correlations are localized at the surface of the nucleus [23], an idea already suggested in Ref. [16] and predicted by more recent theories [24, 25]

In this contribution we recall a quasifree sudden reaction framework to describe proton-target (p, pn) knockout reactions that incorporates the three-body structure of the Borromean core + n + n projectile and the unbound nature of the core + n residue in a consistent way. The method, which was succesfully applied to ^{11}Li, is now extended to ^{14}Be including core excitations. First, the theoretical approach is briefly discussed. Then, some results for the angular and intrinsic momentum distributions of ^{14}Be, as well as the correlation between them, are presented.

2 Theoretical description

As discussed in Refs. [21–23], intermediate-energy proton-target knockout reactions can be described within a quasifree sudden reaction framework. The model assumes a zero-range V_{pn} interaction, and absorption and distortion effects are included through an eikonal S-matrix between the projectile and the proton. In the case of a valence-neutron knockout from a Borromean halo nucleus, the transition amplitude takes the form

$$\mathcal{T} \propto \langle \phi_{\mathrm{c-n}}(\boldsymbol{k}_x, \boldsymbol{x}) \otimes e^{\boldsymbol{k}_y \cdot \boldsymbol{y}} | S(y) \Phi_{g.s.}(\boldsymbol{x}, \boldsymbol{y}) \rangle, \tag{1}$$

where $\{x, y\}$ are the usual Jacobi coordinates depicted in Fig. 1, and $\{k_x, k_y\}$ are the conjugated intrinsic momenta. Here $\Phi_{g.s.}(\boldsymbol{x}, \boldsymbol{y})$ represents the ground state of the Borromean core + n + n projectile, and $\phi_{\mathrm{c-n}}(\boldsymbol{k}_x, \boldsymbol{x})$ is a continuum state describing the unbound core + n subsystem after knockout, which accounts for final state interactions. If no absorption is included (i.e., $S = 1$) and final state interactions are ignored (i.e., replacing $\phi_{\mathrm{c-n}}(\boldsymbol{k}_x, \boldsymbol{x})$ by a plane wave in the x coordinate), the above amplitude $\mathcal{T}$ becomes a simple Fourier transform of the ground-state wavefunction $\Phi_{g.s.}(\boldsymbol{x}, \boldsymbol{y})$. Therefore, Eq. (1) can be understood as a distorted Fourier transform. More details about the derivation of this equation can be found in Ref. [22], where the method was then applied to ^{11}Li(p, pn). Continuum states $\phi_{\mathrm{c-n}}(\boldsymbol{k}_x, \boldsymbol{x})$ are calculated by solving the two-body problem with standard scattering boundary conditions. The three-body ground-state wave function $\Phi_{\mathrm{gs}}(\boldsymbol{x}, \boldsymbol{y})$ can be

*e-mail: jcasal@us.es

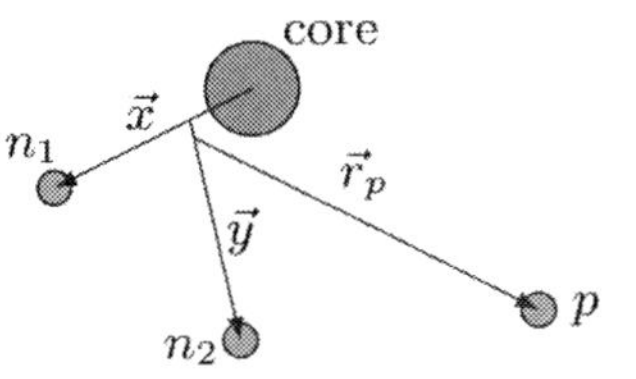

Figure 1. Schematic representation of the proton-target knockout reaction.

obtained within the hyperspherical formalism [1], provided that reliable core-n and n-n interactions are available. In particular, the analytical transformed harmonic oscillator method is adopted [26, 27] to build the wave function in the so-called Jacobi-T representation, where the two valence neutrons are related by the x coordinate. Then, a transformation to the Jacobi-Y coordinates shown in Fig. 1 is performed. For consistency, the same core-n interaction is employed to compute $\phi_{\rm c-n}$ and $\Phi_{\rm gs}$. Following the notation in Ref. [22], the ground state in Jacobi-Y coordinates can be written as

$$\Phi^{j\mu}_{\rm g.s.}(\boldsymbol{x},\boldsymbol{y}) = \frac{1}{xy}\sum_{\eta} \omega_\eta(x,y)\mathcal{Y}^{j\mu}_{\eta}(\widehat{x},\widehat{y}), \tag{2}$$

where $\eta = \left\{l_x, j_x, I, j_1, l_y, j_2\right\}$ is a set of quantum numbers involving all relative angular momenta and spins that follows the coupling scheme

$$\mathcal{Y}^{j\mu}_{\eta}(\widehat{x},\widehat{y}) = \left\{\left[\left(Y_{l_x}(\widehat{x})\otimes\kappa_s\right)_{j_x}\otimes\chi_I\right]_{j_1}\otimes\left[Y_{l_y}(\widehat{y})\otimes\kappa_s\right]_{j_2}\right\}_{j\mu}. \tag{3}$$

Here, $s = 1/2$ is the spin of the neutros and I represents the spin of the core. This means that $\boldsymbol{j}_x = \boldsymbol{l}_x + \boldsymbol{s}$ is the single-particle angular momentum of a neutron with respect to the core, so that $\boldsymbol{j}_x + \boldsymbol{I} = \boldsymbol{j}_1$ is the total angular momentum of the core + n subsystem. With this coupling order, the distorted Fourier transform in Eq. (1) can be given the same form,

$$\begin{aligned}\Psi^{j\mu}(\boldsymbol{k}_x,\boldsymbol{k}_y) &= \left\langle\phi_{\rm c\text{-}n}(\boldsymbol{k}_x,\boldsymbol{x})\otimes e^{i\boldsymbol{k}_y\boldsymbol{y}}\middle|S(y)\Phi^{j\mu}_{\rm g.s.}(\boldsymbol{x},\boldsymbol{y})\right\rangle\\ &= \sum_\eta w_\eta(k_x,k_y)\mathcal{Y}^{j\mu}_\eta(\Omega_x,\Omega_y),\end{aligned} \tag{4}$$

where now the angles refer to that of the conjugate momenta. Here, the w_η functions are obtained as a modified Fourier transform of the different partial waves of the overlaps between $\Phi_{\rm g.s.}$ and $\phi_{\rm c-n}$. It can be shown that

$$w_\eta(k_x,k_y) = (4\pi)^2\frac{i^{-l_x-l_y}}{k_x}\int dy\xi_\eta(k_x,y)j_{l_y}(k_y y)S(y)y, \tag{5}$$

where ξ_η are the corresponding radial overlaps (i.e., obtained by integrating over x). Note that these are formally the same structure overlaps used in DWBA and Transfer-to-the-continuum (TC) calculations presented in Refs. [28] (for transfer) and [29] (for knockout), respectively. It it also worth noting that these overlaps may involve different configurations for a given j_1 of the core + n system. This is particularly important if different core states I are allowed. In this work, core excitations are included by deforming the core + n interaction, in such a way that different core states can be coupled by in a rotational model [10, 30, 31]. From the previous equations, the cross section for the knockout process can be written as

$$\begin{aligned}\sigma &\propto \frac{1}{2j+1}\sum_\mu\int d\boldsymbol{k}_x d\boldsymbol{k}_y\,|\Psi|^2\\ &= \int dk_x dk_y d(\cos\theta)\sum_{\eta\eta'} w_\eta(k_x,k_y)w^*_{\eta'}(k_x,k_y)C_{\eta\eta'}\\ &\times\delta_{II'}\sum_L D^{(L)}_{\eta\eta'}\begin{pmatrix}l_y & l'_y & L\\ 0 & 0 & 0\end{pmatrix}\begin{pmatrix}l_x & l'_x & L\\ 0 & 0 & 0\end{pmatrix}P_L(\cos\theta),\end{aligned} \tag{6}$$

where P_L are Legendre polynomials. The z-axis has been defined in such a way that θ is precisely the angle between $\boldsymbol{k}_x$ and $\boldsymbol{k}_y$, and the quantities $C_{\eta\eta'}$ and $D^{(L)}_{\eta\eta'}$ contain spin factors, phases and 6-j symbols (see the Appendix of Ref. [22]). Equation (6) implies that an asymmetric θ distribution requires mixing between different-parity states in the core-n subsystem, such that odd L values are allowed [17, 23]. It is worth noting that the above cross section can be separated in contributions from different core states to assess the relevance of core excitations in the description of the relevant overlaps.

3 Results

In this work we apply the sudden formalism to describe the ^{14}Be(p,pn) knockout reaction. We use the ^{14}Be three-body wave function corresponding to model P2 in Ref. [10], which contains about $\sim 60\%$ p-wave content mostly linked to a a low-lying p-wave resonance in ^{13}Be. The model includes core excitations in a rotational picture, using an effective quadrupole deformation parameter $\beta_2 = 0.8$ [31], in such a way that $\sim 20\%$ of the norm corresponds to ^{12}Be in its first 2^+ excited state. With a two-neutron separation energy of 1.3 MeV, the computed matter radius is 3.01 fm, which is consistent with Ref. [32]. The ground-state probability density in Jacobi-T representation exhibits a clear maximum at small r_{nn} distances, as shown in Fig. 2, associated to the dineutron configuration. With this structure input, we computed the intrinsic momentum and opening angle distributions by direct integration of Eq. (6). This requires an eikonal S-matrix for the p-core absorption. As in Ref. [22], we employ the $t\rho$ prescription, using a Hartree-Fock density with the SkX interaction [33] for the core. This introduces absorptive effects for small p-core distances, and thus for small p-n distances due to the zero-range approximation for V_{pn}. As a result, the effect of absorption in the calculations is more important at large intrinsic momenta [22].

Figure 3 presents the intrinsic momentum distribution obtained by integrating over all angles θ and specific k_x windows, which correspond to different ^{12}Be + n relative energies $E_{\rm rel} = \hbar^2 k_x^2/(2\mu)$. The main contributions to the distribution are shown with labels $j_1[l_x j_x \otimes I]$, i.e., corresponding to the coupling order given by Eq. (3) for the

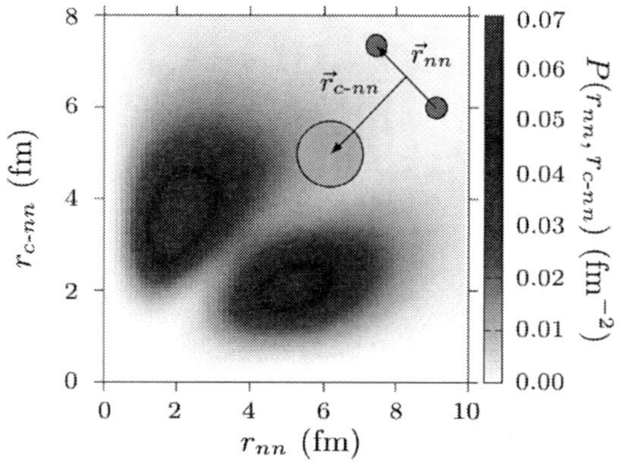

Figure 2. Ground-state probability density of ^{14}Be in Jacobi-*T* coordinates.

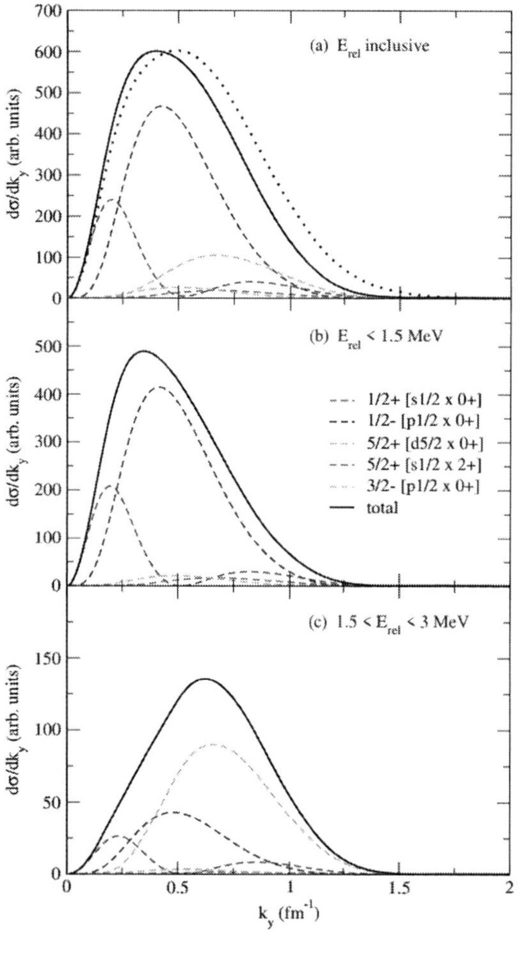

Figure 3. Intrinsic momentum distribution after one-neutron knockout from the ^{14}Be halo, integrated for: a) all core-*n* relative energies; b) E_{rel} < 1.5 MeV; c) 1.5 < E_{rel} < 3.0 MeV. The doted line in panel (a) is the result without including absorption, normalized to the same maximum.

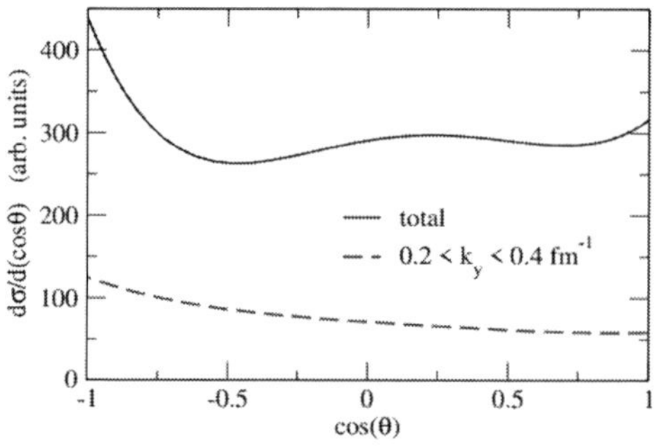

Figure 4. Opening angle distribution. Total (solid line) and restricted to $0.2 < k_y < 0.4$ fm^{-1} (dashed line).

core + *n* subsystem. At small relative energies the distributions are governed by knockout from a $p_{1/2}$ orbital, whereas at higher energies the effect of the $d_{5/2}$ becomes more important. The top panel shows also the total distribution without absorption, indicating that absorption effects tend to narrow the distribution as it occurs at small distances and thus large momenta. Figure 4 shows the opening angle distribution, integrated for all relative energies, and compared with the result in a specific region of intrinsic k_y momenta around the maximum of Fig. 3. The asymmetry in the distribution, associated to the mixing between different-parity components, is more apparent in the restricted k_y window, where the amplitudes of *s*-, *p*- and *d*-waves interfere and lead to $\theta > 90$ deg in momentum space.

In Fig. 5, the average opening angle θ as a function of the intrinsic momentum k_y is presented. The correlation plot exhibits a clear maximum around $k_y \simeq 0.2$ fm^{-1}, corresponding to $\langle\theta\rangle > 90$ deg, where the opposite-parity *s*- and *p*-wave contributions of the momentum distribution in Fig. 3 are comparable. Within the present model, this mixing is also responsible for the maximum of the probability in Fig. 2 at small interneutron distances r_{nn}, which happens at the nuclear surface (not the interior, not the halo tail). Note that, at large intrinsic momenta, the results for $\langle\theta\rangle$ are strongly influenced by absorption effects [22], so the region for $k_y \gtrsim 0.5$ fm^{-1} is not reliable to extract nuclear structure information within the present formalism. The low-k_y maximum, however, is robust under changes in absorption effects. Figure 5 shows also the contribution of each core state (0^+ ground state, 2^+ excited state) to the correlation angle. The weight of each contribution can be easily deduced from the intrinsic momentum distribution. The results indicate that core excitations, while small within the adopted three-body model, produce a reduction of the average angle in momentum space, effectively reducing the dineutron correlations. Therefore, the inclusion of those channels may be crucial to improve the description. The comparison of the present calculations with experimental data is in progress and will be presented elsewhere [34], together with results for ^{11}Li and ^{17}B.

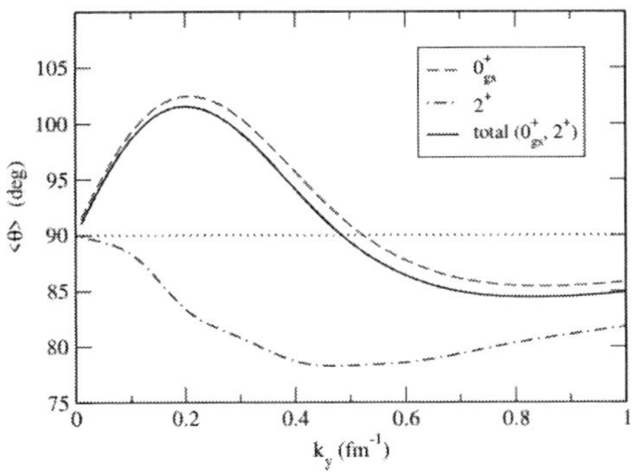

Figure 5. Average opening angle as a function of the intrinsic neutron momentum. The solid line is the complete calculation, while the dashed and dot-dashed lines correspond to the core left in its 0^+ ground state or first 2^+ excited state, respectively.

Acknowledgements

The authors are grateful to A. Corsi and Y. Kubota for useful discussions. This work has been partially funded by MCIN/AEI/10.13039/501100011033 under I+D+i project No. PID2020-114687GB-I00 and under grant IJC2020-043878-I (also funded by "European Union NextGenerationEU/PRTR"), by the European Union's Horizon 2020 research and innovation programme under the Marie Skłodowska-Curie grant agreement No. 101023609, by the Consejería de Economía, Conocimiento, Empresas y Universidad, Junta de Andalucía (Spain) and "ERDF-A Way of Making Europe" under PAIDI 2020 project No. P20_01247, and by the European Social Fund and Junta de Andalucía (PAIDI 2020) under grant number DOC-01006.

References

[1] M. Zhukov, B. Danilin, D. Fedorov, J. Bang, I. Thompson, J. Vaagen, Phys. Rep. **231**, 151 (1993)
[2] E. Nielsen, D. Fedorov, A. Jensen, E. Garrido, Phys. Rep. **347**, 373 (2001)
[3] P.G. Hansen, B. Jonson, Europhys. Lett. **4**, 409 (1987)
[4] K. Riisager, Rev. Mod. Phys. **66**, 1105 (1994)
[5] I. Tanihata, H. Savajols, R. Kanungo, Prog. Part. Nucl. Phys. **68**, 215 (2013)
[6] I. Tanihata et al., Phys. Rev. Lett. **55**, 2676 (1985)
[7] T. Aumann et al., Phys. Rev. C **59**, 1252 (1999)
[8] H. simon et al., Nucl. Phys. A **791**, 267 (2007)
[9] Y. Togano et al., Physics Letters B **761**, 412 (2016)
[10] A. Corsi et al., Phys. Lett. B **797**, 134843 (2019)
[11] K.J. Cook et al., Phys. Rev. Lett. **124**, 212503 (2020)
[12] S. Bagchi et al., Phys. Rev. Lett. **124**, 222504 (2020)
[13] Z.H. Yang et al., Phys. Rev. Lett. **126**, 082501 (2021)
[14] Y.L. Sun et al., Physics Letters B **814**, 136072 (2021)
[15] K. Hagino, H. Sagawa, Phys. Rev. C **72**, 044321 (2005)
[16] A. Migdal, Yad. Fiz. **16**, 427 (1972)
[17] F. Catara, A. Insolia, E. Maglione, A. Vitturi, Phys. Rev. C **29**, 1091 (1984)
[18] T. Nakamura et al., Phys. Rev. Lett. **96**, 252502 (2006)
[19] H. Simon et al., Phys. Rev. Lett. **83**, 496 (1999)
[20] Y. Aksyutina et al., Phys. Lett. B **718**, 1309 (2013)
[21] Y. Kikuchi, K. Ogata, Y. Kubota, M. Sasano, T. Uesaka, Prog. Theor. Exp. Phys. **2016**, 103D03 (2016)
[22] J. Casal, M. Gómez-Ramos, Phys. Rev. C **104**, 024618 (2021)
[23] Y. Kubota et al., Phys. Rev. Lett. **125**, 252501 (2020)
[24] M. Matsuo, Phys. Rev. C **73**, 044309 (2006)
[25] K. Hagino, H. Sagawa, J. Carbonell, P. Schuck, Phys. Rev. Lett. **99**, 022506 (2007)
[26] J. Casal, M. Rodríguez-Gallardo, J.M. Arias, Phys. Rev. C **88**, 014327 (2013)
[27] J. Casal, J. Singh, L. Fortunato, W. Horiuchi, A. Vitturi, Phys. Rev. C **102**, 064627 (2020)
[28] J. Casal, M. Gómez-Ramos, A. Moro, Phys. Lett. B **767**, 307 (2017)
[29] M. Gómez-Ramos, J. Casal, A. Moro, Phys. Lett. B **772**, 115 (2017)
[30] I. Thompson, F. Nunes, B. Danilin, Comput. Phys. Commun. **161**, 87 (2004)
[31] T. Tarutina, I. Thompson, J. Tostevin, Nucl. Phys. A **733**, 53 (2004)
[32] T. Suzuki et al., Nuclear Physics A **658**, 313 (1999)
[33] B. Alex Brown, Phys. Rev. C **58**, 220 (1998)
[34] A. Corsi et al., submitted (2023)

Study of light hypernuclei in Europe: the hypertriton and nnΛ puzzles

C. Rappold[1,*], *P.* Achenbach[2,3], *H.* Alibrahim Alfaki[4], *F.* Amjad[4], *M.* Armstrong[4,5], *K.-H.* Behr[4], *J.* Benlliure[6], *Z.* Brencic[7,8], *T.* Dickel[4,9], *V.* Drozd[4,10], *S.* Dubey[4], *H.* Ekawa[11], *S.* Escrig[1,11], *M.* Feijoo-Fontan[6], *H.* Fujioka[12], *Y.* Gao[11,13,14], *H.* Geissel[4,9], *F.* Goldenbaum[15], *A.* Graña Gonzalez[6], *E.* Haettner[4], *M. N.* Harakeh[10], *Y.* He[11,14], *H.* Heggen[4], *C.* Hornung[4], *N.* Hubbard[4,16], *K.* Itahashi[17,18], *M.* Iwasaki[17,18], *N.* Kalantar-Nayestanaki[10], *A.* Kasagi[11,19], *M.* Kavatsyuk[10], *E.* Kazantseva[4], *A.* Khreptak[20,21], *B.* Kindler[4], *R.* Knoebel[4], *H.* Kollmus[4], *D.* Kostyleva[4], *S.* Kraft-Bermuth[22], *N.* Kurz[4], *E.* Liu[11,13,14], *B.* Lommel[4], *V.* Metag[9], *S.* Minami[4], *D. J.* Morrissey[23], *P.* Moskal[21,24], *I.* Mukha[4], *A.* Muneem[11,25], *M.* Nakagawa[11], *K.* Nakazawa[19], *C.* Nociforo[4], *H. J.* Ong[13,26,27], *S.* Pietri[4], *J.* Pochodzalla[2,3], *S.* Purushothaman[4], *E.* Rocco[4], *J. L.* Rodríguez-Sánchez[6], *P.* Roy[4], *R.* Ruber[28], *T.R.* Saito[4,11,14], *S.* Schadmand[4], *C.* Scheidenberger[4,9], *P.* Schwarz[4], *R.* Sekiya[17,18,29], *V.* Serdyuk[15], *M.* Skurzok[21,24], *B.* Streicher[4], *K.* Suzuki[4,30], *B.* Szczepanczyk[4], *Y. K.* Tanaka[11], *X.* Tang[13], *N.* Tortorelli[4], *M.* Vencelj[7], *H.* Wang[11], *T.* Weber[4], *H.* Weick[4], *M.* Will[4], *K.* Wimmer[4], *A.* Yamamoto[31], *A.* Yanai[11,32], *J.* Yoshida[11,33], *J.* Zhao[4,34], and WASA-FRS/Super-FRS Experiment Collaboration

[1]Instituto de Estructura de la Materia - CSIC, Madrid, Spain
[2]Institute for Nuclear Physics, Johannes Gutenberg University, Mainz, Germany,
[3]Helmholtz Institute Mainz, Johannes Gutenberg University, Mainz, Germany,
[4]GSI Helmholtzzentrum für Schwerionenforschung GmbH, Darmstadt, Germany,
[5]Institut für Kernphysik, Universität Köln, Koln, Germany,
[6]IGFAE, Universidade de Santiago de Compostela, Santiago de Compostela, Spain,
[7]Jozef Stefan Institute, Ljubljana, Slovenia,
[8]University of Ljubljana, Ljubljana, Slovenia,
[9]Justus-Liebig Universität Giessen, Giessen, Germany,
[10]University of Groningen, Groningen, The Netherlands,
[11]High Energy Nuclear Physics Laboratory, RIKEN Cluster for Pioneering Research, RIKEN, Wako, Japan,
[12]Tokyo Institute of Technology, Tokyo, Japan,
[13]Institute of Modern Physics, Chinese Academy of Sciences, Lanzhou, China,
[14]School of Nuclear Science and Technology, University of Chinese Academy of Sciences, Beijing, China,
[15]Institut fur Kernphysik, Forschungszentrum Julich, Julich, Germany,
[16]Institut fur Kernphysik, Technische Universitat Darmstadt, Darmstadt, Germany,
[17]Meson Science Laboratory, Cluster for Pioneering Research, RIKEN, Wako, Japan,
[18]Nishina Center for Accelerator-Based Science, RIKEN, Wako, Japan,
[19]Graduate School of Engineering, Gifu University, Gifu, Japan ,
[20]INFN, Laboratori Nazionali di Frascati, Frascati, Roma, Italy,
[21]Institute of Physics, Jagiellonian University, Krakow, Poland,
[22]TH Mittelhessen University of Applied Sciences, Giessen, Germany,
[23]National Superconducting Cyclotron Laboratory, Michigan State University, East Lansing, USA,
[24]Center for Theranostics, Jagiellonian University, Krakow, Poland,
[25]Faculty of Engineering Sciences, Ghulam Ishaq Khan Institute of Engineering Sciences and Technology, Topi, Pakistan,
[26]Joint Department for Nuclear Physics, Lanzhou University and Institute of Modern Physics, Lanzhou, China,
[27]Research Center for Nuclear Physics, Osaka University, Osaka, Japan,
[28]Uppsala University, Uppsala, Sweden,
[29]Kyoto University, Kyoto, Japan,
[30]Ruhr-Universität Bochum, Institut für Experimentalphysik I, Bochum, Germany,
[31]KEK, Tsukuba, Japan,
[32]Saitama University, Sakura-ku, Japan,
[33]Tohoku University, Sendai, Japan,
[34]Peking University, Beijing, China,

Abstract. The current understanding of light hypernuclei, which are sub-atomic nuclei with strangeness, is being challenged and studied in detail by several European research groups and collaborations. In recent years, studies of hypernuclei using high-energy heavy ion beams have reported unexpected results on the three-body hypernuclear state $^{3}_{\Lambda}$H, named the hypertriton. For some time, reports of a shorter lifetime and larger binding energy than what was previously accepted have created a puzzling situation for its theoretical description; this is known as the "hypertriton puzzle". With the inclusion of the most recent experimental measurements, the current status of the hypertriton puzzle is evolving. Additionally, the possible neutral bound state of a Λ hyperon with two neutrons, nnΛ, has raised questions about our understanding of the formation of light hypernuclei either in bound or resonance states. These results have initiated several ongoing experimental programs all over the world to study these three-body hypernuclear states precisely. We are studying these light hypernuclear states by employing heavy ion beams at 2*A*GeV on a fixed carbon target with the WASA detector system and the Fragment Separator (FRS) at GSI. The WASA-FRS experimental campaign was performed during the first quarter of 2022, and this paper presents a short overview of the campaign and how it seeks to tackle the hypertriton and nnΛ puzzles. Data analysis is ongoing, and several preliminary results will be reported.

1 Introduction

Understanding the baryon-baryon interaction has been a major goal in studying the hypernucleus, which is a bound state of hyperons and nucleons. One key aspect of this research is determining the equation of state of baryonic matter, particularly in the strangeness sector. In recent years, nuclear spectroscopy using heavy-ion beams in collider mode or on fixed targets has emerged as an effective method for investigating hypernuclei. Several European research groups and collaborations are thoroughly investigating and testing our existing knowledge on light hypernuclei [1–5]. In the past ten years, studies of hypernuclear states using high-energy heavy-ion beams have yielded unexpected results regarding the three-body hypernuclear state ${}^{3}_{\Lambda}$H, also known as the hypertriton. The hypertriton is the lightest of the known bound hypernuclei. It is conventionally described as the state of the deuteron nuclear core with which a Λ hyperon is loosely bound. The indication of the weakly bound Λ was first observed in the nuclear emulsions used for hypernuclear studies in the 70s. A Λ binding energy of 130 ± 50 keV was reported [6, 7]. After an extended period, the understanding of the structure of the hypertriton was not challenged, until the first observation of the hypertriton by the STAR collaboration in the Au-Au collisions at $\sqrt{s_{NN}} = 200$ GeV in the ultra-relativistic energy regime [8]. An indication of a possible shorter lifetime than our current understanding was presented. Shortly after the report of the HypHI collaboration of their observation of light hypernuclei, a clear tendency for the experimental hypertriton lifetime leaned towards a shorter value than the different theoretical models predicted at that time [3, 9]. The ALICE collaboration shortly after also observed a hypertriton signal for the first time in Pb-Pb collisions at 2.76 TeV and initially measured a shorter lifetime [5]. Subsequently, more lifetime measurements by the collaborations using heavy-ion beams were obtained with conflicting tendencies [3, 5, 8, 10–13]. Furthermore, the STAR collaboration later published a larger Λ binding energy of 410 ± 120(stat) ± 110(sys) keV [14], while ALICE reported an estimation of a 50 ± 60 ± 100 keV Λ binding energy. All of these experimental observations and comparisons with our original understanding of the hypertriton structure have created a puzzle for its theoretical description, called the "hypertriton puzzle".

Moreover, the potential existence of a neutral bound state involving a Λ hyperon and two neutrons, nnΛ [15], has posed a challenge to the current understanding of the formation of resonance or bound states of light hypernuclei. The HypHI collaboration, besides observing the production of light hypernuclei in the ^{6}Li+^{12}C reaction [16], reported a possible signal in the t+π^- and d+π^- invariant mass spectrum. Those decay channels were associated with the mesonic weak decay of nnΛ $\rightarrow t + \pi^-$ and nnΛ $\rightarrow (d + n) + \pi^-$, where the neutron would not be observed. Extensive theoretical efforts were made to investigate the bound state of the system nnΛ using various approaches and models [17–20]. The results obtained by these efforts did not demonstrate a bound system, while in pionless effective field theory which its existence could not be excluded [21, 22]. Recently, a new experiment labelled E12-17-003 was conducted at JLab to explore the existence of the nnΛ state [20, 23]. The experiment utilized electron beams to bombard a tritium target. Two different interpretations of the results of the data analysis were published by the Hall A collaboration, one indicating possible existence of the nnΛ state with small significance [24], the other suggesting non-existence of nnΛ [25]. Hence, a definitive conclusion has not yet been reached, forming the so-called nnΛ puzzle. Further experimental studies that provide better precision and larger data samples are required to determine the possible existence of the nnΛ state conclusively.

Currently, several experimental efforts are ongoing around the world, particularly in Europe, to address these puzzles, as shown in Figure 1. These ongoing experimental programs across the globe aim at further investigating these three-body hypernuclear states with greater precision.

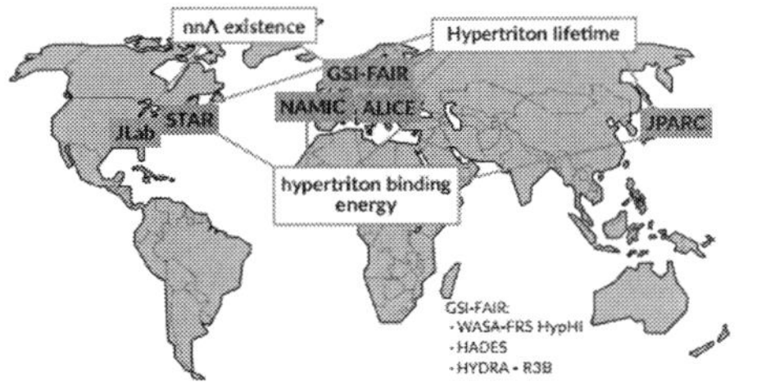

Figure 1. World effort to tackle the hypertriton and nnΛ puzzles.

One such effort in Europe is the WASA-FRS HypHI experiment, which focuses on the precise spectroscopy of light hypernuclei. The experimental objectives include the measurement of the lifetime of ${}^{3}_{\Lambda}$H and ${}^{4}_{\Lambda}$H, followed by the possible observation of the nnΛ state. The study is conducted using heavy ion beams at a kinetic energy of around 2 *A*GeV on a fixed carbon target, with the WASA detector system and the Fragment Separator (FRS) at the GSI-FAIR facility [4]. The WASA-FRS experimental campaign was

*e-mail: christophe.rappold@csic.es

conducted in the first quarter of 2022, with data analysis currently underway. The results of this research will help shed light on the puzzles surrounding three-body hypernuclear states.

2 WASA-FRS HypHI experiment

A new project has been proposed at GSI-FAIR to study hypernuclei using the WASA central detector. This detector, previously located at COSY in Jülich, Germany [26, 27], was transported to GSI-FAIR in 2019. It is used for pion measurement in conjunction with the high-resolution fragment separator (FRS) for measuring decay residues. The objective of this novel experiment, which has been approved for the FAIR-Phase 0 beam-time period, is to investigate the existence of a possible bound state of nnΛ and the hypertriton lifetime puzzle. To assess the feasibility of the experimental method, the experiment was carried out under conditions similar to those of the previously successful HypHI experiment. In particular, the focus is on reconstructing and identifying light hypernuclei, such as $^{3}_{\Lambda}$H, $^{4}_{\Lambda}$H, and nnΛ, using the invariant mass method.

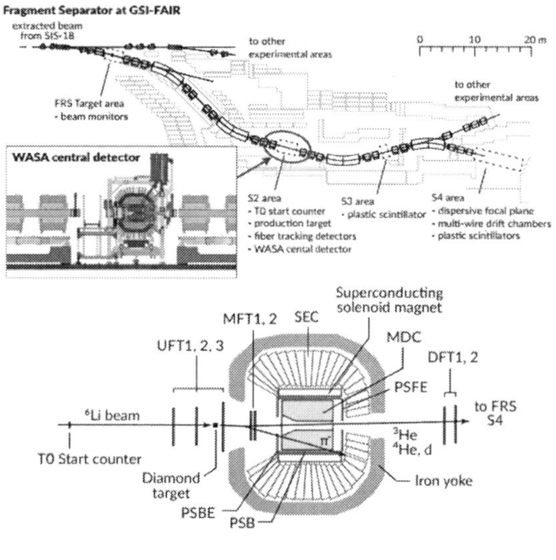

Figure 2. Experimental setup of the WASA-FRS experimental campaign. Top panel: the layout of the FRS, where the beam is directed towards the experimental area S2 for the production of hypernuclei. The FRS segment from S2 to S4 serves as a high-resolution forward spectrometer for the decay fragments. Bottom panel: the cylindrical WASA detector apparatus located in the S2 area, which is used to measure light particles such as pions within the solenoid magnet.

In an exclusive measurement, the mesonic weak decay of $^{3}_{\Lambda}\mathrm{H} \rightarrow {}^{3}\mathrm{He} + \pi^{-}$, $^{4}_{\Lambda}\mathrm{H} \rightarrow {}^{4}\mathrm{He} + \pi^{-}$, and nnΛ → d + n + π^{-} is the focus of the experiment. After the decay, a narrow magnetic rigidity acceptance window is set for the FRS in order to measure precisely the momentum of the outgoing decay fragment (^{3}He, ^{4}He, d). The layout of the FRS is illustrated in the top panel of Figure 2. The detection apparatus at the S2 experimental area is responsible for measuring a large portion of the emitted π^{-}, as depicted in the bottom panel of Figure 2. The experimental setup for our study utilized the WASA central detector, which consists of a superconducting solenoid magnet capable of producing magnetic fields up to 1 T, iron yokes, and a cryogenic system. The WASA detection system includes a calorimeter with CsI crystals (SEC), an inner drift chamber with straw tubes (MDC), and newly developed arrays of plastic barrel hodoscopes (PSB) [28] and end-cap hodoscopes (PSFE and PSBE). In the first quarter of 2022, two experiments encompass a data-taking campaign: S490 to search for η'-nuclei and S447 to study hypernuclei. The experimental setup used for studying hypernuclei (S447, WASA-FRS-HypHI) is illustrated in the bottom panel of Figure 2. In addition to the conventional WASA apparatus, extra scintillating fiber detectors were mounted. Three fiber detector stations, namely UTF1, UTF2, and UTF3, were placed in front of the WASA detectors. Two fiber detector stations, DFT1, and DFT2, were positioned behind the WASA detector. Furthermore, a Mini-fiber detector comprising two fiber complexes, MFT1 and MFT2, was situated inside the iron yoke of the superconducting solenoid magnet of the WASA. Additionally, a small hodoscope, made up of small plastic scintillator fingers referred to as T0 start counter, was mounted in front of the UFT1 to assess the beam particles. A diamond target with a thickness of 9.87 g/cm^2 was located between the tracking station UFT2 and UFT3, where the hypernuclear production would occur during the collision between the ^{6}Li beam at 1.96 *A*GeV directed at the ^{12}C of the target. Negatively charged pions from decays of hypernuclei are to be measured by the WASA detectors with the stations MFT1 and MFT2. Residual nuclei of hypernuclear decays were transferred to and analyzed by the S2-S4 sections of the FRS. Details of the experimental concept and the simulated expectations were discussed in [4]. We evaluated the acceptance and efficiencies and expect the invariant mass resolution to be at least twice as good, with at least an order of magnitude more statistics than collected during the first HypHI experiment.

3 Preliminary outcomes of the S447 data taking campaign

In the experiment, we maintained the counting rate at S4 in the FRS to be below a few kHz. It is essential to note that the standard resolving power of the FRS to measure the momentum of the residual nuclei is 10^4. As such, specific ion-optics settings were developed and used to obtain high acceptance and high resolving power. With the ^{6}Li beam, we have stored 3.3×10^8, 0.9×10^9, and 1.8×10^8 events at S4 of ^{3}He, ^{4}He, and deuterons, respectively. The accumulated data-taking occurred over 40.9, 43.9, and 43.9 hours, respectively. Additionally, we collected data using ^{12}C beams, recording 1.0×10^8 events with ^{3}He at S4 and 2.4×10^5 events with ^{9}C at S4 over 13.5 hours. It will give us a chance to analyze the decay channel $^{9}B \rightarrow \pi^{-} + {}^{9}\mathrm{C}$. The collected data from the experiment is currently undergoing analysis. To identify a fragment that has passed through the FRS beyond the S2 focal plane, the time-of-flight (TOF), and energy deposit measurements obtained from scintillators located at the S3 and S4 focal plane are considered. As shown in Figure 3 top panel, the correlations between the TOF and the energy deposit obtained from the S4 scintillator for ^{3}He, ^{4}He, and deuteron are observed in the left and right panels, respectively. This enables us to clearly identify the fragments. The momentum measurement and analysis in the S2-S4 section of the FRS spectrometer are obtained by ion-optics calibration, and a momentum reconstruction resolution of 5×10^{-4} was preliminarily estimated, as shown in the bottom-left panel of Figure 3. Overall, the identification and analysis performance of fragments in the FRS is reasonably good. Charged particles emitted to the WASA detector system

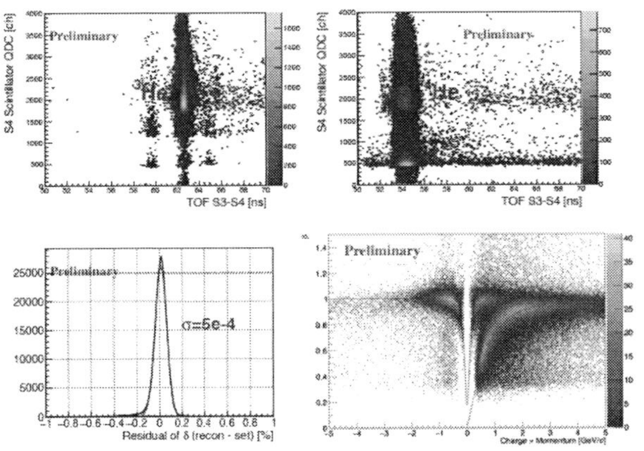

Figure 3. Preliminary results from the calibration are presented. Top panel: the fragment identification at the S2-S4 section of the FRS. The left and right top panels clearly show the identification of ^{3}He, ^{4}He, and deuteron ions, respectively. Bottom left panel: the preliminary momentum resolution achieved after ion-optics analysis of the fragments using the FRS S2-S4 spectrometer is showcased. A resolution of 5×10^{-4} has been achieved. Bottom right panel: the particle identification within the WASA central detection system is presented. Thanks to the β-momentum correlation, light hadrons such as π^-, π^+, and protons can be preliminarily identified.

undergo analysis by the tracking and TOF detectors. The momentum of these particles is estimated using Kalman filter track reconstruction for the hits of MFT, MDC, and PSB. A correlation between velocity β and momentum of the particles is observed, as depicted in the bottom-right panel of Figure 3. Light hadrons, such as π^-, π^+, and proton, can be preliminarily identified. Further data analyses for the hypernuclear reconstruction are in progress and are to be completed.

Experimental activities with the WASA-FRS setup will continue at FAIR Phase 0. Data analysis of the data-taking of 2022 is ongoing, with the hypernuclear reconstruction of the decay channels ^{3}He+π^-, ^{4}He+π^-, d+π^-, contributing to find a solution for the hypertriton and nnΛ puzzle. With the continued experimental efforts at FAIR Phase 0 and 1, we will perform further studies of exotic hypernuclei, including proton-rich [29] and neutron-rich [30] hypernuclei.

4 Acknowledgement

The WASA-FRS project has received funding from the following sources: JSPS KAKENHI (Grant Numbers JP18H01242), JSPS Grant-in-Aid for Early-Career Scientists (Grant No. JP20K14499), JSPS Fostering Joint International Research (B) (Grant No. JP20KK0070), Proyectos I+D+i 2020 ref: PID2020-118009GA-I00, grant 2019-T1/TIC-13194 of the Atracción de Talento Investigador program of the Community of Madrid, the Regional Government of Galicia under the Postdoctoral Fellowship Grant No.ED481D-2021-018, the SciMat and qLife Priority Research Areas budget under the program Excellence Initiative - Research University at the Jagiellonian University, and the European Union's Horizon 2020 research and innovation programme under grant agreement No. 824093.

References

[1] P. Eckert, *et al.*, Proceedings of PANIC 2021, 380, (2022)
[2] S. Simon, EPJ Web of Conf., **259**, 01007 (2022)
[3] C. Rappold *et al.*, Nucl. Phys. A **914**, 519 (2013).
[4] T. R. Saito *et al.*, Nat. Rev. Phys. **3**, 803 (2021).
[5] J. Adam, *et al.*, for ALICE collaboration, Phys. Lett. B **754**, 360 (2016).
[6] G. Bohm, *et al.*, Nucl. Phys. B **4**, 511 (1968)
[7] M. Juric, *et al.*, Nucl. Phys. B **52**, 1 (1973)
[8] The STAR Collaboration, Science **328**, 58 (2010).
[9] C. Rappold, *et al.*, Phys. Lett. B **728** 543 (2014)
[10] Y. Xu for the STAR Collaboration, Proceedings of the HYP2015 021005 (2017).
[11] L. Adamczyk, *et al.*, Phys. Rev. C **97**, 054909 (2018).
[12] J. Chen, *et al.*, Phys. Rep. **760**, 1 (2018).
[13] S. Acharya, *et al.*, Phys. Lett. B **797**, 134905 (2019).
[14] J. Adam, *et al.*, Nat. Phys. **16**, 409 (2020).
[15] C. Rappold, *et al.*, Phys. Rev. C **88**, 041001 (2013).
[16] C. Rappold, *et al.*, Phys. Lett. B **747**, 129 (2015).
[17] E. Hiyama, *et al.*, Phys. Rev. C **89**, 061302 (2014)
[18] A. Gal, *et al.*, Phys. Lett. B **736**, 93 (2014)
[19] H. Garcilazo, *et al.*, Phys. Rev. C **89**, 057001 (2014)
[20] M. Schafer, *et al.*, Phys. Lett. B **808**, 135614 (2020)
[21] S.-I. Ando, *et al.*, Phys. Rev. C **92**, 024325 (2015)
[22] F. Hildenbrand, *et al.*, Phys. Rev. C **100**, 034002 (2019)
[23] I.R. Afnan, *et al.*, Phys. Rev. C **92**, 054608 (2015)
[24] B. Pandey, *et al.*, Phys. Rev. C **105**, L051001 (2022).
[25] K. Itabashi, *et al.*, Few Body Syst. **63**, 16 (2022). K.N. Suzuki, *et al.*, PTEP 2022, Issue 1, 013D01 (2022)
[26] P. Adlarson, *et al.*, Phys. Rev. C **102**, 044322 (2020).
[27] P. Adlarson, *et al.*, Phys. Lett. B **802**, 135205 (2020).
[28] R. Sekiya *et al.*, Nucl. Instruments Methods Phys. Res. A **1034**, 166745 (2022).
[29] C. Rappold *et al.*, Phys. Rev. C **94**, 044616 (2016)
[30] T. R. Saito *et al.*, Eur. Phys. Jour. A **57**, 159 (2021)

Low-lying structure of ^{15}C : information on the N=8 shell gap

J. Lois-Fuentes[1,*], B. Fernández-Domínguez[1], X. Pereira-López[1,2,3], F. Delaunay[2], W.N. Catford[4], A. Matta[2,4], N.A. Orr[2], T. Duguet[5,6], T. Otsuka[7], V. Somà[5], O. Sorlin[8], T. Suzuki[9,10], N.L. Achouri[2], M. Assié[11], S. Bailey[12], B. Bastin[8], Y. Blumenfeld[11], R. Borcea[15], M. Caamaño[1], L. Caceres[8], E. Clément[8], A. Corsi[5], N. Curtis[12], Q. Deshayes[2], F. Farget[8], M. Fisichella[13], G. de France[8], S. Franchoo[11], M. Freer[12], J. Gibelin[2], A. Gillibert[5], G.F. Grinyer[14], F. Hammache[11], O. Kamalou[8], A. Knapton[4], Tz. Kokalova[12], V. Lapoux[5], B. Le Crom[11], S. Leblond[2], F.M. Marqués[2], P. Morfouace[11], J. Pancin[8], L. Perrot[11], J. Piot[8], E. Pollacco[5], D. Ramos[1], D. Regueira-Castro[1], C. Rodríguez-Tajes[1,8], T. Roger[8], F. Rotaru[15], M. Sénoville[2], N. de Séréville[11], R. Smith[12,13], M. Stanoiu[15], I. Stefan[11], C. Stodel[8], D. Suzuki[11], J.C. Thomas[8], N. Timofeyuk[4], M. Vandebrouck[5,8], J. Walshe[12], and C. Wheldon[12]

[1]IGFAE and Dpt. de Física de Partículas, Univ. of Santiago de Compostela, E-15758, Santiago de Compostela, Spain.
[2]Université de Caen Normandie, ENSICAEN, CNRS/IN2P3, LPC Caen UMR6534, F-14000 Caen, France
[3]Center for Exotic Nuclear Studies, Institute for Basic Science (IBS), Daejeon 34126, Republic of Korea
[4]Department of Physics, University of Surrey, Guildford GU2 5XH, UK
[5]IRFU, CEA, Université Paris-Saclay, F-91191 Gif-sur-Yvette, France
[6]KU Leuven, Instituut voor Kern- en Stralingsfysica, 3001 Leuven, Belgium
[7]CNS, University of Tokyo, 7-3-1 Hongo, Bunkyo-ku, Tokyo, Japan
[8]GANIL, CEA/DRF-CNRS/IN2P3, Bd. Henri Becquerel, BP 55027, F-14076 Caen, France
[9]Department of Physics, College of Humanities and Sciences, Nihon University, Sakurajosui 3-25-40, Setagaya-ku, Tokyo, Japan
[10]NAT Research Center, NAT Corporation, 3129-45 Hirahara, Muramatsu, Tokai, Naka, 319-1112, Ibaraki, Japan
[11]Université Paris-Saclay, CNRS/IN2P3, IJCLab, 91405 Orsay, France
[12]School of Physics and Astronomy, University of Birmingham, Birmingham B15 2TT, UK
[13]INFN, Laboratori Nazionali del Sud, Via S. Sofia 44, Catania, Italy
[14]Department of Physics, University of Regina, Regina, SK S4S 0A2, Canada
[15]IFIN-HH, P. O. Box MG-6, 76900 Bucharest-Magurele, Romania

Abstract. The low-lying structure of ^{15}C has been investigated via the neutron-removal d(^{16}C, t) reaction. The experiment was performed at GANIL using a secondary ^{16}C beam produced by fragmentation in the LISE spectrometer at 17.2 MeV/nucleon with an intensity of 5×10^4 pps and 100% purity. The angle and energy of the light ejectile were detected by three MUST2 telescopes. The missing mass technique was used to reconstruct the excitation energy of ^{15}C. In this spectrum, two bound states were observed (gs and the first excited state) and two unbound resonant states above the neutron separation threshold (S_n = 1.218 MeV). From the differential cross sections, information on the angular momentum of the transferred nucleon and spectroscopic factors were deduced.
The excitation energies and the deduced spectroscopic factors of the negative parity states placed above the neutron separation energy are an important measurement of the 2p-1h configurations in ^{15}C. Our results show good agreement with shell-model calculations with the YSOX interaction and show a sensitivity to the N=8 shell gap.

1 Introduction

The study of the evolution of the shell structure when one moves away from the valley of stability has been one of the main tools used in the recent years to learn about fundamental properties of the nuclear interaction.

Nuclei far from the stability with large values of isospin, like the neutron-rich carbon isotopes, are useful to test the theoretical models and interactions. In particular, the shell model calculations with phenomenological interactions in the p-sd shell space (SFO tls [1, 2] and YSOX [3]) manage to reproduce energies and spectroscopic factors in the neutron-rich C isotopes. Still, there is a lack of neutron p-hole states data in this region, which give crucial constraints on the cross-shell part of the interactions. They may be tested by investigating the neutron configurations of the ^{15}C isotope, like the $(p)^{-1}(sd)^2$ which can be studied using the experimental data of cross-shell states measured via the ^{16}C(d,t) reaction.

Here we present the first study on the negative parity states in ^{15}C produced by a single-neutron pick-up reaction, ^{16}C(d,t)^{15}C. ^{16}C can be understood as a ^{14}C core coupled to two neutrons in the sd orbits, therefore the observed bound states in ^{15}C come from the removal of any

*e-mail: juan.lois.fuentes@usc.es

of these two neutrons. In addition a neutron could be also picked from the internal orbits from the core, producing characteristic p-hole states in the unbound part of the spectrum.

Several experiments have been performed in the past to study the structure of ^{15}C. Different probes were used, such us (d,p) [4], (t,p) [5], ^{13}C(^{18}O,^{16}O)^{15}C [6], breakup [7], and knockout experiments [8], but most of them focused on the positive parity states due to the low probability to produce p-hole states with their respective probes.

2 Experimental details

The experiment was performed at the LISE spectrometer in GANIL. It delivered a 17.2 *A* MeV ^{16}C beam produced by fragmentation with an intensity of 5×10^4 pps into a 1.37 mg/cm^2 CD_2 target. The beam direction was determined by a pair of multiwire proportional chambers, the Z of the beam-like particle was identified with a CHARISSA telescope [9] and the outgoing tritium was detected in 3 MUST2 [10] telescopes which covered the forward laboratory angles of the reaction ($10^o - 40^o$). Due to the low energy of the tritons produced in the reaction, only the first layer of MUST2 was used (300 μm of Si). Identification of the mass of the outgoing particle was performed with the E-ToF technique, using the energy deposit from the stopped particle in the Si and the time between the first beam tracking detector and MUST2. The first layer of MUST2 is a X-Y position sensitive DSSSD with a high granularity (strip step 0.8 mm) allowing hit position measurement and track reconstruction (and incident angle on target) of the incident beam trajectory with the upstream beam detector.

3 Data analysis

The kinetic energy of a particle depends both on its velocity and mass, following equation 1. Therefore, when one plots the residual energy vs ToF different lines appear according to the masses of the particles that reach MUST2. The left panel of figure 1 shows the identification matrix were A = 1, 2, 3, 4 and 16 are observed. Using the beam tracking detectors we can determine the hit position on the target and combining this information with the hit position in MUST2 we can determine the total flight length, which allows us to calculate the mass in an absolute way.

$$E \propto A \times \left(\frac{L}{ToF}\right)^2 \quad (1)$$

The experimental limitations due to the energy resolution and energy straggling in the target produces that the selection of the A = 3 cannot be performed directly with a simple gate on figure 1 or a sharp cut on the 1D mass plot shown in the central panel of the same figure. The right panel of figure 1 shows the evolution of the calculated mass with respect to the E_x (assuming ^{15}C), and it can be observed that at low E_x (which corresponds to low energy also in the left panel of figure 1) the lines are much closer. Then, the contamination of A=4 in the selection gate must be taken into account. In order to gate on A=3, $2\sigma_{A=3}$ and $1\sigma_{A=3}$ as lower and upper limits were imposed in order to minimize the main source of contamination, the penetration of α particles (A=4) in our selection cut. The selection was performed taking into account the evolution of A = 3 with the E_x.

Finally a gate was applied on A=2 and 4 in coincidence with the a Z=6 selection of the beam-like particle (which reduces significantly the α contamination). The result was scaled taking into account the percentage of each mass that sinks into the A=3 gate, being this a way of estimating the contamination in the final E_x spectrum.

The ToF vs E technique just selects the events in mass so the gate could be taking tritons or/and ^{3}He particles but different observations confirm the presence of only the first of them. First of all the Q-value of the ^{16}C(d,^{3}He)^{15}B reaction places its ground state at higher energy than any of states shown in this study, naturally separating both reactions. Moreover CHARISSA is used to gate on Z=6, which vetoes ^{15}B. In addition a detail look at the left panel of figure 1 shows that the line corresponding to A=3 abruptly ends around 9 MeV, the punchthrough of the ^{3}H, which confirms the absence of ^{3}He since it would extend up to 20 MeV.

The evolution of the resolution with the E_x was determined using the nptool simulations package [11], which includes the effects of the straggling of the particles in the target and the different layers from MUST2, the resolution of the detectors and the beam energy spread as well as the beam spot and the geometry of the set up. The simulation yields a resolution which decreases with E_x following a polynomial function where $\sigma = 0.90$ MeV is the value obtained for the ground state and 0.55 MeV for the state at 3 MeV. This fast evolution is produced by the smaller straggling of the tritons produced by states at higher energy.

3.1 Results

After the ^{3}H were selected with the A=3 gate described above, in coincidence with Z=6 for the beam-like particle, the E_x spectrum was built using the missing-mass technique. The spectrum displayed on figure 2 shows the E_x in the center of mass angular range 3°-15° and the fit to the present states. The limitations in energy resolution are responsible for the imperfect separation between the well established ground state and first excited state (0.74 MeV) which energies and spin-parity values are well known. The final fit was performed taking into account the evolution of the E_x resolution obtained using comparison to nptool simulations, fitting the two resonances with a Breit-Wigner convoluted with the experimental resolution. In addition the contribution of the background and the one-neutron (1n) phase space are also considered. The first resonance was already observed in previous experiments and J^π=1/2$^-$ was assigned, finding a natural width of ~

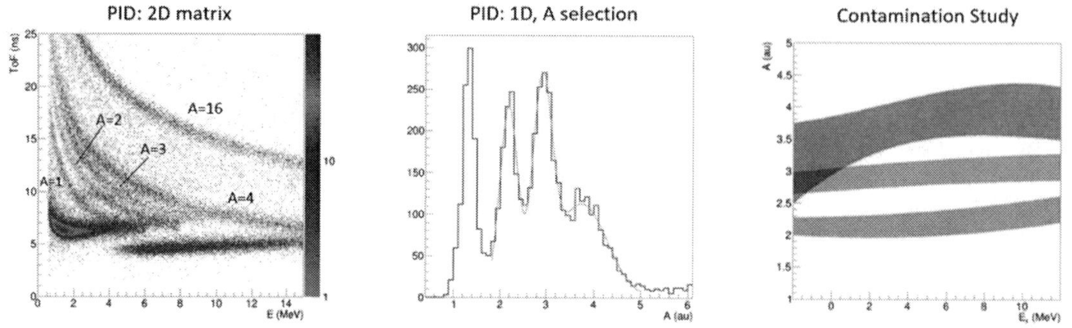

Figure 1. Particle identification plots. The left panel shows the ToF vs the residual energy in the silicon where we can clearly identify the different lines associated to each of the masses that arrive in MUST2. The central plot shows the 1D particle identification showing 4 peaks associated to the light particles. In order to built this plot the E and ToF from the left panel, and the flight length from the target to the hit position in MUST2 were used. The right panel shows evolution of each mass with respect to the E_x of ^{15}C, plotted with its 1σ bands which also evolves with E_x. Studying this plot we can obtain the contamination from A=2 and 4 into our A=3 gate.

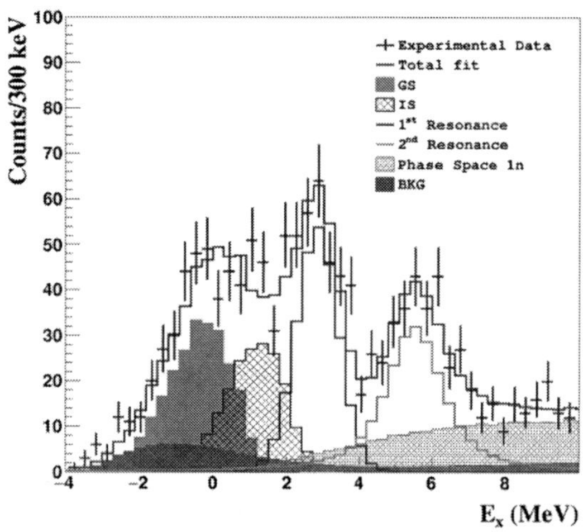

Figure 2. E_x spectrum fitted to the different states observed. In the case of the unbound states a Breit-Wigner fuction convoluted with the experimental resolution is used. The histogram in purple color corresponds to the different sources of background such as the contamination of A=2 and 4 in our A=3 gate. Finally in light blue we also consider the presence of the non-resonant continuum with the 1n phase space.

50 keV. There are works that point out the presence of different states in the region of our second unbound state, but we cannot firmly assign it to any of them. The result from the fitting procedure shows that in addition to the statistical uncertainties there is an overall systematic error in the excitation energy of ±0.35 MeV, which is deduced from the difference of the fit results with respect to the known E_x values of the ground state, first excited state, and first resonance. This error has different origins such as non linearities in the measurement of the energy that could not be properly addressed due to the limited statistics, preventing us from performing an ad-hoc calibration of the states. This systematic error does not have a large impact on our results since our main output from this analysis is the spectroscopic factor of the resonances, which relies on the yield distribution[1] .

Figure 3 presents the angular distributions which were obtained from the individual fits of the spectrum in $\Delta\theta_{CM}$ = 2° slices. They are plotted together with the finite-range Distorted-Wave Born Approximation (DWBA) calculations which are normalized to the data to extract the spectroscopic factors. The DWBA calculations were performed using for the input channel the parameterization of Haixia et al. [12], which was modified to reproduce elastic scattering also measured in the experiment [13, 14]. For the exit channel the optical model potential from Pang et al. [15] was applied. The <d|t> overlap was determined using a potential which reproduced the results obtained from Green's function Monte Carlo calculations [16]. The <^{16}C|^{15}C> overlap was built with a bound neutron wave function using a Wood-Saxon potential with r_0 = 1.25 fm and a = 0.65 fm using a proper depth to reproduce the neutron separation energy.

From the shape of angular distributions the angular momentum of the removed nucleon is deduced, finding consistent results with the previously known assignments for the bound states and the first resonance. In the case of the second resonance both l=1 and l=2 calculations are compared to the data. From the fit, l=2 is more favourable. However, the spectroscopic factor obtained from the fit allows to reject the l=2 since it yields nonphysical occupancy that would exceed the maximum single-particle strength available in the $0d_{5/2}$ orbital. Therefore, an assignment of l=1 was determined for this state. The state can be considered as a hole in the $0p_{3/2}$ orbital since the first resonance almost exhausts the occupation of the $0p_{1/2}$. Previous experiments have already determined the presence of an ℓ=1 state at an energy compatible with our

[1] The impact of the E_x in the single particle cross-section of the second resonance (the only state that we cannot firmly assign to any previous measurement) is orders of magnitude smaller than the quoted uncertainty.

Table 1. Fit results and relative spectroscopic factors compared to theoretical YSOX calculations. The quoted uncertainties are just statistical, the discussion of the systematic ones is included on the text.

E_x (MeV) (t,p) [5]	E_x (MeV)	l	J^π	C^2S	$E_{x,th}$	C^2S_{th}
	(d,t) [This work]				YSOX [3]	
0	-0.30(20)	0	$1/2^+$	0.65(16)	0	0.73
0.744	1.10(25)	2	$5/2^+$	1.35(32)	0.78	1.13
3.105	2.76(7)	1	$1/2^-$	1.74(23)	2.63	1.74
(5.866)	5.41(11)	1	$3/2^-$	1.25(24)	4.97	0.37
		1	$3/2^-$		6.10	0.93

energy measurement, considering the uncertainties, but they assigned it as a $1/2^-$ [5], not being able to exclude $3/2^-$.

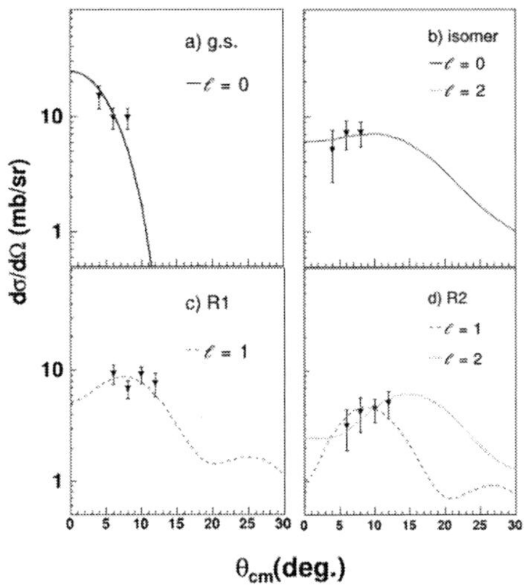

Figure 3. Angular distributions from the different states present in our data. Together with the experimental distributions DWBA calculations are normalized to the data. The shape from the distributions is used to determine the l number, finding consistent results between our data and the previous assignments (a-c). In the case of the second resonance we cannot firmly assign its l value by the shape, since both l=1 and 2 are compatible with our data.

Table 1 contains the compiled results of the fit compared with the states they are assigned and the relative spectroscopic factors which are obtained requiring that the sum of the $1/2^+$ and $5/2^+$ states is equal to 2, which assumes no occupancy of the $0d_{3/2}$. The table also presents the results of shell-model calculations performed with the YSOX interaction [3] which has proven to work well in the CNO neutron-rich region of the nuclear chart.

3.2 Discussion

The results of the spectroscopy of ^{15}C through the (d,t) reaction are presented in table 1. they show valence neutron configurations for the bound states which are in good agreement with previous measurements which used other probes such as knockout [8] and breakup reactions [7]. The C^2S values are also well reproduced by the YSOX interaction.

Above the one-neutron separation threshold the two states assigned l=1 are understood as 2p-1h configurations where two neutrons lay in the sd shell and the hole comes from the $0p_{1/2}$ and the $0p_{3/2}$ orbitals respectively. The spectroscopic factor of the first resonance, which carries most of the single particle strength of the $0p_{1/2}$ orbital, is in very good agreement with the C^2S value predicted by the shell model with the YSOX interaction.

For the second resonance the shell model calculations predict that the strength of the $0p_{3/2}$ orbital is spread over different states, but its two main fragments fall around the position of our resonance. As we observe experimentally the observed width of the second unbound state is compatible with the existence of two states together, an argument that supports that the contributions of the main fragments are being summed in a single state. Theoretical predictions of the shell-model with the YSOX interaction also support this argument since the sum of the C^2S of the two main states of the $0p_{3/2}$ orbital is in good agreement with our experimental result.

References

[1] T. Suzuki, R. Fujimoto and T. Otsuka, Phys. Rev. C 67 (2003) 044302.
[2] T. Suzuki and T. Otsuka, Phys. Rev. C 78 (2008) 061301(R).
[3] C. Yuan et al., Phys. Rev. C 85 (2012) 064324.
[4] J. Goss, Phys. Rev. C 8 (1973) 514-517.
[5] S. Truong, Phys. Rev. C 28 (1983) 977-979.
[6] F. Cappuzzello et al., Phys. Lett. B 711 (2012) 347.
[7] T. Yamaguchi et al., Nuc. Phys. A 724 (2003) 3.
[8] V. Maddalena et al., Phys. Rev. C 63 (2001) 024613.
[9] N. I. Ashwood et al., Phys. Rev. C 70 (2004) 024608.
[10] E. Pollacco et al., Eur. Phys. J. A 25 (2005) 287.
[11] A. Matta et al, J.Phys.G 43 (2016) 4, 045113.
[12] Haixia An et al, Phys. Rev. C 73 (2006) 054605.
[13] X.Pereira-Lopez, PhD Thesis (2016).
[14] X. Pereira-López, Phys. Lett. B 811 (2020) 135939.
[15] D.Y. Pang et al., Phys. Rev. C 79 (2009) 024615.
[16] I. Brida, S.C. Pieper, and R.B. Wiringa, Phys. Rev. C 84 (2011) 024319.

Few-Body Reactions with the Trojan Horse Method

G.G. Rapisarda[1,2,*], *R.* Spartá[3,2], and *A.* Tumino[3,2]
for ASFIN collaboration.

[1]Dipartimento di Fisica e Astronomia "Ettore Majorana," Università degli Studi di Catania, Catania I-95123, Italy
[2]Istituto Nazionale di Fisica Nucleare, Laboratori Nazionali del Sud, I-95123 Catania, Italy
[3]Facoltà di Ingegneria e Architettura, Università degli Studi "Kore", I-94100 Enna, Italy

Abstract. The Trojan Horse method (THM) is a well-established experimental technique to measure nuclear reactions of astrophysical interest avoiding the suppression of the Coulomb barrier affecting experimental direct measurements. In this paper it will describe some of the THM studies involving few-body system of interest for both nuclear physics and nuclear astrophysics, such as the sub-Coulomb proton-proton elastic scattering and the deuteron-deuteron fusion at energies of interest for primordial nucleosynthesis. Moreover, the role of the intercluster motion in nuclei used for THM measurement will be highlight for the discussed physics cases.

1 Introduction

The Trojan Horse Method (THM) (see [1–3] for recent reviews) is a well established experimental technique. Originally introduced for the study of charged-particle-induced nuclear reactions taking place in astrophysical environments, THM has been applied also to investigate the nuclear interaction in few-body systems.

THM, as other indirect methods, has been introduced to overcome the experimental problems related to the study of nuclear reactions relevant for astrophysics. Indeed, in astrophysical environments nuclear reactions take place at very low energies of the order of keV. Such energies are well below the Coulomb barrier of the interacting nuclei, being of the order of MeV. Consequently, the cross section $\sigma(E)$ drops exponentially with decreasing energy, reaching values of 10^{-9} – 10^{-12} barn. In these conditions the experimental evaluation of the cross section is severely hindered and in some cases even beyond present-day technical possibilities. Extrapolation from data at higher energies represents a possible solution, but the exponential decreasing of the cross section could be source of systematic uncertainties. Typically, the cross section is replaced by the astrophysical factor $S(E)$:

$$S(E) = E\sigma(E)\exp(2\pi\eta) \quad (1)$$

because of the weaker energy dependence since the Gamow factor, $\exp(2\pi\eta)$ (η is the Sommerfeld parameter), compensates for the exponential decrease of $\sigma(E)$ [4]. However, extrapolation may lead to considerable uncertainty because contributions to the cross section due to low-energy resonances or tails of broad sub-threshold resonances could be totally missed. Direct reaction measurements that succeeded to perform measurements within the energy range of astrophysical interest [5] have highlighted an enhancement of the cross section at ultra low energies due to atomic electrons surrounding the interaction nuclei referred as *electron screening* effect [6]. The enhancement factor $f_{lab}(E)$ is given by the following equation:

$$f(E)_{lab} = \frac{\sigma_s(E)}{\sigma_b(E)} \sim \exp\left[\pi\eta\left(\frac{U_e}{E}\right)\right] \quad (2)$$

where subscripts s and b indicates respectively the screened and the bare-nucleus cross section, while U_e is the electron screening potential. Since matter in stars is in the form of plasma, a different value of the enhancement factor is obtained with respect to the laboratory case, so the bare nucleus cross section $\sigma_b(E)$ is the experimental parameter required for astrophysical applications. The only way to experimentally get $\sigma_b(E)$ is still through extrapolation procedure, even when experimental data are available down to the astrophysical energy range. In this framework, many indirect methods have been developed with the aim of measuring the bare nucleus cross section for reactions of astrophysical interest, avoiding experimental problems due to Coulomb barrier and electron screening effect.

The THM has been applied in the last decades to the study of charged-particle reactions as well as neutron-induced reactions involved in several astrophysical scenarios (see [7–17] for recent results and reviews). THM has also been employed to study reaction induced by radioactive nuclei [18–21], expanding the THM research field to nuclear processes that take place in explosive astrophysical environments.

In this paper we focus on two THM applications on reactions involving few-body systems. In particular, the study of the *proton-proton* elastic scattering at sub-Coulomb energies to investigate the suppression of the Coulomb field effects, that is, the key feature that makes the THM a unique and powerful experimental approach

*e-mail: grapisarda@lns.infn.it

for the study of nuclear reactions of astrophysical interest. We report also about the study of the ^{2}H(d,p)^{3}H and ^{2}H(d,n)^{3}He channels of two-deuteron fusion process, focusing on the reaction rate evaluation [4], for astrophysical applications, and on the screening potential estimation. In the last section, the results obtained on the study of the inter-cluster motion in nuclei used for THM measurements will be briefly discussed.

2 Trojan Horse Method: basic features

THM is based on the quasi-free (QF) reaction mechanism and allows to extract the cross section of a two-body process $x+A \rightarrow b+B$, at very low center of mass (*cm*) energy performing the measurement of a suitable three-body reaction $a+A \rightarrow b+B+s$ (called Trojan Horse (TH) reaction) in the QF kinematic regime.

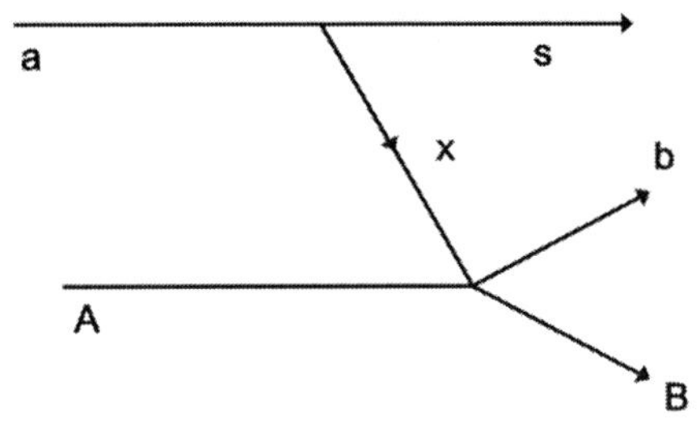

Figure 1. Polar diagram describing the TH reaction $a + A \rightarrow b + B + s$.

The nucleus a, named TH nucleus, is usually a few-body system selected for its high probability to be found in a cluster configuration ($x \oplus s$) and acts as source of virtual projectile/target nucleus. The beam energy is fixed such that the $a + A$ interaction takes place above the $a - A$ Coulomb barrier. In this condition, the TH nucleus is subject to breakup inside the nuclear field of A and the cluster x interacts with A inducing the two-body process of interest, while the cluster s does not perturb the $x - A$ interaction, remaining *spectator* to the binary process. Figure 1 shows the polar diagram describing the TH reaction, where the three-ray vertex refers to the breakup process, while the four-ray vertex to the binary reaction. From an heuristic point of view the hypothesis of breakup within the nuclear field allows the $x - A$ interaction to take place even at zero energy without suffering the Coulomb suppression and the electron screening effect. In a more detailed description, it has to be taken into account that x is a virtual particle, that is $E_x \neq p_x^2/2m_x$. The off-energy shell nature of the $x - A$ interaction will lead to the suppression of the Coulomb field effect [1] in the binary process $x + A \rightarrow b + B$. This is the fundamental feature of the THM for astrophysical applications, moreover, this, as well be shown in sec. 3, opens new perspectives in the investigation of few-body systems interactions.

Despite the high value of the beam energy used in a TH measurament, it is possible to induce the two-body reaction at low energy down to zero. In QF conditions the two-body reaction takes place at the energy E_{QF} given by the following equation:

$$E_{QF} = \frac{m_x}{m_x + m_A} E_A - B_{xs} \tag{3}$$

where the $x-s$ binding energy B_{xs} plays a key role in compensating for the high beam energy required to overcome the $a - A$ Coulomb barrier. E_{QF} is uniquely determined from equation 3, once that the projectile energy is fixed. Actually, the QF process contribution is still high as long as the relative $x-s$ momentum p_{xs} is small compared with the bound state wave number $k_{xs} = \sqrt{2\mu_{xs}B_{xs}}$ [22]. Thus, selecting p_{xs} value within the range $0 \leq p_{xs} < k_{sx}$, it is possible to measure the two body process cross section in a wide E_{cm} range using a single beam energy. From the experimental point of view, this feature represents one of the most important advantages of the THM.

The simplest theoretical approach for THM is given by the plane-wave impulse approximation (PWIA). In PWIA the three-body cross section is factorized as in the following equation:

$$\frac{d^3\sigma}{dE_{cm}d\Omega_b d\Omega_B} \propto \mathrm{KF}|\phi(\mathbf{p}_{xs})|^2 \left(\frac{d\sigma_{xA\rightarrow bB}}{d\Omega}\right)^{HOES} \tag{4}$$

where the first term KF is the *kinematical factor*, which depends on the masses, momenta and angles of the outgoing particles [1], $|\phi(\mathbf{p}_{xs})|^2$ is the momentum distribution for the $x - s$ system given by the squared module of the Fourier transform of the radial wave function describing the $x - s$ relative motion inside a. The last factor is the differential cross section for the two-body process in the *cm* system. The *cm* energy E_{cm} is calculated following the post-collision prescription as:

$$E_{cm} = E_{bB} - Q_{2b}, \tag{5}$$

where E_{bB} is the $b - B$ relative energy and Q_{2b} is the Q-value of the binary reaction.

From equation 4 the two-body cross section can be obtained dividing the experimental three-body cross section by a Monte Carlo simulation of the $\mathrm{KF}|\phi(\mathbf{p}_{xs})|^2$ factor as shown by the following equation:

$$\left(\frac{d\sigma}{d\Omega}\right)_{bB}^{HOES} \propto \frac{d^3\sigma}{dE_{cm}d\Omega_b d\Omega_B} / \mathrm{KF}|\phi(\mathbf{p}_{xs})|^2 \tag{6}$$

The superscript *HOES* underlines that the two-body process $x + A \rightarrow b + B$ is *Half-Off-Energy-Shell*, since the entry channel is off-energy-shell due to the virtual nature of the transferred particle x. At low energies the HOES cross section reproduces the on-energy-shell (OES) one if the correction for the penetrability of the Coulomb barrier is introduced. Both experimental [23] and theoretical [24] works have demonstrated that the only consequence of the HOES nature of binary cross section consists in the suppression of the Coulomb barrier effects.

3 Proton-proton sub-Coulomb elastic scattering via THM

It is known that the $p-p$ scattering at low energy is characterized by an interference effect between the Coulomb scattering amplitude and the nuclear one [25]. In particular, the repulsive Coulomb field and the attractive nuclear one produce a destructive interference. Because of this effect, excitation function for $\vartheta_{cm} = 90°$ is characterized by a deep minimum at E_{cm} =191.2 keV (for more theoretical details refer to [25]). Several experimental measurements within the interference region confirm the theoretical expectations ([26] and references therein). The low-energy proton-proton scattering peculiarity represents an unique opportunity to further test the key THM feature, that is, the suppression of the Coulomb effect in the TH binary process. Indeed, a suppression of the Coulomb amplitude should make the interference minimum disappear and the $p-p$ excitation function should approach the $p-n$ or $n-n$ one. In this framework, $p-p$ elastic scattering around the interference minimum energy has been indirectly measured applying the THM to the three-body reaction ^{2}H(p,pp)n.

Four experimental runs [27–30] have been performed where the ^{2}H(p,pp)n reaction was induced by a proton beams of 5-6 MeV delivered on a CD_2 target, where deuteron was used as source of the virtual proton target. The two outgoing protons were detected in coincidence. The kinematic conditions of each run were changed in order to increase the QF mechanism contribution and to induce the HOES proton-proton elastic scattering at lower energy. In particular, the last two experimental runs [29, 30] allowed to fully cover the interference energy range. More experimental details can be found in [27–30].

Following the THM prescription [1], data analysis required the selection of the events corresponding to the reaction channel of interest produced by QF breakup mechanism. The contribution of the Final State Interaction (FSI) was evaluated to be less than 10% and was subtracted from experimental data [30]. Moreover, data were selected in order to have the cm angle ϑ_{cm} of the two-body process in the range of 90° ± 10°, where the Coulomb-nuclear interference produces the maximum effects. The experimental HOES $p-p$ scattering cross section was obtained using equation 6, by dividing the three-body coincidence yield by the factor $KF|\phi(\mathbf{p}_{xs})|^2$ obtained through a Monte Carlo calculation. $|\phi(\mathbf{p}_{xs})|^2$ was evaluated using the Fourier transform of the s-wave radial Hultén function for the deuteron bound state, where the expected experimental FWHM of the momentum distribution was taken into account [30, 31] (see also sec. 5). The TH $p-p$ excitation function is reported in figure 2 (black triangles) as the weighted average of the experimental data from the four THM runs.

TH data are compared with the OES $p-p$ excitation function integrated over $80° \le \vartheta_{cm} \le 100°$ (black line in figure 2). Experimental data were normalized to the OES cross section close to the $p-p$ Coulomb barrier (500 keV). Impressively, the TH excitation function does not show any interference minimum, being in total disagreement with the OES $p-p$ one below the Coulomb barrier. On the other hand, the trends are comparable above the Coulomb barrier where the process is dominated by the nuclear field. TH data are further compared with the nucleon-nucleon nuclear $l = 0$ cross section described in terms of scattering length a and effective radius r_0. Three different calculations were performed, taking into account a and r_0 values for $p-n$, $n-n$ [32], and nuclear $p-p$ scattering [25]. Results are reported in figure 3 where the green line represents $p-n$ scattering, the red line represents $n-n$ scattering and the blue line the nuclear $p-p$ scattering. The TH cross section, within the experimental errors, is in agreement with the nucleon-nucleon nuclear cross section. This result gives a strong qualitative indication that the TH $p-p$ excitation function is dominated by the nuclear field and that the Coulomb field ef-

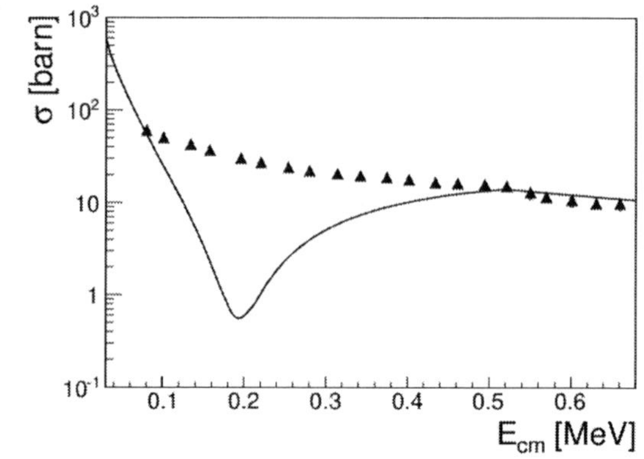

Figure 2. Black triangles represent TH $p-p$ excitation function (weighted average of the four runs) [30]. Black line is the theoretical OES $p-p$ excitation function [25].

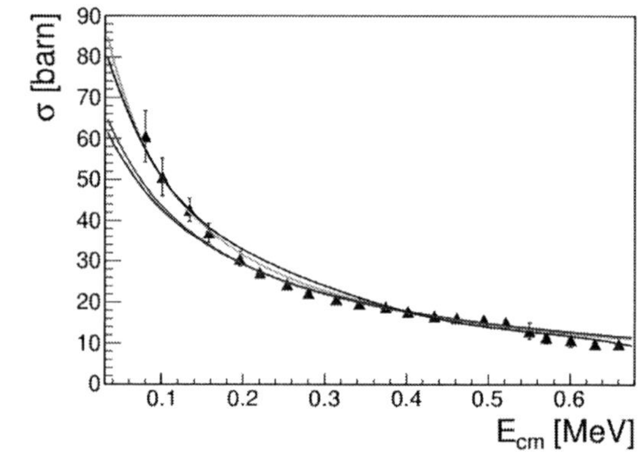

Figure 3. Black triangles represent TH $p-p$ excitation function (weighted average of the four runs). TH data are compared with low energy nucleon-nucleon cross section: green line represents $p-n$ scattering, red line represents $n-n$ scattering and the blue line the nuclear $p-p$ scattering. Black line is the theoretical HOES cross section for the low-energy $p-p$ scattering [29, 30].

fects are suppressed. TH data are further compared with an independent calculation of the theoretical low-energy $p-p$ scattering HOES cross section (black line in figure 3) [29, 30] resulting in very good agreement with the THM data. The obtained result represents a fundamental experimental test that clearly confirms the key THM feature, that is the suppression of the Coulomb effects. Moreover, it provides an interesting result from the nuclear physics point of view since, for the first time, the nuclear part of the $p-p$ interaction has been experimentally investigate at sub-Coulomb energies.

4 The *d+d* reaction studied via the THM

A great experimental effort has been devoted to the study of $p-{}^3H$ and $n-{}^3He$ channels of the $d+d$ reaction induced at sub-Coulomb energies, because of their crucial role for astrophysics and for applied physics. Indeed, ${}^2H(d,p){}^3H$ and ${}^2H(d,n){}^3He$ reactions are in the party of twelve of the Standard Big Bang Nucleosynthesis (SBBN) reactions, being strongly influential for all the primordial abundances [33], and are among the deuterium burning channels in the Pre Main Sequence (PMS) stars [4]. Besides, these reactions are among the few possible to be used in the future fusion power plants, under research at the moment [34, 35]. About the former topic, the energy range of interest spans from 0 to 1.5 MeV in order to have an accurate reaction rate calculation [4], while in the latter case the range of interest is from 0 to 30 keV.
The available data sets obtained from direct measurement show important discrepancies. Furthermore, experimental data are missing in the energy range from 200 keV and 1 MeV. Theoretical calculations deviates each other up to 15%. All these experimental and theoretical uncertainties have an important impact on the evaluation of the reaction rate. In this framework the $d+d$ interaction was studied applying the THM to get a coherent measurement of the cross sections from 30 keV to at least 1.5 MeV, to reduce the experimental errors and to obtain an accurate calculation of the reaction rates.
THM was applied to ${}^2H({}^3He,p{}^3H){}^1H$ and ${}^2H({}^3He,n{}^3He){}^1H$ three-body processes to get the astrophysical factor of the ${}^2H(d,p){}^3H$ and ${}^2H(d,n){}^3He$ respectively [36]. Two experimental runs were performed with a ${}^3He$ beam (at 17 and 18 MeV beam energy) impinging on CD_2 target (with thickness 150 $\mu g/\ cm^2$). ${}^3He$ was used as TH nucleus for its $d \oplus p$ internal structure, as a source of virtual deuteron, while the proton is the spectator. Outgoing particles have been detected in coincidence by means of two couples of telescopes made up of two stages of silicon detectors, allowing also for particles identifications [36]. Experimental data have been analyzed using a more sophisticated theoretical approach, that made use of MPWBA [37]. The desired two-body cross sections can be considered as made of two non-resonant components for the $l = 0$ and $l = 1$ partial waves. The energy dependence of the cross section for the two components can be described by their penetrability factors [38], so that, writing the two channels $p-{}^3H$ and $n-{}^3He$ as $C+c$, the two-body cross section was assumed as:

$$\frac{d\sigma}{dEd\Omega}_{(d+d\to C+c)} = \frac{1}{E_{dd}} \sum_{l=0,1} C_l P_l^2 k_{dd} R T_l(k_{dd}R), \quad (7)$$

where: the relative energy E_{dd} is related to the momentum k_{dd}, to the reduced mass μ_{dd} and to the binding energy B_{dp} of ${}^3He$ by the relation $E_{dd} = \frac{\hbar k_{dd}^2}{2\mu_{dd}} - B_{dp}$; P_l is the Legendre polynomial; R is the channel radius; $T_l(k_{dd}R)$ is the penetrability factor [36].

To apply the MPWBA approach it is necessary to know the values of the scaling factors $C_{l=0,1}$, thus equation 7 entered in equation 6 to fit the experimental three-body coincidence yield, using C_l and the channel radius R as free parameters. The obtained values together with more details on the procedure are reported in [36].

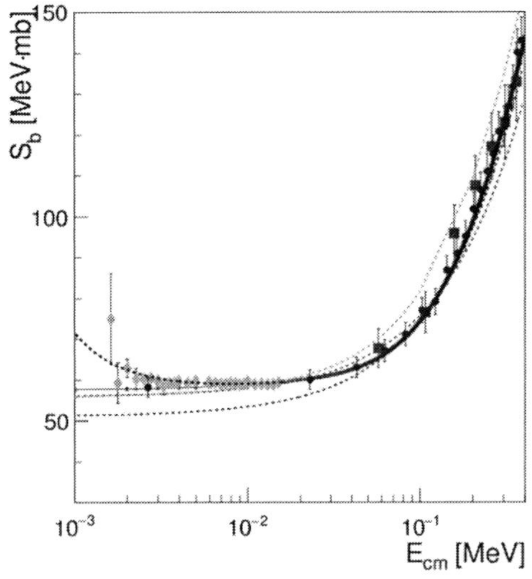

Figure 4. Bare nucleus TH S(E) for $p-{}^3H$ channel. Black dots and blue squares represent TH data from the 18 MeV and 17 MeV runs respectively. Black solid line represents the parameterization of the THM S(E) factor (taking into account the weighted sum of two THM data sets). Low-energy data by [39] are shown as green diamonds. The black dashed line represents the result of the fit to the green diamonds. The red and blue lines are the polynomial fit by [40] and [41], while the green line is the *ab initio* calculation by [42].

Figure 4 and 5 show the bare nucleus TH S(E) factor for both $p-{}^3H$ and $n-{}^3He$ channels compared with low-energy data by [39], and with polynomial fits by [40] and [41] (red and blue dashed lines). In particular, black dots represent the TH astrophysical factor for both channels, obtained from the experiment with beam energy at 18 MeV. Blue squares in figure 4 represent the TH S(E) factor for $p-{}^3H$ channel from the 17 MeV run. Moreover, the black solid lines show the parameterization of the THM S(E) factor (see [36] for details). The two-deuterons interaction is suitable for accurate *ab initio* calculation [42] shown as the green dashed line in figure 4 and 5.

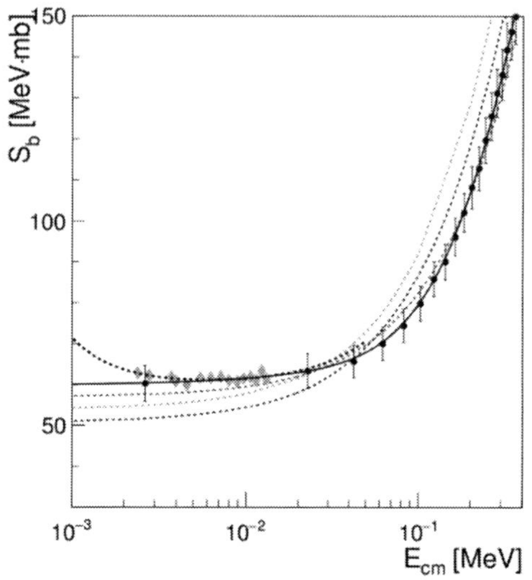

Figure 5. Bare nucleus THM S(E) for n–^{3}He channel. Black dots represent TH data. Black solid line represents the parameterization of the THM S(E) factor. Low-energy data by [39] are shown as green diamonds. The black dashed line represents the result of the fit to the green diamonds. The red and blue lines are the polynomial fit by [40] and [41], while the green line is the *ab initio* calculation by [42].

THM experiments allowed for the first time to measure the S(E) factor in a very wide energy range from 0 to 1.5 MeV. This result enabled to evaluate a very precise reaction rate, with less than 5% error, overall the region of interest for astrophysical scenarios, differing up to 20% with respect to other calculations [38, 40, 41].

Since, as mentioned before, nuclear reactions performed in the laboratory are affected by a different electron screening than in plasma, the obtained bare nucleus cross section below 30 keV provides a very important result also in the framework of study carried on the fusion power prototype plants.

Furthermore, a new estimation of the electron screening potential has been obtained. For both p–^{3}H and n–^{3}He channels, U_e has been evaluated by comparison with low energy data by Greife [39]. For the p–^{3}H channel, U_e has been evaluated in 13.4±0.6 eV, consistent with the adiabatic limit 14 eV. For n–^{3}He channel U_e =11.7±1.6 eV was evaluated, in agreement, as expected, with the result obtained for the p–^{3}H channel, but very discordant with the value given in [39] (U_e= 25±5 eV). More details about the procedure of extraction of the electron screening potential can be found in [36].

5 Study of the inter-cluster motion in nuclei used for THM measurements

A THM measurement gives access to the study of the TH nucleus inter-cluster motion. Indeed, taking into account the PWIA and the equation 4, it is possible to measure the momentum distribution of the TH nucleus. A systematic study has shown a correlation between the FWHM of the TH nucleus experimental momentum distribution and the transferred momentum from particle A to the system $F = b+B$ (see figure 1) [31]. This study considered several nuclei used as TH nucleus, here as an example the case of deuteron is discussed. In figure 6 the FWHM (indicated as W in figure and in equation 8) of the deuteron momentum distribution from different experiments (see [31] for the used data sets) is reported as a function of the transferred momentum q_t [31]. A reduction of the FWHM with respect to the theoretical value given by the Hultén function, that is 58 MeV/c, is evident at low transferred momentum. On the other hand at higher transferred momentum the FWHM approaches the theoretical value. The experimental trend have been reproduced by equation 8

$$W(q_t) = f_0[1 - \exp(-q_t/q_0)] \tag{8}$$

where f_0 = 58 MeV/c is the asymptotic theoretical value of the FWHM and q_0 = 48 ± 2 MeV/c is obtained by fitting several experimental data from literature as reported in [31]. Similar results have been obtained for the ^{3}He case as for other nuclei used in THM measurements as TH nuclei [31].

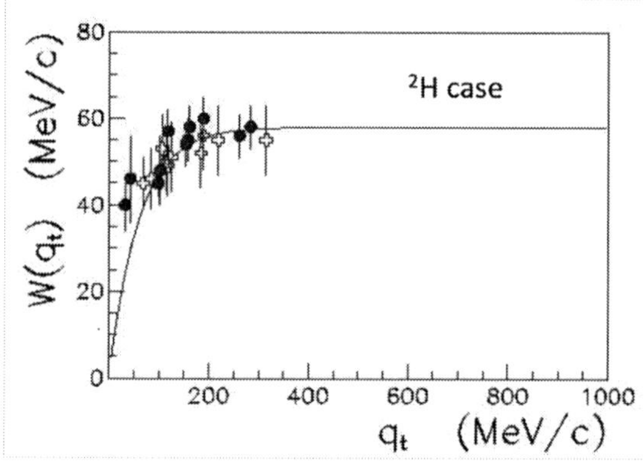

Figure 6. Deuteron experimental momentum distribution FWHM (W) as a function of the transferred momentum q_t [31].

This effect has been explain in a qualitatively way taking into account the onset of the interaction between the spectator and the final-state fragments b and B that produce a destructive interference with the pole diagram that describe the quasi-free mechanism (see figure 1). While for $q_t >> 5k_{xs}$ (with $k_{xs} = \sqrt{(2\mu B_{xs})}$ the contribution of the spectator interaction is negligible, at low energies for $q_t \approx k_{xs}$ this contribution produce a destructive interference reducing the total amplitude of the TH process. This distortion effect is reflected in the reduction of the FWHM of the momentum distribution [31]. Further theoretical analysis are required to better understand this effects. Nevertheless the applicability of the PWIA is still possible once that the experimental FWHM is taken into account.

6 Conclusions

The THM is an indirect experimental technique mainly applied for the study of nuclear reaction of astrophysical interest. On the other hand, the results described in this paper underline the possibility to apply the THM for investigation of few-body system. In particular, it has been shown how a THM measurement gives access to the nuclear part of a nuclear reaction even at sub-Coulomb energies. Further studies are ongoing in order to obtain from THM data the Coulomb free $p-p$ scattering length. Moreover, in the next future THM could be applied to the quasi-free $n+d \rightarrow n+n+p$ reaction, in order to provide a better estimation of the $n-n$ scattering length.

References

[1] C. Spitaleri, M. La Cognata, L. Lamia, R.G. Pizzone, A. Tumino, Eur. Phys. J. A **55**, 161 (2019)
[2] R.G. Pizzone et al., Eur. Phys. J. A, **56**, 283 (2020)
[3] A. Tumino, et al., Ann. Rev. Mod. Phys. **71(1)**, 345–376.(2021)
[4] C.E. Rolfs and W.S. Rodney, *Cauldrons in the Cosmos*, The University of Chicago Press, (1988)
[5] R. Bonetti, et al., Phys. Rev. Lett.**82**, 5205 (1999)
[6] H.J. Assenbaum, K. Langanke, C. Rolfs, Z. Phys. A **327**, 461 (1987)
[7] G.G. Rapisarda et al., Eur. Phys. J. A, **54**, 189, (2018)
[8] A. Cvetinovic et al., Phys. Rev. C, **97**, 065801 (2018)
[9] S. Palmerini et al., Eur. Phys. J Plus, **136**, 898 (2021)
[10] M. La Cognata et al., Astrophys. J., **941**, 96 (2022)
[11] G.L. Guardo et al., Eur. Phys. J. A, **55**, 211 (2019)
[12] R.G. Pizzone et al., Eur. Phys. J. A, **56**, 199 (2020)
[13] R. Spartá et al., Eur. Phys. J. A, **57**, 170 (2021)
[14] R. Spartá et al., Front. Astron. Space Sci., **7**, 560149, (2020)
[15] G.G. Rapisarda et al., Front. Astron. Space Sci., **7**, 589240, (2021)
[16] M.L. Sergi et al., Universe, **8**, 128 (2022)
[17] R. Spartá et al., Frontiers in Physics, **10**, 896011 (2022)
[18] R.G. Pizzone, et al., Eur. Phys. J. A, **52**, 24 (2016)
[19] M. La Cognata, et al., Astrophys. J., **846**, 65 (2017)
[20] L. Lamia, et al., Astrophys. J., **879**, 23 (2019)
[21] S. Hayakawa, et al., Astrophys. J. Lett.,**915**, L13 (2021)
[22] I.S. Shapiro, *Interaction of high-energy particles with nuclei.* International School of Physics Enrico Fermi, Course 38, edited by E.O. Ericson, p. 210, Academitices, New York (1967)
[23] M. Gulino, et al., J. Phys. G, **37**, 125105 (2010)
[24] A. M. Mukhamedzhanov, Phys. Rev. C, **84**, 044616 (2011)
[25] J. D. Jackson, J. M. Blatt, Rev. Mod. Phys.**22**, 77, (1950)
[26] H. Dombrowski, A. Khoukaz, R. Santo, Nucl. Phys. A, **619**, 97, (1997)
[27] M. Pellegriti, et al., Prog. Theor. Phys. Supp., **154**, 349, (2004)
[28] A. Tumino, et al., Nucl. Phys. A, **787**, 337, (2007)
[29] A. Tumino, et al., Phys. Rev. Lett., **98**, 252502, (2007)
[30] A. Tumino, et al., Phys. Rev. C, **78**, 064001 (2008)
[31] R.G. Pizzone, et al., Phys. Rev. C, **80**, 025807, (2009)
[32] G. A. Miller, B. M. K. Nefkens, I. Šlaus,Phys. Rep., **194**, 1, (1990)
[33] G. Steigman, **15**, 01, pp. 1-35 (2006)
[34] J. Lindl, Physics of Plasmas **2**, 3933 (1995)
[35] C. Yamanaka, *Inertial confinement fusion: The quest for ignition and energy gain using indirect drive*,, Nuclear Fusion, **39**, 6 (1999)
[36] A. Tumino et al., Astrophys. J.**785**, 96, (2014)
[37] S. Typel and G. Baur, AnPhy **203**, 228 (2003)
[38] P. Descouvemont et al., ADNDT **88**, 203 (2004)
[39] U. Greife et al., Z. Phys. A **351**, 107-112 (1995)
[40] C. Angulo et al., Nucl. Phys. A**656**, 3-183 (1999)
[41] R. H. Cyburt, Phys. Rev. D **70**, 023505 (2004)
[42] K. Arai et al., Phys. Rev. Letters **107**, 132502 (2011)

EPJ Web of Conferences **290**, 10001 (2023) https://doi.org/10.1051/epjconf/202329010001
EuNPC 2022

Neutrino oscillation anomalies

Christoph Andreas Ternes[1,2]*

[1]Istituto Nazionale di Fisica Nucleare (INFN), Sezione di Torino, Via P. Giuria 1, I–10125 Torino, Italy
[2]Dipartimento di Fisica, Università di Torino, via P. Giuria 1, I–10125 Torino, Italy

Abstract. I give an overview of some of the several anomalies appearing in neutrino oscillation experiments, setting particular focus to the reactor antineutrino anomaly and the Gallium anomaly. I will discuss these two anomalies in some detail and, in particular, compare their explanation due to neutrino oscillations in presence of a light sterile neutrino among each other and also with the bounds from the analyses of reactor spectral ratio data, β-decay data, and solar neutrino data.

1 Introduction

The possible existence of light sterile neutrinos is a hot topic of current research in high-energy physics. It was motivated by anomalies found in short-baseline neutrino oscillation experiments: the Gallium Anomaly, the Reactor Antineutrino Anomaly, and the LSND and MiniBooNE anomalies (see the reviews in Refs. [1–4]). Most puzzling is the large Gallium Anomaly, which has been recently revived by the results of the BEST experiment [5, 6]. Here, we discuss the status of short-baseline ν_e and $\bar{\nu}_e$ disappearance and we compare the neutrino oscillation explanation of the Gallium Anomaly with the constraints from other experiments.

The standard paradigm in the phenomenology of massive neutrinos is the three-neutrino mixing scheme in which the three well known active flavor neutrinos ν_e, ν_μ, and ν_τ take part in the weak interactions of the Standard Model and are unitary superpositions of three massive neutrinos ν_1, ν_2, and ν_3 with respective masses m_1, m_2, and m_3. The two independent squared-mass differences $\Delta m^2_{21} \approx 7.4 \times 10^{-5}\,\text{eV}^2$ and $|\Delta m^2_{31}| \approx 2.5 \times 10^{-3}\,\text{eV}^2$ (with $\Delta m^2_{kj} \equiv m^2_k - m^2_j$) generate the oscillations observed in solar, atmospheric and long-baseline neutrino oscillation experiments (see for example Ref. [7]). Short-baseline (SBL) oscillations that could explain the Reactor Antineutrino Anomaly and the Gallium Anomaly require the existence of at least one additional squared-mass difference $\Delta m^2_{\text{SBL}} \gtrsim 1\,\text{eV}^2$. In the minimal 3+1 scenario that we consider here there is a non-standard massive neutrino ν_4 with mass $m_4 \gtrsim 1\,\text{eV}$ which generates the short-baseline squared-mass difference $\Delta m^2_{\text{SBL}} \simeq \Delta m^2_{41}$. In the flavor basis, the new neutrino corresponds to a sterile neutrino ν_s which does not take part in the weak interactions of the Standard Model (see the reviews in Refs. [1–4]). In this framework, the effective short-baseline survival probability of electron neutrinos and antineutrinos relevant for reactor and Gallium experiments is given by $P_{ee} \simeq 1 - \sin^2 2\vartheta_{ee} \sin^2\left(\frac{\Delta m^2_{41} L}{4E}\right)$. The effective mixing angle ϑ_{ee} depends on the element U_{e4} of the 4×4 mixing matrix U through the relation $\sin^2 2\vartheta_{ee} = 4|U_{e4}|^2(1 - |U_{e4}|^2)$.

Currently, the Reactor Antineutrino Anomaly is regarded to be resolved or, at least, diminished with the new refinements of reactor flux models [8, 9], but the Gallium Anomaly is reinforced by the new measurements of the BEST experiment [5, 6]. Therefore, it is desirable to pay special attention to the Gallium Anomaly, and look for possible viable solutions.

2 The Gallium Anomaly

The Gallium Anomaly was originally [10–12] a deficit of events observed in the GALLEX [13–15] and SAGE [10, 16–18] source experiments aimed at testing the solar neutrino detection done in these experiments through the process $\nu_e + {}^{71}\text{Ga} \to e^- + {}^{71}\text{Ge}$. Two source experiments have been done by the GALLEX collaboration using an intense artificial ${}^{51}\text{Cr}$ radioactive source placed inside the detector. This source emitted electron neutrinos through the electron capture (EC) process $e^- + {}^{51}\text{Cr} \to {}^{51}\text{V} + \nu_e$. The SAGE collaboration performed a source experiment with a ${}^{51}\text{Cr}$ radioactive source and another with a ${}^{37}\text{Ar}$ radioactive source, which emitted electron neutrinos through the EC process $e^- + {}^{37}\text{Ar} \to {}^{37}\text{Cl} + \nu_e$. The deficits of the observed rates with respect to the rates calculated from the well-measured activity of the sources and different cross sections for the detection process have been discussed in many papers. The measurements from GALLEX and SAGE have been recently confirmed in the BEST experiment [5, 6], which was using also a ${}^{51}\text{Cr}$ radioactive source.

The determination of the significance of the Gallium anomaly depends on the theoretical prediction of the detection cross section for which there are several model calculations. The difference between these cross section models is the contribution to the cross section coming

*ternes@to.infn.it

from the transitions from the ground state of ^{71}Ga to excited states of ^{71}Ge. The ground-state to ground-state cross section is known with a very small uncertainty from the measured lifetime of ^{71}Ge [19], whereas the transitions to the excited states are obtained from theoretical calculations. These relative contributions are expected to be at the few % level, with the exact value depending on the cross section model. The cross section models that we consider here are a "Ground-state model" where no contribution from the excited states is considered and the calculations from Bahcall [19], Kostensalo et al. [20] and Semenov [21]. It has been shown that the statistical significance of the Gallium Anomaly is large, about 5-6σ, for all the cross section models [22, 23]. The Gallium Anomaly can be explained by short-baseline neutrino oscillations in the 3+1 active-sterile neutrino mixing framework described briefly in Section 1. Fig. 1 shows the 2σ allowed regions in the ($\sin^2 2\vartheta_{ee}, \Delta m^2_{41}$) plane obtained from the analyses of the Gallium data with the different cross section models. One can see that in all cases there is a clear indication of a relatively large value of $\sin^2 2\vartheta_{ee}$ for $\Delta m^2_{41} \gtrsim 0.6\,\mathrm{eV}^2$. The minimum 2$\sigma$ value of $\sin^2 2\vartheta_{ee}$ is about 0.14 in the case of the Ground State model. For the other cross section models the allowed regions lie at even larger values of $\sin^2 2\vartheta_{ee}$. The Kostensalo Shell Model cross section requires values of $\sin^2 2\vartheta_{ee}$ that are only slightly larger than those of the Ground State cross section, $\sin^2 2\vartheta_{ee} \gtrsim 0.17$, because of the small contribution of the transitions to the excited states of ^{71}Ge in this cross section model. The Bahcall cross section implies $\sin^2 2\vartheta_{ee} \gtrsim 0.17$, and the Semenov cross sections give $\sin^2 2\vartheta_{ee} \gtrsim 0.23$.

3 The reactor rates constraints

The phenomenology of reactor neutrinos focused on the Reactor Antineutrino Anomaly in 2011 [24], after the reevaluations of the predicted reactor antineutrino fluxes by Mueller *et al.* [25] and Huber [26] (HM model), which led to a deficit of about 2.5σ of observed event rates in short-baseline reactor neutrino experiments with respect to the predictions. Recent theoretical and experimental developments led to new reactor flux models with enhanced (HKSS model of Hayen, Kostensalo, Severijns, and Suhonen [27]) or diminished (EF model of Estienne, Fallot *et al* [28] and KI model of Kopeikin *et al.* [29]) Reactor Antineutrino Anomaly. In particular, the EF and KI flux models give average ratios of observed and predicted events that are only about 1σ smaller than unity, hinting at the demise of the Reactor Antineutrino Anomaly [8, 9]. Here we use the results of the analysis in Ref. [9], where it was shown that all the reactor flux models imply upper bounds for the 3+1 active-sterile neutrino mixing parameter $\sin^2 2\vartheta_{ee}$ that are smaller than about 0.25 at 2σ. This bound is in tension with the large values of $\sin^2 2\vartheta_{ee}$ which are required to explain the Gallium Anomaly, as discussed in Section 2.

Figure 1 shows a comparison of the 2σ allowed regions in the ($\sin^2 2\vartheta_{ee}, \Delta m^2_{41}$) plane obtained from the analyses of reactor rates and Gallium data. One can see that

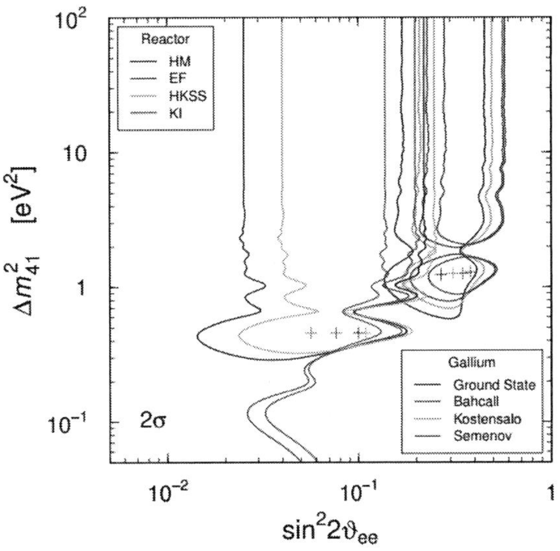

Figure 1. Contours delimiting the 2σ allowed regions in the ($\sin^2 2\vartheta_{ee}, \Delta m^2_{41}$) plane obtained from the Gallium data with different cross sections compared with those obtained in Ref. [9] from the analysis of short-baseline reactor rate data. The best-fit points are indicated by crosses.

there is only a small overlap of the reactor and Gallium regions for some models.

Table 2 of Ref. [22] shows the χ^2 difference $\Delta\chi^2_{\mathrm{PG}}$ of the parameter goodness of fit test [30] between the reactor rates and the Gallium data corresponding to the models in Fig. 1. Also shown are the corresponding values of the parameter goodness of fit $\mathrm{GoF_{PG}}$, which quantifies the tension between the fits of reactor rates and Gallium data in the different models. Considering the extreme case of Ground State Gallium cross section model, the largest tension is obtained with the KI reactor flux model ($\mathrm{GoF_{PG}} = 0.26\%$). Indeed, one can see from Fig. 1 that the corresponding 2σ allowed regions have only a very marginal overlap for $\Delta m^2_{41} \simeq 1\,\mathrm{eV}^2$.

The more realistic Bahcall, Kostensalo, and Semenov cross section models give larger tensions between the reactor rates and Gallium data. If we consider as severe a tension with a $\mathrm{GoF_{PG}}$ smaller than 1%, for the Bahcall and Kostensalo cross section models there is a severe tension between Gallium data and reactor rates for all reactor flux models, except the HKSS model, while for the Semenov cross section model there is a severe tension for all of the reactor flux models. One can also notice that the KI reactor flux model gives the maximal tension for all Gallium cross section models and it is always severe. The EF reactor flux model gives a severe tension for all the Gallium cross section models, except for the Ground State model, where it is a marginal 1.1%.

Since the EF and KI reactor flux models are those that may have solved the Reactor Antineutrino Anomaly and are currently believed to represent reliable replacements of the standard HM model, we conclude that the tension between Gallium data and reactor rates is a serious issue in the framework of 3+1 active-sterile neutrino oscillations.

4 Short-baseline reactor spectral ratios

Short-baseline oscillations of reactor antineutrinos can be probed in a model-independent way by comparing the rates or spectra measured at different distances from the reactor antineutrino source. This is the approach adopted by the recent DANSS [31, 32], PROSPECT [33, 34], and STEREO [35, 36] experiments. Important results have been obtained also by the NEOS collaboration [37] and by a joint analysis of the RENO and NEOS collaborations [38]. In 2017 [37] the NEOS collaboration published the results of the comparison of NEOS data at about 24 m from the reactor source with the prediction obtained from the neutrino flux measured in the Daya Bay experiment [39] at a distance of about 550 m from the reactor source, where the short-baseline oscillations are averaged. Recently, the RENO and NEOS collaborations published a joint paper [38] with the results of the comparison of NEOS data with the prediction obtained from the neutrino flux measured in the RENO experiment at a distance of about 419 m from the reactor source, which is in the same reactor complex of NEOS. This comparison allowed the RENO and NEOS collaborations to reduce the systematic uncertainties with respect to the NEOS/Daya Bay analysis.

In the following we consider both the NEOS/Daya Bay and NEOS/RENO data, because it is obviously unknown which one of the Daya Bay and RENO neutrino spectra is more accurate. Note that in this analysis we do not consider the controversial results of the Neutrino-4 experiment: the Neutrino-4 collaboration claimed a 2.9σ evidence of short-baseline neutrino oscillations with large mixing ($\sin^2 2\vartheta_{ee} = 0.36 \pm 0.12$) at $\Delta m^2_{41} = 7.3 \pm 1.17$ [40]. However, these results were criticized in Refs. [41–44].

In the following we discuss the compatibility of the results of the fits of reactor spectral ratio data with the neutrino oscillation interpretation of the Gallium data and we perform combined fits with other experimental results. For convenience, we introduce the following notation:

RSRF(N/DB) combined reactor spectral ratio fit of the NEOS/Daya Bay, DANSS, PROSPECT, STEREO, and Bugey-3 data.

RSRF(N/R) combined reactor spectral ratio fit of the NEOS/RENO, DANSS, PROSPECT, STEREO, and Bugey-3 data.

As shown in Fig. 2, these combined fits of the data of reactor spectral ratio experiments favor short-baseline neutrino oscillations at values of Δm^2_{41} which are compatible with the neutrino oscillation interpretation of the Gallium data discussed in Section 2, but the required values of the mixing angle are smaller. Therefore, there is again a tension between the results of reactor spectral ratio experiments and those of the Gallium experiments. For all the four Gallium detection cross section models, the parameter goodness of fit is well below 1% for RSRF(N/DB),

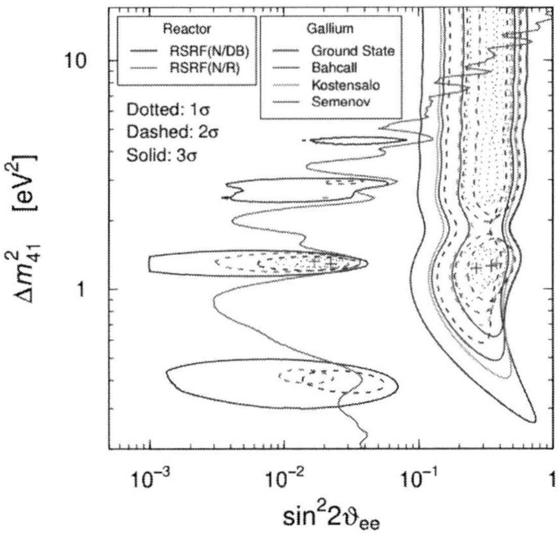

Figure 2. Contours delimiting the 1σ, 2σ, and 3σ allowed regions in the $(\sin^2 2\vartheta_{ee}, \Delta m^2_{41})$ plane obtained from the Gallium data with different cross sections compared with those obtained from the two analyses of reactor spectral ratio data discussed in the text. The best-fit points are indicated by crosses.

whereas it is slightly above 1% for RSRF(N/R) as indicated in Tab. 3 of Ref. [22]. The larger compatibility of RSRF(N/R) with the Gallium data may seem contradictory with the smaller statistical significance of short-baseline neutrino oscillations of RSRF(N/R) (2.6σ) with respect to RSRF(N/DB) (3.1σ). However, one can see from Fig. 2 that the 3σ allowed region of RSRF(N/R) extends to large values of $\sin^2 2\vartheta_{ee}$ for large values of Δm^2_{41}, leading to a relative compatibility with the Gallium allowed regions. On the other hand, the 3σ allowed regions of RSRF(N/DB) are closed and lie at values of $\sin^2 2\vartheta_{ee}$ that are incompatible with the Gallium allowed regions.

5 Global ν_e and $\bar{\nu}_e$ disappearance analysis

In this Section we present the results of the global analysis of the ν_e and $\bar{\nu}_e$ disappearance data in the framework of 3+1 active-sterile neutrino mixing. The data that we consider are the reactor rates and the reactor spectral ratio data discussed above. We include also the Tritium limits and the solar bound discussed in detail in Ref. [22]. We will discuss the global tension between the data that we consider and the Gallium data if the Gallium Anomaly is considered as due to 3+1 active-sterile neutrino mixing. Since the tension is very strong, we cannot include the Gallium data in a global fit in the framework of 3+1 active-sterile neutrino mixing, and we presume that the Gallium Anomaly is due to other reasons.

The results of the global fits that we obtained with different data sets (NEOS/Daya Bay or NEOS/RENO) and different reactor flux models are listed in Tab. 1. One can see that the goodness of fit is high. There is a 3.1–3.3σ

	Global RSRF(N/DB) Fit			
	HM	HKSS	EF	KI
χ^2_{min}	393.5	395.2	391.2	391.4
GoF	43%	40%	46%	46%
$(\sin^2 2\vartheta_{ee})_{\text{b.f.}}$	0.022	0.022	0.022	0.022
$(\Delta m^2_{41})_{\text{b.f.}}/\text{eV}^2$	1.29	1.29	1.29	1.29
$\Delta\chi^2_{4\nu-3\nu}$	13.8	14.1	12.6	12.9
$n\sigma_{4\nu-3\nu}$	3.3	3.3	3.1	3.2
	Global RSRF(N/R) Fit			
	HM	HKSS	EF	KI
χ^2_{min}	386.5	388.3	384.0	384.2
GoF	53%	50%	56%	56%
$(\sin^2 2\vartheta_{ee})_{\text{b.f.}}$	0.017	0.019	0.017	0.017
$(\Delta m^2_{41})_{\text{b.f.}}/\text{eV}^2$	1.32	1.32	1.32	1.32
$\Delta\chi^2_{4\nu-3\nu}$	10.1	10.3	9.1	9.3
$n\sigma_{4\nu-3\nu}$	2.7	2.8	2.6	2.6

Table 1. Results of the global ν_e and $\bar{\nu}_e$ disappearance fits obtained using NEOS/Daya Bay (RSRF(N/DB)) or NEOS/RENO (RSRF(N/R)) data. The column titles HM, HKSS, EF, and KI refer to the four reactor neutrino fluxes discussed in Section 3. The tables show the minimum value χ^2_{min} of χ^2, the goodness of fit (GoF), the best-fit values of $\sin^2 2\vartheta_{ee}$ and Δm^2_{41}, the χ^2 difference $\Delta\chi^2_{4\nu-3\nu}$ between the 3+1 4ν fit and the 3ν fit, and the statistical significance ($n\sigma_{4\nu-3\nu}$) of the corresponding indication in favor of 3+1 4ν mixing.

indication in favor of 3+1 active-sterile neutrino mixing in the global fits with the NEOS/Daya Bay data. The indication decreases to 2.6–2.8σ if the NEOS/RENO are used. The values of the best-fit points are in all cases around $\sin^2 2\vartheta_{ee} \simeq 0.02$ and $\Delta m^2_{41} \simeq 1.3\ \text{eV}^2$.

Figure 3 shows the 2σ and 3σ allowed regions in the $(\sin^2 2\vartheta_{ee}, \Delta m^2_{41})$ plane obtained from the global fits with different neutrino flux models and considering either the NEOS/Daya Bay [37] spectral ratio (Figs. 3(a) and 3(c)) or the NEOS/RENO [38] spectral ratio (Figs. 3(b) and 3(d)).

The restrictions of the solar neutrino constraint increase the tension with the results of the Gallium experiments in the framework of 3+1 active-sterile neutrino mixing. From Tab. 8 of Ref. [22], one can see that the global fit gives values of parameter goodness of fit that are well below 1% for all the fits with different data combinations.

In Fig. 3 one can notice the curious approximate coincidence of the values of Δm^2_{41} for the global best-fit points ($(\Delta m^2_{41})_{\text{b.f.}} = 1.3\,\text{eV}^2$) and the Gallium best-fit points ($(\Delta m^2_{41})_{\text{b.f.}} = 1.3\,\text{eV}^2$ for Bahcall, Kostensalo and Semenov, and $(\Delta m^2_{41})_{\text{b.f.}} = 1.2\,\text{eV}^2$ for Ground State). This coincidence has no meaning in the framework of 3+1 active-sterile neutrino mixing, because the best-fit values of the unique mixing parameter $\sin^2 2\vartheta_{ee}$ are too different and the 3σ allowed regions are well separated for $\Delta m^2_{41} \approx 1.3\,\text{eV}^2$.

6 Summary and conclusions

We have discussed in a systematic way the results of ν_e and $\bar{\nu}_e$ disappearance experiments which are relevant for the hypothetical existence of short-baseline neutrino oscillations due to 3+1 active-sterile neutrino mixing. We started in Section 2 with the analysis of the results of the Gallium source experiments, motivated by the recent results of the BEST experiment [5, 6], which revived the Gallium Anomaly confirming the results of the GALLEX [13–15] and SAGE [10, 16–18] experiments. We have shown that the explanation of the Gallium Anomaly in the framework of 3+1 active-sterile neutrino mixing requires rather large values of the mixing between ν_e and the new massive neutrino ν_4 for all the cross section models of neutrino detection in the Gallium source experiments. This means that ν_4 is not almost entirely sterile, as it would be required for considering 3+1 active-sterile mixing as a perturbation of standard three-neutrino mixing. This is required for the explanation of the neutrino oscillations observed in solar, atmospheric and long-baseline neutrino oscillation experiments.

We also considered the results of reactor neutrino experiments and we presented the results of 3+1 fits of the measured rates and the ratios of spectra measured at different distances. In Section 3, we have shown that the measured reactor rates imply upper bounds for active-sterile mixing that are in tension with the results of the Gallium experiments for all the cross section models of neutrino detection in the Gallium source experiments and all the reactor neutrino flux models.

In Section 4, we presented the results of a global fit of the most recent available data of the reactor neutrino experiments which measured the ratio of the neutrino spectra at different distances in order to probe neutrino oscillations independently from the absolute values of the neutrino fluxes. We have shown that the results depend significantly on the choice to consider either the NEOS/Daya Bay [37] or NEOS/RENO [38] spectral ratio data. In the case of the NEOS/Daya Bay, there is a 3.1σ indication in favor of short-baseline oscillations with best-fit parameter values $\sin^2 2\vartheta_{ee} = 0.022$ and $\Delta m^2_{41} = 1.29\,\text{eV}^2$, which is driven by the overlap of the surrounding allowed regions of NEOS/Daya Bay and DANSS [32]. On the other hand, the overlap of the NEOS/RENO and DANSS allowed regions is smaller and leads to an indication in favor of short-baseline oscillations of only 2.6σ with best-fit parameters values $\sin^2 2\vartheta_{ee} = 0.017$ and $\Delta m^2_{41} = 1.32\,\text{eV}^2$. Although the NEOS/RENO comparison may be favored by the smaller systematic uncertainties which were estimated by the NEOS and RENO collaborations using the fact that the two experiments detect neutrino fluxes from similar reactors in the same complex, we think that we cannot dismiss the NEOS/Daya Bay data, which should be considered as a different measurement based on the Daya Bay neutrino flux measurement. Since we cannot combine the two measurements because the NEOS data would be double counted, we remain with the ambiguity of the two different results which hopefully will be solved by future measurements.

We have also shown that the results of the reactor spectral ratio experiments are in tension with the neutrino oscillation explanation of the Gallium Anomaly and the tension is stronger when the NEOS/Daya Bay data are con-

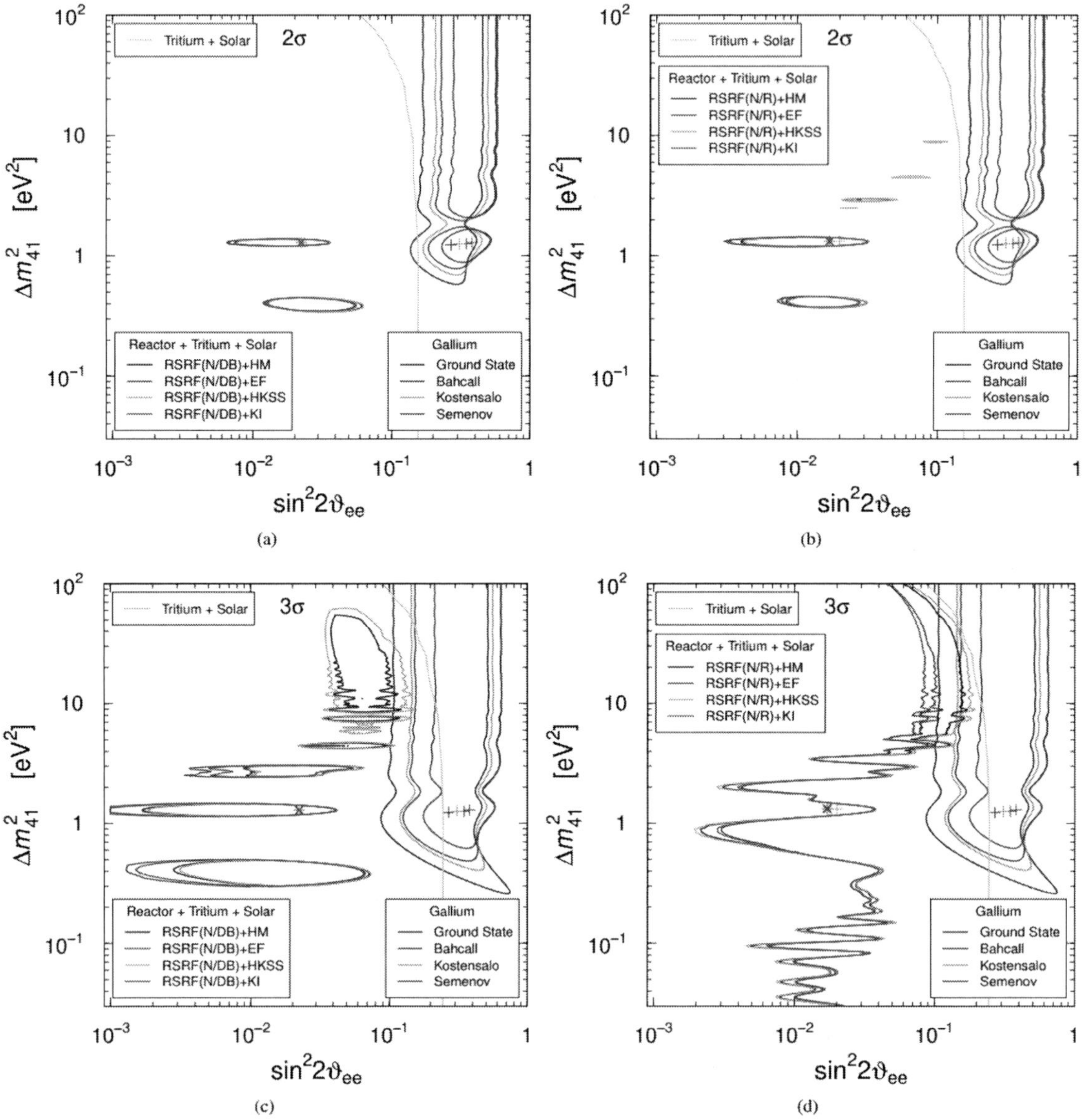

Figure 3. Comparison of the contours delimiting the [(a) and (b)] 2σ and [(c) and (d)] 3σ allowed regions in the $(\sin^2 2\vartheta_{ee}, \Delta m^2_{41})$ plane obtained from the combined analysis of the data of the reactor rate experiments with different flux models, the spectral ratio experiments,reactor the Tritium experiments, and the solar bound with those obtained from the Gallium data with different cross sections. Also shown is the 3σ bound obtained from the combination of the Tritium and solar bounds. The figures differ by the use of [(a) and (c)] NEOS/Daya Bay [37] or [(b) and (d)] NEOS/RENO [38] spectral ratio data. The best-fit points are indicated by crosses.

sidered, with about 0.15% parameter goodness of fit for all the Gallium detection cross section models, whereas considering the NEOS/RENO data we obtain about 1.3%. Finally, we discussed in Section 5 the results of the global fit of the ν_e and $\bar{\nu}_e$ disappearance data. We have shown that the tension is dramatic in the case of the global fit, with values of the parameter goodness of fit well below 1% for all the cases with different data choices (NEOS/Daya Bay or NEOS/RENO), different reactor flux models, and different Gallium detection cross sections.

In conclusion, we think that the results presented in Ref. [22] show the present status of our knowledge on short-baseline ν_e and $\bar{\nu}_e$ disappearance in the framework of 3+1 active-sterile neutrino mixing and the dramatic tension between the the neutrino oscillation explanation of the Gallium Anomaly and the results of the other experiments. We conclude that it is very likely that the Gallium Anomaly is not due to neutrino oscillations and some other explanation must be found.

Acknowledgments

Work supported by the research grant "The Dark Universe: A Synergic Multimessenger Approach" number 2017X7X85K under the program "PRIN 2017" funded by the Italian Ministero dell'Istruzione, Università e della Ricerca (MIUR). C.A.T. also acknowledges support from *Departments of Excellence* grant awarded by MIUR and the research grant *TAsP (Theoretical Astroparticle Physics)* funded by Istituto Nazionale di Fisica Nucleare (INFN).

References

[1] C. Giunti, T. Lasserre, Ann. Rev. Nucl. Part. Sci. **69**, 163 (2019), arXiv:1901.08330
[2] A. Diaz, C. Arguelles, G. Collin, J. Conrad, M. Shaevitz, Phys.Rept. **884**, 1 (2020), arXiv:1906.00045
[3] S. Boser, C. Buck, C. Giunti, J. Lesgourgues, L. Ludhova, S. Mertens, A. Schukraft, M. Wurm, Prog.Part.Nucl.Phys. **111**, 103736 (2020), arXiv:1906.01739
[4] B. Dasgupta, J. Kopp, Phys.Rept. **928**, 63 (2021), arXiv:2106.05913
[5] V. Barinov et al. (BEST), Phys.Rev.Lett. **128**, 232501 (2022), arXiv:2109.11482
[6] V.V. Barinov et al. (2022), 2201.07364
[7] P.F. de Salas, D.V. Forero, S. Gariazzo, P. Martinez-Mirave, O. Mena, C.A. Ternes, M. Tortola, J.W.F. Valle, JHEP **2021**, 071 (2020), arXiv:2006.11237
[8] J.M. Berryman, P. Huber, JHEP **2101**, 167 (2021), arXiv:2005.01756
[9] C. Giunti, Y. Li, C. Ternes, Z. Xin, Phys.Lett.B **829**, 137054 (2022), arXiv:2110.06820
[10] J.N. Abdurashitov et al. (SAGE), Phys. Rev. **C73**, 045805 (2006), nucl-ex/0512041
[11] M. Laveder, Nucl. Phys. Proc. Suppl. **168**, 344 (2007)
[12] C. Giunti, M. Laveder, Mod. Phys. Lett. **A22**, 2499 (2007), hep-ph/0610352
[13] P. Anselmann et al. (GALLEX), Phys. Lett. **B342**, 440 (1995)
[14] W. Hampel et al. (GALLEX), Phys. Lett. **B420**, 114 (1998)
[15] F. Kaether, W. Hampel, G. Heusser, J. Kiko, T. Kirsten, Phys. Lett. B **685**, 47 (2010), 1001.2731
[16] J.N. Abdurashitov et al. (SAGE), Phys. Rev. Lett. **77**, 4708 (1996)
[17] J.N. Abdurashitov et al. (SAGE), Phys. Rev. **C59**, 2246 (1999), hep-ph/9803418
[18] J.N. Abdurashitov et al. (SAGE), Phys. Rev. C **80**, 015807 (2009), 0901.2200
[19] J.N. Bahcall, Phys. Rev. **C56**, 3391 (1997), hep-ph/9710491
[20] J. Kostensalo, J. Suhonen, C. Giunti, P.C. Srivastava, Phys.Lett. **B795**, 542 (2019), arXiv:1906.10980
[21] S.V. Semenov, Phys. Atom. Nucl. **83**, 1549 (2020)
[22] C. Giunti, Y.F. Li, C.A. Ternes, O. Tyagi, Z. Xin, JHEP **10**, 164 (2022), 2209.00916
[23] C. Giunti, Y.F. Li, C.A. Ternes, Z. Xin (2022), 2212.09722
[24] G. Mention et al., Phys. Rev. **D83**, 073006 (2011), arXiv:1101.2755
[25] T.A. Mueller et al., Phys. Rev. **C83**, 054615 (2011), arXiv:1101.2663
[26] P. Huber, Phys. Rev. **C84**, 024617 (2011), arXiv:1106.0687
[27] L. Hayen, J. Kostensalo, N. Severijns, J. Suhonen, Phys.Rev. **C100**, 054323 (2019), arXiv:1908.08302
[28] M. Estienne, M. Fallot et al., Phys. Rev. Lett. **123**, 022502 (2019), arXiv:1904.09358
[29] V. Kopeikin, M. Skorokhvatov, O. Titov, Phys. Rev. D **104**, L071301 (2021), 2103.01684
[30] M. Maltoni, T. Schwetz, Phys. Rev. **D68**, 033020 (2003), hep-ph/0304176
[31] I. Alekseev et al. (DANSS), Phys.Lett. **B787**, 56 (2018), arXiv:1804.04046
[32] M. Danilov (2022), talk presented at ICHEP 2022, 41th International Conference on High Energy Physics, 6-13 July 2022, Bologna, Italy
[33] J. Ashenfelter et al. (PROSPECT), Phys.Rev.Lett. **121**, 251802 (2018), arXiv:1806.02784
[34] M. Andriamirado et al. (PROSPECT), Phys.Rev. **D103**, 032001 (2021), arXiv:2006.11210
[35] H. Almazan et al. (STEREO), Phys.Rev.Lett. **121**, 161801 (2018), arXiv:1806.02096
[36] H. Almazan Molina et al. (STEREO), Phys.Rev. **D102**, 052002 (2020), arXiv:1912.06582
[37] Y. Ko et al. (NEOS), Phys.Rev.Lett. **118**, 121802 (2017), arXiv:1610.05134
[38] Z. Atif et al. (RENO, NEOS), Phys. Rev. D **105**, L111101 (2022), arXiv:2011.00896
[39] F. An et al. (Daya Bay), Chin.Phys. **C41**, 013002 (2017), arXiv:1607.05378
[40] A. Serebrov et al., Phys.Rev.D **104**, 032003 (2021), arXiv:2005.05301
[41] M. Danilov, J.Phys.Conf.Ser. **1390**, 012049 (2019), arXiv:1812.04085
[42] M. Andriamirado et al. (PROSPECT, STEREO) (2020), 2006.13147
[43] M.V. Danilov, N.A. Skrobova, JETP Lett. **112**, 452 (2020)
[44] C. Giunti, Y.F. Li, C.A. Ternes, Y.Y. Zhang, Phys. Lett. B **816**, 136214 (2021), 2101.06785

Charmed Galaxies

Julia Becker Tjus[1,2,3,*] and *Wolfgang* Rhode[1,3,4,**]

[1]Lehrstuhl für Theoretische Physik IV, insb. Plasma-Astroteilchenphysik, Fakultät für Physik & Astronomie, Ruhr-Universität Bochum, 44780 Bochum, Germany
[2]Department of Space, Earth and Environment, Chalmers University of Technology, 412 96 Gothenburg, Sweden
[3]Ruhr Astroparticle and Plasma Physics Center (RAPP Center), Germany
[4]Fakultät für Physik, Technische Universität Dortmund, 44221 Dortmund, Germany

Abstract. The quest for finding the origins of cosmic rays has been going on for many decades. Cosmic rays as charged particles react to cosmic magnetic fields and typically travel in diffusive motion through the Universe. Their imprint on Earth therefore holds little information on their origin, and finding the sources of cosmic rays is a major challenge. The question of their origins has been a leading questions in physics and astrophysics in the past decades. To solve this riddle, a multimessenger approach is used, including cosmic-ray interaction products in the searches, specifically gamma-rays and neutrinos produced in the resulting particle showers. In this multimessenger picture, the detection of high-energy neutrinos from the cosmos by IceCube - as a unique tracer of cosmic rays - is an important piece of the puzzle. First evidence for neutrino emission from the active galaxies TXS0506+056 and NGC1068 indicates that a significant fraction comes from such sources. In this paper, the intriguing fact that gamma-rays seem to be absorbed in these sources discussed. The possibility of neutrinos being produced in regions of high photon or gas densities, together with the possibility of in the future revealing neutrinos from the decay of charmed particles, will be investigated, possibly opening a window to *Charmed Galaxies*.

1 Introduction

The existence of high-energy cosmic rays is still one of the biggest riddles in physics and astrophysics. More than 100 years after their first detection, it is still mostly unclear how the Universe can manage to accelerate particles up to 10^{20} eV energies. The search for the specific acceleration mechanism is complicated by the fact that the origin of cosmic rays is not deducible from their arrival directions at Earth: these charged particles react to large- and small-scale magnetic fields in extragalactic space and in the Milky Way, leaving a close-to isotropic arrival pattern as their imprint at Earth. A way around this problem is the search for neutral particles that are produced in cosmic-ray interactions in the sources. Hadronic interactions lead to the production of high-energy gamma-rays and neutrinos, dominantly from pion and Kaon decays:

$$p\,p \;\rightarrow\; \#(\pi^{\pm}/\pi^{0}) + \#(K^{\pm}/K_L^{0}) + X \qquad (1)$$

$$p\,\gamma \;\rightarrow\; \left\{ \begin{array}{l} \Delta^{+} \rightarrow n\pi^{+}/p\pi^{0} \\ \#(\pi^{\pm}/\pi^{0}) + \#(K^{\pm}/K_L^{0}) + X \end{array} \right. , \qquad (2)$$

Here, multi-pion and multi-Kaon production is leading the production of high-energy neutrinos and gamma-rays, X stands for everything else that is produced in the particle showers. The decay processes of pions and Kaons are given in Table 1 and lead to the production of gamma-rays and neutrinos as shown. Rare processes from the decay of charmed mesons like $D^{\pm}/D^{0}$ are typically subdominant, but could become important in regions of extreme densities as will be discussed below.

The above equations are formulated for protons, which make up around 90% of the cosmic-ray spectrum at GeV energies. Significant contributions from helium up to iron are present at higher energies, see e.g. [11] for a review.The proton-proton cross-section can in principle be used to estimate the cross sections of the heavier nuclei (mass-number times the proton cross-section at 1/mass-number of the total energy per nucleus). Alternatively, the interactions can be treated in more detail by including the full interaction networks, as it is done for instance in [8, 16, 22, 24, 26]. Effects of rare particle decays can be taken into account by using event generators like SYBILL, EPOS, or QGSJET, as we will see that these can play a role in the dense cores of galaxies (see Section 3.2). At this point, more than 150 astrophysical sources have been detected in TeV gamma-rays [14]. These are, however, ambiguous signs of particle acceleration, as these photons can arise from the acceration of electrons as well, via bremsstrahlung or inverse Compton scattering. A clear sign of hadronic acceleration are neutrinos, as these are not produced by electrons. These, on the other hand, are difficult to detect. With the instrumentation of one cubic kilometer of Antarctic Ice, it was possible for the first time to establish the existence of a diffuse flux of high-energy

*e-mail: julia.tjus@rub.de
**e-mail: wolfgang.rhode@tu-dortmund.de

Table 1. Processes relevant in cosmic-ray air showers, after [18]. The mean life time in the rest frame of the particle is given as $\overline{\tau}_0$. Data from [28].

decay channel	valence quark contribution	branching ratio	$\overline{\tau}_0$/s
$\mu^{\pm} \rightarrow e^{\mp} + \overline{\nu}_e/\nu_e + \nu_\mu$	-	100%	$2.2 \cdot 10^{-6}$
$\pi^{\pm} \rightarrow \mu^{\pm} + \nu_\mu/\overline{\nu_\mu}$	u/d	99.99%	$2.6 \cdot 10^{-8}$
$\pi^0 \rightarrow \gamma\gamma$	u/d	98.8%	$8.4 \cdot 10^{-17}$
$K^{\pm}\mu^{\pm} + \nu_\mu/\overline{\nu}_\mu$	u/s	63.55%	$1.2 \cdot 10^{-8}$
$K_L^0 \rightarrow (\pi^+ + e^- + \overline{\nu}_e)/(\pi^- + e^+ + \nu_e)$	d/s	40.55%	$5.1 \cdot 10^{-8}$
$K_L^0 \rightarrow (\pi^+ + \mu^- + \overline{\nu}_\mu)/(\pi^- + \mu^+ + \nu_\mu)$	d/s	27.04%	$5.1 \cdot 10^{-8}$
$K^{\pm} \rightarrow \pi^0 + e^{\pm} + \nu_e/\overline{\nu}_e$	u/s	5.07%	$1.2 \cdot 10^{-8}$
$K^{\pm} \rightarrow \pi^0 + \mu^{\pm} + \nu_\mu/\overline{\nu}_\mu$	u/s	3.35%	$1.2 \cdot 10^{-8}$
$D^{\pm} \rightarrow \overline{K}^0 + \mu^{+/-} + \nu_\mu/\overline{\nu}_\mu$	d/c	9.2%	$1.0 \cdot 10^{-12}$
$D^0 \rightarrow \overline{K}^- + \mu^+ + \nu_\mu$	u/c	3.3%	$4.1 \cdot 10^{-13}$
$D_s^{+/-} \rightarrow \tau^{\pm} + \nu_\tau/\overline{\nu}_\tau$	s/c	5.5%	$5.0 \cdot 10^{-13}$
$\tau^{\pm} \rightarrow (e^+ + \nu_e + \nu_\tau)/(e^- + \overline{\nu}_e + \overline{\nu}_\tau)$	-	17.83%	$2.9 \cdot 10^{-13}$
$\tau^{\pm} \rightarrow (\mu^+ + \nu_\mu + \nu_\tau)/(\mu^- + \overline{\nu}_\mu + \overline{\nu}_\tau)$	-	17.41%	$2.9 \cdot 10^{-13}$
$\eta \rightarrow \mu^+\mu^+\gamma$	u/d	0.031%	$5.0 \cdot 10^{-19}$

neutrinos from the cosmos [1]. More recent measurements show first evidence that active galaxies are among the sources, with concrete evidence of emission from the blazar TXS 0506+056 [2] and the Seyfert-Starburst composite NGC 1068 [4]. Here, the current knowledge on these galaxies and their features of hadronic emission will be summarized, and prospects will be given to study rare particle decays in dense cores of active galaxies.

2 Neutrino and gamma-ray emission from active galaxies

The class of active galaxies describes all galaxies that host an active core, very abstractly meaning that there is activity connected to the supermassive black hole (SMBH) in the center. This very general class is subdivided into different sub-catagories and the objects can look very different to the observer at Earth. Very distinct features are an accretion disk around the black hole, a torus of dust that is at a few tens of parsecs distance from the SMBH, and a jet of relativistic plasma that can reach up to Megaparsec in scale and typically radiates brightly at radio wavelengths. This description of active Galaxies is very simplistic, as there are many more features that divide up this general class in an entire zoo of subclasses. In the context of cosmic-ray acceleration and interaction, active Galaxies represent interesting candidates to reach particle energies up to the highest energies of 10^{20} eV observed at Earth. There is a simple argument that sources have to fulfill the Hillas criterion [20],

$$E_{\max} = 10^{18}\ \mathrm{eV} \cdot Z \cdot (B/\mu\mathrm{G}) \cdot (R/\mathrm{kpc}) . \qquad (3)$$

This equation is valid for nuclei of charge Z and velocity $v \sim c$, in a magnetic field B that is present in a source with radius R, see [11] for a review. There are only a few object classes in the Universe who can provide conditions to reach $E_{\max} = 10^{20}$ eV, among them are the cores of active Galaxies with their extreme magnetic fields of up to several Gauss at radii on parsec-scale, as well as their jets with strong magnetic fields and hundreds of parsec extension in the inner jet, and lower fields (μG to mG), but significantly larger extensions (R up to kpc-Mpc) in the outer jets. But what is the mechanism to accelerate particles in these environments? As large-scale electric fields that are needed for linear acceleration cannot be created on longer time scales in natural conditions (the plasma would try to go back to neutrality), stochastic acceleration via fluctuations in the magnetic field, called Fermi acceleration, is a possibility. This can happen at shock fronts, with an efficiency of first order with respect to the shock velocity, or otherwise on moving magnetic turbulence, which is a second order effect in terms of the velocity of the magnetic turbulence. Strong shock fronts do exist in jets of active galaxies, and when the jet is pointed directly toward Earth, the relativistic plasma is boosted by the Lorentz factor γ, even boosting the particle energies. These objects are called blazars. The variability time scales of the emission points to the existence of small (parsec-scale) plasmoids that move along the jet axis with relativistic speed. One possibility for particle accelaration is thus via second order Fermi acceleration in which the plasmoids work as the moving magnetic turbulence needed for this process. All of these scenarios, however, do not solve the so-called *injection problem*: in simulations, particles are not picked up by the process of Fermi acceleration unless they are already out of equilibrium, so they must be pre-accelerated somewhere. In active Galaxies, one option to pre-accelerate particles is in events of magnetic reconnection that happen at the launching site of the jet and/or on the accretion disk that surrounds the black hole [27]. With such a scenario at hand, even galaxies that do not host strong radio jets - so-called Seyfert galaxies - can be sources of cosmic rays and neutrinos. Here, the total electromagnetic luminosity is rather dominated by the emissions connected to the accretion disk. While these might not be able to accelerate all the way up to 10^{20} eV, they could be a source of PeV cosmic-rays and this way even TeV-PeV neutrinos. This idea has started to become interesting since the first strong evidence of neutrino emission from the Seyfert-starburst composite galaxy NGC1068. Evidence for a neutrino signal was observed in 9 years of data (2011-2020) in the TeV energy range, with an indication of a steep neutrino spectrum, i.e. around E^{-3} [4]. The most fascinating fact about this measurement is that the estimated neutrino luminosity is significantly larger than the measured gamma-ray luminosity. This is to first order in strong contradiction to the fact that the production of gamma-rays and neutrinos via pion decay happens at about the same flux level. The only way to reduce gamma-rays to the level observed at GeV-TeV energies is to absorb these gamma-rays. Cores of active galaxies are ideal places for such absorption: the strong Corona that is produced in connection with the accretion disk in fact leaves little room for gamma-ray emission above GeV energies [23]. In fact, other emission regions from farther outside are needed to fully explain the Fermi observations. One possibility is the starburst ring aound the black hole at around 1 kpc distance [15], or an ultra-fast outflow launched from the core of these galaxies [25].

The fact that NGC1068 seems to be a gamma-ray absorbed neutrino source does not come as a surprise: the luminosity of the diffuse neutrino flux is also significantly higher than one of the diffuse gamma-ray flux, so that this flux in general must be dominated by gamma-ray absorbed sources [6]. Another piece of evidence that fits well into this scheme is the potential correlation of high-energy neutrinos with the position of blazars. In the sample of 9 years of data, the blazars TXS 0506+056 and PKS 1424+240 show local significances above 3σ [4]. It is interesting to note that these two sources are even correlated to events that are made public via the IceCube alert system. A high-energy neutrino event from TXS 0506+056 in particular was seen in coincidence with a gamma-ray flare (Fermi), giving a 3σ significance of evidence that this is a source of high-energy neutrinos on its own [2]. In [21], four neutrinos with high probability of being of astrophysical origin and are at the same time in spatial coincidence with blazars. This concerns the blazars PKS B1424-418, TXS 0506+056 and PKS 1502+106. Fig. 1 shows the lightcurve of the Fermi emission at GeV photon energy at the time of the neutrino arrival, which are indicated for each panel as purple, vertical lines. What has been shown in [21] is that the gamma-ray blazars show a local (TSX0506+056) or even global (PKS B1424) minimum in their gamma-ray light-curves at the time of neutrino detection. This can imply that even in the case of blazars, the neutrinos are emitted preferably when the emission region is opaque to high-energy gamma-rays. Since the same target that absorbs the gamma-rays creates the neutrinos to begin with, such a scenario is in fact quite plausible. For TXS 0506+056, the interaction of two jets has been discussed [13], and would correspond to an environment that all of the sudden becomes opaque to the gamma-rays through the enhanced density present during the interaction of the jets, which at the same time would enhance the neutrino production.

So, to summarize, to produce sources of high-energy neutrinos and at the same time reduce the gamma-ray flux at above GeV energies significantly, very dense environments are needed that host a cosmic-ray acceleration region. In the following chapter, we discuss how high gas densities could be an alternative explanation to the scenario of high photon fields that is discussed above, and what effects this could have on the neutrino flux.

3 Charmed galaxies

In this section, the possibility of hadronic interactions in dense cores of galaxies is discussed and how these could contribute to the flux of high-energy neutrinos in the Universe. We will first discuss in more detail how hadronic and photohadronic interactions can take place in cores and jets of galaxies (Section 3.1), then discuss the details of the particle showers and how rare decays could emerge toward high energies (Section 3.2), and finally, where such signatures of Charmed Galaxies could be revealed in the future (Section 3.3)

3.1 Hadronic interactions in dense cores of galaxies

In general, hadron-hadron interactions are well-suited for the production of neutrinos as discussed in Section 1. In fact, the production of neutrinos in hadron-hadron interactions has a quite low kinematic threshold close to the mass of the respective pions that are responsible for their production, so the spectra rise at a few tens of MeV. As the proton-proton cross section is almost constant at $\sigma_{\rm pp} \sim 3 \cdot 10^{-26}\ {\rm cm}^{-2}$, the secondary spectra basically follow the one of the initial protons above this threshold, i.e. E^{-p}, see e.g. [10] for a discussion. Photohadronic interactions, on the other hand, have a threshold that is determined by the production of the mass of the Δ^+ resonance, so 1.232 GeV beam energy which translates to a threshold that depends on the average energy of particles in the target photon field $< \epsilon >$ (see e.g. [9]) as

$$E_{\rm th}^{p} = \frac{m_\Delta^2 - m_p^2}{4 < \epsilon >}\,. \tag{4}$$

Taking for instance the Corona of an AGN like NGC1068 as a target photon field, results in a threshold of

$$E_{\rm th}^{p} \approx 14\,{\rm TeV}\left(\frac{< \epsilon_{\rm photon} >}{0.01\ {\rm MeV}}\right)^{-1}\,. \tag{5}$$

Thus, an interaction with the coronal X-ray field - which has a temperature of around 0.01 MeV - only happens above TeV cosmic-ray energies. Around 1/20th of the proton energy is transferred to neutrinos and thus, the neutrino spectra will exist at around a few hundred GeV upward only. The same is true for the gamma-rays without considering absorption. Thus, proton-proton interactions will happen at significantly lower energies already, where the flux of cosmic rays is significantly higher, at least when assuming a power-law shape of the energy spectrum.

In the environment of an accretion disk of an active Galaxy, the neutrinos can be produced by hadron-hadron interactions with the Coronal gas, which is expected to

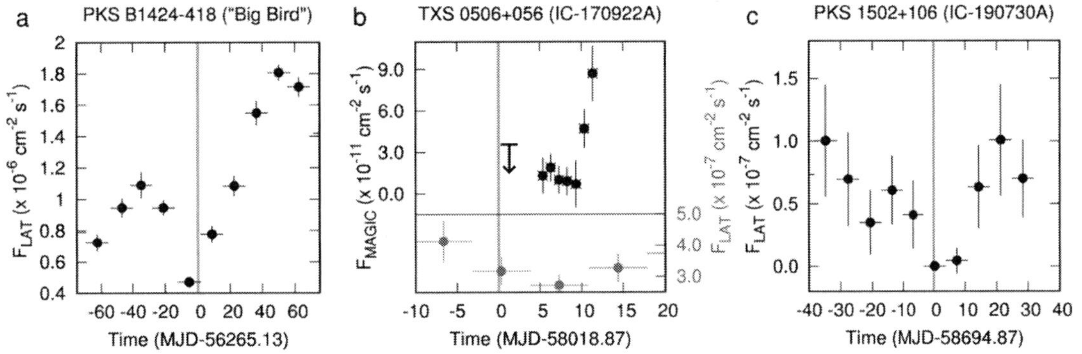

Figure 1. Fermi lightcurves of three gamma-ray blazars (from left: PKS B1424-418, TXS 0506+056, and PKS 1502+106) Neutrinos from the direction of gamma-ray blazars. The arrival of a high-energy neutrino from the direction of these is indicated by the purpel, vertical line in each of the panels. For TXS 0506+056, MAGIC data are shown in black (left y-axis), Fermi data in blue/grey (right y-axis). Figure from [21], reproduced with permission of the authors and IOP Publishing.

have a density of $n \sim 10^{9\pm1}$ cm^{-3} with an extesion of a few tens Schwarzschild-radii (max $100\,r_s$). Another option is that the interaction happens with the photon field of the Corona, but this process would only start to work efficiently above TeV energy.

Now, this is a very interesting environment for shower production from the perspective of particle physics: at high densities, the charged pions will start to interact before they can decay, possibly revealing the neutrino flux from promptly-decaying neutrinos as we discuss below.

3.2 Charmed neutrinos

The observation of rare decays in particle showers opens the window to studying the fundamental properties of matter in the form of the strong and weak interactions, it helps to make precision measurements in particle physics, as well as it provides clues of how to move to beyond-standard-model physics. Earth-bound accelerators like CERN can study the interaction directly at the vertex and have increased our understanding of the standard model of particle physics and beyond significantly. But even the observation of particle showers in the Earth's atmosphere has contributed to our understanding of the microsopic world of particle physics. It is the result of its roughly exponentially increasing density toward the ground that makes the Earth's atmosphere a most interesting test laboratory for studying the evolution of particle showers in dense targets. At a height of approximately ~ 30 km above ground, cosmic-ray nuclei hit atmospheric ones up to the highest energies at $\sim 10^{20}$ eV. In these nucleus-nucleus interactions, an energy-dependent multiplicity of mesons of different masses is produced. Their interaction and decay products are detected (among others) by the detectors of astroparticle physics. In underground experiments like the IceCube Neutrino Observatory, the predominant signal consists of muons from the first interaction above the detector and from all meson decays below ground. The Pierre Auger Observatory also detects muons from secondary interactions, but with a lower energy threshold. In order to draw solid conclusions from the measured signal about the primary cosmic rays, these interactions must be known. Previous studies could be used to quantify the proton-air cross section with Auger [5], and the neutrino-nucleon cross section with IceCube [3]. In the current discussion of these processes, two relevant problems arise in the general context of astrophysical interaction setups. On the one hand, there are tensions between the measured multiplicity of the final generated muons and the current theoretical predictions, also indicating uncertainties in the calculation of the neutrino flux created in interactions in the Earth's atmosphere [7]. On the other hand, high-density environments that are tested by measurements in the Earth's atmosphere reveal a change in the energy spectra of muons and neutrinos: low-energetic and heavy mesons decay before they interact and therefore inherit the primary cosmic-ray power-law energy spectrum. Light or high-energetic mesons, however, interact before they decay, because of their signifiantly longer life times. Table 1 show those processes relevant for neutrino and muon production in an air shower, and thus also in particle showers in astrophysical environments. Pions and Kaons in their rest frames decay after around 10^{-8} s, while heavy measons like D^+ decay after 10^{-12} s already. Thus, if the density of a target is increased at constant energy and the limit is reached at which the interaction length of the light mesons becomes shorter than the decay length, the decay products of the heavy mesons are filtered out. The condition needed for the pions (and Kaons) to interact before decay is an optical depth for pion-gas (Kaon-gas) interactions that is larger than one:

$$\tau_{\pi+p/K+p} = \frac{L}{\lambda_{\rm mfp}^{\pi/K}} \geq 1\,. \tag{6}$$

Here, the mean free path of the pions/Kaons is determined as $\lambda_{\rm mfp} = (\rho\,\sigma_{\pi p/Kp})^{-1}$ and the propagation length is given by the decay length in the observer's frame, $L = c \cdot \gamma \cdot \tau^0_{\pi/K}$. Using $\gamma = E/(m\,c^2)$, we receive a relation between the density of the source and the energy of the neutrinos resulting from the decaying pions and Kaons (we assume that $E_\nu = E_\pi/4$):

$$E_\nu \gtrsim 10^5\,\mathrm{PeV}\left(\frac{\rho}{10^{10}\mathrm{cm}^{-3}}\right)^{-1} . \qquad (7)$$

This is a limit for the production of neutrinos via pion and Kaon decay. This shows that at the highest energies, this flux will disappear and the flux of promptly decaying particles is predicted to emerge. It is interesting that it is the absolute density that is relevant for this effect rather than the column depth. So, inhomogenieties in the plasma might enhance the effect and shift it toward lower energies. The effect of charm production has been discussed previously in astrophysical sources [19] and more specifically in supernova explosions with slow jets [12]. Here, we argue that even the dense cores of galaxies might how this effect.

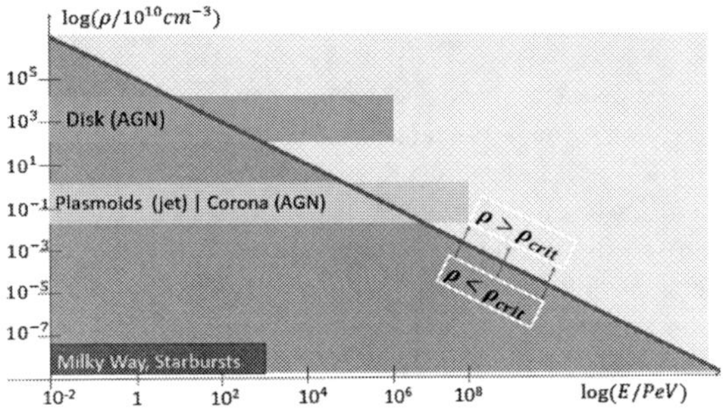

Figure 2. Regions of gas density (ρ) and cosmic-ray energies (E) of the different astrophysical systems. The solid black line shows the case where the decay length of the charged pions equals the time scale for interactions of pions with the gas, assuming a constant density medium.

3.3 Neutrino emission from Compact Obscured Nuclei (CONs)

The fact that the detection of the diffuse flux of neutrinos by IceCube must be connected to gamma-ray absorbed sources together with the strong evidence that NGC1068 is among the sources shows that neutrino production must happen in very dense environments as discussed above. This specifically makes galaxies with a dense and active core outstanding candidates. The production of high-energy protons in this scenario is believed to happen very close to the black hole, possibly via magnetic reconnection in connection to the accretion disk, or via a combination of reconnection and the plasmoid-instability at the foot of an AGN jet. Interaction can then happen with the local gas target in the Corona or jet, or with the photon field. The absorption of gamma-rays is then typically discussed as being due to interactions of the high-energy photons from hadronic interactions with the Coronal photon field. Another option would be the interaction via Bethe-Heitler pair production, i.e. $p\gamma \to pe^+e^-$, or more generally the interaction of photons in the field of the nuclei of the gas. This type of absorption could also happen outside of the Corona and no strong background photon fields for interaction are needed to absorb the photons. The existence of compact obscured nuclei (CONs) has been established in recent year, see e.g.[17] for a review. Compact obscured nuclei (CONs) have been shown to be present in close to 40% of ULIRGs and more than 20% of LIRGs [17]. While the existence of CONs in lower luminosity galaxies is still unclear, these measurements reveal that there is a large number of (active) core out in the Universe that have extreme densities. Due to their huge column densities of $10^{25} - 10^{26}$ cm^2, these are ideal environments for the production of neutrinos. At the same time, these compact structures are more extended than the very innermost region of the active galaxy, i.e. typically have a size of $10 - 100$ pc, thus leaving a larger radius for particle acceleration to happen.

4 Summary, Conclusion & Outlook

The measurements of high-energy neutrinos with IceCube in the past decades has taught us quite a lot concerning the properties of potential neutrino emitters: the detection of a diffuse flux from the cosmos shows that there is no obvious correlation with the Galactic Plane, implying that a large fraction of the flux must be of extragalactic nature. As the neutrino flux is significantly larger than the signal detected in gamma-rays, this means that we are looking at gamma-ray absorbed sources. First pieces of evidence point to emission from the blazars TXS 0506+056 and PKS 1424+240, as well as the Seyfert-starburst composite NGC1068. These potential sources match well with the idea of gamma-ray absorbed sources and start to establish dense cores of galaxies as neutrino emitters. Such a scenario also opens the window to with the next generation neutrino telescopes IceCube-Gen2 and KM3NeT search for a contribution from charmed neutrinos, thus opening the window to studying *Charmed Galaxies*.

Acknowledgements

JBT and WR acknowledge support from the Collaborative Research Center SFB1491 *Cosmic Interacting Matters - From Source to Signal*, funded by the Deutsche Forschungsgemeinschaft DFG under the grant number 445052434. JBT also acknowledges DFG funding via the MICRO project (No. 445990517). JBT would like to thank Chalmers University of Technology for awarding the Jubilee Professorship for the year 2023. Part of this work has been performed at Chalmers in the context of the Jubilee Professorship. JBT and WR would like to thank in particular Susanne Aalto and Francis Halzen for valuable discussions on the topic.

References

[1] Aartsen et al. (IceCube Coll.), Evidence for High-Energy Extraterrestrial Neutrinos at the IceCube Detector, Science 342, 1242856 (2013)

[2] Aartsen et al. (IceCube Coll.), Multimessenger observations of a flaring blazar coincident with high-energy neutrino IceCube-170922A, Science 361, eaat1378 (2018)

[3] Aartsen et al. (IceCube Coll.), Measurement of the multi-TeV neutrino interaction cross-section with IceCube using Earth absorption, Nature 551, 596 (2017)

[4] Abbasi et al. (IceCube Coll.), Evidence for neutrino emission from the nearby active galaxy NGC 1068, Science 378, 538 (2022) doi:10.1126/science.abg3395

[5] Abreu, P., et al. (Auger Coll.), Measurement of the Proton-Air Cross Section at s=57TeV with the Pierre Auger Observatory, Physical Review Letters 109, 062002 (2012)

[6] Ahlers & Halzen, High-energy cosmic neutrino puzzle: a review, Reports on Progress in Physics 78, 126901 (2015)

[7] Albrecht, Cazon, Dembinski, Fedynitch, Kampert, Pierog, Rhode, Soldin, Spaan, Ulrich, & Unger, The Muon Puzzle in cosmic-ray induced air showers and its connection to the Large Hadron Collider, Astrophysics and Space Science 367, 27 (2022)

[8] Alves Batista et al., CRPropa 3.2 - an advanced framework for high-energy particle propagation in extragalactic and galactic spaces, Journal of Cosmology and Astroparticle Physics 2022, 035 (2022)

[9] Becker, High-energy neutrinos in the context of multimessenger astrophysics, Physics Reports 458, 173 (2008)

[10] Becker Tjus, Eichmann, Halzen, Kheirandish, & Saba, High-energy neutrinos from radio galaxies, Physical Review D 89, 123005 (2014)

[11] Becker Tjus & Merten, Closing in on the origin of Galactic cosmic rays using multimessenger information, Physics Reports 872, 1 (2020)

[12] Bhattacharya, Enberg, Reno, & Sarcevic, Charm decay in slow-jet supernovae as the origin of the IceCube ultra-high energy neutrino events, Journal of Cosmology and Astroparticle Physics 2015, 034 (2015)

[13] Britzen et al., A cosmic collider: Was the IceCube neutrino generated in a precessing jet-jet interaction in TXS 0506+056?, Astronomy and Astrophysics 630, A103 (2019)

[14] Chadwick, 35 Years of Ground-Based Gamma-ray Astronomy, Universe 7, 432 (2021)

[15] Eichmann, Oikonomou, Salvatore, Dettmar and Becker Tjus, Solving the Multimessenger Puzzle of the AGN-starburst Composite Galaxy NGC 1068, The Astrophysical Journal **939**, 43 (2022)

[16] Evoli, Gaggero, Vittino, Di Mauro, Grasso, & Mazziotta, Cosmic-ray propagation with DRAGON2: II. Nuclear interactions with the interstellar gas, Journal of Cosmology and Astroparticle Physics 2018, 006 (2018)

[17] Falstad et al., CON-quest. Searching for the most obscured galaxy nuclei, Astronomy and Astrophysics 649, A105 (2021)

[18] Fedynitch, Cascade equations and hadronic interactions at high energies, Ph.D. Thesis (2015)

[19] Enberg, Reno, & Sarcevic, High energy neutrinos from charm in astrophysical sources, Physical Review D 79, 053006 (2009)

[20] Hillas, The Origin of Ultra-High-Energy Cosmic Rays, Annual Review of Astronomy and Astrophysics 22, 425 (1984)

[21] Kun, Bartos, Becker Tjus, Biermann, Halzen and Mező, Cosmic Neutrinos from Temporarily Gamma-suppressed Blazars, The Astrophysical Journal 911, L18 (2021) doi:10.3847/2041-8213/abf1ec

[22] Merten, Becker Tjus, Fichtner, Eichmann, & Sigl, CRPropa 3.1—a low energy extension based on stochastic differential equations, Journal of Cosmology and Astroparticle Physics 2017, 046 (2017)

[23] Murase, Kimura, & Mészáros, Hidden Cores of Active Galactic Nuclei as the Origin of Medium-Energy Neutrinos: Critical Tests with the MeV Gamma-Ray Connection, Physical Review Letters 125, 011101 (2020)

[24] Orlando, Johannesson, Moskalenko, Porter, & Strong, GALPROP cosmic-ray propagation code: recent results and updates, Nuclear and Particle Physics Proceedings 297-299, 129 (2018) doi:10.1016/j.nuclphysbps.2018.07.020

[25] Peretti, Lamastra, Saturni, Ahlers, Blasi, Morlino, & Cristofari, Diffusive shock acceleration at EeV and associated multimessenger flux from ultra-fast outflows driven by Active Galactic Nuclei, arXiv e-prints arXiv:2301.13689 (2023)

[26] Strong & Moskalenko, The GALPROP program for cosmic-ray propagation: new developments, 26th International Cosmic Ray Conference (ICRC26), Volume 4 4, 255 (1999)

[27] Sironi & Beloborodov, Kinetic Simulations of Radiative Magnetic Reconnection in the Coronae of Accreting Black Holes, The Astrophysical Journal 899, 52 (2020)

[28] Zyla et al. (Particle Data Group), Review of Particle Physics, Progress of Theoretical and Experimental Physics 2020, 083C01 (2020)

Constraints on the nuclear Equation of State from heavy ion reaction dynamics

Maria Colonna[1,*]

[1]INFN-LNS, *Laboratori Nazionali del Sud*, 95123 Catania, Italy

Abstract. Heavy ion collisions allow one to create and explore, in laboratory experiments, transient states of nuclear matter under several conditions of density, temperature and charge asymmetry. The complex many-body dynamics is suitably described by transport theories which employ effective interactions, thus connecting fundamental nuclear ingredients to sensitive observables of experimental interest. We review here recent analyses which allow to constrain, in particular, the controversial density behavior of the symmetry energy of the nuclear Equation of State, also outlining future perspectives.

1 Introduction

The understanding of several phenomena occurring at different scales, ranging from nuclear systems to astrophysical compact objects, is governed by the nuclear matter Equation of State (EOS). Its features can be explored by mean of heavy ion collision experiments, where transient states of nuclear matter spanning a large variety of regimes are created. More precisely, the reaction dynamics obeys to the properties of the nuclear effective interaction, which provides the Equation of State in the equilibrium limit. Over the past years, measurements of experimental observables, like isoscalar collective vibrations, collective flows, rapidity distribution and meson production, have contributed to constrain the EOS of symmetric matter for densities up to five time the saturation value [1]. More recently, the availability of exotic beams has made it possible to explore, in laboratory conditions, new aspects of nuclear structure and dynamics up to extreme ratios of neutron (N) to proton (Z) numbers, also opening the way to the investigation of the EOS of asymmetric matter. Indeed, the isovector part of the nuclear effective interaction and the corresponding density dependence of the nuclear symmetry energy (Asy-EOS) are still quite debated. Nevertheless, this information is crucial in the astrophysical context, e.g. for the understanding of the properties of compact objects such as neutron stars, whose crust behaves as low-density asymmetric nuclear matter [2, 3] and whose core may touch extreme values of density and asymmetry. The interpretaion of gravitation wave emission signals also rely on a good knowledge of the Asy-EOS.

Moreover, the low-density behavior of the symmetry energy also affects the structure of exotic nuclei, their masses and the appearance of features involving the neutron skin, such as the dipole polarizability, which are currently under intense investigation [4, 5].

From the theoretical point of view, the study of heavy ion collisions requires considerable efforts to tackle the dynamics of the nuclear many-body system. Transport theories appear as a suitable tool, though several approaches are available on the market and, for a safe comparison to experimental data, it is very important to reduce, as much as possible, the model dependence of the results. To this purpose, the Transport Model Evaluation Project (TMEP) was initiated a few years ago, aiming at comparing the results of different models under controlled situations and establishing a kind of systematic theoretical error [6].

Several observables which are sensitive to the Asy-EOS and testable experimentally, have been suggested [7–10]. After a brief description of transport models, in this article we will review recent results on dissipative collisions in a wide range of beam energies, from reactions at Fermi energies up to the $AGeV$ range. Low to Fermi energies bring information on the Asy-EOS around (below) normal density, while intermediate energies will probe high density regions.

2 Transport models and effective interactions

Nuclear reactions at Fermi/intermediate energies can be modeled by solving transport equations based on mean field theories, with short range (2p-2h) correlations included via hard nucleon-nucleon (elastic or inelastic) collisions and with stochastic terms, accounting for the effect of unknown correlations, selfconsistently evaluated from the mean phase-space trajectory, see [7, 11–13]. Fluctuations are essential in order to get distributions as well as to allow for the growth of dynamical instabilities.

In the energy range up to a few hundred $AMeV$, an appropriate tool is the so-called Boltzmann-Langevin equa-

*e-mail: colonna@lns.infn.it

tion (BLE) [11, 14]:

$$\frac{df}{dt} = \frac{\partial f}{\partial t} + \{f, H\} = I_{coll}[f] + \delta I[f], \quad (1)$$

where $f(\mathbf{r}, \mathbf{p}, t)$ is the one-body distribution function, the semi-classical analog of the Wigner transform of the one-body density matrix, $H(\mathbf{r}, \mathbf{p}, t)$ the mean field Hamiltonian, I_{coll} the two-body collision term incorporating the Fermi statistics of the particles, and $\delta I[f]$ its fluctuating part. The so-called BUU approaches (see [8] and refs. therein) neglect the latter term. Among extended BUU-like models, which incorporate fluctuation terms, we mention the approximate treatment of the BLE introduced in Ref.[15], namely the Stochastic Mean Field (SMF) model, and the BLOB model [16], which implements the full stochastic collision integral. The numerical procedure to integrate the transport equations is generally based on the test-particle method.

Molecular dynamics (QMD) approaches represent an alternative avenue to solve the semi-classical transport equations [17–19]. Within such a family of models, the nuclear many-body state is written as a product (antisymmetrized only in the case of the AMD [13, 20, 21] approach) of single particle states, represented by Gaussian wave packets. Equations of motion are derived for the centroids, according to the mean-field potential. Two-body collisions between nucleon packets are treated stochastically. Moreover, the localization of the wave packets induces some additional correlations and fluctuations also in the mean-field propagation.

From the above discussion, we expect mean-field (BUU-like) theories to better describe genuine mean-field effects, owing to the better mapping of the one-body distribution function in phase space. On the other hand, the effect of fluctuations is emphasized in QMD-like models. In the following we will discuss results obtained with both classes of models. Recent studies performed within the TMEP project have shown that BUU and QMD models can give convergent results for the main features of the reaction dynamics under controlled situations, as far as numerical details are suitably adjusted [6].

A common ingredient is certainly the nuclear effective interaction, associated with a given EOS. Thus from the comparison of reaction observables evaluated in simulated events with experimental data one can finally get some hints on nuclear matter properties. Our discussion will be mainly concentrated on the symmetry energy E_{sym}, which enters the energy density functional, ϵ, adopted to represent the nuclear interaction. In its simplest form (i.e., at zero temperature and neglecting momentum-dependent terms), the latter quantity can be written as: $\epsilon(\rho, \rho_i) \equiv \epsilon(\rho) + \rho \frac{E_{sym}}{A}(\rho_i/\rho)^2 + O(\rho_i/\rho)^4 + ..$, expressed in terms of total ($\rho = \rho_p + \rho_n$) and isospin ($\rho_i = \rho_p - \rho_n$) densities. E_{sym} gets a kinetic contribution directly from basic Pauli correlations and a potential part, $C_{pot}(\rho)$, from the highly controversial isospin dependence of the effective interactions:

$$\frac{E_{sym}}{A} = \frac{E_{sym}}{A}(kin) + \frac{E_{sym}}{A}(pot) \equiv \frac{\epsilon_F}{3} + \frac{C_{pot}(\rho)}{2\rho_0}\rho \quad (2)$$

(ρ_0 denotes the saturation density and A the nucleon number). The nuclear mean-field potential is consistently derived from the energy functional ϵ. The isoscalar section is fixed requiring that the saturation properties of symmetric nuclear matter, with a compressibility modulus around $K \approx 200 - 250 MeV$, are reproduced.

Fig.1b illustrates three different parameterizations of $C(\rho) \equiv E_{sym}/A$: the asysoft, the asystiff and asysuperstiff respectively, see [7] for a detailed description.

The sensitivity of the simulation results can be tested against these different choices.

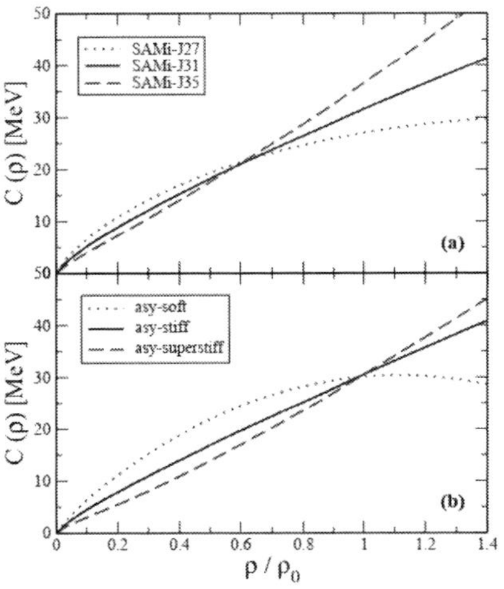

Figure 1. (Color online) (a) Three effective parameterizations of the symmetry energy, as given by the Skyrme SAMi-J family [22]. (b) Parametrizations corresponding to momentum independent interactions: asystiff (dotted line), asysoft (full line) and asysuperstiff (dashed line).

Momentum-dependent (MD) effective interactions may also be implemented into transport equation, allowing to also discuss effects related to the isovector channel of the interaction, leading to the splitting of neutron and proton effective masses. In Fig.1a the symmetry energy behavior is shown for three (momentum dependent) Skyrme-like interactions, whose parameters are chosen to fit structure properties of selected nuclei, see Refs.[4, 22].

3 Fragmentation mechanisms at Fermi energies

The occurrence of surface instabilities (or meta-stabilities) is particularly important for the description of reaction dynamics of low-energy heavy ion collisions and one could envisage a smooth transition between a fast fission mechanism, leading to the emission of two sizable fragments and the disassembly of the system into several pieces. In the Fermi energy regime (30-60 $AMeV$), heavy ion collisions are characterized by the manifestation of volume instabilities as well. Depending on the reaction centrality, one may explore a variety of reaction mechanisms, ranging from (incomplete) fusion and deep-inelastic binary processes, up to neck fragmentation and multifragment production. Focusing on multifragmentation in central collisions, simulations based on stochastic approaches have allowed, through the comparison with experimental data, to shed light on the compression-expansion reaction dynamics, yielding independent information on the nuclear matter compressibility [23]. An example of comparison between theory and experiment [24, 25] is given in Fig. 2 [26], where SMF and BLOB models are tested around the multifragmentation threshold (32 *AMeV*) and in full multifragmentation regime (45 *AMeV*). One can observe that, thanks to the more accurate treatment of the fluctuation dynamics, the BLOB model, employing a soft EOS (compressibility K = 200 MeV), is suitable to give a good reproduction of the fragment multiplicity detected by INDRA [24, 25].

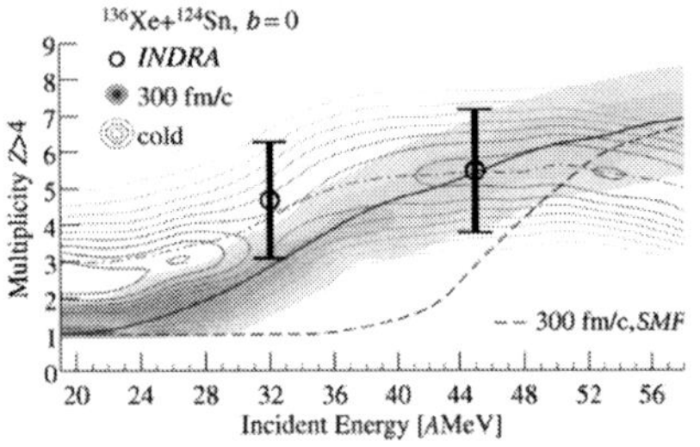

Figure 2. (Color online) BLOB simulation: evolution of IMF ($Z > 4$) multiplicity, as a function of incident energy for the system $^{136}Xe+^{124}Sn$ and a selection of central collisions at 300fm/c (colour shades) and for the cold system (grey contours). Corresponding mean values are indicated for the BLOB simulation (full blue line for primary fragments and dot-dashed line for final fragments) and for a SMF calculation (black dashed line, primary fragments). Corresponding experimental data from Indra experiments at GANIL [24, 25] are added for comparison, with average (symbols) and variance (bars) of the multiplicity distributions.

4 Isospin transport: diffusion and migration effects

In the Fermi energy range, reactions between systems having different $N//Z$ are charecterized by a direct isospin transport in binary events (isospin diffusion). This process involves the low density neck region and is sensitive to the low density behavior of E_{sym}, see Refs.[27–30]. Moreover, it is now quite well established that a large part of the reaction cross section for dissipative collisions at Fermi energies goes through the neck fragmentation channel, with intermediate mass fragments (IMF) directly produced in the interacting zone in semiperipheral collisions on short time scales [31, 32]. It is possible to predict interesting isospin transport effects also for this fragmentation mechanism. Clusters emerging from the neck are formed in a dilute asymmetric matter but always in contact with the regions of the projectile-like and target-like remnants almost at normal densities, thus favoring the neutron enrichment of the neck region (isospin migration).

An easy understanding of the role of the isospin degree of freedom is gained in the hydrodynamical limit, considering the behavior of the *proton/neutron* currents as a function of density (ρ) and asymmetry ($\beta = \rho_i/\rho$) gradients [33]. The latter can be expressed as

$$\mathbf{j}_{p/n} = D^{\rho}_{p/n}\nabla\rho - D^{\beta}_{p/n}\nabla\beta, \tag{3}$$

with $D^{\rho}_{p/n}$ the drift, and $D^{\beta}_{p/n}$ the diffusion coefficients for transport [33]. Of interest for the study of isospin effects are the differences of currents between protons and neutrons which have a simple relation to the density dependence of the symmetry energy

$$\begin{aligned} D^{\rho}_{n} - D^{\rho}_{p} &\propto 4\beta\frac{\partial E_{sym}}{\partial\rho}, \\ D^{\beta}_{n} - D^{\beta}_{p} &\propto 4\rho E_{sym}. \end{aligned} \tag{4}$$

Thus the isospin transport due to density gradients (isospin migration) depends on the slope, $L = \frac{\partial E_{sym}}{\partial\rho}$, of the symmetry energy, or the symmetry pressure, while the transport due to isospin concentration gradients (isospin diffusion) depends on the absolute value of the symmetry energy. Hence transport phenomena in nuclear reactions appear directly linked to the Asy-EOS properties. Recent experiments have been dedicated to a comprehensive study of isospin transport effects, within the same set of data [34].

4.1 The isospin transport ratio

In peripheral and semi-peripheral reactions, it is interesting to look at the asymmetries of the various parts of the interacting system in the exit channel: emitted particles, projectile-like (PLF) and target-like fragments (TLF), and in the case of ternary (or higher multiplicity) events, IMF's. In particular, one can study the so-called isospin transport ratio, which is defined as

$$R^{x}_{P,T} = \frac{2(x^M - x^{eq})}{(x^H - x^L)}, \tag{5}$$

with $x^{eq} = \frac{1}{2}(x^H + x^L)$. Here, x is an isospin sensitive quantity that has to be investigated with respect to equilibration. We consider primarily the asymmetry $I = (N - Z)/A$, but also other quantities, such as isoscaling coefficients, ratios of production of light fragments, etc, can be of interest [29, 35, 36]. The indices H and L refer to symmetric reactions between a heavy (n-rich) and a light (n-poor) system, while M refers to the mixed reaction. P, T denote the rapidity region, in which this quantity is measured, in particular the PLF and TLF rapidity regions. Clearly, this ratio is ± 1 in the projectile and target regions, respectively, for complete transparency, and oppositely for complete rebound, while it is zero for complete equilibration (isospin diffusion).

The centrality dependence of the isospin ratio, for instance in the case of Sn + Sn collisions at 35 and 50 *AMeV*, has been investigated in experiments as well as in theory [27–29]. Both BUU-like and QMD-like codes exhibit a nice sensitivity to the symmetry energy value at low-density, yielding a larger degree of equilibration in the "soft" case (i.e., for a higher symmetry energy value at low density, as expected). Information about the stiffness of the symmetry energy, pointing to a L value in the range of 40-80 Mev (for a symmetry energy coefficient around 30 MeV at saturation) has been extracted from the analysis of the MSU data presented in [29, 30], based on the ImQMD transport model [19].

4.2 Isospin effects in neck emission

In presence of density gradients, as the ones occurring when a low-density neck region is formed between the two reaction partners in semi-peripheral collisions, the isospin transport is mainly ruled by the density derivative of the symmetry energy (see Eqs.(3-4)) and so we expect a larger neutron flow to the neck clusters for a stiffer symmetry energy around saturation [7]. This mechanism leads to the neutron enrichment of the neck region (isospin migration). This effect is nicely illustrated by BLOB simulations of Xe + Sn reactions at 32 $AMeV$, see Fig.3 [26], which represent the isotopic content of the neck primary fragments as a function of their transverse velocity. The calculations predict that slower fragments, emitted on longer time scales, can become rather neutron rich in the asystiff case, owing to isospin migration effects.

In order to build observables less affected by secondary decay effects, one can consider the ratio of the asymmetries of the IMF's (β_{IMF}) to those of the PLF/TLF residues (β_{res}). This quantity can be roughly estimated on the basis of simple energy balance considerations. By imposing to get a maximum (negative) variation of E_{sym} when transfering the neutron richness from PLF and TLF towards the neck region, one obtains [28]:

$$\frac{\beta_{IMF}}{\beta_{res}} = \frac{E_{sym}(\rho_R)}{E_{sym}(\rho_I)} \tag{6}$$

From this simple argument the ratio between the IMF and residue asymmetries should depend only on symmetry energy properties and, in particular, on the symmetry energy values corresponding to the residue and neck densities (ρ_R and ρ_I), as appropriate for isospin migration. It should also be larger than one, more so for the asystiff than for the asysoft EOS. Results obtained with SMF, for stiff and soft Asy-EOS (see Fig.1), are represented in Fig.4, for Sn + Sn reactions with different initial charge asymmetry, β_{in}, at 50 $AMeV$. One can see that this ratio is nicely dependent on the Asy-EOS only (being larger in the asystiff case) and not on the system considered. If final asymmetries were affected in the same way by secondary evaporation in the case of neck and PLF fragments, then one could directly compare the results of Fig.4 to data. However, due to the different size and temperature of the neck region with respect to PLF or TLF fragments, de-excitation effects should be carefully checked with the help of suitable decay codes.

In recent experimental analyses performed on Sn+Ni data at 35 $AMeV$ by the Chimera Collab.[32] a strong correlation between neutron enrichement and fragment alignement with respect to the PLF-TLF direction (when the short emission time selection is enforced) has been evidenced, that points to a stiff behavior of the symmetry energy ($L \approx 75$ MeV), for which a larger neutron enrichment of neck fragments is seen (see Fig.3), compatible with the constraints extracted from the analysis of isospin diffusion discussed above. Other experimental analyses have been dedicated to investigate the isotopic content of the fragments emitted at mid-rapidity, in connection to kinematical features and emission time scales [37]. Very recent experiments have opened the possibility to investigate both isospin transport phenomena discussed above (isospin equilibration and migration) within the same set of data [34, 38]. Forthcoming analyses, including the comparison with transport simulations, are thus expected to reduce the uncertainties and provide more stringent constraints about the symmetry energy behavior at low density.

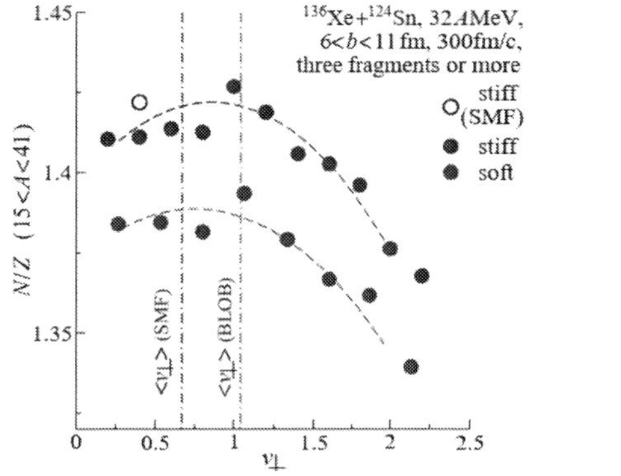

Figure 3. (Color online) BLOB simulations: Isotopic content of neck fragments at 300fm/c, identified in the impact parameter region of maximum production, as a function of the transverse velocity component with respect to the PLF-TLF axis for a stiff and a soft form of the symmetry energy potential.

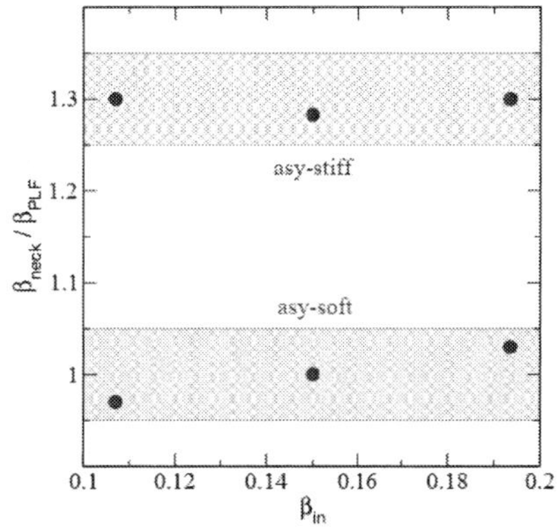

Figure 4. (Color online) Ratio between the neck IMF and the PLF asymmetries, in Sn + Sn reactions at 50 $AMeV$, as a function of the system initial asymmetry. The bands indicate the uncertainty in the calculations [9].

5 Symmetry energy constraints at supra-saturation density

Reactions with neutron-rich systems at intermediate energies (100-500 $AMeV$) are of interest in order to have high momentum particles and to test regions of high baryon and isovector density during the reaction dynamics. In such a context, it is particularly important to consider momentum dependent effective interactions, which essentially lead to the concept of effective masses.

Also in this energy regime it is very interesting to study the collective response of the system. Collective flows are very good candidates to probe the Asy-EOS, since they are expected to be rather sensitive to the features of the nuclear effective interaction, including the momentum dependence of the mean field, see [7, 39]. The transverse flow, $V_1(y, p_t) = \langle \frac{p_x}{p_t} \rangle$, where $p_t = \sqrt{p_x^2 + p_y^2}$ is the transverse momentum and y the rapidity along the beam direction, provides information on the anisotropy of nucleon emission on the reaction plane. Very important for the reaction dynamics is also the elliptic flow, $V_2(y, p_t) = \langle \frac{p_x^2 - p_y^2}{p_t^2} \rangle$. The sign of V_2 indicates the azimuthal anisotropy of the emission: on the reaction plane ($V_2 > 0$) or out-of-plane ($squeeze - out$, $V_2 < 0$) [39]. In particular, the difference between neutron and proton flows is mainly ruled by the balance between the Coulomb repulsion (acting for protons) and the symmetry potential (repulsive for neutrons in neutron-rich systems), and hence it appears as a suitable observable to investigate the supra-saturation behavior of the symmetry energy [40–43].

Let us consider semicentral ($b/b_{max} = 0.5$) collisions of ^{197}Au+^{197}Au at 400 $AMeV$. In the interacting zone baryon densities about $1.7 - 1.8\rho_0$ can be reached in a transient time of the order of 15-20 fm/c. The system is quickly expanding and the freeze-out time is around 50 fm/c [43]. A rather abundant particle emission is observed over this time scale. The comparison of transport calculations (QMD-like codes [41, 42, 44]) to FOPI-LAND and ASY-EOS data [45, 46] has recently allowed to extract new constraints on the symmetry energy slope: L = 72 ± 13 MeV.

However, in order to conclude on the properties of the effective interaction (asystiffness and MD), it would be desirable to combine the analysis of several obervables. For instance, in the considered beam energy range, the N/Z content of the particle emission looks particularly sensitive just to the sign of neutron/proton mass splitting, rather than to the asystiffness [43]. Recent experimental analyses look very promising in this direction [30, 45].

A quite interesting observable is meson production, arising from inelastic nucleon collisions. As first pointed out in Ref.[47], pion ratios, such as the π^-/π^+ ratio, are expected to reflect the N/Z content of the high density regions reached along the reaction dynamics, which in turn depends on the symmetry energy trend (the N/Z ratio being larger in the asysoft case, because more neutrons are kept inside the system). However, more recent calculations show that emission threshold effects (which are different for π^+ and π^-) may reverse the trend of the pion ratio with respect to the effects associated with the N/Z ratio of the high density regions [48–51]. From the analysis discussed in [51], the following contraints was extracted on the symmetry energy slope: L = 79.9 ± 37.6 MeV. Open problems are related to possible effects associated with the width of the Δ resonance (including off-shell effects), Δ and pion potentials in the nuclear medium [52–56]. Moreover, recent AMD calculations have shown that light particle emission (clustering effects) may also affect pion ratios [21]. There is presently an intense activity in the field, also triggered by new devoted experiments [51, 57, 58].

6 Summary and perspectives

Heavy ion collisions, from low up to relativistic energies, offer a quite rich phenomelogy to explore the behavior of the nuclear Equation of State. Recent results have been reviewed here, focusing on isospin effects and the possibility to learn about the nuclear symmetry energy. From the theoretical point of view, transport theories are an essential tool to handle the description of the reaction dynamics and to connect the features of the nuclear effective interaction (and nuclear EOS) to predictions of observables of experimental interest, directly comparable to data. Here we have discussed selected observables and corresponding constraints which have been extracted over the past recent years. In several cases, it would be highly desirable to further reduce the model dependence of the results, in view of a more robust combined analysis of the available experimental data. A dedicated program (TMEP), aiming at comparing the predictions of different transport models under controlled situations, to understand and eliminate possible sources of discrepancies, is running since a few years [6] and new developments are currently being considered.

Within such a phenomenological approach to investigate the nuclear EOS, the synergy between theory and experiments is of crucial importance. From the experimental point of view, new detection systems and analyses, which allow to investigate several reaction mechanisms and observables within the same data set, look extremely promising.

7 Acknowledgements

Useful discussions with P.Napolitani and H.H.Wolter are gratefully acknowledged. Discussions within the TMEP collaboration are also acknowledged.

References

[1] P. Danielewicz, R. Lacey and W.G. Lynch, Science **298**, 1592 (2002).

[2] J.M. Lattimer and M. Prakash, Phys. Rep. **442**, 109 (2007).

[3] S. Burrello *et al.*, Phys. Rev. C **92**, 055804 (2015).

[4] S. Burrello *et al.*, Phys. Rev. C **99**, 054314 (2019).

[5] P-G Reinhard, X. Roca-Maza and W. Nazarewicz, Phys. Rev. Lett. **129**, 232501 (2022).

[6] J.Xu *et al.*, Phys. Rev. C **93**, 044609 (2016); Y.X. Zhang *et al.*, Phys. Rev. C **97**, 034625 (2018); A. Ono *et al.*, Phys. Rev. C **100**, 044617 (2019); M. Colonna *et al.*, Phys. Rev. C **104**, 024603 (2021); H. Wolter *et al.*, Prog. Part. Nucl. Phys. **125**, 103962 (2022).
[7] V. Baran, M. Colonna, V. Greco and M. Di Toro, Phys. Rep. **410**, 335 (2005).
[8] B.A. Li, L.W. Chen and C.M. Ko, Phys. Rep. **465**, 113 (2008).
[9] M. Colonna, V. Baran and M. Di Toro, Eur. Phys. Jou. A **50**, 30 (2014); M. Colonna, Prog. Part. Nucl. Phys. **113**, 103775 (2020).
[10] C.J. Horowitz *et al.*, Jou. Phys. G **41**, 093001 (2014).
[11] P. Chomaz, M. Colonna and J. Randrup, Phys. Rep. **389**, 263 (2004).
[12] J. Xu, Prog. Part. Nucl. Phys. **106**, 312 (2019).
[13] A. Ono, Prog. Part. Nucl. Phys. **105**, 139–179 (2019).
[14] Y. Abe *et al.*, Phys. Rep. **275**, 49 (1996).
[15] M. Colonna *et al.*, Nucl. Phys. A **642**, 449 (1998); J. Rizzo, Ph. Chomaz, M. Colonna, Nucl. Phys. A **806**, 40 (2008).
[16] P. Napolitani, M. Colonna, Phys. Lett. B **726**, 382 (2013).
[17] J. Aichelin, Phys. Rep.**202**, 233 (1991).
[18] M. Papa *et al.*, Phys. Rev. C **64**, 024612 (2001).
[19] Y.X. Zhang and Z. Li, Phys. Rev. C **74**, 014602 (2006).
[20] A. Ono, Phys. Rev.C **59**, 853 (1999).
[21] N. Ikeno *et al.*, Phys. ReV. C **93**, 044612 (2016).
[22] X. Roca-Maza *et al.*, Phys. Rev. C **87**, 034301 (2013).
[23] P. Napolitani, M. Colonna, Eur. Phys. Jou. A, **117**, 07008 (2016).
[24] F. Gagnon-Moisan *et al.* (INDRA Collaboration), Phys. Rev. C **86**,044617(2012).
[25] G. Ademard *et al.* (INDRA Collaboration), Eur. Phys. J. A **50**,33(2014).
[26] M. Colonna, P. Napolitani, V. Baran, *Nuclear Particle Correlations and Cluster Physics*, pag.403 (Ed. W-U Schroeder, World Scientific, Singapore, 2017).
[27] M.B. Tsang *et al.*, Phys. Rev. Lett. **92**, 062701 (2004); B-A Li, L-W Chen, Phys. Rev. C **72**, 064611 (2005).
[28] J. Rizzo *et al.*, Nucl. Phys. A **806**, 79 (2008); E. Galichet *et al.*, Phys. Rev. C **79**, 064615 (2009).
[29] M.B. Tsang *et al.*, Phys. Rev. Lett. **102**, 122701 (2009); Z.Y. Sun *et al.*, Phys. Rev. C **82**, 051603 (2010).
[30] W.G. Lynch and M.B. Tsang, Phys. Lett. B **830**, 137098 (2022).
[31] M. Di Toro, A. Olmi and R. Roy, Eur. Phys. Jou. A **30**, 65 (2006).
[32] E. De Filippo *et al.* (Chimera Collab.), Phys. Rev. C **86**, 014610 (2012); A. Pagano *et al.*, Eur. Phys. Jou. A **56**, 102 (2020).
[33] V. Baran *et al.*, Phys. Rev. C **72**, 064620 (2005).
[34] S. Piantelli *et al.*, Phys. Rev. C **103**, 014603 (2021).
[35] M. Colonna *et al.*, Phys. Rev. C **78**, 064618 (2008).
[36] M. Colonna, Phys. Rev. Lett. **110**, 042701 (2013).
[37] A. Jedele *et al.*, Phys. Rev. Lett. **118**, 062501 (2017); A. Rodriguez Manso *et al.*, Phys. Rev. C **95**, 044604 (2017).
[38] S. Valdré *et al.*, Nucl. Instr. and Meth. A **930**, 27 (2019).
[39] P. Danielewicz, Nucl. Phys. A **673**, 375 (2000).
[40] B-A Li *et al.*, Phys. Rev. Lett. **88**, 192701 (2002).
[41] P. Russotto *et al.*, Phys. Lett. B **697**, 471 (2011).
[42] M.D. Cozma, Phys. Lett. B **700**, 139 (2011).
[43] V. Giordano *et al.*, Phys. Rev. C **81**, 044611 (2010).
[44] M.D. Cozma *et al.*, Phys. Rev. C **88**, 044912 (2013).
[45] P. Russotto *et al.*, Phys. Rev. C **94**, 034608 (2016).
[46] S. Huth *et al.*, Nature, **606**, 276 (2022).
[47] Z. Xiao *et al.*, Phys. Rev. Lett. **102**, 062502 (2009).
[48] G.Ferini *et al.*, Phys. Rev. Lett. **97**, 202301 (2006).
[49] M.D. Cozma, Phys. Lett. B **753**, 166 (2016).
[50] T. Song, C.M. Ko, Phys. Rev. C **91**, 014901 (2015).
[51] J. Estee *et al.* , Phys. Rev. Lett. **126**, 162701 (2021); M.D. Cozma, M.B. Tsang, Eur. Phys. J. A **57** (11), 309 (2021).
[52] J. Hong and P. Danielewicz, Phys. Rev. C **90**, 024605 (2014).
[53] C. Guo *et al.*, Phys. Rev. C **91**, 054615 (2015).
[54] Z. Feng, Eur.Phys.Jou. A **53**, 30 (2017).
[55] Z. Zhang and C-M Ko, Phys. Rev. C **95**, 064604 (2017).
[56] M.D. Cozma *et al.*, Phys. Rev. C **95**, 014601 (2017).
[57] M.B. Tsang *et al.*, Phys. Rev. C **95**, 044614 (2017).
[58] G. Jhang *et al.*, Phys. Lett. B **813**, 136016 (2021).

Long range plans to study the nuclear equation-of-state from sub- to supra-saturation densities with heavy-ion collisions.

A. Le Fèvre[1], *M. Colonna*[2], *G. Verde*[3], *K. Agarwal*[4], *J. Aichelin*[5], *T. Aumann*[6], *A. Chbihi*[7], *M.D. Cozma*[8], *E. De Filippo*[3], *Hannah Elfner*[1], *T. Galatyuk*[6,1], *Ch. Hartnack*[5], *Y. Leifels*[1], *S. Pirrone*[3], *P. Senger*[1], *J. Stroth*[1], *P. Russotto*[2], *H.R. Schmidt*[4], *A. Schwenk*[6], *H. Wolter*[9]

[1]GSI Helmholtzzentrum für Schwerionenforschung GmbH, Darmstadt, Germany.
[2]Istituto Nazionale di Fisica Nucleare, Laboratori Nazionali del Sud, Catania, Italy
[3]Istituto Nazionale di Fisica Nucleare, Sezione di Catania, Catania, Italy
[4]Physikalisches Institut, Eberhard Karls Universität Tübingen, Tübingen, Germany.
[5]IMT Atlantique, Université de Nantes, France
[6]Department of Physics, Technische Universität Darmstadt, Darmstadt, Germany.
[7]GANIL, Caen, France
[8]IFIN-HH Bucharest, Romania
[9]University of Munich, Munich, Germany

Abstract. We cover here the present state-of-the-art in constraining the nuclear equation-of-state (EoS) and the symmetry energy using heavy-ion collisions (HIC), from sub- to supra- saturation densities, from Fermi to (ultra-) relativistic beam energies. We also discuss how HIC constraints on the EoS contribute to the knowledge of thermodynamical properties of neutron star matter. Necessary improvements and challenges are outlined, in particular in the perspective, for HICs, of staying competitive with future astrophysical multimessenger observations.

1 Introduction

Constraining the nuclear equation-of-state (EoS) of nuclear matter, from sub- to supra-saturation densities has given rise, since two decades, to a very intense experimental and theoretical activity worldwide. Important heavy-ion collision (HIC) projects are being carried-out at European heavy-ion accelerator facilities, from Fermi energies (GANIL, INFN-LNS) to (ultra-) relativistic energies (GSI, CERN). An early strong involvement in these studies has also been established in the USA, specifically at the Lawrence Berkeley National Laboratory (LBNL), at Brookhaven National Laboratory (BNL) and at the Michigan State University (MSU). Apart from a deeper knowledge of the EoS, which is a fundamental nuclear matter property, one of the main aims is to understand astrophysical processes like core-collapse supernovae explosions, neutron star properties like mass, radii and deformability, binary mergers and their associated phenomena such as gravitational wave emissions and kilonova signals. In a pioneering work [1], constraints on the pressure of symmetric nuclear matter (SNM) at large densities could be obtained from transport model calculations (pBUU) by analysing experimental data from LBNL and BNL on ^{197}Au+^{197}Au collisions at energies up to 10 GeV per nucleon. These collisions probe nuclear matter at 2–4.5 times saturation density ($n_{sat}\approx0.17$ fm^{-3}). Similarly, around the same time and at later years, several experimental HIC campaigns at GSI Darmstadt have been performed to study the EoS of isospin symmetric matter at lower densities, ranging from 0.7 to 3·n_{sat}. First, comparing the yields of kaons [2] produced in C+C and Au+Au collisions at sub-threshold incident energy, the KaoS collaboration, by using various transport models (IQMD [3] and RQMD [4]), has concluded that the EoS of SNM is soft (with a nuclear incompressibility $K_0\approx200$ MeV) provided that the right momentum dependence of the nucleon-nucleon potential is taken into account, as demonstrated before in pA collision experiments. Thereafter, a complete excitation function campaign was performed by the FOPI collaboration with Au+Au collisions at energies between 90 AMeV and 1.5 AGeV. By studying the elliptic flow of light charged particles, the results of the KaoS collaboration could be confirmed. These new measurements with FOPI were performed with an unprecedented precision [5]. Both IQMD and UrQMD, used to extract EoS contraints from the data, agreed on KaoS conclusions, leading to incompressibility values of $K_0=190 \pm 30$ MeV and 220 ± 40 MeV, respectively [6]. In Fig. 1, the present constraints from HIC experiments on symmetric nuclear matter EoS are summarized from intermediate energies (GSI data, red band) to ultra-relativistic energies (AGS data, green band). As a comparison, recent predictions from chiral effective field theory are displayed as well (2-body N^{2}LO (light blue band) and 3-body N^{3}LO (dark blue band) interactions), showing a fair agreement with FOPI results. Another complementary valuable observable to constrain the EoS at high density is the continuum dilepton radiation, as addressed in White Papers by CBM and HADES, the authors believe that adding this observable will help reducing model uncertainties, which is a big issue in this field [7].

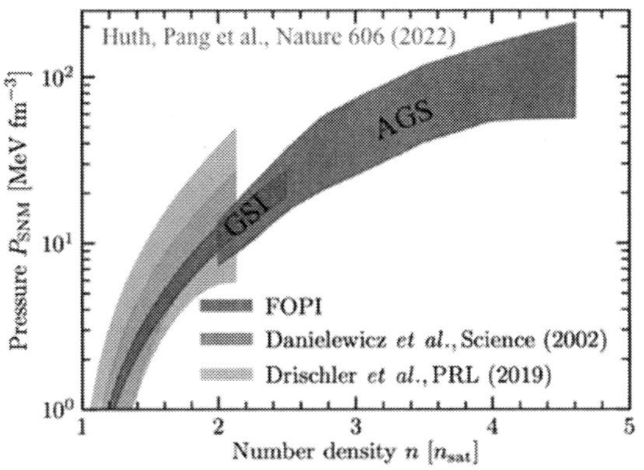

Fig. 1. Constraints from HIC experiments on symmetric nuclear matter EoS. See text for details.

2 Symmetry Energy

With the growing interest in properties of exotic nuclei and neutron star matter, a strong international activity aiming at constraining the nuclear symmetry energy has arisen. Numerous experimental on-earth methods based on heavy-ion collision studies have been developed. Sub-saturation densities have been probed on the one hand with measuring neutron-skin thicknesses of nuclei, collective resonances, short range correlations, isotope yields from the break-up of colliding heavy-ions, nuclear masses, and isobaric analog states, providing a more and more precise description of the density evolution of the symmetry energy as depicted on Figs. 2 and 5. On the other hand, at supra-saturation density, the symmetry energy has been constrained with the best so far achieved accuracy through the comparison of the elliptic flow of neutrons and light charged particles: this was obtained from FOPI-LAND re-analyzed data [8] and later improved by the ASY-EOS experiment at GSI [9]. The successful S394 experiment held by the ASY-EOS Collaboration at the GSI [9] has pushed further the boundaries of our knowledge of the symmetry energy of the nuclear equation-of-state (EoS), up to two times the saturation density. The symmetry energy is found to be moderately soft with a nearly linear behaviour. After this achieved accuracy of the methods used by the ASY-EOS collaboration, the astrophysics community recently recognized (see the NEOS conferences, the Rußbach (Austria) School of Astrophysics and the Hirschegg 2020 Workshop) that laboratory-based HIC are also a powerful tool to constrain neutron star matter properties. Recently, an attempt has been done at RIKEN in Japan with the SπIRIT experiment [10], to contribute to this constraint at medium-high densities by measuring the yield ratio of π^+ over π^-. This result is also shown in Fig. 2. It is consistent with the results obtained at GSI, though less accurate, mainly due to the omission of non-resonant pion production channels. Confirmation of this result using other transport models is required and foreseen for the near future.

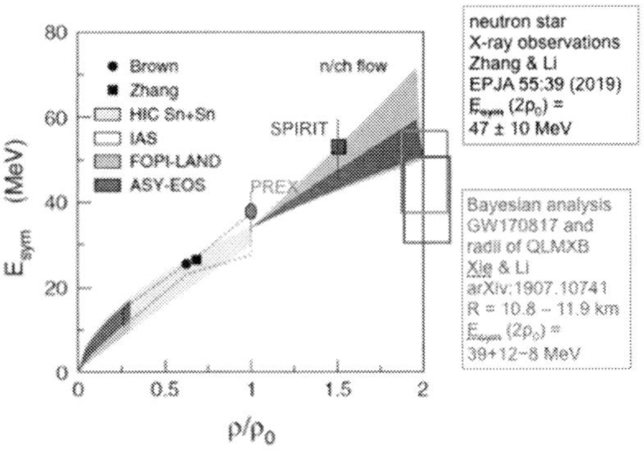

Fig. 2. Compilation of nuclear symmetry energy constraints obtained with heavy ion experiments, compared with recent astrophysical constraints. Taken from Ref. [9] and complemented with results from Horowitz et al., J. Phys. G (2014). Recent PREX [11] and SπRIT data have been added.

The most accurate experimental result on the neutron-skin is presently provided by the PREX experiment [11], but it still remains in the present situation, with very large 1 sigma errors, far not as accurate as e.g. the flow method mentioned before or the chiral effective field theory to constrain the symmetry energy. In the future, developments of methods applicable to neutron-rich nuclei are needed, in order to provide precise and accurate data. At FAIR, proton elastic scattering of radioactive beams can be performed at the storage ring, after the collector ring will be implemented. In addition, an electron-heavy ion collider for radioactive beams at GSI/FAIR is proposed for the future as an outside experiment built at GSI. Then neutron density distributions and corresponding neutron skins can be extracted precisely. This also points to the need of bench-marking nuclear transport models, which is compulsory to draw robust constraints on the EoS. The new constraints coming from HIC have stimulated, since 2017, a renewed and intense activity on the theoretical side, including a world network initiative of code comparison TMEP (Transport Model Evaluation Project). This long-term international transport code comparison project has been pursued [12], and is still on-going, presently focusing on the pion production for better interpreting S π RIT and similar experimental data sets. A very close collaboration between theory and experiment is of utmost importance to obtain reliable information from HIC experiments.

In a pioneering attempt to extract a symmetry energy constraint at large density, the FOPI collaboration measured K^+/K^0 yields from Ru+Ru and Zr+Zr collisions [13]. The results remained inconclusive due to low experimental statistics, the difficulties of measuring kaon ratios for one collision system rather than double ratios, and uncertainties in describing kaon production. The advantage of using kaons relies on the fact that they are expected to be produced at the early stages of the collisions, when nuclear matter is in a high-density state due to compression, and do not interact significantly with the dynamically evolving system before being detected. The EoS is from the kaon multiplicity and kaon participant dependence [14]. They may thus

https://doi.org/10.1051/epjconf/202329010004

represent a more sensitive probe of the supra-saturation density dependence of the symmetry energy, as compared to the pion ratios [15].

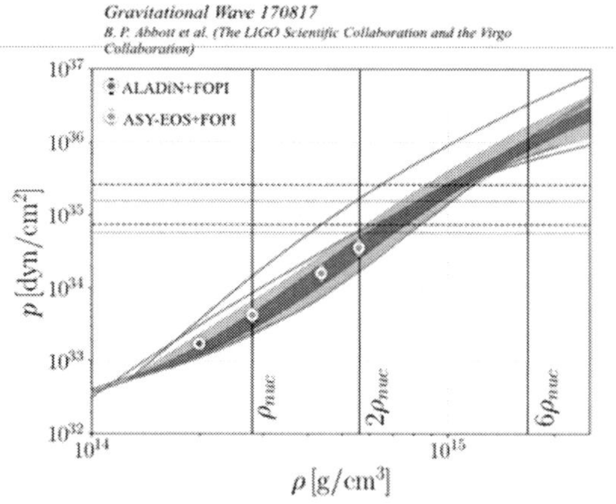

Fig. 3. Superimposition of pressure reconstruction in a neutron star from HIC constraints at several densities (colored dots) on gravitational wave 170817 (B.P. Abbot et al., PRL 119, 161101 (2017)) constraint (blue area). It shows the remarkable agreement between both macroscopic (GW) and microscopic (HIC) approaches.

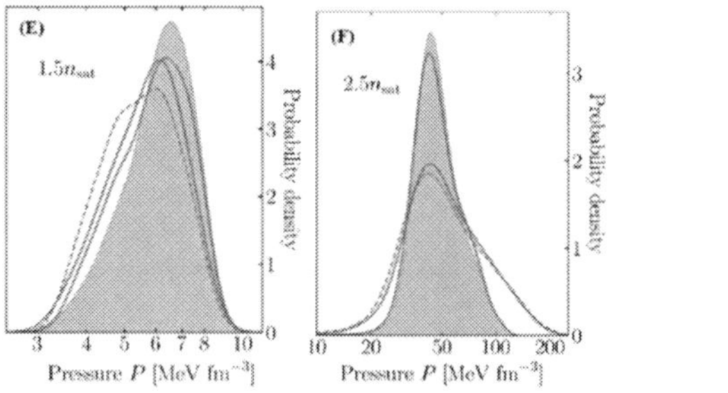

Fig. 4. Posterior constraints on the EoS of neutron-star matter, obtained from the Bayesian analysis on distributions for the pressure at 1.5 nsat (left) and 2.5nsat (right) at different stages of the analysis, with respective constraints coming from HIC's (orange curve), astrophysical multi-messenger observations (green curve), and with combining both (light blue shaded region) respectively. The dashed curve is the prior distribution. From Ref. (S. Huth at al., Nature 606, 276 (2022)).

Recently, motivated by the increasing precision obtained by astrophysical multi-messenger constraints of the EoS of neutron star matter (with the measurement of gravitational waves, kilonovas, X-ray emissions of binary pulsars), an interdisciplinary research group gathering nuclear theorists, heavy-ion collision experimentalists, and astrophysicists has quantitatively demonstrated the impact of HICs on the knowledge of neutron star matter properties, that is remarkably consistent with astrophysical measurements, and therefore very complementary: "The simultaneous analysis of the HIC data and the full suite of astrophysical data on neutron stars within a self-consistent Bayesian framework is a novel contribution to the literature » (S. Huth et al., Nature 606, 276 (2022)). Several conclusions have been drawn from this study (as it is illustrated by Fig. 3 and 4):

1. Constraints from HIC data favor higher pressures, similarly to what is deduced from NICER explorations, with an overall remarkable consistency with chiral EFT and astrophysical constraints

2. Up to densities of the order of $1.5\rho_0$, HICs constrain the EoS of neutron stars with an accuracy comparable to that of most recent astrophysical findings, favoring a somehow stiffer EoS (higher pressure).

3. Above $1.5\rho_0$, astrophysical measurements are still more accurate, and drive the neutron star EoS, but with lower statistics.

4. The most significant densities for constraining NS radii are as it follows: for 1.4 $M_\odot$, $\rho \approx 1.6\ n_{sat}$; for 2 $M_\odot$: $\rho \approx (2\text{–}2.5)\ n_{sat}$

One of the main conclusions of this innovative work is to significantly contribute to constrain neutrons star radii and the density dependence of pressure in neutron stars, HIC should not only improve the accuracy of the symmetry energy constraints just above saturation density, but also probe it at larger densities, as compared to the so far limited reachable regime of about 1.5 n_{sat}.

Various experimental HIC projects, focused especially on constraining the symmetry energy below and close to 3 n_{sat} are planned. Among them, we mention: a) a second ASY-EOS experiment at GSI at higher incident energies, aiming at performing new measurements of elliptic flow with improved accuracy, probing higher densities with respect to the first ASY-EOS experiment, and profiting of the improved capabilities of the new NeuLAND detector; b) pion production measurements close to production threshold, to be performed at RIKEN.

Other experiments such as HADES at GSI, aiming at measuring kaon yields close to the production threshold, also promise to possess the potential of constraining the symmetry energy at larger densities. But the pre-requisite would be to better bench-mark transport models to get more robust symmetry energy constraints, considering the complexity of mechanisms and dynamics ruling the kaon emission in the fireball. A robust description of pion production is a prerequisite for kaon studies [16]. This concerns models like IQMD, dcQMD [17], UrQMD, PHQMD [18], SMASH [19].

At high densities, complementary information can be provided by investigating the isospin dependence of short-range correlations. A first experiment has been performed at R3B FAIR Phase-0. A full program with an upgraded setup will provide data on a wide isospin range along isotopes. In the farer future, short range correlations could be studied at a multi-GeV radioactive ion beam facility which is proposed as a later moderate upgrade of FAIR.

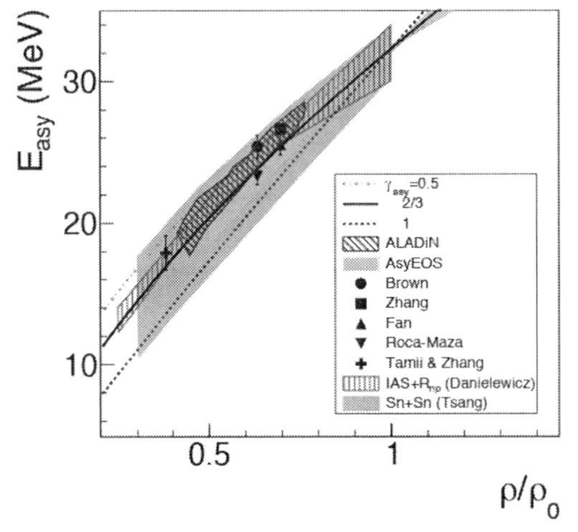

Fig. 5. Compilation of constraints on symmetry energy obtained with heavy-ion studies. The blue hatched area are the most recent (preliminary) results obtained from the analysis of the spectator decay products of various projectiles as measured by the ALADiN@GSI S254 experiment [23]. To guide the eye, density dependence functionals used in IQMD and FRIGA clustering algorithm [24] are shown as lines for various values of the power exponent of the symmetry potential used in IQMD-FRIGA (related to the stiffness of the symmetry energy dependence on density): 0.5 (green dashed-dotted), 2/3 (black solid) and 1 (blue dashed). Various recent findings of the literature are displayed for comparison: AsyEOS (from Russotto et al. (2016) re-analysed with IQMD) as orange area, HIC (Sn+Sn) determined from isospin diffusion observables measured in mid-peripheral collisions of Sn isotopes [25] (grey area), IAS + Rnp from the analysis of isobaric analog states (IAS) supplemented with additional constraints from neutron-skin data [26] (magenta hatched area), along with results (as black markers) extracted from nuclear structure information by Brown [27], Zhang and Chen [28], Fan et al. [29], Roca-Maza et al. [30], Zhang and Chen [31] using dipole polarisability data by Tamii et al. [32].

As shown in the interdisciplinary study of S. Huth et al., Nature **606**, 276 (2022), the combination of HIC constraints on the nuclear EoS with effective field theory of strong interactions is very successful in determining with an enhanced accuracy the thermodynamical properties of neutron star matter. This points out that powerful many-body theories must be further developed to support such joint efforts. In parallel to the chiral effective field theory which is commonly used in Bayesian analysis of neutrons-star multi-messenger studies, constraining strongly the sub- to near saturation region, one should note that HIC may also provide a valuable contribution, because they are increasingly accurate in constraining the symmetry energy at low densities. This is demonstrated by numerous recent experimental constraints such as the ones illustrated on Fig. 5, with a remarkable increasing accuracy and consistency between results using different observables and methods. A similar study has been recently published by Lynch et al. (Phys. Lett. B 830, 137098 (2022)), where also the attempt is made to determine the sensitive density of a particular observable more systematically. Similar efforts are on-going using the isospin migration/distillation effect as a probe of the strength of the symmetry energy in neutron rich or isospin asymmetric colliding systems at the Fermi energy regime (see for instance Phys. Rev. C 78, 064618 (2008) and Nucl. Phys. A 730 (2004) 329), with e.g. the INDRA-VAMOS [20], INDRA-FAZIA [21] and CHIMERA-FARCOS [22] experiments. However, even at these energies there is a strong need for transport model benchmarking in order to provide robust probes, concerning in particular a more accurate description of the cluster formation in HIC's. Adding to the constraining power of probes, new opportunities with radioactive ion beams are presently emerging, with the need of neutron detection, at both INFN-LNS and GANIL.

3 Beyond densities of 3 n_{sat}

In the next decade, astrophysical multi-messenger observations are foreseen to deliver new constraints on the neutron matter EoS, with unprecedented accuracy, thanks to more precise apparata and larger statistics, being able to detect events coming from larger distances in the universe. In the context of HIC explorations, in order to maintain a competitive contribution to our understanding of neutron star physics, it is compulsory to probe densities that are higher than $3{\cdot}n_{sat}$. Such task will concern both the symmetric part of the EoS, which is still not accurately constrained (see Fig. 1, green area) at these densities, and the symmetry energy. Such a program could be performed using the already existing HADES set-up, as well as the future CBM [33] and R3B detector ensembles at GSI/FAIR, with HIC's at bombarding energies above 2 A·GeV. In particular, the precise measurement of the collective flow of protons and light fragments in HIC at energies up to 10A GeV by the CBM experiment will significantly improve our knowledge on the high-density EOS of symmetric nuclear matter. On the theoretical side, a strong effort of the community working on transport models for relativistic energy collisions needs to be sustained in order to deliver observables that are both robust and sensitive to the nuclear EoS (for both isospin symmetric and asymmetric matter). Indeed, elliptic flow phenomena, that are very successful to constrain the EoS at lower energies, will be significantly reduced above 2-3 A·GeV incident energies. The reason is the very fast escape of spectators with respect to the expanding evolution of the fireball, which eliminates the spectator shadowing effect that is needed to generate a negative elliptic flow of particles emitted from the participant region, whose amplitude informs on the expansion velocity – hence on the compressibility – of this initially compressed region [34]. Due to the highly relativistic nature of the dynamics at such bombarding energies, a new generation of relativistic transport models needs to be developed (within the scientific context of SMASH, PHQMD and other approaches), using the state-of-the-art description of in-medium potentials, momentum/effective mass dependencies, short-range correlations, off-shell transport with spectral functions for all particles, and possibly sub-nuclear degrees-of-freedom, benchmarked with e.g. data taken at SIS18 at the highest available beam energies (FOPI, ASY-EOS, HADES).

4 Conclusions for a long-range plan

According to the above-mentioned physical motivations, it will be important to support the following actions:

- High precision measurements of cluster production at low density, of isospin diffusion and migration effects using HICs, and exotic systems at GANIL, INFN-LNS and GSI;

- High precision measurements of elliptic flow of neutrons and protons at relativistic energies to probe supra-saturation densities up to around 3·n_{sat} (i.e. up to about 2 A·GeV) at GSI (R3B);

- Perform high precision measurements of collective flow of protons and light fragments in HICs up to 10 A·GeV (up to 5·n_{sat}) at FAIR;

- Stimulate significant efforts in developing new neutron detectors to be used in various energy regimes, characterized by good angular and energy resolutions with a high granularity and efficiency;

- A strong collaboration to develop and use transport models in Europe (IQMD, UrQMD, dcQMD, SMASH, BUU, PHQMD, and others) aiming at providing new and robust observables sensitive to the nuclear EoS and to the symmetry energy.

- The opportunity/challenge to apply Bayesian statistics just for the model-to-data comparison to extract the nuclear equation of state from heavy-ion collisions with transport codes.

- Develop alternative methods to PREX for neutron-rich nuclei, in order to provide precise and accurate data

References

1. P. Danielewicz, R. Lacey & W.G. Lynch, Science **298**, 1592 (2002)
2. C. Sturm et al., Phys. Rev. Lett. **86**, 39 (2001)
3. C. Hartnack et al., Phys. Rev. Lett. **96**, 012302 (2006)
4. C. Fuchs et al., Phys. Rev. Lett. **86**, 1974 (2001); C. Fuchs, Prog. Part. Nucl. Phys. **56**, 1 (2006).
5. W. Reisdorf et al., Nucl. Phys. A**876**, 1 (2012); A. Le Fèvre at al., Nucl. Phys. A **945**, 112 (2016) ; Y. Wang et al., Phys. Lett. B **778**, 207 (2018)
6. S. A. Bass et al., Progr. Part. Nucl. Phys. **41**, 255 (1998); Q. Li and M. Bleicher, J. Phys. G **36**, 015111 (2009).
7. HADES Collaboration, Nat. Phys. **15** (2019) 10
8. P. Russotto et al., Phys. Lett. B **697**, 471 (2011).
9. P. Russotto et al., Phys. Rev. C **94**, 034608 (2016).
10. J. Estee et al., Phys. Rev. Lett. **126**, 162701 (2021)
11. D. Adhikari et al., Phys. Rev. Lett. **126**, 172502 (2021)
12. H. Wolter et al., Prog. Part. Nucl. Phys. **125** (2022) 103962
13. X. Lopez et al., Phys. Rev. C **75**, 011901(R) (2007)
14. C Hartnack et al., Phys. Rep. **510**, 119 (2012)
15. G. Ferini et al., Phys. Rev. Lett. **97**, 202301 (2006)
16. J. Adamczewski-Musch et al., Eur. Phys. J. A **56**, 259 (2020)
17. M.D. Cozma: Eur. Phys. J. A54(3) and Refs. therein
18. J. Aichelin et al., Phys. Rev. C **101**, 044905 (2020)
19. J. Weil et al., Phys. Rev. C **94**, 054905 (2016)
20. Q. Fable et al., Phys. Rev. C **106**, 024605 (2022)
21. C. Ciampi et al., Phys. Rev. C **106**, 024603 (2022)
22. E. De Filippo, Eur. Phys. J A **50**, 32 (2014)
23. C. Sfienti et al., Nucl. Phys. A **749**, 83c (2005)
24. A. Le Fèvre et al., Phys. Rev. C **100**, 034904 (2019)
25. M.B. Tsang et al., Phys. Rev. Lett. **102** (2009) 122701.
26. P. Danielewicz and J. Lee, Nucl. Phys. A **922**, 1 (2014)
27. B. A. Brown, Phys. Rev. Lett. **111**, 232502 (2013)
28. Z. Zhang et al., Phys. Lett. B **726**, 234 (2013)
29. X. Fan et al., Phys. Rev. C **89**, 017305 (2014)
30. X. Roca-Maza et al., Phys. Rev. C **87**, 034301 (2013).
31. Zhang et al., Phys. Rev. C **92**, 031301(R) (2015)
32. A. Tamii et al., Phys. Rev. Lett. **107**, 062502 (2011)
33. CBM Collaboration, Eur. Phys. J.A **53**, 3 (2017)
34. A. Le Fèvre et al., Phys. Rev. C **98**, 034901 (2018)
35. C. Drischler, K. Hebeler, and A. Schwenk, Phys. Rev. Lett. **122**, 042501 (2019).
36. C. Drischler, R.J. Furnstahl, J.A. Melendez, and D.R. Phillips, Phys. Rev. Lett. **125**, 202702 (2020).
37. S. Huth, C. Wellenhofer an A. Schwenk, Phys. Rev. C **103**, 025803 (2021).
38. G. Raaijmakers, et al., Astrophys. J. Lett. **918**, L29 (2021).
39. E. Annala, T. Gorda, A. Kurkela, J. Nättilä and A. Vuorinen, Nat. Phys. **16**, 907–910 (2020).
40. A. Bauswein, O. Just, H.-T. Janka and N. Stergioulas, Astrophys. *J.* **850**, L34 (2017).
41. E. Annala, T. Gorda, A. Kurkela and A. Vuorinen, Phys. Rev. Lett. **120**, 172703 (2018).
42. E.R. Most, L.R. Weih, L. Rezzolla and J. Schaffner-Bielich, Phys. Rev. Lett. **120**, 261103 (2018).
43. B.P. Abbott, et al., Phys. Rev. Lett. **121**, 161101 (2018).

44. D. Radice and L. Dai, Eur. Phys. J. **A55**, 50 (2019).
45. C.D. Capano, et al., Nat. Astron. **4**, 625–632 (2020).
46. I. Legred, K. Chatziioannou, R. Essick, S. Han and P. Landry, Phys. Rev. D **104**, 063003 (2021).
47. P. Danielewicz, R. Lacey, and W.G. Lynch, Science **298**, 1592–1596 (2002).
48. M.B. Tsang, W.G. Lynch, P. Danielewicz, and C.Y. Tsang, Phys. Lett. B **795**, 533–536 (2019).
49. I. Tews, J. Carlson, S. Gandolfi and S. Reddy, Astrophys. J. **860**, 149 (2018).
50. E. Epelbaum, H.-W. Hammer and U.-G. Meissner, Rev. Mod. Phys. **81**, 1773–1825 (2009).
51. R. Machleidt, and D.R. Entem, Phys. Rep. **503**, 1–75 (2011).
52. E. Epelbaum, H. Krebs and U.-G. Meißner, Eur. Phys. J. A **51**, 53 (2015).
53. J. Carlson, et al., Rev. Mod. Phys. **87**, 1067 (2015).
54. R. Essick, I. Tews, P. Landry, S. Reddy, and D.E. Holz, Phys. Rev. C **102**, 055803 (2020).
55. I. Tews, J. Margueron, and S. Reddy, Phys. Rev. **C98**, 045804 (2018).
56. J. Antoniadis, et al., Science **340**, 1233232 (2013).
57. Z. Arzoumanian, al., Astrophys. J. Suppl. Ser. **235**, 37 (2018).
58. E. Fonseca, et al., Astrophys. J. Lett. 915, L12 (2021).
59. P.T.H. Pang, et al., Astrophys. J. 922, 14 (2021).
60. B. Margalit, and B.D. Metzger, Astrophys. J. 850, L19 (2017).
61. L. Rezzolla, E.R. Most, and L.R. Weih, Astrophys. J. 852, L25 (2018).
62. T. Dietrich, et al., Phys. Rev. D 100, 044003 (2019).
63. B.P. Abbott, et al., Phys. Rev. X 9, 011001 (2019).
64. S. Huth, P.T.H. Pang, et al., Nature 606 (2022).

EPJ Web of Conferences **290**, 10005 (2023) https://doi.org/10.1051/epjconf/202329010005
EuNPC 2022

Nuclear Astrophysics at the Low-Energy Frontiers: Updates from underground laboratories

Francesca Cavanna[1,*]

[1]Istituto Nazionale di Fisica Nucleare, sezione di Torino, Via Pietro Giuria 1, 10125 Torino

Abstract. Nuclear fusion reactions are the heart of nuclear astrophysics: they sensitively influence the nucleosynthesis of the elements in the earliest stages of the Universe and in all the objects formed thereafter; control the associated energy generation and neutrino luminosity; influence the evolution of stars. Unfortunately, measuring reaction cross sections at astrophysically relevant energies is exceptionally challenging due to Coulomb repulsion between nuclei, resulting in cross section values as low as fbar. Laboratorial measurements of these cross sections are often unfeasible due to overwhelming cosmic-ray-induced backgrounds. One effective solution to this problem is to conduct experiments in underground laboratories. The Laboratory for Underground Nuclear Astrophysics (LUNA) is an experimental approach based on an underground accelerator focusing on studying nuclear fusion reactions. Its primary objective is to accomplish direct measurement of cross sections for nuclear reactions that have significance in stellar and primordial nucleosynthesis. This article will present the latest findings and future objectives.

1 Introduction

Nuclear astrophysics is as a multifarious field, interrelated with astronomy, cosmology, neutrino physics, and nuclear physics. Nuclear fusion reactions serve as the centerpiece of this discipline. The lifespan of stars can be depicted as a series of phases in which successively heavier elements incinerate at the core. With primordial hydrogen and helium as a starting point, various material is produced, and theoretical models endeavor to align anticipated elemental ratios with cosmic observations. Since interacting nuclei have positive charges, they repel each other, proportional to their nuclear charge, meaning that a temperature of approximately 10^7 K is necessary for hydrogen burning so that the projectiles can break through the Coulomb barrier. Fusion is only possible if the Coulomb barrier is surpassed. Classically, the interaction energy must be higher than the effective height of the Coulomb barrier, while in quantum mechanics, particles with an energy $E<E_C$ possess a slim probability to penetrate the Coulomb barrier (Tunnel effect). This outcome coincides with the Gamow peak [1]. At these energies nuclear cross sections can significantly decrease (pbar or fbar), thereby resulting in a count rate being much lesser than the environmental background of a detector.

One possible solution to reduce natural and cosmic background is to establish an accelerator facility deep underground, such as the LUNA experiment under the Gran Sasso Mountain with a thickness of 1400 m of rock — equivalent to 3800 m of water — to suppress muon and neutron fluxes by six and three orders of magnitude, respectively. The experimental apparatus (see figure 1) includes a 400 kV electrostatic accelerator which provides a high-intensity proton or alpha beam. The beam energy at the exit of the accelerator has a spread of less than 100 eV, and the energy drift is less than 5 eV/h. The beam energy uncertainty is kept to 0.3 keV, and the calibration is frequently rechecked and found consistent with new measurements [2]. The beam can be delivered to either a solid or gas target [3]. In the case of a solid target, the proton beam can be guided and focused to the target station using an extremely stable analysing magnet and a copper pipe that is cooled to liquid nitrogen temperature to serve as a cold trap and prevent carbon buildup. A negative voltage can be applied to the cold trap to suppress secondary electrons. The isolated target holder also acts as a Faraday cup for beam integration, and the target is typically water-cooled with deionized water. The use of a gas target provides the advantage of isotopical purity and high stability over prolonged periods. For nuclear astrophysics research, a windowless gas target is typically preferred to avoid energy straggling that can occur with a solid containment along the beam path. At the low proton energies required for astrophysics, a gas target separated from the accelerator by a solid wall would be unsuitable due to the difficulty of constructing a thin enough window to minimize beam energy straggling to less than 1 keV while withstanding the pressure gradient between the target and accelerator, typically ranging a few mbar. Consequently, a vacuum system with differential pumping stages has been implemented at LUNA to contain the gas inside the target chamber. This setup consists of an analyzing magnet, a vertical steering magnet, and vacuum pumps required to

*e-mail: francesca.cavanna@to.infn.it

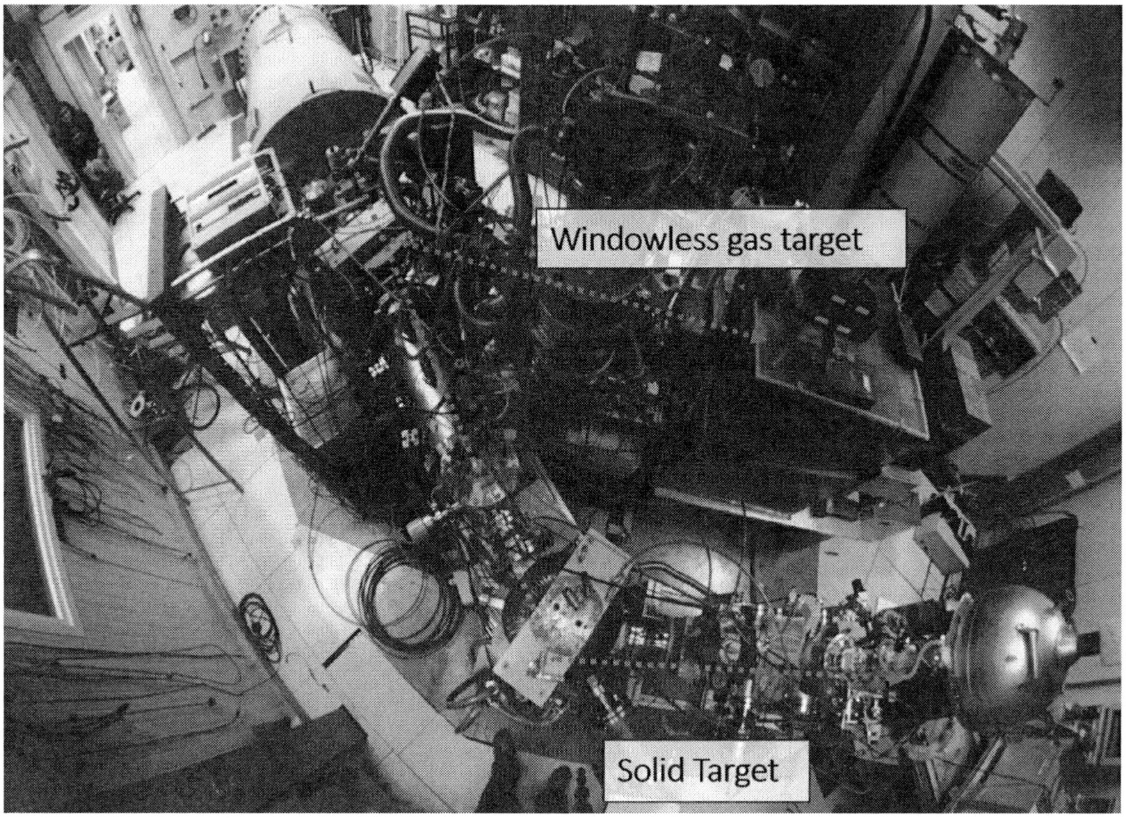

Figure 1. LUNA experimental setup. The two beam lines, for solid and gas targets, are visible.

maintain a pressure of 10^{-7} mbar inside the accelerating tube. The beam is transmitted through several apertures of decreasing diameter to collimate the beam into the target chamber [4, 5]. Thanks to the described experimental apparatus, crucial hydrogen [6] and helium-burning [7, 8] and Big Bang Nucleosynthesis [9] reaction cross sections have been measured in the past down to astrophysical energy levels. Few selected scientific cases will be discussed in the following.

2 The D(p,γ)^{3}He cross section measurement

Big Bang Nucleosynthesis (BBN) occurs during the first minutes of cosmological time in a rapidly expanding hot and dense Universe, where a fraction of protons and nearly all free neutrons end up bound in ^{4}He, while D, ^{3}H, ^{3}He, ^{6}Li, ^{7}Li and ^{7}Be nuclei form in trace quantities. The D(p,γ)^{3}He is responsible for deuterium burning during BBN and it is of primary importance in cosmology because it affects the primordial deuterium abundance, that in turn is very sensitive to fundamental cosmological parameters such as the baryon density and the amount of relativistic species permeating the early Universe [10]. This reaction is also of a particular interest in theoretical nuclear physics because it offers a unique opportunity to test *ab – initio* calculations. Up to few years ago, the D(p,γ)^{3}He reaction was the most uncertain among the deuterium destruction channels. In the low-energy range ($E_{\rm cm} \simeq 2-20$ keV), mostly relevant to hydrogen burning in the Sun and in protostars, cross sections were obtained with a systematic error of at most 5.3% with the 50 kV LUNA accelerator[11]. In the BBN energy range and beyond ($E_{\rm cm} \simeq 30-700$ keV), several data sets were available, however not with the required accuracy, with systematic errors of 9% or higher [12–15]. The situation was further compounded by the fact that a recent *ab initio* calculation [16] disagrees at the 8% level with the best fit of experimental data reported in Iliadis et al. [17]. These large uncertainties had a significant impact on the comparison between predicted and observed primordial abundance of deuterium. As a consequence of the poor experimental data for the D(p,γ)^{3}He reaction, firm conclusions on the cosmological model could not be drawn. For all these reasons, a new experimental campaign started at LUNA in 2016.

The experiment was performed using the gas target filled with molecular deuterium; γ-rays were detected with a 137% HPGe detector in close geometry with the interaction chamber. The cross-section was measured with an unprecedented high precision in 30-50 keV steps inside the energy range of interest for BBN (30 keV < E_{cm}

< 300 keV) [9, 18]; the corresponding S-facor is shown in figure 2. Beam-induced background contribution and possible deuterium implantation were evaluated with measurements with inhert ^{4}He gas in the target chamber.

3 The ^{17}O(p,γ)^{18}F cross section measurement

The ^{17}O(p,γ)^{18}F reaction plays a crucial role in the hydrogen burning phases of different stellar scenarios, as Asymptotic Giant Branch Stars (AGB) [19]. At temperature of 20 MK < T < 80 MK the main contribution to the astrophysical reaction rate comes from the poorly constrained 65 keV resonance [20], whose strength is presently determined only through indirect measurements. Considering the typical experimental values in a laboratory for beam current, isotopic enrichment and detection efficiency, this strength yields to an expected count rate of less than one count per Coulomb, making the direct measurement of this resonance extremely challenging. A new high sensitivity setup has been installed at LUNA; the residual background underground was further reduced by a devoted shielding of lead and borated (5%) polyethylene. Gamma-rays were detected using a 4π BGO [21, 22] detector whose efficiency was optimized installing aluminum target chamber and holder. With about 400 C accumulated charge on Ta_2O_5 targets, with nominal ^{17}O enrichment of 90%, the LUNA collaboration has performed the first direct measurement of the 65 keV resonance strength.

4 The LUNA-MV project

The activity of the LUNA experiment will continue in the next 20 years with the LUNA-MV project which aims at studying the phases of stellar evolution successive to the quiescent Hydrogen burning, in particular the helium and carbon burning phases. To face the challenge, a new 3.5 MV single-ended accelerator has been installed in hall B of Gran Sasso Laboratories and is now part of a facility called: Bellotti Ion Beam Facility. The new accelerator can deliver hydrogen, helium and carbon high-intensity beams[23]. The direct measurement of the cross section of the ^{12}C+^{12}C reaction at the relevant astrophysical energies is the main goal of the first part of the project. The neutron sources relevant for stellar nucleosynthesis beyond iron i.e. ^{13}C(α,n)^{16}O and ^{22}Ne(α,n)^{25}Mg will be studied as well. The first reaction which will be measured and used also for the commissioning of the accelerator is the ^{14}N(p,γ)^{15}O, of crucial importance for the CNO solar neutrinos [24]. The accelerator is deployed inside a new room with concrete walls. The concrete walls and the room roof (thickness of 80 cm) serve as neutron shielding to prevent any possible background to the other experiments installed in the same site. The shielding guarantees that, in the worst operating conditions (i.e.: maximum energy and beam intensity), the neutron flux just outside the accelerator room will be about one tenth of the laboratory natural background. The accelerator is a custom Inline Cockcroft Walton accelerator, manufactured at High Voltage Engineering Europe (HVEE). The Terminal Voltage will range from 0.2 to 3.5 MV. As with the present LUNA400 accelerator, ion beams of H+, ^{4}He+, ^{12}C+ and ^{12}C++ in the energy range from 0.350 to 7 MeV could be sent to two different beam lines.

Thanks to the success of the LUNA collaboration showing the potential of underground nuclear astrophysics, now other laboratories around the world are presently under construction or just in operation, in particular one in the Republic of China (JUNA) and another one in the United States (CASPAR). In the close future it will possible to compare experimental results of the different laboratories.

References

[1] C. Iliadis, *Nuclear Physics of Stars* (Wiley-VCH, Weinheim, 2007)

[2] A. Formicola, G. Imbriani, M. Junker, D. Bemmerer, R. Bonetti, C. Broggini, C. Casella, P. Corvisiero, H. Costantini, G. Gervino et al., Nucl. Inst. Meth. A **507**, 609 (2003)

[3] F. Cavanna, P. Prati, Int. J. Mod. Phys. A **33**, 1843010 (2018)

[4] F. Cavanna, R. Depalo, M.L. Menzel, M. Aliotta, M. Anders, D. Bemmerer, C. Broggini, C.G. Bruno, A. Caciolli, P. Corvisiero et al., Eur. Phys. J. A **50**, 179 (2014), 1411.2888

[5] F. Ferraro, M. Takács, D. Piatti, V. Mossa, M. Aliotta, D. Bemmerer, A. Best, A. Boeltzig, C. Broggini, C. Bruno et al., Eur. Phys. J. A **54**, 44 (2018)

[6] A. Boeltzig, C.G. Bruno, F. Cavanna, S. Cristallo, T. Davinson, R. Depalo, R.J. deBoer, A. Di Leva, F. Ferraro, G. Imbriani et al., The European Physical Journal A **52** (2016)

[7] G.F. Ciani, L. Csedreki, D. Rapagnani, M. Aliotta, J. Balibrea-Correa, F. Barile, D. Bemmerer, A. Best, A. Boeltzig, C. Broggini et al. (LUNA Collaboration), Phys. Rev. Lett. **127**, 152701 (2021)

[8] D. Piatti, T. Chillery, R. Depalo, M. Aliotta, D. Bemmerer, A. Best, A. Boeltzig, C. Broggini, C.G. Bruno, A. Caciolli et al. (LUNA Collaboration), Phys. Rev. C **102**, 052802 (2020)

[9] V. Mossa, K. Stöckel, F. Cavanna, F. Ferraro, M. Aliotta, F. Barile, D. Bemmerer, A. Best, A. Boeltzig, C. Broggini et al., Nature **587**, 210 (2020)

[10] B.D. Fields, Annu. Rev. Nucl. Part. Sci. **61**, 47 (2011)

[11] C. Casella, H. Costantini, A. Lemut, B. Limata, R. Bonetti, C. Broggini, L. Campajola, P. Corvisiero, J. Cruz, A. D'Onofrio et al., Nucl. Phys. A **706**, 203 (2002)

[12] L. Ma, H. Karwowski, C. Brune, Z. Ayer, T. Black, J. Blackmon, E. Ludwig, M. Viviani, A. Kievsky, R. Schiavilla, Phys. Rev. C **55**, 588 (1997)

[13] G. Schmid, B. Rice, R. Chasteler, M. Godwin, G. Kiang, L. Kiang, C. Laymon, R. Prior, D. Tilley, H. Weller, Phys. Rev. C **56**, 2565 (1997)

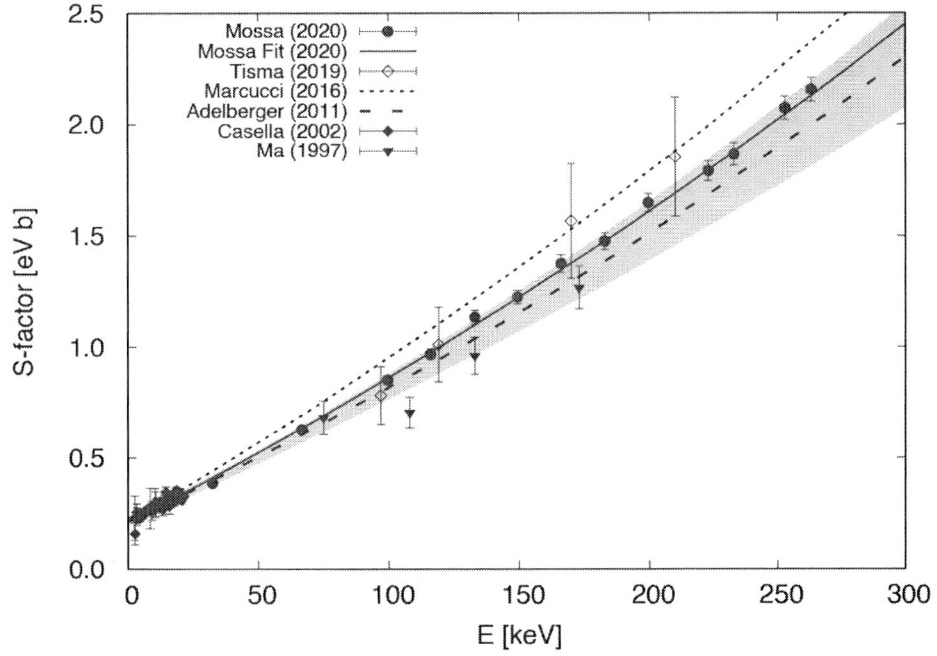

Figure 2. S-factor of the D(p,γ)^{3}He reaction; LUNA results are represented with filled red circles. Other experimental data are also shown. The best fit (red solid line) include all the reported experimental data. Band represents the 68% confidence level.

[14] G. Griffiths, M. Lal, C. Scarfe, Can. J. Phys. **41**, 724 (1963)

[15] I. Tišma, M. Lipoglavšek, M. Mihovilovič, S. Markelj, M. Vencelj, J. Vesić, Eur. Phys. J. A **55**, 137 (2019)

[16] L. Marcucci, G. Mangano, A. Kievsky, M. Viviani, Phys. Rev. Lett. **116**, 102501 (2016)

[17] C. Iliadis, K.S. Anderson, A. Coc, F.X. Timmes, S. Starrfield, Astrophys. J. **831**, 107 (2016)

[18] V. Mossa, K. Stöckel, F. Cavanna, F. Ferraro, M. Aliotta, F. Barile, D. Bemmerer, A. Best, A. Boeltzig, C. Broggini et al., Eur. Phys. J. A **56**, 144 (2020), 2005.00002

[19] C. Ananna, F. Barile, A. Boeltzig, C.G. Bruno, F. Cavanna, G.F. Ciani, A. Compagnucci, L. Csedreki, R. Depalo, F. Ferraro et al., Universe **8** (2022)

[20] C.G. Bruno, D.A. Scott, M. Aliotta, A. Formicola, A. Best, A. Boeltzig, D. Bemmerer, C. Broggini, A. Caciolli, F. Cavanna et al. (LUNA Collaboration), Phys. Rev. Lett. **117**, 142502 (2016)

[21] A. Boeltzig, A. Best, G. Imbriani, M. Junker, M. Aliotta, D. Bemmerer, C. Broggini, C.G. Bruno, R. Buompane, A. Caciolli et al., Journal of Physics G: Nuclear and Particle Physics **45**, 025203 (2018)

[22] J. Skowronski, R.M. Gesue, A. Boeltzig, G.F. Ciani, D. Piatti, D. Rapagnani, M. Aliotta, C. Ananna, F. Barile, D. Bemmerer et al., Journal of Physics G: Nuclear and Particle Physics **50**, 045201 (2023)

[23] A. Sen, G. Domínguez-Cañizares, N.C. Podaru, D.J.W. Mous, M. Junker, G. Imbriani, V. Rigato, **450**, 390 (2019)

[24] F. Ferraro, G.F. Ciani, A. Boeltzig, F. Cavanna, S. Zavatarelli, Frontiers in Astronomy and Space Sciences **7**, 119 (2021)

Upgrade to the focal-plane detector system of the S800 spectrograph at FRIB

Marco Cortesi[1*], *Jorge* Pereira[1], *Sean* Dziubinski[1,2], *Steven* Lidia[1], *Remco* Zegers[1,2], *Daniel* Bazin[1,2], *Giordano* Cerizza[1], *Joseph* Asciutto[1], and *Alexandra* Gade[1,2]

[1]Facility for Rare Isotope Beams, Michigan state University, East Lansing, MI 48824, USA
[2]Department of Physics and Astronomy, Michigan State University, East Lansing, MI 48824, USA

Abstract. The present work describes the ongoing upgrade of the focal-plane detector system and the associated data acquisition system of the large-acceptance S800 magnetic spectrograph. It includes the development of an innovative micro-pattern gaseous detector readout for the tracking system, a novel energy-loss detector concept for Z-identification, and a high resolution optical readout for the timing scintillator detector. The upgraded focal-plane detector system will allow for improved particle identification for heavy-ion beams above Z=100, at higher rate capability (> 25 kHz).

1 Introduction

The S800 is a large-acceptance, high-resolution spectrograph [1] used for experiments with radioactive beams (RIBs) produced by projectile fragmentation at the Facility for Rare Isotope Beams (FRIB) [2]. The S800 started its operation at the end of the '90s at the National Superconducting Cyclotron Laboratory (NSCL), and it was a key instrument to capitalize on the high-luminosity and energy (up to ~100 MeV/u) RIBs experiments through the use of thick targets. It has been an important experimental device for the international low-energy nuclear experimental user community. The S800 will also be a crucial tool also at FRIB, with about 40% of the experiments expected to be performed on the S800 until the completion of the High Rigidity Spectrometer [3].

The typical experimental scheme is based on a fast, exotic beam (v/c >0.4) delivered by the FRIB fragment separator, impinging on a reaction target located at the S800 pivot point. The emerging reaction products are identified and characterized by the spectrograph, using its focal-plane detection system (FPD) [4], via the simultaneous measurement of magnetic rigidity (Bρ), time-of-flight (ToF) and energy loss (dE).

The basic configuration of the S800 FPD comprises [5]: two drift chambers (DC) that provide tracking capability; an energy-loss measurement device used to identify the atomic number (Z); a large timing plastic scintillator, which combined with a small scintillator placed in the beamline prior to the target, or combined with the accelerator radiofrequency phase information, yields ToF information; and a CsI(Na) hodoscope readout by a 2D array of photomultiplier tubes (PMTs) [6], used to measure either the total kinetic energy of implanted nuclei (allowing the identification of different charge states) or gamma rays emitted from isomers (isomer tagging). All of the particle detectors described above have an effective area of 30x60 cm^2 and are capable of event-by-event particle detection at maximum rate up to 5 kHz, with 100% transmission efficiency, with the exception of the hodoscope, which absorbs all particle energy.

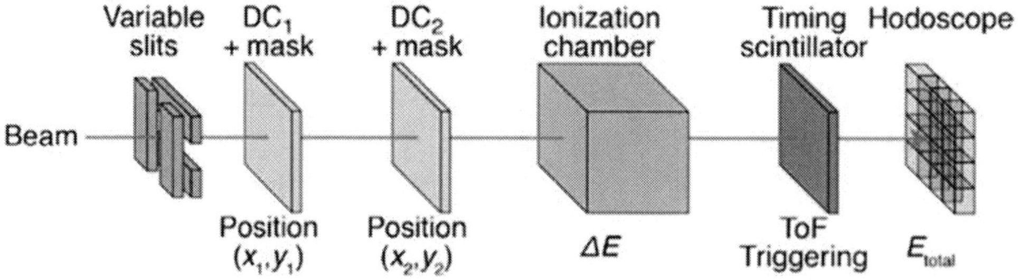

Fig. 1. Schematic drawing of the focal-plane detector system used for particle identification at the FRIB S800 spectrograph.

* Corresponding author: cortesi@frib.msu.edu

A focal-plane detector system upgrade, with innovative readout technologies and improved detector designs, has been initiated for the S800 to cope with beams of the high-intensity heavy-ion beams planned for the FRIB science program. In this work, the plan for the upgrade of the S800 FDP will be presented and described in detail. This includes the implementation of a novel position-sensitive gas-avalanche readout for the tracking system [7], an innovative high-rate energy-loss detector concept for Z-identification [8], and a revised high-resolution optical-readout for the timing plastic scintillator. The improved FPD is being designed to improve high-rate tracking capabilities, identification performance, and overall reliability. Furthermore, a new, high-performance front-end electronics and data acquisition system is being upgraded and implemented in order to take advantage of the new capabilities of the novel detector systems.

Through the development of a high-performance tracking and particle identification detector system for the S800 spectrometer as well as the optimization of the data acquisition system hardware, FRIB will be able to continue its rich program of nuclear physics with exotic beams. In addition, it would also serve as a benchmark for the development of similar detector systems to be used at FRIB. This includes, for instance, the Sweeper Magnet [9] and the future High Rigidity Spectrometer [3], [10].

2 The tracking system

The tracking system comprises two drift chambers (DCs), 1 meter apart, for the measurement of the transversal position and the angle of the charged particles that reach the focal-plane. The design of the DC is based on the old cathode readout drift chamber (CRDC) [5], with the exception of the single wire avalanche architecture, being replaced by a more advanced micro-pattern gaseous avalanche structure. The latter consists of a two-layer M-THGEM [11], which serves as a pre-amplification stage, and is suspended over a position-sensitive micromegas detector [12] – see Fig. 2.

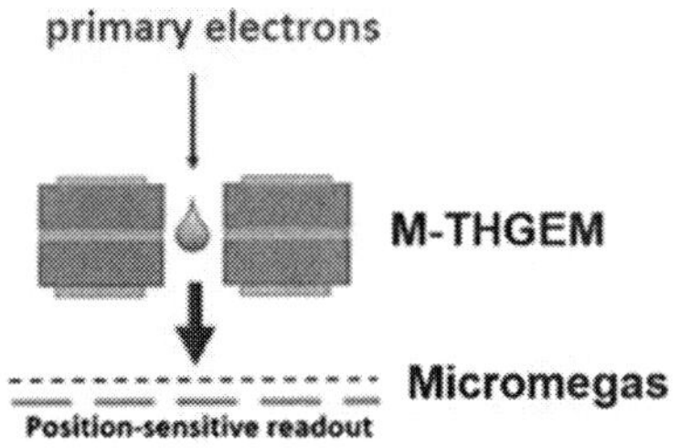

Fig. 2. Updated readout structure of the drift chamber for the S800 tracking system, comprising a segmented micromegas board preceded by an M-THGEM as a pre-amplification stage.

This novel hybrid readout configuration has the advantages of stable, high-gain operation at low pressure (typically 40 Torr). For reducing heavy-ion energy/angle straggling, low pressure operation is necessary to reduce the material budget. In addition, it provides excellent position resolution (0.25 mm RMS in both directions), and a better reduction of ion backflow. The latter is crucial for an overall improvement of the operational stability of the detector at high rates.

Furthermore, in the conventional CRDC, operated in the slow $CF_4/20\%iC_4H_{10}$ gas to delay aging problems, the typical drift times of the electrons to the anode wires are from a few hundred ns to around 20 μs, depending on the position of the track in the non-dispersive direction. The relatively long drift times limit the maximum rate that the detector can process properly to around 5 kHz. With the newly developed detector configuration, the operation in a faster counting gas is possible (i.e., pure CF_4), resulting in an increase in electron drift velocity, and thus a better rate capability (>20 KHz).

The localization of the charged particle along the dispersive direction is achieved by computing the center of gravity of the avalanche charge induced on the pads (1.25 mm wide pitch) of the segmented micromegas board – a total of 480 pads along the 60 cm long direction. The localization along the non-dispersive coordinate is given by measuring the drift time of the primary ionization electrons to reach the micromegas mesh from the point of their origin in the drift region, measured with respect to an external trigger (e.g. the trigger signal from the ToF plastic scintillator).

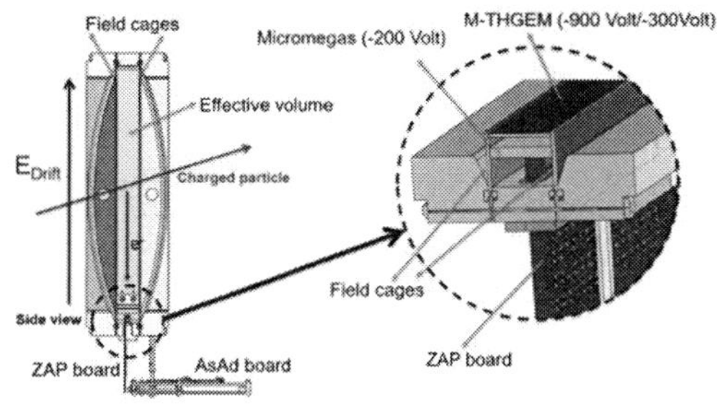

Fig. 3. Details of the mechanical design of the drift chamber and of the gas avalanche readout scheme.

The 3 cm wide effective volume (drift region) is sandwiched between two printed circuit boards (PCB) and two 12-μm PPTA windows mounted on frames (Fig. 3). Each PCB is made of un-masked G-10, and includes a field shaping foil (70 μg/cm^2 polypropylene) to ensure a uniform field in the active region of the detector. The shaping foils are made of 1.9 mm pitch of evaporated (0.1 μm thick) gold strips perpendicularly oriented to the electric field.

The signals from the micromegas pads are read out and processed by a Generic Electronics for TPCs (GET) electronics based DAQ system [13], characterized by low noise, high data transfer capabilities and multi-hit capability for efficient digital rejection of pile-up events.

The updated tracking system for the S800 will be commissioned in early spring 2023.

3 The energy-loss (dE/E) measurement

An ionization chamber with optical readout, denoted as an energy-loss optical scintillator system (ELOSS) [8], will be installed downstream of the DCs and it will be used to identify the atomic number (Z) of the transmitted isotopes from their energy loss. ELOSS will replace the conventional "charge" ionization chamber (IC) [5] used for the identification of the atomic number (Z) in the last three decades at the S800. It has an active area that is the same as that of the DC and an effective depth of about 100 mm.

The ELOSS vessel is filled with high scintillation-yield gas, such as Xenon (Xe). The scintillation light emitted along the track of a beam particle that cross ELOSS is intense (approximately 75 photons/keV [14], [15]) with prompt photons are emitted within a few nanoseconds. The scintillation light is read out by an array of vacuum photomultiplier tubes (PMTs) that are placed around the detector volume (Fig. 4).

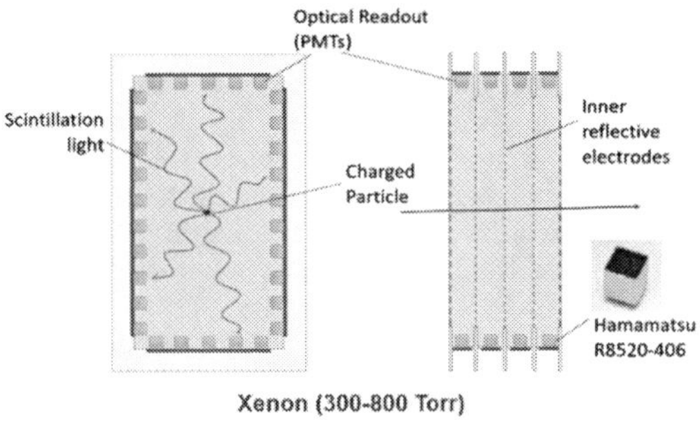

Fig. 4. Schematic drawing of the ELOSS detector.

The optical readout comprises 120 Hamamatsu PMTs (Model R8520-406 [16]) arranged in four sectors (30 PMTs uniformly spaced per sector). The sectors are sandwiched between aluminized thin foils (1.5 um thick), which allow scintillation photons to be directed towards the optical readout for increased photon collection efficiency. Alternatively, the aluminized foil can be biased to create a uniform electric field with a strength comparable to the end of the ionization region of Xe (proportional scintillation mode). Stimulated electroluminescence produces a higher overall light yield and thus better photon statistics. The PMT signals are processed by a Mesytec QDC-MDPP-32 digitizer [17], which provides high resolution timing and amplitude on an event-by-event basis.

The ELOSS detector is a cutting-edge experimental device that offers several noteworthy technical advantages compared to traditional charge readout systems. Unlike conventional ICs, the detector readout of ELOSS is completely decoupled from the drift region, hence the small detector capacitance from an optical readout, together with the low limit of the photo-sensor threshold (down to single-photon detection efficiency), allow ELOSS to reach larger signal-to-noise ratios, and consequently a better resolving power. ELOSS is expected to achieve a 0.4% (σ) energy resolution compared to the 1.2% (σ) resolution [5] achieved by conventional IC. The better intrinsic energy resolution of ELOSS will be able to extend Z-identification to elements with an atomic number Z>50.

ELOSS is based on recording electroluminescence emission, and it is therefore a much faster detector than a conventional IC. In the case of Xe gas, more than 90% of the scintillation photons are emitted within 100 ns providing an extremely short signal rise time. This results in a rate capability (up to 50 kHz) that is several times higher compared to the traditional IC (limited to 5 kHz).

Finally, ELOSS may also provide good timing information (<100 ps) and moderate localization capability (a few mm) compared to virtually no usable timing/position information provided by the traditional IC.

The ELOSS detector and interfaces (gas handling system and DAQ) will be commissioned at the S800 in early fall 2023.

4 The timing plastic scintillator

A large-area scintillation detector is placed behind ELOSS to provide timing information, as well as energy loss (at a moderate resolution). The upgraded version of the timing detector features a large (30 cm × 60 cm) plastic scintillator sheet (Eljen EJ-230), with thickness between 1-3 mm, depending on the nuclear charge of the fragments to be detected. The EJ-230 is designed to handle very fast timing (rise time of 0.5 ns) over large areas, with light output of about 11,000 ph/MeV and a long light-attenuation length. The light is readout by a series of very fast Hamamatsu PMTs (model H6533), positioned around the scintillator area. The PMTs are directly coupled to the scintillator through an optical pad, with no need for a light guide.

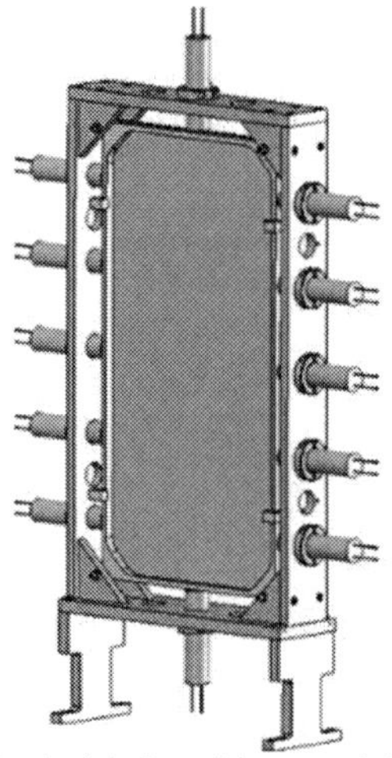

Fig. 5. Mechanical design of the upgraded scintillator detector for timing measurements.

A single PMT is mounted on each of the horizontal sides of the plastic scintillator – i.e. along the non-dispersed beam direction, and five PMTs along each of the two vertical sides – i.e. along the dispersed beam direction (Fig. 5). However, the design of the detector frame allows one to include up to 20 PMTs. The number and arrangement of the PMTs around the plastic scintillator was defined through a systematic Monte

Carlo simulation and measurements performed with a prototype unit, with the aim of attaining the highest photon-collection efficiency and a uniform fast timing response across the full detector area. Details of the procedure and the results of the readout optimization will be reported in a future publication [18].

The PMT signals of the scintillator detector are processed by a Mesytec MCFD16 (selectable constant fraction discriminator (CFD) or leading edge (LE) [19], followed by a multi-hit Mesytec MTDC32 [20] for timing. A second branch of the detector signals is processed by a Mesytec digitizer MDPP-32 [17] for extracting the integrated charge.

Different time-of-flight paths can be selected by combining the timing signals from the S800 scintillator and other upstream timing detectors along the beam line, the fragment separator, and the RF signal. The timing resolution for a point-like beam spot in the focal-plane is expected to be <35 ps (σ). However, the resolution worsens significantly (up to 1 ns) when the whole focal-plane is illuminated, because of the different path lengths of the fragments along the beam line and the spectrometer. Good time resolution can be recovered by including in the calculation the trajectories of the fragments using the S800 tracking system. The plastic scintillators can withstand maximum rates of up to 10^6 particles per second.

The new upgraded timing plastic scintillator, which replaced the old configuration based on only two PMTs, will be commissioned in early spring 2023.

5 The FRIB data acquisition system

The focal-plane detector systems of the S800 spectrometer is read out via a set of different Data Acquisition Systems (DAQs), specific of the various detector technologies. FRIB takes advantage of the experience accumulated over many years of experiments running at the NSCL by adopting a software suite that provides a flexible and extensible framework known as FRIBDAQ (previously NSCLDAQ). FRIBDAQ features a generalized data flow architecture that supports data transfer from an arbitrary set of source processes to an arbitrary set of destination processes (consumers) reachable over the network. To construct the data stream, also known as the data pipeline, consumers can, in turn, become sources for successive stages to build more complex architectures.

The backbone of the data flow in FRIBDAQ is the ring buffer, a memory location that acts as simple proxy object used to perform traffic aggregation and accessible to multiple processes. Access is however controlled in such a way that only one process can contribute data to it, while many processes can read from it. FRIBDAQ features a timestamped generic event builder. It can accept event fragments from segments of the dataflow system, order these data by timestamp and emit events that consist of fragments that are within a user specified coincidence window. This data is then submitted back to the FRIBDAQ data flow system where they can be used in later stages of the event builder. This design allows for the combination of detectors with independent data acquisition systems without modification to each individual component, a routine procedure at FRIB. A collection of standalone executables has been provided to readout data from what have become standard data source types. These include support for VMUSB and CCUSB modules from Wiener/JTEC as well as a generic framework that is used both for legacy SBS/Bit3 PCI/VME bus bridges, XIA Digital systems based on the PIXIE-16 module, and General Electronics for TPCs (GET), which will be used to readout the new S800 tracking detectors. FRIBDAQ is continuously improved and adapted for new hardware by the FRIB Laboratory staff, in coordination and collaboration with the users of the facility.

A very important aspect of FRIBDAQ is to be designed to be open both in terms of ability to extend its functionality through plugins and to attach and analyze to the data pipeline with a multiplicity of data analysis framework, including CERN ROOT system [21] and Python scientific libraries for analysis. FRIBSpecTcl is for data analysis what FRIBDAQ is for data flow. It is a C++ framework that allows experimenters to write custom analysis software that seamlessly integrates with a histogramming engine and viewer. The Graphical User Interface (GUI) allows users to quickly create histograms from the data and to draw and apply on the fly one- and two-dimensional gates for more restrictive event selection. FRIBSpecTcl has integrated support for both ROOT and Python libraries for more advance data analysis tools and it is maintained and continuously improved by the Laboratory staff.

6 Summary

The upgraded configuration chosen for the S800 focal-plane detector system has been presented and discussed. Through the newly developed readout scheme based on micro-pattern gaseous detectors for the tracking system and high-resolution particle identification, it is possible to detect isotopes with a wide range of energies and stopping powers. In addition, the rate capability is increased by about 2-3 fold when compared to the old detector configuration. The upgraded detector systems will also serve as models for other applications, including experiments with the Sweeper Magnet and the future High Rigidity Spectrometer.

Acknowledgement

This article is based on work supported by the U.S. Department of Energy (DOE), Office of Science, Office of Nuclear Physics and used resources of FRIB, which is a DOE Office of Science User Facility, under Award Number DE-SC0000661. Michigan State University operates the Facility for Rare Isotope Beams as a user facility for the U.S. Department of Energy Office of Science (DOE-SC), supporting the mission of the DOE-SC Office of Nuclear Physics. Support from the U.S.

References

[1] D. Bazin, *et al.*, *NIM B* **204** (2003), pp. 629–633
[2] J. Wei *et al.*, *Mod. Phys. Lett. A* **37** (2022), 2230006.
[3] R. Zegers, *Nuclear Physics News* **32** (2022), pp. 16–19.
[3] M. Hausmann *et al.*, *NIM B* **317** (2013), pp. 349–353.
[4] J. Yurkon *et al.*, *NIM A* **422** (1999), pp. 291–295.
[5] K. Meierbachtol *et al.*, *NIM A* **652** (2010), pp. 668–670.
[6] M. Cortesi *et al.*, *J. Inst.* **15** (2020), pp. P03025–P03025.
[7] M. Cortesi *et al.*, *Review of Scientific Instruments* **93** (2022), p. 123305.
[8] J. Toth *et al.*, *IEEE Trans. on Applied Supercond.* **12** (2002), pp. 341–344.
[9] S. Noji *et al.*, *NIM A* **1045** (2023), pp. 167548.
[10] M. Cortesi *et al.*, *Review of Scientific Instruments* **88** (2017), p. 013303.
[11] G. Charpak *et al.*, *NIM A* **478** (2002), pp. 26–36.
[12] E. C. Pollacco *et al., NIM A* **887** (2018), pp. 81–93.
[13] T. Doke *et al.*, *Jpn. J. Appl. Phys.* **41** (2002), p. 1538.
[14] M. Tanaka *et al.*, *NIM A* **457** (2001), pp. 454–463.
[16] "Hamamatsu Photomultiplier tube R8520-406 Datasheet (https://www.hamamatsu.com/)."
[17] E. Aprile, A. E. Bolotnikov, A. I. Bolozdynya, and T. Doke, *Noble Gas Detectors*. John Wiley & Sons, 2007. ISBN-10: 3527405976.
[18] J. Pereira *et al.*, "A new high resolution ToF detector for the S800 spectrograph.", *in preparation*.
[19] "Mesytec 32 channels fast high resolution time an amplitude digitizer MDPP-32 Datasheet (https://www.mesytec.com)."
[20] "Mesytec 16 channels constant fraction discriminator MCFD-16 Datasheet (https://www.mesytec.com)."
[21] "Mesytec 32 channels fast high resolution time digitizer MTDC-32 Datasheet (https://www.mesytec.com)."
[22] "ROOT open-source data analysis framework (https://root.cern.ch/)."

Status and Perspectives of Silicon Detectors

Alberto Cervelli[1]*

[1]Istituto Nazionale di Fisica Nucleare - Sezione di Bologna, Viale Berti Pichat 6/2, 40138 Bologna, Italy

Abstract. In the following a review of the present status of silicon tracking and vertexing systems and their future developments will be presented. We will show the modern detectors used in present day experiments both in nuclear and elementary particle physics, and their achieved performances. Later we present a review of the near-future systems which are now being designed, built, or commissioned together with an outlook on the future developments for next-generation silicon detectors.

1 Introduction

Silicon detectors have been used in tracking since the start of the 80s. The development of silicon detectors in high energy physics followed closely the one in the microelectronic industry, which share its technology and design with the detectors used nowadays for tracking and vertexing.

What should be kept in mind is that for tracking purposes, and especially for reconstructing secondary vertexes, silicon trackers have to be placed very close to the interaction point, in order to achieve the best results, should it be on a fixed target or near colliding beams. Their operating environment is always characterized by a large number of particles traversing the detector. This results in mainly two requirements for any silicon tracker: the need for a finely segmented detector, to cope with large number of hits, without reducing the performances due to occupancy or dead-time, and the possibility to withstand incredibly high radiation doses. These two requirements need to be tailored to the particular need of each experiment, both in terms of its physics case, and depending on the environment provided by the accelerator in use. For example hadron colliders have higher dose-rates compared to electron-positron colliders due to the production of baryons, conversely electron-positron colliders provide harsher conditions for the on-detector electronics caused by high fluxes of photons produced by syncrotron radiation.

The first silicon trackers had active elements connected via thin wires to peripheral readout, and then, from 1985 onwards, larger active silicon sensor were produced thanks to very large-scale integration (VLSI).

During the 90s and early 2000s many new and interesting results were obtained thanks to the new development in silicon trackers: the results obtained by the experiments taking data at LEP[1], with precision measurement at the Z resonance, the top discovery by CDF[2], or the measurement of indirect and direct CP violation in the B-sector by BaBar and Belle experiments[3]. Today the currently operating detectors in the LHC experiments [4–7], and B-factories[8] reached unprecedented precision, complexity, the possibility to continue working without a reduction in performances in environments with tremendous amount of radiation, and to record a large amount of data with great speed.

Silicon detectors for tracking evolved in complexity, and reduced their production cost together with the development of smaller pitches used in the microelectronic industry: this allowed to have an exponential increase in the surface covered and with the increase in density in the electronics it allowed for a larger number of channels, high read-out speeds, and always better performances. The evolution of the covered surface of silicon trackers in time is shown in figure 1.

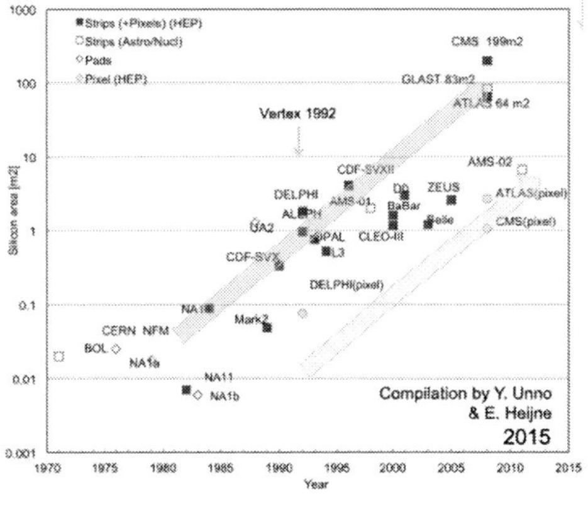

Figure 1. Evolution of silicon trackers surface over the last 4 decades, both silicon strips and pixels had an exponential increase in covered areas and number of channels.

The evolution on both the sensor and their electronics is ever evolving, with many new improvement foreseen for silicon trackers in the near future: after almost 50 years,

*e-mail: alberto.cervelli@bo.infn.it

https://doi.org/10.1051/epjconf/202329010007

silicon tracker systems are still one of the most advanced detectors present in many high energy physics detectors.

In the following discussion we will focus on pixel detectors for tracking, as they represent one of the most and active fields of research in tracker detector development.

2 Different Designs

To better understand how each detector, and so its design, should be tailored to the environment it is bound to work in a good example is provided by the different designs of the Belle2 tracker, and the ongoing design of the EIC tracking system.

Despite a very different physic goal, both EIC and Belle2 need to measure the $D^* \rightarrow D^0\pi_s^+ \rightarrow (K_S\pi^0)\pi_s^+$: however in the case of EIC, an electron-ion collider, the main backgrounds arises from collision products,resulting in high charged particles multiplicity, leading to the requirement for high granularity. On the other hand, in the case of Belle2 the background is dominated by beam induced signal, which instead requires a better timing resolution, rather than better granularity.

A similar situation arises from the design of the beam line and the spectrometer magnet: EIC foresee a larger beam pipe and higher B-Field, requiring a better spatial resolution for tracking especially at low transverse momentum, Belle2 conversely has a smaller beam pipe, weaker magnet so the momentum resolution depends less strongly on the spatial resolution, on the other hand its electronics has to withstand a large amount of radiation due to the closeness of the system to the beam.

In conclusion, even the optimization for the detection of the same process will need different approaches in the design and the research of a particular tracking system, so in sections 5-7 we will provide a description of the building blocks for the future generation of pixel tracking systems.

3 The LHC-era

The LHC experiments provided one of the best benchmarks for tracking systems working in high radiation and high charged track multiplicity environments, the three 4-π experiments at LHC (ALICE, ATLAS, CMS) shared a similar approach in their design for the first runs of data-taking. These detectors have tracking systems with an innermost part instrumented with pixels and an outermost part with silicon strip sensor, all designed to survive doses in excess of 10^{12}n.eq. Due to the differences in their respective physics case ATLAS and CMS have systems able to reach a resolution of $O(10\mu\text{m})$, while ALICE system has a poorer position resolution $O(10\text{mm})$, but can cover a wider range of transverse momenta.

In the coming years these three systems will undergo an upgrade to copewith the higher beam intensities, and the resulting higher luminosity provided by the LHC. For the pixel sub-detectors of ATLAS and CMS the upgrade foresee a similar approach: the pseudorapidity coverage of the system will be extended, with a detector surface increase following the increase of coverage, the technology for the pixel sensors will not change, relying on hybrid pixels with new design for both sensor and front-end electronics. The goal is to achieve better performances, or at least on-par with previous run of data-taking in harsher pile-up conditions[9]. ALICE, on the other hand, will undergo a major change in its tracking system, replacing its innermost tracker with a monolithic pixel detector. The new detector will be the first to be used in such a large scale in a high energy experiment at a collider[10], increasing both the position resolution performances. Standalone tracking at low momenta will be possible thanks to a lower material budget provided by the thin monolithic sensors.

4 Pixel architectures

In order to understand the status and the research and development of silicon pixels for the future generations of high energy physics experiments it is important to have a look at the two possible approaches leading to a sensor and front-end assembly. The two broad families constituting pixels are the hybrid and monolithic architectures for pixel design.

Hybrid pixels are characterized by having the sensor built on a different silicon layer than the front-end electronics. The two silicon substrates are usually interconnected trough the use of bump-bonding (i.e. the deposition of small amounts of conducting material between contact pads on the two substrates). Monolithic pixel instead take advantage of the modern lithographic processes allowing to make deep wells in the silicon bulk, where the front-end electronic is placed. A schematic view of both Hybrid and monolithic sensors is shown in figure 2.

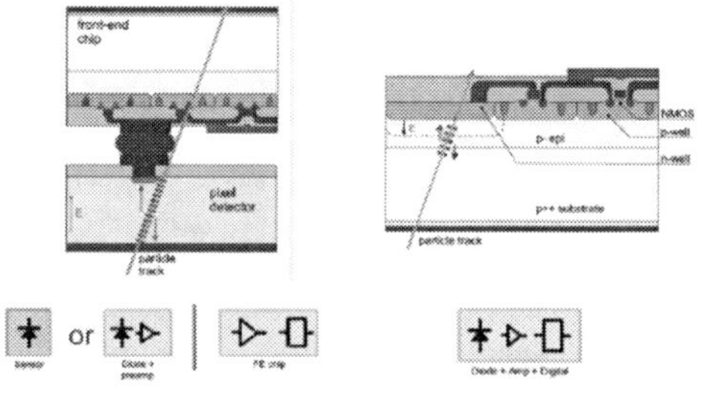

Figure 2. Schetch design for Hybrid pixel (left) where the second substrate harboring the FE is visible on top, and monolithic pixels (right), where the electronics is placed on the same substrate as the sensor.

The Hybrid architecture allows to perform the design and optimization of sensor and front-end separately, where no technology commonality is needed and different processes can be used in the two components with minimal interplay in their performances. The main drawback comes from the need to connect the sensors to the front-end electronics: the bump-bonding process is expensive and the soldering material needed between the two layers of an hybrid module limit the minimum pitch achievable by the pixels. The presence of two layers of silicon increases also the budget material of the detector requiring O(500μm) of silicon for each module, increasing the multiple scattering, so making the detector less resolute for low momenta particles. On the other hand, hybrid modules allow for the use of high resistivity bulk in the sensor layer, largely increasing the radiation hardness of the sensor, without affecting the front-end performances, for the lack of technology commonality also the size of the full module is not strictly limited in surface, as it would happen for instance in a fully CMOS technology sensor, allowing for large area coverage.

In Monolithic sensors front-end and sensor are designed on the same substrate: this requires a commonality between the design of the two components, usually CMOS technology: the price to pay to avoid interconnection, and reduce the sensor thickness thanks to the lack of a second layer of silicon harboring the front-end, is represented by the need of a shared design and production process. Overall monolithic pixels have typically better performances in environment requiring low material budget (i.e. low momentum tracks), with lower (with respect to hybrid pixels) level of radiations: this limitation comes from the usually lower resistivity of monolithic sensor bulks, limiting the charge collection efficiency after absorbing high doses of radiation. The other main limitation comes from the small module sizes allowed for monolithic processes, which lead to a larger number of sensors to cover a given instrumented surface.

5 Hybrid R&D

As stated in the previous section, Hybrid pixels provide the best solution for high radiation environment where large areas should be instrumented: the main lines of research go in the direction of having even radiation-harder sensors, and being able to operate properly after large accumulated doses. On the other hand there is a will to overcome the drawbacks of hybrid pixels, trying to reduce the interconnection spacing due to the bump-bonding, or thinning the sensor itself.

HV protection at edges

Nowadays sensors built to operate in high radiation environment, such as those in use at LHC detectors, are built with n-in-p technology. This choice is made so to have sensors and guard rings on the same side of the pixel, a cheaper solution than n-in-n, but more importantly so that the bulk does not experience a type inversion during the operations. While a p substrate does not experience type inversion, nevertheless the operating voltage will increase exponentially with the accumulated dose. During most of their operations Hybrid pixel operate at voltages up to 600V, while the front-end electronics, placed just few microns above, operates at low voltages. The difference in operating voltages is critical at the edges of the sensors, where a high voltage difference can cause discharges between the edges of the sensor and the one of the front-end, causing irreversible damage to the latter.

In recent times two paths have been explored with good results yo mitigate this effect, so allowing for even higher sensor voltages to allow operations following higher accumulated doses. The first consist in the deposition of an insulator in the lithographic process of the sensor, this can be done without the need to protect the front-end structures, but increase the complexity of the sensor production process making it more expensive. An alternative consists in the deposition of parylene over the module assembly, after it has been bump-bonded, while cheaper, this procedure need the front-end to be protected during the deposition, making it riskier. Both procedures have been tested with good results, with the latter going to possibly be the standard procedure for the construction of the new ATLAS tracker.

3D technology

One of the main effects of radiation damage for silicon trackers is the increase of defects in the sensor bulk. These defects reduce the charge collection efficiency, and strongly depend on the drift length of the charge inside the sensor itself. In the last decade it became possible to produce 3D structures inside the silicon bulk: using these processes the sensor electrodes are deposited in columns inside the sensor, opposed to the usual planar configuration with electrodes deposited on the faces of the sensors. A graphic comparison between 3D sensors and planar sensor geometry is shown in figure 3.

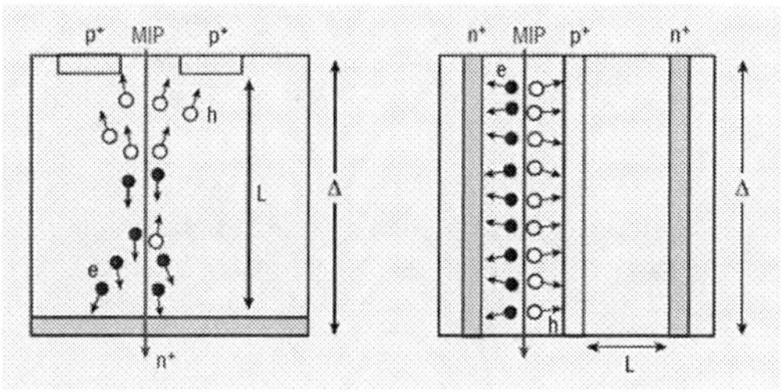

Figure 3. Comparison of planar (left) and 3D (right) electrode geometry, highlighting the charge drift between electrodes.

The 3D geometry decouples the charge drift length and the sensor thickness, providing a better radiation hardness by reducing charge drift inside the detector. While the first prototypes suffered from inefficiencies along the electrode columns, that reduced the available signal for collection, the problem has been mitigated by tilting the sensor with respect to the most probable directions of the incoming particles. The complex structure of the electrodes require an expensive and rather low yield production processes, with yields of O(80%), however more complex structures allow also for reducing the pitch between neighboring electrodes: the latest prototypes achieved a 50μm×50μm sensor cells. The first use in large scale of 3D sensors in a collider experiment was the ATLAS insertable B-Layer [11], which is operating since 2015.

https://doi.org/10.1051/epjconf/202329010007

Thin sensors

Since they rely on two silicon substrates, hybrid detectors tend to be inherently thick, their high material budget limiting their precision in tracking low momentum tracks. Moreover a thin sensor is easier to operate after irradiation: a lower voltage is needed to fully deplete the sensor after irradiation, due to the reduced volume of the bulk, the shorter collection distance reduce drift lenght; it allow for larger electric fields, high speed collection, and lower power dissipation. Thinner sensors on the other hand are more mechanically fragile, and the thinning process can resultsin lower yields with respect to thicker sensors.

Recent results [12] have shown that thin sensors, with thickness down to 75μm can operate with high efficiencies after irradiation and at very high voltages, without incurring in breakdown or charge collection losses, as shown in figure. 4. These sensors can then operate in high radiation environment for long time and good reliability, and they are also a great platform for timing measurements for 4D tracking, as will be described in section 7.

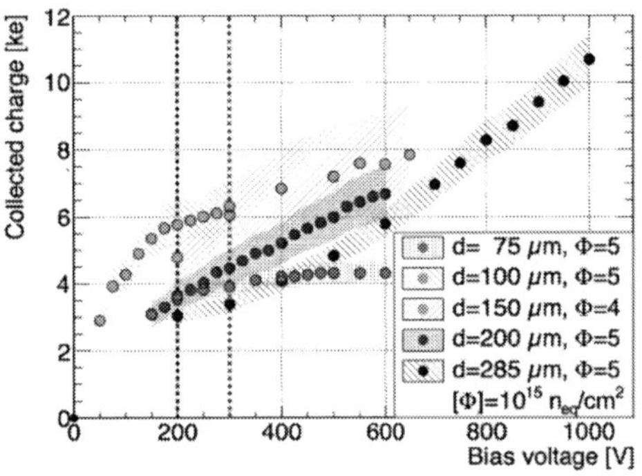

Figure 4. Collected charge as a function of bias voltage for irradiated sensors with thicknesses ranging from 75 μm to 285 μm [12].

Small pitches

Bump-bonding process drives the interconnection pitch dimension: this has a direct effect on the minimum pitch between two neighboring pixel sensor elements. Over the years new technologies and techniques reduced the pitch dimension driven by the bonding processes, as shown in figure 5.

The physical dimension of the soldering item can achieve a pitch between soldering of the order of 10μm: new prototypes are testing Hybrid modules electrically connected through the use of anisotropic conductive films (ACF)[13]. ACFs provide a cheap solution to the expensive flip-chipping processes and decouples the sensor pitch and the bump-bonding size by providing a thin film with conducting micro-particles in an insulating medium: in fact only the micro-particles in contact to both the sensor electrode and the front-end provide a viable electrical connection, while cross talk and shortings between neighboring pixels is reduced due to the insulating nature of the ACF medium.

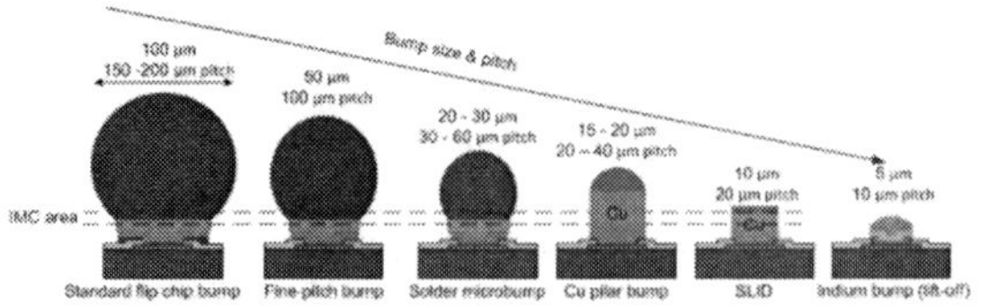

Figure 5. Soldering dimension for different bonding techniques ordered in decreasing size, from the ordinary flip-chipping to indium bumps.

6 Monolithic R&D

Monolithic active pixels sensors (MAPS) are characterized by the use of a single silicon substrate hosting both the sensor and the electronics: the need for a common sensor and front-end technology can be achieved predominantly using CMOS technology. CMOS processes are cheap, widely available, and allow for a large number of detector elements to be produced, thanks to the common availability of producers. The front-end electronics are usually placed in a deep n-well while the sensor is formed by a depleted p-substrate with high resistivity.

The MAPS sensors can be very thin, resulting in detectors with $0.1\% X_0$ thickness per layer, their thinness is paired with a thin bulk, which can be fully or partially depleted. In both cases the depletion region is rather shallow, with respect to hybrid pixels, with depleted region depths ranging between 10-50μm. The drawback of their thinness is given by the small signal produced in the shallow depletion region, a limited radiation tolerance, due to long collection times, which is particularly true if the sensor collect charge by diffusion.

In order to mitigate some of the drawbacks and exploit to the best the potential of each MAPS design, one of the following three approaches for electrode placement can be followed: large electrodes, small electrodes, or buried electrodes. A schematic view of the three designs can be found in figure 6

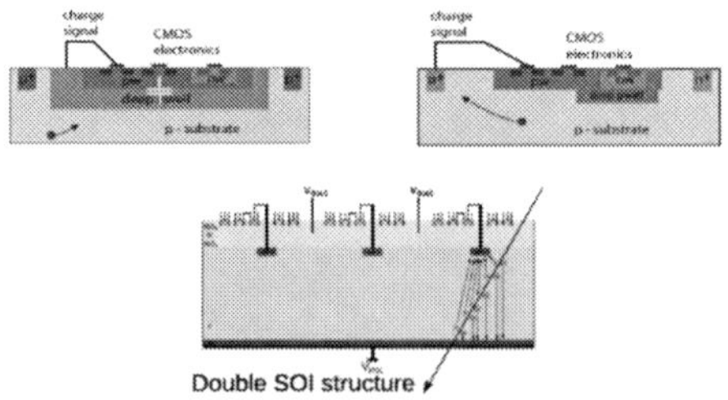

Figure 6. Different electrode design for MAPS, from top left, clockwise: large electrode design, small electrode design, buried electrodes.

Large electrode design uses the well holding the electronics as the collection electrode for the charge released in the bulk by traversing particles. With this approach there are little zones of low field in the collection region, and large areas for collection allow short drift for the charges to be collected, allowing for better radiation hardness. On the other hand large electrodes result higher capacitance for the sensor, and hence higher noise. Moreover the closeness between the collection electrode and the electronics are a source of potential cross-talk between the digital and analog electronic component, interlocking even more the need fora common approach in the design of the sensor and front-end.

Some of the shortcomings of large electrodes design can be overcome through the use of smaller electrodes. In this way the electrode is decoupled from the front end, as charge is collected on an implant different from the deep well holding the analog circuitry. This design does not suffer from cross talk, and provides smaller sensor capacitances, allowing for higher signal-to-noise ratios and faster response. However the longer drift collection length makes it less radiation hard, and need adjustments to CMOS processes to maintain good performances after irradiation.

Another way of decoupling the electrodes and the circuitry is the use of buried electrodes: in this design schema, the electronics and the collection electrodes are separated by a silicon oxide layer, and the connection between the electrode and the overlaying electronics is provided through conductive vias. This approach decouples the design of sensor and electronics, and at the same time offer the possibility to use lower or higher resistivity substrates for the depletion region. The main liability of the design come from the charge build-up in the oxide layer after irradiation.

Diffusion MAPS

MAPS relying on diffusion for charge collection are coming to be used in large apparatuses in the coming years, in particular the ALICE upgrade foresee the use of monolithic sensors for its innermost detector. The high radiation doses it has to survive high doses and so a moderate reverse bias voltage will be applied to help the charge collection. The detector will be closer to the interaction point with respect to the present one, moving from 39mm to 21mm and will reduce the total budget for the first tracker layer from 1.14% X_0 to 0.3% X_0 together with a reduction of the pixel size to 30μm × 30μm pixels. ALICE MAPS will use large n-type collection electrodes placed in high resistivity epitaxial layer. And the new systems will provide standalone tracking for low momenta particle, which was hardly possible with the system used at present.

Depleted MAPS

Fully depleted MAPS are yet to be used for the first time in large apparatuses, however Mu3e experiment is going to employ fully depleted radiation hard CMOS pixels for its operations, starting in less than 5 years. Depleted MAPS have clear advantages over non-depleted ones: while sharing the same light weight characteristic, they are inherently faster with a collection time reduction proportional to the bias voltage (V_B), which makes them suitable for their use also in high repetition environments or where track multiplicity is very high (i.e. at LHC-like experiments). The depletion voltage scales with $\sqrt{V_B}$ and so makes it possible to fully deplete thin detectors without having to apply voltages as high as in hybrid pixels, even after irradiation.

In order to achieve the performance and operation goals for fully depleted maps, some technological challenges need to be overcome: in particular the need for radiation hard sensors require deep submicron technologies and multiple well process are needed to decouple front-end electronics from the sensitive region. This would be allowed by 130-18'μm feature size. Beside very fine features radiation hard can be achieved only if they are coupled with high resistivity substrates, and processes allowing for high voltage usage in the sensors.

With these technology challenges tackled Mu3e experiment will be able to reach unprecedented precision in the measurement of lepton flavor violating process $\mu^+ \rightarrow e^+e^-e^+$, thanks to a detector with a 80μm×80μm pixel size, a time resolution better than 20ns, and a material budget lower than 0.1 X_0 per layer. This configuration will allow for reliable and precise tracking of low momentum electrons, in an environment densely populated by tracks from decaying muons.

7 4D detectors

Silicon trackers are inherently fast, with collection times of the order of tenths of nanoseconds, however they are not fast enough for resolving hits from particles produced during a bunch-crossing in a highly luminous collider. In the past, the ghost-hits problems in strip detector through the pixellization of the detector. By making the tracker having a time resolution of the order of tenths of picoseconds we could solve the ambiguity in track reconstruction each hit to the track based on the time of charge release in the detector, possibly using radiation-hard detectors. This process is usually called 4D tracking, as it adds time information to the spatial one.

Timing resolution can be written as:

$$\sigma_t = \left(\frac{Noise}{dV/dt}\right)^2 + \Delta^2_{Ionization} + \Delta^2_{shape} + \sigma^2_{TDC} \qquad (1)$$

if we focus on the first term, which is the leading term in our resolution, it becomes clear that a large slew rate for the signal is needed. As ratio between the peak current (S) and the time raise is constant (i.e. Thicker detectors have longer signals, not higher signals), the consequence is silicon detectors should be thinner to have better time resolution, as shown in figure 7[14].

In order to maximize the slew rate of the signal it is important to increase the peak current for it: in order to do so, and reach the goal timing resolutions a gain layer for the sensor can be beneficial. In order to do so a thin (few nm) p++-type doping is added below the n-type collected electrode: in this way a local very large field is put in place allowing for large gains (factor of 10 to 30 before

 https://doi.org/10.1051/epjconf/202329010007

irradiation). This design is usually referred to as low gain avalanche diods (LGAD), and are one of the most promising designs for 4D tracking. Their main drawback comes from their gain layer itself: as the gain strongly depends on the doping density in the gain layer: the doping densities are strongly affected by radiation damage, resulting in lower gains after irradiation, and hence a lower timing resolution,as the gain is one of the leading terms for the slew rate of the signal sensors.

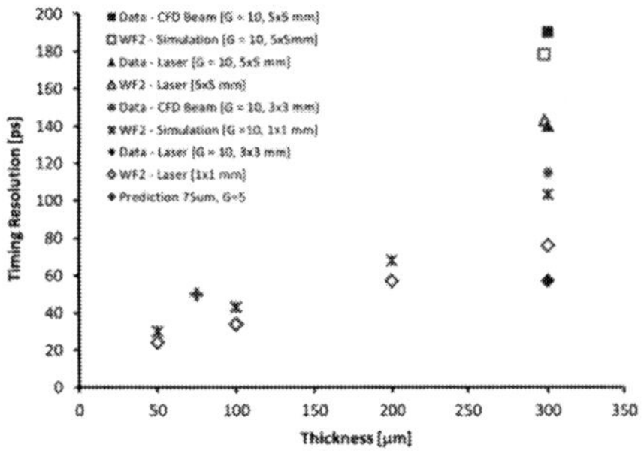

Figure 7. Timing resolution on LGAD detectors for different thicknesses and production processes.

8 Outlook

Silicon trackers have been a staple in high energy physics experiments since the eighties, and followed closely the development happened in the microelectronics industry and they evolved to ever bigger and more complex apparatuses. Since the nineties they have been crucial for some of the most important discoveries and measurements in our field, and they continue to do so today.

Different technical solutions have been devised for different physics cases and different operating conditions, and in particular the development of pixelated detectors improved the already high precision in position measurement achieved by silicon strips detector.

Even with all the achieved goals and the improvements reached in the last decade, silicon pixel trackers are pushing their limits by the day: new technologies to reach higher radiation tolerance, extending the lifetime and reliability, and improving the system resolution have been developed for the next phase of experiment upgrades at LHC and for near-future experiments such as EIC or Mu3e, pushing the boundaries for both the consolidated Hybrid pixel modules, and paving the road for large-scale use of monolithic pixel detectors.

Hopefully the next generation of trackers will also be able to perform time measurement with unprecedented precision, allowing the trackers to match both the space and time coordinates for the track, to make them even more better performing in high track density conditions.

To conclude, silicon trackers are the most advanced, complex, and precise detectors for tracking charged particles, and probably still be in the coming decade, and their advancements are starting to open new paradigms in track reconstruction, and operation in harsh environments.

References

[1] J. Drees, International Journal of Modern Physics A **17** (2023) 3259.

[2] CDF Collaboration, Phys. Rev. Lett. **74** (1995) 2626.

[3] A. Bevan et al., *The Physics of the B Factories* (Springer Berlin, Heidelberg, 2015) 920.

[4] ATLAS Collaboration, JINST **3** (2008) S08003.

[5] CMS Collaboration, JINST **3** (2008) S08004.

[6] LHCb Collaboration, JINST **3** (2008) S08005.

[7] ALICE Collaboration, JINST **3** (2008) S08002.

[8] G. Dujany et al, JINST **17** (2022) C08014.

[9] ATLAS Collaboration, ATL-PHYS-PUB-2021-024 (2021).

[10] ALICE Collaboration, ALICE-PUBLIC-2018-013 (2019).

[11] B. Abbott et al, JINST **13** (2018) T05008.

[12] J.C. Beyer, arXiv:1612.01281 (2016).

[13] J.V. Schmidt et al. JINST **18** (2023) C01040.

[14] N. Cartiglia et al. J.NIMA **979** (2020) 164383.

https://doi.org/10.1051/epjconf/202329010008

From correlations to universal behavior in few-nucleon systems

Alejandro Kievsky[1,*], *Mario* Gattobigio[2,**], *Luca* Girlanda[3,***], and *Michele* Viviani[1,****]

[1]Istituto Nazionale di Fisica Nucleare, Sezione di Pisa, Largo B. Pontecorvo, 56127 Pisa, Italy
[2]Université Côte d'Azur, CNRS, Institut de Physique de Nice, 1361 route des Lucioles, 06560 Valbonne, France
[3]Dipartimento di Matematica e Fisica "E. De Giorgi", Università del Salento, I-73100 Lecce, Italy

Abstract. The deuteron, the only two-nucleon bound state, has a shallow character: its binding energy is strictly related to zero-energy parameters, the triplet scattering length ${}^3a_{np}$ and triplet effective range ${}^3r_{np}$. This fact places the deuteron inside the universal window, a region in which systems having a large value of the two-body scattering length are located. When the scattering length is large compared to the interaction range certain types of correlations can be observed. Increasing the number of nucleons these correlations are responsible for the particular spectrum of light nuclei as for example the lack of excited states in three- and four-nucleon systems. In this presentation some constraints imposed by the large values of the singlet and triplet scattering lengths in the spectrum of light nuclear systems are discussed.

1 Introduction

When the kinetic and potential energies are such that their difference is almost zero the system is close to support a bound state. The scattering length is large and negative, and a virtual state is located close to the zero-energy threshold. With a small change of the system parameters, the scattering lengths jumps from $-\infty$ to $+\infty$ and the virtual state transforms into a bound state. In this particular region, when the scattering length is large with respect to the interaction range, the bound (or virtual) state has a shallow character. It is worth to note that weakly bound systems define a class of universality. Due to the large tail of the wave function, the particles stay most of the time outside the interaction range and many of their properties can be explained in terms of the probability to be in that region. This probability is well estimated by the relation $1 - r_e/a$, with r_e the effective range and a the scattering length. Close to the unitary limit, $a \to \infty$, the ratio r_e/a has a small value. Moreover these two quantities determine the s-wave phase-shift δ at low energies through the effective range expansion

$$k\cot\delta = -\frac{1}{a} + \frac{1}{2}r_e k^2 \ldots \quad (1)$$

with the energy $E = \hbar^2k^2/m$. From the experimental point of view, these two quantities are extracted from the phases analyzing their behavior at very low energies, when the effective range function, $S_k = k\cot\delta$, reaches a linear dependence in terms of the energy. Accepted values in the case of two nucleons are ${}^3a_{np} = 5.419(7)$ fm and ${}^3r_{np} = 1.753(8)$ fm for the spin $S = 1$ state and ${}^1a_{np} = -23.740(20)$ fm and ${}^1r_{np} = 2.77(5)$ fm for the spin $S = 0$ state, see Ref.[1] and references therein. In the case of two neutrons the quantities in the $S = 0$ state are ${}^1a_{nn} = -18.90(40)$ fm and ${}^1r_{nn} = 2.75(11)$ fm. For two protons the values are ${}^1a_{pp} = -7.8063$ fm and ${}^1r_{pp} = 2.773(14)$ fm. However after subtracting the electromagnetic contribution, the scattering length results ${}^1a_{pp} = -17.137$ fm. Though there is some model dependence in subtracting the electromagnetic contribution, all values of these quantities are such that the two-nucleon system in both spin states is located close to the unitary limit.

A model independent description of the physics inside the unitary window can be given by an effective field theory (EFT) based on the separation of scales between the typical momenta $Q \sim 1/a$ of the system and the underlying high momentum scale $\sim 1/\ell$, with ℓ a typical range of the system [2–5]. This condition is well fulfilled in nuclear physics and this approach is known as pionless-EFT. Using EFT one can systematically improve the prediction of the observables. For instance the effective range expansion can be reproduced by such an expansion. At leading order (LO) the information encoded in the scattering length a is introduced, whereas the finite-range nature of the interaction, represented by the effective range r_e, is found at the next-to-the-leading order (NLO). In addition, inside the unitary window, there is an energy pole close to the two-particle threshold. The extension of the effective range function to the negative energy pole results in

$$k_d = \frac{1}{a} + \frac{1}{2}r_e k_d^2 \ldots \quad (2)$$

where $E_2 = -\hbar^2k_d^2/m$ is the energy of the virtual or bound state.

*e-mail: kievsky@pi.infn.it
**e-mail: mario.gattobigio@univ-cotedazur.fr
***e-mail: girlanda@le.infn.it
****e-mail: viviani@pi.infn.it

The most remarkable property of systems at the unitary limit appears in the three-body system where the Efimov effect can be observed [6, 7]. When the strength of the two-body interaction is such that there is a bound state at zero-energy, an infinite tower of geometrically distributed energy states appears in the three-body system with the energy threshold $E_3 = 0$ as an accumulating point. The energy ratio of successive levels $E_3^{n+1}/E_3^n = e^{-2\pi/s_0}$ is a universal constant, with s_0 depending on the mass ratio of the constituents; for three equal bosons $s_0 \simeq 1.00624$ so that $e^{-2\pi/s_0} \simeq (1/22.7)^2$. This effect, predicted by V. Efimov around 50 years ago, was observed 35 years after its prediction by the group of R. Grimm [8]. An enormous amount of work, experimental as well as theoretical, has been, and still is, dedicated to study this phenomenon. An introduction to this sector of research can be found in the following reviews [9–16] and references therein.

In the present work we discuss a description of systems inside the unitary window based on a gaussian potential. We call this a gaussian characterization of the window and use the characterization to link different systems along the window and to study the appearance of correlations. As a difference to the standard use of the EFT framework, the gaussian characterization treats at the same level the scattering length and effective range [17–22], so many times the obtained results have to be compared to the NLO of the EFT framework.

2 Gaussian characterization of the two-nucleon system

The dynamics of two-body systems inside the universal window is highly independent of the details of their mutual interaction. This is motivated by Eq.(1) and (2): the presence of the shallow state allows for a second order expansion of its energy in r_e/a introducing a strict correlation between the low energy parameters a, k_d and r_e

$$k_d^{-1} = \frac{a}{2}\left(1 + \sqrt{1 - 2r_e/a}\right) . \tag{3}$$

When this relation is fulfilled with an error at the level of $(r_e/a)^2$, the bound or virtual state can be considered shallow. Moreover, these quantities completely determine the S-matrix of systems having one bound or virtual state

$$S(k) = \frac{k + ik_d}{k - ik_d}\frac{k + i/r_B}{k - i/r_B}, \tag{4}$$

where we have introduced the length $r_B = a - 1/k_d$.

To highlight the universal properties of systems belonging to the unitary window, we propose to study the two-nucleon system using of a two-parameter short-range potential. We consider this potential a minimal low-energy representation of the two-particle interaction fixed by two low-energy data, a (or k_d) and r_e (or r_B). Specifically we use a Gaussian potential

$$V(r) = V_0 e^{-r^2/r_0^2} , \tag{5}$$

where r is the interparticle distance, while the strength V_0 and the range r_0 are parameters useful to explore the low-energy dynamics associated with the existence of one (bound or virtual) state close to threshold. In Fig.1 we show how we proceed to characterize the unitary window using the gaussian potential. In the upper panel the solid circles indicate the location of the NN system inside the unitary window. The circles are placed using the corresponding experimental data (in the pp case electromagnetic effects have been substracted). The solid line encompasses the solutions of the Schrödinger equation using the gaussian potential with variable strength and range. In fact, the only condition imposed to find the gaussian curve is to limit the gaussian potential to support one bound state for positive values of a and to have no bound states for negative values of a. Once these conditions are verified the results of a generic gaussian potential lie on the solid curve. The specific values of the gaussian parameters necessary to describe a particular system are given in the bottom panel from the position of each system on the x-axis. As a example a gaussian potential with a range $r_0 = 1.55$ fm is able to describe simultaneously the triplet np scattering length, the effective range and the deuteron binding energy. In the caso of the 0^+ states, the np, nn and pp systems are well described by a gaussian of range $r_0 = 1.85$ fm.

Different points along the gaussian curve are related trough the running of the gaussian strength

$$\frac{V_0 m r_0^2}{\hbar^2} = C_0 \left(1 + \alpha_1 \frac{r_0}{a} + \alpha_2 \left(\frac{r_0}{a}\right)^2 + \ldots\right) \tag{6}$$

where $C_0 = 2.6840$ is the value at which a bound state starts to be supported by the potential and α_1, α_2 are constants, the same for all gaussian potentials [23, 24]. The above equation can be used to connect different systems lying on the gaussian curve. For example the gaussian curve can be designed using a gaussian potential with constant range and variable strength. The complete curve can be explored in this way and the above formula relates the strength of the gaussian potential to the different values of the scattering length. Constructing the curve in this way the values of the ratio r_e/a at which specific systems are located (those indicated by the symbols in Fig.1) will be reached. Therefore a second possible movement on the curve would be to vary the strength V_0 and the range r_0 maintaining fixed the ratio r_e/a. The above equation relates the pairs of values r_0, V_0 needed to perform this action. It should be noted that systems having the same ratio r_e/a are connected by a scale transformation: $a \to \lambda a$ and $k_d \to k_d/\lambda$ implies $r_B \to \lambda r_B$ and $r_e \to \lambda r_e$.

When the interaction between two particles is strongly repulsive at short distances the two-body system is, as a consequence, highly correlated. For bound systems the probability to be inside the repulsive core is very small. Accordingly, the wave function in that region is almost zero and increases rapidly towards the attractive region. Therefore the total energy results from a big cancellation between the kinetic and potential energy. Systems such as the deuteron are examples of this kind of correlation. It is interesting to analyze the description of these systems in terms of the low-energy parameters. Outside the in-

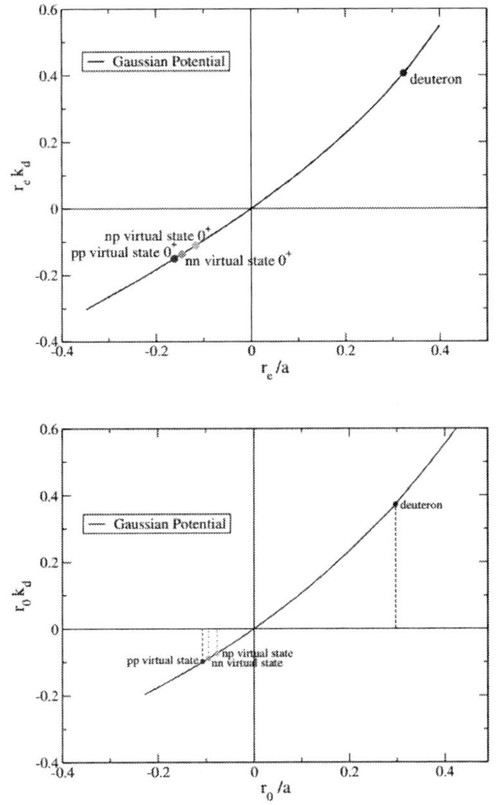

Figure 1. Upper panel: the binding momentum as a function of the scattering length (both in units of the effective range) for a gaussian potential. Specific systems are shown by colored symbols. Bottom panel: the binding momentum as a function of the scattering length (both in units of the gaussian range). The position onto the x-axis of the indicated systems is given.

teraction region the s-wave reduced wave function of the system is

$$\psi_B(r \to \infty) = C_a e^{-k_d r}\,, \tag{7}$$

where C_a is the asymptotic normalization constant well described, inside the unitary window, by the relations

$$C_a^2 = 2k_d \frac{1}{1-k_d r_e} = 2k_d e^{2k_d r_B}\,. \tag{8}$$

These relations are fulfilled up to second order in the ratio r_e/a. At the same order the mean square radius results

$$\langle r^2 \rangle = \frac{a^2}{8}\left[1 + \left(\frac{r_B}{a}\right)^2\right] = \frac{1}{8k_d^2} e^{2k_d r_B}\,. \tag{9}$$

Both observables are governed by the scaling function $f_{sc} = e^{2k_d r_B}$. At the same order, the probability P_e of the particles to be in the classical forbiden region is defined as

$$P_e = C_a^2 \int_{2r_B}^{\infty} e^{-2k_d r} dr = \frac{C_a^2}{2k_d} e^{-4k_d r_B} = \frac{1-k_d r_e}{e^{-4k_d r_B}} = e^{-2k_d r_B}\,, \tag{10}$$

where we have identified $2r_B = r_e a k_d$ as the lower limit for two particles to be considered outside the interaction region [21].Very close to the unitary limit $P_e \approx 1 - r_e/a$. For weakly bound systems P_e is governed by the ratio $2k_d r_B$, therefore we consider the systems inside the unitary window as strongly correlated. In the Table 1 we give the experimental values for the deuteron observables (second column) and the calculated values using the scaling function f_{sc} (third column). As can be seen the experimental values are estimated well below a 1% accuracy. Better than the expected error given by $(r_e/a)^2$ which for the $S = 1$ state is about 0.1.

E (MeV)	2.224575(9)	2.223
C_a (fm$^{-1/2}$)	0.8781(44)	0.8786
$\sqrt{\langle r^2 \rangle}$ (fm)	1.97535(85)	1.971
P_e		0.601

Table 1. Experimental values (second column) and calculated values (third column) of the indicated observables.

3 The universal window for $A \leq 6$

The gaussian characterization introduced at the level of two particles can be extended to describe a general number of particles. In this case the potential energy is

$$V = V_0 \sum_{i<j} e^{-r_{ij}^2/r_0^2}\,, \tag{11}$$

and the unitary window can be explored by varying the parameters of the gaussian potential. Though the above potential applies well for equal boson systems, in the case of nucleons the potential could be different in the different spin-isospin states. Accordingly we extend the above definition to

$$V = V_0 \sum_{i<j} e^{-r_{ij}^2/r_0^2} \mathcal{P}_{01} + V_1 \sum_{i<j} e^{-r_{ij}^2/r_1^2} \mathcal{P}_{10}\,, \tag{12}$$

where $\mathcal{P}_{01}$ projects on spin-isospin channels $S = 0, T = 1$ and $\mathcal{P}_{10}$ projects on spin-isospin channels $S = 1, T = 0$. Due to antisymmetrization these two channels correspond to an s-wave interaction, therefore the gaussian parameters can be fixed by the low energy s-wave parameters. In order to maintain low the number of parameters in the following we explore the unitary window using the same range values, $r_0 = r_1$, in both spin states. It is interesting to notice that in this case and when the two terms are at the unitary limit, the potential results [23]

$$V = 2.6840 \frac{\hbar^2}{m r_0^2} \sum_{i<j} e^{-r_{ij}^2/r_0^2}\,, \tag{13}$$

and the binding momentum for $A = 3, 4$ are [19]

$$K_3 = \frac{0.4883}{r_0}\,, \tag{14}$$

$$K_4 = \frac{1.1847}{r_0}\,. \tag{15}$$

To determine the binding energies at the unitary limit of the three- and four-nucleon systems, the range of the gaussian potential has to be known. In order to determine the range, we proceed in the following way. In Fig.2 we show the gaussian characterization of the universal window for three- and four nucleons obtained using a gaussian potential having equal range in both spin channels and the two strengths, V_0 and V_1, related to follow the nuclear cut, defined as the path in which the singlet and triplet np scattering lengths have a fixed ratio $^1a_{np}/^3a_{np} = -4.3$ (see Refs.[19, 25]). In the figure the solid green square and solid red square show the position of ^{3}H and ^{4}He on the gaussian curve. In fact at those positions the ratio $\kappa_N/k_d = \sqrt{E_N/E_2}$ correspond to the experimental ratios. For $N = 3$, $\sqrt{E_3/E_2} = 3.81$, whereas for $N = 4$, $\sqrt{E_4/E_2} = 13.13$. The corresponding values at the x-axis are $k_d r_0 = 0.457$, for $N = 3$, and $k_d r_0 = 0.481$, for $N = 4$. From the deuteron binding momentum $k_d = 0.231$ fm^{-1} we can determine the ranges $r_0^{(3)}$ and $r_0^{(4)}$. With these ranges the following gaussian potentials

$$V = V_0 \sum_{i<j} e^{-(r_{ij}/r_0^{(N)})^2} \mathcal{P}_{01} + V_1 \sum_{i<j} e^{-(r_{ij}/r_0^{(N)})^2} \mathcal{P}_{10} \quad (16)$$

describe simultaneously the experimental values of E_N and E_2 at the position of the gaussian curve indicated by the green square ($N = 3$) and red square ($N = 4$) in the figure. From the obtained ranges we can determine the binding energies at the unitary limit.

$$E_3 = 0.4883 \frac{\hbar^2}{m[r_0^{(3)}]^2} = 2.54\,\text{MeV}\,, \quad (17)$$

$$E_4 = 1.1847 \frac{\hbar^2}{m[r_0^{(4)}]^2} = 13.5\,\text{MeV}\,. \quad (18)$$

The above values are the values at unitary if the systems follow the nuclear cut moving from the physical point to the unitary limit. To check to which extend the gaussian path encodes movements along the unitary window we can scale one of the widely used NN potentials, the Argonne AV14 potential [26]. The AV14 potential can be cast in the following way

$$V(i,j) = \sum_{p=1,14} V_p(r_{ij}) O_{ij}^p = \sum_{STv} V_v^{ST}(r_{ij}) O_v^{ST}(ij) \quad (19)$$

with O_{ij}^p the 14 operators that characterize the potential. In the last equality the potential is given in the four spin-isospin channels $ST = 01, 10, 00, 11$. To bring the potential to the unitary limit the spin-isospin channels $ST = 01, 10$ are scaled by 1.0633 and 0.8 respectively. With this calibration the two scattering lengths, $^1a_{np}$ and $^3a_{np}$, are close to infinity. The binding energy of ^{3}H calculated with the scaled potential results $B(^3\text{H}) = 2.4$ MeV, very close to the value predicted by the gaussian characterization. This numerical observation confirms the low sensitivity to the interaction details this particular region has.

Figure 2 shows another interesting question regarding the existence of excited states in ^{3}H and ^{4}He. At unitarity the spectrum of ^{3}H has the Efimov structure, an infinite tower of geometrically distributed states accumulating at $E = 0$. In the figure the first state is shown with

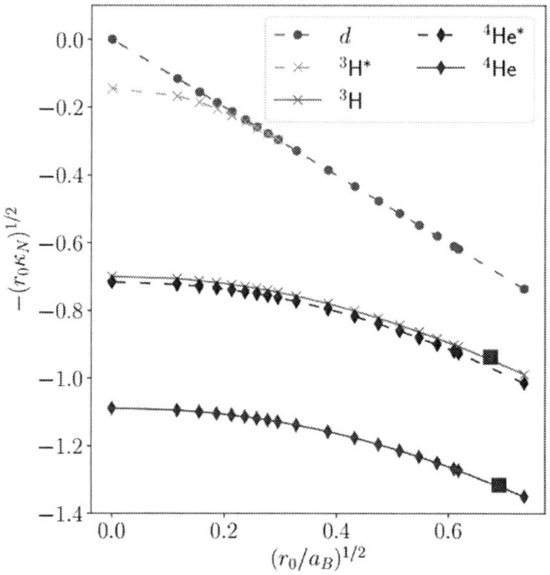

Figure 2. The binding momentum for $N = 2, 3, 4$ as a function of the two-body binding momentum $k_d = 1/a_B$, both in units of the gaussian range, calculated using the spin-dependent gaussian potential.^{3}H* and ^{4}He* are the first excited state for the three- and four-nucleon system, respectively.

the orange dashed line. As the parameters of the gaussian potential are varied to follow the nuclear cut, the excited state disappears very soon crossing the two-body threshold and becoming a virtual state [20]. The excited state disappears around a value of the triplet scattering length of 18 fm, very far from the physical point. Therefore we can conclude that the absence of excited states in the three-nucleon system is related to the position of the deuteron in the universal window and can be explained using the gaussian characterization of the window. In the case of ^{4}He, Fig.2 shows that there is an excited state at unitarity and that the excited state remains along the path (blue diamonds). However in the figure the Coulomb interaction has not been considered. When it is taken into account, the three-nucleon systems, ^{3}H and ^{3}He, split and at the same time the excited state of ^{4}He move into the continuum [19].

We now comment on the formation of the $A = 6$ nuclei as emerging from the unitary limit. This is illustrated in Fig.3. At unitarity, the potential of Eq.(13) is used to solve the Schrödinger equation for $A \geq 3$. Due to antisymmetrization of the wave function, the $A = 6$ nuclei result unbound below the ^{4}He threshold. Differently from what happens at the physical point, at unitarity the np system has zero energy and, therefore ^{6}He and ^{6}Li have the same threshold represented by the ^{4}He binding energy. As the strength of the potential increases along the nuclear cut the two nuclei appear, first ^{6}Li (violet squares) and then ^{6}He (red stars). The exact position of these nuclei on top of the gaussian characterization is discussed in Ref.[19]. In the figure the energies are given in units of the triton energy

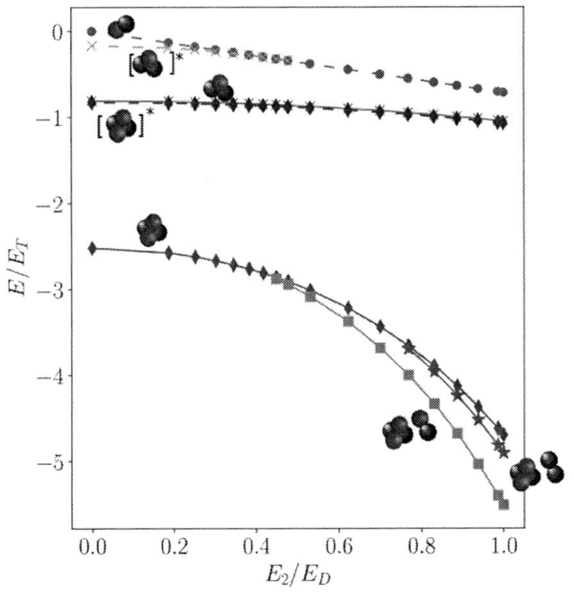

Figure 3. The energies of the two-, three-, four- and six-nucleon systems, in units of the triton binding energy E_T, as a function of the two-body energy, in units of the deuteron energy. Symbols indicate the different nuclei (red and blue circles representing protons and neutrons respectively). The '*' indicates excited states.

E_T as a function of the two-body energy in units of the deuteron energy.

4 Conclusions

In the present work we have discussed the gaussian characterization of the unitary window in the case of few-nucleon systems. Systems belonging to this window show low sensitivity to the interaction details. The dynamics is governed by the scattering length and by the location on the unitary window given by the ratio r_e/a. The gaussian paths are useful to connect different systems, for example atomic and nuclear systems, or the same system in different spin channels, as the NN system in $S = 0$ or $S = 1$. They are also useful to follow real (or ideal) movements of the systems inside the window. Though in nuclear systems at present it is not possible to perform experiments in which the nuclear interaction is changed, this can be achieved for atomic systems in traps using Feshbach resonances. For these particular systems, trapped cold atoms, it was shown that a particular class of universal properties appears known as van der Waals universality. In Ref. [21] this property has been discussed showing that this class of universality is well described by the gaussian characterization. This is a manifestation of two facts; from one side, due to the low importance on the details of the interaction, the real interaction between two-particles can be effectively represented by a two-parameter potential as a gaussian. At that point, movements of the systems inside the unitary window can be followed by varying the parameters of that effective potential. On the other hand, movements along the window can be followed by scaling the original potential with the result that the scaled potential follows closely the gaussian path. In the present discussion we have shown explicitly this fact by constructing a two-body gaussian potential able to describe simultaneously the triton and deuteron biding energies and, with a different range, the alpha particle and the deuteron. Then, using the range of those gaussian potentials, we have predicted the corresponding binding energies at unitarity. To verify the quality of this procedure we have proposed an ideal experiment by scaling the spin-isospin channels $S, T = 0, 1$ and $1, 0$ of the AV14 potential in order to bring the system at the unitary limit. Then we have calculated the ^{3}H binding energy at unitarity, using the scaled potential, and surprisingly the energy value resulted very close to that one predicted by the gaussian potential showing that the scaled potential follows the gaussian path.

One important property of the gaussian characterization is the possibility of connecting the physical points to the unitary limit allowing the study of the nuclear levels at this limit. In this respect the unitary limit can be seen as a mirror of the nuclear levels in which the scale symmetry is better realized. First of all we have shown the evolution of the excited states. At unitarity the three-nucleon system described by the potential of Eq.(13) has an infinite number of excited states geometrically distributed (Efimov specrum). By varying the strength of the interaction following the nuclear cut, the system moves away from that limit and the excited states disappear one by one, the last one when the np scattering length is around 18 fm. In the case of ^{4}He, the excited state is present along the gaussian path and its evolution onto a resonance state depends on the inclusion of the Coulomb interaction. Interestingly we have shown that ^{6}He and ^{6}Li are not bound at the unitary limit, their binding strongly depends on the deuteron one. They emerge as stable states close to the point at which the excited state of the triton disappears. We can conclude that the particular spectrum of the light nuclei is strongly determined by the position of the deuteron and the 0^+ virtual states inside the unitary window. With the corresponding scattering lengths and effective ranges as the only input parameters, the $A \leq 6$ spectrum can be predicted and continuously linked to the unitary point. The extension of the present analysis up to ^{12}C is at present underway.

References

[1] M. Piarulli, L. Girlanda, R. Schiavilla, R. Navarro Pérez, J.E. Amaro, and E. Ruiz Arriola, Phys. Rev. C **91**, 024003 (2015)
[2] H.W. Hammer, S. König S, and U. van Kolck, *Rev. Mod. Phys.* 92:025004 (2020)
[3] U. van Kolck, *Nucl. Phys. A* 645:273 (1999)
[4] P. Bedaque, H.W. Hammer, and U. van Kolck, *Phys. Rev. Lett.* 82:463 (1999)

[5] P. Bedaque, H.W. Hammer, and U. van Kolck, *Nucl. Phys. A* 646:444 (1999)

[6] V. Efimov, *Phys. Lett. B* 33:563 (1970)

[7] V. Efimov, *Sov. J. Nucl. Phys.* 12:589 (1971)

[8] T. Kraemer, et al., *Nature* 440:315 (2006)

[9] E. Braaten, and H.W. Hammer, *Phys. Rep.* 428:259 (2006)

[10] H.W. Hammer, and L. Platter, *Eur. Phys. J. A* 32:113 (2007)

[11] L. Platter, *Few-Body Syst.* 46:139 (2009)

[12] F. Ferlaino, and R. Grimm, *Physics* 3:9 (2010)

[13] F. Ferlaino, et al., *Few-Body Syst.* 51:113 (2011)

[14] T. Frederico, et al., *Few-Body Syst.* 51:87 (2011)

[15] C.H. Greene, P. Giannakeas, J. Pérez-Ríos, *Rev. Mod. Phys.* 89:035006 (2017)

[16] P. Naidon, S. Endo, *Rep. Prog. Phys.* 80:056001 (2017)

[17] A. Kievsky, A. Polls, B. Juliá Díaz, and N.K. Timofeyuk, Phys. Rev A **96**, 040501(R) (2017)

[18] A. Kievsky, A. Polls, B. Juliá Díaz, N.K. Timofeyuk, and M. Gattobigio, Phys. Rev A **102**, 063320 (2020)

[19] M. Gattobigio, A. Kievsky, and M. Viviani, Phys. Rev. C **100**, 034004 (2019).

[20] A. Deltuva, M. Gattobigio, A. Kievsky, and M. Viviani, Phys. Rev. C **102**, 064001 (2020).

[21] A. Kievsky, M. Gattobigio, L. Girlanda, and M. Viviani, Annu.Rev.Nucl.Part.Sci. 2021.71:465-90

[22] P. Recchia, A. Kievsky, L. Girlanda, and M. Gattobigio, Phys. Rev. A **106**, 022812 (2022)

[23] R. Álvarez-Rodríguez, A. Deltuva, M. Gattobigio, A. Kievsky, *Phys. Rev. A***93**, 062701 (2016)

[24] Bazak B, Eliyahu M, van Kolck U. *Phys. Rev. A* **94**, 052502 (2016)

[25] A. Kievsky, M. Gattobigio, *Few-Body Syst.* 57:217 (2016)

[26] R.B. Wiringa, R.A. Smith, and T.L. Ainsworth, Phys. Rev. C **29**, 1207 (1984)

AUTHOR INDEX